PHOTONICS
OPTOELECTRONICS

for

MTech, BTech, MSc (Phys), BSc (Hons/Pass), GATE, NET/SET

PHOTONICS
OPTOELECTRONICS

for
MTech, BTech, MSc (Phys), BSc (Hons/Pass), GATE, NET/SET

SL Kakani
MSc (Physics), PhD
Former Executive Director
Institute of Technology and Management
(Affiliated to Rajasthan Technical University, Kota)
Bhilwara 311001, Rajasthan, India
E-mail: slkakani28@gmail.com

Shubhra Kakani
MSc (Physics), PhD
MLV Government College
Bhilwara 311001, Rajasthan, India

CBS Publishers & Distributors Pvt Ltd
New Delhi • Bengaluru • Chennai • Kochi • Kolkata • Mumbai
Hyderabad • Nagpur • Patna • Pune

Disclaimer

Science and technology are constantly changing fields. New research and experience broaden the scope of information and knowledge. The authors have tried their best in giving information available to them while preparing the material for this book. Although, all efforts have been made to ensure optimum accuracy of the material, yet it is quite possible some errors might have been left uncorrected. The publisher, the printer and the authors will not be held responsible for any inadvertent errors, omissions or inaccuracies.

PHOTONICS
OPTOELECTRONICS

ISBN: 978-93-86478-22-1

Copyright © Authors and Publisher

First Edition: 2017

All rights reserved. No part of this book may be reproduced or transmitted in any form or by any means, electronic or mechanical, including photocopying, recording, or any information storage and retrieval system without permission, in writing, from the authors and the publisher.

Published by Satish Kumar Jain and produced by Varun Jain for
CBS Publishers & Distributors Pvt Ltd
4819/XI Prahlad Street, 24 Ansari Road, Daryaganj, New Delhi 110 002, India.
Ph: 23289259, 23266861, 23266867   Website: www.cbspd.com
Fax: 011-23243014   e-mail: delhi@cbspd.com; cbspubs@airtelmail.in.

Corporate Office: 204 FIE, Industrial Area, Patparganj, Delhi-110092
Ph: 4934 4934   Fax: 4934 4935   e-mail: publishing@cbspd.com; publicity@cbspd.com

Branches

- **Bengaluru:** Seema House 2975, 17th Cross, K.R. Road,
  Banasankari 2nd Stage, Bengaluru 560 070, Karnataka
  Ph: +91-80-26771678/79   Fax: +91-80-26771680   e-mail: bangalore@cbspd.com
- **Chennai:** 7, Subbaraya Street, Shenoy Nagar, Chennai 600 030, Tamil Nadu
  Ph: +91-44-26680620, 26681366   Fax: +91-44-42032115   e-mail: chennai@cbspd.com
- **Kochi:** Ashana House, No. 39/1904, AM Thomas Road, Valanjambalam,
  Ernakulam 682 018, Kochi, Kerala
  Ph: +91-484-4059061-65   Fax: +91-484-4059065   e-mail: kochi@cbspd.com
- **Kolkata:** 6/B, Ground Floor, Rameswar Shaw Road, Kolkata-700 014, West Bengal
  Ph: +91-33-22891126, 22891127, 22891128   e-mail: kolkata@cbspd.com
- **Mumbai:** 83-C, Dr E Moses Road, Worli, Mumbai-400018, Maharashtra
  Ph: +91-22-24902340/41   Fax: +91-22-24902342   e-mail: mumbai@cbspd.com

Representatives

- **Hyderabad**   0-9885175004
- **Patna**   0-9334159340
- **Nagpur**   0-9021734563
- **Pune**   0-9623451994

Printed at:
Swastik Packagings, Patparganj Industrial Area, Delhi, India

Preface

In the modern era of information technology, the efficient transfer of information is highly relevant to our well-being. Photonics and optical fiber systems, owing to their better suitability, are poised to form the very means of many segments of industry and academic research. The book, supplemented with detailed illustrations, examples and references, provides a comprehensive source of information for students, engineers and scientists in this field. Mainly, the book fulfills the requirements of a suitable text for MTech, BTech, MSc and BSc (Hons) electronics students of all Indian and foreign universities.

The book is divided into sixteen chapters covering photonics, optoelectronics materials, devices and systems. A special chapter on metamaterials is also given. Each topic is discussed in detail and systematic manner, in an easily understandable mode as well as language. Topics are supplemented with worked out problems, glimpses, short-answer questions, multiple choice questions with answers, which further help in better understanding of the subject.

While preparing the manuscript, we consulted various books, research papers and review articles and got their help. We are grateful to publishers and authors of those books and research papers. We would like to thank all our friends and students who forwarded their useful suggestions.

Salient features of the book:
- Extensive coverage of topics related to photonics | optoelectronics
- Special chapters on optical switches, laser-based systems, metamaterials, etc.
- Each chapter is followed by a good number of worked out problems, suggested readings, glimpses, review questions, problems, short-answer questions and multiple choice questions with answers

We hope that this student-friendly text will definitely cater to the needs of students, teachers and researchers.

We are highly thankful to CBS Publishers & Distributors, New Delhi. We would like to put on record the sincere efforts of Mr YN Arjuna and his team comprising Ms Ritu Chawla, Ms Poonam Kapoor Bhatia, Ms Sanjubala, Mr Manish Raj and Mr Kuldeep for bringing out the book in the present form.

We welcome comments, suggestions and corrections from the readers.

SL Kanani
Shubhra Kakani

Contents

1. Introduction   1–7

    1.1. Introduction *1*
        Glimpses *4*
        Review Questions *6*
        Short Question Answers *6*

2. Nature of Light and Foundation of Electromagnetic Theory   8–53

    2.1 Introduction *8*
    2.2 Polarization of Light *11*
    2.3 Polarization of Sensitive Materials *13*
    2.4 Quantum Nature of Light *15*
    2.5 Electromagnetic Waves *17*
        Illustrative Examples *36*
        Suggested Readings *41*
        Glimpses *41*
        Review Questions *45*
        Problems *46*
        Short Answer Questions *47*
        Multiple Choice Questions *49*
        Answers *53*

3. Optical Fiber Waveguides   54–112

    3.1 Introduction *54*
    3.2 Concept of Electromagnetic Waves *54*
    3.3 Solution in an Inhomogeneous Medium *56*
    3.4 Planar Optical Waveguide *60*
    3.5 The Modes of a Symmetric Step Index Fiber *62*
    3.6 Power Distribution and Confinement Factor *68*
    3.7 Cylindrical Waveguides *70*
    3.8 Photonic Crystal Fibers *84*
        Suggested Readings *95*
        Glimpses *95*
        Review Questions *99*

Problems *100*
Short Answer Questions *104*
Multiple Choice Questions *108*
Answers *112*

4. Optical Fibers   113–192

4.1 Introduction *113*
4.2 Optical Fiber *116*
4.3 Optical Fiber as Waveguide: Principle of Propagation *118*
4.4 Principle of Optical Fiber: Total Internal Reflection *118*
4.5 Modes of Propagation: V-Parameter *125*
4.6 Index Profile or Refractive Index Profile *129*
4.7 Types of Optical Fibers *130*
4.8 Ray Propagation in Graded-Index Fibers *139*
4.9 Effect of Material Dispersion *142*
4.10 The Combined Effect of Multipath and Material Dispersion *144*
4.11 Calculation of RMS Pulse Width *145*
4.12 Advantages of Optical Fiber *145*
4.13 Signal Degradation *147*
4.14 Fiber Losses *147*
4.15 Optical Fiber Cables *150*
4.16 Fiber Fabrication *153*
4.17 Polarization *159*
4.18 Optical Fiber Amplification *162*
4.19 Optical Nonlinearities in Fibers *164*
4.20 Solitons in Optical Fiber *165*
4.21 Applications of Fiber Optic Cables *167*
Suggested Readings *180*
Glimpses *181*
Review Questions *183*
Problems *184*
Short Answer Questions *186*
Multiple Choice Questions *189*
Answers *192*

5. Fiber Optic Cables and Connectors   193–253

5.1 Introduction *193*
5.2 Fiber Materials *194*
5.3 Fiber Fabrication Methods *196*
5.4 Optical Fiber Cables *202*
5.5 Fiber-Optic Connections and Related Losses *206*
5.6 Connectors and Splices *211*
5.7 Applications of Connectors and Splices *211*
5.8 Requirements of Connectors and Splices *212*
5.9 Fiber Connectors *212*
5.10 Mechanical Considerations *215*
5.11 Fiber-optic Connector Types *215*

5.12 Adapters for Different Fiber-Optic Connector Types 216
5.13 Fiber-Optic Connector Structures 217
5.14 Fiber-Optic Connector Assembly Techniques 217
5.15 Fiber Splicing 223
5.16 Connectors Verses Splices 227
5.17 Characterization of Optical Fibers 228
5.18 Applications of the Fiber Optic Cables 235
    Glimpses 244
    Suggested Readings 245
    Review Questions 246
    Problems 247
    Short Answer Questions 249
    Multiple Choice Questions 251
    Answers 253

6. Optical Sources   254–327

6.1 Introduction 254
6.2 Spontaneous and Stimulated Emission 261
6.3 Relation between Einstein's Coefficients 262
6.4 Laser Components 264
6.5 Population Inversion 266
6.6 Characteristics of Laser Radiation 268
6.7 Types of Lasers 269
6.8 Laser Oscillations and Resonant Modes 270
6.9 Semiconductor Lasers 271
6.10 Rate Equations 288
6.11 Analysis Based on Rate Equations 294
6.12 Applications of Lasers 304
    Suggested Readings 315
    Glimpses 315
    Review Questions 317
    Problems 318
    Short Answer Questions 319
    Multiple Choice Questions 324
    Answers 327

7. Optoelectronic Devices   328–419

7.1 Introduction 328
7.2 Photoelectric Effect 329
7.3 Photodetectors 333
7.4 Photodiode 342
7.5 PIN Photodiodes 354
7.6 Response Time of Photodiodes 359
7.7 Avalanche Photodiodes 361
7.8 Schottky Photodiode 367
7.9 Light Emitting Diodes (LEDs) 374
7.10 Laser Diode 388

Suggested Readings *411*
Glimpses *412*
Review Questions *413*
Problems *415*
Short Answer Questions *416*
Multiple Choice Questions *417*
Answers *419*

8. Photovoltaic Devices   420–448

8.1 Introduction *420*
8.2 Principle of Photovoltaic Cell *423*
8.3 Basic Model of a Solar Cell *428*
8.4 Multijunctions *430*
8.5 Temperature Effects *434*
8.6 Other Technologies: Dye-Sensitized Cells *435*
8.7 Applications of Photovoltaic System *437*
Glimpses *443*
Suggested Readings *444*
Review Questions *444*
Short Answer Questions *444*
Multiple Choice Questions *447*
Answers *448*

9. Optical Receivers   449

9.1 Introduction *449*
9.2 Principle of Optoelectronic Detection *451*
9.3 Optical Receiver *452*
9.4 Photodetectors *453*
9.5 Principles of Photodetection *454*
9.6 Properties of Semiconductor Photodetectors *456*
9.7 Performance Parameters of Photodetectors *461*
9.8 Types of Optical Detectors *462*
9.9 Photodetector Noise *476*
9.10 Receiver Analysis *480*
9.11 BER of an Ideal Optical Receiver *481*
Suggested Readings *489*
Glimpses *489*
Review Questions *491*
Problems *492*
Short Answer Questions *495*
Multiple Choice Questions *499*
Answers *501*

10. Optoelectronic Modulation   502–522

10.1 Introduction *502*
10.2 Electro-optic Effect *503*
10.3 Linear Electro-optic Modulator *503*

Contents xi

    10.4 Transverse Electro-optic Modulator *509*
    10.5 Acousto-optic Effect and Acousto-optic Modulators *511*
        Suggested Readings *517*
        Glimpses *517*
        Review Questions *518*
        Problems *518*
        Short Answer Questions *519*
        Multiple Choice Questions *520*
        Answers *522*

11. Optical Amplifiers   523–606

    11.1 Introduction *523*
    11.2 Semiconductor Optical Amplifiers *524*
    11.3 Rare Earth Doped Fiber Amplifiers *548*
    11.4 Raman Fiber Optical Amplifiers or Fiber Raman Amplifiers (FRA) *560*
    11.5 Planer Waveguide Optical Amplifiers *564*
    11.6 Linear Optical Amplifiers *554*
    11.7 Basic Applications of Optical Amplifiers *564*
        Suggested Readings *571*
        Glimpses *571*
        Review Questions *574*
        Problems *575*
        Short Answer Questions *576*
        Multiple Choice Questions *578*
        Answers *580*

    Appendix—Optical Switches
    A11.1 Introduction *581*
    A11.2 Opto-mechanical Switches *581*
    A11.3 Electro-optic Switches *592*
    A11.4 Thermo-optic Switches *595*
    A11.5 Acousto-optic Switches *600*
    A11.6 Micro Electromechanical System (MEMS) *601*
    A11.7 3D MEMS Based Optical Switches *604*
    A11.8 Micro-optic Mechanical System (MOMS) *606*

12. Wavelength Division Multiplexing (WDM) Technology   607–662

    12.1 Introduction *607*
    12.2 Time Division Multiplexing *612*
    12.3 Frequency Division Multiplexing (FDM) *612*
    12.4 Dense Wavelength Division Multiplexing (DWDM) *613*
    12.5 Coarse Wavelength Division Multiplexing (CWDM) *613*
    12.6 Passive Components *613*
    12.7 Multiplexers and Demultiplexers *624*
    12.8 Active Components *633*
    12.9 Topologies and Architectures *639*
        Suggested Readings *652*

xii Photonics | Optoelectronics

    Glimpses *652*
    Review Questions *655*
    Problems *655*
    Short Answer Questions *657*
    Multiple Choice Questions *660*
    Answers *662*

13. Fiber Optic Communication   663–761

    13.1 Introduction *663*
    13.2 Essential Components of Fiber Communication System *665*
    13.3 Basic Communication System *680*
    13.4 Types of Topologies *681*
    13.5 Types of Networks *684*
    13.6 Submarine Cables *687*
    13.7 Open System Interconnection (OSI) *689*
    13.8 System Architectures *690*
    13.9 Line Coding in Optical Links *694*
    13.10 Error Control or Correction *696*
    13.11 Performance of Passive Linear Optical Networks *697*
    13.12 Performance of Star Optical Networks *699*
    13.13 SONET and SDH *700*
    13.14 Multiplexing Terminology and Signaling Hierarchy *703*
    13.15 SONET and SDH Transmission Rates *707*
    13.16 SONET Systems *707*
    13.17 Metro and Long-Haul Optical Networks *708*
    13.18 Network Configuration *709*
    13.19 Nonlinear Effects *715*
    13.20 Dispersion *720*
    13.21 Solitons *728*
    Suggested Readings *749*
    Glimpses *750*
    Review Questions *753*
    Problems *753*
    Short Answer Questions *755*
    Multiple Choice Questions *759*
    Answers *761*

14. Optical Fiber Sensors   762–787

    14.1 Introduction *762*
    14.2 Fiber-optic Sensor (FOS) *763*
    14.3 Classification of Fiber-optic Sensors *764*
    14.4 Intensity-modulated Sensors *765*
    14.5 Phase-modulated Sensors *768*
    14.6 Spectrally Modulated Sensors *772*
    14.7 Distributed Fiber-optic Sensors (DFOSs) *775*
    14.8 Fiber-optic Smart Structures *777*
    14.9 Industrial Applications of Fiber-optic Sensors *779*

Contents **xiii**

    Suggested Readings *781*
    Glimpses *781*
    Review Questions *782*
    Problems *783*
    Short Answer Questions *783*
    Multiple Choice Questions *785*
    Answers *787*

15. Laser Based Systems   788–832

    15.1 Introduction *788*
    15.2 Solid State Lasers *789*
    15.3 Gas Lasers *794*
    15.4 Dye or Liquid Lasers *798*
    15.5 Chemical Lasers *800*
    15.6 Color Center Laser *802*
    15.7 Special Lasers *803*
    15.8 X-ray Laser *805*
    15.9 Q-Switching *806*
    15.10 Mode Locking *809*
    15.11 Application of Lasers *812*
    Suggested Readings *825*
    Glimpses *826*
    Review Questions *827*
    Problems *827*
    Short Answer Questions *829*
    Multiple Choice Questions *830*
    Answers *832*

16. Metamaterials   833–876

    16.1 Introduction *833*
    16.2 Prehistory of Metamaterials *835*
    16.3 Electromagnetic Metamaterials *838*
    16.4 General Equations for the RH and TLS *839*
    16.5 Metamaterial Lenses and Superlenses *847*
    16.6 Electromagnetic and Photonic Crystals *848*
    16.7 Hyperlens *848*
    16.8 Cloaking *851*
    16.9 Metamaterial Transmission Lines *851*
    16.10 Controllable Metamaterials *852*
    16.11 Optoelectronic Control of Metamaterials *852*
    16.12 Extraordinary Transmission *854*
    16.13 Photonic Crystals (PhCs) *855*
    16.14 Fabrication of Metamaterials *855*
    16.15 How to Create Metamaterials? *856*
    16.16 Some Applications of Metamaterials *863*
    16.17 Metamaterials with an Active Element *868*
    Glimpses *868*

xiv Photonics | Optoelectronics

       Suggested Readings *870*
       Review Questions *870*
       Short Answer Questions *871*
       Fill in the Blanks *874*
       Multiple Choice Questions *875*
       Answers *876*

Index   **877–881**

1

Introduction

1.1 INTRODUCTION

Photonics (also known as optoelectronics) is the technology of creation, transmission, detection, control and application of light. Basically, photonics is the study of interaction of photons and electrons. Photonics has several applications in the following areas:

a. Ecology
   i. Solar cell energy generation
   ii. Air quality and pollution monitoring

b. Imaging
   i. Camcorders
   ii. Satellite weather pictures
   iii. Digital cameras
   iv. Night vision
   v. Military surveillance

c. Information Display
   i. Computer terminals
   ii. Traffic signals
   iii. Operating displays in automobiles and appliances

d. Information Storage
   i. CD-ROM
   ii. DVD

e. Life Sciences
   i. Identification of molecules and proteins
   ii. Lighting

2 Photonics | Optoelectronics

f. Medicine
   i. Minimally invasive diagnostics
   ii. Photodynamic chemotherapy

g. Telecommunication
   i. Lasers
   ii. Photodetectors
   iii. Light modulators

Fibre optic communication is an important part of photonics. The growth of internet and its capacity to transmit both images and sound has been made possible only because of vast improvements in speed and capacity of fibre optic communications. At the heart of this revolution are the semiconductor lasers, fast light modulators, photodiodes, and communications grade optical fibre. The growth of telecommunication systems got a big jolt with the deployment of optical fibres in 1980, creating the first system of optical telecommunications networks. There was another big jolt in 1990, when *optical amplifiers* were rediscovered and adapted to optical fibre telecommunications. This implemented multiple wavelength transmission (wavelength-division multiplexing) and made it possible to develop internet (Fig. 1.1).

Photonics is based on the physics and devices of conventional optics. But its main topic is the physics of *nonlinear optical processes*. Thus, it is necessary to analyse possible nonlinear processes and investigate suitable materials with methods of *nonlinear spectroscopy*.

Light is built out of *photons* which are *quantum mechanical* and *relativistic* particles. Thus, light shows *particle* and *wave* properties in the sense of our macroscopic understanding. Light moves with the maximum speed (3×10^8 ms^{-1} in vacuum). Light from *lasers* shows new statistical properties.

In *nonlinear optics* all these properties are much more important than in conventional optics. Light-induced changes in the materials responsible for nonlinear effects are mostly functions of these light properties and the superposition of different light beams leads to complicated effects.

Nonlinear effects can be differentiated in interactions with absorption, called *resonant interactions*, and interactions with nonabsoring, transparent materials, called *nonresonant interactions*.

Applications of these nonlinear effects demand sufficient knowledge of the *nonlinear properties of possible materials*. One of the main topics is the invention of new useful materials with high nonlinear coefficients. Most of today's known materials demand intensities of more than 10^{20} photons cm$^{-2}\cdot$s^{-1}.

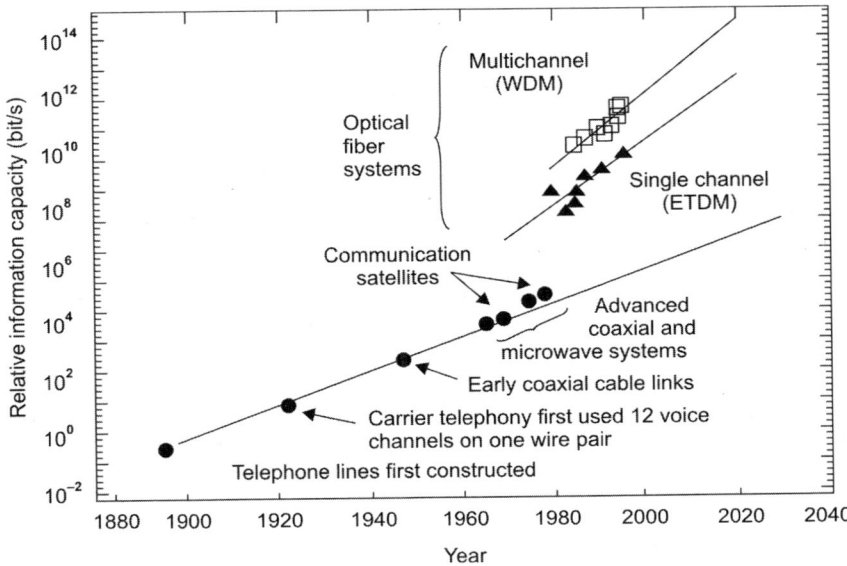

Fig. 1.1. The growth of telecommunication systems

New materials may become available by *femtochemistry* using short light pulses for well-timed ignition of molecular chemical reactions.

New materials are needed for the invention of new *solar energy technology*. Nonlinear optical investigations may be necessary for their exploration.

One of the fastest growing areas in photonics is *laser medicine*. Almost all parts of the body are treatable. This includes, in particular, eye lens and retina, skin, vessels and stone demolition. The possibility through thin fibres into the body enables new techniques with minimal invasion to be developed.

Photonic sensors have been the subject of intensive research over the last two decades. These are emerging as an important branch of photonics. Their development is based on analysing different components of optical signal like intensity polarization, pulse shape or arrival time and also other phenomena like interference. Utilizing those various possibilities leads to different types of sensors. There are endless possibilities for their use. The best known applications are in civil and military environments for detection of a wide variety of biological, chemical and nuclear agents.

An important medium for sensing is the optical fibre and its recent variations which found their own ways of applications in sensing. These includes: photonic band gap (PBG) fibres, microstructure official fibres and hybrid random hole fibres (HRHF). They offer higher resolution, lower cost and/or expanded detection range capability for sources and detection schemes. There are some of the photonic devices whose properties can be directly applied to sensors, e.g. *light emitting diodes* (LED), high finesse, wide-spreading-gallery-mode disc resonators, etc.

Intensive work on the application of *silicon* in photonics is in progress. Silicon is a material of choice in electronics and it is also economical to build photonic devices on this material. However, silicon is not a *direct-gap* material and therefore cannot efficiently generate light. Work on hybrid silicon laser is in progress.

There is another area, associated with *computer technology*, the continuation of Moore's law and progress which it reflects is becoming increasingly dependent on ultra-fast data transfer between and within microprocessor. The existing electrical interconnects are expected to be replaced by high speed *optical interconnects*. They are seen as a promising way forward, and *silicon phonics* is seen as particularly useful, due to the ability to integrate electronics and optical components on the same silicon chip.

Significant progress has been seen in advances of *optical modulators* based on silicon. In designing and optimizing those devices *simulations* play a very important role. Recent advances in modelling and simulations of silicon phonic devices, e.g. recently developed simulation methods for submicrometre innovative *silicon-on-insulator* (SOI) guiding structures and phonic devices is commendable.

A new emerging direction within photonics is related to *quantum communications* and *quantum information processing*.

Photons can potentially play the crucial role in the quantum information processing due to their low-noise properties and ease of manipulation at the single *orbit* level. In recent years, several *quantum gates* have been implemented using integrated photonics circuits.

In all topics of photonics, a close connection of basic knowledge about the physical principles on one hand and technical possibilities on the other side is typical.

GLIMPSES

- *Photon* is a particle representing one quantum of light. It has energy $E = h\nu$, where ν is the frequency of the radiated energy. The rest mass of photon is zero. *Electron* is a stable negatively charged elementary particle having a mass of 9.10939×10^{-31} kg, a charge of 1.602177×10^{-19} C and a spin of $1/2$.
- *Electrons* have been used, e.g. for long distance communication since the discovery of first telegraph. After the invention of electronic devices such as tubes, e.g. the triode, radiocommunication became possible. Later semiconductor devices such as transistors, and

computer chips were used for this purpose. These electronic devices allow telecommunication with about 10^9 bit/s.
- The *electromagnetic field of photons* oscillates much faster than is possible for electrons. Using light for communication purposes became possible after the invention of laser (**L**ight **A**mplification by **S**timulated **E**mission of **R**adiation). The commonly known fibers for transmitting light over long distances of many hundred kilometers demand several devices for generating, switching and amplifying the light. Thus, the engineers created the word *photonics* to describe the combination of light technologies and electronics in communication.
- The term *photonics* now covers almost all processes using laser light in science, medicine and technology, but it does not include illumination and simple conventional optical techniques. Thus, photonics includes mainly the nonlinear interactions of light with matter. In this case, the characteristic effects are a nonlinear function of the intensity of applied light. Obviously, the term has close relations to *nonlinear optics* and *quantum optics*.
- The nonlinear optical processes demand, with a few exceptions, such as photosynthesis, lasers for providing a sufficiently large number of photons per area and time. Typically more than 10^{18} photons cm^{-2} s^{-1} are needed to reach nonlinearity in materials with fast reaction times.
- An *optical fiber* is generally a thin strand of glass that is used to carry a beam of light. Once the light is introduced in the fiber, by using a lens, e.g., it can only escape by propagating to the other end of the fibre. The light beam is prevented from leaking out of the sidewalls by an effect called *total internal reflection*. Thus, the fiber acts as a *guide* for photons.
- *Photonic devices* are used to convert photons to electrons and *vice versa*. Photons and electrons are two of the basics quantum-mechanical particles. Like all quantum mechanical particles, electrons and photons also exhibit *dual nature*, i.e. sometimes as particles and sometimes as waves.
- *Solar cell* is a device that uses solar radiation to drive electric current. One type, used to power equipment in spacecraft and artificial satellites is essentially a semiconductor in which a voltage is set up across the *pn* junction when photons from the sun fall on the surface (*photovoltaic cell*). Other types of solar cells, used in desert regions, water heating systems, etc. comprises complex *thermopiles*, one set of junctions being illuminated by solar radiation. A *solar battery* consists of several solar cells. A *solar panel* is a large flat array of solar cells attached to the outside of a spacecraft or satellite and oriented to receive the maximum solar radiations.

6 Photonics | Optoelectronics

REVIEW QUESTIONS

1. What is photonics?
2. How inventions of lasers and optical fibers helped the phenomenal growth of telecommunication systems?
3. What are electrons and photons?
4. What is nonlinear optics?
5. What are photonic sensors?

SHORT ANSWER QUESTIONS

1. Mention two simple principles that support almost all the science of photonic devices.

Ans. There are two simple principles that support almost all the science of photonic devices. First, Boltzmann relationship expressing the relationship between molecules at energy E_2 relative to those at energy E_1 as:

$$\frac{n(E_2)}{n(E_1)} = \exp[-(E_2 - E_1)/k_B T] \quad (1)$$

where $k_B = 1.380658 \times 10^{-23}$ JK^{-1}

is Boltzmann's constant,

and second is Planck's equation, relating the energy of a photon ($h\nu$) to the frequency of light wave (ν) associated with the photon, i.e.:

$$E = h\nu \quad (2)$$

where h is Planck's constant ($h = 6.625 \times 10^{-34}$ Js)

2. Mention the important properties of photons.

Ans.   i. Photon is a quantum of electromagnetic radiation.
  ii. Photon has an energy of $h\nu$ where h is Planck's constant and ν the frequency of radiation.
  iii. For some purposes photons can be considered as *elementary particles* travelling at the speed of light ($c = 3 \times 10^8$ m/s) and having a momentum $h\nu/c$ or h/λ (where λ is the wavelength of light).
  iv. Mass of photons is almost zero.
  v. Photons can cause excitation of atoms and molecules and more energetic one can cause *ionization*.
  vi. Photons fulfill at least following four uncertainty conditions:
    • Uncertainty of position and momentum

$$\Delta x \Delta p_x \geq \frac{h}{\pi} \text{ and } \Delta y \Delta p_y \geq \frac{h}{\pi}$$

- *Uncertainty of position and angle*
  With the relation
  $$\Delta x = w_c,\ \Delta p = \Delta k h/2\pi \text{ and } \Delta k = k\theta,$$
  one obtains, Position-angle uncertainty
  $$w_0\, \theta \geq \frac{\lambda}{\pi}$$

- *Uncertainty of energy and time*
  $$\Delta E\, \Delta t \geq \frac{h}{2\pi}$$

- *Frequency-time uncertainty*
  $$\Delta v\, \Delta t \geq \frac{1}{2\pi}$$

3. Mention some of the important applications of photonics.

Ans.
  i. Opto-optical switches
  ii. Development of quantum computer
  iii. Preparation of new materials by femtochemistry using short light pulses for well-timed ignition of molecular-chemical reactions.
  iv. Detection of environmental pollution
  v. New materials for the invention of advanced solar energy technologies
  vi. New nonlinear devices
  vii. Laser medicine
  viii. Development of new quantum information technologies

4. What is photonics?

Ans. Photonics is the field which involves electromagnetic energy, such as light, where the fundamental object is a photon. In some sense photonics is parallel to electronics which involves electrons. Photonics is often referred to as optoelectronics or as electro-optics to indicate that both fields a lot in common. In fact, there is a lot of interplay of photonics and electronics, e.g. a laser is driven by electricity to produce light or to modulate that light to transmit data.

2
Nature of Light and Foundation of Electromagnetic Theory

2.1 INTRODUCTION

Light is a form of energy which evokes visual sensation of objects. However, nowadays radiations like ultraviolet, infrared are also included in this form of energy; which do not excite visual sensation but show similar effects.

Geometrical optics or ray optics describes the light propagation in terms of ray. The ray is an abstraction, which can be used as an approximate model to describe the propagation of light. Light rays tend to bend at the interface between two dissimilar media, and may curve in a medium where the refractive index changes. The path taken by the rays indicate how the actual ray will propagate. However, this simple ray model of optics fails to account for optical effects such as *diffraction* and *polarization*.

The correct explanation of diffraction was given by Fresnel in 1815. Fresnel showed that the approximately rectilinear propagation character of light could be interpreted on the assumption that light is a *wave motion*, and that the diffraction fringes could thus be accounted for in detail. Later, the work of Maxwell in 1864 theorized that light waves must be *electromagnetic in nature*. Furthermore, observation of *polarization* effects indicated that light waves are *transverse* in nature, i.e. the wave motion is perpendicular to the direction in which the wave travels. In this *wave* or *physical optics* view point, the electromagnetic waves radiated by a small optical source can be represented by a spherical wavefront with the source at the centre (Fig. 2.1a). A *wavefront* or *phase front* is defined as the locus of all points in the wave train which have the same phase. Generally, one draws wavefronts passing either the maxima or minima of the wave, such that the peak or trough of a sine wave, for example. Obviously, wavefronts are separated by one wavelength.

When the wavelength of light is much smaller than the object (or opening) which it encounters, the wavefront appears as straight lines to this object or opening. In this case, the light wave can be represented as a *plane wave*, and its direction of travel can be shown by a *light ray* which is drawn perpendicular to the phase front (Fig. 2.1b). The light rays concept allows large-scale optical effects, e.g. reflection and refraction to be analyzed by the simple geometrical process of *ray tracing*. The concept of light ray is very useful because the rays show the direction of energy flow in the light beam.

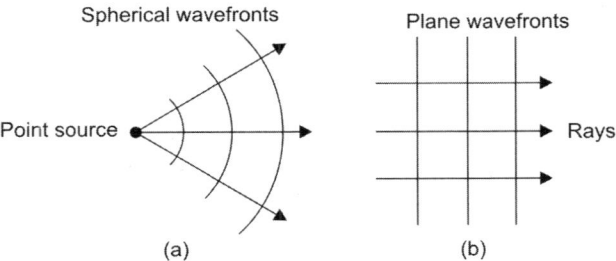

Fig. 2.1. Representations of spherical and plane wavefronts and their associated rays

These light rays obey following rules:
i. In a medium, light rays travel with velocity v given by:

$$v = \frac{c}{n} \quad (2.1)$$

where $c = 3 \times 10^8$ ms^{-1} is the velocity of light in vacuum and n is known as refractive index of the medium. Table 2.1 shows some typical values of refractive indices. As v is always less than c, n is always greater than 1. For air $n = n_a \approx 1$.

Table 2.1. Refractive index for some materials

Material	Refractive index (n)
Air	1.0
Water	1.33
Silica glass	1.5
GaAs	3.35
Silicon	3.5
Germanium	4.0

ii. In a uniform medium, rays travel in a straight path.

iii. The phenomenon of *refraction* of light at the interface between two transparent media of uniform indices of refraction is governed by **Snell's law**. Consider a ray of light passing from a medium of refractive index n_1 into a medium of refractive index n_2 (Fig. 2.2a). Assume that

$n_1 > n_2$ and that the angles of incidence and refraction with respect to the normal to the interface are ϕ_1 and ϕ_2 respectively. Then, according to Snell's law:

$$n_1 \sin \phi_1 = n_2 \sin \phi_2 \qquad (2.2)$$

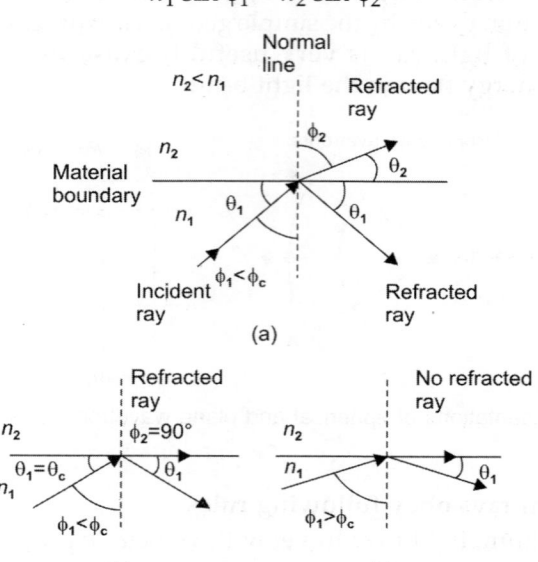

Fig. 2.2. Representation of the critical angle and total internal reflection at a glass–air interface (a) Refraction of a ray of light. (b) Critical ray incident at $\phi_1 = \phi_c$ and refracted at $\phi_2 = \pi/2$. (c) Total internal reflection ($\phi_1 > \phi_c$)

As $n_1 > n_2$, if we increase the angle of incidence ϕ_1, the angle of refraction ϕ_2 will go on increasing until a critical situation is reached, when for a certain value of $\phi_1 = \phi_c$, ϕ_2 becomes $\pi/2$, and the refracted ray passes along the interface. This angle $\phi_1 = \phi_c$ is called the *critical angle*. Substituting the values of $\phi_1 = \phi_c$ and $\phi_2 = \pi/2$, in Eq. (2.2), we find,

$$n_1 = \sin \phi_c = n_2 \sin (\pi/2) = n_2.$$

Thus,

$$\sin \phi_c = \frac{n_2}{n_1} \qquad (2.3)$$

If the angle of incidence ϕ_1 is further increased beyond ϕ_c, the ray is no longer refracted but is reflected back into the same medium (Fig. 2.2c), this is ideally expected. In practice, however, there is always some tunneling of optical energy through this interface. The wave carrying this energy in called *evanescent wave*. One can explain this in terms of electromagnetic theory). This phenomenon is called total *internal reflection*, which is responsible for the propagation of light through *optical fibre*.

In addition, when light is totally internally reflected, a phase change δ occurs in the reflected wave. This phase change depends on the angle $\theta_1 < \frac{\pi}{2} - \phi_c$ according to relationships:

$$\tan\frac{\delta_N}{2} = \frac{\sqrt{n^2 \cos^2\theta_1 - 1}}{n \sin\theta_1} \quad (2.4)$$

$$\tan\frac{\delta_P}{2} = n\frac{\sqrt{n^2 \cos^2\theta_1 - 1}}{\sin\theta_1} \quad (2.5)$$

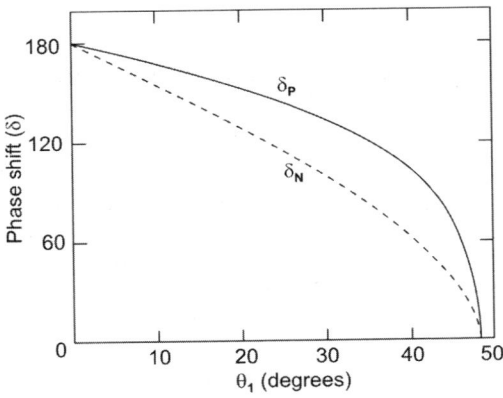

Fig. 2.3. Phase shifts occurring from the reflection of wave components normal (δ_N) and parallel (δ_P) to the plane of incidence

Here, δ_N and δ_P are the phase shifts of the electric-field wave components normal and parallel to the phase of incidence, respectively, and $n = \frac{n_1}{n_2}$. Fig. 2.3 shows these phase shifts for a glass-air interface ($n = 1.5$ and $f_c = 42.5°$). The values range from zero immediately at the critical angle to $\frac{\pi}{2} - \phi_c$, when $\phi_c = 90°$.

2.2 POLARIZATION OF LIGHT

Polarization of light waves demonstrates that they are *transverse*. An ordinary light wave consists of many transverse electromagnetic waves that vibrate in a variety of directions, i.e. in more than one plane is called *unpolarized light*. One can pictorically represent any arbitrary direction of vibration as a combination of a parallel vibration and a perpendicular vibration (Fig. 2.4).

In an unpolarized light, all directions of vibration at right angles to that of propagation of light as consisting of two orthogonal plane polarization components, one that lies in the plane of incidence (the

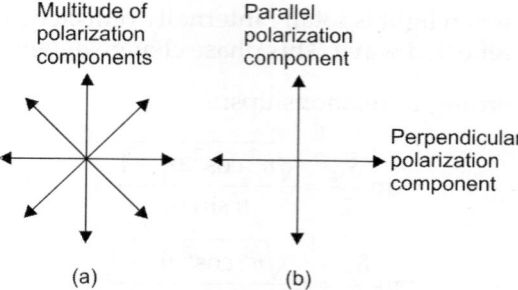

Fig. 2.4. Polarization of light represented as a combination of a parallel vibration and a perpendicular vibration

plane containing the incident ray and reflected ray) and the other of which lies in a plane perpedicular to the plane of incidence. These are termed as the *parallel polarization* and the *perpendicular polarization* components, respectively.

Unpolarized light can be split into separate polarization component either by reflection of nonmetallic surface or by refraction when the light beam passes from one material to another. Fig. 2.5 shows that when an unpolarized light beam travelling in air impinges on a nonmetallic surface such as glass, part of the beam is reflected and part is refracted into a glass. The reflected beam is partially polarized and at a specific angle, i.e. *Brewster's angle* [Brewster discovered that for a certain angle of incidence, monochromatic light was 100% polarized upon reflection. The refracted beam is partially polarized, while the reflected beam is completely polarized parallel to the reflecting surface. Furthermore; at this angle of incidence, the reflected and refracted beams

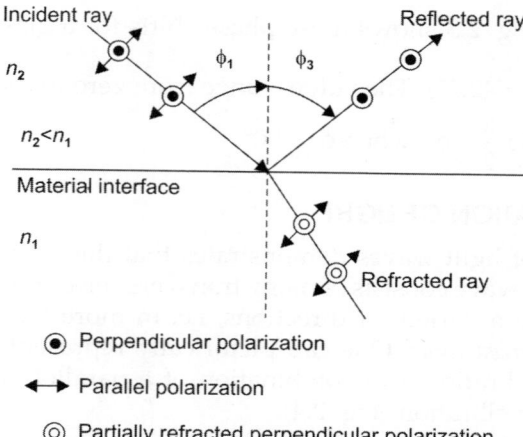

Fig. 2.5. Behaviour of an unpolarized light beam at the interface between air and nonmetallic surface

are perpendicular. The value of polarizing angle ϕ_p depends upon the refractive index of the refractive medium, i.e. $n = \tan \phi_p$] the reflected light is completely perpendicularly polarized. The parallel component of the refracted beam is transmitted entirely into the glass, whereas the perpendicular component is only partially refracted. How much of the refracted light is polarized depends on the angle at which the light approaches the surface and on the material composition.

2.3 POLARIZATION OF SENSITIVE MATERIALS

The polarization characteristics of light are important when one examining the behaviour of components such as *optical isolators* and *light filters*. We shall restrict to the study of three polarization – sensitive materials or devices that are used in such components. These devices are *polarizers*, *Faraday rotators*, and *birefringent* crystals.

(i) *Polarizer*: This is a material or device that transmits only one polarization component and blocks the other, e.g. when unpolarized light enters a polarizer that has a vertical transmission axis (Fig. 2.6), only the vertical polarization component passes through the polarizer (device). This concept is used in polarizing sunglasses to reduce the glare of partially polarized sunlight reflections from the road or water surfaces. In order to see the polarization property of sunglasses, tilt your head sideways, you will see a number of glare spots. The polarization filters in the sunglasses block out the polarized light coming from these glare spots when you held your head normally.

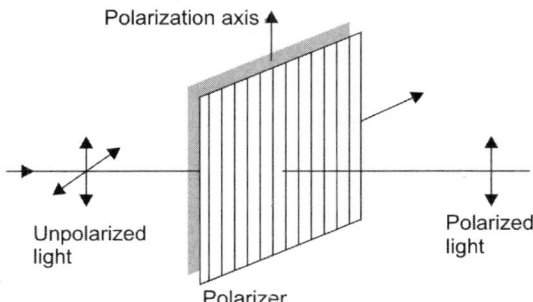

Fig. 2.6. Only the vertical polarization component passes through a vertically oriented polarizer

(ii) *A Faraday rotator*: This is a device that rotates the *state of polarization* (SOP) of light passing through it by a specific angle. A popular device rotates the SOP clockwise by 45° or $\lambda/4$ (Fig. 2.7).

We may note that this rotation is independent of the SOP of input light, but the rotation angle is different depending on the direction in which the light passes through the Faraday rotator, e.g. the rotation process is not reciprocal. In this process, the SOP of the input light is

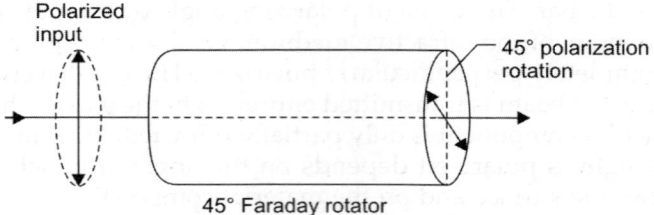

Fig. 2.7. A Faraday rotator rotates the state of polarization clockwise by 45° or a quarter of a wavelength ($\lambda/4$)

maintained after the rotation, e.g. if the input light to a 45° Faraday rotator is linearly polarized in a vertical direction, then the rotated light existing the crystal is also linearly polarized at 45° angle. The Faraday rotator material is usually an asymmetric crystal, e.g. yttrium iron garnet (YIG) and the degree of angular rotation is proportional to the thickness of the device.

(iii) *Double-refractive crystals or Birefrigent*: These crystals have a property called double refraction, or birefrigent which is defined as the double refraction of light in a transparent, molecularly ordered material, which is manifested by the existence of orientation-dependent difference in refractive index.

The phenomenon of double refraction (Fig. 2.8a) is based on the laws of electromagnetism due to Maxwell. One of the most familiar example of double refraction occurs with calcium carbonate (calcite) crystals. Birefrigent splits the light signals entering it into two orthogonally (perpendicularly) polarized beams (Fig. 2.8b). One of the beams is called an *ordinary* (o) – ray, since it obeys Snell's law of refraction at the surface of crystals. The second ray is called the *extraordinary* (e) ray, since it refracts at an angle that deviates from the prediction of the standard form of Snell's law.

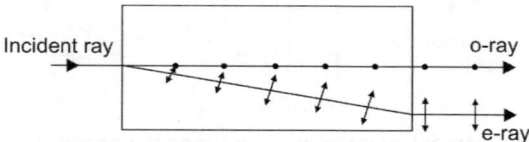

Fig. 2.8a. Phenomenon of double refraction exhibited by a birefringent crystal. The electric-field vectors of the *o*-ray and *e*-ray are shown to vibrate perpendicular and parallel to the plane of the figure, respectively

As shown in Fig. 2.8b, each of the two orthogonal polarization components refracted at different angle, e.g. if the incident unpolarized light arrives at an angle perpendicular to the surface of the device, the *o*-ray can pass straight through the device whereas the *e*-ray component, is deflected at a slight angle and hence it follows a different path through the material.

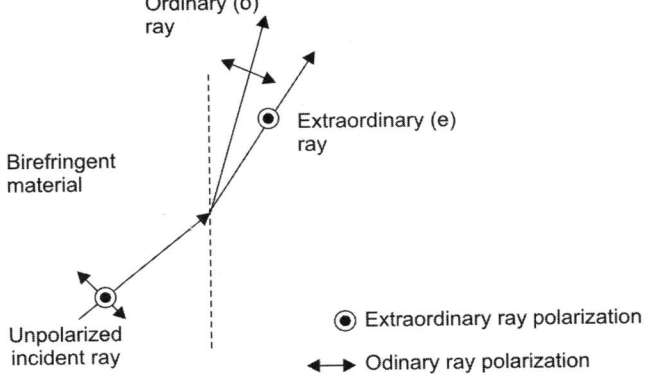

Fig. 2.8b. A birefringent crystal splits the light signal entering it into two perpendicularly polarized beam

The difference in refractive indices or birefringence, between the e and o rays travelling through an anisotropic crystal is given by:

$$\text{Birefringence } (B) = [n_e - n_o] \qquad (2.6)$$

where n_e and n_o are the refractive indices experienced by the e and o ray respectively.

The optical path difference (Δ) is defined by the relative phase shift between the ordinary and extraordinary rays as they emerge from an anisotropic material. Thus,

$$\Delta = (n_1 - n_2)\, t \qquad (2.7)$$

where n_1 and n_2 are refractive index values for a system having two values and t is thickness of the system.

There are some birefringent crystals that are used in optical communication components. Table 2.2 lists n_o and n_e of some of these birefringent crystals along with their applications.

2.4 QUANTUM NATURE OF LIGHT

We have read that the distribution of light past aperatures and obstacles could be explained only by attributing the wave nature of it. Thus, the wave theory of light adequately accounts for all phenomena related to the transmission of light. However, the wave theory of light is found inadequate in dealing with the interaction of light and matter, e.g. dispersion, emission and absorption of light. The particle nature of light arises from the observation that light energy is always emitted or absorbed in discrete units called as *quanta* or *photons*. In all experiments used to show the existence of photons, reveal that photon energy is related only to the frequency ν, i.e.:

$$E = h\nu \qquad (2.8)$$

where $h = 6.625 \times 10^{-34}$ is the Planck's constant. This frequency ν, in turn, must be measured by observing a wave property of light.

Table 2.2. Common birefringent crystals and some of their applications

Crystal name	Symbol	n_o (ordinary index)	n_e (extraordinary index)	Applications
Calcite	$CaCO_3$	1.658	1.486	Polarization controllers and beam splitters
Lithium niobate	$LiNbO_3$	2.286	2.200	Light signal modulators
Rutile	TiO_2	2.616	2.903	Optical isolators and circulator
Yttrium vanadate	YVO_4	1.945	2.149	Optical isolators, circulators, and beam displacers

When light is incident on an atom, a photon can transfer its energy to an electron within this atom, thereby exciting it to a higher energy level. In this process either all or none of the photon energy is imparted to the electron. The energy *absorbed* by the electron must be exactly equal to that required to excite the electron to higher energy level E_2 from lower energy level E_1, i.e.:

$$E_2 - E_1 = h\nu \qquad (2.9)$$

Conversely, an electron in an excited state can fall to a lower energy state separated from it by an energy $h\nu$ by *emitting* a photon of exactly this energy.

In view of these developments, light must be regarded as having a *dual nature*. Light exhibits the characteristics of a wave in some situations and characteristics of a particle in other situations.

Now, it is well established that *light is an energy-carrying electromagnetic wave that emanates from vibrating electrons in atoms*. When light is transmitted through matter, some of the electrons in the matter are forced into vibration. In this way, vibrations in the emitter are transmitted to vibrations in the receiver.

2.5 ELECTROMAGNETIC WAVES

Although, we are not always aware of the presence of electromagnetic (em) waves, but these waves permeate our environment regularly. In the form of *visible light*, they enable us to view the world around us with our eyes. Infrared waves from the surface of the earth warm our environment, radio-frequency waves carry our favourite radio entertainment, microwaves cook our food and are used in communication systems and the list goes on and on.

The fundamental laws of electricity and magnetism—*Maxwell's equations* form the basis of electromagnetic phenomena. One of these

Nature of Light and Foundation of Electromagnetic Theory 17

equations predicts that a time-varying electric field produces a magnetic field just as a time-varying magnetic field produces an electric field. From this generalization, Maxwell provided the final important link between electric and magnetic fields. The most *dramatic prediction of his equations is the existence of electromagnetic waves that propagate through empty space with the speed of light* ($c = 3 \times 10^8$ ms^{-1}). This discovery led to many practical applications, such as radio and television, and to the realization that light is one form of electromagnetic radiations. Figure 2.9 shows the propagation of electromagnetic wave at one instant.

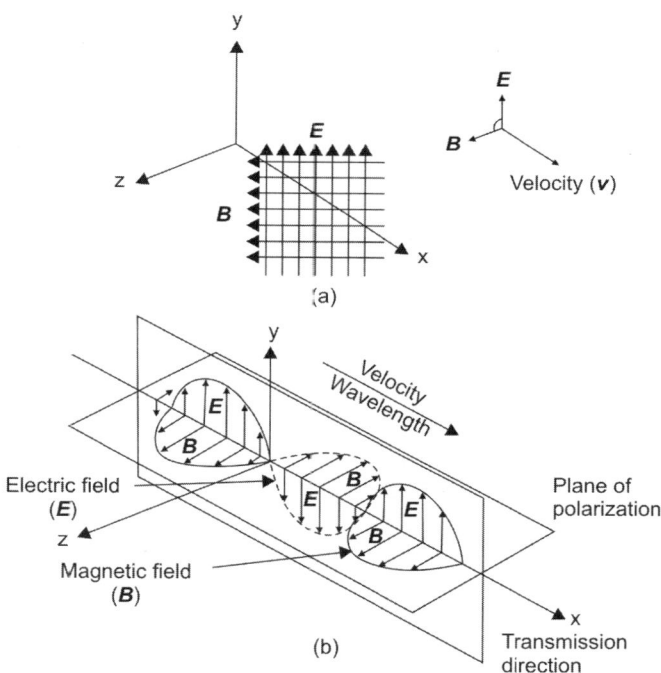

Fig. 2.9. Propagation of a plane wave, **E** and **B** vectors

We may note that:
(i) the two fields **E** and **B** are perpendicular to each other, and
(ii) both the fields **E** and **B** are perpendicular to the direction of propagation of the wave.

The second fact is a property characteristic of *transverse wave*. This means an electromagnetic wave is a transverse wave.

The following properties are associated with an electromagnetic wave travelling through *free space*:
(i) Electromagnetic waves travel with the speed of light
   ($c = 3 \times 10^8$ ms^{-1} in vacuum).

(ii) Electromagnetic waves are transverse waves.
(iii) The ratio of electric (**E**) to magnetic field (**B**) in an electromagnetic wave equals the speed of light ($E/B = c$).
(iv) Electromagnetic waves carry both energy and momentum which can be delivered to a surface. Figure 2.10 shows the electromagnetic radiation spectrum.

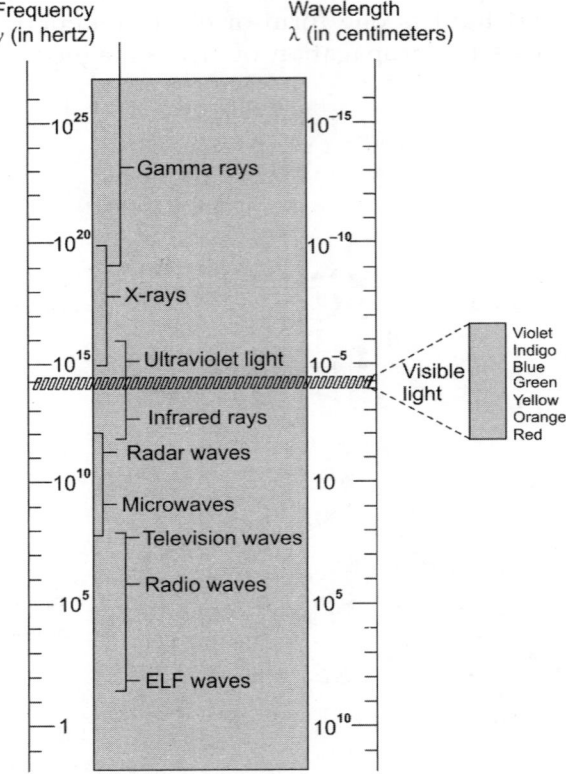

Fig. 2.10. Spectrum of electromagnetic radiation

2.5.1 Maxwell's Equations and Boundary Conditions

The space variations of electric and magnetic field components are related to time variations of magnetic and electric field components respectively. This interdependence gives rise to the phenomenon of electromagnetic wave propagation.

Maxwell in 1873 published his original finding of the electromagnetic theory of light. This theory has led to many important discoveries, including the existence of electromagnetic waves. Based on his theory, all electric, magnetic, electromagnetic, and optical phenomena are governed by the same fundamental laws of electromagnetism. These laws are written mathematically in terms of *Maxwell equations* (in MKS units):

$$\nabla \times E = -\frac{\partial B}{\partial t} \qquad (2.10)$$

$$\nabla \times H = J + \frac{\partial D}{\partial t} \qquad (2.11)$$

$$\nabla \cdot D = \rho \qquad (2.12)$$

$$\nabla \cdot B = O \qquad (2.13)$$

where
  E is electric field intensity (Vm^{-1} or NC^{-1}),
  B is magnetic field (T or tesla),
  J is electric current density (Am^{-2}),
  ρ is volume charge density (Cm^{-3}).
where $D = P + \varepsilon_o E$.

The operator ∇ in cartesian coordinates is:

$$\nabla = \hat{i}\frac{\partial}{\partial x} + \hat{j}\frac{\partial}{\partial y} + \hat{k}\frac{\partial}{\partial y} \qquad (2.14)$$

The above relations are supplemented with constitutive relations:

$$D = \varepsilon E \qquad (2.15)$$

$$B = \mu H \qquad (2.16)$$

$$J = \sigma E \qquad (2.17)$$

where $\varepsilon = \varepsilon_o \varepsilon_r$ is the dielectric permittivity (Fm^{-1}), $\mu = \mu_o \mu_r$ is permeability (Hm^{-1}), σ is electrical conductivity and ε_r is relative dielectric constant. For optical problems, one may take $\mu_r = 1$. The permittivity of a vacuum, $\varepsilon_o = 8.854 \times 10^{-12}$ Fm^{-1} and permeability of a vacuum, $\mu_o = 4\pi \times 10^{-7}$ Hm^{-1}.

First, we recall two mathematical theorems, Gauss's theorem:

$$\oint_S F \cdot dS = \int_V \nabla \cdot F dV \qquad (2.18)$$

where S is the closed surface area defining volume V, and Stokes's theorem:

$$\oint_L F \cdot dl = \oint_S \nabla \times F \cdot dS \qquad (2.19)$$

where contour L define surface area A. With the help of the above theorem, Maxwell's equations in *differential form can be transformed into an integral form*. Integral forms of Maxwell's equation are:

$$\oint_S D \cdot dS = \int_V \rho dV \qquad (2.20)$$

$$\oint_L B \cdot dS = 0 \qquad (2.21)$$

$$\oint_L E \cdot dl = -\int_S \frac{\partial B}{\partial t} \cdot dS \qquad (2.22)$$

$$\oint_L H \cdot dl = \oint_S \left(J + \frac{\partial D}{\partial t}\right) \cdot dS \qquad (2.23)$$

Properties of the medium are mostly determined by ε, μ and σ. Further, for the dielectric medium the main role is played by ε. The medium is known as *linear*, if ε is independent of E; otherwise it is non-linear. If it does not depend on position in space, the medium is said to be *homogeneous*; otherwise it is *inhomogeneous*. If properties are independent of direction, the medium is *isotropic*; otherwise it is *anisotropic*.

Boundary Conditions

Boundary conditions are derived from an integral form of *Maxwell's equations*. For this purpose we separate all vectors into two components, one parallel to the interface and other normal to the interface. The derivation of boundary conditions is facilitated by using the contour and cylindrical shapes as shown in Fig. 2.11. It will be done independently for the electric and magnetic fields.

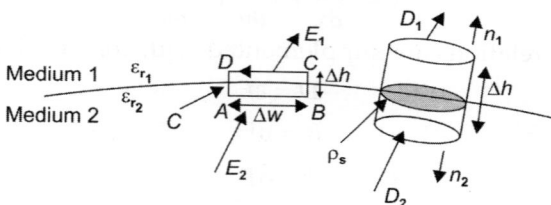

Fig. 2.11. An interface between two media. Contour and volume used to derive boundary conditions for fields between two different dielectrics are shown

(i) Electric Boundary Conditions

First, analyse transversal components. Integrate Eq. (2.22) over closed loop C (Fig. 2.11) and then set $\Delta h \to 0$:

$$\int_{ABCDA} E \cdot dl = -E_1 \cdot dl + E_2 \cdot dl$$
$$= -E_{1t}\Delta W + E_{2t}\Delta W$$
$$= 0$$

Therefore:

$$E_{1t} = E_{1t} \qquad (2.24)$$

The tangenital component of the electric field is therefore continuous across the boundary between any two dielectric media. Using general relation [Eq. (2.15)], one obtains the boundary conditions for tangenital component of vector D.

$$\frac{D_{1t}}{\varepsilon_{r1}} = \frac{D_{2t}}{\varepsilon_{r2}} \qquad (2.25)$$

In order to obtain conditions for normal components, consider the cylinder shown in Fig. 2.11. Here n_1 and n_2 are normal vectors pointing outwards of the top and bottom surfaces into corresponding dielectrics. Apply Gauss's law with integration over surface S of the cylinder.

$$\int_S \mathbf{D} \cdot d\mathbf{s} = \int_{op} \mathbf{D}_1 \cdot n_1 dS + \int_{bottom} \mathbf{D}_2 \cdot n_2 dS = \rho_s \Delta S \quad (2.26)$$

The contribution from the outer surface of the cylinder vanishes in the limit $\Delta h \to 0$. Use the face that $n_2 = -n_1$ and have:

$$n_1 \cdot (\mathbf{D}_1 - \mathbf{D}_2) = \rho_s \quad (2.27)$$

or

$$\varepsilon_{r1} E_{2n} - \varepsilon_{r2} E_{1n} = \rho_s \quad (2.28)$$

Here we have used general relation [Eq. (2.15)]. Normal component of vector $\mathbf{D}$ is not continuous across boundary (unless $\rho_s = 0$).

(ii) Magnetic Boundary Conditions

When deriving boundary conditions for magnetic field, we use approach similar as for electric fields, we use Eq. (2.21) where integration is over the cylinder.

$$\int_S \mathbf{B} \cdot d\mathbf{S} = 0$$

or

$$B_{1n} = B_{2n} \quad (2.29)$$

Using general relation [Eq. (2.16)], one obtains conditions for magnetic field $\mathbf{H}$:

$$\mu_{r1} H_{1n} = \mu_{r2} H_{2n} \quad (2.30)$$

The above relations tell us that normal component $\mathbf{B}$ is continuous across the boundary.

To obtain how transversal magnetic components behave across interface, apply Ampere's law for contour C and then let $\Delta h \to 0$, one obtains:

$$\int_S \mathbf{H} \cdot d\mathbf{l} = \int_A^B \mathbf{H}_2 \cdot d\mathbf{l} - \int_C^D \mathbf{H}_1 \cdot d\mathbf{l} = I$$

Here I is the net current crossing the surface of the loop. Let Δh approach zero, the surface of the loop approaches a thin line of length ΔW. Hence, the total current flowing through this thin line is $I = J_s \cdot \Delta W$, where J_s is the magnitude of the normal component of the surface current density traversing the loop. We can therefore express the above equation as:

$$H_{2t} - H_{1t} = J_s \quad (2.31)$$

Utilizing unit vector $\hat{n}_2$, the above relation can be written as:

$$n_2 \times (\mathbf{H}_1 - \mathbf{H}_2) = J_s \quad (2.32)$$

where n_2 is the normal vector pointing away from medium 2 (Fig. 2.11). J_s is the surface current.

In Table 2.3 we summarize boundary conditions between two dielectrics for the electric and magnetic fields. The behaviour of various field components is shown schematically in Fig. 2.12.

Table 2.3. Summary of boundary conditions for electric and magnetic fields

Field components	General form	Specific form
Tangential E	$n_2 \times (E_1 - E_2) = 0$	$E_{1t} = E_{2t}$
Normal D	$n_2 \cdot (D_1 - D_2) = \rho_s$	$D_{1t} = D_{2t} = \rho$
Tangential H	$n_2 \times (H_1 - H_2) = J_s$	$H_{2t} = H_{1t}$
Normal B	$n_2 \times (B_1 - B_2) = 0$	$B_{1n} = E_{2n}$

2.5.2 Wave Equation

Here, we will derive the wave equation. The wave equation for a source-free medium where $\rho_v = 0$ and $J = 0$. The wave equation is obtained by applying curl operation to both sides of Eq. (2.10).

$$\nabla \times \nabla \times E = -\frac{\partial}{\partial t}(\nabla \times B)$$

$$= -\mu \frac{\partial}{\partial t}(\nabla \times H)$$

$$= -\mu \frac{\partial}{\partial t}\frac{\partial D}{\partial t} = -\mu\varepsilon \frac{\partial}{\partial t}\frac{\partial^2 D}{\partial t^2}$$

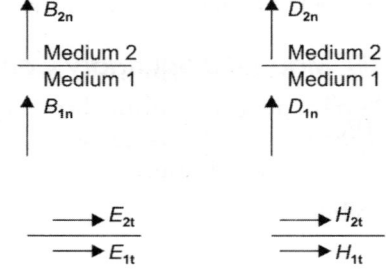

Fig. 2.12. Boundary conditions for fields

where we have used Maxwell Eq. (2.11) and relation (2.15). Next apply the following mathematical formula:

$$\nabla \times \nabla \times E = \nabla(\nabla \cdot E) - \nabla^2 E \qquad (2.33)$$

With the help of Maxwell Eq. (2.12), one finds finally:

$$\nabla^2 E = \mu\varepsilon \frac{\partial^2}{\partial t^2} E \qquad (2.34)$$

which is the *desired wave equation*. The quantity $\mu\varepsilon$ is related to velocity of light (c) in a vacuum as (assuming $\mu_r = 1$).

$$\mu\varepsilon = \mu_0 \varepsilon_0 \varepsilon_r = \frac{n^2}{c^2} \qquad (2.35)$$

where n is the refractive index of the medium.

Similarly, we can get general form of wave equation of H as:

$$\nabla^2 H = \mu\varepsilon \frac{\partial^2 H}{\partial t^2} \qquad (2.35a)$$

2.5.3 Time Harmonic Fields

In many practical situations fields have sinusoidal time dependence and are known as *time-harmonic*. This fact is expressed as:

$$E(r, t) = Re\{E(r)\, e^{j\omega t}\} \quad (2.36)$$

where $E(r)$ is the phasor form of $E(r, t)$ and is in general complex. $Re\{....\}$ indicates "taking the real part in" quantity in brackets. Finally, ω is the angular frequency in rad/s. In what follows, all fields will be represented in phasor notation.

Applying the time-harmonic assumption (2.36) to source – free Maxwell's equations results in:

$$\nabla \times E = -j\omega\mu H \quad (2.37)$$

$$\nabla \times H = -j\omega\varepsilon E \quad (2.38)$$

Applying the time harmonic assumption again to wave equation [Eq. (2.34)] gives:

$$\nabla^2 E + k^2 E = 0 \quad (2.39)$$

where $k = \omega\sqrt{\mu\varepsilon}$. Explicitly, the above wave equation is:

$$\left(\frac{\partial^2}{\partial x^2} + \frac{\partial^2}{\partial y^2} + \frac{\partial^2}{\partial z^2} + k^2\right) E_i = 0 \quad (2.40)$$

with $i = x, y, z$.

As an example, let us consider propagation of a uniform plane wave characterized by uniform electric field with nonzero component E_x. Assume also:

$$\frac{\partial^2 E_x}{\partial x^2} = 0,\; \frac{\partial^2 E_x}{\partial y^2} = 0 \quad (2.41)$$

The wave equation reduces to:

$$\frac{\partial^2 E_x}{\partial z^2} + k^2 E_x = 0 \quad (2.42)$$

and has the following forward propagating solution:

$$E_x(z) = E_0 e^{jkz} \quad (2.43)$$

Magnetic field is determined from the Maxwell's Equation (2.37):

$$\nabla \times E = \begin{vmatrix} \hat{a}_x & \hat{a}_y & \hat{a}_z \\ 0 & 0 & \frac{\partial}{\partial z} \\ E_x(z) & 0 & 0 \end{vmatrix} \quad (2.44)$$

$$= -j\omega\mu\, (\hat{a}_x H_x + \hat{a}_y H_y + \hat{a}_z H_z)$$

Here $\hat{a}_x, \hat{a}_y, \hat{a}_z$ are unit vectors along x, y, z axes, respectively. From the above equation, one finds:

$$H_x = 0 \tag{2.45}$$

$$H_y = \frac{1}{j\omega\mu}\frac{\partial E_x(z)}{\partial z}$$

$$H_z = 0$$

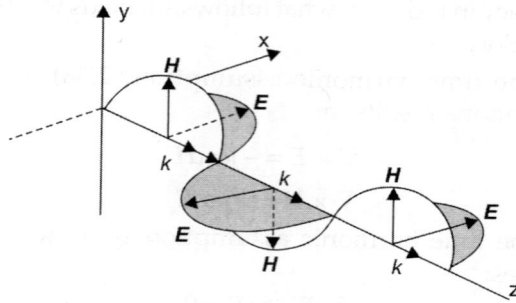

Fig. 2.13. Visualization of the electric and magnetic fields

Using the solution for $E_x(z)$, one finally obtains the expression for magnetic field:

$$H_x = \hat{a}_y H_y(z) \tag{2.46}$$

or

$$H_x = \frac{1}{-j\omega\mu}(-jkE_x)\hat{a}_y \tag{2.47}$$

$$= \hat{a}_y \frac{k}{\omega\mu} E_x(z)$$

$$= \hat{a}_y \frac{1}{Z} E_x(z)$$

where the impedance Z of the medium is defined as $Z = \sqrt{\frac{\mu}{\varepsilon}}$. The propagating wave is shown in Fig. 2.13.

We finish this section by providing a useful relation between electric and magnetic fields. Assuming time-harmonic and plane-wave dependence of both fields as:

$$E \sim \exp(i\omega t - i\mathbf{k}\cdot\mathbf{r})$$

and using constitutive relation [Eq. (2.16)], Maxwell equation [Eq. (2.10)] takes the form

$$\mathbf{k} \times \mathbf{E} = \omega\mu_0 \mathbf{H}$$

Introducing unit vector $\hat{k}$ along wave vector $\mathbf{k}$ and the expression for wave number:

$$k = \omega n\sqrt{\mu_0\varepsilon_0}, \text{ one finds}$$

$$n\hat{k} \times \mathbf{E} = Z_0 \mathbf{H}$$

where Z_0 is the *impedance* of the free-space.

2.5.4 Polarized Waves

Let us now discuss the concept of polarization of electromagnetic waves. Polarization characterizes the curve which E vector makes (in the plane orthogonal to the direction of propagation) at a given point in space as a function of time. In the most general case, the curve produced is an ellipse and, accordingly, the wave is called elliptically polarized. Under certain conditions, the ellipse may be reduced to a circle or a segment of a straight line. In those cases it is said that the wave's polarization is circular or linear, respectively. Since the magnetic field vector is related to the electric field vector, it does not need separate discussion. First consider a single electromagnetic wave.

2.5.4.1 Linearly Polarized Waves

Consider an electromagnetic wave characterized by electric field vector E directed along the x-axis.

$$E = \hat{a}_x \, E_0 \cos(\omega t - kz + \phi) \quad (2.48)$$

It is known as a linearly polarized plane wave with the electric field vector oscillating in the x direction. Wave propagates in the $+z$ direction. In Eq. (2.48), $\omega = 2\pi\nu$ is the angular frequency and k is the propagation constant defined as:

$$k = \frac{\omega}{v}$$

where $v = c/n$ is the velocity of the electromagnetic wave in the medium having refractive index n, ϕ is known as a phase of electromagnetic wave.

The electromagnetic wave can also be written in the complex representation as:

$$E = \hat{a}_x E_0 e^{i(\omega t - kz + \phi)} \quad (2.49)$$

The actual field as described by Eq. (2.48) is obtained from Eq. (2.49) by taking *real* part. A more general expression for the electromagnetic wave is:

$$E = \hat{e} E_0 e^{i(\omega t - k \cdot r + \phi)}$$

which is known as the *plane polarized wave*. Here unit vector $\hat{e}$ lies in the plane known as plane of polarization. It is perpendicular to vector k which describe direction of propagation:

$$k \cdot \hat{e} = 0$$

2.5.4.2 Circularly and Elliptically Polarized Waves

In general, when we have an arbitrary number of plane waves propagating in the same direction they add up to a complicated wave.

In the simplest case, one has only two such plane waves. To be specific, consider two plane waves oscillating along orthogonal direction. They are linearly polarized having the same frequencies and propagating in the same direction:

$$E_1 = E_x \hat{a}_x = \hat{a}_x E_{0x} \cos(\omega t - kz) \quad (2.50)$$

$$E_2 = E_y \hat{a}_y = \hat{a}_y E_{0y} \cos(\omega t - kz) + \phi) \quad (2.51)$$

We want to know the type of the resulting wave and the curve traced by the tip of the total electric vector $E = E_1 + E_2$

$$E = E_1 + E_2$$
$$= E_0 \{\cos(\omega t - kz) + \cos(\omega t - kz + \phi)\}$$

First, eliminate $\cos(\omega t - kz)$ term. From Eq. (2.50), we have

$$\cos(\omega t - kz) = \frac{E_x}{E_{0x}} \quad (2.52)$$

Using trigonometric indentify

$$\cos(\alpha - \beta) = \cos\alpha \cos\beta + \sin\alpha \sin\beta$$

We can express Eq. (2.51) as follows

$$E_y = E_{0y} \{\cos(\omega t - kz) \cos\delta + [1 - \cos^2(\omega t - kz)]^{1/2} \sin\delta\}$$

Substitute Eq. (2.52) in the above and have:

$$\frac{E_y}{E_{0y}} = \frac{E_x}{E_{0x}} \cos\phi + \left(1 - \frac{E_x^2}{E_{0x}^2}\right)^{1/2} \sin\phi$$

Squaring both sides gives

$$\left(\frac{E_y}{E_{0y}} = \frac{E_x}{E_{0x}} \cos\phi\right)^2 = \left(1 - \frac{E_x^2}{E_{0x}^2}\right)^{1/2} \sin\phi$$

or $\quad \dfrac{E_y^2}{E_{0y}^2} - 2\cos\phi \dfrac{E_y}{E_{0y}} \dfrac{E_x}{E_{0x}} + \dfrac{E_x^2}{E_{0x}^2} \cos^2\phi + \dfrac{E_x^2}{E_{0x}^2} \sin^2\phi = \sin^2\phi$

Finally, the above equation gives:

$$\left(\frac{E_y}{E_{0y}}\right)^2 + \left(\frac{E_x}{E_{0x}}\right)^2 - 2\left(\frac{E_y}{E_{0y}}\right)\left(\frac{E_x}{E_{0x}}\right) \cos\phi = \sin^2\phi \quad (2.53)$$

This is *general equation of an ellipse*. Thus the endpoint of $E(z, t)$ will trace an ellipse at a given point in space. It is said that the wave is *elliptically polarized*.

Nature of Light and Foundation of Electromagnetic Theory

When phase $\phi = \dfrac{\pi}{2}$, the resultant total electric field is:

$$\left(\frac{E_y}{E_{0y}}\right)^2 + \left(\frac{E_x}{E_{0x}}\right)^2 = 1$$

which describes *right elliptically polarized wave* since as time increases the end of electric vector E rotates clockwise on the circumference of an ellipse. Typical situations are illustrated in Fig. 2.14 where we show elliptic, circular and linear polarizations.

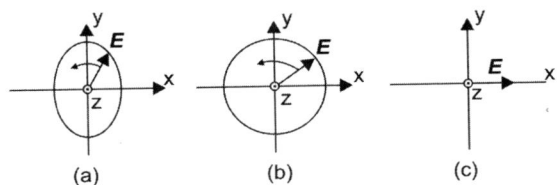

Fig. 2.14. Typical states of polarization: (a) elliptic, (b) circular, and (c) linear

2.5.5 Fresnel Coefficients and Phases

In this section, we discuss electromagnetic wave undergoing reflection at the boundary between two dielectrics (see Fig. 2.15). The plane of incidence is defined as the plane formed by unit vector $\hat{n}$ normal to the interface between two media and the directions of propagation of the incident and reflected waves.

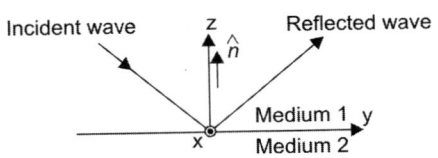

Fig. 2.15. Plane of incidence

We will derive so-called *Fresnel coefficients* which are reflection and transmission coefficients expressed in terms of the angle of incidence and material properties (ε dielectric constants) of the two dielectrics. Further, Fresnel phases are determined from Fresnel coefficients using the following definition:

$$r = e^{-2j\phi} \qquad (2.54)$$

Both reflection coefficients r and phases ϕ will be be calculated for both types of modes, TE and TM.

2.5.5.1 TE Polarization

Referring to Fig. 2.16 the E_{1i}, E_{1r}, E_{2t} complete values of incident, reflected and transmitted electric fields in medium 1 and 2. The incident electric field in medium 1(E_{1i}) is parallel to the interface between both media. Such orientation is known as TE polarization. It is often said that electric field vector **E** is normal to the plane of incidence. Such configuration is also known as *s-polarization*.

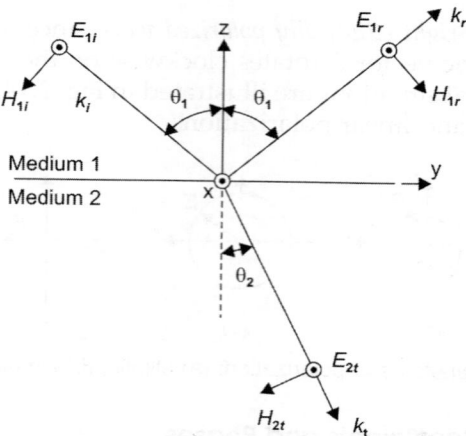

Fig. 2.16. Fresnel reflection for TE polarization. Directions of vector for TE polarized wave

Boundary conditions require that the tangential component of the total field **E** and the total field **H** on both sides of the interface be equal. These conditions result in the following equations:

$$E_{1i} + E_{1r} = E_{2t}$$

$$-H_{1i} \cos \phi_1 + H_{1r} \cos \theta_1 = -H_{2t} \cos \theta_2 \quad (2.55)$$

We also have the following relations involving impedances in both media:

$$\frac{E_{1i}}{H_{1i}} = Z_1, \frac{E_{1r}}{H_{1r}} = Z_1, \frac{E_{2t}}{H_{2t}} = Z_2 \quad (2.56)$$

For a non-magnetic medium ($\mu_r = 1$) the impedance can be written as $Z_0 = \sqrt{\frac{Z_0}{n}}$, with n being the refractive index of the medium and $Z_0 = \sqrt{\frac{\mu_0}{\varepsilon_0}}$, is the impedance of a free space. Replacing magnetic field in the second equation of Eq. (2.55) by the electric fields using Eq. (2.56) gives:

$$E_{1i} + E_{1r} = E_{2t} \quad (2.57)$$

$$-\frac{E_{1i}}{Z_1} \cos \theta_1 + \frac{E_{1r}}{Z_1} \cos \theta_1 = -\frac{E_{2t}}{Z_2} \cos \theta_2$$

Reflection coefficient is defined as:

$$r_{TE} = \frac{E_{1r}}{E_{1i}} \qquad (2.58)$$

From Eq. (2.57) by eliminating E_{2t} and using definition (2.58), one obtains:

$$r_{TE} = \frac{Z_2 \cos\theta_1 - Z_1 \cos\theta_2}{Z_2 \cos\theta_1 + Z_1 \cos\theta_2}$$

Replacing impedances Z by refractive indices n, we finally have:

$$r_{TE} = \frac{n_1 \cos\theta_1 - n_2 \cos\theta_2}{n_1 \cos\theta_1 + n_2 \cos\theta_2}$$

$$= \frac{n_1 \cos\theta_1 - \sqrt{n_2^2 - n_1^2 \cos\theta_2}}{n_1 \cos\theta_1 + \sqrt{n_2^2 - n_1^2 \cos\theta_2}} \qquad (2.59)$$

using Snell's law. For angles θ_1 such that $n_2^2 - n_1^2 \cos\theta_2 < 0$, the reflection becomes complex. For such cases, we write it as:

$$r_{TE} = \frac{n_1 \cos\theta_1 - j\sqrt{n_1^2 \cos\theta_1 - n_2^2}}{n_1 \cos\theta_1 + j\sqrt{n_1^2 \cos\theta_1 - n_2^2}} \equiv \frac{a - jb}{a + jb} \qquad (2.59a)$$

Such complex number can be expressed as:

$$r_{TE} = \frac{e^{-j\phi_{TE}}}{e^{j\phi_{TE}}} = e^{-2j\phi_{TE}} \qquad (2.59b)$$

where we have defined $a + jb = e^{-j\phi_{TE}}$. Finally using the definition of Fresnel phase, Eq. (2.54) one obtains:

$$\tan\phi_{TE} = \frac{\sqrt{n_1^2 \cos\theta_1 - n_2^2}}{n_1 \cos\theta_1} \qquad (2.60)$$

The above equation represents phase shift during reflection for TE polarized electromagnetic wave.

2.5.5.2 TM Polarization

Field configuration used to analyse reflection for TM polarization (magnetic field parallel to the interface) is shown in Fig. 2.17. Here, electric field vector E is parallel to the plane of incidence. Such configuration is also known as *p*-polarization.

We will drive coefficient of reflection r_{TM} for TM mode. As before E_{1i}, E_{1r}, E_{2t} are (complex) values of incident, reflected and transmitted

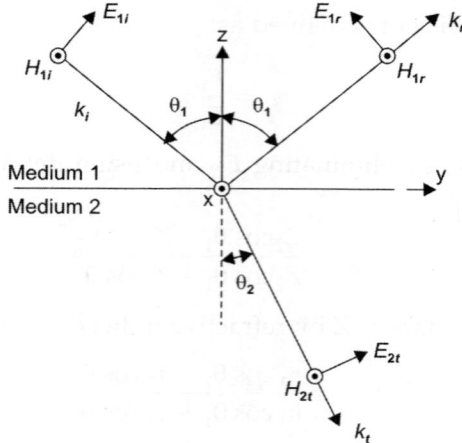

Fig. 2.17. Fresnel reflection for TM polarization. Directions of vectors for a TM polarized wave

electric fields in medium 1 and 2. Similar notation holds for magnetic vectors. Boundary conditions require that tangenital components are continuous across interface. The relevant conditions are

$$E_{1j} \cos \theta_1 - E_{1r} \cos \theta_1 = E_{2t} \cos \theta_2 \tag{2.61}$$

Using Eq. (2.56) for impedances to eliminate magnetic field in the previous equations results in the following:

$$E_{1i} \cos \theta_1 - E_{1r} \cos \theta_1 = E_{2t} \cos \theta_2 \tag{2.62}$$

$$\frac{E_{1i}}{Z_1} + \frac{E_{1r}}{Z_1} = \frac{E_{2t}}{Z_2} \tag{2.63}$$

Reflection coefficient is defined as:

$$r_{TM} = \frac{E_{1r}}{E_{1i}} \tag{2.64}$$

From Eq. (2.62):

$$r_{TM} = \frac{Z_1 \cos \theta_1 - Z_2 \cos \theta_2}{Z_1 \cos \theta_1 + Z_2 \cos \theta_2}$$

It can be also expressed in terms of refractive index:

$$r_{TM} = \frac{n_2 \cos \theta_1 - n_1 \cos \theta_2}{n_2 \cos \theta_1 + n_1 \cos \theta_2}$$

$$= \frac{n_2 \cos \theta_1 - \sqrt{n_1^2 - n_2^2} \cos \theta_1}{n_2 \cos \theta_1 + \sqrt{n_1^2 - n_2^2} \cos \theta_1} \tag{2.65}$$

TM phase is obtained in the same way as for TE polarization. Final result is:

$$\tan\phi_{TM} = \frac{n_1^2}{n_2^2} \times \frac{\sqrt{n_1^2 \cos\theta_1 - n_2^2}}{n_1 \cos\theta_1} \quad (2.66)$$

2.5.6 Polarization by Reflection from Dielectric Surfaces

In interpreting the formulas for r_{TE} and r_{TM}, we distinguish between two situations:

(1) for $n_1 < n_2$ or $n = \dfrac{n_2}{n_1} > 1$, one defines so-called *external reflection*.

(2) for $n_1 < n_2$ or $n = \dfrac{n_2}{n_1} < 1$, one defines so-called *internal reflection*.

Condition (1) is air-to-glass reflection and (2) is the glass-to-air reflection.

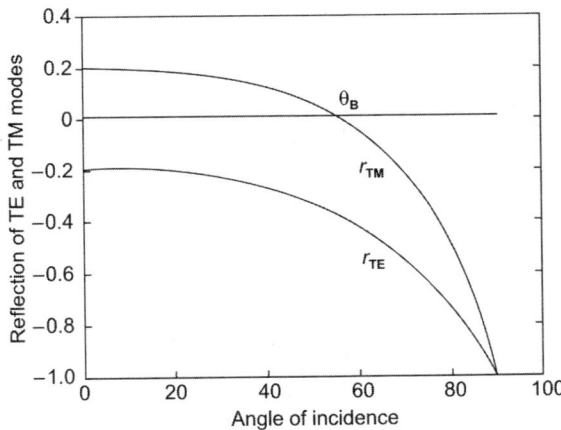

Fig. 2.18. Reflection of TE and TM modes for external reflection with $n = 1.50$ (Brewster's angle is also shown)

A plot of coefficients of reflection for $n = 1.50$ is shown in Fig. 2.18. The so-called *Brewster's angle* for which $r_{TM} = 0$ is also shown. One can obtain polarization by reflection at Brewster's angle, see Fig. 2.19. In the figure we illustrate properties of the reflected and transmitted light incident at the surface at Brewster's angle.

Unpolarized light is incident at the interface at Brewster's angle θ_B. Upon reflection, one obtains polarized TM light. At the same time, the refracted light is only partially polarized since r_{TE} is nonzero.

At Brewster's angle θ_B of incidence, coefficient of reflection r_{TM} (also known as r_{11} since E is parallel to the plane of incidence) is zero, i.e.

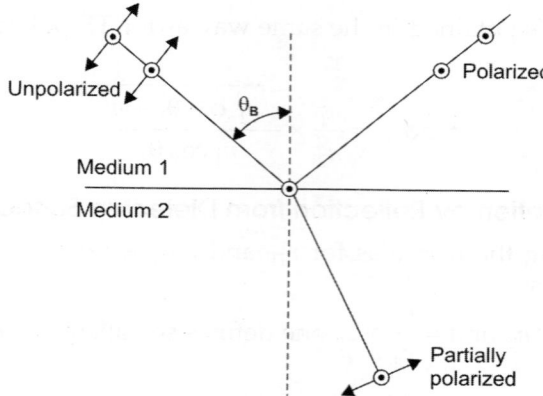

Fig. 2.19. Polarization at Brewster's angle

$r_{11} = 0$. The above happens when the sum of the angles of incidence and refraction is equal to $\pi/2$, i.e.:

$$\theta_1 + \theta_2 = \frac{\pi}{2}$$

No such angle exists for the s polarization (TE). Thus if an unpolarized light is incident at Brewster's angle, the reflected light will be linearly polarized, in fact s-polarized.

2.5.6.1 Expression for Brewster's Angle

Expression for r_{TM} is:

$$r_{TM} = \frac{n_2 \cos\theta_1 - \frac{n_1}{n_2}\sqrt{n_2^2 - n_1^2 \sin^2\theta_1}}{n_2 \cos\theta_1 + \frac{n_1}{n_2}\sqrt{n_2^2 - n_1^2 \sin^2\theta_1}} \qquad (2.67)$$

Vanishing for r_{TM} corresponds to $\theta_1 = \theta_B$. We have

$$n_2 \cos\theta_B - \frac{n_1}{n_2}\sqrt{n_2^2 - n_1^2 \sin^2\theta_B} = 0$$

from which we find an expression for Brewster's angle

$$\tan\theta_B = \frac{n_2}{n_1} \qquad (2.68)$$

2.5.7 Antireflection Coating

There is a need to reduce (or completely eliminate) the effect of reflections. Therefore, the practical question is how to eliminate reflections at the interface between two dielectrics.

Nature of Light and Foundation of Electromagnetic Theory

As seen before, light passing through a boundary between two dielectrics is lost due to reflection. The effect depends on the difference of the values of refractive indices between neighbouring layers. One can observe that for an equal refractive indices there will be no reflection but also no refraction, which is not a very interesting possibility.

A practical method of reducing reflections is to use several layers with properly selected values of refractive indices. For a single layer, interference of two reflected waves can lead to elimination of reflection, but only at a particular wavelength.

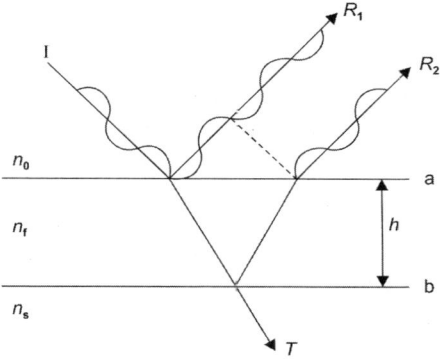

Fig. 2.20. Illustration of reflection from single dielectric film layer (case of destructive interference is shown)

Let us analyse this situation in more detail. Consider reflections of two waves R_1 and R_2 (Fig. 2.20). One of the conditions for destructive interference is that the amplitudes of both reflected waves should be equal. The reflection of ray R_1 at interface 'a' is described by reflection coefficient.

$$R = \left(\frac{n_0 - n_f}{n_0 + n_f}\right)^2 \tag{2.69}$$

whereas reflection of ray R_2 at interface 'b' is described by coefficient:

$$R = \left(\frac{n_s - n_f}{n_s + n_f}\right)^2 \tag{2.70}$$

Both coefficient should be equal which gives:

$$\left(\frac{n_0 - n_f}{n_0 + n_f}\right)^2 = \left(\frac{n_s - n_f}{n_s + n_f}\right)^2$$

Assuming that the upper medium is air, i.e. $n_0 = 1$, from the above one finds expression for the value of refractive index for a layer as:

$$n_f = \sqrt{n_s} \tag{2.71}$$

34 Photonics | Optoelectronics

Therefore, at a particular wavelength there will be no reflection once the refractive index of coating layer is given by Eq. (2.67). If there is a need to design structures producing no reflection over some frequency band, one must design multilayer structure consisting of layers with different refractive indices. In order to design such structures a more realistic description is necessary.

2.5.8 Bragg Mirrors

The *Bragg mirror*, also known as the *Bragg reflector*, consists of identical layers of dielectrics with high and low values of refractive indices as shown in Fig. 2.21. The main interest in fabricating such structures is that they have extremely high reflectivities at optical and infrared frequencies. They are important elements of VCSELs where high reflectivity and bandwidth are required. A typical structure forming Bragg mirror consists of N layers of dielectrics with refractive indices n_L (low refractive index) and n_H (high refractive index). The ratio of those values, the so-called *contrast ratio*, plays an important role.

The structure is known as a quarter-wave dielectric stack, which means that the optical thicknesses are quarter-wavelength long; that is $n_H.a_H = n_L.a_L = \lambda_0/4$ at some wavelength λ_0. The structure consists of an odd number of layers with the high index layer being the first and the last layers.

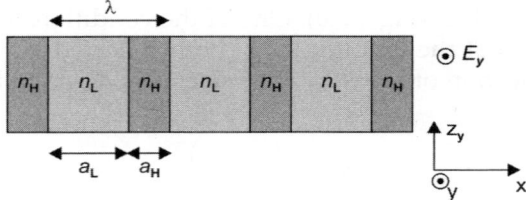

Fig. 2.21. Schematic of a seven-layer dielectric (Bragg) mirror; shown are $N = 3$ periods

Spectrum of reflectivity of a Bragg mirror is shown in Fig. 2.22 for $N = 10$ periods of Bragg reflector.

2.5.9 POYNTING THEOREM

Electromagnetic waves carry with them electromagnetic power. We will now derive a relation between the rate of such energy transfer and the electric and magnetic field intensities. We start with Maxwell's equations [Eqs (2.11) and (2.12)].

$$\nabla \times E = -\frac{\partial B}{\partial t} \qquad (2.72)$$

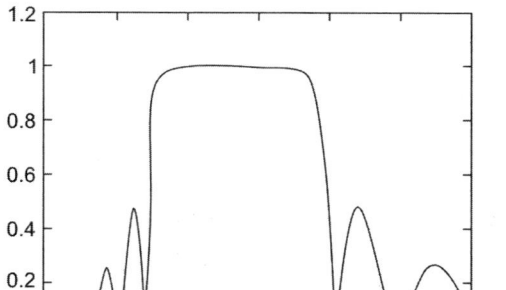

Fig. 2.22. Spectrum of reflectivity of Bragg mirror for $N = 10$ periods of Bragg reflector

$$\nabla \times H = \frac{\partial D}{\partial t} + J \qquad (2.73)$$

Using the following mathematical indentify:

$$\nabla \cdot (E \times H) = H \cdot (\nabla \times E) - E (\nabla \times H) \qquad (2.74)$$

Substituting Maxwell's equations into (2.74) and using constitutive relations, one obtains:

$$\begin{aligned}
\nabla \cdot (E \times H) &= -H \cdot \mu \frac{\partial H}{\partial t} - E \cdot J - E \cdot \mu \frac{\partial E}{\partial t} \\
&= -\mu \frac{1}{2} \frac{\partial H \cdot H}{\partial t} - \sigma E^2 - \varepsilon \frac{1}{2} \frac{\partial E \cdot E}{\partial t} \\
&= -\frac{\partial}{\partial t}\left(\frac{1}{2}\varepsilon E^2 + \frac{1}{2}\mu H^2\right) - \sigma E^2
\end{aligned} \qquad (2.75)$$

Integrate (2.75) over volume V and apply Gauss's theorem, one finds:

$$\int_V \nabla \cdot (E \cdot H) dV = -\frac{\partial}{\partial t} \int_V \left(\frac{1}{2}\varepsilon E^2 + \frac{1}{2}\mu H^2\right) - dV - \int_V \sigma E^2 dV \qquad (2.76)$$

The Poynting vector P is defined as:

$$P = E \times H \qquad (2.77)$$

Eq. (2.76) can be written as:

$$\oint_S P \cdot dS = \frac{\partial}{\partial t} \int_V (w_e + w_m) dV + \int_V p_\sigma dV \qquad (2.78)$$

where

$w_e = \frac{1}{2}\varepsilon E^2$ is electric energy density

$w_m = \frac{1}{2}\mu H^2$ is magnetic energy density

$p_\sigma = \sigma E^2$ is Ohmic power density

From Eq. (2.78), one can interpret vector P as representing the power flow unit area.

The Poynting vector is applied to time constant fields. The direction of the vector P indicates the direction of the instantaneous power flow at the point. Many of us think that Poynting vector P is a vector. This homogeneous is accidental, but correct. The Poynting vector P has the dimensions of *power per unit area* and its unit is Wm^{-2}. It is a vector, because it indicates not only the magnitude of energy but also its direction. It is Poynting theorem that the vector product or cross product $P = E \times H$ at any point is a measure of the time rate of energy flow per unit area of that point.

The direction of flow is perpendicular to E and H and is the direction of Poynting vector $P = E \times H$.

ILLUSTRATIVE EXAMPLES

Example 1

(a) Consider normal incidence of light on an air-silica interface. Compute the fraction of reflected and transmitted power. Also, express the transmitted loss in decibels. Assume refractive index of silica to be 1.45. (b) Repeat the calculation for Si which has $n = 3.50$. (c) Consider coupling of GaAs optical source with a refractive index of 3.6 to a silica fibre which has refractive index of 1.48. Assume close physical contact of the fibre end and the source.

Solution

(a) The corresponding coefficient known as reflectance

$$R = |r|^2 = \left(\frac{n_1 - n_2}{n_1 + n_2}\right)^2 \qquad (i)$$

Substituting values for air and silica, one obtains:

$$R = \left(\frac{1.45 - 1.00}{1.45 + 1.00}\right)^2 = 0.033 \qquad (ii)$$

so about 3% of the light is reflected. The remainder, 97% is transmitted.

The transmission loss in dB is:

$$\log_{10} 0.97 = 0.13 \text{ [dB]}$$

This result shows that there is about 0.2 dB loss when light enters glass from air.

(b) For Si, using the same procedure we obtain:

$$R = \left(\frac{1.00 - 3.50}{1.00 + 3.50}\right)^2 = 0.309$$

This means that about 31% of light is reflected.

(c) The Fresnel reflection at the interface is:

$$R = \left(\frac{3.60 - 1.48}{3.60 + 1.48}\right)^2 = 1.704$$

Therefore, about 17.4% of the created optical power is emitted back into the source. The power coupled into optical fibre is:

$$P_{coupled} = (1 - R) P_{emitted}$$

The power loss, α in decibels is:

$$\alpha = -10 \log_{10} \frac{P_{coupled}}{P_{emitted}}$$

$$= -10 \log_{10} (1 - R)$$

$$= -10 \log_{10} (1 - 0.174) - 10 \log_{10} (0.826)$$

$$= 0.83 \text{ dB}$$

Example 2

A lossless dielectric medium has $\sigma = 0$, $\mu_r = 1$ and $\varepsilon_r = 4$. An electromagnetic wave has magnetic field components expressed as:

$$H = -0.1 \cos(\omega t - z)\hat{i} + 0.5 \sin(\omega t - z)\hat{j} \text{ A/m}$$

Determine:

(a) the phase constant β,
(b) the angular velocity,
(c) the wave impedance, and
(d) the components of the electric field intensity of the wave.

Solution

Here:

$\sigma = 0$, $\mu_r = 1$

$\varepsilon_r = 4$, $\alpha = 0$ for lossless dielectric

(a) Looking at the equations of H, we can find β directly. $\beta = 1$ rad/m [as $H = H_m \cos(\omega t - \beta_z)$ type].

(b) $\beta = \omega\sqrt{\mu\varepsilon} = \omega\sqrt{\mu_0\varepsilon_0}\sqrt{\varepsilon_r\mu_r} = \dfrac{\omega}{c}\sqrt{4} = \dfrac{2\omega}{c}$

$\therefore \omega = \dfrac{\beta c}{2} = \dfrac{1\times 3\times 10^8}{2} = 1.5\times 10^8$ rad/s

(c) Wave impedance:

$$\eta = \sqrt{\dfrac{\mu}{\varepsilon}} = \sqrt{\dfrac{\mu_0}{\varepsilon_0}}\sqrt{\dfrac{\mu_r}{\varepsilon_r}} = 377\times\sqrt{\dfrac{1}{4}} = \dfrac{377}{2} \approx 60\pi\,\Omega$$

(d) $\mathbf{H} = -0.1\cos(\omega t - z)\,\hat{j}$ Am^{-1}

The wave is travelling in z-direction and has components of **H** in $\hat{i}$ and $\hat{j}$-direction, H_x and H_y, respectively, varying with respect to z. For finding **E**, we use Maxwell's equation for a lossless medium.

$$\nabla\times\mathbf{H} = \varepsilon\dfrac{\partial \mathbf{E}}{\partial t}$$

$$\nabla\times\mathbf{H} = \begin{vmatrix} \hat{i} & \hat{j} & \hat{k} \\ \dfrac{\partial}{\partial x} & \dfrac{\partial}{\partial y} & \dfrac{\partial}{\partial z} \\ H_x & H_y & H_z \end{vmatrix} = -\dfrac{\partial H_y}{\partial z}\hat{i} + \dfrac{\partial H_y}{\partial z}\hat{j}$$

$$\mathbf{E} = \dfrac{1}{\varepsilon}\int(\nabla\times\mathbf{H})\,dt = \dfrac{0.5}{\varepsilon\omega}\sin(\omega t - z)\hat{i} + \dfrac{0.1}{\varepsilon\omega}\cos(\omega t - z)\hat{j}$$

or $\mathbf{E} = 94.12\sin(\omega t - z)\hat{i} + 18.83\cos(\omega t - z)\hat{j}$ V/m

Example 3

The electric field intensity of a uniform plane wave in air is 7500 Vm^{-1} in the $\hat{j}$ direction.

The wave is propagating in the $\hat{i}$ direction at a frequency of 2×10^9 rad/s. Determine: (i) the wavelength of the wave, (ii) the frequency, (iii) the time period, and (iv) the amplitude of **H**. (Take $v = 3\times 10^8$ ms^{-1}).

Solution

Here:

$E_y = 7500\cos(2\times 10^9\, t - \beta x)$ in air
$\varepsilon_0 = 8.854\times 10^{-12}$ F/m, $\mu_0 = 4\pi\times 10^{-7}$ Hm^{-1}

(i) $\lambda = \dfrac{v}{\nu} = \dfrac{3\times 10^8}{(2\times 10^9/2\pi)} = 0.943$ m

(ii) $v = \dfrac{\omega}{2\pi} = \dfrac{2 \times 10^9}{2\pi} = 318.3$ MHz

(iii) $T = \dfrac{1}{v} = \dfrac{10^{-6}}{318.3} = 3.142$ ns

(iv) $\dfrac{E}{H} = \eta = \sqrt{\dfrac{\mu_0}{\varepsilon_0}} \cong 377\,\Omega$

E_y and H_z components shall exist.

$|H| = \dfrac{7500}{377} = 19.9$ A/m

This gives $H_z = 19.9 \cos(2 \times 10^9 t - \beta x)$.

Example 4

The electromagnetic wave intensity received on the surface of the earth from the sun is found to be 1.33 kW/m². Find the amplitude of electric field vector associated with sunlight as received on earth surface. Assume sun's light to be monochromatic ($\lambda = 6000$Å).

Solution

The energy transported by an electromagnetic wave per unit area per second during propagation is represented by Poynting vector $\mathbf{P}$ as:

$$\mathbf{P} = \mathbf{E} \times \mathbf{H}$$

The energy flux per unit area per second is:

$$|\mathbf{P}| = |\mathbf{E} \times \mathbf{H}| = EH \sin 90° = EH$$

The energy flux per unit area per second at the earth surface:

$$|\mathbf{P}| = 1.33 \text{ kW/m}^2$$
$$= 1.33 \times 10^3 \text{ Jm}^{-2}\text{ s}^{-1}$$

or $\quad EH = 1330$ Jm^{-2} s^{-1} \hfill (i)

We know that:

$$\eta = \dfrac{E}{H} = \sqrt{\dfrac{\mu_0}{\varepsilon_0}} = \sqrt{\dfrac{4\pi \times 10^{-7} \text{ Wb/A·m}}{8.854 \times 10^{-12} \text{ C}^2/\text{Nm}^2}}$$

or $\quad \dfrac{E}{H} = 376.72\,\Omega \cong 377\,\Omega$ \hfill (ii)

Multiplying Eqs (i) and (ii), one obtains:

$$EH \times \dfrac{E}{H} = 1330 \times 37$$

or $\quad E^2 = 501410$ or $E = 708.1$ VM^{-1}

Substituting this value in Eq. (i), one obtains:

$$H = \frac{1330}{708.1} = 1.878 \text{ Am}^{-1}$$

∴ The amplitude of electric and magnetic fields of radiation are:

$$E_0 = E\sqrt{2} = 708.1\sqrt{2} = 1001.4 \approx 1001 \text{ Vm}^{-1}$$

and $$H_0 = H\sqrt{2} = 1.878\sqrt{2} = 2.65 \text{ Vm}^{-1}$$

Example 5

For a lossy dielectric material having $\mu_r = 1$, $\varepsilon_r = 48$, $\sigma = 20$ S/m, calculate the attenuation constant, phase constant and intrinsic impedance at a frequency of 16 GHz.

Solution

We have:

$$\frac{\sigma}{\omega\varepsilon} = \frac{20 \times 10^{12}}{2\pi \times 16 \times 10^9 \times 48 \times 8.856} = 0.47$$

$$\gamma = i\omega\sqrt{\mu\varepsilon}\sqrt{1 - i\frac{\sigma}{\omega\varepsilon}}$$

$$= i(2\pi) 16 \times 10^9 \sqrt{4\pi \times 10^{-7} \times 48 \times 8.854 \times 10^{-12}} \times \sqrt{1 - i(0.47)}$$

$$= i2323.25\sqrt{1.0966} \angle -24.23°$$

$$= 2432.88 \angle 77.89° = 510.4 + i2378.7 \text{ m}^{-1}$$

$\alpha = 510.4$ Np/m

$\beta = 2378.7$ rad/m

The intrinsic impedance:

$$\eta = \sqrt{\frac{i\omega\mu}{\sigma + i\omega\varepsilon}} = \sqrt{\frac{\mu}{\varepsilon\left(1 + \frac{\sigma}{i\omega\varepsilon}\right)}}$$

$$= \sqrt{\frac{4\pi \times 10^{-7}}{48 \times 8.854 \times 10^{-12}}} \times \sqrt{\frac{1}{1 - j(0.47)}}$$

$$= \frac{54.377}{\sqrt{1.0966} \angle -24.23°}$$

$$= 51.93 \angle 12.12° \, \Omega$$

The electric field (E_y) leads the magnetic field (H_z) by 12.12° at every point.

Example 6
Find the skin depth δ of an electromagnetic wave in copper at $\nu = 60$ Hz. For copper $\sigma = 5.8 \times 10^7$ mho/m, and $\mu_r = 1$.

Solution
For copper at $\nu = 60$ Hz:

$$\frac{\sigma}{\omega\varepsilon} = \frac{5.8 \times 10^7}{2\pi \times 60 \times 8.854 \times 10^{-12}}$$

$$= 1.74 \times 10^{14} >> 1$$

Therefore, at $\eta = 60$ Hz, copper is a very good conductor. The depth of penetration

$$\delta = \frac{1}{\beta} = \sqrt{\frac{2}{\omega\mu\sigma}}$$

$$= \sqrt{\frac{2}{2\pi \times 60 \times 4\pi \times 10^{-7} \times 5.8 \times 10^7}}$$

$$= 8.53 \times 10^{-3} \text{ m}.$$

SUGGESTED READINGS
1. Kakani SL and Hemrajani C, 'Electromagnetics', CBS Publications, New Delhi-2 (2016).
2. Sadiku MNO, 'Elements of Electromagnetics', Oxford University Press, Newyork, 2nd Ed. (1995).
3. Cullwick EG, 'The Fundamentals of Electromagnetism, 3rd Ed., Cambridge University Press (1966).
4. Kakani SL and Bhandari KC, 'Optics', Sultan Chand & Sons, New Delhi-2, 2nd Ed. (2015).
5. de Groot SR, The Maxwell's Equations, North Holland, Amsterdam (1969).

GLIMPSES
- **Light:** The form of electromagnetic radiation to which the human eye is sensitive and on which our visual awareness of the universe and its contents relies (colour).

  Light may be regarded as a form of *electromagnetic radiation* consisting of interdependent mutually perpendicular transverse oscillations of an electric and magnetic field. It forms a narrow section of the electromagnetic spectrum, the wavelength range (for normal vision) being approximately 390 nm (violet) to 740 nm (red). According to *quantum theory*, light is absorbed in packets of light quanta or photons (energy = $h\nu$).

The existence of two theories, i.e. wave theory and quantum theory to account for light is referred to as the *wave particle duality*. The wave theory and quantum theory of light are regarded as complementary; phenomena involving the propagation of light can be interpreted adequately on the wave basis, but when interactions of light (e.g. photoelectric effect, etc.) are under consideration, the quantum theory has to be employed.

During the course of evolution of wave mechanics it has become evident that *electrons* and other *elementary particles* have dual wave and quantum properties.

- The observed behaviour of light as it travels from one place to another, and its interaction with matter had led to the development of a lot of technological tools, e.g. microscope, telescope, high speed cameras, etc. All these instruments use the intricate behaviour of light as it travels from one place to another, and as it interacts with material objects.

- *Electromagnetic (em) spectrum*: An orderly arrangement of radiation according to wavelength (λ), frequency (ν), or energy is known as spectrum. The em spectrum is such a spectrum. In it, the components of radiation are arranged in order of frequency and wavelength.

  All the radiations constituting the e.m. spectrum travel at the same speed in vacuum ($c = 3 \times 10^8$ ms^{-1}).

- According to *Maxwell's* electromagnetic theory, the changing fields produced by the oscillating charges result in electromagnetic disturbances that travel through space as waves. The waves sent out by oscillating charges are viewed as fluctuating electric and magnetic fields, and hence, they are called em waves. These waves travel with the speed of light ($c = 3 \times 10^8$ ms^{-1}) in vacuum.

- EM waves are transverse waves.

  $\dfrac{E}{B} = c$ (velocity of light in vacuum).

- EM waves carry both energy and momentum, which can be delivered to the surface.

- *Maxwell's equations*: A series of classical equations that govern the behaviour of em waves in all practical situations. They connect vector quantities applying to any point in a varying electric or magnetic field. The equations are:

$$\nabla \cdot D = \rho \tag{1}$$

$$\nabla \cdot B = 0 \tag{2}$$

$$\nabla \times E = -\frac{\partial B}{\partial t} \tag{3}$$

$$\nabla \times \mathbf{H} = \mathbf{J} + \frac{\partial \mathbf{B}}{\partial t} \qquad (4)$$

where
$D = \epsilon E$
$B = \mu H$
$J = \sigma E$

From these equations, Maxwell demonstrated that each field vector E or B obeys a wave equation, he showed that when a varying electric field exists, it is accompanied by a varying magnetic field induced at right angle, and *vice versa*, and the two form an electromagnetic field that could propagate as a transverse wave. He showed that in vacuum, the speed of the wave, $c = \dfrac{1}{\sqrt{\varepsilon_0 \mu_0}}$, where ε_0 and μ_0 are the permittivity and permeability of vacuum respectively.

Maxwell's equation [Eq. (1)] represents Coulomb's law; Eq. (2) represents the absence of *magnetic monopoles*; Eq. (3) represents Faraday's laws of electromagnetic induction and Eq. (4) represents a generalization of Ampere's law.

- Integral forms of Maxwell's equations:

$$\oint_S \mathbf{D}.d\mathbf{S} = \int_V \rho dV$$

$$\oint_S \mathbf{B}.d\mathbf{S} = 0$$

$$\oint_L \mathbf{E}.d\mathbf{l} = -\int_S \frac{\partial \mathbf{B}}{\partial t}$$

$$\oint_L \mathbf{H}.d\mathbf{l} = -\int_S \left(\mathbf{J} + \frac{\partial \mathbf{D}}{\partial t} \right) d\mathbf{S}$$

- Solving Maxwell's equation, one obtains wave equations of E and H as:

$$\nabla^2 E = \mu\varepsilon \frac{\partial^2 E}{dt^2}$$

$$\nabla^2 H = \mu\varepsilon \frac{\partial^2 H}{dt^2}$$

In general case, em wave propagation involves electric and magnetic fields having more than one components, each dependent on all three coordinates.

- The ratio of E and H in a wave is denoted by η, and is called *intrinsic impedance*.

$$\eta = \frac{E}{H} = \sqrt{\frac{\mu}{\varepsilon}}$$

For free space:

$$\eta = \sqrt{\frac{\mu_0}{\varepsilon_0}} \cong 377 \text{ or } 120\pi$$

- *Maxwell's equations when the material is lossy*, i.e. it will have finite conductivity σ:

$\nabla \cdot D = 0$

$\nabla \cdot E = 0$

$\nabla \times E = -i\omega\mu H$

$\nabla \times H = (\sigma + i\omega\varepsilon)E$

and $\nabla \times (\nabla \times E) = \nabla \cdot (\nabla \cdot E) - D^2 E$

On solving, one gets intrinsic impedance:

$$\eta = \frac{E_y}{H_z} = \frac{i\omega\mu}{\gamma}$$

where $\gamma = i\omega\sqrt{\mu\varepsilon}\sqrt{1 - \frac{i\sigma}{\omega\varepsilon}}$

$\therefore \quad \eta = \sqrt{1 - \frac{i\omega\mu}{(\sigma + i\omega\varepsilon)}}$

Obviously, η is a complex quantity.

In the loss dielectrics, the electric field leads the magnetic field in time phase.

- In electromagnetics, materials are roughly divided into two classes: (i) *conductors*, and (ii) *dielectrics* or *insulators*. We may note that the dividing line between two classes of materials in not sharp and some media are considered as conductors in one part of radio frequency range, and as dielectric (with loss) in another part of the range.

- *Skin death*: In a medium of high conductivity, the wave is attenuated as its progress due to those losses which occur in the medium. In a good conductor, the rate of attenuation is very great and the wave may penetrate only a very short distance before being reduced to a negligible small percentage of its original strength. The *depth of penetration* or *skin depth* (δ) is defined as the depth in which the wave has been attenuated to $1/e$ or approximately 37% of its original value.

$$\delta = \frac{1}{\omega\sqrt{\frac{\mu\varepsilon}{2}\left[1 + \frac{\sigma^2}{\omega^2\varepsilon^2} - 1\right]}}$$

For good conductor

$$\frac{\sigma}{\omega\varepsilon} \gg 1, \text{ so}$$

$$\delta = \sqrt{\frac{2}{\omega\mu\sigma}} = \sqrt{\frac{1}{\pi\nu\mu\sigma}}$$

For copper $\sigma = 5.8 \times 10^7$ Sm^{-1}, $\mu = \mu_0$ at $f = 1$ MHz:

$$\therefore \delta = \frac{1}{\sqrt{\pi \times 10^6 \times 4\pi \times 10^{-7} \times 5.8 \times 10^7}} = 0.0661 \text{ mm}$$

Value at 50 Hz

$$\delta = \frac{1}{\pi \times 50 \times 4\pi \times 10^{-7} \times 5.8 \times 10^7} = 9.35 \text{ mm}$$

At *microwave* frequency 10,000 MHz, $\delta = 6.61 \times 10^{-4}$ mm, which is about 1/8th of the wavelength of visible light.

- *Poynting vector*: The vector P giving the direction and magnitude of energy flow in an electromagnetic field
  $P = E \times H$,
  where E and H are mutually orthogonal vectors of the electric and magnetic fields, respectively. If energy dissipation takes place during the propagation, this vector becomes complex with a real part equal to the average energy flow.

REVIEW QUESTIONS

1. What is the nature of light? Explain briefly.
2. What do you understand by dual nature of light? Explain with examples.
3. Write Maxwell's equations in differential and integral forms. Derive the wave equation from these equations for free space and charge free region.
4. Show that for uniform plane waves in a perfect dielectric medium, E and H are normal to each other and also show that the ratio of their magnitude is constant of the medium. Write the name of the constant and explain its significance.
5. Define skin depth (δ). Show that in case of a semifinite solid conductor δ is given by:

$$\delta = \sqrt{\frac{2}{\omega\mu\sigma}}$$

Symbols have their usual meanings.

6. State Poynting theorem and show that Poynting vector:
   $P = E \times H$

7. Using Maxwell's equations, show that in free space, the electromagnetic wave propages with the speed of light, i.e.:
$$v = \frac{1}{\sqrt{\mu_0 \varepsilon_0}} = 3 \times 10^8 \text{ ms}^{-1} = c$$

8. Enumerate Maxwell's equations and show that they predict existence of electromagnetic waves.

9. What are polarization sensitive materials?
Give examples of few common birefrigent crystals with their applications.

PROBLEMS

1. If the earth receives 2 cal min^{-1} cm^{-2} solar energy, calculate the amplitudes of the electric and magnetic fields of radiation?
   [**Ans.** $E_0 = E/\sqrt{2} = 1024.3$ Vm^{-1},
   $H_0 = H/\sqrt{2} = 2.717$ Am^{-1}]

2. Show that a linearly polarized plane wave can be decomposed into a right hand and a left-hand circularly polarized waves of equal amplitudes.

3. Assume that for some materials refractive index at a particular wavelength is negative. Discuss the consequences of such an assumption. Consider modification of Snell's law.

4. A uniform plane wave in a medium having $\sigma = 10^{-3}$ Sm^{-1}, $\varepsilon_r = 80\ \varepsilon_0$ and $\mu_r = \mu_0$ is having a frequency 10 kHz. Calculate the various parameters of the wave.
   [**Ans.** $\alpha = 2\pi \times 10^{-3}$ Np m^{-1}
   $\beta = 2\pi \times 10^{-3}$ rad m^{-1}, $\eta = 2\pi\ (1 + j)\Omega$
   $\lambda = 1000$ m, $v = 10^7$ ms^{-1}]

5. The relative permeability of distilled water is 81. Show that the refractive index and the velocity of light in it are $n = 9.0$ and $v = 3.33 \times 10^7$ ms^{-1} respectively.

6. Calculate the skin depth (δ) for an e.m. wave of frequency 1 MHz travelling through copper. Given $\alpha = 5.8 \times 10^7$ Sm^{-1} and $n = 4\pi \times 10^{-7}$.
   [**Ans.** $\delta = 66\ \mu$m]

7. A plane e.m. wave is travelling in the positive Z direction in an unbounded lossless dielectric medium with relative permeability $\mu_r = 1$ and relative permittivity $\varepsilon_r = 3$ has an electric field intensity $E = 6$ Vm^{-1}. Calculate.
   (i) the speed of e.m. waves in the given medium, and
   (ii) the impedance of the medium.
   [**Ans.** (i) $v = 1.732 \times 10^8$ ms^{-1}
   (ii) $\eta = 2.17 \times 10^2\ \Omega$]

SHORT ANSWER QUESTIONS

1. What is the basis of Maxwell's em theory?

Ans. The theory developed by Maxwell is based on the following statements:
   (i) Electric fields originate on positive charges and terminate on negative charges. The electric field due to point charges can be determined at a location by applying Coulomb's force law to positive test charge placed at that location.
   (ii) Magnetic field line always form closed loops, i.e. they do not begin or end anywhere.
   (iii) A varying magnetic field induces an e.m.f. and, hence, an electric field. This is Faraday's law.
   (iv) Magnetic fields are generated by moving charges (or currents), as summarized in Ampere's law.

2. What is quantum nature of light?

Ans. Light energy is always emitted or absorbed in discrete units called *quanta* or *photons*. Energy of photon depend only on frequency ν of a photon and is given by:
$$E = h\nu$$

3. Write basic optical laws.

Ans. (i) *Refractive index (n)*: The ratio of the speed of light in a vacuum to that in matter is the index of refraction of material and is given by:
$$n = c/v$$
(ii) *Snell's law*: The relationship at the interface is known as Snell's law and is given by:
$$n_1 \sin\phi_1 = n_2 \sin\phi_2$$
where n_1 and n_2 are refractive indices of medium 1 and medium 2 respectively. ϕ_1 is the angle of incidence between the incident ray and the normal to the surface in medium 1 and angle ϕ_2 is angle of refraction between the normal and refracted ray in medium 2.

4. Write the important properties of e.m. waves propagating through free space.

Ans. (i) EM waves travel with speed of light in vacuum ($c = 3 \times 10^8$ ms^{-1})
   (ii) EM waves are transverse in nature
   (iii) $E/B = c$ (velocity of light)
   (iv) EM waves carry both energy and momentum.

5. What is intrinsic impedance? What is its value for free space?

Ans. $\eta = \dfrac{E}{H} = \sqrt{\dfrac{\mu}{\varepsilon}}$

For free space $\eta = \sqrt{\dfrac{\mu_0}{\varepsilon_0}} \cong 377$ or 120π

6. In electromagnetics, how materials are divided?

Ans. (i) Conductors, (ii) Dielectrics or insulators.

The dividing line between two classes is not sharp and some media are considered as conductors in one part of radio frequency range, and as dielectric (with loss) in another part of the range.

7. What is Poynting vector?

Ans. $P = E \times H$

P is interpreted as an instantaneous power density that is measured in Wm^{-2}, P can be used to determine total power crossing the surface in an outward sense.

The Poynting vector is applied to time constant fields.

8. Draw a figure of plane polarized em wave and mention its important features.

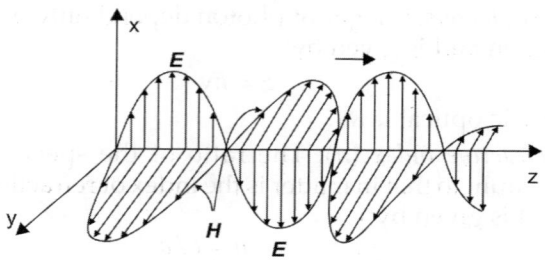

Fig. 2.23. An instantaneous snapshot of a plane-polarized transverse wave showing the electric and magnetic field vectors **E** and **H**, with the wave moving to the right

Ans. Figure 2.23 shows plane-polarized electromagnetic wave. We observe that:

(i) Vibrations of its electric field vector E are parallel to each other for all points in the wave.

(ii) At all these points (i), the vibrating E vector and the direction of propagation form a plane called the *plane of vibration* or the *plane of polarization*. All such planes are parallel in a plane-polarized wave.

(iii) In this case the resultant electric-field vector may be resolved in two components which have a random phase difference. If the wave is propagating along the z-axis and the two components are along the x and y axes and have the same amplitudes with a phase difference of $\pi/2$, the wave is said to be circularly polarized. However, if the amplitudes of the two components are not same, but their phase difference is still $\pi/2$, the resultant wave is said to be elliptically polarized.

Nature of Light and Foundation of Electromagnetic Theory

9. **How velocity of propagation of light is affected in anisotropic materials, e.g. KDP, etc.?**

Ans. In optically anisotropic materials, the velocity of propagation of light depends on the direction as well as the state of polarization. Thus, there are two principal refractive indices: One corresponding to the ordinary (o) ray (which follows Snell's law) and other corresponding to the extraordinary (e) ray (which does not follow Snell's law).

10. **What are uniaxial crystals? Give two examples.**

Ans. Crystals, which have two principal refractive indices and one optic axis (the direction along which the velocity of both the rays is same) are termed *uniaxial crystals*. Example are calcite and quartz.

11. **What is Pockel's electro-optic effect?**

Ans. When an electric field is applied across a birefringent crystal, e.g. KDP, this may change its refractive indices. When this change is linearly proportional to the electric field, the effect is called the *Pockel's electro-optic effect*. This can be used for phase modulation.

12. **What happens when a voltage V is applied along the optic axis of KDP crystal?**

Ans. KDP is a birefringent crystal. When a voltage V is applied along the optic axis of this crystal, the incident plane-polarized light splits into two components and the net phase shift between them is given by:

$$\phi = \frac{2\pi}{\lambda} r_{63} n_0^3 V$$

MULTIPLE CHOICE QUESTIONS

1. The mass of photon of wavelength λ is given by:
   (a) $hc\lambda$
   (b) $h/\lambda c$
   (c) $\dfrac{hc}{\lambda}$
   (d) $\dfrac{h\lambda}{c}$

   [**Hint.** We have $hc = mc^2$ or $\dfrac{hc}{\lambda} = mc^2$ ($\because c = \nu\lambda$). Hence $m = h/\lambda c$)

2. A photon has energy $E = h\nu$ and momentum $p = \dfrac{h}{\lambda}$. The speed of light in terms of E and p is:
   (a) Ep
   (b) $\sqrt{EP}$
   (c) p/E
   (d) E/p

   [**Hint.** $\dfrac{E}{p} = \dfrac{h\nu}{h/\lambda} = \nu\lambda = c$]

3. The momentum of a photon of wavelength λ is:
   (a) $\dfrac{h}{c\lambda}$
   (b) $\dfrac{hc}{\lambda}$
   (c) $h\lambda$
   (d) h/λ

   [**Hint.** $E = h\nu = mc^2 \therefore m = \dfrac{h\nu}{c^2} = \dfrac{h}{c\lambda}$
   $\therefore$ Momentum of photon, $mc = h/\lambda$]

4. The momentum of a particle of mass m and charge q is equal to that of photon of wavelength λ. The speed of particle is given by:
   (a) $qh\lambda$
   (b) $\dfrac{mh}{\lambda}$
   (c) $\dfrac{h}{m\lambda}$
   (d) $\dfrac{h\lambda}{m}$

   [**Hint.** We have $mv = \dfrac{h}{\lambda} \therefore v = \dfrac{h}{m\lambda}$]

5. The concept that a changing electric field in a conductor produces induced magnetic field was proposed by:
   (a) Faraday
   (b) Biot–Savart
   (c) Maxwell
   (d) Oersted

6. The concept of displacement current was proposed by:
   (a) Maxwell
   (b) Faraday
   (c) Ampere
   (d) Gauss

7. Maxwell's modified Ampere's law is valid:
   (a) only when electric field does not change with time
   (b) only when electric field varies with time
   (c) in the above situations
   (d) none of the above

8. The dimensions of Planck's constant (h) are:
   (a) ML^2T^{-2}
   (b) $ML^{-3}T^{-5}$
   (c) ML^2T^{-1}
   (d) being a constant, h has no dimension

   [**Hint.** Dimensions of h = dimensions of energy $\times$ dimensions of time = $ML^2T^{-2}T = ML^2T^{-1}$]

9. Which is the incorrect statement about the electromagnetic waves:
   (a) the electromagnetic field vector E and B are mutually perpendicular and they are also perpendicular to the direction of propagation of the electromagnetic wave

Nature of Light and Foundation of Electromagnetic Theory 51

  (b) the field vector E and H are in same phase
  (c) the field vector E and H are along the same direction
  (d) electromagnetic waves are transverse in nature
10. Which one is the incorrect statement about the electromagnetic waves:
  (a) the energy density associated with the electromagnetic wave in free space propagates with a speed less than the speed of light
  (b) in free space, the electromagnetic waves travel with the speed of light
  (c) the direction of flow of electromagnetic energy along the direction of propagation of wave
  (d) the electrostatic energy density is equal to magnetic energy density
11. Poynting vector P is:
  (a) $E \cdot H$
  (b) $E \times H$
  (c) E/H
  (d) $\oint (E \times H) \cdot dS$
12. Intrinsic impedance of dielectric with $\sigma = 0$ is:
  (a) μ/ε
  (b) $\sqrt{\dfrac{\mu}{\varepsilon}}$
  (c) $\sqrt{\mu\varepsilon}$
  (d) $1\sqrt{\mu\varepsilon}$
13. For good dielectrics:
  (a) $\dfrac{\sigma}{\omega\varepsilon} = 0$
  (b) $\dfrac{\sigma}{\omega\varepsilon} \gg 1$
  (c) $\dfrac{\sigma}{\omega\varepsilon} = 1$
  (d) $\dfrac{\sigma}{\omega\varepsilon} \ll 1$
14. The skin depth for a good conductor is:
  (a) $\sqrt{\dfrac{2}{\omega\mu\sigma}}$
  (b) $\sqrt{2\omega\mu\sigma}$
  (c) $\sqrt{\dfrac{\omega\mu\sigma}{2}}$
  (d) $\dfrac{2}{\sqrt{\omega\mu\sigma}}$
15. The intrinsic impedance of good conductor is given by:
  (a) $\sqrt{\dfrac{\mu}{\varepsilon}}\left(1 + \dfrac{i\sigma}{\sqrt{2\omega\varepsilon}}\right)$
  (b) $\sqrt{\dfrac{i\omega\mu}{\sigma}}$
  (c) $\sqrt{\dfrac{i\omega\mu}{\sigma + i\omega\mu}}$
  (d) $\sqrt{i\omega\mu\sigma\left(1 + i\dfrac{\omega\mu}{\sigma}\right)}$

16. A monochromatic electromagnetic wave means that:
    (a) the field strength at a point varies with time according to sine and cosine function
    (b) the wave always travels in the same direction
    (c) electric field vector E lies in one direction only
    (d) magnetic field vector B must be perpendicular to the direction of propagation
17. "The net power flowing out of a given volume V is equal to the time rate of decrease in energy stored within V minus the ohmic losses" is the statement of:
    (a) Gauss's theorem   (b) Stoke's theorem
    (c) Poynting theorem   (d) none of these
18. The depth of penetration is the depth in which the electromagnetic wave has been attenuated to:
    (a) e of the original value   (b) $\frac{1}{e}$ of the original value
    (c) 50% of the original value (d) 100% of the original value
19. A half-wave plate will introduce a path difference (between two emerging beams) of:
    (a) $\lambda/4$ (b) $\lambda/2$
    (c) $3\lambda/4$ (d) λ
20. In a longitudinal electro-optic modulator, half-wave voltage is that voltage which introduces the following phase shift between two polarization components:
    (a) $\pi/4$ (b) $\pi/2$
    (c) π (d) 2π
21. In a transverse electro-optic modulator:
    (a) V_π is independent of the length l and width d of the modulator crystal
    (b) V_π is dependent on the length l but not on the width d of the crystal
    (c) V_π is dependent on the width d but not on the length l of the crystal
    (d) V_π is dependent on the ratio d/l
22. In a Raman-Nath modulator, the acousto-optic grating is:
    (a) so thin that it behaves almost like a plane transmission grating
    (b) so thick that it behaves almost like Bragg's crystal grating
    (c) analogous to a concave Rowland's grating
    (d) quite complicated
23. Consider an electromagnetic wave travelling in the z-direction. The orthogonal components of its resultant E-vector are along

the x- and y-direction. Assume that the two components have same amplitudes but a phase difference of $\pi/2$. The resultant E-vector at any point in space:
(a) is constant in amplitude but rotates with an angular frequency ω
(b) changes in amplitude but rotates with an angular frequency ω
(c) remains stationary
(d) varies randomly

24. The electromagnetic wave of Question 23, is said to be:
(a) unpolarized
(b) plane-polarized
(c) circularly polarized
(d) elliptically polarized

25. In a birefringent crystal:
(a) the o-ray follows Snell's law but the e-ray does not
(b) the e-ray follows Snell's law but the o-ray does not
(c) both o-ray and e-ray follow Snell's law
(d) both the o-ray and e-ray do not follow Snell's law

26. In a doubly refracting crystal, the optic axis is the direction in which:
(a) v_0 is greater than v_e
(b) v_0 is equal to v_e
(c) v_0 is less than v_e
(d) v_0 and v_e vary randomly

27. A uniaxial crystal has:
(a) one principal refractive index and no optic axis
(b) one principal refractive index and one optic axis
(c) two principal refractive indices and one optic axis
(d) three principal refractive indices and two optic axes

28. A quarter-wave plate will introduce a phase difference (between two emerging beams) of:
(a) $\pi/4$
(b) $\pi/2$
(c) π
(d) $3\pi/2$

ANSWERS

1. (b) 2. (d) 3. (d) 4. (c) 5. (c) 6. (a) 7. (c)
8. (c) 9. (c) 10. (a) 11. (b) 12. (b) 13. (d) 14. (a)
15. (b) 16. (a) 17. (c) 18. (b) 19. (b) 20. (c) 21. (d)
22. (a) 23. (a) 24. (c) 25. (a) 26. (b) 27. (c) 28. (b)

3

Optical Fiber Waveguides

3.1 INTRODUCTION

In Chapter 2, we have read that a uniform plane electromagnetic (em) wave propagates in an unbounded medium as a transverse electromagnetic wave (TEM) having E and H (or B) vectors both perpendicular to each other and to the direction of propagation. We have also examined the boundary effects on the propagation of an em wave. We shall now discuss the behaviour of such waves in the vicinity of the boundaries so configured as to guide that wave along a certain path. Such systems are called *waveguides*. The most efficient way of transmitting energy over short distances is by using wave guides. Wave guides were first evolved and used in practical electronics and now assumed great importance in fiber optic communication – *optoelectronics*.

In waveguide, both electric and magnetic fields are confined to space within the waveguide. The dielectric loss within the waveguide is negligibly small because the guides are normally air filled. Within a waveguide several modes of electromagnetic waves could be propagated. The mode of propagation of a waveguide is determined from solutions of Maxwell's equations.

3.2 CONCEPT OF ELECTROMAGNETIC WAVES

We will briefly summarize the results of the electromagnetic theory discussed in Chapter 2. Maxwell's set of four equations are:

$$\nabla \times E = -\frac{\partial B}{\partial t} \tag{3.1}$$

$$\nabla \times H = J + \frac{\partial D}{\partial t} \tag{3.2}$$

$$\nabla \cdot D = \rho \tag{3.3}$$

$$\nabla \cdot B = 0 \tag{3.4}$$

where, E is the electric field vector in $V \cdot m^{-1}$, H is the magnetic field vector in $A \cdot m^{-1}$, D is electric displacement vector in $C \cdot m^{-2}$, it is related to E by the relation $D = \varepsilon E$, where E is the dielectric permittivity of the medium. B is the magnetic induction vector in H·m^{-1}, it is related to H by the relation $B = \mu H$, where μ is the magnetic permeability of the medium. ρ is the charge density of the medium in C·m^{-3}. J is the current density in A·m^{-2}, it is related to E by the relation $J = \sigma E$, where σ is the conductivity of the medium in A·V^{-1}·m^{-1}. ∇ is 'del' operator defined as:

$$\nabla = \hat{i}\frac{\partial}{\partial x} + \hat{j}\frac{\partial}{\partial y} + \hat{k}\frac{\partial}{\partial z}$$

We have $\mu = \mu_0 \mu_r$, $\mu_0 = 4\pi \times 10^{-7}$ N·A^{-2} (or H·m^{-1}) is the permeability of free space, and μ_r is the relative permeability of the material and very close to unity. For most dielectrics, $\varepsilon = \varepsilon_0 \varepsilon_r$, where $\varepsilon_0 = 8.854 \times 10^{-12}$ C^2·N^{-1}·m^{-2} is the permittivity of free space, and ε_r is the relative permittivity of the material.

Inside an ideal dielectric material, ρ (free charge density) = 0 and $\sigma = 0$. Therefore, *Maxwell's equations* for a dielectric medium reduces to:

$$\nabla \times E = \frac{\partial B}{\partial t} \tag{3.5}$$

$$\nabla \times H = \frac{\partial D}{\partial t} \tag{3.6}$$

$$\nabla \cdot D = 0 \tag{3.7}$$

$$\nabla \cdot B = 0 \tag{3.8}$$

Substituting for D and B and taking the curl of Eqs (3.5) and (3.6) gives:

$$\nabla \times (\nabla \times E) = -\mu\varepsilon \frac{\partial^2 E}{\partial t^2} \tag{3.9}$$

$$\nabla \times (\nabla \times H) = -\mu\varepsilon \frac{\partial^2 H}{\partial t^2} \tag{3.10}$$

In obtaining these equations we have used:

$$\nabla \cdot E = \nabla \cdot (D/\varepsilon) = 0 \tag{3.11}$$

Using the vector identity:

$$\nabla \times (\nabla \times E) = \nabla \cdot (\nabla \cdot E) - \nabla^2 E \tag{3.12}$$

We obtain the nondispersive wave equations:

$$\nabla^2 E - \mu\varepsilon \frac{\partial^2 E}{\partial t^2} = 0 \tag{3.13}$$

$$\nabla^2 H - \mu\varepsilon \frac{\partial^2 H}{\partial t^2} = 0 \tag{3.14}$$

where, ∇^2 is the Laplacian operator. For rectangular cartesian and cylindrical polar coordinates, the above wave equations hold for each component of the field vector, each component satisfying the scalar wave equation:

$$\nabla^2 \psi - \mu\varepsilon \frac{\partial^2 \psi}{\partial t^2} = 0 \qquad (3.15a)$$

or

$$\nabla^2 \psi - \frac{1}{v_p^2} \frac{\partial^2 \psi}{\partial t^2} = 0 \qquad (3.15b)$$

where ψ may represent a component of E or H field and v_p is the phase velocity (velocity of propagation of a point of constant phase in the wave) in the dielectric medium. It follows that:

$$v_p = \frac{1}{(\mu\varepsilon)} = \frac{1}{\sqrt{\mu\varepsilon}} = \frac{1}{\sqrt{\mu_r \mu_0 \varepsilon_r \varepsilon_0}} \qquad (3.16)$$

where μ_r and ε_r are the relative permeability and permittivity of free space, respectively. The velocity of light in free space c is therefore:

$$c = \frac{1}{\sqrt{\mu_o \varepsilon_o}} \qquad (3.17)$$

For an *isotropic medium*, the refractive index n is related to ε by relation to ε by the relation $n = \sqrt{\varepsilon/\varepsilon_0} = \sqrt{\varepsilon_r}$, with $\mu_r = 1$. Therefore, $v_p = c/n$. Eq. (3.15a) then becomes:

$$\nabla^2 \psi - \frac{n^2}{c^2} \frac{\partial^2 \psi}{\partial t^2} = 0 \qquad (3.18)$$

The solution of wave equation (3.18) can be verified by substituting:
$$\psi = \psi_0 \exp\left[i\left(\omega t - \beta z\right)\right] \qquad (3.19)$$

This represents a uniform plane wave propagating in Z direction with a phase velocity $v_p = \omega/\beta$, where ω is the angular frequency of the field, t is the time, β is the propagation phase constant, and $i = \sqrt{-1}$.

3.3 SOLUTION IN AN INHOMOGENEOUS MEDIUM

We now find the solution in an isotropic, linear, nonconducting, non-magnetic, but inhomogeneous medium. The divergence of electric displacement vector D becomes:

$$\nabla \cdot D = \nabla \cdot (\varepsilon E) = \varepsilon_0 \nabla \cdot (\varepsilon_r E) = 0$$
$$= \varepsilon_0 \left[\nabla(\varepsilon_r) \cdot E + \varepsilon_r (\nabla \cdot E)\right] = 0 \qquad (3.20)$$

This gives:

$$\nabla \cdot E = -\left(\frac{1}{\varepsilon_r}\right) \nabla(\varepsilon_r) E \qquad (3.21)$$

Optical Fiber Waveguides

From Eq. (3.9), one finds

$$\nabla \times \nabla \times E = -\mu \frac{\partial^2 D}{\partial t^2} = -\mu \varepsilon_0 \varepsilon_r \frac{\partial^2 E}{\partial t^2} \qquad (3.22)$$

since $\mu = \mu_0 \mu_r = \mu_0$ ($\because \mu_r = 1$) and $D = \varepsilon_0 E = \varepsilon_0 \varepsilon_r E$.

From Eqs (3.12) and (3.22), one obtains:

$$\nabla(\nabla \cdot E) - \nabla^2 E = -\mu \varepsilon_0 \varepsilon_r \frac{\partial^2 E}{\partial t^2}$$

Rearranging, one obtains:

$$\nabla^2 E - \nabla(\nabla \cdot E) - \mu_0 \varepsilon_0 \varepsilon_r \frac{\partial^2 E}{\partial t^2} = 0 \qquad (3.23)$$

Substituing for $\nabla \cdot E$ from Eq. (3.21) in Eq. (3.23), one obtains:

$$\nabla^2 E - \nabla \left\{ \frac{1}{\varepsilon_r} \nabla(\varepsilon_r) \cdot E \right\} - \mu \varepsilon_0 \varepsilon_r \frac{\partial^2 E}{\partial t^2} = 0 \qquad (3.24)$$

Equation (3.24) shows that for an inhomogeneous medium, the equation for the Cartesion components of E, i.e., E_x, E_y, and E_z are coupled. Now, for homogeneous medium, the second term on the L.H.S. of Eq. (3.24) will become zero ($\because \nabla \cdot E = 0$ for a homogeneous medium). In this case, the Cartesian components of E will satisfy the scalar wave equation represented by Eq. (3.18).

One can obtain a similar equation to Eq. (3.24) for H also. Taking the curl of Eq. (3.6), one obtains:

$$\nabla \times \nabla \times H = \nabla \times \left(\frac{\partial D}{\partial t} \right) = \frac{\partial}{\partial y}(\nabla \times D)$$

$$= \frac{\partial}{\partial t}(\nabla \times \varepsilon E) = \varepsilon_0 \frac{\partial}{\partial t}(\nabla \times \varepsilon_r E) \qquad (3.25)$$

But, we have:

$$\nabla \times \nabla \times H = \nabla(\nabla \cdot H) - \nabla^2 H$$

From Eq. (3.8), $\nabla \cdot B = \nabla \cdot (\mu_0 H) = 0$

$$\therefore \qquad \nabla \times \nabla \times H = -\nabla^2 H \qquad (3.26)$$

Equating Eqs (3.25) and (3.26), one obtains:

$$\nabla^2 H + \varepsilon_0 \frac{\partial}{\partial t}(\nabla \times \varepsilon_r E) = 0$$

or

$$\nabla^2 H + \varepsilon_0 \frac{\partial}{\partial t}[(\nabla \varepsilon_r) \times E + \varepsilon_r (\nabla \times E)] = 0$$

or

$$\nabla^2 H + \varepsilon_0 (\nabla \varepsilon_r) \times \frac{\partial E}{\partial t} + \varepsilon_0 \varepsilon_r \frac{\partial}{\partial t}(\nabla \times E) = 0$$

Using Eqs (3.5) and (3.6) and rearranging, one obtains:

$$\nabla^2 H + \frac{1}{\varepsilon_r}\{(\nabla \varepsilon_r) \times (\nabla \times H)\} - \mu_0 \varepsilon_0 \varepsilon_r \frac{\partial^2 H}{\partial t^2} = 0 \qquad (3.27)$$

We again see that the equations for Cartesian components of H, i.e. H_x, H_y and H_z, are coupled. To simplify Eqs (3.24) and (3.27), let us set the z-coordinates along the direction of propagation of the wave represented by Eqs (3.24) and (3.27). Let us assume that $n = \sqrt{\varepsilon_r}$ does not vary with y and z. Then, one finds:

$$\varepsilon_r = n^2 = n^2(x) \qquad (3.28)$$

This reveals that the y- and z-dependence of the fields, in general will be of the form $\exp[-i(\gamma y + \beta z)]$. However, we put $\gamma = 0$ wihout any loss of generality. Now, the equations governing the modes of propagation of E_j or H_j, where $J = x, y, z$, may be written as:

$$E_j = E_j(x)\, e^{i(\omega t - \beta z)} \qquad (3.29)$$

and
$$H_j = H_j(x)\, e^{i(\omega t - \beta z)} \qquad (3.30)$$

Here $E_j(x)$ and $H_j(x)$ represents the transverse field distributions that do not change as the field propagates through the medium.

Now, we write Eqs (3.5) and (3.6) in Cartesian coordinates. From Eqs (3.5), i.e.:

$$\nabla \times E = -\frac{\partial B}{\partial t} = -\mu_0 \frac{\partial H}{\partial t}$$

one obtains:

$$\hat{i}\left(\frac{\partial E_z}{\partial y} - \frac{\partial E_y}{\partial z}\right) + \hat{j}\left(\frac{\partial E_x}{\partial z} - \frac{\partial E_z}{\partial x}\right) + \hat{k}\left(\frac{\partial E_y}{\partial x} - \frac{\partial E_x}{\partial y}\right)$$
$$= -\mu_0 \frac{\partial}{\partial t}\{\hat{i}\, H_x + \hat{j}\, H_y + \hat{k}\, H_z\} \qquad (3.31)$$

and from Eq. (3.6), i.e.

$$\nabla \times H = -\frac{\partial D}{\partial t} = \varepsilon_0 \varepsilon_r \frac{\partial E}{\partial t} = \varepsilon_0 n^2 \frac{\partial E}{\partial t}$$

one obtains:

$$\hat{i}\left(\frac{\partial H_z}{\partial y} - \frac{\partial H_y}{\partial z}\right) + \hat{j}\left(\frac{\partial H_x}{\partial z} - \frac{\partial H_z}{\partial x}\right) + \hat{k}\left(\frac{\partial H_y}{\partial x} - \frac{\partial H_x}{\partial y}\right)$$
$$= -\varepsilon_0 n^2 \frac{\partial}{\partial t}\{\hat{i}\, E_x + \hat{j}\, E_y + \hat{k}\, E_z\}] \qquad (3.32)$$

The fields E and H do not vary with y, we may take the $\partial/\partial y$ terms equal to zero. Now, substituting the values of E_j and H_j from Eqs (3.29) and (3.30) into Eq. (3.31), one obtains:

$$-\hat{i}\frac{\partial}{\partial z}\{E_y\,e^{i(\omega t-\beta z)}\} + \hat{j}\frac{\partial}{\partial z}\{E_x\,e^{i(\omega t-\beta z)}\} + \hat{k}\frac{\partial}{\partial x}\{E_x\,e^{i(\omega t-\beta z)}\}$$

$$= -\hat{i}\,\mu_0\frac{\partial}{\partial t}H_x\,e^{i(\omega t-\beta z)} - \hat{j}\,\mu_0\frac{\partial}{\partial t}H_y\,e^{i(\omega t-\beta z)} - \hat{k}\,\mu_0\frac{\partial}{\partial t}H_z\,e^{i(\omega t-\beta z)}$$

or

$$\hat{i}(i\beta)E_y - \hat{j}\left[i\beta E_x + \frac{\partial E_z}{\partial x}\right] + \hat{k}\frac{\partial E_y}{\partial x}$$

$$= -\hat{i}\,\mu_0(i\omega)H_x - \hat{j}\,\mu_0(i\omega)H_y - \hat{k}\,\mu_0(i\omega)H_z$$

Comparing the corresponding components on both sides, one obtains the following set of equations:

$$i\beta E_y = -i\omega\mu_0 H_x \tag{3.33a}$$

or

$$\beta E_y = -\mu_0\omega H_x$$

$$i\beta E_x + \frac{\partial E_z}{\partial x} = i\mu_0\omega H_y \tag{3.33b}$$

$$\frac{\partial E_y}{\partial x} = -i\mu_0\omega H_z \tag{3.33c}$$

Now substituting the values of E_j and H_j from Eqs (3.29) and (3.30) into Eq. (3.32) and setting to zero all terms containing $\partial/\partial y$, one obtains:

$$-\hat{i}\frac{\partial}{\partial z}H_y\,e^{i(\omega t-\beta z)} + \hat{j}\left[\frac{\partial}{\partial t}\{H_x\,e^{i(\omega t-\beta z)}\} - \frac{\partial}{\partial t}H_z\,e^{i(\omega t-\beta z)}\right]$$

$$+ \hat{k}\frac{\partial}{\partial t}\{H_x\,e^{i(\omega t-\beta z)}\}$$

$$= \varepsilon_0 n^2\,\hat{i}\frac{\partial}{\partial t}\{E_x e^{i(\omega t-\beta z)}\} + \varepsilon_0 n^2\,\hat{j}\frac{\partial}{\partial t}\{E_y e^{i(\omega t-\beta z)}\} + \varepsilon_0 n^2\,\hat{k}\frac{\partial}{\partial t}\{E_z\,e^{i(\omega t-\beta z)}\}$$

or

$$\hat{i}(i\beta)H_y - \hat{j}\left\{-(i\beta H_x) - \frac{\partial H_z}{\partial x}\right\} + \hat{k}\frac{\partial H_y}{\partial x}$$

$$= \hat{i}(i\omega\varepsilon_0 n^2)E_x + \hat{j}(i\omega\varepsilon_0 n^2)E_y + \hat{k}(i\omega\varepsilon_0 n^2)E_z$$

Comparing the respective components, one obtains:

$$i\beta H_y = -i\omega\mu_0 n^2 E_x \tag{3.34a}$$

or

$$\beta H_y = -\varepsilon_0 n^2 \omega E_x$$

$$-i\beta H_x - \frac{\partial H_z}{\partial x} = i\varepsilon_0 n^2 \omega E_y \tag{3.34b}$$

$$\frac{\partial H_y}{\partial x} = i\varepsilon_0 n^2 \omega E_z \tag{3.34c}$$

where $n^2 = n^2(x)$.

The above six equations, i.e., (3.33a–3.33c, 3.34a–3.34c) form two independent sets. Thus, Eqs (3.33a), (3.33c), and (3.34b) involve E_y, H_x, and H_z. Clearly, the field components E_x, E_z, and H_y are zero. The modes described by these equations are called *transverse electric (TE) modes* as the electric field has only the transverse component E_y.

Renumbering these equations as follows, we find:

$$\beta E_y = -\mu_0 \omega H_x \qquad (3.35)$$

$$\frac{\partial E_y}{\partial x} = -i\mu_0 \omega H_z \quad \bigg| \text{TE modes} \qquad (3.36)$$

$$-i\beta H_x - \frac{\partial H_z}{\partial x} = i\varepsilon_0 \omega n^2(x) E_y \qquad (3.37)$$

The second set of equations is formed by Eqs (3.34a), (3.34c) and (3.33b), which involve only H_y, E_x, and E_z, and the field components E_y, H_x and H_z are zero. These are called *transverse magnetic (TM) modes* because the magnetic field herein has only a transverse component H_y. The second set of equations is also renumbered, we find:

$$\beta H_y = -\varepsilon_0 \omega n^2(x) E_x \qquad (3.38)$$

$$\frac{\partial H_y}{\partial x} = -i\varepsilon_0 \omega n^2(x) E_z \quad \bigg| \text{TE modes} \qquad (3.39)$$

$$i\beta H_x + \frac{\partial E_z}{\partial x} = -i\mu_0 \omega H_y \qquad (3.40)$$

3.4 PLANAR OPTICAL WAVEGUIDE

The planar optical waveguide is the simplest form of the optical waveguide and these are important components in integrated optical devices.

Figure 3.1 shows the geometrical configuration of simplest optical waveguide. We may assume that it consists of a thin dielectric slab of refractive index n_1 and sandwiched between two symmetrical dielectric slabs of refractive index n_2 and infinite thickness ($n_2 < n_1$). The waveguide is oriented such that the wave propagates along the z-direction. The y and z dimensions of the guide are assumed to extend to infinity. The thickness of the slabs is along the x-direction (Fig. 3.1). A ray of light launched into the guide slab or layer would progress by multiple reflections as shown in Fig. 3.2. One may assume that such a ray represents a *plane TEM wave* travelling at an angle θ with the z-axis. As the refractive index within the guide layer is n_1, the wavelength of light in the layer is reduced to $\lambda_m = \lambda/n_1$, where λ is the wavelength of light in vacuum and the propagation constant is increased to:

$$\beta_2 = \frac{2\pi}{\lambda_m} = \frac{2\pi n_1}{\lambda} = k n_1$$

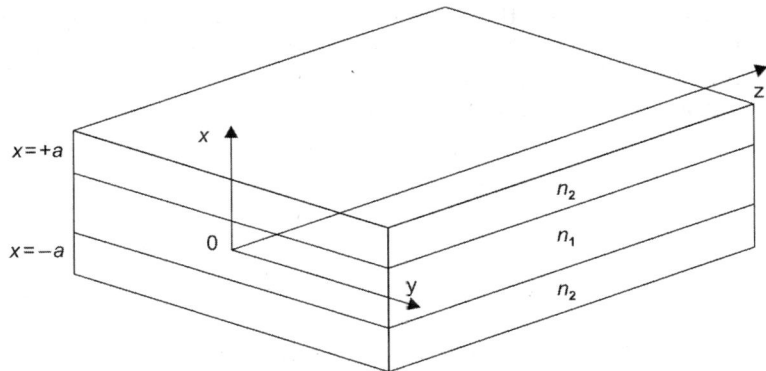

Fig. 3.1. Structure of a planar optical waveguide

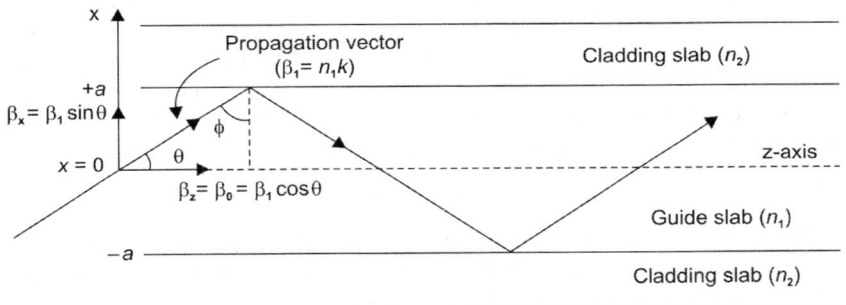

Fig. 3.2. Ray propagation in a planar waveguide

where $k = 2\pi/\lambda$ is the vacuum propagation constant or the propagation vector.

Let the propagation vector β_1 makes an angle θ with the z-axis (which is the same as the guide axis), the plane wave may be resolved into two component plane waves propagating in the z and x directions as shown in Fig. 3.2. The component of the propagation vector in the z-direction or, in other words, the effective propagation constant of the guided wave (along the z-direction) will be as follows:

$$\beta = \beta_2 = \beta_1 \cos\theta \qquad (3.41)$$

The limiting value of θ, i.e. θ_m is related to the critical angle ϕ_c at the interface of the guide layer and the cladding slabs, and is given by the following relation:

$$\sin\phi_c = \cos\theta_m = n_2/n_1 \qquad (3.42)$$

Clearly, the minimum value of β in the z-direction, i.e. β_{min}, will be determined by the maximum value of θ, i.e. θ_m or:

$$\beta_{min} = \beta_1 \cos\theta_m = \beta_1 \frac{n_2}{n_1} = \beta_2 \qquad (3.43)$$

62 Photonics | Optoelectronics

The maximum value that β can have is β_1, which corresponds to $\theta = 0$, i.e. the plane TEM waves travelling parallel to the guide axis: $\beta_{max} \approx \beta_1$. One, therefore, expect β to lie between β_1 and β_2, $\beta_2 < \beta < \beta_1$. The component of the propagation vector β_1 in the x-direction is given by

$$\beta_x = \beta_1 \sin\theta = n_1 k \sin\theta$$

where

$$k = 2\pi/\lambda$$

or

$$\beta_x = \frac{2\pi}{\lambda_m} = \sin\theta \tag{3.44}$$

We see that this component of the plane wave is reflected at the interface between the guide layer and the cladding slabs. When the total phase change after two successive reflections at the upper and lower interfaces is equal to $2i\pi$ radians, where i is an integer (0, 1, 2, 3, ...), constructive interference will occur and a standing-wave pattern will be formed in the x-direction. This stable field pattern in the x-direction with only a periodic z-dependence is known as a mode. Thus, only a finite number of discrete modes which satisfy the above condition will propagate through the guide, i.e., $4a\beta_x = 2i\pi$:

or

$$4a \sin\theta_i = i\lambda_m \tag{3.45}$$

We see that each value of θ_i corresponds to a particular mode with its own characteristic field pattern and its own propagation constant β_1 in the z-direction. Obviously, β_i lies between β_1 and β_2. Since the maximum value that θ_i can take is θ_m, the number of guided modes is limited to:

$$M = i_{max} = \frac{4a \sin\theta_m}{\lambda_m} = \frac{4an_1 \sin\theta_m}{\lambda} = \frac{4a}{\lambda} = (n_1^2 - n_2^2)^{1/2} \tag{3.46}$$

We may note that the requirement for the ith mode to be propagated is that $i \leq (4a/\lambda)(n_1^2 - n_2^2)^{1/2}$. The mode corresponding to the highest value of i, i.e., i_{max}, does not meet the condition for total internal reflection, as the value of θ_m corresponds exactly to the critical angle ϕ_c, and is refracted. Clearly, it can propagate freely in the cladding slabs and is said to be a *radiation mode*.

3.5 THE MODES OF A SYMMETRIC STEP INDEX FIBER

For the symmetric waveguide structure (Fig. 3.1), we have $n^2(-x) = n^2(x)$. Moreover, the structure is step index type, as the refractive index of guide layer is n_1 and the refractive index of the cladding slabs is n_2. Further, both n_1 and n_2 are constants and $n_1 > n_2$. We first take up the discussion of TE modes.

Substituting the values of H_x and H_z from Eqs (3.35) and (3.36) into Eq. (3.37), one obtains:

Optical Fiber Waveguides 63

$$-i\beta\left(\frac{\beta}{\mu_0\omega}\right)E_y - \frac{\partial}{\partial x}\left(\frac{1}{-i\mu_0\omega}\right)\frac{\partial E_y}{\partial x} = i\varepsilon_0\omega n^2(x) E_y$$

As $E_y = E_y(x)$, the partial derivative involving E_y may as well be written as a full derivative. Thus, rearranging the above equation, one may write:

$$\frac{d^2E_y}{dx^2} - \beta^2 E_y + \mu_0\varepsilon_0\omega^2 n^2(x) E_y = 0$$

or

$$\frac{d^2E_y}{dx^2} + [k^2 n^2(x) - \beta^2] E_y = 0 \qquad (3.46a)$$

as $\mu_0\omega_0 = 1/c^2$ and $\omega/c = k$

In the waveguide of Fig. 3.1, we have:

$$n(x) = \begin{cases} n_1 \text{ for } |x| < a \\ n_2 \text{ for } |x| < a \end{cases} \qquad (3.47)$$

Moreover, E_y and H_z (and hence $\partial E_y/\partial x$) are continuous at $x = \pm a$ because E_y and H_z are tangenital components to the planes represented by $x = \pm a$, and H_z is proportional to $\partial E_y/\partial x$.

Substituting for $n(x)$ from Eq. (3.47) in Eq. (3.46), one obtains in the guide layer:

$$\frac{d^2E_y}{dx^2} + [k^2 n_1^2 - \beta^2] E_y = 0 \, (|x| < a) \qquad (3.48)$$

and in the cladding layer:

$$\frac{d^2E_y}{dx^2} + [k^2 n_2^2 - \beta^2] E_y = 0 \, (|x| < a) \qquad (3.49)$$

Les us put

$$u^2 = k^2 n_1^2 - \beta^2 = \beta_1^2 - \beta^2 \qquad (3.50)$$

and

$$w^2 = \beta^2 - k^2 n_2^2 = \beta^2 - \beta_2^2 \qquad (3.51)$$

Eqs (3.48) and (3.49) take the forms:

$$\frac{d^2E_y}{dx^2} + u^2 E_y = 0 \, (|x| < a) \qquad (3.52)$$

and

$$\frac{d^2E_y}{dx^2} - w^2 E_y = 0 \, (|x| > a) \qquad (3.53)$$

For the wave to be guided through the layer, both parameters u and w have to be real. Thus

$$\beta_1^2 \, (= k^2 n_1^2) > \beta^2 > \beta_2^2 \, (= k^2 n_2^2) \qquad (3.54)$$

With these conditions, the solutions in the guide layer are oscillatory, while those in the cladding layers decay exponentially. This is what we exactly expected. Thus, for a guided-wave solution, the propagation constant β must lie between β_1 and β_2. The same inference is also obtained from ray analysis.

Since the refractive index $n(x)$ is symmetrically distributed about $x = 0$, the solutions are either symmetric or antisymmetric functions of x. Therefore, one must have:

$$E_y(-x) = E_y(x) \text{ symmetric modes} \quad (3.55)$$
$$E_y(-x) = E_y(x) \text{ antisymmetric modes} \quad (3.56)$$

For symmetric modes, the electric field distribution takes the form:

$$E_y(x) = \begin{cases} A \cos ux, & |x| < a \\ C \exp(-w|x|), & |x| > a \end{cases} \quad (3.57, 3.58)$$

where A and C are constants. The continuity of $E_y(x)$ and dE_y/dx at $x = \pm a$ gives the following equations

$$A \cos(ua) = Ce^{-wa} \quad (3.59)$$
and
$$-uA \sin(ua) = -wCe^{-wa} \quad (3.60)$$

Dividing Eq. (3.60) by Eq. (3.59), one obtains:

$$u \tan(ua) = w$$
or
$$ua \tan(ua) = wa \quad (3.61)$$

Let us now, define a new dimensionless waveguide parameter called the *normalized frequency parameter* V. From Eqs (3.48) and (3.49), one finds:

$$u^2 + w^2 = k^2 n_1^2 - \beta^2 + \beta^2 - k^2 n_2^2 = k^2(n_1^2 - n_2^2)$$

or
$$(ua)^2 + (wa^2) = k^2 a^2 (n_1^2 - n_2^2) = \left(\frac{2\pi}{\lambda}\right)^2 a^2 (n_1^2 - n_2^2)$$

Thus, V can be defined as:

$$V = \{(ua)^2 + (wa^2)\}^{1/2} = \frac{2\pi}{\lambda}(n_1^2 - n_2^2)^{1/2} \quad (3.62)$$

In terms of V, Eq. (3.61) taken the form:

$$ua \tan(ua) = \{V^2 - (ua)^2\}^{1/2} \quad (3.63)$$

For antisymmetric modes, the solution take the form:

$$E_y(x) = \begin{cases} B \sin ux, & |x| < a \\ \dfrac{x}{|x|} D \exp(-w|x|), & |x| > a \end{cases} \quad (3.64, 3.65)$$

where B and D are constants. Following exactly the above procedure, one obtains:

$$-ua \cot(ua) = wa \quad (3.66)$$

or, in terms of the parameter V, one obtains:

$$-ua \cot(ua) = \{V^2 - (ua)^2\}^{1/2} \quad (3.67)$$

Eqs (3.61) to (3.66) are transcendisal equations and one can find their solution by plotting ua as a function of wa for $ua \tan(ua) = wa$ and $-ua \cot(ua) = wa$. Eq. (3.62) is plotted as arcs of circles for constant

Optical Fiber Waveguides 65

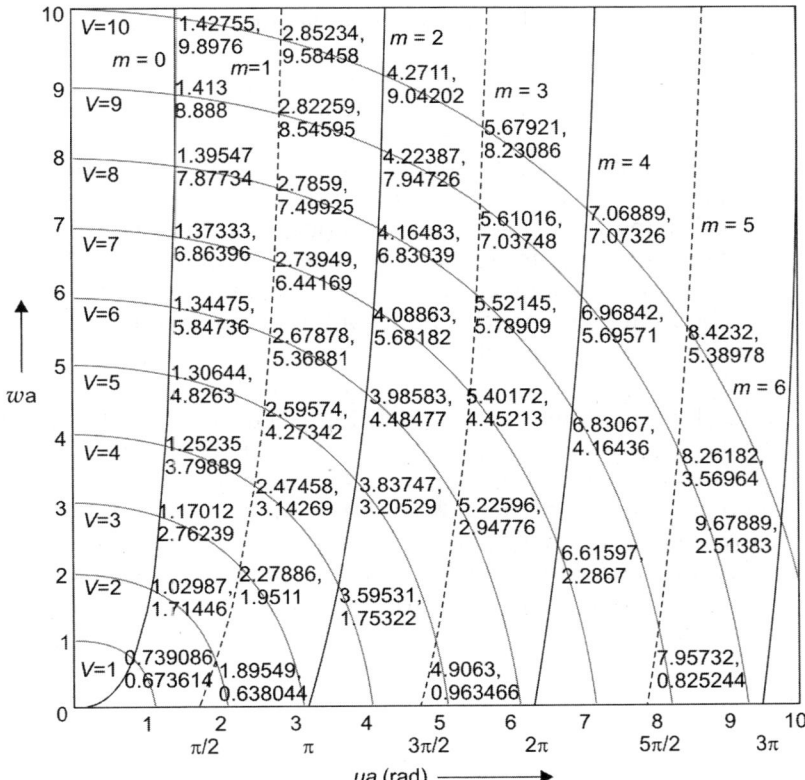

Fig. 3.3. Graphical solution of Eqs (3.61) and (3.66) for obtaining the propagation parameters *ua* and *wa* (dashed lines). $V^2 = [(ua)^2 + (wa)^2]$ = const. is plotted as arcs of circles for constant *V*-values (light face solid lines)

V-values. Fig. 3.3 shows the graphical solution of Eqs (3.61) and (3.62) for obtaining the propagation parameters *ua* and *wa* of TE modes for constant values of *V* in a planar waveguide.

One can derive following information from Fig. 3.3

(i) For $0 \leq V \leq \pi/2$ (i.e., for an arc of a circle of radius corresponding to $V < \pi/2$), there is only one intersection with the bold solid curve marked $m = 0$. This is the only solution for the guided TE mode, i.e. the waveguide supports only one discrete TE mode and this mode is symmetric in *x*.

(ii) For $\pi/2 \leq V \leq \pi$ (i.e., for an arc of circle of radius corresponding to $\pi/2 < V < \pi$), the arc intersects at two points; one on the bold solid line $m = 0$ and the other on the dashed line $m = 1$. This means that, therefore, two TE modes, one symmetric and the other antisymmetric. In general, if:

$$(2m)\frac{\pi}{2} \leq V \leq (2m+1)\frac{\pi}{2} \tag{3.68}$$

we will have $m + 1$ symmetric and m antisymmetric modes, and if:

$$(2m+1)\frac{\pi}{2} \leq V \leq (2m+2)\frac{\pi}{2} \quad (3.69)$$

we will have $m + 1$ symmetric and $m + 1$ antisymmetric modes, where $m = 0, 1, 2, ..., m = 0, 2, 4, ...$ correspond to symmetric modes and $m = 1, 3, 5, ...$ correspond to antisymmetric modes. The maximum number of TE modes, M, supported by the guide would be an integer close to or greater than $2V/\pi$. This is found to be in agreement with Eq. (3.46), which was obtained on the basis of the ray model.

(iii) We may note in Fig. 3.3 that for the fundamental mode ($m = 0$), ua always lies between 0 and $\pi/2$ and the corresponding electric field $E_y(x)$ for $|x| < a$ will have no zeros. For the next mode ($m = 1$), which is antisymmetric in x, ua lies between $\pi/2$ and π and the corresponding field distribution has one zero (at $x = 0$). One can extend the analysis to prove that the electric-field distribution $E_y(x)$ for the mth mode will have m zeros between $x = -a$ and $x = +a$. The mode patterns for the first few modes are as shown in Fig. 3.4.

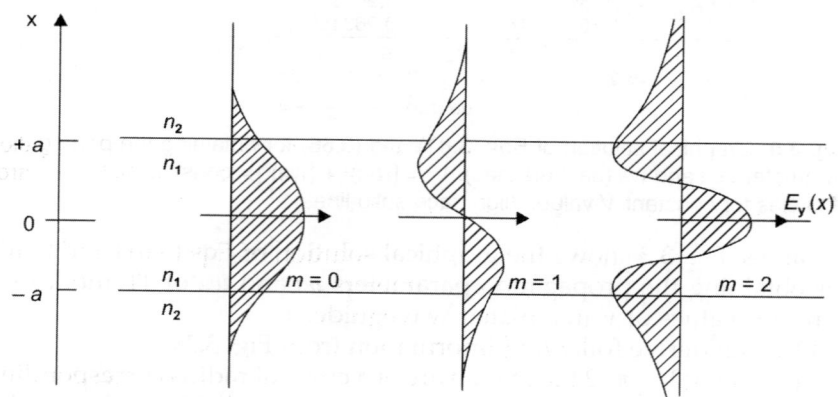

Fig. 3.4. Field distribution for the first three TE modes ($m = 0$, 1, and 2) of a planar waveguide

There are some more important points about these modes:

(i) A relevant parameter is the normalized propagation constant denoted by b, defined by the following relation:

$$b = \frac{\beta^2 - \beta_2^2}{\beta_1^2 - \beta_2^2} = 1 - \left(\frac{ua}{V}\right)^2 = \left(\frac{wa}{V}\right)^2 \quad (3.70)$$

For a given guided mode, the value of b lies between 0 and 1. The variation of b with V for first few modes is shown in Fig. 3.5.

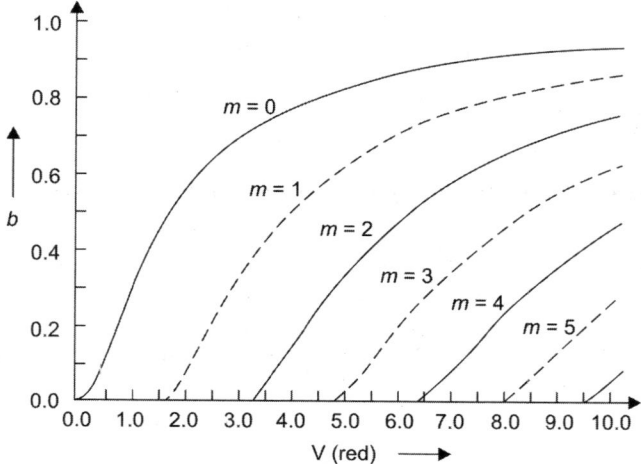

Fig. 3.5. Variation of the normalized propagation parameter b for TE modes with the V parameter for a planar waveguide

When β becomes equal to β_2, $b = 0$ and the mode is said to have reached the 'cut-off'. Clearly, at cut-off, $ua = V = V_c$ and $wa = 0$. This occurs at:

$$wa = V_c = \frac{m\pi}{2} \tag{3.71}$$

As $\quad V = \dfrac{2\pi a}{\lambda}(n_1^2 - n_2^2)^{1/2} \quad$ (from Eq. 3.62)

equating Eqs (3.62) and (3.71), one can obtain the thicknss of the guide layer necessary to support m modes:

$$\frac{m\pi}{2} = \frac{2\pi a}{\lambda}(n_1^2 - n_2^2)^{1/2}$$

or
$$2a = \frac{m\lambda}{2(n_1^2 - n_2^2)^{1/2}} \tag{3.72}$$

where $m = 0, 1, 2, \ldots$. We may note that fundamental mode has no cut-off frequency.

(ii) The modes for which $\beta^2 < \beta_2^2$ are called *radiation modes*. These modes are continuous. In terms of the ray model, these modes correspond to the rays for which total internal reflection does not occur, and get refracted into the cladding.

(iii) All the modes are orthogonal. If the field $E_y(x)$ corresponding to a guided mode with a propagation constant β_m is represened by $\omega_{m(x)}$ and its complex conjugate by $\omega_m(x)$, then one can show that:

$$\int_{-\infty}^{\infty} \psi_m^*(x)\,\psi_m(x)\,dx = 0 \text{ for } m \neq n$$

This is the condition of *orthogonality*.

After determining the electric-field distribution $E_y(x)$, it is possible to calculate the H_x and H_z components of the magnetic field by substituting the value of E_y in Eq. (3.35) and $\partial E_y/\partial x$ in Eq. (3.36). A similar analysis may be performed for the TM modes of a planar waveguide.

3.6 POWER DISTRIBUTION AND CONFINEMENT FACTOR

We have seen (Fig. 3.4) that the electric-field distribution of the guided modes extends beyond the boundary of the guide layer, and the extent of penetration into the cladding layer depends on the thickness and the mode number. It is important for us to know, in many situations, the fraction of guided optical power confined with the guide layer. This fraction is called the *confinement factor*.

The power flow is given by Poynting vector defined by:

$$\mathbf{S} = \mathbf{E} \times \mathbf{H}$$

where $\mathbf{E}$ and $\mathbf{H}$ are normally expressed in the complex form. However, the actual fields are the real part of the complex form. Now, taking the time average of the Poynting vector, one obtains:

$$\langle \mathbf{S} \rangle = \langle \operatorname{Re} \mathbf{E} \times \operatorname{Re} \mathbf{H} \rangle$$
$$= \frac{1}{2} \operatorname{Re} \langle \mathbf{E} \times \mathbf{H}^* \rangle$$

where $\mathbf{H}^*$ is the complex conjugate of $\mathbf{H}$. Obviously, the time average of S along the z-direction will be given by:

$$\langle S_z \rangle = \frac{1}{2} \langle E_x H_y - H_x E_y \rangle \qquad (3.73)$$

We have already seen that, for TE modes, the field components E_x and E_z vanish and:

$$H_x = \frac{\beta}{\mu_0 \omega} E_y$$

Therefore
$$\langle S_z \rangle = \frac{1}{2} \frac{\beta}{\omega \mu_0} |E_y^2| \qquad (3.74)$$

For a particular mode, the power associated per unit area per unit length in the y-direction is given by:

$$P = \frac{1}{2} \frac{\beta}{\omega \mu_0} \int_{x=-\infty}^{\infty} |E_y^2| \, dx \qquad (3.75)$$

The power inside the guide layer is obtained as:

$$P_{in} = \frac{1}{2} \frac{\beta}{\omega \mu_0} \int_{-a}^{a} E_y^2 \, dx \qquad (3.76)$$

and the power outside the guide layer is obtained as:

$$P_{out} = \frac{1}{2} \frac{\beta}{\omega \mu_0} \left[\int_{-\infty}^{a} E_y^2 \, dx + \int_{a}^{\infty} E_y^2 \, dx \right] \qquad (3.77)$$

For symmetric TE modes, substituting for E_y and Eq. (3.57), one obtains:

$$P_{in} = \frac{1}{2}\frac{\beta}{\omega\mu_0} 2\int_0^a (A\cos ux)^2 \, dx = \frac{\beta}{\omega\mu_0} A^2 \int_0^a \frac{1}{2}(1+\cos 2ux)\,dx$$

$$P_{in} = \frac{\beta}{2\omega\mu_0} A^2 \left[x + \frac{1}{2u}\sin 2ux \right]_0^a = \frac{\beta}{2\omega\mu_0} A^2 \left[a + \frac{1}{2u}\sin 2ua \right] \quad (3.78)$$

Substituting the value of E_y from Eq. (58), one obtains:

$$P_{out} = \frac{1}{2}\frac{\beta}{\omega\mu_0} 2\int_a^\infty (Ce^{-wx})^2 \, dx = \frac{\beta}{\omega\mu_0} C^2 \left[-\frac{1}{2w}e^{-2wx} \right]_a^\infty$$

or
$$P_{out} = \frac{\beta}{\omega\mu_0} C^2 \left[\frac{1}{2w}e^{-2wx} \right] \quad (3.79)$$

From Eqs (3.78) and (3.79), one can derive the final expression for P as:

$$P = P_{in} + P_{out}$$

$$= \frac{\beta}{2\omega\mu_0} A^2 \left[a + \frac{1}{2u}\sin 2ua \right] + \frac{\beta}{\omega\mu_0} C^2 \left[\frac{1}{2w}e^{-2wa} \right]$$

$$= \frac{\beta A^2}{2\omega\mu_0} \left[a + \frac{1}{2u}\sin 2wa + \left(\frac{C}{A}\right)^2 \frac{1}{w}e^{-2wa} \right]$$

Putting the value of $C/A = e^{wa}\cos ua$ from Eq. (3.59), one obtains:

$$P = \frac{\beta A^2}{2\omega\mu_0} \left[a + \frac{1}{2u}\sin 2ua + \frac{1}{w}\cos^2 ua \right]$$

$$= \frac{\beta A^2}{4\omega\mu_0} \left[2a + \frac{2}{w} + \frac{2\sin(ua)\cos(ua)}{u} - \frac{2\sin^2 ua}{w} \right]$$

$$= \frac{\beta A^2}{4\omega\mu_0} \left[\left(2a + \frac{2}{w}\right) + \frac{2\sin(ua)\cos(ua)}{uwa} - \{wa - ua\tan ua\} \right]$$

From Eq. (3.61), the term within the braces becomes zero. This yield the following expression for P:

$$P = \frac{\beta A^2}{4\omega\mu_0}\left(a + \frac{1}{w}\right) \quad (3.80)$$

A similar expression may also be derived for power carried by antisymmetric modes.

The fraction of power confined within the guide layer, called the *confinement factor G*, is given by:

$$G = \frac{P_{in}}{P_{in} + P_{out}} \quad (3.81)$$

$$= \frac{a + \frac{1}{2}\sin 2ua}{\left[a + \frac{1}{2u}\sin 2ua\right] + \left[\frac{1}{w}e^{-2wa}\right]\left(\frac{C}{A}\right)^2} \quad (3.82)$$

Putting the value of $C/A = e^{wa}\cos ua$, from Eq. (3.59), in Eq. (3.82), one obtains:

$$G = \frac{\left[a + \frac{1}{2u}\sin 2ua\right]}{\left[a + \frac{1}{2u}\sin 2ua\right] + \left[\frac{1}{w}e^{-2wa}(e^{wa}\cos ua)^2\right]}$$

$$= \left[\frac{a + \frac{1}{2u}\sin 2ua + \frac{1}{w}\cos^2 ua}{a + \frac{1}{2u}\sin 2ua}\right]$$

$$= \left[1 + \frac{\cos^2 ua}{wa\{1 + [\sin(ua)\cos(ua)/ua]\}}\right]^{-1} \quad (3.83)$$

A similar expression may also be derived for antisymmetric modes. Some salient features of Eq. (3.83) are as follows:

(i) As $a \to 0$, $G \to 0$; that is, for a very thin guide layer, there is no guidance of light, i.e., almost all the power is lost in the cladding.

(ii) As the G-factor depends on u and w and both these parameters in turn depend on m, G will vary with the mode number.

(iii) One can show that G increases, first rapidly and then slowly, with increase in the thickness of the guide layer for each mode.

3.7 CYLINDRICAL WAVEGUIDES

3.7.1 Modal Analysis of an Ideal Step-Index Optical Fiber

We now study the wave propagations in a cylindrical homogeneous core dielectric waveguide. This type of waveguide with a constant refractive index core is known as a *step index fiber*.

Let us consider a step-index fiber consisting of a uniform cylindrical dielectric core of radius a and refractive index n, surrounded by an infinitely thick uniform dielectric cladding of refractive index n_2. Let us find a solution of wave equation for electric and magnetic fields for the modes propagated by such a fiber.

Eqs (3.24) and (3.27) shows that different components of **E** as well as **H** are coupled.

In the present case $n = \sqrt{\varepsilon_r}$ is constant $= n_1$ upto $r \leq a$ and equal to n_2 for points $r > a$ but there is a discontinuity at $r = a$. We assume that this discontinuity is small, i.e. $n_1 \approx n_2$. This is termed as a weakly guiding approximation. With this approximation, one may neglect the second term on the LHS of Eqs (3.24) and (3.27) and each Cartesian component of **E** and **H** satisfies the scalar wave equation; putting $\varepsilon_r = n^2$, we have

$$\nabla^2 \psi = \varepsilon_0 \mu_0 n^2 \frac{\partial^2 \psi}{dt^2} \tag{3.84}$$

where ψ represents scalar E and H field. We may note that the optical fiber boundary conditions have cylindrical symmetry and we assume that the direction of propagation of the electromagnetic waves is along the axis of the optic fiber., which is taken as z-axis. In the scalar wave approximation, the modes may be assumed to be nearly transverse and they may possess an arbitrary state of polarization. These linearly polarized modes are referred to as *LP modes*. The propagation constants of the TE and TM modes are nearly equal.

In cylindrical coordinates (r, ϕ, z), Eq. (3.84) may be expressed as

$$\nabla^2 \psi = \varepsilon_0 \mu_0 n^2 \frac{\partial^2 \psi}{\partial t^2} = \frac{\partial^2 \psi}{\partial r^2} + \frac{1}{r} \frac{\partial \psi}{\partial r} + \frac{1}{r^2} \frac{\partial^2 \psi}{\partial \phi^2}$$

$$+ \frac{\partial^2 \psi}{\partial z^2} - \varepsilon_0 \mu_0 n^2 \frac{\partial^2 \psi}{\partial t^2} = 0 \tag{3.85}$$

Since n may depend on the transverse coordinates (r, ϕ), though it usually depends only on r, and the wave is propagating along the z-direction, one may write the solution of Eq. (3.85) as

$$\psi(r, \phi, z, t) = \psi(r, \phi) \, e^{-i(\omega t \times \beta z)} \tag{3.86}$$

Substituting the value of ψ from Eq. (3.86) in Eq. (3.85), we obtains:

$$e^{i(\omega t - \beta z)} \frac{\partial^2 \psi}{\partial r^2} + \frac{1}{r} e^{i(\omega t - \beta z)} \frac{\partial \psi}{\partial r} + \frac{1}{r^2} e^{i(\omega t - \beta z)} \frac{\partial^2 \psi}{\partial \phi^2}$$

$$+ \psi(-\beta^2) e^{i(\omega t - \beta z)} - \varepsilon_0 \mu_0 n^2 (-\omega^2) \psi \, e^{i(\omega t - \beta z)} = 0$$

or

$$\frac{\partial^2 \psi}{\partial r^2} + \frac{1}{r} \frac{\partial \psi}{\partial r} + \frac{1}{r^2} \frac{\partial^2 \psi}{\partial \phi^2} + -[\varepsilon_0 \mu_0 \omega^2 n^2 - \beta^2] \psi = 0$$

Putting $\varepsilon_0 \mu_0 = 1/c^2$ and $\omega/c = k$, the free-space wave number, in the above equation, one obtains:

$$\frac{\partial^2 \psi}{\partial r^2} + \frac{1}{r} \frac{\partial \psi}{\partial r} + \frac{1}{r^2} \frac{\partial^2 \psi}{\partial \phi^2} + [n^2 k^2 - \beta^2] \psi = 0 \tag{3.87}$$

Since the fiber under consideration has cylindrical symmetry, the variables can be separated:

$$\psi(r, \phi) = R(r)\,\Phi(\phi) \qquad (3.88)$$

where R is a function of only r and Φ is a function of only ϕ. Substituting ψ from Eq. (3.88) in Eq. (3.87), one obtains:

$$\Phi \frac{\partial^2 R}{\partial r^2} + \frac{1}{Rr}\Phi \frac{dR}{dr} + \frac{R}{r^2}\frac{\partial^2 \Phi}{\partial \phi^2} + [n^2 k^2 - \beta^2]\,R\Phi = 0$$

Since the derivatives involved are dependent either on r or ϕ only, the partial derivatives may be replaced by full derivatives. Further, on dividing the entire LHS by $R\Phi$, one obtains:

$$\frac{1}{R}\frac{d^2 R}{dr^2} + \frac{1}{Rr}\frac{dR}{dr} + \frac{1}{r^2}\frac{1}{\Phi}\frac{d^2 \Phi}{d\phi^2} + [n^2 k^2 - \beta^2] = 0$$

or

$$\frac{r^2}{R}\left(\frac{d^2 R}{dr^2} + \frac{1}{Rr}\frac{dR}{dr}\right) + r^2[n^2 k^2 - \beta^2] = -\frac{1}{\Phi}\frac{d^2\Phi}{d\phi^2} = l^2 \quad \text{(say)} \quad (3.89)$$

where l is a constant, known as an *azimuthal eigenvalue*.

The dependence of Φ on ϕ will be of the form $e^{il\phi}$. For the function to be single-valued, i.e., $\Phi(\phi + 2\pi) = \Phi(\phi)$, the constant l is required to be an integer, i.e.:

$$l = 0, 1, 2, 3, \ldots \qquad (3.90)$$

Therefore the complete transverse field will be given by:

$$\psi(r, \phi, z, t) = R(r)\,e^{il\phi}\,e^{i(\omega t - \beta z)} \qquad (3.91)$$

The radial part of Eq. (3.89) may be written as:

$$\frac{r^2}{R}\left(\frac{d^2 R}{dr^2} + \frac{1}{Rr}\frac{dR}{dr}\right) + r^2(n^2 k^2 - \beta^2) = l^2$$

which after *rearrangement* gives:

$$r^2 \frac{d^2 R}{dr^2} + r\frac{dR}{dr} + [r^2(n^2 k^2 - \beta^2) - l^2]\,R = 0 \qquad (3.92)$$

We know that $n = n_1$ for $r \leq a$ and $n = n_2$ for $r > a$. Thus, substituting the value of n in Eq. (3.92), we obtain for the case of a step-index fiber:

$$r^2 \frac{d^2 R}{dr^2} + r\frac{dR}{dr} + [r^2(k^2 n_1^2 - \beta^2) - l^2]\,R = 0, \; r \leq a \qquad (3.93)$$

and

$$r^2 \frac{d^2 R}{dr^2} + r\frac{dR}{dr} + [r^2(k^2 n_2^2 - \beta^2) - l^2]\,R = 0, \; r > a \qquad (3.94)$$

In order to simplify the above equations, one may put:

$$u^2 \equiv (k^2 n_1^2 - \beta^2)\,a^2 \qquad (3.95)$$

and

$$w^2 \equiv (\beta^2 - k^2 n_1^2)\,a^2 \qquad (3.96)$$

The normalized waveguide parameter V for the fiber can be defined by:

$$V = (u^2 + w^2)^{1/2} = ka(n_1^2 - n_2^2)^{1/2} = \frac{2\pi a}{\lambda}(n_1^2 - n_2^2)^{1/2} \quad (3.97)$$

Substituting the values of u and w in Eqs (3.93) and (3.94), one obtains:

$$r^2 \frac{d^2 R}{dr^2} + r \frac{dR}{dr} + \left(\frac{u^2 r^2}{a^2} - l^2 \right) R = 0, r \leq a \quad (3.98)$$

and

$$r^2 \frac{d^2 R}{dr^2} + r \frac{dR}{dr} + \left(\frac{w^2 r^2}{a^2} - l^2 \right) R = 0, r > a \quad (3.99)$$

Equations (3.98) and (3.99) are second-order equations and hence should possess two independent solutions. The solutions corresponding to Eq. (3.98) are the *Bessel function* of the first kind. The solutions corresponding to Eq. (3.99) are the Bessel function of the second kind and the modified *Bessel function* of the *second kind*. The modified Bessel function of the first kind has a discontinuity at the origin and the Bessel function of the second kind has an asymptotic form. Hence these are discarded in the solutions for fiber modes. For the solutions to be well behaved, that is, be finite at $r = 0$ and tend to zero as $r \to \infty$, it is essential that both u and ω are real. Therefore, a valid solution of Eq. (3.98) would be given by the first kind of Bessel function of order l and that of Eq. (3.99) would be given by the second kind of modified Bessel function of order l. Thus, one finds:

$$R(r) = A J_l \left(\frac{ur}{a} \right), r < a \quad (3.100)$$

$$R(r) = B K_l \left(\frac{wr}{a} \right), r > a \quad (3.101)$$

The Bessel function of the first kind of order l and argument x, denoted by $J_l(x)$, is defined in terms of an infinite series as follows:

$$J_l(x) = \sum_{n=0}^{\infty} \frac{(-1)^n}{n! \Gamma(n + l + 1)} \left(\frac{x}{2} \right)^{2n+1} \quad (3.102)$$

where $x = ur/a$ and the gamma function $\Gamma(n + l + 1) = (n + l)!$ For $l = 0$:

$$J_0(x) = 1 - \frac{\left(\frac{1}{4}x^2\right)}{(1!)^2} + \frac{\left(\frac{1}{4}x^2\right)^2}{(2!)^2} + \frac{\left(\frac{1}{4}x^2\right)^3}{(3!)^2} + \ldots$$

and for $l = 1$:

$$J_1(x) = \frac{1}{2}x - \frac{\left(\frac{1}{2}x\right)^3}{2!} - \frac{\left(\frac{1}{2}x\right)^5}{2!3!} - \ldots$$

and so on the higher values of l.

The second kind of modified Bessel function of order l is given by:

$$K_l(\tilde{x}) = (\pi/2)\, i^{-(l+1)}\, H_l(-i\tilde{x}), \quad \tilde{x} = \frac{wr}{a} \qquad (3.103)$$

where $H_l(-i\tilde{x})$ is a Hankel function, which is a linear combination of Bessel functions of the first (J_l) and second (Y_l) kind. These functions have been chosen for $x \ll 1$:

$$J_l(x) = \frac{1}{l!}\left(\frac{x}{2}\right)^l, \quad l = 0, 1, \ldots \qquad (3.104)$$

$$K_l(\tilde{x}) = (l-1)!\, 2^{l-1}\, \tilde{x}^{-1}, \quad l \geq 1 \qquad (3.105)$$

and for $x \gg 1$:

$$J_l(\tilde{x}) = \sqrt{\frac{2}{\pi x}}\, \cos\left[x - \frac{\pi(2l+1)}{4}\right] \qquad (3.106)$$

and

$$K_l(\tilde{x}) = \sqrt{\frac{2}{\pi x}}\, e^{-x}\left[1 + \frac{4l^2 - 1}{8\tilde{x}}\right] \qquad (3.107)$$

Thus $J_1(x)$ is a well behaved function for $r < a$ and $K_l(\tilde{x})$ is well behaved for $r > a$.

For further analysis, some recurrence relations for these functions (with argument x) and some asymptotic forms are given as follows:

$$J_{-l} = (-1)^l\, J_l \qquad (3.108)$$

$$J_l' = \frac{1}{2}(J_{l-1} - J_{l+1}) = \pm J_{l\mp 1} \mp \frac{lJ_l}{x} \qquad (3.109)$$

$$J_{l\mp 1} = \frac{2lJ_l}{x} - J_{l\mp 1} \qquad (3.110)$$

$$J_{l\mp 2} = \frac{2(l \mp 1)J_{l\mp 1}}{x} - J_l \qquad (3.111)$$

$$K_l = K_{-l} \qquad (3.112)$$

$$K_l' = \frac{1}{2}(K_{l-1} - K_{l+1}) = \mp \frac{lK_l}{x} - K_{l\mp 1} \qquad (3.113)$$

$$K_{l\mp 1} = \mp \frac{2lK_l}{x} + K_{l\pm 1} \qquad (3.114)$$

$$K_{l\mp 2} = \mp \frac{2(l \mp 1)K_{l\mp 1}}{x} + K_l \qquad (3.115)$$

Here, prime denotes the first derivative. For $\tilde{x} \ll 1$, we have:

$$\frac{K_0}{K_1} = \tilde{x} \ln \frac{2}{1.782\,\tilde{x}} \tag{3.116}$$

$$\frac{K_{l-1}}{K_l} = \frac{\tilde{x}}{2(l-1)}, \; l \geq 2 \tag{3.117}$$

$$\frac{K_{l+1}}{K_l} = \frac{2l}{\tilde{x}}, \; l \geq 1 \tag{3.118}$$

For $\tilde{x} \gg 1$, we have

$$\frac{K_{l \mp 1}}{K_l} = 1 + \frac{1 \mp 2l}{2\tilde{x}} \tag{3.119}$$

Since ψ is continuous at $r = a$, $R(r)$ must be continuous at $r = a$. Imposing this condition, one finds the values of constants A and B. Thus from Eqs (100) and (101), we have:

$$A = \frac{R(a)}{J_l(u)} \tag{3.120}$$

$$B = \frac{R(a)}{K_l(w)} \tag{3.121}$$

Substituting the values of R and Φ in Eq. (3.88), one obtains the transverse dependence of the modal fields as follows:

$$\psi(r, \phi) = A J_l\left(\frac{ur}{a}\right) \begin{bmatrix} \cos(l\phi); \, r < a \\ \sin(l\phi) \end{bmatrix} \tag{3.122}$$

and

$$\psi(r, \phi) = B K_l\left(\frac{wr}{a}\right) \begin{bmatrix} \cos(l\phi); \, r > a \\ \sin(l\phi) \end{bmatrix} \tag{3.123}$$

Now $\partial\psi/\partial r$ is also continuous at $r = a$:

$$\left.\frac{\partial\psi}{\partial r}\right|_{r=a} = A\frac{u}{a} J_l'\left(\frac{ur}{a}\right)\cos l\phi \bigg|_{r=a} = A\frac{u}{a} J_l'(u) \cos l\phi$$

$$\left.\frac{\partial\psi}{\partial r}\right|_{r=a} = B\frac{w}{a} K_l'\left(\frac{wr}{a}\right)\cos l\phi \bigg|_{r=a} = B\frac{w}{a} K_l'(w) \cos l\phi$$

Thus, substituting the values of A and B from Eqs (3.120) and (3.121), one obtains

$$\frac{R(a)}{J_l(u)} \frac{u}{a} J_l'(u) \cos l\phi = \frac{R(a)}{K_l(\omega)} \frac{w}{a} K_l'(w) \cos l\phi$$

or

$$\frac{u J_l'(u)}{J_l(u)} = \frac{w K_l'(w)}{K_l(w)}$$

Thus the continuity of ψ and $d\psi/dr$ at the core-cladding interface ($r = a$) leads to an eigenvalue equation of the form:

$$\frac{uJ_l'(u)}{J_l(u)} = \frac{wK_l'(w)}{K_l(w)} \tag{3.124}$$

Substituting the the values of J_l' and K_l' from Eqs (3.109) and (3.110), respectively, in Eq. (3.124), one obtains:

$$\frac{u}{J_l(u)}\left[\pm J_{l\mp 1}(u) \mp \frac{lJ_l(u)}{u}\right] = \frac{w}{K_l(w)}\left[\mp \frac{lK_l(w)}{w} - K_{l\mp 1}(w)\right]$$

or

$$\pm u\frac{J_{l\mp 1}(u)}{J_l(u)} \mp l = \mp l - \frac{wK_{l\mp 1}(w)}{K_l(w)}$$

One can write this equation in either of the following two forms:

$$\frac{uJ_{l+1}(u)}{J_l(u)} = \frac{wK_{l+1}(w)}{K_l(w)} \tag{3.125}$$

or

$$\frac{uJ_{l+1}(u)}{J_l(u)} = -w\frac{K_{l+1}(w)}{K_l(w)} \tag{3.126}$$

One can obtain, from Eqs (3.125) and (3.126), the values of u and w for various values of l and the corresponding values of the propagation constant β. For β-values lying within the range:

$$\beta_2^2 (= n_2^2 k^2) < \beta^2 < \beta_1^2 (= n_1^2 k^2) \tag{3.127}$$

the radial part of the field, $R(r)$, in the core is given by the Bessel function $J_l(x)$, $x = ur/a$, which is oscillatory in nature. Hence there exist m allowed solutions for β for each value of l. Therefore, each allowed value of β is characterized by two integers l and m. The first integer l is associated with two circular functions $\cos l\phi$ and $\sin l\phi$ corresponding to the azimuthal part of the solution, and the second integer m is associated with the mth root of the eigenvalue equation corresponding to the radial part of the solution. These are known as guided modes.

From Eq. (3.97), we have $V^2 = u^2 + w^2$. Thus the solution of the transcendental equations (for given values of l and V) will give universal curves describing the dependence of u or w on V. The value of β can be calculated by substituting the values of u (or w) in the defining equations. Alternatively, one can define the normalized propagation constant b as follows:

$$b = \frac{\beta^2 - \beta_2^2}{\beta_1^2 - \beta_2^2} = \frac{\beta^2 - n_2^2 k^2}{n_1^2 k^2 - n_2^2 k^2} = \frac{w^2}{V^2} = 1 - \frac{u^2}{V^2} \tag{3.128}$$

Since β lies between β_1 (= $n_1 k$) and β_2 (= $n_2 k$) for the guided nodes, the value of b will lie between 0 (for $\beta = \beta_2$) and 1 (for $\beta = \beta_1$). The plots of b as a function of V, shown in Fig. 3.6, form universal curves.

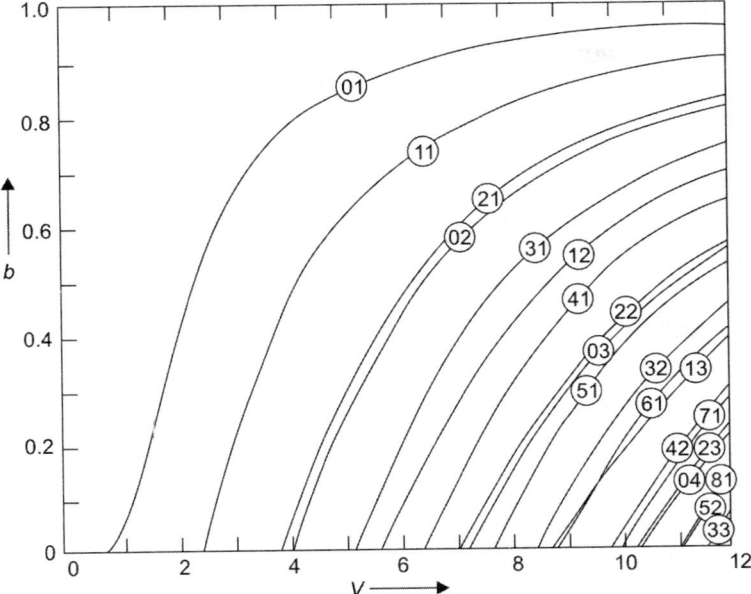

Fig. 3.6. Plots of the normalized propagation constant *b* as a function of the normalized frequency parameter *V* for a silicon (Si) fiber

A mode ceases to be guided when $\beta^2 < \beta_2^2$. Such modes are called *radiation modes*. In terms of the ray model, these modes correspond to rays that undergo refraction, rather than total internal reflection, at the core-cladding interface. When such modes are excited, they quickly leak away from the core. The condition $\beta = \beta_2$ corresponds to what is known as the *cut-off of a mode*. Thus, at $\beta = \beta_2$; $b = 0$, $w = 0$, and $u = V = V_c$, where V_c is the cut-off frequency. We may note that:

$$\lim_{w \to 0} w \frac{K_{l-1}(w)}{K_l(w)} \to 0, l = 0, 1, 2, 3, \ldots$$

Therefore, the RHS of Eq. (3.125) vanishes as $w \to 0$. Thus, Eq. (3.125) may be used for finding the cut-off frequencies of different modes.

For $l = 0$, one obtains from Eq. (3.125)

$$u \frac{J_{-1}(u)}{J_0(u)} = -w \frac{K_{-1}(w)}{K_0(w)}$$

Using Eqs (3.108) and (3.112), one obtains:
$$J_{-1}(u) = -J_1(u) \text{ and } K_{-1}(w) = -K_1(w)$$

Substituting the value of $J_{-1}(w)$ and $K_{-1}(w)$ in the above equation, one obtains:

$$\frac{uJ_1(u)}{J_0(u)} = w \frac{K_1(w)}{K_0(w)} \tag{3.129}$$

All the cuf-off frequency, $w = 0$, $u = V = V_c$, the RHS = 0, and Eq. (3.129) is transformed into:

$$\frac{V_c J_1(V_c)}{J_0(V_c)} = 0$$

or

$$J_1(V_c) = 0 \tag{3.130}$$

The roots of Eq. (3.130) give the value of the cut-off frequency for $l = 0$ and $m = 1, 2, 3, \ldots$.

Similarly, for $l = 1$, one obtains from Eq. (3.126):

$$\frac{u J_0(u)}{J_1(u)} = -w \frac{K_0(w)}{K_1(w)} \tag{3.131}$$

At the cut-off frequency, $w = 0$, $u = V = V_c$, the RHS = 0, and we obtain:

$$J_0(V_c) = 0 \tag{3.132}$$

The roots of Eq. (3.132) gives V_c for $l = 1$ and $m = 1, 2, 3, \ldots$.

For $l \geq 2$ modes, the following equation gives the value of V_c.

$$J_{l-1}(V_c) = 0, \quad V_c \neq 0$$

We may note here that, for $l > 2$, the root $V_c = 0$ must be included because:

$$\lim_{V \to 0} V \frac{J_{l-1}(V)}{J_l(V)} \neq 0 \text{ for } l \geq 2$$

The values of the cut-off frequencies for the first few LP modes are given in Table 3.1. Figure 3.7 shows the oscillatory nature of Bessel functions J_0 and J_1 and their roots.

Table 3.1. Cut-off frequencies of the first few lower order LP_{lm} modes in a step index (Si) fiber

l	m			
	1	2	3	4
0	0	3.832	7.106	10.173
1	2.405	5.520	8.654	11.790
2	3.832	7.016	10.173	13.324
3	5.136	8.417	11.620	14.796

Prior to proceeding further, we must remember few important points about the modes:

(i) The $l = 0$ modes have twofold degeneracy corresponding to two orthogonal linearly polarized states.

(ii) The $l \geq 1$ modes have fourfold degeneracy as each polarization state may have ϕ-dependence of the $\cos l\phi$ type or the $\sin l\phi$ type.

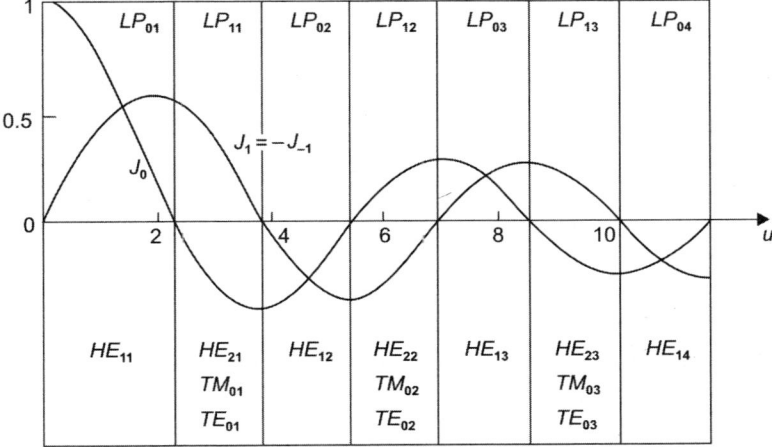

Fig. 3.7. Plot of Bessel functions J_0 and J_1, indicating the range of allowed values of u for lower order modes

(iii) Total number of modes (when $V \gg 1$) guided along a step-index fiber is given approximately by $M = V^2/2$. As we know from Eq. (3.97) that the V-value of the fiber is dependent on the dimensions, the numerical aperture of the fiber, and the wavelength of the light signal; the total number of guided modes in a particular fiber at a specific wavelength is fixed.

3.7.2 Fraction Modal Power Distribution

Now, we calculate the fractional modal power distribution in the core and cladding of a Si fiber using a scalar approximation. In the core, power is given by:

$$P_{core} = (\text{constant}) \int_{r=0}^{a} \int_{\phi=0}^{2\pi} |\psi(r,\phi)|^2 \, r \, dr \, d\phi$$

$$= (\text{constant}) \{R(a)\}^2 \int_{r=0}^{a} \frac{J_l^2(ur/a) \, r \, dr}{J_l^2(u)} \int_{\phi=0}^{2\pi} \cos^2 l\phi \, d\phi$$

Here, we have substituted the value of $\psi(r,\phi)$ from Eq. (3.122) and taken the ϕ-dependence to be of the form $\cos(l\phi)$. The solution of the above integral can be shown to be:

$$P_{core} = C\pi a^2 \left[1 - \frac{J_{l-1}(u) J_{l+1}(u)}{J_l^2(u)} \right] \tag{3.133}$$

where C is a constant.

Similarly, one can obtain the power distribution in the cladding by solving the integral:

$$P_{\text{cladding}} = (\text{constant}) \int_{r=a}^{\infty} \int_{\phi=0}^{2\pi} |\psi(r,\phi)|^2 \, r \, dr \, d\phi$$

Substituting the value of $\psi(r,\phi)$ from Eq. (3.123) in the above relation and again taking the $\cos(l\phi)$ form, one obtains:

$$P_{\text{cladding}} = (\text{constant}) \{R(a)\}^2 \int_{r=a}^{\infty} \frac{K_l^2(wr/a)}{K_l^2(w)} r \, dr \int_{\phi=0}^{2\pi} \cos^2(l\phi) d\phi$$

$$= C\pi a^2 \left[\frac{K_{l-1}(w) K_{l+1}(w)}{K_l^2(w)} - 1 \right] \tag{3.134}$$

Adding Eqs (3.133) and (3.134), one can obtain an expression for the total power P_T as follows:

$$P_T = P_{\text{core}} + P_{\text{cladding}}$$

$$= C\pi a^2 \left[1 - \frac{J_{l-1}(u) J_{l+1}(u)}{J_l^2(u)} \right] + C\pi a^2 \left[\frac{K_{l-1}(w) K_{l+1}(w)}{K_l^2(w)} - 1 \right]$$

$$= C\pi a^2 \left[\frac{K_{l-1}(w) K_{l+1}(w)}{K_l^2(w)} - \frac{J_{l-1}(u) J_{l+1}(u)}{J_l^2(u)} \right] \tag{3.135}$$

Multiplying the eigenvalue equation (3.125) and (3.124), one obtains:

$$\frac{u^2 J_{l-1}(u) J_{l+1}(u)}{J_l^2(u)} = -\frac{w^2 K_{l-1}(u) K_{l+1}(u)}{K_l^2(u)} \tag{3.136}$$

Using Eq. (3.136), Eq. (3.135) may be expressed as:

$$P_T = C\pi a^2 \left(\frac{K_{l-1}(w) K_{l+1}(w)}{K_l^2(w)} \right) \left(1 + \frac{w^2}{u^2} \right)$$

$$= C\pi a^2 \frac{V^2}{u^2} \left[\frac{K_{l-1}(w) K_{l+1}(w)}{K_l^2(w)} \right] \tag{3.137}$$

Now, using Eq. (3.136), Eq. (3.133) may be expressed as:

$$P_{\text{core}} = C\pi a^2 \left[1 + \frac{w^2}{u^2} \frac{K_{l-1}(w) K_{l+1}(w)}{K_l^2(w)} \right] \tag{3.138}$$

Dividing Eq. (3.138) by Eq. (3.137), one obtains the fractional power propagating in the core as:

$$\frac{P_{\text{core}}}{P_T} = \frac{C\pi a^2 \left[1 + \dfrac{w^2}{u^2} \dfrac{K_{l-1}(w) K_{l+1}(w)}{K_l^2(w)} \right]}{C\pi a^2 \dfrac{V^2}{u^2} \left[\dfrac{K_{l-1}(w) K_{l+1}(w)}{K_l^2(w)} \right]}$$

$$= \left[\frac{u^2}{V^2} \frac{K_l^2(w)}{K_{l-1}(w) K_{l+1}(w)} + \frac{w^2}{V^2}\right] \quad (3.139)$$

The functional power propagating in the cladding is expressed as:

$$\frac{P_{\text{cladding}}}{P_T} = 1 - \frac{P_{\text{core}}}{P_T} = \frac{u^2}{V^2}\left[1 - \frac{K_l^2(w)}{K_{l-1}(w) K_{l+1}(w)}\right] \quad (3.140)$$

Fig. 3.8 shows the plots of fractional power propagating in the core and the cladding for some lower order LP modes. It is interesting to note that for the first two lower order modes, the power flow is mostly in the cladding near cut-off. Using Eqs (3.116) to (3.118), one can show that as $V \to V_c$, $w \to 0$, and $u \to V_c$, we obtain

$$\frac{P_{\text{core}}}{P_T} \to \begin{cases} 0 & \text{for } l = 0 \text{ and } 1 \\ \frac{(l-1)}{l} & \text{for } l \geq 2 \end{cases} \quad (3.141)$$

However, for larger values of l, the power remains in the core even at or just beyond cut-off. Another point to be mentioned is that the power associated with a particular mode is mostly confined in the core for large values of V.

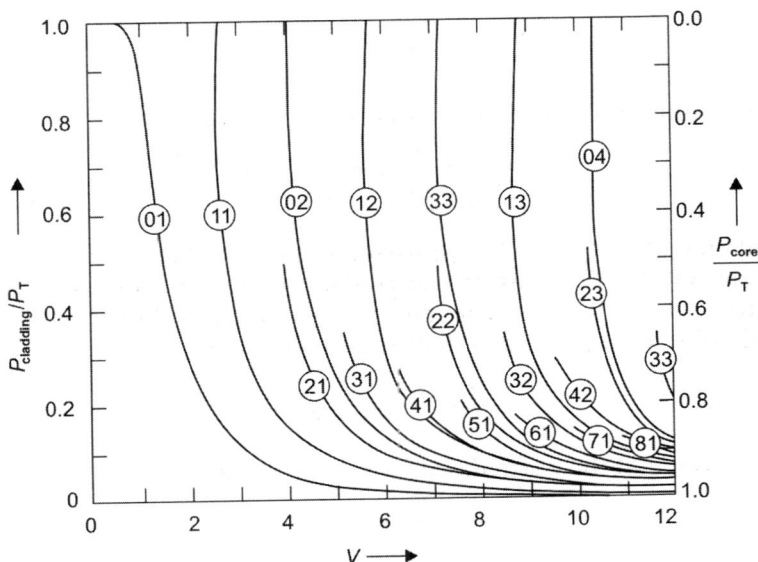

Fig. 3.8. Schematic of plots of fractional power contained in the core and cladding of a Si fiber as a function of V

3.7.3 Graded Index (GI) Fibers

The refractive index profile for a multimode graded index (GI) fiber is given by:

$$n(r) = n_0\{1 - 2\Delta (r/a)^\alpha\}^{1/2}, \quad r \le a \tag{3.142}$$

$$= n_0\{1 - 2\Delta\}^{1/2} = n_c, \quad r \ge a \tag{3.143}$$

where $\Delta = (n_0^2 - n_c^2)/2n_0^2 = \dfrac{(n_0 - n_c)}{n_0}$

when $\Delta \ll 1$, n_0 is the refractive index at $r = 0$, n_c is the refractive index of the cladding, and α is called the profile parameter.

In general, for a cylindrical dielectric waveguide, e.g. an optical fiber, the electric and magnetic field are governed by Eqs (3.24) and (3.27) respectively. The exact solution of these equations for graded-index (GI) fibers are difficult to obtain. However, using the WKB (Wentzel-Kramers-Brillouin) approximation, one can study the propagation characteristics of these fibers. The propagation constant b_p of the pth mode in a GI fiber with an α-profile may be given by:

$$\beta_p \approx \beta_0 \left[1 - 2\Delta\left(\frac{p}{M_g}\right)\right]^{1/2} \tag{3.144}$$

where $p = 1, 2, 3, \ldots, M_g$ and $\beta_0 = kn_0$. Here, M_g represents the total number of guided modes given by:

$$M_g = k^2 n_0^2 a^2 \, \Delta \frac{1}{2}\left(\frac{\alpha}{\alpha+2}\right) \tag{3.145}$$

Substituting the values of Δ and k, one obtains:

$$M_g = \frac{1}{2}\left(\frac{\alpha}{\alpha+2}\right)\left[\frac{2\pi a}{\lambda}(n_0^2 - n_c^2)^{1/2}\right]^2 \tag{3.146}$$

Equation (3.146) gives the approximate modal volume or the number of modes guided by the α-profile graded-index fiber. For a step-index fiber ($\alpha = \infty$), $n = n_1$ in the core and $n = n_2$ in the cladding. Clearly, the modal volume M_s for such a fiber will be given by:

$$M_s \approx \frac{1}{2}\left[\frac{2\pi a}{\lambda}(n_1^2 - n_2^2)^{1/2}\right]^2 \tag{3.147}$$

We know that the normalized frequency parameter is given by:

$$V = \frac{2\pi a}{\lambda}(n_1^2 - n_2^2)^{1/2}$$

Therefore

$$M_s \approx \frac{V^2}{2} \tag{3.148}$$

For a GI fiber, the numerical aperture (NA) for the guided rays varies with r as follows:

$$NA(r) = \begin{cases} \{n^2(r) - n_c^2\}^{1/2} & \text{for } r < a \\ 0 & \text{for } r \geq a \end{cases}$$

However, for small variation of $n(r)$ with r:

$$NA \approx \{n_0^2 - n_c^2\}^{1/2}$$

and

$$V \approx \frac{2\pi a}{\lambda} (n_0^2 - n_c^2)^{1/2}$$

With this approximation, one finds:

$$M_g \cong \frac{\alpha}{\alpha + 2} \frac{V^2}{2} \tag{3.149}$$

We must take care when using Eqs (3.144)–(3.149) because the WKB is valid for highly multimoded fibers with V much greater than 1.

For a parabolic profile, $\alpha = 2$, and:

$$M_g \approx \frac{V^2}{4}$$

One can show that the cut-off value of the normalized frequency, V_c, to support a single mode in a graded-index fiber is given by:

$$V_c = 2.405 \left(1 + \frac{2}{\alpha}\right)^{1/2} \tag{3.150}$$

Clearly, it is possible to determine the structural and/or operational parameters of the fiber which give single-mode operation.

3.7.4 Limitations of Multimode Fibers

We have read that multimode fibers support many modes. The higher order modes (corresponding to oblique rays in terms of the ray model) travel slower and hence arrive at the other end of the fiber later than the lower order modes (corresponding to axial rays). This means that different modes travel with different group velocities. Thus, a light pulse propagating through such a fiber will get broadened. This is called *multipath dispersion* or *intermodal dispersion*. The pulse dispersion per unit length for a step-index fiber ($\alpha = \infty$) is given by:

$$\frac{\Delta T}{L} = \frac{n_1}{n_2}\left(\frac{n_1 - n_2}{c}\right) \approx \frac{n_1 \Delta}{c} \tag{3.151}$$

Similarly, one can show that the pulse dispersion per unit length for a graded-index fiber with a parabolic profile ($\alpha = 2$) may be expressed as:

$$\frac{\Delta T}{L} = \frac{n_0}{2c} \Delta^2 \tag{3.152}$$

and that for a GI fiber with an optimum profile ($\alpha = 2 - 2\Delta$) may be expressed as:

$$\frac{\Delta T}{L} \approx \frac{n_0}{8c} \Delta^2 \qquad (3.153)$$

Thus, pulse broadening due to intermodal dispersion, varying from about 0.05 nm/km to 80 nm/km, depending on the value of α and the core index n_0, has been observed in multimode fibers. This severely restricts the use of such fibers in long-haul communications.

Moreover, a light pulse has a number of spectral components (of different frequencies), and the group velocity of a mode varies with frequency. Therefore different spectral components in the pulse propagate with slightly different group velocities, resulting in pulse broadening. This phenomenon is called *group velocity dispersion* (GVD) or *intramodal dispersion*.

We may note that in spite of these limitations, multimode fibers are used in local area network and short-haul communication links. However, prior to using them, the link power budget and the rise-time budget have to be analysed.

3.8 PHOTONIC CRYSTAL FIBERS

So far we have concentrated on optical fibers comprising solid silica core and cladding regions in which the light is guided by a small increase in refractive index in the core facilitated through doping the silicon with germanium. More recently, however, a new class of microstructured optical fiber containing a fine array of air holes running longitudinally down the fiber cladding has been developed. Since the microstructure within the fiber is often highly periodic due to the fabrication process, these fibers are usually referred to as *photonic crystal fibers* (PCFs), or sometimes just as holey fibers. Whereas in conventional optical fibers electromagnetic modes are guided by total internal reflection in the core region, which has a slightly raised refractive index, in PCFs two distinct guidance mechanisms arise.

Although the guided modes can be trapped in a fiber core which exhibits a higher average index than the cladding containing the air holes by an effect similar to total internal reflection, alternatively they may be trapped in a core of either higher, or indeed lower, average index by a photonic bandgap effect. In the former case the effect is often termed modified total internal reflection and the fibers are referred to as index guided, while in the latter they are called photonic bandgap fibers. Furthermore, the existence of two different guidance mechanisms makes PCFs versatile in their range of potential applications. For example, PCFs have been used to realize various optical components and devices including long period gratings, multimode interference power splitters, tunable coupled cavity fiber lasers, fiber amplifiers,

Optical Fiber Waveguides 85

multichannel add/drop filters, wavelength converters and wavelength demultiplexers. As with conventional optical fibers, however, a crucial issue with PCFs has been the reduction in overall transmission losses which were initially several hundred decibels per kilometer even with the most straightforward designs. Increased control over the homogeneity of the fiber structure together with the use of highly purified silicon as the base material has now lowered these losses to a level of a very few decibels per kilometer for most PCF types, with a loss of just 0.3dB km^{-1} at 1.55 μm for a 100 km span being recently reported.

3.8.1 Index-Guided Microstructures

Although the principles of guidance and the characteristics of index-guided PCFs are similar to those of conventional fiber, there is greater index contrast since the cladding contains air holes with a refractive index of 1 in comparison with the normal silica cladding index of 1.457 which is close to the germanium-doped core index of 1.462. A fundamental physical difference, however, between index-guided PCFs and conventional fibers arises from the manner in which the guided mode interacts with the cladding region. Whereas in a conventional fiber this interaction is largely first order and independent of wavelength, the large index contrast combined with the small structure dimensions cause the effective cladding index to be a strong function of wavelength. For short wavelengths the effective cladding index is only slightly lower than the core index and hence they remain tightly confined to the core. At longer wavelengths, however, the mode samples more of the cladding and the effective index contrast is larger. This wavelength dependence results in a large number of unusual optical properties which can be tailored. For example, the high index contrast enables the PCF core to be reduced from around 8 μm in conventional fiber to less than 1 μm, which increases the intensity of the light in the core and enhances the nonlinear effects.

Two common index-guided PCF designs are shown diagrammatically in Fig. 3.9. In both cases a solid-core region is surrounded by a cladding region containing air holes. The cladding region in Fig. 3.9(a) comprises a hexagonal array of air holes while in Fig. 3.9(b) the cladding air holes are not uniform in size and do not extend too far from the core. It should be noted that the hole diameter d and hole to hole spacing or pitch Λ are critical design parameters used to specify the structure of the PCF. For example, in a silica PCF with the structure depicted in Fig. 3.9(a) when the air fill fraction is low (i.e. $d/\Lambda < 0.4$), then the fiber can be single-moded at all wavelength. This property, which cannot be attained in conventional fibers, is particularly significant for broadband applications such as *wavelength division multiplexed transmission*.

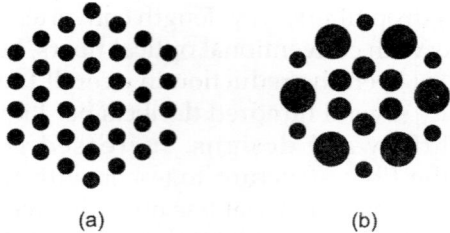

Fig. 3.9. Two index-guided photonic crystal fiber structures. The dark areas are air holes while the white areas are silica

As PCFs have a wider range of optical properties in comparison with standard optical fibers, they provide for the possibility of new and technologically important fiber devices. When the hole region covers more than 20% of the fiber cross-section, for instance index-guided PCFs display an interesting range of dispersive properties which could find application as *dispersion-compensating* or *dispersion-controlling fiber components*. In such fibers it is possible to produce very high optical nonlinearity per unit length in which modest light intensities can induce substantial nonlinear effect. For example, while several kilometers of conventional fiber are normally required to achieve 2R data regeneration, it was obtained with just 3.3 m of large air-filling fraction PCF. In addition, filling the cladding holes with polymers or liquid crystals allows external fields to be used to dynamically vary the fiber properties. The temperature sensitivity of a polymer within the cladding holes may be employed to tune a Bragg grating written into the core. By contrast, index-guided PCFs with small holes and large hole spacings provide very large mode area (and hence low optical nonlinearities) and have potential applications in high-power delivery (e.g. laser welding and machining) as well as high-power fiber lasers and amplifiers. Furthermore, the large index contrast between silica and air enables production of such PCFs with large multimoded cores which also have very high numerical aperture values (greater than 0.7). Hence these fibers are useful for the collection and transmission of high optical powers in situations where signal distortion is not an issue. Finally, it is apparent that PCFs can be readily spliced to conventional fibers, thus enabling their integration with existing components and subsystems.

3.8.2 Photonic Bandgap Fibers

Photonic bandgap (PBG) fibers are a class of microstructured fiber in which a periodic arrangement of air holes is required to ensure guidance. This periodic arrangement of cladding air holes provides for the formation of a photonic bandgap in the transverse plane of the fiber. As a PBG fiber exhibits a two-dimensional bandgap, then wavelength within this bandgap cannot propagate perpendicular to the fiber axis (i.e. in the cladding) and they can therefore be confined to propagate within a region in which the refractive index is lower than the surrounding material. Hence utilizing the photonic bandgap effect light can, for example, be guided within a low-index, air-filled core region creating fiber properties quite different from those obtained without the bandgap. Although, as with index-guided PCFs, PBG fibers can also guide light in regions with higher refractive index, it is the lower index region guidance feature which is of particular interest. In addition, a further distinctive feature is that while index-guiding fibers usually have a guided mode at all wavelengths, PBG fibers only guide in certain wavelength bands, and furthermore, it is possible to have wavelengths at which higher order modes are guided while the fundamental mode is not.

Two important PBG fiber structures are displayed in Fig. 3.10. The honeycomb fiber design shown in Fig. 3.10(a) was the first PBG fiber to be experimentally realized in 1998 and adaptations of this structure continue to be pursued. A triangular array of air holes of sufficient size as displayed in Fig. 3.10(b), however, provides for the possibility, unique to PBG fibers, of guiding electromagnetic modes in air. In this case a large hollow core has been defined by removing the silica around seven air holes in the center of the structure. These fibers, which are termed *air-guiding* or *hollow-core PBG fibers*, enable more than 98% of the guided mode field energy to propagate in the air regions. Such air-guiding fibers have attracted attention because they potentially provide an environment in which optical propagation can take place with little attenuation as the localization of light in the air core removes the limitations caused by material absorption losses. The fabrication of hollow-core fiber with low propagation losses, however, has proved to be quite difficult, with losses of the order of 13 dB km^{-1}. Moreover, the fibers tend to be highly dispersive with narrow transmission windows and while single-mode operation is possible, it is not as straightforward to achieve in comparison with index-guiding PCFs.

More recently, the fabrication and characterization of a new type of solid silica-based photonic crystal fiber which guides light using the

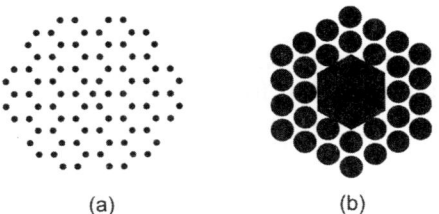

Fig. 3.10. Photonic bandgap (PBG) fiber structures in which the dark areas are air (lower refractive index) and the lighter area is the higher refractive index: (a) honeycomb PBG fiber; (b) air-guiding PBG fiber

PBG mechanism has been reported in the literature. The fiber employed a two-dimensional periodic array of germanium-doped rods in the core region. It was therefore referred to as a nanostructure core fiber and exhibited a minimum attenuation of 2.6 dB km^{-1} at a wavelength of 1.59 µm. Furthermore, the fiber displayed greater bending sensitivity than conventional single-mode fiber as a result of the much smaller index difference between the core and the leaky modes which could provide for potential applications in the optical sensing of curvature and stress. In addition, it is indicated that the all-solid silica structure would facilitate fiber fabrication using existing technology and birefringence of the order of 10^{-4} is easily achievable with a large mode field diameter up to 10 µm, thus enabling its use within fiber lasers and gyroscope applications.

Example 1

A silica optical fiber with a core diameter large enough to be considered by ray theory analysis has a core refractive index of 1.50 and a cladding refractive index of 1.47. Calculate (a) the critical angle at the core-cladding interface; (b) the NA for the fiber; (c) the acceptance angle in air for the fiber

Solution

(a) $\theta_c = \sin^{-1} \dfrac{n_2}{n_1}$

$= \sin^{-1} \dfrac{1.47}{1.50} = 78.5°$

(b) $NA = (n_1 - n_2)^{½} = [(1.50)^2 - (1.47)^2]^{½}$
$= (2.25 - 2.16)^{½}$
$= 0.30$

(c) The acceptance angle:
$\theta_a = \sin^{-1}(NA) = \sin^{-1} 0.30$
$= 17.4°$

Example 2

An optical fiber in air has an NA of 0.4. Compare the acceptance angle for meridional rays with that for skew rays which change by 100° at each reflection.

Solution

The acceptance angle for meridional ray is given by:

$$NA = n_0 \sin \theta_a = (n_1^2 - n_2^2)^{½}$$

With $n_0 = 1$, we have
$$\theta_a = \sin^{-1} NA$$
$$= \sin^{-1} 0.4$$
$$= 23.6°$$

The skew rays change direction by 100° at each reflection, therefore $\gamma = 50°$. Hence:

$$\theta_{as} = \sin^{-1}\left(\frac{NA}{\cos \gamma}\right)$$

$$= \sin^{-1}\left(\frac{0.4}{\cos 50°}\right)$$

$$= 38.5°$$

In this example, the acceptance angle for the skew rays is about 15° greater than the corresponding angle for meridional rays. However, we must note that we have compared the θ_a of one particular skew ray path. When the light input to the fiber is at an angle to the fiber axis, it is possible that γ will vary from zero for meridional rays for 90° for rays which enter the fiber at the core-cladding interface giving acceptance of skew rays over a conical half angle of $\pi/2$ radians.

Example 3

A symmetric step-index (SI) planar waveguide is made of glass with $n_1 = 1.5$ and $n_2 = 1.49$. The thickness of the guide layer is 9.83 μm and the guide is excited by a source of wavelength $\lambda = 0.85$ μm. What is the range of the propagation constants? What is the maximum number of modes supported by the guide? [B.Tech.]

Solution

The phase propagation constant β lies between β_1 and β_2. Here:

$$\beta_1 = kn_1 = \frac{2\pi n_1}{\lambda} = \frac{2\pi}{(0.85 \times 10^{-6})} \times 1.50 = 11.0082 \times 10^6 \text{ m}^{-1}$$

and $\quad \beta_2 = kn_2 = \dfrac{2\pi n_2}{\lambda} = \dfrac{2\pi}{(0.85 \times 10^{-6})} \times 1.49 = 11.008 \times 10^6 \text{ m}^{-1}$

The maximum number of modes that the guide can support is given by Eq. (3.43), i.e.:

$$M = \frac{4a}{\lambda}(n_1^2 - n_2^2)^{1/2} = \frac{2 \times (9.83)}{0.85}[(1.5)^2 - (1.49)^2]^{1/2} \approx 4$$

Example 4

A multimode step index fiber with a core diameter of 80 μm and a relative index difference of 1.5% is operating at a wavelength of 0.85

μm. If the core refractive index is 1.48, estimate (a) the normalized frequency for the fiber; (b) the number of guided modes.

Solution

(a) The normalized frequency is given by:

$$V \simeq \frac{2\pi}{\lambda} a n_1 (2\Delta)^{1/2}$$

$$= \frac{2\pi \times 40 \times 10^{-6} \times 1.48}{0.85 \times 10^{-6}} (2 \times 0.015)^{1/2}$$

$$= 75.8$$

(b) Total number of guided modes is given by:

$$M_s \simeq \frac{V^2}{2} = \frac{5745.6}{2} = 2873$$

Hence this fiber has a V number of approximately 76, giving nearly 3000 guided modes.

Example 5

What should be the maximum thickness of the guide slab of a symmetrical SI planar waveguide so that it supports only the fundamental TE mode? Take $n_1 = 3.6$, $n_2 = 3.56$, and $\lambda = 0.85$ μm. [B.Tech.]

Solution

For the waveguide to support only the fundamental mode, V should be less than $\pi/2$, i.e.:

$$\frac{2\pi a}{\lambda} (n_1^2 - n_2^2)^{1/2} < \pi/2$$

or

$$2a < \frac{\lambda}{(n_1^2 - n_2^2)^{1/2}}$$

$$< \frac{0.85 \ \mu m}{2\{(3.6)^2 - (3.56)^2\}^{1/2}}$$

$$< 0.793 \ \mu m$$

Clearly, the thickness of the guide slab should not be more than 0.793 μm.

Example 6

The cutoff wavelength of a step index fiber is quoted as $\lambda_c = 1.20$ μm. If the fiber is operated at wavelength $\lambda = 1.55$ μm, what is V?

Solution

$$V = 2.405 \frac{\lambda_c}{\lambda}$$

$$= 2.405 \left(\frac{1.20}{1.55}\right) = 1.86$$

Example 7

Calculate the G-factor for the fundamental TE mode supported by the waveguide having $n_1 = 3.6$, $n_2 = 3.56$ and $\lambda = 0.85$ μm. Given $2a = 0.793$ μm. [B.Tech.]

Solution

For thickness $2a = 0.793$ μm, $V = \pi/2$. The point of intersection of an arc of $V = \pi/2$ with the curve $ua \tan ua = wa$ for $m = 0$ yield

$$ua = 0.934 \text{ and } wa = 1.262$$

Using Eq. (3.83), the confinement factor G may be calculated as follows:

$$G = \left[1 + \frac{\cos^2(0.934)}{1.262\{1 + (\sin 0.934)(\cos 0.934)/0.934\}}\right]^{-1}$$

$$= 0.8436$$

This means that only 84.36% of the total power carried by fundamental mode is confined within the guide layer; the remaining 15.64% extends into the cladding slabs.

Example 8

Graded index fiber with a parabolic refractive index profile core has a refractive index at the core axis of 1.5 and a relative index difference of 1%. Estimate the maximum possible core diameter which allows single mode operation at a wavelength of 1.3 μm.

Solution

The maximum value of normalized frequency for single mode operation is given by:

$$V = 2.405 \left(1 + \frac{2}{\alpha}\right)^{1/2}$$

Here $\alpha = 2$:

∴

$$V = 2.405 \left(1 + \frac{2}{2}\right)^{1/2}$$

$$= 2.405 \sqrt{2} = 2.4\sqrt{2}$$

The maximum core radius may be obtained from:

$$V = \frac{2\pi}{\lambda} a n_1 (2\Delta)^{1/2}$$

or

$$a = \frac{V\lambda}{2\pi n_1 (2\Delta)^{1/2}}$$

$$= \frac{2.4\sqrt{2} \times 1.3 \times 10^{-6}}{2\pi \times 1.5 \times (0.02)^{1/2}}$$

$$= 3.3 \, \mu m$$

Hence the maximum core diameter which allows single mode operation is approximately 6.6 µm.

Example 9

A multimode step-index fiber has a relative refractive index difference of 2% and a core refractive index of 1.5. The number of modes propagating at a wavelength of 1.3 µm is 1000. Calculate the diameter of the fiber core. [B.Tech.]

Solution

$$M_s = \frac{V^2}{2} = \frac{1}{2}\left[\frac{2\pi a}{\lambda} n_1 \sqrt{2\Delta}\right]^2 = \frac{1}{2}\frac{4\pi^2 a^2 n_1^2 2\Delta}{\lambda^2}$$

$$\therefore \quad 2a = \frac{\lambda}{\pi n_1}\left(\frac{M_s}{\Delta}\right)^{1/2} = \frac{1.3}{\pi \times 1.5}\sqrt{\frac{1000}{0.02}} = 62 \, \mu m$$

Example 10

Determine the cutoff wavelength (λ_c) for a step index fiber to exhibit single mode operation when the core refractive index and radius are 1.46 and 4.5 µm, respectivity, with the relative index difference being 0.25%.

Solution

The cutoff wavelength is given by

$$\lambda_c = \frac{2\pi a n_1}{V_c}(2\Delta)^{1/2}$$

$$= \frac{2\pi a n_1}{2.405}(2\Delta)^{1/2}$$

Substituting the values in the above relation, one obtains:

$$\lambda_c = \frac{2\pi \times 4.5 \times 1.46 \times (0.005)^{1/2}}{2.405} \, \mu m$$

$$= 1.214 \, \mu m$$

$$= 1214 \, nm$$

Hence the fiber is a single-mode *d* to *a* wavelength of 1214 nm.

Example 11

A step-index fiber has a numerical aperture of 0.17 and a core diameter of 100 µm. Determine the normalized frequency parameter of the fiber when light of wavelength 0.85 µm is transmitted through it. Also estimate the number of guided modes propagating in the fiber.
[B.Tech.]

Solution
We have:
$$V = \frac{2\pi a}{\lambda}(n_1^2 - n_2^2)^{1/2} = \frac{2\pi a(\text{NA})}{\lambda} = \frac{\lambda(100\ \mu m)(0.17)}{(0.85\ \mu m)}$$
$$= 62.83$$

Therefore $M_s = \frac{V^2}{2} = 1974$

Example 12

A graded-index fiber with a parabolic profile supports the propagation of 700 guided modes. The fiber has a relative refractive index difference of 2%, a core refractive index of 1.45 and a core diameter of 75 µm. Calculate the wavelength of light propagating in the fiber. Further, estimate the maximum diameter of the fiber core which can give single-mode operation at the same length.
[B.Tech.]

Solution

We have $M_g \cong \left(\frac{\alpha}{\alpha+2}\right)\frac{V^2}{2}$ for the mode volume to evaluate V. With $\alpha = 2$ for the parabolic profile, we have:
$$V = \sqrt{4M_g} = \sqrt{4 \times 700} = 52.91$$
and
$$\lambda = \frac{2\pi a}{V}n_1\sqrt{2\Delta} = \frac{\pi \times (75\ \pi m)}{52.91} \times 1.45 \times (2 \times 0.02)^{1/2}$$
$$= 1.3\ \mu m$$

The cutoff value of the normalized frequency V_c for single-mode operation in a graded-index fiber is given by the relation:
$$V_c = 2.405\left(1+\frac{2}{\alpha}\right)^{1/2}.$$

Thus, with $\alpha = 2$:
$$V_c = 2.405\sqrt{2}$$

The maximum core diameter may be obtained as follows:
$$2a = \frac{V_c \lambda}{\pi n_1 \sqrt{(2\Delta)}} = \frac{2.405\sqrt{2} \times (1.3\ \mu m)}{\pi \times 1.45 \times \sqrt{(2 \times 0.02)}} = 4.85\ \mu m$$

Example 13

Given that a useful approximation for the eigenvalue of the single-mode step index fiber cladding W is given by:
$$W(V) \simeq 1.1428\,V - 0.9960$$

Deduce an approximation for the normalized propagation constant $b(v)$. [B.Tech.]

Solution

The normalized propagation constant is given by:
$$b(V) = 1 - \frac{(V^2 - W^2)}{V^2} = \frac{W^2}{V^2}$$

Now, substituting the approximation given above, one obtains:
$$b(V) \simeq \left(\frac{(1.1428V - 0.9960)}{V^2}\right)^2$$
$$= \left(1.1428 - \frac{0.9960}{V}\right)^2$$

The relative error on this approximation for $b(V)$ is less than 0.2% for $1.5 \le V \le 2.5$ and less than 2% for $1 \le V \le 3$.

Example 14

A parabolic profile graded index single-mode fiber designed for operation at a wavelength of 1.30 µm has a cutoff wavelength of 1.08 µm. From experimental measurement it is established that the first minimum in the diffraction pattern occurs at an angle of 12°. Determine the spot size at the operating wavelength (based on the ESI technique). [B.Tech.]

Solution

The effective core radius is given by:
$$a_{\text{eff}} = 3.832/(k \sin \theta_{\min})$$
where
$$k = 2\pi/\lambda.$$

Substituting the given values, we obtain:
$$a_{\text{eff}} = \frac{3.832\,\lambda}{2\pi \sin \theta_{\min}}$$
$$= \frac{3.3832 \times 1.30 \times 10^{-6}}{2\pi \sin 12°} = 3.81 \ \mu m$$

The effective normalized frequency is given by:
$$V_{\text{eff}} = 2.405\left(\frac{V}{V_c}\right) = 2.405\left(\frac{\lambda_c}{\lambda}\right)$$
$$= 2.405\left(\frac{1.08}{1.30}\right) = 2.00$$

∴ The spot size can be obtained from:

$$\omega_0 = a_{eff}\left[0.6043 + 1.755\, V_{eff}^{-1/2} + 2.78\, V_{eff}^{-6}\right]$$
$$= 3.81 \times 10^{-6}\,[0.6043 + 1.755\,(2.00)^{-1/2} + 2.78\,(2.00)^{-6}]$$
$$= 4.83\ \mu m$$

SUGGESTED READINGS

1. Pollock CE and Lipson M, 'Integrated Photonics', Kluwer Academic Publishers (2003).
2. Kogelnik H, Theory of Optical Waveguides. In Tamir T, Ed. Springer Series in Electronics and Photonics. Guided-wave Optoelectronics, Vol. 26, pp. 7–88, Springer (1988).
3. Midwinter TE, 'Optical Fibers for Transmission', Wiley (1979).
4. Okamoto K, Fundamentals of Optical Waveguides' (2nd ed.), Academic Press (2006).
5. Yeh CW, 'Optical Waveguide Theory', IEEE Trans. Circuit Syst. CAS-26(3.12), pp. 1011–1019 (1979).
6. Marcuse A, 'Theory of Dielectric Optical Waveguides, Academic Press (1974).
7. Degiorgio V and Cristiani I, 'Photonics': A Short Course, Springer (2014).
8. Kumar S and Jamel Deen M, 'Fiber Optic Communications', Wiley (2014).

GLIMPSES

- Light is an electromagnetic wave. It travels in vacuum with a speed, $c = 3 \times 10^8$ ms^{-1}.
- Maxwell's electromagnetic (em) theory is based on a set of four equations. One can derive a wave equation describing the propagation of em waves in any medium.
- In a isotropic and homogeneous dielectric medium, the wave equation is represented by:

$$\nabla^2 \psi - \frac{n^2}{c^2}\frac{\partial^2 \psi}{\partial t^2} = 0$$

where ψ represents the scalar field component of **E** and **H**. The solution of this equation is of the form:

$$\psi = \psi_0 \exp\{i\,(\omega t - \beta z)\}$$

where the symbols have their usual meanings.
- A waveguide is used to guide electromagnetic waves through it. In waveguide, both electric and magnetic fields are confined to space within the waveguide. The dielectric loss within the waveguide is

negligibly small because the guides are normally air filled. Within a waveguide several *modes* could be propagated. The mode of propagation of em waves of a waveguide is determined from solution to Maxwell's equations. If the frequency of a particular signal is above the cutoff frequency of the particular mode, then that particular signal will be passed through it or else it gets attenuated.

- A *planar guide* is the simplest form of optical waveguide. One may assume it consists of a slab of dielectric with refractive index n_1, sandwiched between two regions of lower refractive index n_2.
- A *mode* is a stable (electric or magnetic) field pattern in the transverse direction (e.g., the x-direction) with only a periodic z-dependence. In an isotropic and inhomogeneous dielectric medium, the modes of wave propagation can be described by the following sets of equations:

$$\left. \begin{array}{l} \beta E_y = -\mu_0 \omega H_x \\ \dfrac{\partial E_y}{\partial x} = -i\mu_0 \omega H_z \\ -i\beta H_x - \dfrac{\partial H_z}{\partial x} = i\varepsilon_0 \omega n^2(x) E_y \end{array} \right\} \text{TE modes}$$

$$\left. \begin{array}{l} \beta H_y = -\varepsilon_0 \omega n^2(x) E_x \\ \dfrac{\partial H_y}{\partial x} = i\varepsilon_0 \omega n^2(x) E_z \\ -i\beta E_x + \dfrac{\partial E_z}{\partial x} = i\mu_0 \omega H_y \end{array} \right\} \text{TE modes}$$

- The maximum number of TE modes, M, supported by a symmetrical step-index planar waveguide is an integer close to or greater than $2V/\pi$, where V is normalized frequency given by:

$$V = \frac{2\pi}{\lambda} a \sqrt{n_1^2 - n_2^2}$$

- The normalized propagation constant b for a fiber is given by:

$$b = \frac{\beta^2 - \beta_1^2}{\beta_1^2 - \beta_2^2} = 1 - \left(\frac{ua}{V}\right)^2 = \left(\frac{wa}{V}\right)^2$$

- For a guided mode b must lie between 0 and 1, and also β^2 must lie between $\beta_1^2 - \beta_2^2$.
- The cutoff of a mode occurs at $b = 0$, $ua = V_c = m\pi/2$ and $wa = 0$. However, the fundamental mode has no cutoff frequency.

- The thickness of a guide layer that can support m modes is given by:
$$2a = \frac{m\lambda}{2(n_1^2 - n_2^2)^{1/2}}$$
- The thickness of a guide layer required to support only the fundamental mode is given by:
$$2a \leq \frac{\lambda}{2(n_1^2 - n_2^2)^{1/2}}$$
- The fraction of guided optical power that is confined within a guide layer is called the *confinement factor*. The confinement factor depends on the thickness of the guide layer and the mode number.
- The optical fiber with a core of constant refractive index n_1 and a cladding of a slightly lower refractive index n_2 is known as *step index* fiber. This is because the refractive index profile for this type of fiber makes a step change at the core cladding index. The refractive index profile may be defined as:
$$n(r) = \begin{cases} n_1 & r \leq a \text{ (core)} \\ n_2 & r \geq a \text{ (cladding)} \end{cases}$$

in both cases.
- The wave equation, togther with the boundary conditions for cylindrical wave guides, describes the propagation of electromagnetic waves in *step-index* and *graded-index* optical fibers. Solving these equations for an ideal step-index fiber, under the weakly guiding approximation, gives a set of solutions:
$$\psi(r, \phi, z, t) = R(r) e^{il\phi} e^{i(\omega t - \beta z)}$$

where
$$R(r) = \begin{cases} AJ_1\left(\frac{ur}{a}\right) & r < a \\ BK_1\left(\frac{wr}{a}\right) & r > a \end{cases}$$

The Bessel function $J_l(ur/a)$ are oscillatory in nature, and hence there exist m allowed solutions (corresponding to m roots of J_l) for each value of l. Thus, the propagation phase constant β is characterized by two integers l and m.
- The number of modes for smaller V and the propagation parameters for different modes can be found from the universal b versus V curves. However, for $V \gg 1$, the modal volumes for graded-index and step-index fibers may be calculated using the following approximate relations:
$$M_g = \frac{1}{2}\left(\frac{\alpha}{\alpha + 2}\right)\left[\frac{2\pi a}{\lambda}(n_0^2 - n_c^2)^{1/2}\right]^2$$

$$\approx \left(\frac{\alpha}{\alpha+2}\right)\frac{V^2}{2} \qquad (1)$$

and
$$M_s = \frac{1}{2}\left[\frac{2\pi a}{\lambda}(n_1^2 - n_2^2)^{1/2}\right]^2 \approx \frac{V^2}{2} \qquad (2)$$

M_s allows an estimate of the number of guided modes propagating in a particular multimode step index fiber.

- The optical power is launched into a large number of guided modes, each having different spatial field distributions, propagation constants, etc. In an ideal multimode step index fiber with properties (i.e. relative index difference, core diameter) which are independent of distance, there is no mode coupling and the optical power launched into a particular mode remains in that mode and travels independently of the power launched into the other guided modes.
- *Graded index fibers* do not have a constant refractive index in the core but a decreasing core index $n(r)$ with radial distance from a maximum value of n_1 at the axis to a constant value n_2 beyond the core radius a in the cladding. This index variation may be represented as:

$$n(r) = \begin{cases} n_1(1 - 2\Delta(r/a)^\alpha)^{1/2} & r < a \text{ (core)} \\ n_1(1 - 2\Delta)^{1/2} & r \geq a \text{ (cladding)} \end{cases}$$

where Δ is the relative refractive index difference and α is the profile parameter which gives the characteristic refractive index profile of the fiber core.

- For a parabolic refractive index profile fiber ($\alpha = 2$), total number of guided modes or volume $M_g \approx V^2/4$, which is half the number supported by a step index fiber ($\alpha = \infty$) with the same V value.
- The cutoff value of the normalized frequency V_c to support a single mode in a graded-index fiber is given by:

$$V_c = 2.405\left(1 + \frac{2}{\alpha}\right)^{1/2}$$

Therefore, as in the step index case, it is possible to determine the fiber parameters which give single mode operation.

- For a step index fiber $V_c = 2.405$, the cutoff wavelength is given by:

$$\lambda_c = \frac{V\lambda}{2.405}$$

- Intermodal dispersion in multimode fibers restricts their use in long-haul communications.

REVIEW QUESTIONS

1. Explain the concept of electromagnetic modes in relation to a planar optical waveguide. Discuss the modification that may be made to electromagnetic mode theory in a planar waveguide in order to describe optical propagation in a cylindrical fiber.
2. Explain a plane TEM wave. How the wavelength of this wave change with the medium while propagating?
3. Write Maxwell's equations for the metallic and dielectric media and explain their significance.
4. Explain phase velocity and group velocity. Write expressions for these and explain the difference between them.
5. What is difference between propagation constant k, β and b? Find the interrelationship between them.
6. What do you understand by modes? Distinguish between symmetric and antisymmetric modes of a planar SI waveguide.
7. Derive the expression for the confinement factor,

$$G = \left[1 + \frac{\sin^2(ua)}{wa\left\langle 1 - \frac{\sin(ua)\cos(ua)}{ua}\right\rangle}\right]$$

for the antisymmetric TE modes of a symmetrical SI planar wavelength.
8. Mention the boundary conditions and assumptions made while solving the wave equation for an ideal step-index fiber. Also explain, how these conditions and assumptions are satisfied in a real optical fiber?
9. Define the normalized frequency for an optical fiber and explain its use in determination of the number of guided modes propagating within a step index fiber.
10. Explain, what is meant by graded index fiber, giving the expression for the possible refractive index profile?
11. What is difference between multimode step-index and graded index fibers and also between multimode and single mode fibers?
12. Explain the difference between the propagation phase constant β and the normalized propagation parameter b? Find the relation between two.
13. Derive the necessary expressions for the pulse broadening per unit length due to intermodal dispersion for a GI fiber with parabolic profile ($\alpha = 2$) and GI fiber with an optimum profile.
14. The transverse dependence of the modals:

$$\psi_z(r, \phi) = AJ_l\left(\frac{ur}{a}\right)\begin{bmatrix}\cos(l\phi) \\ \sin(l\phi)\end{bmatrix} : r < a$$

and $\psi_z(r, \phi) = BK_l\left(\dfrac{wr}{a}\right)\begin{bmatrix}\cos(l\phi): r > a \\ \sin(l\phi)\end{bmatrix}$

give the axial components of E or H, i.e. E_z or H_z. Using the following relations, find the transverse components E_r, E_ϕ, H_r, and H_ϕ:

$$E_r = -\dfrac{i}{k_{r^2}}\left(\beta\dfrac{\partial E_z}{\partial r} + \omega\mu\dfrac{1}{r}\dfrac{\partial H_z}{\partial \phi}\right)$$

$$E_\phi = -\dfrac{i}{k_{r^2}}\left(\dfrac{\beta}{r}\dfrac{\partial E_z}{\partial \phi} - \omega\mu\dfrac{\partial H_z}{\partial r}\right)$$

$$H_r = -\dfrac{i}{k_{r^2}}\left(\beta\dfrac{\partial H_z}{\partial r} - \omega\varepsilon\dfrac{1}{r}\dfrac{\partial E_z}{\partial \phi}\right)$$

$$H_\phi = -\dfrac{i}{k_{r^2}}\left(\dfrac{\beta}{r}\dfrac{\partial H_z}{\partial \phi} + \omega\varepsilon\dfrac{\partial E_z}{\partial r}\right)$$

where K_r is the radial components of the propagation vector K in the waveguide and has an amplitude given by:

$$K_r = \sqrt{\omega^2\varepsilon\mu - \beta^2}$$

15. Define the terms cutoff wavelength and dominant mode as applied to waveguides.

16. If the local numerical aperature (NA) of a graded-index fiber at a radial distance r is given by:

$$NA(r) = \{n^2(r) - (n_c^2)\}^{1/2} \quad \text{for } r < a$$

then show that the *rms* value of the NA of the fiber for an α-profile (taken over the core area is given by the following relation:

$$(NA)_{rms} = \left[\left(\dfrac{\alpha}{\alpha+2}\right)(n_0^2 - n_c^2)\right]^{1/2}$$

17. Explain the photonic crystal fiber (PCF) and also explain the guidance mechanisms for electromagnetic modes in such optical fibers.

18. Compare and contrast the performance attributors, potential drawbacks and possible applications of index-guided PCFs and photonic bandgap fibers.

PROBLEMS

1. A step index fiber has a solid acceptance angle in air of 0.115 radians and a relative refractive index difference of 0.9%. Calculate the speed of light in the fiber core.
   [**Ans.** 2.11×10^8 ms^{-1}]

2. A step index fiber in air has a $NA = 0.16$, a core refractive index of 1.45 and a core diameter of 60 μm. Find the normalized frequency for the fiber when light as a wavelength of 0.9 μm is transmitted. Also estimate the number of guided modes propagating in the fiber.
   [Ans. 33.5, 561]

3. Perform a detailed analysis of the TM modes (which are characterized by the field components E_x, E_z, and H_y) of a symmetrical SI planar waveguide. [B.Tech.]
   [Hint:
   Step (a): Substituting for E_x and E_z from Eqs (3.1) and (3.2) respectively, in Eq. (3.3), we obtains:
   $$\beta \mu_y = -\varepsilon_0 \omega n^2(x) E_x \tag{1}$$
   $$\frac{\partial H_y}{\partial x} = i\varepsilon_0 \omega n^2(x) E_z \tag{2}$$
   $$i\beta E_x + \frac{\partial E_z}{\partial x} = i\mu_0 \omega H_y \tag{3}$$
   $$\frac{d^2 H_y}{dx^2} - \left[\frac{1}{n^2(x)} \frac{dn^2(x)}{dx}\right] \frac{dH_y}{dx} + [k^2 n^2(x) - \beta^2] H_y = 0$$
   Step (b): Substitute for $n(x)$ in this equation in the region $|x| < a$ and $|x| > a$, to get two differential equations for the two regions.
   Step (c): Apply the boundary conditions. Here H_y and E_z are tangenital components to the planes $x = \pm a$. Thus H_y and $[1/n^2(x)] (dH_y/dx)$ must be continuous at $x \pm a$.
   Step (d): Write the solutions for symmetric and antisymmetric modes and arrive at the following transcendental equations:
   $$ua \tan ua = \left(\frac{n_1}{n_2}\right)^2 wa \text{ for symmetric modes}$$
   $$-ua \cot ua = \left(\frac{n_1}{n_2}\right)^2 wa \text{ for antisymmetric modes}$$
   where the symbols have their usual meaning.
   Step (e): Now, solve these equations graphically and discuss the results.]

4. A multimode graded index fiber has an acceptance angle in air of 8°. Estimate the relative refractive index difference between the core axis and the cladding when the refractive index at the core axis is 1.52.
   [Ans. 0.42%]

5. Calculate the maximum thickness of the guide slab of a symmetrical planar waveguide so that it supports the first 10 modes. Take $n_1 =$

3.6, $n_2 = 3.598$, and $\lambda = 0.90$ μm. Calculate also the maximum and minimum values of the propagation constant β. [B.Tech.]
[Ans. $2a = 11.875$ μm, $\beta_1 = 2.513 \times 10^5$ m^{-1}, $\beta_2 = 2.499 \times 10^5$ m^{-1}]

6. Calculate the G-factors for the modes supported by a planar waveguide whose layer has a thickness 6.523 μm, $n_1 = 1.50$, and $n_2 = 1.48$. The guide is excited by a light of wavelength $\lambda = 1.0$ mm. [B.Tech.]
[Ans. $G_0 = 0.98828$, $G_1 = 0.94889$, $G_2 = 0.85992$, $G_3 = 0.50960$]

7. A planar waveguide is formed from a 14 μm thick film of dielectric material of refractive index 1.46 sandwiched between two infinite dielectric slabs of refractive index 1.455. (a) Calculate the number of TE modes of propagation that the guide supports at a wavelength of 1.3 μm. (b) Estimate the propagation parameters u_m and w_m for different values of m and hence estimate b_m and β_m for each. [B.Tech.]

[Ans. (a) Three modes corresponding to $m = 0, 1$ and 2.
(b) Since $V = 4$, the abscissae of the intersecting points of the ua versus wa curves (see Fig. 3.3) for $m = 0, 1$ and 2 with the quadrant of a circle of radius $V = 4$ give the values of $u_m a$, and the corresponding ordinates give the values of $w_m a$. Similarly, from Fig. 3.5, one can get the values of b_m corresponding to $V = 4$. β_m can be obtained from:

$\beta_m = \{\beta_2^2 + b_m (\beta_1^2 - \beta_2^2)\}^{1/2}$, where $\beta_1 = 2\pi n_1 / \lambda$ and $\beta_2 = 2\pi n_2 / \lambda$.]

8. A multimode Si fiber has a core diameter of 50 μm, a core index of 1.46, and a relative refractive index difference of 1%. It is operating at a wavelength of 1.3 μm. Calculate (i) the refractive index of the cladding, (ii) the normalized frequency parameter V, and (iii) the total number of modes guided by the fiber. [B.Tech.]
[Ans. (i) 1.445, (ii) 25, (iii) 312]

9. The cutoff wavelength of a step index fiber is quoted as $\lambda_c = 1.20$ μm. If the fiber is operated at wavelength $\lambda = 1.55$ μm, what is normalized frequency or V number? [B.Tech.]

[Hint: $V = 2.405 \dfrac{\lambda_c}{\lambda} = 2.405 \dfrac{1.20}{1.55} = 1.86$]

11. A graded-index single-mode fiber has a core axis refractive index of 1.5, a triangular index profile ($\alpha = 1$) in the core, and a relative index difference of 1.3%. Calculate the core diameter of the fiber if it has to transmit (i) $\lambda = 1.3$ μm and (ii) l = 1.55 μm.
[Ans. (i) 7.1 μm, (ii) 8.5 μm]

12. It is reported that the effective number of modes guided by a curved multimode GI fiber of radiu a is given by:

$$(M_g)_{eff} = (M_g) \left[1 - \frac{(\alpha+2)}{2\alpha\Delta} \left\{ \frac{2a}{R} + \left(\frac{3}{2n_c kR} \right)^{2/3} \right\} \right]$$

where α is the profile parameter, Δ is the relative refractive index difference, n_c is the refractive index of the cladding, $k = 2\pi/\lambda$, and M_g is the total number of guided modes in a single fiber are given by:

$$M_g \cong \frac{\alpha}{\alpha+2} \frac{V^2}{2}$$

Calculate the radius of curvature R such that the effective number of guided model reduces to half its maximum value. Take $\alpha = 2$, $n_c = 1.48$, $\Delta = 0.01$, $a = 50$ μm, and $\lambda = 0.85$ μm. [M.Tech.]

[Ans. $R \approx 1.66$ cm]

13. The refractive indices of the core and cladding of a SI fiber are 1.48 and 1.465, respectively. Light of wavelength 0.85 μm is guided through it. Calculate the minimum and maximum values of the propagation phase constant β. [B.Tech.]

[Ans. 10.82×10^6 m^{-1}, 10.93×10^6 m^{-1}]

14. The velocity of light in the core of a step index fiber is 2.01×10^8 ms^{-1}, and the critical angle at the core-cladding interface is 80°. Find the NA and the acceptance angle for the fiber in air. Assume that it has a core diameter suitable for consideration by ray analysis. Take $c = 3 \times 10^8$ ms^{-1} in vacuum.

[Ans. 0.263, 15.2°]

15. A graded-index fiber with a triangular profile supports the propagation of 500 modes. The core axis refractive index is 1.46 and the core diameter is 75 μm. If the wavelength of light propagating through the fiber is 1.3 μm, calculate (a) the relative refractive index difference Δ of the fiber and (b) the maximum diameter of the fiber core which would give single-mode operation at the same wavelength. [B.Tech.]

[Ans. (a) 0.021, (b) 5.76 μm]

16. A graded-index fiber has a core diameter of 40 μm, $\alpha = 2$, $n_0 = 1.460$, and $n_c = 1.445$. If it is excited by a source of $\lambda = 1.3$ μm, calculate (a) (NA)$_{rms}$; (b) β_0, Δ, and V; and (c) the total number of propagating modes.

[Ans. (a) 0.1476, (b) 7056 mm^{-1}, 0.0102, 25.23, (c) 159]

17. A multimode step index fiber has a relative refractive index difference of 1% and a core refractive index of 1.5. The number of modes propagating at a wavelength of 1.3 μm is 1100. Find the diameter of the fiber core. [B.Tech.]

[Ans. 92 μm]

18. The relative refractive index difference between the core axis and the cladding of a graded index fiber is 7% when the refractive index at the core axis is 1.45. Calculate NA of the fiber when:
    (a) the index profile is not taken into account; and
    (b) the index profile is taken to be triangular.
    Comment on the results. [M.Sc. (Ele.)]
    [**Ans.** (a) 0.172, (b) 0.171]

SHORT ANSWER QUESTIONS

1. Write the Maxwell's relations for a medium with zero conductivity.

Ans. $\nabla \times E = -\dfrac{\partial B}{\partial t}$

$\nabla \times H = -\dfrac{\partial D}{\partial t}$

$\nabla \cdot D = 0$ (no free charges)

$\nabla \cdot B = 0$ (no free poles)

2. What is a planar guide?

Ans. A planar guide is the simplest form of optical waveguide. One may assume it consists of a slab of dielectric with refractive index n_1 sandwitched between two regions of lower refractive index n_2.

3. Write the wave equation in an isotropic and homogeneous dielectric medium.

Ans. $\nabla^2 \psi - \dfrac{n^2}{c^2} \dfrac{\partial^2 \psi}{\partial t^2} = 0$

where ψ represents the scalar field components of **E** and **H**. The solution of above equation is of the form:

$$\psi = \psi_0 \exp[i(\omega t - \beta z)]$$

where the symbols have usual meaning.

4. What are phase and group velocities?

Ans. Within all electromagnetic waves, whether plane or otherwise, there are points of constant phase. For plane waves these constant phase points form a surface which is referred to as a wavefront. As a monochromatic light wave propagates along a waveguide in the z-direction these points of constant phase travel at a phase velocity V_p given by:

$$V_p = \omega/\beta$$

where ω is the angular frequency of the wave. However, it is impossible in practice to produce perfectly monochromatic light waves, and light energy is generally composed of a sum of plane wave components of different frequencies. Often the situation exists where a group of waves with closely similar frequencies propagate so that their resultant forms a *packet of waves*. This wave packet does not travel at the phase velocity of the individual waves but is observed to move at a group velocity of V_g given by:

$$V_g = \frac{\delta\omega}{\delta\beta}$$

The group velocity is of greatest importance in the study of transmission characteristics of optical fibers as it relates the propogation characteristics of observable wave groups or packets of light.

If propagation in an infinite medium of refractive index n_1 is considered, then the propagation constant may be written as:

$$\beta = n_1 \frac{2\pi}{\lambda} = n_1 \frac{\omega}{c}$$

Now, $\quad V_p = \frac{c}{n_1}$

Further, in the limit $\delta\omega/\delta\beta$ becomes, the group velocity:

$$V_g = \frac{d\lambda}{d\beta} \cdot \frac{d\omega}{d\lambda} = \frac{d}{d\lambda}\left(n_1 \frac{2\pi}{\lambda}\right)^{-1}\left(-\frac{\omega}{\lambda}\right)$$

$$= -\frac{\omega}{2\pi\lambda}\left(\frac{1}{\lambda}\frac{dn_1}{d\lambda} - \frac{n_1}{\lambda^2}\right)^{-1}$$

$$= \frac{c}{\left(n_1 - n\dfrac{dn_1}{dl}\right)} = \frac{c}{N_g}$$

where N_g is group index of the guide.

5. Explain the importance of the choice of cladding material.

Ans. (i) The cladding should be transparent to light at the wavelength over which the guide is to operate.

(ii) Ideally, the cladding should consist of a solid material in order to avoid both damage to the guide and accumulation of foreign matter on the guide walls. These effects degrade the reflection process by interaction with the evanescent field. This in part explains the poor performance (high losses) of early optical waveguides with air cladding.

(iii) The cladding thickness must be sufficient to allow the evanescent field to decay to a low value or losses from the penetrating energy may be encountered. In many cases, however, the magnitude of the field falls off rapidly with distance from the guide-cladding interface. This may occur within distances equivalent to a few wavelengths of the transmitted light.

Therefore, the most widely used optical fibers consist of a core and cladding, both made of glass. The cladding refractive index is thus higher than would be the case with liquid or gaseous cladding giving a lower NA for the fiber, but it provides a far more practical solution.

6. What is *Goos-Haenchen* shift?

Ans. The phase change incurred with the total internal reflection of light beam on a planar dielectric interface may be understood from physical observation. Careful examination shows that the reflected beam is shifted laterally from the trajectory predicted by simple ray theory analysis. This lateral displacement is known as the Goos-Haenchen shift.

7. What is physical interpretation of mode number *m*?

Ans. The mode number m denotes the number of half-cycles of the electric field (for TE modes) or magnetic field (for TM modes) that occur over the transverse dimension. The lowest order mode ($m = 1$) is seen to have no cutoff – it will propagate from zero frequency on up. One will thus achieve single-mode operation (actually a single pair of TE and TM modes) if we can assure that $m = 2$ modes are below cutoff.

8. Write the equations for the modes of wave equation in an isotropic and inhomogeneous dielectric medium.

Ans.
$$\left.\begin{array}{l} \beta E_y = \mu_0 \omega H_x \\ \dfrac{\partial E_y}{\partial x} = -i\mu_0 \omega H_z \\ -i\beta H_x - \dfrac{\partial H_z}{\partial x} = i\varepsilon_0 \omega n^2(x) E_y \end{array}\right\} \text{TE modes}$$

$$\left.\begin{array}{l} \beta H_y = -\varepsilon_0 \omega n^2(x) E_x \\ \dfrac{\partial H_y}{\partial x} = -i\varepsilon_0 \omega n^2(x) E_z \\ i\beta E_x + \dfrac{\partial E_z}{\partial x} = i\mu_0 \omega H_y \end{array}\right\} \text{TM modes}$$

9. What is normalized frequency?

Ans. $V = ka(n_1^2 - n_2^2)^{1/2}$

$\quad\quad = \dfrac{2\pi}{\lambda} a(n_1^2 - n_2^2)^{1/2}$

$$= \frac{2\pi}{\lambda} a\,(NA) = \frac{2\pi}{\lambda} a n_1 \sqrt{(2\Delta)}$$

The normalized frequency is a dimensionless parameter and hence is also sometimes simply called the V number or value of the fiber. It combines in a very useful manner the information about three important design variables for the fiber: namely, the core radius a, the relative refractive index difference Δ and the operating wavelength λ.

10. How waveguide perturbations affect the propagation characteristics of the fiber?

Ans. Waveguide perturbations such as deviations of the fiber axis from straightness, variations in the core diameter, irregularities at the core-cladding interface and refractive index variations may change the propagation characteristics of the fiber. These will have the effect of coupling energy travelling in one mode to another depending on the specific perturbation.

11. Write the advantages of multimode fibers over single mode fibers for lower band applications.

Ans. The advantages are:
   (i) the use of spatially incoherent optical sources (e.g. most light emitting diodes) which cannot be efficiently coupled to single-mode fibers;
   (ii) larger NAs, as well as core diameters, facilitating coupling to optical sources.
   (iii) lower tolerance requirements on fiber connectors.

12. What are graded index fibers?

Ans. These do not have a constant refractive index in the core (this is why they are sometimes referred to as inhomogeneous core fibers) but a decreasing core index $n(r)$ with radial distance from a maximum value of n_1 at the axis to a constant value n_2 beyond the core radius a in the cladding. This index variation may be represented as:

$$n(r) = \begin{cases} n_1 \left(1 - 2\Delta \left(\frac{r}{a}\right)^\alpha\right)^{1/2} & r < a \text{ (core)} \\ n_1 (1 - 2\Delta)^{1/2} = n_2 & r \geq a \text{ (cladding)} \end{cases}$$

where Δ is the relative refractive index difference and α is the profile parameter which gives the characteristic refractive index profile of the fiber core.

13. What is the advantage of the propagation of a single mode within an optical fiber?

Ans. The signal dispersion caused by the delay differences between different modes in a multimode fiber may be avoided.

14. **What are the reasons that single-mode fibers emerge as a viable communication medium and most widely used fiber type within telecommunications?**

Ans.   (i) They exhibit the greatest transmission bandwidths and the lowest losses of the fiber transmission media.
  (ii) They have a superior transmission quality over other fiber types because of the absence of the modal noice.
  (iii) They offer a substantial upgrade cabpability (i.e. future proofing) for future wide bandwidth services using either faster optical transmitters and receivers or advanced transmission techniques (e.g. coherent technology).
  (iv) They are compatible with the developing integrated optics technology.

15. **What is cutoff wavelength?**

Ans. The single-mode operation only occurs above a theoretical cutoff wavelength λ_c given by:

$$\lambda_c = \frac{2\pi a n_1}{V_c} (2\Delta)^{1/2}$$

where V_c is the cutoff normalized frequency. Hence λ_c is the wavelength above which a particular fiber become single-moded. Thus, we have:

$$\frac{\lambda_c}{\lambda} = \frac{V}{V_c}$$

For step index fiber $V_c = 2.405$, the cutoff wavelength is given by:

$$\lambda_c = \frac{V\lambda}{2.405}$$

MULTIPLE CHOICE QUESTIONS

1. In a wave guide, suffix mn of the modes TE/TM denoe:
   (a) Half wavelength of E field and full wavelength of H field
   (b) Half wavelengths of E and H fields in directions other than guide axis
   (c) Full wavelength of E field and half wavelength of H field
   (d) Half wavelength of E and H fields

2. In a symmetrical SI planar waveguide, the refractive index of the guide layer is $n_1 = 1.5$ and $\Delta = 0.001$. If the thickness of the guide layer is 8.5 µm, for what wavelength will the guide support only a fundamental mode?
   (a) 0.85 µm (b) 1.14 µm
   (c) 1.36 µm (d) 1.55 µm [B.Tech.]

3. A symmetrical SI planar waveguide is to be excited by a source of central wavelength $\lambda = 0.85$ μm. Assume that $n_1 = 1.5$ and $\Delta = 0.01$. What should be the thickness of the guide layer so that it supports one symmetric and one antisymmetric TE mode?
   (a) 4.00 μm (b) 6.00 μm
   (c) 10.00 μm (d) 12.00 μm [B.Tech.]

4. For a given guided mode, the normalized propagation constant lies between:
   (a) $-\infty$ and $+\infty$ (b) 0 and ∞
   (c) 0 and 1 (d) –1 and +1.5

5. The range of the phase propagation constant ($\beta_2 < \beta < \beta_1$) for a wave guided by a planar waveguide can be obtained if the following is known:
   (a) Wavelength of the excitation source
   (b) Refractive index of the guide layer
   (c) Refractive index of the cladding slabs
   (d) All of the abve

6. Which of the following relation for NA is correct:
   (a) $NA = n_1 (2\Delta)$ (b) $NA = n_1 (2\Delta)^{1/2}$
   (c) $NA = n_1^{1/2} (2\Delta)$ (d) $NA = n_1^2 (2\Delta)^{1/2}$

7. Inside an ideal dielectric medium:
   (a) the free charge density ρ is zero but the conductivity σ is nonzero
   (b) ρ is nonzero and σ is zero
   (c) both ρ and σ are zero (d) both ρ and σ are nonzero

8. For symmetric TE modes supported by symmetrical SI planar waveguide, the stable electric-field distribution inside the guide layer takes:
   (a) a sinusoidal form (b) a cosinusoidal form
   (c) an exponential form (d) a tangential form [B.Tech.]

9. The maximum number of TE modes supported by a symmetrical SI planar waveguide is an integer greater than:
   (a) $2V/\pi$ (b) $V/2\pi$
   (c) $2\pi V$ (d) zero

10. The relation between the group velocity V_g and group index (N_g) of the guide is:
    (a) $V_g = \dfrac{c}{N_g}$ (b) $V_g = \dfrac{c}{\sqrt{N_g}}$
    (c) $V_g = c N_g$ (d) $V_g = \dfrac{c}{N_g^{3/2}}$

11. Mode volume or total number of guided modes M_s for a step index fiber is related to V value for the fiber by the approximate expression:
    (a) $M_s \simeq \dfrac{V}{2}$
    (b) $M_s \simeq \dfrac{V^3}{2}$
    (c) $M_s \simeq \dfrac{V^2}{2}$
    (d) $M_s = 2V^2$

12. For a typical symmetrical SI planar waveguide, the value of the V-parameter is 5. The guide supports the following TE modes.
    (a) Two symmetric and three antisymmetric
    (b) Three symmetric and two antisymmetric
    (c) Two symmetric and two antisymmetric
    (d) One symmetric and one antisymmetric [B.Tech.]

13. The electric-field distribution $E_y(x)$ in a planar waveguide corresponding to the $m = 3$ TE mode will have the following number of zeros between $x = -a$ and $x = +a$.
    (a) Three
    (b) Two
    (c) One
    (d) Five [B.Tech.]

14. In a planar waveguide, for a typical mode, 70% of its total power remains in the guide layer and the remaining 30% extends into the cladding slabs. Its confinement factor for the mode is:
    (a) 0.40
    (b) 0.60
    (c) 0.70
    (d) 1.00

15. The effective refractive index (n_{eff}) or phase index or normalized phase change coefficient for single-mode fiber bear the following relation with the propagation constant of the fundamental mode (β) and vacuum propagation constant (k):
    (a) $n_{eff} = \beta k$
    (b) $n_{eff} = \dfrac{k}{\beta}$
    (c) $n_{eff} = \dfrac{\beta}{k}$
    (d) $n_{eff} = \sqrt{\dfrac{\beta}{k}}$

16. A step-index fiber has a core of refractive index 1.5 and a cladding of refractive index 1.49. The core diameter is 100 µm. How many guided modes are supported by the fiber if the wavelength of light is 0.85 µm?
    (a) 380
    (b) 870
    (c) 1160
    (d) 2040

17. A graded-index fiber has a triangular profile with $n_0 = 1.48$ and $\Delta = 0.02$. If it is excited by a source of $\lambda = 1.0$ µm, what is the range of phase propagation constants for the modes supported by the fiber?

(a) 3.438–5.327 μm^{-1} (b) 5.289–7.142 μm^{-1}
(c) 6.315–8.342 μm^{-1} (d) 8.620–9.299 μm^{-1} [B.Tech.]
18. If the core diameter of the fiber of Question 17 is 50 μm, what is the value of the normalized frequency parameter?
(a) 19.61 (b) 26.53
(c) 31.41 (d) 50.72 [B.Tech.]
19. For any multimode optical fiber, what is the range of the normalized propagation parameter?
(a) 0–1 (b) 1–10
(c) Cannot be calculated unless λ is known
(d) Cannot be calculated unless the profile parameter is known
20. Single mode operation only occurs above a theoretical cutoff wavelength λ_c given by $\lambda_c = \dfrac{2\pi a n_1}{V_c}(2\Delta)^{1/2}$. For step index fiber V_c is equal to:
(a) 1 (b) 1.5
(c) 2.405 (d) 4
21. The relationship between the effective refractive index (n_{eff}) and the normalized propagation constant (b) (b is a dimensionless parameter varies between 0 and 1) is:
(a) $b \simeq n_{eff}$ (b) $b \simeq (n_{eff} - n_2)$
(c) $b \simeq \dfrac{n_{eff} - n_2}{n_1 - n_2}$ (d) $b \simeq \dfrac{n_{eff} + n_2}{n_1 + n_2}$
22. For the optical fiber of Question 18, what is the total number of modes supported by the fiber?
(a) 94 (b) 164
(c) 208 (d) 500
23. In a step-index fiber, what is the cut-off frequency of the LP$_{11}$ mode?
(a) 0.0 (b) 2.405
(c) 5.832 (d) 8.520
24. A GI fiber with a parabolic profile has an axial refractive index of 1.46 and Δ of 0.5%. What is the pulse broadening per unit length due to intermodal dispersion?
(a) 50.4 mm/km (b) 60.8 ns/km
(c) 60.8 ps/km (d) Zero [B.Tech.]
25. In a multimode SI fiber, the higher order modes propagate within the fiber with:
(a) lower group velocity than the lower order modes
(b) higher group velocity than the lower order modes
(c) same group velocity as that of lower order modes
(d) random group velocity

26. An unclad fiber with a core refractive index of 1.46 and core diameter of 60 μm is placed in air. What is the normalized frequency for the fiber when light of wavelength 0.85 Δm is transmitted through it?
    (a) 70.74
    (b) 108.23
    (c) 221.65
    (d) 375.02

27. A 0.5 mm thick slab of glass ($n_1 = 1.45$) is surrounded by air ($n_2 = 1$). The slab guides infrared light at wavelength $\lambda = 1.0$ μm. TE and TM modes propagte will be:
    (a) 1100
    (b) 1550
    (c) 2102
    (d) 2505

28. The cutoff wavelength of a step index fiber is quoted as $\lambda_c = 1.20$ μm. If the fiber is operated at wavelength $\lambda = 1.55$ μm, then V is:
    (a) 1.1
    (b) 1.4
    (c) 1.86
    (d) 2.1

    [Hint: $V = 2.405 \dfrac{\lambda_c}{\lambda} = 2.405 \left(\dfrac{1.20}{1.55}\right) = 1.86$)

29. The condition for single-mode operation in a step index fiber is:
    (a) $\lambda > \lambda_c = \dfrac{2\pi a}{2.405}(n_1 - n_2)^{1/2}$
    (b) $\lambda > \lambda_c = \dfrac{2\pi a}{2.405}(n_1^2 - n_2^2)^{1/3}$
    (c) $\lambda > \lambda_c = \dfrac{2\pi a}{2.405}(n_1^2 - n_2^2)^{1/2}$
    (d) $\lambda > \lambda_c = \dfrac{2\pi a}{2.405}(n_1 + n_2)$

ANSWERS

1. (b)	2. (b)	3. (b)	4. (c)	5. (d)	6. (b)	7. (c)
8. (b)	9. (a)	10. (a)	11. (c)	12. (c)	13. (a)	14. (c)
15. (c)	16. (d)	17. (d)	18. (c)	19. (a)	20. (c)	21. (c)
22. (b)	23. (b)	24. (c)	25. (a)	26. (c)	27. (c)	28. (c)
29. (c)						

4

Optical Fibers

4.1 INTRODUCTION

It has been known and understood by the time of Isaac Newton that light beams could be trapped and guided in a medium of higher index of refraction material surrounded by lower index of refraction material (Fig. 4.1). Three hundred years later, we figured out how to use this observation and revolutionized the telecommunication industry.

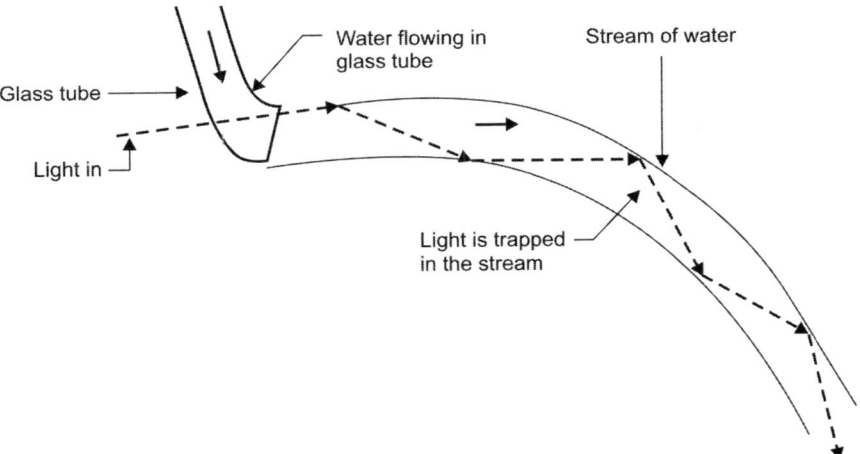

Fig. 4.1. Schematic diagram of the experimental demonstration by Newton that light can be guided in a stream of water

In 1870, Tyndall, first studied transmission in a confined path. In 1930, H. Lamm conducted experiments on image transmission through quartz fibres. In 1936, G.C. Southworth studied theoretically and experimentally, the behaviour of electromagnetic waves in cylindrical nonconductive materials; similar to today's optical fibre laser light transmission. In the early 1960s, Bell Laboratories (BS) and Standard Telephone Laboratories (UK) demonstrated optical transmission through low loss glass fibres useful for long distane communications.

In 1966, Charles Kao and Charles Hockham proposed that an *optical fibre* might be used as a means of telecommunication, provided the signal loss could be made less than 20 decibels per kilometer (dB/kM). At that time optical fibres exhibited losses of the order of 1000 dB/km.

At this point, it is important to know why the need for optical fibres as a transmission medium was felt. Truly speaking, the transfer of information from one point to another, i.e., communication is achieved by superimposing (or modulating) the information onto an electromagnetic wave, which acts as a carrier for the information signal. The modulation carrier is then transmitted through the information channel (open or guided) to the receiver, where it is demodulated and the original information sent to the destination. Now, the carrier frequencies present certain limitations in handling the volume and speed of information transfer. These limitations generated the need for *increased carrier frequency*. In fibre optic systems, the carrier frequencies are selected from the optical range (particularly the infrared part of the spectrum (Fig. 4.2).

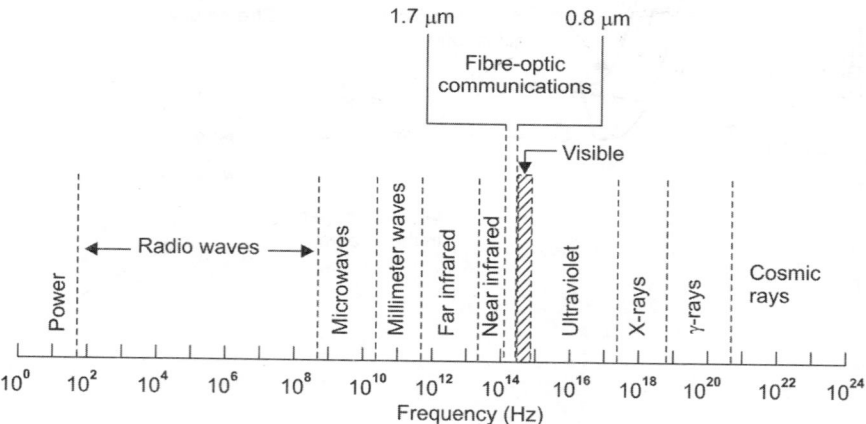

Fig. 4.2. Electromagnetic spectrum

We can see that the typical frequencies are of the order of 10^{14} Hz, which is 10,000 times greater than that of microwaves. Obviously, optical fibres are the most suitable medium for transmitting these frequencies, and hence optical fibres present theoretically unlimited possibilities. Thus, we can say that:

(i) Optical fibres deals with study of propagation of light through transparent dielectric waveguides. The optical fibres are used for transmission of data from point to point location. Currently, the optical fibre systems are extensively used as the transmission line between terrestrial hardwired systems.

(ii) The carrier frequencies used in conventional systems had the limitations in handling the volume and rate of data transmission. The greater the carrier frequency, larger the available bandwidth and information carrying capacity.

The more general term used for an optical fibre is *light wave guide* and for fibre optics communication as *light wave communication*. However, the name *photonics* for fibre optic communications is becoming popular day to day. *Photonics* is very much helping the technology for fibre optic communication to enhance its performance and become more cost effective. Today, optical fibres are finding use in large number of areas, e.g. *sensors, display systems, copying machines, defence services* due to high privacy, etc. In telecommunications, modulating a beam of light (LED or LASER) as a source of light and using optical fibre waveguides have the inherent advantage of very large bandwidth, small dimensions, low losses bandwidth and insensitivity to electrical interference.

Optical fibres are dielectric waveguides, which are fabricated from glass or plastic. A fibre is as thin as human hair designed to guide light wave along their length. They are operated on optical frequencies. Usually an optical fibre is a cylindrical waveguide system through which the optical wave can propagate. The principle by which light wave travels through the fibre is **'total internal reflection'** without loss of incident intensity.

An important structure used in optical systems is layered structure or waveguide structure. These structures are used to confine optical waves in a well defined region and guide their propagation. One can make layered structures from non-crystalline or crystalline materials. For example, glass is used to produce optical fibres for use in optical communication, whereas semiconductor waveguides are used in semiconductor lasers.

In short, we can say that fibre optics have revolutionised the modern world. Two areas in which optical fibres are used extensively are *communications* and *imaging*. Optical fibres have made it possible to perform internal medical examinations with minimal intrusion, an example is imaging. For example, an *endoscope* is used for inspecting different parts of the body. A bundle of optical fibres is used to carry, light from a source into the patient's body, where it illuminates the area of interest. Another bundle, i.e. the output bundle, carries the image light back to a camera mounted externally. Each output fibre turns into a pixel in the final image, so it is important to have a coherent bundle in order to reconstruct the picture properly. Obviously, Endoscopes have the profound advantage that they allow visual examinations of internal organs without requiring surgery, or even an incision, and the procedure is quite safe, because no electrical equipment enters the body.

In this chapter, we will review the fundamental properties of optical fibers. This will include the material and fabrication properties and optical propagation fundamentals. Propagation in the optical fibers used in today's communication systems is best described using the concept of *guided modes*. The *fibre-optic amplifier* represents a major advance in fiber-optic technology. Commercialization of this device is in the initial stages. The promise of optical communication using many wavelengths and exploiting *nonlinearities*, such as *soliton* propagation is indeed exciting.

4.2 OPTICAL FIBER

Optical fibers are the dielectric waveguides, which are fabricated from glass or plastic and are operated on optical frequencies. These are normally cylindrical in form. The optical fiber cable (Fig. 4.3) has three principal sections: (i) Core, (ii) Cladding, and (iii) Jacket or Sheath.

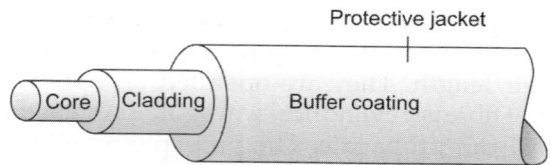

Fig. 4.3. A single fibre structure

(i) **Core:** Core is the innermost region of the fiber. Core has specific property of conducting an optical beam. Core is usually made of glass ($n_1 = 1.5$) or plastic. Glass is used as core material for low and medium loss fibers, whereas plastic core is used for relatively higher loss fibers. Quartz ($n_1 = 1.46$) is also used for core material. The diameter of core varies from 8 to 200 µm. We may note that the core is the actual working structure of the optical fiber, which is covered with another layer of glass or plastic having slightly different chemical composition known as cladding.

(ii) **Cladding:** The core is surrounded by a solid dielectric cladding having refractive index n_2 which is less than n_1 of core ($n_2 < n_1$). The cladding material is either glass or plastic. Although cladding is not mandatory for light propagation along the core, however, it has the following advantages:

    (a) It reduces scattering loss arising due to dielectric discontinuities at the boundary of core surface and air ($n = I$).

    (b) It adds mechanical strength to the fiber.

    (c) It protects the core from absorbing surface contaminants.

We can have three types of core and cladding: (i) Plastic core and cladding, (ii) Glass (silica) core and cladding, and (iii) Glass core with plastic cladding.

Although glass has low loss, plastic provides more flexibility and ruggedness to fiber, less expensive, also 60% less heavy than glass and is stress tolerant. Hence plastic fiber is preferred for small distance communication. The optical fiber may have an abrupt boundary core between cladding, i.e. they may exhibit a gradual change in material between them.

(iii) **Jacket or Sheat:** The outermost section of optical fiber is made up of plastic or special kind of polymer and other material usually opaque in nature. This is known as jacket or sheat. A typical single fiber cable is shown in Fig. 4.4.

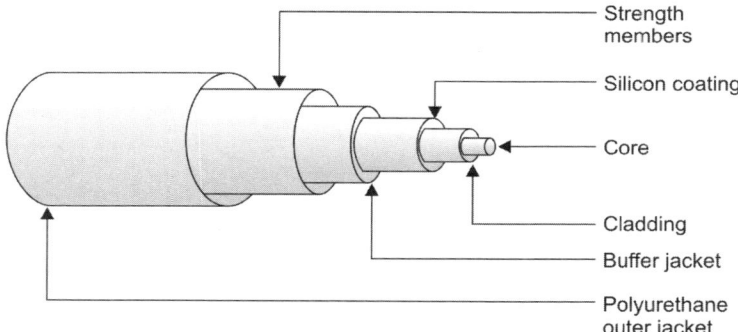

Fig. 4.4. A schematic representation of typical fiber cable

Jacket protects the core from abrasion, interaction with environment, crushing, absorption, moisture and all other adversities of the terrestrial atmosphere. Obviously, jacket enhances tensile strength of fiber.

Apart from its use as communication channel, optical fibers find wide uses in many other areas. Sensors for detecting electrical, mechanical and thermal energies are made using optical fibers. Copier machines, simple display systems, medical diagnostic, fibroscopes, etc. also utilize fiber optics.

The optical properties of cladding are different from those of the core. Jacket protects the fiber structure from moisture, abrasion, mechanical shocks and other environmental hazards.

The core forms the actual working structure of optical fiber. An incident (wave) ray of light entering the core suffers a number of total internal reflections at the core-cladding interface. Obviously the interface between core and cladding acts as a mirror at which total internal reflection of the transmitted light takes place.

4.3 OPTICAL FIBER AS WAVEGUIDE: PRINCIPLE OF PROPAGATION

A waveguide is a tubular structure through which energy of sort could be guided in the form of waves. As *light* waves can be guided through a fiber, it is called *light guide*. Usually, it is also called as *fiber wave guide* or *fiber light wave*.

As stated earlier, the cladding in an optical fiber always has a lower refractive index than that of core. The light signals which enters into the core can strike the interface of the core and cladding only at large angles of incidence, because of the ray geometry (Fig. 4.5). We can see that the light signal undergoes reflection after reflection within the fiber core. Since each reflection is a total internal reflection, the signal sustains its strength and also confines itself completely within the core during propagation. Clearly the optical fiber functions as a waveguide.

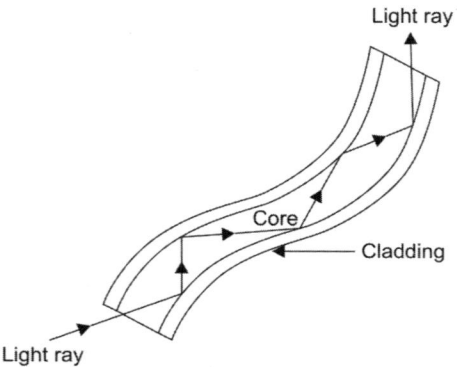

Fig. 4.5. Optical fiber as a waveguide

The propagation of light continues as long as the fiber is not bent too sharply, since for sharp bends, the light fails to undergo total internal reflection because of which the signal strength comes down drastically. Obviously, care is taken to avoid very short bends in the signal carrying fiber. We may make that, for all analysis, it is primary made use of.

4.4 PRINCIPLE OF OPTICAL FIBER: TOTAL INTERNAL REFLECTION

When light travels from denser to rarer medium, it moves away from normal. The angle of refraction is larger than angle of incidence. When angle of incidence increases, angle of refraction also increases. At one specific angle of incidence, angle of refraction becomes 90° and it is the limiting case. When light travels from denser to rarer medium, angle of incidence at which angle of refraction becomes 90°, is called *critical angle* (θ_c).

According to Snell's Law:

$$n_1 \cdot \sin\theta_1 = n_2 K \cdot \sin\theta_2 \quad \text{or} \quad \frac{\sin\theta_1}{\sin\theta_2} = \frac{n_2}{n_1} \quad (4.1a)$$

where θ_1 is the angle of incidence in the medium of refractive index n_1 and θ_2 is the angle of refraction in the medium of refractive index n_2. In this case $n_1 > n_2$. When $\theta_1 = \theta_c$, $\theta_2 = 90°$, therefore $\sin\theta_c = n_2/n_1$.

When the angle of incidence is greater than critical angle, the angle of refraction becomes greater than 90° and it is no more refraction. The light comes back in the same (original) medium and it is called *total internal reflection*. This principle is used in optical fiber ray transmission.

As stated earlier, optical fiber consists of *core* of refractive index n_1 and cladding of refractive index n_2 ($n_1 > n_2$). At the interface of core-cladding (Fig. 4.3), there occurs total internal reflection and ray travels through core of the fiber. Fig. 4.6 shows the geometry of launching a light ray into an optical fiber.

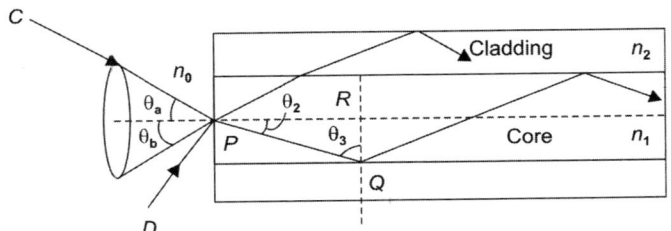

Fig. 4.6. Ray transmission through an optical fiber

The incident rays with angle of incidence greater than θ_c at the core cladding surface are transmitted by total internal reflection. A ray travelling parallel to axis (or making relatively small angle with axis) is called *meridional* ray. For example, CP is the meridional ray. The ray CP makes an angle of incidence θ_a at air-core interface. The rays making an angle greater than θ_a at air-core interface, make angle of incidence of core-cladding interface, less than θ_c and they are lost as they do not undergo total internal reflection. For example, ray DP. The rays entering the fiber core within the *acceptance cone* specified by conical half-angle θ_a are transmitted and others are lost. The maximum angle of incidence at air-core (or any other medium and core) interface, specified by the conical half-angle, for the light rays to be transmitted through the optical fiber is called *acceptance angle* for the fiber.

In optical fiber ray transmission, generally a term *numerical aperture* (NA) is used. This term is used in optics: where it *defines the light-gathering power of lens or express the angle of view of lens. The NA is the product of the refractive index of the surrounding medium and sine of half the angle of view of the lens.*

In optical fiber, NA gives the relation between the acceptance angle and refractive index of three media involved, viz. core, cladding and air (or any other medium). Suppose θ_a is the *acceptance angle*, or the acceptance cone half angle refractive index of air or any other medium present at core interface is n_0 of core is n_1 and that of cladding is n_2 ($n_1 > n_2$).

From Snell's law:
$$n_0 \sin \theta_1 = n_1 \sin \theta_2 \quad (4.1b)$$

In triangle PQR, $\theta_3 = [(\pi/2) - \theta_2]$. Here $\theta_3 > \theta_c$, therefore:
$$n_0 \sin \theta_1 = n_1 \cos \theta_3$$
$$= n_1 \sqrt{1 - \sin^2 \theta_3}$$

For limiting case, $\theta_1 \to \theta_a$ and $\theta_3 \to \theta_c$, hence:
$$n_0 \sin \theta_a = n_1 \sqrt{1 - \sin^2 \theta_c}$$
$$= n_1 \sqrt{1 - \left(\frac{n_2}{n_1}\right)^2}$$
$$= \sqrt{n_1^2 - n_2^2}$$

$$NA = n_0 \sin \theta_a = \sqrt{n_1^2 - n_2^2} \text{ or } \sin \theta_a = \sqrt{\frac{n_1^2 - n_2^2}{n_0}} \quad (4.2)$$

we have:
$$\sin \theta_a = \sin(\theta_1)_{\max}$$
$$= \left(\frac{n_1}{n_0}\right) \cos \phi_c$$

We know that:
$$\theta_c = \frac{n_2}{n_1}$$

If surrounding medium is air, $n_0 = 1$ and $NA = \sin \theta_a = \sqrt{n_1^2 - n_2^2}$ or $\theta_a = \sin^{-1} \sqrt{n_1^2 - n_2^2}$.

Sometimes the NA is expressed in terms of relative refractive index Δ, between core and cladding. Δ is also called as index difference:
$$\Delta = \frac{n_1 - n_2}{n_1}$$
$$n_1^2 - n_2^2 = (n_1 + n_2)(n_1 - n_2)$$
$$= \left(\frac{n_1 + n_2}{2}\right)\left(\frac{n_1 - n_2}{n_1}\right) 2n_1 \quad \left(\because \frac{n_1 + n_2}{2} \approx n_1\right)$$
$$= 2n_1^2 \cdot \Delta$$

Thus, $$NA = n_1\sqrt{2\Delta} = \sqrt{n_1^2 - n_2^2} \qquad (4.3)$$

Clearly, NA represents a figure of merit used to find the light gathering capability of a fibre. Higher the NA, higher is θ_a and higher the amount of light coupled to the fiber. Obviously, NA is a measure of the ability of the fiber to accept light for transmission. NA is effectively depending only on the refractive indices of the core and cladding of materials and is not a function of fiber dimension.

NA is a *dimension-less quantity* lying typically between 0.14 and 0.50. The maximum value of NA for step index fiber occure for air cladding, i.e. without any solid cladding, which can be determined from Fig. 4.7.

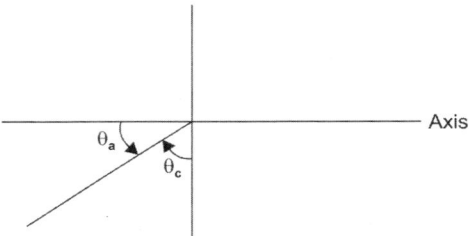

Fig. 4.7. Acceptance angle for air cladding

For air cladding:
Acceptance angle, $\theta_a = 90° - \theta_c$
$$NA = \sin\theta_a = \sin(90° - \theta_c) \qquad (4.4)$$
Acceptance angle θ_a for grade index fibre is given as:
$$NA = \sin\theta_c \qquad (4.5)$$

Variation of NA with acceptance angle (θ_a) is shown in Fig. 4.8a. We may note that:

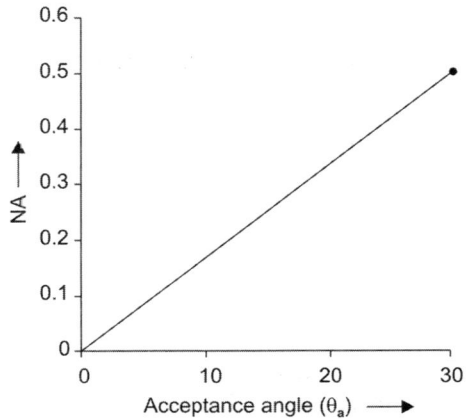

Fig. 4.8a. Variation of NA with acceptance angle

(i) The larger the diameter of the core, the larger is the acceptance angle.
(ii) The larger the difference in the refractive indices of the core and cladding, the large is the acceptance angle.
(iii) If $\theta_1 > (\theta_1)_{max}$ (or $\theta_1 > \theta_0$), the ray refracts through the cladding and the corresponding optical signal is lost.

So far, we have considered only the propagation of a single ray of light. However, a pulse of light consists of several rays, which may propagate at all values of α (Fig. 4.8(b)). varying from 0 to α_m.

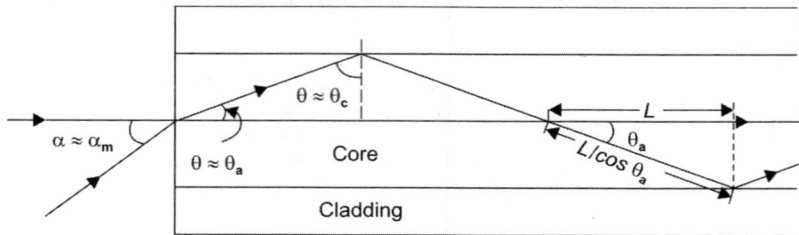

Fig. 4.8b. Trajectories of two extreme rays inside the core of a step-index fiber

The paths transversed by two extreme rays, one corresponding to $\alpha = 0$ and other ray corresponding to α very nearly equal to (but less than) α_m, are shown in Fig. 4.8(b).

We see that an axial ray travels a distance L inside the core of index n_1 with velocity v in time:

$$t_1 = \frac{L}{v} = \frac{Ln_1}{c}$$

where $n_1 = c/v$, c is velocity of light.

The most oblique ray corresponding to $\alpha \approx \alpha_m$ will cover the same length of fiber (axial length L, but actual distance $L/\cos\theta_a$) in time t_2 given by:

$$t_2 = \frac{L/\cos\theta_a}{v} = \frac{Ln_1}{c\cos\theta_a} = \frac{Ln_1}{c\sin\theta_c}$$

$$= \frac{Ln_1}{c\left(\frac{n_2}{n_1}\right)} = \frac{Ln_1^2}{cn^2}$$

We should remember that the two rays are launched at the same time, but will be separated by a time interval ΔT after travelling the length L of the fiber, given by:

$$\Delta T = t_2 - t_1 = \frac{Ln_1^2}{cn_2} - \frac{Ln_1}{c} = \frac{Ln_1}{c}\left(\frac{n_1 - n_2}{n_2}\right)$$

Thus, a light pulse consisting of rays spread over $\alpha = 0$ to $\alpha = \alpha_m$ will be broadened as it propagates through the fiber, and the pulse broadening per unit length of traversal will be given by:

$$\frac{\Delta T}{L} = \frac{n_1}{n_2}\left(\frac{n_1 - n_2}{c}\right)$$

Note 1: θ_a or $(\theta_1)_{max}$, i.e. acceptance angle or acceptance cone half angle [Eq. (4.2)] defines the maximum angle in which external light rays couple at air-fiber interface and travel down the fiber with a response maximum 10 dB down the peak value. Rotating the acceptance angle around the fiber axis, we get the acceptance cone of the fiber as shown in Fig. 4.9.

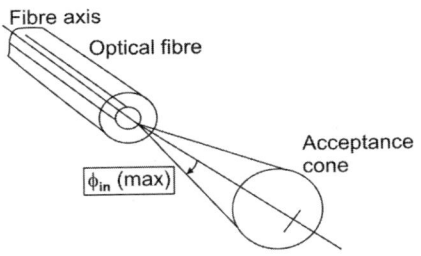

Fig. 4.9. Acceptance cone of a fiber cable

Note 2: *Acceptance angle is defined as the maximum angle that a ray can have relative to the axis of the fiber so that is may propagate down the fiber.* It may also be defined as *the maximum angle from the fiber axis at which light may enter the fiber so that it will propagate in the core through total internal reflection.* We may note that:
  (i) The larger is the diameter of the core, the larger is the acceptance angle.
  (ii) The large the difference in the refractive indices of the core and cladding, the larger will be the acceptance angle.

Note 3: The refractive index (n) of a dielectic material depends on its dielectric constant (ε), i.e. $n = \sqrt{\varepsilon}$. The refractive index of same common materials are given in Table 4.1.

Note 4: Fractional refractive index change or relative refractive index difference when $n_1 \neq n_2$:

$$\Delta = \frac{n_1 - n_2}{n_1}$$

For the internal reflection $n_1 > n_2$. Obviously, to guide light ray effectively through a fiber:

Table 4.1. Typical indices of refraction (at 589 nm)

Medium	Index of refraction (n)
Vacuum	1.0
Air	1.0003 (= 1.0)
Water	1.33
Ethyl alcohol	1.36
Fused quarts	1.46
Glass fiber	1.5–1.9
Diamond	2.0–2.42
Silicon	3.4
Gallium arsenide	3.6

(i) $\Delta \ll 1$.

(ii) Δ is always positive

Typically Δ is of the order of 0.01.

Note 5: NA determines the *light gathering ability* of the fiber. This means NA is a measure of the amount of light that can be accepted by the fiber. When fibers used in short distance communication, NA ranges from 0.1 to 0.3. When NA is too small, it is difficult to launch power into the fiber.

Larger the value of NA means more is the amount of light accepted from the fiber.

NA is not a function of fiber dimensions and NA effectively depends only on refractive indices of core and cladding.

Note 6: Decibels (dB): The relative power level between two points along a fiber-optic communication link is measured in decibels (dB). For a particular wavelength λ, if P_0 is the power launched at one end of the link and P is the power received at the other end, then the efficiency (η) of transmission of the link is:

$$\eta = P/P_0$$

When the measure P and P_0 both in the same units, then P/P_0 ratio is expressed in decibels as:

$$dB = 10 \log_{10}(P/P_0)$$

We may note that there is always some loss in the communication link. This means that P/P_0 is always less than 1 and $\log_{10}(P/P_0)$ is always negative.

In order to make absolute measurements, P_0 is assigned a reference value, normally 1 mW. The value of power (say, P) relative to P_0 (= 1 mW) is denoted by dB_m. Obviously:

$$dB_m = 10 \log_{10} \frac{P \, (mW)}{P_0 \, (1 \, mW)} = 10 \log_{10} P \qquad (4.6)$$

Note 7: To prevent energy losses via absorption and scattering, the cladding should be at least a few wavelengths thick. One can see that a typical value for n_2/n_1 is 0.09. For this value, one finds the critical angle as 0.142 radius or 8.11°. Thus, rays travelling at an angle less than 8.11° relative to the reference axis will be totally internally reflected and guided by the fiber. Higher angle rays will enter the cladding and be lost due to high levels of scattering and absorption.

We may note that dielectric cladding on glass core reduces scattering loss, protects core from absorbing external optical disturbances and provides mechanical strength to main core glass fiber. Sometimes, there is buffer coating over cladding which adds further strength to main fiber and protects it from mechanical vibrations and impact.

If ray AB is rotated around the fiber axis keeping θ_a same, it describes a conical surface as shown in Fig. 4.10. Now only those rays that are funneled into the fiber within this cone having a full angle $2\theta_a$ will only be total internally reflected and thus confined within the fiber propagation, i.e., only rays within the cone are accepted. Therefore, the cone is called acceptance cone. Light incident at an angle beyond θ_a will be refracted though the cladding and the corresponding optical energy is lost.

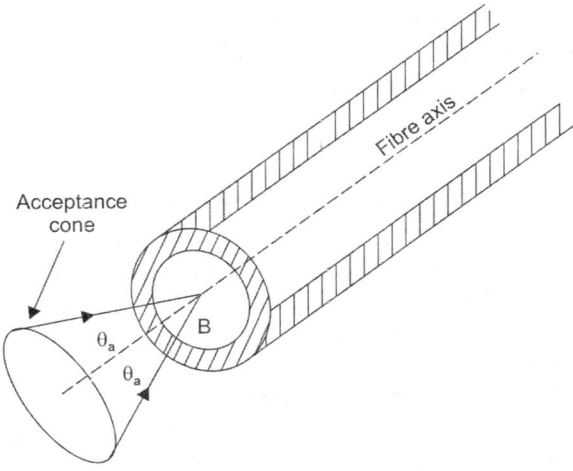

Fig. 4.10: Conical surface described by a ray due to rotation around the fiber axis

4.5 MODES OF PROPAGATION: V-PARAMETER

Light transmitted through one end of the fiber propagate down the fiber in the form of transverse magnetic (TM) and transverse electric (TE) modes. Let us first understand how the modes are formed. A ray of transmitted light propagates down the fiber as shown in Fig. 4.11.

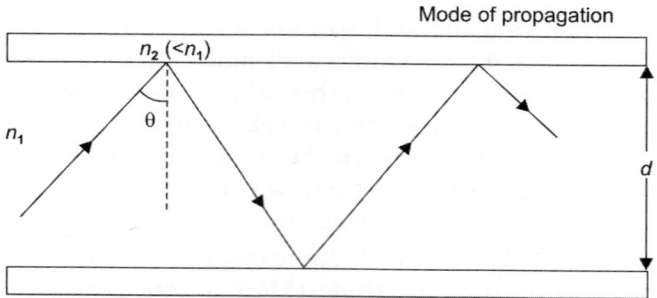

Fig. 4.11. Path of a light ray down a planar dielectric waveguide that results when the angle of incidence at the boundary is $\theta_1 > \theta_c$

Due to the total internal reflection provided the angle of incidence at the core-cladding interface is greater than the critical angle ($\theta > \theta_c$). A ray indicated in reality represents infinite number of such parallel rays, all are very close to each other since the beam transmitted has some cross-section.

If a line is drawn normally as shown in Fig. 4.12a, it represents plane wavefront and all points along the same wavefront must be in phase with each other. The PQ after reflection at Q travels to R and then travels parallel to PQ from R, i.e. $UTPR$ represents a plane wavefront and, hence, the points P and R must be in phase with each other. If they are out of phase with each other, destructive interference takes place and propagation is not possible. Now, moving from P to R along the ray the phase change is given by:

$$\text{Phase change} = (PQ + QR)\frac{2\pi n_1}{\lambda_0} - 2\psi \quad (4.6a)$$

where λ_0 is the wavelength of light in vacuum and ψ is the phase change due to reflection. It should be remembered that whenever light is reflected at the interface of denser to rarer media, additional phase change occurs and the magnitude of the change ψ depends on the angle of incidence θ. Such change is absent if reflection takes place at the interface of rarer to denser medium. Since reflection takes place both at Q and R, 2ψ appears in Eq. (4.6a). From triangle PQR:

$$\cos 2\theta = \frac{PQ}{QR}$$

$$PQ = QR \cos 2\theta$$

or
$$PQ + QR = QR(1 + \cos 2\theta)$$

∴
$$= 2QR \cos 2\theta$$

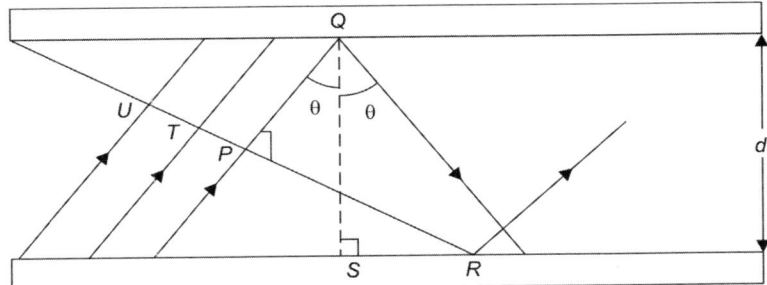

Fig. 4.12a. Ray paths having the same internal angle θ within a planar waveguide. A wavefront is shown connecting the points *UTPR* must, therefore, have the same phase

From triangle QSR:

$$\cos\theta = \frac{QS}{QR}$$

$$QS = d = QR\cos\theta$$

∴ $$PQ + QR = 2d\cos\theta$$

Hence, Eq. (4.6a) can be written as:

$$\text{Phase change} = \frac{4\pi n_1 d \cos\theta}{\lambda_0} - 2\psi = 2\pi m$$

or

$$\frac{2\pi n_1 d \cos\theta}{\lambda_0} - \psi = 4\pi m \quad (4.6b)$$

where m is an integer.

For each value of m, there will be a corresponding value of θ, i.e. θ_a. Rearranging Eq. (4.6b), we have:

$$\cos\theta_a = \frac{(m\pi + \psi)\lambda_0}{2\pi d n_1} \quad (4.7)$$

We may note that it is not possible to obtain an explicit expression for θ in terms of m (m can have only few specific integer values, since θ_a can take values in the range θ_c to $\pi/2$.

Now, rearranging Eq. (4.6b), one obtains:

$$m = \frac{2d n_1 \cos\theta_a}{\lambda_0} - \frac{\psi}{\pi} \quad (4.8)$$

Further, we know that $\sin\theta_c = \frac{n_2}{n_1}$. Since $\theta_a \geq \theta_c$ to maintain total internal reflection (TIR) condition:

$$\sin\theta_a \geq \frac{n_2}{n_1}$$

and
$$\cos\theta_a = \sqrt{(1-\sin^2\theta_a)}$$
$$\leq \sqrt{\left(1-\frac{n_2^2}{n_1^2}\right)}$$
$$\leq \sqrt{\frac{n_1^2-n_2^2}{n_1}}$$

Now, one can write Eq. (4.8) as:

$$m \leq \frac{2dn_1}{\lambda_0}\frac{\sqrt{n_1^2-n_2^2}}{n_1} - \frac{\psi}{\pi} \qquad (4.9)$$

or
$$m \leq \frac{2V}{\pi} - \frac{\theta}{\pi} \qquad (4.10)$$

where
$$V = \frac{\pi d}{\lambda_0}\sqrt{n_1^2-n_2^2}$$

or
$$V = \frac{\pi d}{\lambda_0} NA \qquad (4.11)$$

where V is called as the *normalized frequency* or *cutoff parameter* or *V parameter*.

V parameter depend on:
 (i) Characteristics of optical fiber, and
 (ii) The wavelength of light propagating.

We see that each value of m is associated with a distinct wave pattern or mode within the waveguide. When $m > \left(\frac{2V}{\pi} - \frac{\psi}{\pi}\right)$, the condition for TIR will not be satisfied, the mode is said to be beyond "cutoff".

From Eq. (4.11), one finds that if $2V < \psi$, no mode can be propagated. However, for any value of V, one finds that it is always possible to find an angle θ such that the corresponding value of ψ is less than $2V$, consequently, at least one mode can always be propagated. In general, one finds that the light launched at the fiber and within, the acceptance cone meets the core cladding interface at an angle θ varying between θ_c and $\pi/2$. Of all the three rays only those which satisfy Eq. (4.10) alone propagate forming modes as shown in Fig. 4.12(a).

We can write Eq. (4.11) as:

$$V = \frac{2\pi r}{\lambda_0}\sqrt{n_1^2-n_2^2} \quad (\because d=2r) \qquad (4.12)$$

where r is the radius of the core.

Eq. (4.12) can also be expressed in terms of fractional index change (Δ) as:
$$V = n_1 \sqrt{2\Delta}\, kr \qquad (4.13)$$
where $k = 2\pi/\lambda_0$ is the radial wave vector.

The total number of modes can be obtained as:
$$N = \int V dV = \frac{V^2}{2}$$
$$= \frac{1}{2}\left(NA \times \frac{2\pi r}{\lambda_0}\right)^2 \qquad (4.14)$$

We may note that V parameter actually determines the number of modes supported by optical fiber.

We may also note that *mode* refers to the number of paths for the light rays in the fibre cable. In single mode follows single path through core. In multimode, the light takes many paths through the core (Fig. 4.12(b)).

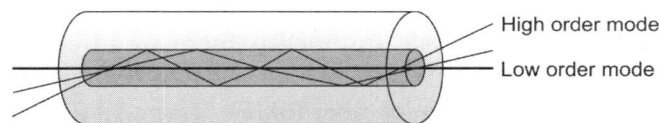

Fig. 4.12b. Modes in optical fibers

4.6 INDEX PROFILE OR REFRACTIVE INDEX PROFILE

An *index profile* for an optical fiber is a plot of refractive index (n) drawn on horizontal axis versus the distance from the core axis drawn on the vertical axis as shown in Fig. 4.13.

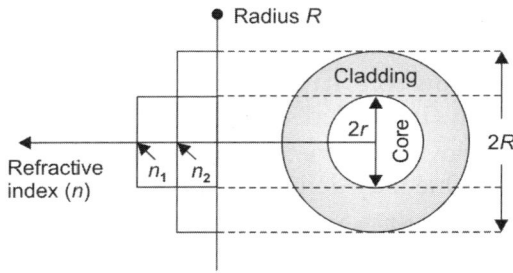

Fig. 4.13. Index profile for an optical fiber

The index profile of multimode fiber can be either a step index type of graded index type. The index profile of a single mode fibre is usually a step index type.

4.7 TYPES OF OPTICAL FIBERS

There are *two basic ways of classifying* fiber optic cables. The first way is how the refractive index varies across the cross-section of the cable. The second way by its mode. The mode refers to the various paths that rays can take in passing through the fiber.

There are two basic ways of defining the index of refraction variation across a cable, i.e. their step index and graded index. Step index refers to the fact that there is a sharply defined step in the index of refraction (n) where the fiber core and the cladding interface. It means that the core has one constant refractive index n_1 while the cladding has another constant refractive index n_2. In other type of cable that has a graded index, the refractive index of core is not constant. Instead, the refractive index varies smoothly and continuously over the diameter of the core. As we get closer to the centre of the core, the refractive index gradually increases, reaching a maximum at the centre of the core and declining as the other outer edge of the core is reached. The refractive index of the cladding is constant.

Each type of optical fiber cable is classified by one of these methods of *rating*, i.e. the *index* or *mode*. In practice, there are three commonly used types of fiber optic cable:

(i) Single mode step index fiber (SMF)
(ii) Multimode step index fiber (MMF)
(iii) Multimode graded index fiber (GRIN)

(a) Single Mode Step Index Fiber (SMF)

A single mode fiber has a *core material of uniform refractive index* value. Similarly, cladding also has a material of uniform index but of lesser value. This result in a sudden increase in the value of refractive index profile takes the shape of a step. The diameter value of core is about 2 µm to 15 µm and external diameter of cladding is 60 µm to 70 µm.

Because of its narrow core, it can guide just a *single mode* as shown in Fig. 4.14. This is why, it is called *single mode fiber*.

The refractive index profile for SMF makes a step change at the core-cladding interface. One may define the refractive index as:

$$n(r) = n_1 r < a \text{ (core)}$$

$$n_2 r \geq \text{ (cladding)}$$

Single mode fibres are the most extensively used ones and they constitute 80% of all the fibers that are manufactured world today. They need lasers as the source of light. Though less expensive, it is very difficult to splice them. They find particular application in submarine cable system.

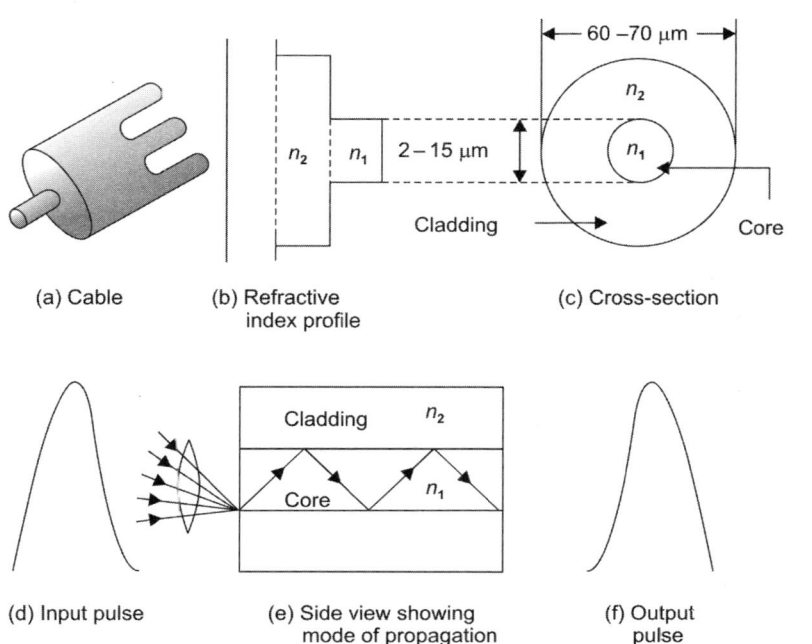

Fig. 4.14. Single mode step-index fiber

(b) Multimode Step Index Fiber (MMF)

The geometry of a multimode step index fiber is shown in Fig. 4.15. Its construction is similar to that of the single mode step index fiber, but differs in its core that has a much larger diameter, by virtue of which it will be able to support propagation of large number of modes as shown in Fig. 4.15. Its refractive index profile is also similar to that of a single mode fiber, but with a larger plane region for the core.

The multimode step index fiber can accept either a laser or a light emitting diode (LED) as source of light. It is least expensive of all. Its typical application is in the *data links*, which has lower bandwidth requirements.

Multimode step index fibers allow the propagation of a finite number of guided modes along the channel. The number of guided modes is dependent upon the physical parameter (i.e. relative refractive index difference, core radius) of the fiber and the wavelengths of the transmitted light which are included in the normalized frequency v for the fiber. There is a cutoff value of normalized frequency v_c for guided modes below which they cannot exist. However, mode propagation does not entirely cease below cutoff. Modes may propagate as unguided or leaky modes which can travel considerable distances along the fiber. Nevertheless, it is the guided modes which are of paramount importance in optical fiber communication as these are confined to the

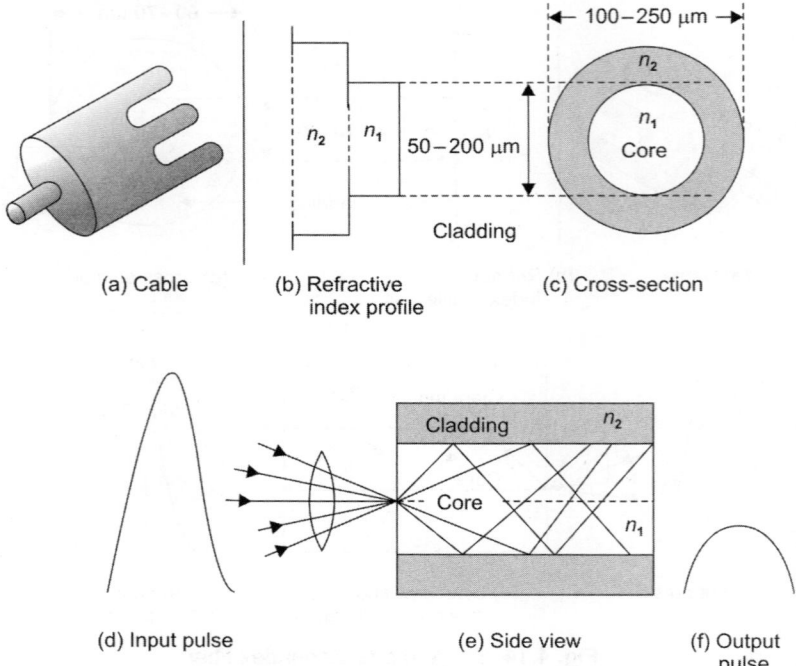

Fig. 4.15. Multimode step index fiber

fiber over its full length. The total number of guided modes or mode volume M_s for a step index fiber is related to the V value for the fiber by the expression:

$$M_s = \frac{V^2}{2} \tag{4.15}$$

Which allows an estimate of the number of guided modes propagating in a particular multimode step index fiber.

The total number of modes M entering the fiber depends on wavelength (λ), radius of fiber (r) and refractive indices (n_1, n_2) and is given by the relation:

$$M = \frac{2\pi^2 r^2}{\lambda^2}(n_1^2 - n_2^2) = \frac{1}{2}\left[\frac{\pi d}{\lambda} \text{NA}\right]^2 \tag{4.15a}$$

where d is the core diameter ($d = 2r$ or $2a$). An important parameter connected with cut off conditions for fiber modes is normalized frequency V (or V-number or V-parameter) given by the relation

$$V^2 = 2M \tag{4.15b}$$

A model is referred to as being cutoff when it is no longer bound to the core of the fiber. V-number is a dimensionless number that determines how many modes a fiber can support. The percentage of power flow in cladding depends on M and hence V as given by the relation:

$$\frac{P_{\text{clad}}}{P} = \frac{4}{3} M^{1/2} \tag{4.15c}$$

Therefore, the optical power is launched into a large number of guided modes, each having different spatial field distribution, propagation constants etc. In an ideal multimode step index fiber with properties (i.e. relative index difference, core diameter) which are independent of distance, there is no mode coupling, and the optical power launched into a particular mode remains in that mode and travels independently of the power launched into the other guided modes. Also, the majority of these guided modes operates far from the cutoff and is well confined to the fiber core. Thus most of the optical power is carried in the core region and not in the cladding. The properties of the cladding (e.g. thickness) do not, therefore, signiticantly affect the propagation of these modes.

(c) Multimode Graded Index Fiber (GRIN)

Multimode graded index fiber is also denoted as GRIN. The geometry of the GRIN multimode fiber is same as that of the MMF. Its core material has a special feature that its refractive index value decreases in the radially outward direction from the axis, and becomes equal to that of the cladding at the interface. But the refractive index of the cladding remains uniform. Its refractive index profile is shown in Fig. 4.16. Either

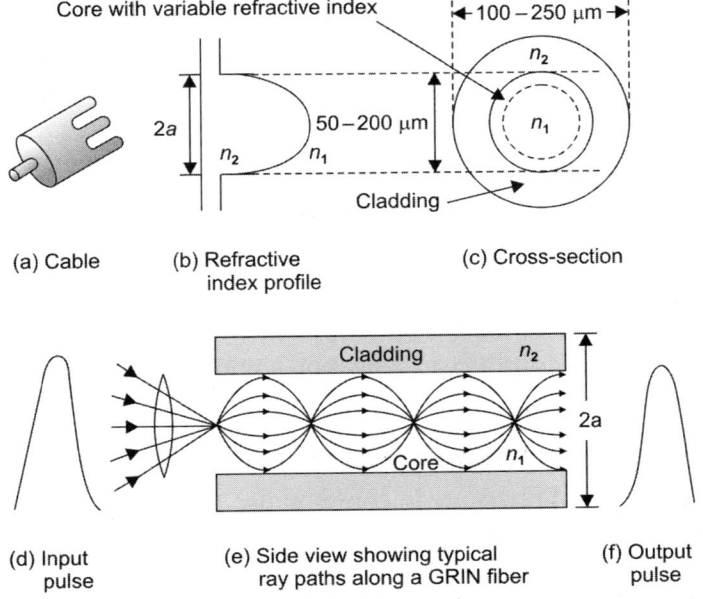

Fig. 4.16. Multimode graded index fibre

a laser or light emitting diode (LED) can be the source for the GRIN multimode fiber. It is the most expensive of all. Its splicing could be done with some difficulty. Its typical application is the *telephone trunk* between central offices.

In the graded index fiber, the acceptance angle and the numerical aperture decease with radial distance from the axis. The numerical aperture of the graded index multimode fiber is given by:

$$NA = n_1 \sqrt{2\Delta \left[1 - \left(\frac{r}{a}\right)^2\right]} \qquad (4.15d)$$

where a is the total radius of the core and r is the varying radius of the core.

The index variation in these fibers may be represented as:

$$n(r) = n_1 \left[1 - 2\Delta (r/a)^\alpha\right]^{1/2} \qquad r < a \text{ (core)}$$

$$= n_1 \left[(1 - 2\Delta)^{1/2}\right] = n_2 \qquad r \geq a \text{ (cladding)}$$

where Δ is relative refractive index difference and α is the profile parameter which gives the characteristic refractive index profile of the fiber core (Fig. 4.17).

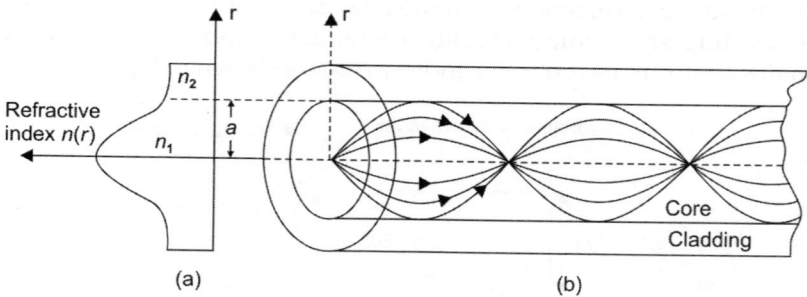

Fig. 4.17. The refractive index profile and ray transmission in a multimode graded index fiber

When $\alpha = \infty$ the fiber represents a step index profile, a parabolic profile when $\alpha = 2$ and a triangular profile when $\alpha = 1$.

Fibers with $\alpha = 2$ produce propagation having a near parabolic profile (Fig. 4.18).

4.7.1 Multimode and Single Mode Fibers

These two types of fibers are in common use. Both these fibers are 125 microns in outside diameter.

Multimode fiber optic cable has a large diameter core that is much larger than the wavelength of light transmitted, and therefore has

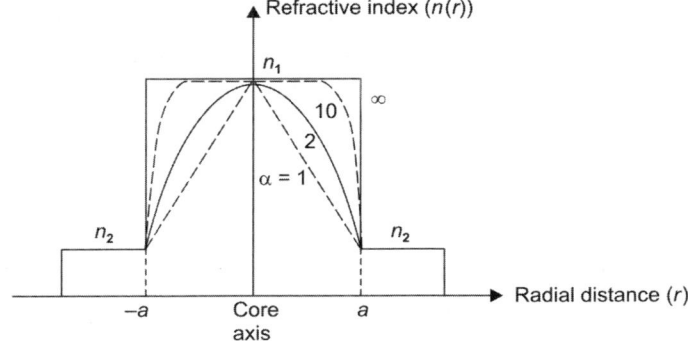

Fig. 4.18. Fiber refractive index profile for different values of α

multiple pathways of light. Several wavelengths of light are used in the fiber core.

Multimode fiber has light travelling in the core in many rays, called modes. It has a bigger core, almost always 62.5 microns, but sometimes 50 microns and is used with LED source at wavelengths of 850 and 1300 mm for slower local area networks (LANs) and lasers at 850 and 1310 mm for network running at gigabits per second or more.

Singlemode fiber has a much smaller core, only about 9 microns, so that the light travels in only one ray. It is used for telephone and CATV with laser sources at 1300 nm and 1550 nm.

Plastic optical fiber (POF) is large core (about 1 mm) fiber that can only be used for short, low speed networks. Step index multimode was the first fiber design but is too slow for most uses, due to the dispersion caused by the different path lengths of the various modes. Step index fiber is rare. Only POF uses a step index design today.

Graded index multimode fiber uses variations in the composition of the glass in the core to compensate for the different path lengths of the modes. It offers hundreds of times more bandwidth than step index fiber i.e., up to about 2 gigahertz over 1 km.

Singlemode fiber shrinks the core down so small that the light can only travel in one ray. This increases the bandwidth to almost infinitely. But it is particularly limited to about 100,000 gigahertz.

The size of an optic fiber matters. Fiber comes in two types, singlemode and multimode. Except for fibers used in specialty applications, singlemode fiber can be considered as one size and type. When dealing with long haul telecom or submarine cases, specially single mode fibers are used.

Multimode fibers originally came in several sizes, optimized for various networks and sources, but the data industry standardized on 62.5 core fibers. 62.5/125 fiber has a 62.5 micron core and a 125 micron cladding. Multimode fiber optic cable can be used for most general fiber applications. Multimode cable comes with two different core sizes: 50 micron or 62.5 micron. Although 50-micron fiber features a smaller core, which is the light-carrying portion of the fiber, both 62.5- and 50-micron cable feature the same glass cladding diameter of 125 microns. Both can be used in the same types of networks, although 50-micron cable is used more often for premise applications such as backbone, horizontal, and intrabuilding connections. Both types can use either LED or laser light sources.

The main difference between 50 micron and 62.5 micron cable is in band-width. 50 micron cable features three times the bandwidth of standard 62.5 micron cable, particularly at 850 nm. The 850 nm wavelength is becoming more important as lasers are being used more frequently as a light source.

Other differences are distance and speed. 50 micron cable provides longer link lengths and higher speeds in the 850 nm wavelength.

Fiber type	Bandwidth (minimum)	at 850 nm	at 1310 nm
50/125 μm	50 MHz/km	500 m	500 m
62.5/125 μm	160 MHz/km	220 m	500 m

Singlemode fiber optic cable has a small core and only one pathway of light. With only a single wavelength of light passing through its core, singlemode realigns the light toward the centre of the core instead of simply bouncing it off the edge of the core as with multimode. Singlemode is typically used in long-haul network connections spread out over extended areas, longer than a few miles. For example, they can be used for connections between switching offices. Singlemode cable features a 9-micron glass core.

Duplex cable consists of two fibers, usually in a zipcord (side-by-side) style. Use duplex multimode or singlemode fiber optic cable for applications that require simultaneous, bi-directional data transfer. Workstations, fiber switches and servers, fiber modems, and similar hardware require duplex cable. Duplex fiber is available in singlemode and multimode.

Simplex fiber optic cable consists of a single fiber, and is used in applications that only require one-way data transfer. For instance, an interstate trucking scale that sends the weight of the truck to a monitoring station or an oil line monitor that sends data about oil flow to a central location. Simplex fiber is available in singlemode and multi-mode.

Although it may seem from what we have discussed about total internal reflection that any ray of light can travel down the fiber, in fact, because of the wave nature of light, only certain ray directions can actually travel down the fiber. These are called the *fiber mode*. In a multimode fiber many different modes are supported by the fiber.

Because its core is so narrow that singlemode fiber can support only one mode. This is called the *lowest order mode*. Singlemode fiber has some advantages over multimode fiber.

Graded index fiber has a different core structure from single mode and multimode fiber. Whereas in a step-index fiber fiber the refractive index of the core is constant throughout the core, in a graded index fiber the value of the refractive index changes from the centre of the core onwards. It has a quadratic profile. This means that the *refractive index of the core is proportional to the square of the distance from the centre of the fiber.*

The two main types of fiber in use today are *step-index multimode* and *step-index singlemode* fiber. The step-index part of the name can be understood from Fig. 4.19 which shows the cross-section of the fiber. Step-index refers to the abrupt change in refractive index between the core and cladding materials in contrast to graded-index fibers where refractive index changes gradually over the diameter of the fiber (Fig. 4.20).

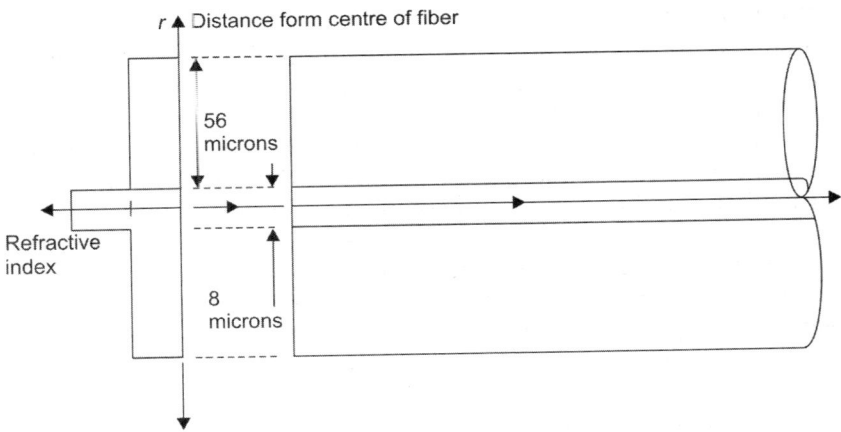

Fig. 4.19. Single mode optical fiber

Multimode fibers have cores of around 50 μm and outside diameters of about 125 μm. Singlemode fiber has a core reduced to below 10 μm to allow only one mode of propagation to be supported.

Multimode fiber can capture light from the light source and pass light to the receiver with high efficiency, so can be used with low-cost light emitting diodes (LEDs). High precision connectors are not required because the large core diameter allows wide-tolerance on mechanics. Multimode modal dispersion severely limits the usable bandwidth. Multimode fibers suffer from higher losses than singlemode fibers.

Multimode fiber has found some application in *cost-sensitive areas* such as *LAN* but even here, it is too costly compared to copper solutions and local-loop applications. But its poor bandwidth and high-loss

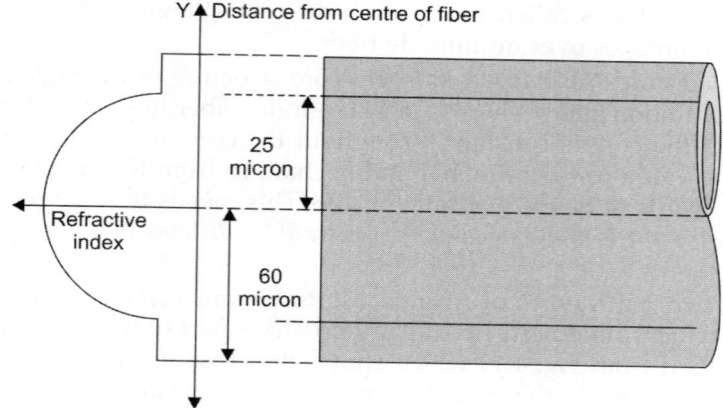

Fig. 4.20. Graded index optical fiber

characteristics means that its application in high-data rate links has been very limited.

Singlemode fiber exhibits lower attenuation. Attenuation of singlemode fiber is specified at 0.37 dB/km at 1310 nm, in effect allowing a non-repeated run to be increased by a factor of two over multimode fiber. The use of singlemode fiber completely eliminates modal-dispersion which is the key cause of bandwidth limitation in multimode optical fiber, but this does not mean that it has infinite bandwidth. The dispersion left is called chromatic dispersion (so called as it is wavelength dependent). Chromatic dispersion is caused by the core material itself and is actually negative at short wavelength and moves positive at longer wavelengths. This creates a *'magic' wavelength* at which dispersion is actually zero.

This is, interestingly enough, at about 1310 nm which explains the wide use of this particular wavelength. If 1310 nm is used on a singlemode fiber, it is easy to achieve a bandwidth of several Gbit/s with losses of around 37 dB/km. Thus, in a singlemode fiber, attenuation is the limiting factor for long-distance transmission.

The sizes of these two main types of fibers are shown in Fig. 4.21.

The characteristics of singlemode fiber: Bandwidth can be in the order of many Gbit/s with very low attenuation. This allows long-distance unrepeated transmission up to around 50 km.

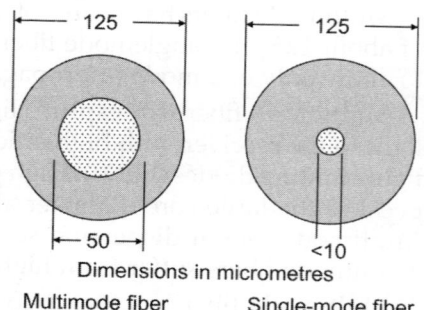

Fig. 4.21. Sizes of the two main types of fiber

The small diameter (10 µm) of the core necessitates the use of expensive laser diodes to enable efficient light coupling and pass sufficient light into the fiber.

The small core diameter needs extremely precise connectors e.g. if two fibers are misaligned by only 1 µm the overlap area is reduced by about 15% or attenuation equivalent to several km of fiber. Singlemode connectors are thus more expensive.

The performance of singlemode fiber is so good that it is the only type of fiber used for long distance links.

With copper cables larger size means less resistance and therefore more current, but with fiber the opposite is true.

So a smaller core size means higher bandwidth and greater distances.

A comparison between single mode step index fiber and multimode graded index fiber is given in Table 4.2.

Table 4.2. Comparison between single mode step index fiber and multimode graded index fiber

S. No.	Single mode step index fiber	Multimode graded index fiber
1.	In single mode fibre, the diameter of core is very small. Only one mode is allowed to propagate through it. Core diameter ~ 2–15 µm, cladding diameter ~ 60–70 cm	In multimode fiber, the diameter of core is comparatively large so that more mode can propagate through it. Core diameter ~50–200 µm cladding diameter ~100–250 µm
2.	Difference in refractive indices of core and cladding is usually very small	Difference in refractive indices of core and cladding is larger
3.	A very narrow source of light, i.e. either laser or LED can only be used to launch light in the fiber	Light source should not necessarily be very narrow
4.	Numerical aperture NA of single mode fiber is usually small	Numerical aperture of multimode fiber is large
5.	Refractive index of core is constant and changes abruptly at core-cladding interface	Refractive index of core decreases parabolically from the centre of core-cladding interface
6.	Single mode fiber are expensive but more efficient	Multimode fibers are comparatively cheap, but have low information carrying capability

4.8 RAY PROPAGATION IN GRADED-INDEX FIBERS

In a step-index fiber, the refractive index of the core is constant, n_1, and that of the cladding is also constant, n_2; n_1 being greater than n_2. The refractive index n is a step function of the radial distance. A pulse of light launched in such a fiber will get broadened as it propagates through it due to multipath time dispersion. Therefore, such fibers cannot be used for long-haul applications. In order to overcome this difficulty, another class of fibers is made, in which the core index is not constant but varies with radius r according to the relation

$$n(r) = \begin{cases} n_1 = n_0 \left[1 - 2\Delta\left(\dfrac{r}{a}\right)^\alpha\right]^{1/2} & \text{for } r \leq a \\ n_2 = n_0 \left[1 - 2\Delta\right]^{1/2} = n_c & \text{for } b \geq r \geq a \end{cases} \qquad (4.16)$$

where $n(r)$ is the refractive index at radius r, a is the core radius, b is the radius of the cladding, n_0 is the maximum value of the refractive index along the axis of the core, Δ is the relative refractive index difference, and α is called the *profile parameter*. Such a fiber is called a *graded-index* (GI) fiber. For $\alpha = 1$, the index profile is triangular; for $\alpha = 2$, the profile is parabolic; and for $\alpha = \infty$, the profile is that of a Si fiber (Fig. 4.22).

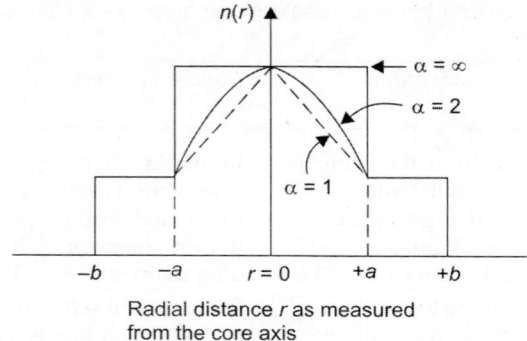

Fig. 4.22. Variation of $n(r)$ with r for different refractive index profiles

For a parabolic profile, which reduces the modal dispersion considerably, the expression for NA can be obtained as follows:

$$\begin{aligned} \text{NA} &= (n_1^2 - n_2^2)^{1/2} \\ &= \left[n_0^2\left\{1 - 2\Delta\, 1 - 2\Delta\left(\dfrac{r}{a}\right)^2\right\} - n_0^2\,(1 - 2\Delta)\right]^{1/2} \\ &= n_0\left[2\Delta\left(1 - \dfrac{r^2}{a^2}\right)\right]^{1/2} \qquad (4.17) \end{aligned}$$

Therefore axial $\text{NA} = n_0\sqrt{2\Delta}$. This means that the NA decreases with increasing r and becomes zero at $r = a$.

In order to apreciate ray propagation through a GI fiber, let us first visualize the core of this fiber as having been made up of several coaxial cylindrical layers (Fig. 4.23).

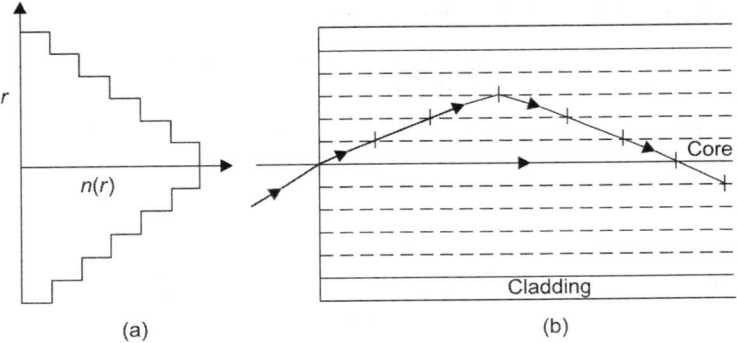

Fig. 4.23. (a) Variation of *n* with *r*, (b) ray traversal through different layers of the core

The refractive index of the central cylinder is the highest, and it goes on decreasing in the successive cylindrical layers. Thus, the meridional ray shown takes on a curved path, as it suffers multiple refractions at the successive interfaces of high to low refractive indices. The angle of incidence for this ray goes on increasing until the condition for total internal reflection is met; the ray then travels back towards the core axis. On the other hand, the axial ray travels uninterrupted.

In this configuration, the multipath time dispersion will be less than that in SI fibers. This is because the rays near the core axis have to travel shorter paths compared to those near the core-cladding interface. However, the velocity of the rays near the axis will be less than that of the meridional rays because the former have to travel through a region of high refractive index ($v = c/n$). Thus, both the rays will reach the other end of the fiber almost simultaneously, thereby reducing the multipath dispersion. If the refractive index profile is such that the time taken for the axial and the most oblique ray is same, the multipath dispersion will be zero. In practice, a parabolic profile ($\alpha = 2$), (Fig. 4.24) reduces this type of dispersion considerably.

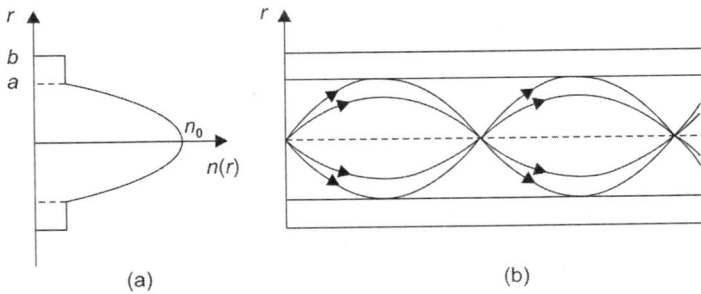

Fig. 4.24. (a) The parabolic profile of a GI fiber, (b) ray path in such a fiber

4.9 EFFECT OF MATERIAL DISPERSION

The refractive index n of any transparent medium is given by $n = c/v$, where c is the velocity of light in vacuum and v is its velocity in the medium. In terms of wave theory, v is called the *phase velocity* v_p of the wave in the medium. Thus, we have:

$$v = v_p = c/n \tag{4.18}$$

If $\omega (= 2\pi f)$ is the angular frequency of the wave (in rad/s), f being the frequency in hertz, and $\beta\, (= 2\pi/\lambda_m)$ is the propagation constant, λ_m being the wavelength of light in the medium (which is equal to λ/n; λ is the free-space wavelength), then the phase velocity of the wave is also expressed as:

$$v_p = \omega/\beta \tag{4.19}$$

Using Eqs (4.18) and (4.19), one obtains:

$$v_p = c/n = \omega/\beta$$

or

$$n = \frac{c\beta}{\omega} \tag{4.20}$$

At this point, it is important to know that any signal superposed onto a wave does not propagate with the phase v_p of the wave, but travels with a group velocity v_g given by the followign expression:

$$v_g = \frac{d\omega}{d\beta} = \frac{1}{d\beta/d\omega} \tag{4.21}$$

In a non-dispersive medium, v_p and v_g are same, as v_p is independent of the frequency ω; but in a dispersive medium, where v_p is a function of ω:

$$v_g = \frac{1}{d\beta/d\omega} = \frac{v_p}{1-(\omega/v_p)(dv_p/d\omega)} \tag{4.22}$$

Thus a signal, which is normally a light pulse, will travel through a dispersive medium (e.g., the core of the optical fiber) with speed v_g. Therefore, for such a pulse, we may define a group index n_g such that:

$$n_g = c/v_g \tag{4.23}$$

Let us subtitute for v_g from Eq. (4.21) in Eq. (4.23) to get:

$$n_g = c\frac{d\beta}{d\omega} = c\frac{d}{d\omega}\left(\frac{\omega n}{c}\right) = \frac{d}{d\omega}(n\omega)$$

$$= n + \omega\frac{dn}{d\omega} \tag{4.24}$$

This is an *important expression*, relating the group index n_x with the ordinary refractive index or the phase index n.

Since

$$\frac{dn}{d\omega} = \frac{dn}{d\lambda} \cdot \frac{d\lambda}{d\omega}$$

and
$$\omega = \frac{2\pi c}{\lambda}$$

$$\frac{d\omega}{d\lambda} = -\frac{2\pi c}{\lambda^2}$$

We have from Eq. (4.24):

$$n_g = n + \frac{2\pi c}{\lambda}\frac{dn}{d\lambda}\left(-\frac{\lambda^2}{2\pi c}\right) = n - \lambda\frac{dn}{d\lambda} \qquad (4.25)$$

Thus
$$v_g = \frac{c}{n_g} = \frac{c}{(n - \lambda\, dn/d\lambda)} \qquad (4.26)$$

Obviously, a light pulse, therefore, will travel through the core of an optical fiber of length L in time t given by:

$$t = \frac{L}{v_g} = \left[n - \lambda\frac{dn}{d\lambda}\right]\frac{L}{c} \qquad (4.27)$$

If the spectrum of the light source has a spread of wavelength $\Delta\lambda$ about λ and if the medium of the core is dispersive, the pulse will spread out as it propagates and will arrive at the other end of length L, over a spread of time Δt. If $\Delta\lambda$ is much smaller than the central wavelength λ, we can write:

$$\Delta t = \frac{dt}{d\lambda}\Delta\lambda = \frac{L}{c}\left[\frac{dn}{d\lambda} - \frac{dn}{d\lambda} - \lambda\frac{d^2n}{d\lambda^2}\right]\Delta\lambda$$

$$= -\frac{L}{c}\lambda\frac{d^2n}{d\lambda^2}\Delta\lambda \qquad (4.28)$$

If $\Delta\lambda$ is the full width at half maximum (FWHM) of the peak spectral power of the optical source at λ, then its relative spectral width γ is given by:

$$\gamma = \left|\frac{\Delta\lambda}{\lambda}\right| \qquad (4.29)$$

If an impulse of negligible width is launched into the fiber, then Δt will be the half power width τ of the output (or the broadened pulse). Thus, the pulse broadening due to material dispersion may be given by:

$$\tau = \frac{L}{c}\gamma\left|\lambda^2\frac{d^2n}{d\lambda^2}\right|$$

or
$$\frac{\tau}{L} = \frac{\gamma}{c}\left|\lambda^2\frac{d^2n}{d\lambda^2}\right| \qquad (4.30)$$

The material dispersion of optical fibers is quoted in terms of the material dispersion parameter D_m given by:

$$D_m = \frac{1}{L}\frac{\tau}{\Delta\lambda} = \frac{\lambda}{c}\left|\frac{d^2n}{d\lambda^2}\right| \qquad (4.31)$$

D_m has the units of ps nm^{-1} km^{-1}. Variation of D_m with wavelength (λ) for pure silica is shown in Fig. 4.25. We may note that majority of fibers are manufactured using silica as the host material.

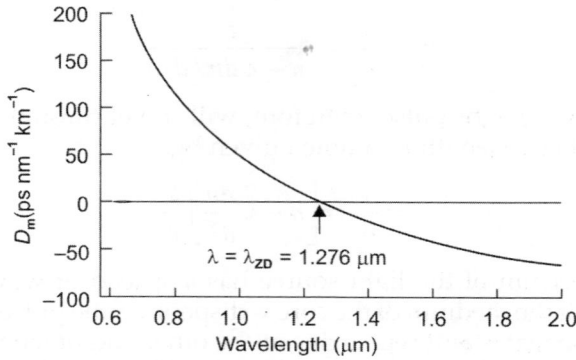

Fig. 4.25. Material dispersion parameter D_m as a function of λ for pure silica

We may note that D_m changes sign at $\lambda = \lambda_{ZD} = 1.276$ μm (for pure silica). This point has been frequently referred to as the *wavelengh of zero material dispersion*. This wavelength can be changed slightly by adding other dopants to silica. Clearly, the use of an optical source with a narrow spectral width (e.g., injection laser) around λ_{ZD} would substantially reduce the pulse broadening due to material dispersion.

4.10 THE COMBINED EFFECT OF MULTIPATH AND MATERIAL DISPERSION

In a fiber-optic communication system, the optical power generated by the optical source, e.g., LED, is proportional to the input current to the transmitter. The optical power received by the detector is proportional to the power launched into and propagated by the optical fiber. This, in turn, gives rise to a proportional current at the receiver end, thus giving an overal linearity to the system.

We have alreay seen that multipath and material dispersion lead to the broadening of the pulse launched into a fiber. Thus, the received pulse represents the impulse response of the fiber. If we assume that the FWHM of the transmitted pulse is τ_0, and the impulse responses due to multipath and material dispersion lead to approximately Gaussian pulses of FWHM τ_1 and τ_2, respectively, as shown in Fig. 4.26, and that the two mechanisms are almost independent of each other, then the received pulse width at half maximum power, τ, will be given by:

$$\tau = \left[\tau_0^2 + \tau_1^2 + \tau_2^2\right]^{1/2} \tag{4.32}$$

$$\frac{\tau}{L} = \left[\left(\frac{\tau_0}{L}\right)^2 + \left(\frac{\tau_1}{L}\right)^2 + \left(\frac{\tau_2}{L}\right)^2\right]^{1/2} \tag{4.33}$$

where L is the total length of the fiber.

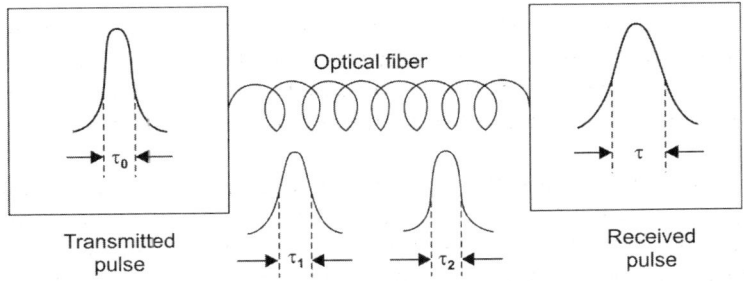

Fig. 4.26. The combined effect of multipath and material dispersion

It is also possible to express the same result in terms of the root mean square (rms) pulse widths. The calculation of the rms pulse width is given in the next section. Thus, if the transmitted pulse is Gaussian in shape and has an rms width σ_0 and this pulse is broadened by both multipath and material dispersion leading to nearly Gaussian pulses of rms width σ_1 and σ_2, respectively, then the rms width σ of the received pulse is given by:

$$\frac{\sigma}{L} = \left[\left(\frac{\sigma_0}{L}\right)^2 + \left(\frac{\sigma_1}{L}\right)^2 + \left(\frac{\sigma_2}{L}\right)^2\right]^{1/2} \tag{4.34}$$

4.11 CALCULATION OF RMS PULSE WIDTH

The rms width σ of a pulse is defined as follows. If $p(t)$ is the power distribution in the pulse as a function of time t and the total energy in the pulse is:

$$\varepsilon = \int_{-\infty}^{\infty} p(t)\, dt \tag{4.35}$$

then its rms width σ is given by:

$$\sigma^2 = \frac{1}{\varepsilon}\int_{-\infty}^{\infty} t^2 p(t)\, dt - \left[\frac{1}{\varepsilon}\int_{-\infty}^{\infty} tp(t)\, dt\right]^2 \tag{4.36}$$

4.12 ADVANTAGES OF OPTICAL FIBER

The transmission and reception of large amount of data/information at the fastest rate is a fundamental requirement of communication technology. This data may be in any form, viz. text, sound, pictures etc.

The information-carrying capacity is directly related to the bandwidth (or frequency extent) of the modulated carrier, which is generally limited to a fixed fraction of the carrier frequency. Theoretically, greater the carrier frequency, larger is the available transmission bandwidth and thus the information-carrying capacity of the system. Therefore, radio frequency communication was developed at VHF and UHF. The communication at optical frequency, 10^{14} Hz, offers an increase in the bandwidth by a factor of 10^4, over high frequency microwave transmission of 10^{10} Hz.

The invention of laser in 1960 provided a powerful coherent light source together with the possibility of modulation at high frequency. The semiconductor lasers (1962) and low loss optical fibre waveguides boosted the development of optical fiber communication in 1960s and 1970s.

The theoretical transmission loss of a silica fiber at a wavelength of 1.55 μm is 0.13 dB/km. It is limited by Rayleigh scattering and vibrational absorption of Si–O bonds. The best transmission reported for infrared fibers is 0.73 dB/km using a fluoride glass. Two major factors are associated with this; one, if the losses are less, efficiency increases and two, the spacing of repeaters also increases, which reduces cost and complexity of system. For conventional metal cables, losses are around 5 dB/km.*

The diameter of optical fiber is very small, of the order of μm. Even including protective coating, size is small compared to conventional metal cables. Due to small size and low weight, they are used in aircrafts, ships, satellites, automobiles etc. for iternal communications.

The fibers are made of glass or plastic polymers, therefore, they are electrical insulators. There are no problems of short circuit, sparks, earth loop and interface, electromagnetic interferences, switching transients resulting from electromagnetic pulses, radio frequency interference, susceptibility to lightning stroke, cross talk, etc. The signl transmission of optical fiber is safe because it does not radiate much and it is almost impossible to tap the signal in non-invasive manner (non-invasive means without drawing optical power from the fiber). Tapping can be immediately detected. Therefore, this type of communication is very useful in military or bank or defence or high security related communications.

The technological progress made it possible to manufacture fiber with high tensile strengths. They can be twisted or bent into small radii without damage. Hence, these cables are superior to copper cables in terms of size, weight, storage, handling, transportation and installation. The cost of manufacturing glass fiber cables is also low compared to metal cables. Due to low loss, line amplifiers or repeaters are widely

---

* Signal attenuation in number of decibels per unit length $\alpha_{dB} = 10 \log_{10} (P_I/P_o) \div L$, where P_i is the input transmitted optical power, P_o is the output optical power and L is fibre length.

separated. It reduces the cost of equipments, installation, maintenance etc. The overall cost and benefits of optical fiber communications are better than metal cable or radio wave communications.

There are some problems also. For example, fragility of bare and small size create problems in splicing (means forming permanent joints), connectors, losses in couplers, complex testing procedure, stress corrosion etc. But engineering and technological advancements are expected to overcome these problems.

4.13 SIGNAL DEGRADATION

When a light signal modulated by analogue or digital information propagates through a fiber, it is degraded at other end in the sense that power loss and shape distortions takes place in the signal. Loss or signal attenuation weakens the strength of light signal after travelling a certain distance. A repeater is required at a distance where the signal is not severely attenuated. Similarly, due to distortion, optical signal pulses (in digital communication) are broadened as they travel along the fiber. The excessive broadening may cause overlapping with neighbouring pulses, thereby, creating error ('1' may be replaced by '0' or *vice versa*) in the receiver output. Obviously, signal degradation is mainly due to two reasons: (i) attenuation (loss) and (ii) distortion (broadening of pulses).

Fiber loss and distortion depends on the material used and structural defects during fiber manufacturing. Materials used in making the fiber are glasses and plastics. The most common optically transparent glasses from which fiber is made is silica which has refractive index $n_1 = 1.458$ at 850 nm. The cladding material (having slightly lower index (n_2) is made by adding 'dopants' such as B_2O_3, GeO_2 or P_2O_5 in silica.

4.14 FIBER LOSSES

The basic mechanisms responsible for fiber losses are:

(i) *Absorption:* Absorption depends on fiber material and wavelength. Glass introduces less absorption loss as compared to plastic. Absorption of energy may be caused by atomic defects, impurity atom or basic constituent atoms of fiber material. This loss includes ultraviolet absorption, infrared absorption and ion-resonance absorption. Absorption loss due to impurities is the major source of loss in optical fiber. There are two important types of impurities: (i) the transition metal ions and hydroxyl (OH) ions. The transient metal impurities, i.e. Cu, Fe, Co, V, Cr, Ni and Mn absorb strongly in the region of interest. This means that the fiber be free from these impurities in the electronic absorption. The trapped OH^- ions in the fiber material absorb at 0.95 µm and 1.39 µm. These regions corresponds to region of interest and hence the presence of OH^- ion should be minimized and as far as possible should be kept below 0.01 part per million (ppm). There are certain wavelength ranges at which attenuation is minimum. Such a range of wavelength is called as **optical window or transmission window**. These windows are quite suitable for transmission of information.

(ii) *Scattering losses:* Light scattering (Rayleigh scattering) is caused by structural imperfection in the guided mode besides the fiber material. The order of losses introduced per kilometer due to scattering is 2.5 dB at 8.20 nm, 0.24 dB at 1300 nm, and 0.012 dB at 1550 nm. This reveals that loss reduces with increasing wavelength. This loss arises due to microscope irregularities in material density, structural inhomogeneities or defects during manufacture process. Light is refracted by imperfection and some of it is spread out to escape through the cladding (Fig. 4.27). For Rayleigh scattering proportional to λ^{-4}, this sets a lower limit on wavelength that can be transmitted through a glass fiber to about 0.8 μm. Below this wavelength, scattering loss is quite high.

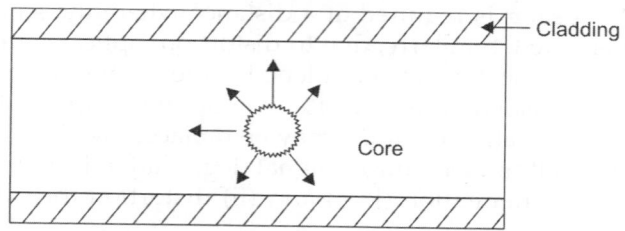

Fig. 4.27. Scattering losses

(iii) *Radiative losses:* These originate from peturbations in fiber, geometry, e.g. bends (microscopic or macroscopic) of finite radius. Bends can arise when cable turns a corner or when fiber is incorporated in cable.

(iv) *Coupling losses (splicing losses):* These losses occurs at junctions connecting source to fiber, fiber to fiber and fiber to photodetector. Losses at the junction of two fiber sections are caused mainly due to misalignment. There are four types of misalignment, namely lateral misalignment, gap displacement, angular misalignment and imperfect finish of surface.

(v) *Loss due to geometrical effects:* During manufacture and/or installation, bending (microscopic or macroscopic) causes loss of power (Fig. 4.28(a)). Due to small cracks in glass, microbending is caused. When the cracks extend to a large distance along the length of the optical fiber, microbending occurs. However, coupling can occur due to microbending but microbending obstructs more propagation.

4.14.1 Attenuation

In addition to the requirement of minimum pulse broadening during pulse propagation, minimum loss in optical power is also required. The optical fiber fabrication process minimize the introduction of common transition-metal impurities, such as iron, copper, cobalt, etc.,

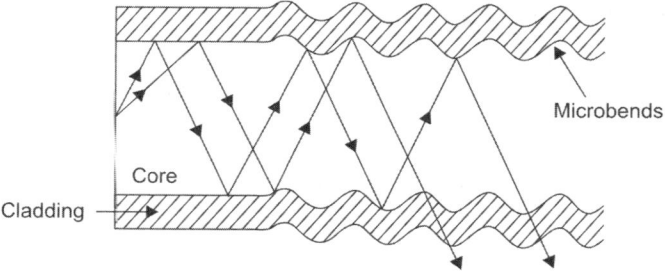

Fig. 4.28a. Losses due to tight bending of fiber

into the glass. The important source of attenuation in fiber can be classified as due to molecular vibrational resonance, vibrational resonances of the OH ion, and Rayleigh scattering. The molecular vibrational resonance in optical fiber glasses occurs in the infrared region beyond 5 μms, but these bands have tails that extend into the region at which optical signals propagate (0.8–1.6 μms). Although small, these bands contribute a loss to optical propagation on the order of 0.5 dB/km and less. Although small, these bands contribute a loss to optical propagation on the order of 0.5 dB/km and less. Vibrational resonances in fibers can also result from the inclusion of the OH ions in the silica matrix. An important vibrational resonance of the OH ion occurs at 2.71 μms. There are overtones and combination bands at 0.95, 1.25, and 1.38 μms due to the 2.71-μm resonance. Optical fiber fabrication processes limit the amount of OH ions through proper doping procedures, and it is possible to fabricate fibers with OH ion inclusions on the order of parts per billion. The window of transmission of 0.8, 1.3 and 1.55 μm are chosen so that losses due to OH ion vibrational resonances are limited.

Rayleigh scattering of light is strongly wavelength-dependent and varies inversely with the fourth power of the wavelength. This source of optical loss results from concentration fluctuations at high temperatures that are frozen in place as the glass cools through the transformation region. One can calculate an effective loss coefficient due to Rayleigh scattering by the expression.

$$\alpha_R = \frac{8\mu^3}{3\lambda^4}(n^2 - 1) kTB \qquad (4.37)$$

Here, B represents the isothermal compressibility of the glass, T is the fictive temperature and k is Boltzman's constant. At 1550 nms, losses resulting from Rayleigh scattering are on the order of 0.13 dB/km. This window is minimum with respect to optical attenuation as it is optimized with respect to the losses resulting from Rayleigh scattering (which increase with lower wavelength). Losses in this window are as low as 0.2 dB/km, whereas in the 1310 nm window, they are on the order of 0.35 dB/km.

4.14.2 Dispersion-Shifted Fiber

The attenuation of silica-based fibers is a minimum near 1.55 µms. For this reason, there is considerable interest in transmitting information at this wavelength. Also, with the recent availability of erbium-doped fiber amplifiers at 1.55 µms, this window of operation is most advantageous for long distance communication. Step-index, single-mode fibers, however, exhibit considerable pulse broadening in this window, approximately 18 ps/(nm·km). Figure 4.28(b) shows the segmented core profile of a dispersion-shifted fiber and that of the platform profile. The segmented core profile was the first dispersion-shifted profile capable of shifting the λ_0 while maintaining other important parameters.

Studies shows that the dispersion minimum can be easily shifted to the 1.55 µm window., while at the same time maintaining other important properties of the fiber. Other important parameters include both bend loss, spot size (r_0), and cutoff wavelength (λ_c). The cutoff wavelength refers to the wavelength below which multimode propagation exists. These dispersion-shifted profiles are now finding application in *transoceanic communication* and the *trunk lines*. The use of dispersion-shifted fibers with erbium-doped amplifiers requires a minimum dispersion, on the order of 0.2 ps. (nm·km) because optical nonlinearities distort the signal at the zero dispersion wavelength λ_0. Of particular concern is four-wave mixing.

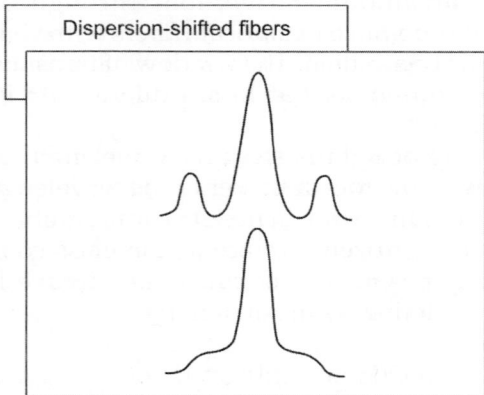

Fig. 4.28b. Dispersion-shifted profiles

Some standard optical fibers are listed in Table 4.3.

4.15 OPTICAL FIBER CABLES

Where optical fibers are to be installed in a working environment their *mechanical properties* are of prime importance. Bare glass fibers are brittle and have small sectional areas which make them susceptible to damages

Optical Fibers

Table 4.3. Standard optical fibers

Optical fiber (type)	Cladding diameter (μm)	Core cladding (μm)	D	Applications
Single mode (8/125)	125	8	0.1% to 0.2%	(i) Long distance (ii) High data rate
Multimode (50/125)	125	50	1% to 2%	(i) Short distance (ii) Low data rate
Multimode (62.5/125)	125	62.5	1% to 2%	LAN
Multimode (100/140)	140	100	1% to 2%	LAN

when employing normal transmission line handling procedures. It is, therefore, necessary to cover the fibers to improve their tensile strength and to protect them against external influences. This is done by surrounding the fiber with protective layers referred to as coating and cabling. A plastic coating with high elastic modulus is applied directly to the fiber cladding. This coated fiber is incorporated to an optical cable to increase its resistance to mechanical strain, stress and environmental conditions.

The optical cable gives:
  (i) *Fiber protection*: The optical cable protects the fiber against damage and breakage during installation and also throughout the life of the fiber.
  (ii) *Stability of the fiber* must have good stable transmission characteristic such that attenuation due to cabling is minimized.
  (iii) *Cable strength*: The mechanical properties such as tension, torsion, and compression, etc. of optical cables must be similar to electrical transmission cables.
  (iv) *Identification and joining of the fibers within the cable*: If the fibers are arranged in a suitable geometry, it may be possible to use multiple jointing techniques rather than jointing each fiber individually.

One or more structural members are usually included in the optical fiber cable to serve as a core foundation around which the buffered fibers may be wrapped. The structural members may also be strength member. In certain cases, a central steel wire acts as both a structural and a strength member. Structural members may be non-metallic with plastic, fiberglass. Strength members are preferred to have a high Young's modulus, high strain capability, flexibility and low weight per unit length.

The cable is usually covered with an outer plastic sheath to reduce abrasion and to provide extra protection against external mechanical effects such as crushing.

4.15.1 Fiber Strength and Durability

Optical fibers are fabricated from silica or a compound of glasses that are brittle and exhibit perfect elasticity until their breaking point is reached. The bulk material strength of glass is high and is given by the following expression:

$$S_t = \left(\frac{\gamma_p E}{4la}\right)^{1/2} \tag{4.38}$$

where S_t is the cohesive strength, γ_p is the surface energy of the material, E the Young's modulus and la is the atomic spacing or bond distance.

4.15.2 Fiber-Optic Connectors

The interconnection of optical components is a vital part of an optical system, having a major effect on performance. *Interconnection* between two fiber-optic cables is achieved either by *connectors* or *splices* which link the ends of the fiber cables optically and mechanically.

Connector are devices used to connect a fiber-optic cable to an optical fiber device, such as a detector, optical amplifier, optical light power meter, or link to another fiber cable. They are designed to be easily and reliably connected and disconnected. The connector create an intimate contact between the mated halves to minimise the power loss across the junction. They are appropriate for *in-door* applications.

Splices are used to permanently connect one fiber-optic cable to another. Splices are suitable for outdoor and indoor applications. Some types of splices are used to temporarily connect for *quick testing* purposes.

The key to a fiber-optic interconnection is precise alignment of the mated fiber cable cores so that the couples from one fiber, across the junction, into the other fiber. This precise alignment creates a challenge for designers. There are many applications for fiber connectors and splices in fiber systems, such as:

(i) Connecting between a pair of fiber cables, using connectors or a splice, is an essential part of any fiber system.

(ii) Interfacing devices to local area network.

(iii) Connecting and disconnecting fiber cables to patch panels where signals can be checked and routed in a fiber system.

(iv) Connecting and splicing may be required on short fiber cables for wiring, testing devices, and at other intermediate points between transmitters and receivers.

(v) Dividing a fiber system into subsystems which simplifies the selection, installation, and maintenance of fiber systems.

(vi) Temporarily connecting remote mobile systems and recording equipment in many fiber systems.

4.16 FIBER FABRICATION

From the consideration of *optical wave guiding*, it is clear that a variation of refractive index inside the optical fiber (i.e. between the core and the cladding) is a fundamental necessity in the fabrication of fibers for light transmission. Hence, at least two different materials which are transparent to light over the current operating wavelength range (0.8 to 1.6 µm) are required. In practice, these materials must exhibit relatively low optical attenuation and they must, therefore, havge low intrinsic absorption and scattering losses. A number of organic and inorganic substances meet these conditions in the visible and near infrared regions of the spectrum. In order to avoid scattering losses in excess of the fundamental intrinsic losses, scattering centers such as bubbles, strains and grain boundaries must be eradicated. This limits the choice of suitable materials for the fabrication of optical fibers to either glasses or certain plastics.

In case of graded index fibers, it is essential that the refractive index of the material may be varied by suitable doping with another compatible material. Hence, these two materials should have mutual solubility over a relatively wide range of concentrations that can be only achieved in glasses and glass-like materials. Glass exhibit the best material characteristics for use in low loss optical fibers. Plastic fibers find some use in short-haul, low band width applications.

Generally, there are two methods of preparing optical glasses:
(1) Liquid phase melting techniques
(2) Vapour phase deposition methods

4.16.1 Liquid Phase Melting Techniques

In the liquid phase melting technique, the glass is processed in the molten state producing a multicomponent glass structure.

The first stage is the preparation of material powders usually in oxides or carbonates of the required constituents. These include oxides such as SiO_2, B_2O_2, Al_2O_3, and carbonates such as Na_2CO_3, $CaCO_3$ and $BaCO_3$ which will decompose into oxides during the glass melting. Very high purity is essential which is attained by fine filtration and co-precipitation, followed by solvent extraction before recrysttalization and final drying in a vacuum to remove any residual OH ions.

These high purity powdered glass materials are melted to form a homogeneous, bubble free multicomponent glass.

The refractive index may be varied by either a change in the composition of the various constituents or by ion exchange when materials are in the molten phase. Communication may arise during melting from several sources e.g. furnace environment and the crucible.

A technique for avoiding this contamination involves melting the glass directly into a radio frequency (~5 MHz) induction furnace while cooling the silica by gas or water flow.

The glass is homogenized and dried by bubbling pure gases through the melt, while protecting against any airborne dust particles, the melt is then cooled and formed into long rods of multicomponent rods.

To produce fine optical fiber waveguides, a preform is made using the rod in tube process. A rod of core glass is inserted into a tube of cladding glass and the preform is drawn in a vertical muffle furnace (Fig. 4.29). This technique is useful for the production of step index fibers with large core and cladding diameters.

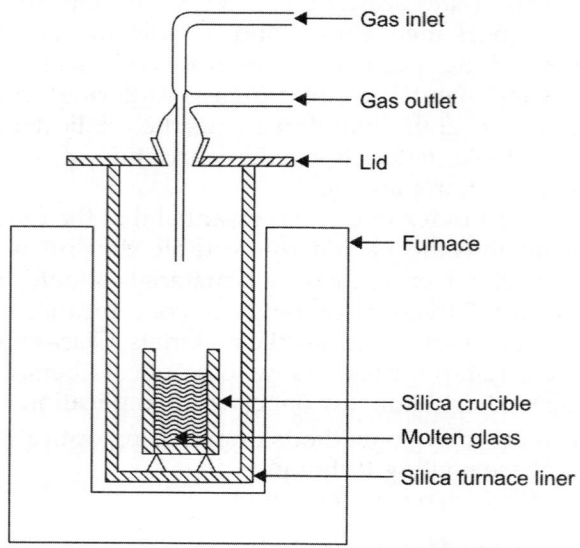

Fig. 4.29. Schematic of glassmaking furnace for the production of high purity glasses

Another technique suitable for the production of large core diameter step index fibers is called the *stratified melt process*. This process involves pouring a layer of cladding glass over the core glass in a platinum crucible. A bait glass rod is dipped slowly into the molten combination and slowly withdrawn giving a composite core-clad preform which may then be drawn into a fiber.

The double crucible method is used in the drawing of optical fibers.

The core and cladding glass in the form of separate rods is fed into two concentric platinum crucible m which is located in a muffle furnace capable of heating the crucible content to a temperature of between 800° and 1200°C. The crucibles have nozzles in their bases from which the clad fiber is drawn directly from the melt. Index grading may be achieved through the diffusion of mobile ions across the core cladding interface within the molten glass. It is possible to produce fibers with a reasonable refractive index profile using this process, hence graded index fibers produced by this technique are subsequently less dispersive than step index but do not have the bandwidth-length products of optimum profile fibers.

Using very high purity melting techniques and the *double crucible drawing method* (Fig. 4.30), step index and graded index fibers with attenuation as low as 3.4 dβ·km^{-1} and 1.1 dβ·km^{-1} respectively have been produced. Liquid-phase techniques have the inherent disadvantage of obtaining and maintaining extremely pure glass which limits their ability to produce low loss fibers, the advantage of this technique is the possibility of continuous production of optical fibers.

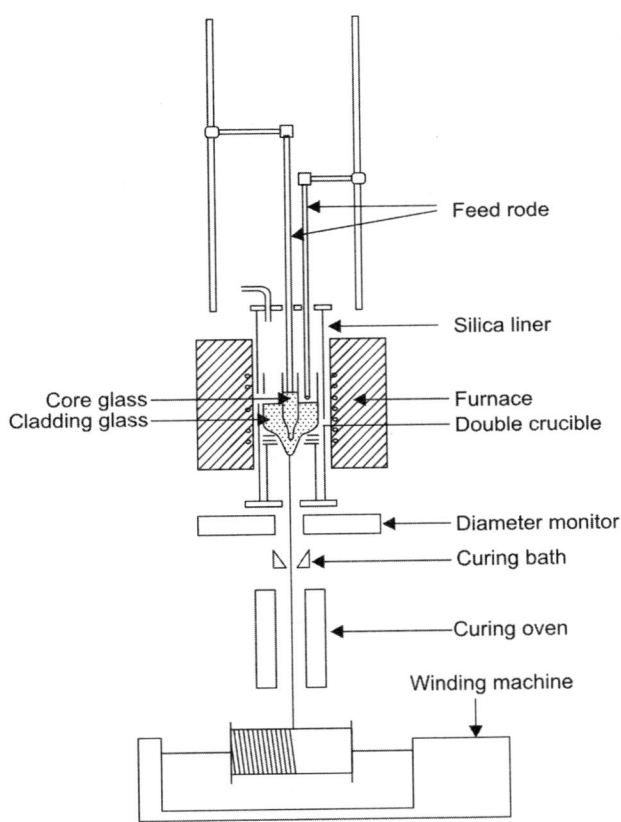

Fig. 4.30. Schematic of double crucible method for fiber drawing

4.16.2 Vapour Phase Deposition Technique

Vapour phase deposition technique is used to produce silica rich gases of the highest transparency and with the optimal optical properties. The starting materials are volatile compounds such as $SiCl_4$, SiF_4, BCl_3, BBr_3 and $POCl_3$ which may be distilled to reduce the concentration of most transition metal impurities to below one part in 10^9, giving negligible absorption losses from these elements. Refractive index modification is achieved through the formation of dopants such as TiO_2, GeO_2, P_2O_5 and F. Gases mixtures of the silica containing compounds, the doping oxidation reaction where the deposition is usually on a stack of successive layers. Hence, the dopant concentration may be varied gradually to produce a graded index profile or maintained to give a step index profile. In case of subtractive results in a solid rod or perform whereas the hollow tube must be collapsed to give a solid perform from which the fiber may be drawn.

There are various techniques in vapour-phase deposition which produce low loss fiber.

The vapour-phase deposition techniques are broadly categorized into two categories (Fig. 4.31):
(1) Flame hydrolysis
(2) Chemical vapour deposition

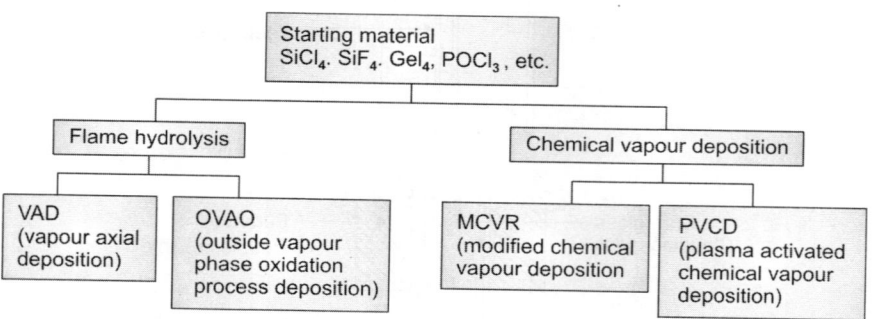

Fig. 4.31. Flame hydrolysis and chemical vapour deposition techniques

These techniques have all demonstrated relatively similar performance for the fabrication of both multimode and single mode fiber of stand step and graded index designs.

4.16.3 Modern Fiber-Optic Processes

Modern fiber-optic processes enable the fabrication of low-cost fibers of excellent quality. In all glass-forming processes, silica tetrachloride and the dopants, such as germanium tetrachloride, are delivered to the reaction region as vapors, where silica and germania are formed. The byproduct is chlorine gas, Cl_2. In addition to glass blank fabrication, high-speed draw (greater than one meter per second) are used to draw and coat the fiber with an organic material that protects the fiber from handling and from the environment.

There are basically three methods used to form the glass blank. The modified chemical vapor deposition process (MCVD), the outside vapor deposition process (OVD), and the vapor-axial deposition process (VAD). In the MCVD process, successive layers of SiO_2 and dopants, which include germania, phosphorous, and fluorine, are deposited on the inside of a fused silica tube by mixing the chloride vapors and oxygen at a temperature on the order of 1800°C. The temperature is maintained using a burner which traverses the outside of the tube. After the dopants are deposited, the temperature of the tube is raised, via the burner, to collapse the tube. In the OVD process, the core and cladding layers are deposited on a rotating mandrel by the flame hydrolysis process. Upon completing the deposition process, the mandrel is removed from the preform, and the preform is sintered in a furnace to form the fiber blank. The VAD process is also a flame hydrolysis process, but, in this technique, the soot is deposited axially. Fig. 4.32 is a schematic of the three blank fabrication processes.

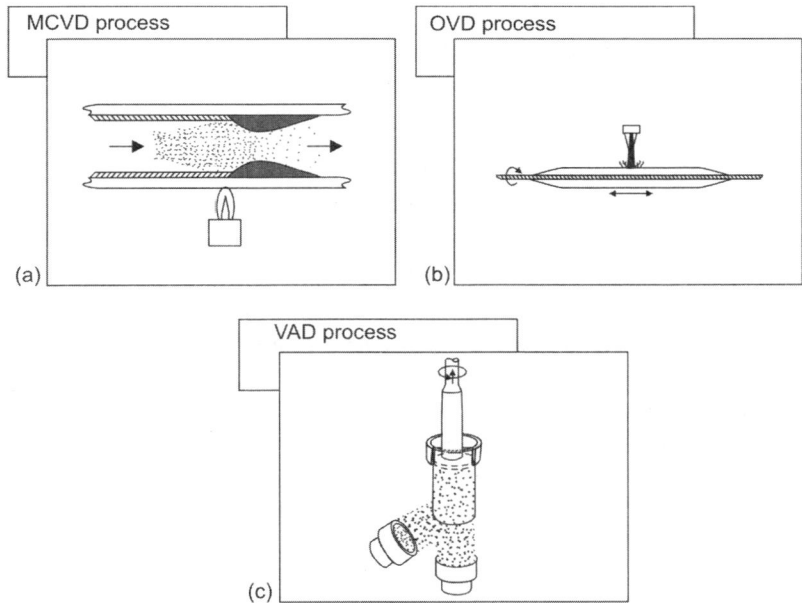

Fig. 4.32. Schematic of three blank fabrication processes: (a) MCVD; (b) OVD; (c) VAD

In the drawing process, the blank is fed from above into the drawing portion of the furnace while being drawn from the bottom using tractors. The fiber is then wound onto a drum while being monitored for tensile strength. The temperature during draw is on the order of 2000°C. After exciting the furnace, the fiber is coated with a UV-curable coating before winding on the drum.

4.16.4 Advanced Fiber Optic Cables

The scaling down of fiber light sources brings the benefits of fiber technology to wide range of applications. These developments have led to a surge of interest in advanced fiber cables for both military and industrial applications. Advanced fiber cables are also used for transmitting high volumes of data in communication systems over long distances for getting *very clear images*, and in building many sophisticated instruments for a variety of applications. By creating new core designs, adding dopants to the fiber core and cladding, and developing advanced fiber-optic cable technology. For example, the core of the *holey fiber* consists of many air holes acts as a single fiber. This fiber enables a high data transmission rate and capacity, and consequently, reduces the cost of network. The holey fibers have tubes or spaces in the core along the fiber's length. Some types of these advanced fiber-optic cables are listed below:

(i) Dual-core fiber for high power laser
(ii) Fiber-Bragg gratings
(iii) Chirped fiber Bragg gratings
(iv) Blazed fiber Bragg gratings
(v) Nonzero-dispersion fiber-optic cables
(vi) Photonic crystal fiber cables
(vii) Polymer-Holey-fiber cables
(viii) Liquid crystal photonic bandgap fiber cables
(ix) Lenses and trapped fiber cables
(x) Bend-insensitive fiber cables
(xi) Nanoribbon fiber optic cables

4.16.5 Fiber Optic Cables vs Copper Cables

There is still a place for fiber optic and copper cables in communication systems, but the shrinking price gap, coupled with increasing bandwidth demands, make fiber cables worth using in more situations than ever before. The customers of small and large communication providers are dispersed all over the world, and they need to send and receive lots of data. Customers require large amounts of bandwidth, and fiber cables are the only medium that can support this. Some of those customer's bandwidth requirements have been growing exponentially since the beginning of the twenty-first century. In reality, all fiber networks have a lot of room for future growth. The ultimate choice is whether to use fiber optic cables. As mentioned above, the fiber optic technology is moving forward to create high-capacity fibers with low production and installation costs and increasing bandwidths. The overall cost difference between optical fiber and twisted-pair copper cabling has been reduced. Now the choice between optical fiber and twisted-pair copper cabling has shifted in favour of the fiber cable.

Desktop computers require very high bandwidths. One way to meet this need is to wire them with fiber optic cables. However, copper cable has continued to prove more capable than expected; every time that new, higher-speed network standards appear to be forcing a move to fiber cable, someone has found a way to pump more data through the old copper cables.

Still, fiber cables can be made even more economically attractive by rethinking the way the network is physically laid out. Because fiber cable can be run for longer distance than copper, the network could be laid out without the wiring closets full of additional gear that are common in copper-based networks. Instead, fiber might run directly from the desktop back to the server or to the backbone connecting floors of a building, and the savings on the intermediate gear might more than cancel out the higher cost of installing the fiber cable. Running fiber cable to small enclosures close to users, and then covering the last short distances with copper, provides an economical alternative that minimizes the length of the twisted-pair cable used.

The increasing use of wireless networking also opens up a variation on the fiber cable network layout. Fiber can be run to a wireless access point that can then be used to serve a group. Thus, the copper cable can be eliminated altogether without actually taking fiber cable to every machine.

Many companies are removing existing copper cables and replacing them with fiber cables. When facilities are built or refurnished, fiber cables are installed. The choice between copper and fiber cables depends on several factors, including the applications being run on the network, the company's future plans, and the demands of costumers.

In particular, there are some specific situations in which fiber has advantages over copper cable. First, fiber cable is immune to electrical interference and tapping. It also carries high data capacity over long distances and is small and lightweight. When it comes to testing, fiber may still require some fairly sophisticated equipments—but the newer standards for copper cabling present the same issues.

In the end, the choice between fiber and copper cables comes down to the company's networking requirements, the needs of individual users, and the budget. Table 4.4 presents a side-by-side comparison of the important differences between fiber and copper cables.

4.17 POLARIZATION

Single-mode fibers capable of maintaining an input linear polarization are known as *polarization-preserving fibers*. These fibers employ either an elliptical core or stress rods placed 180° from each other and outside the core. The elliptical core and/or stress rods introduce a birefringence that removes the degeneracy of the orthogonally polarized mode. The *birefringence* is defined as the difference between the effective indices of the two orthogonal modes and is on the order of 10^{-4}. Fig. 4.33 shows the index profile for stress rod and elliptical core polarization-maintaining fibers. Polarization-preserving fibers are not used at present in telecommunication system but rather find application in the area of fiber-optic sensors. As an example, fiber-optic gyroscopes use

160 Photonics | Optoelectronics

Table 4.4. Fiber optic cables vs Copper cables

S.No.	Fiber optic cables	Copper cables
1.	Fiber-based systems are more expensive to buy and install	Copper-based systems are less expensive to buy and install
2.	Fiber is clearly the superior technologically. Installing fiber ensures performance, as even higher speed networks will emerge in the future	Installing copper cable ensures performance for low-speed networks
3.	Carry high data capacity over long distances	Carry low data capacity over short distances
4.	Wide bandwidth	Limited bandwidth
5.	Low loss per cable length	Conventional loss per cable length
6.	Immune to electrical interference and tapping	Not immune to electrical interference and tapping
7.	Small size and lightweight	Large size and heavyweight
8.	New technology reduces installation time	Conventional technology keeps the same installation time
9.	High safety	Low safety
10.	Fast-developing technology	Steady-state developing technology

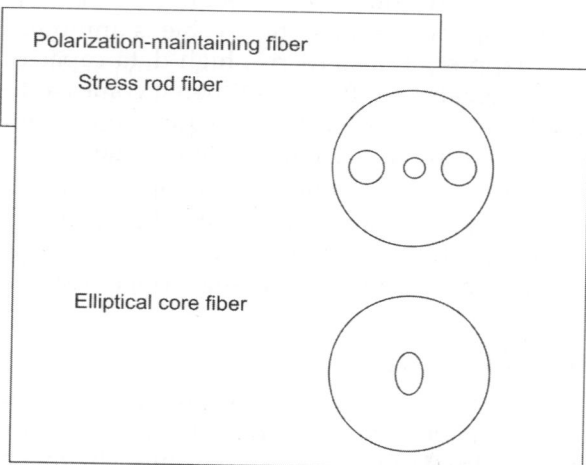

Fig. 4.33. Polarization-maintaining fibers, index profiles

polarization-preserving fiber to prevent coupling from one polarization to the other in the gyroscope coil. Such coupling leads to inaccurate sensing and is eliminated by using only one polarization. An important parameter for these fibers is the *beat length*, defined as the wavelength divided by the birefringence (δn). This parameter describes the beating which results from the fact that the propagation speed along the fast axis is considerably larger than that along the slow axis. Highly polarized fiber exhibits beat lengths on the order of less than a centimeter. Standard fibers have very small birefringence and exhibit beat lengths greater than ten meters.

The slight differences in propagation speeds for the two polarization modes in standard single-mode fiber lead to PMD. Polarization-mode dispersion for standard fiber is, typically, significantly less than 1 ps/sqrt (km). The inverse square root length dependence results from the fact that the polarization randomly couples as the light propagates through the fiber. Transoceanic systems using optical amplifiers eliminate the need for expensive electronic regenerators. Without signal regeneration, polarization-mode dispersion can become significant over hundreds of kilometers. Also, CATV analog transmission systems can be sensitive to PMD when component polarization-dependent loss and significant laser chirping occur. In this situation, composite second-order distortion (CSO) can significantly degrade the signal.

As mentioned, small-core eccentricities, on the order of a few percent, can be the source of PMD. Significant literature exists on the effect of core ellipticity on polarization dispersion for step-index fibers. However, these perturbation models predict that the polarization dispersion increases linearly with length. The square root length dependence that is experimentally observed results from random coupling of the polarization modes. In step-index fibers, polarization dispersion results from both form *birefringence* and *stress birefringence*, both of which depend on core ellipticity and are approximated in the low ellipticity limit as:

$$e = \frac{A - B}{A + B} \tag{4.39}$$

and

$$E = 1 - \frac{B^2}{A^2}, \tag{4.40}$$

where A and B are the major and minor diameters. The stress birefringence depends on the expansion differences of the core and cladding, $\Delta\alpha$, which, in turn, depend on the dopant concentrations. The stress birefringence also depends on the fictive temperature T which is used to approximate a point below which the glass structure cannot change on the same time scale as the cooling rate, a stress optic coefficient C_s, and Poisson's ratio σ.

The measured value of PMD can be sensitive to fiber deployment. This is because the amount of mode coupling varies with deployment and with the environment. The effect of mode coupling on PMD is modeled using the statistics of random occurrences. This analysis explains the square root length dependence of PMD and the distribution of PMD values with changes in the environment and the fact that repeated measurements of PMD show that the observed values obey a Maxwellian process:

$$P(\Delta\tau) = \frac{2\tau}{\sqrt{2\pi}\,q^3} e^{-(\Delta\tau^2/2q^2)} \tag{4.41}$$

In Equation (4.41), q^2 is the variance. Another aspect of the random coupling of the polarization modes is a significant wavelength dependence of PMD.

One final and important aspect of polarization modes in fibers is the *evolution of the polarization states as light propagates through a fiber*. Light linearly polarized as it enters the fiber not only evolves into circularly or elliptically polarized light, as it propagates, but significant mode coupling occurs. This complicates the analysis and measurement of PMD. Techniques to minimize the effects of mode coupling on the stability of PMD have been described in 1994 by Judy.

4.18 OPTICAL FIBER AMPLIFICATION

An important aspect of optical fiber research is rare-earth doping for amplification and lasing. Amplification in optical fiber has not only renewed interest in the materials and propagative aspects of fiber research, but it has also significantly affected the systems aspects of optical communication. Doping of fibers for optical amplification has been under study since the 1960s, when neodymium was used as a dopant. Interest in rare-earth doping was renewed in the 1980s when research scientists showed amplification in the 1.55 µm region, which coincidentally is the low-loss transmission window. The fact that erbium-doping enables amplification in the transmission window of lowest loss has attracted much interest in this aspect of optical fiber research.

In the erbium-doped fiber amplifier (in its simplest form), Fig. 4.34, an erbium-doped fiber with lengths on the order of meters and dopant levels on the order of 2 ppm, is spliced to a wavelength-dependent, fiber-optic coupler. The coupler enables one to continuously pump the erbium-doped fiber with light emitted from a high-power semiconductor laser diode at 980 or 1480 nm. Filters and optical isolators are often included to minimize spontaneous emission noise and reflections. The pump light is used to excite ions from the ground state to an excited state. Signal light entering the fiber initiates stimulated emission and is coherently amplified. Years of research at many laboratories has led to the development of *erbium amplifiers*. Such technical issues as wavelength dependence of gain, gain saturation, polarization dependence and spontaneous emission, among others, have been carefully studied. Spontaneous emission occurs when ions

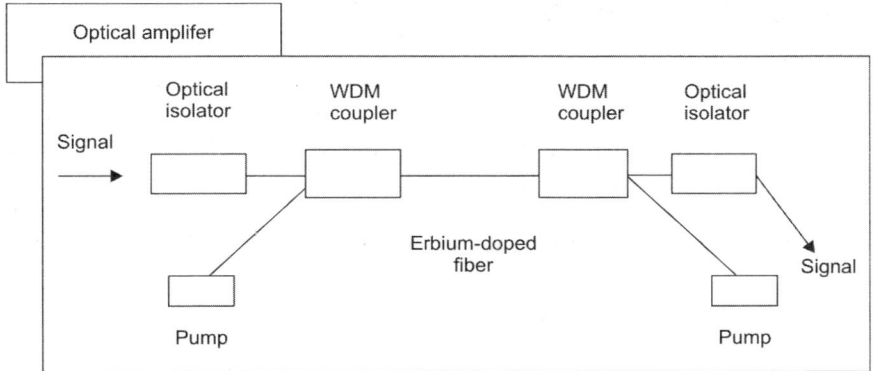

Fig. 4.34. Schematic (simple) of the optical amplifier

in the excited state spontaneously relax to the ground state contributing to noise. This phenomenon in itself significantly affects the signal-to-noise ratio of an amplifier-based communication system. Another important parameter of the optical amplifier is the concentration of erbium ions. An optimum concentration of erbium ions avoids clustering which alters the excited states and results in elevating one ion to a higher state and emission to the ground of neighbouring ions.

Research in the area of rare-earth doped optical fibers is far from complete. Issues, such as multiple wavelength amplification, need to be further addressed. Research into the possibilities of using erbium-doped fibers in a lasing configuration is useful for picosecond pulse sources. Another aspect of rare-earth doping is the interest in amplification at 1.3 µms. Amplification in the low-loss transmission windows requires dopants other than erbium (praseodymium, co-doped with neodymium, for example), and also requires th use of a fluoride glass host. In a silica glass, the phonon vibrational spectrum affects the amplification process. In a fluoride-based glass, however, the phonon edge is shifted to higher frequencies, thereby, enabling reasonable amplification in the 1.3 µm window. Other issues remain, however, including the amount of gain and the wavelength dependence of the gain.

Fluoride fibers are melted at temperatures far below that of silica and, therefore, cannot be fused to silica fibers. The index is not well matched to silica and, therefore, leads to strong back reflections. More importantly, the fabrication technology of fluoride-based fibers is less advanced than that of silica. Nonetheless, interest in amplification at 1.3 µms remains because there is a huge installed base of optical fiber that is optimized for minimum pulse distortion at 1.3 µms rather than 1.55 µms.

4.19 OPTICAL NONLINEARITIES IN FIBERS

Nonlinear effects in silica fibers have been considered unimportant. However, the use of optical amplifiers and the promise of dense wavelength-division multiplexing imposes a number of limitations on the ultimate bandwidth of optical communication. Optical nonlinearities will be discussed in detail in other chapters, but it is important to mention the limitation that these nonlinearities impose on communication bit rates. These nonlinearities are stimulated Brillouin scattering, four-wave mixing, cross-phase modulation, and stimulated Raman scattering.

Multiple wavelength mixing was first observed in optical fibers by Hill in 1978. The output power of light generated at a fourth wavelength can be written as:

$$P_F(L) = \frac{1024\pi^6}{n^4\lambda^2 c^2}(D_x)^2 \frac{P_i(0)P_j(0)P_k(0)}{A_{\text{eff}}^2} e^{-\alpha L} \frac{(1-e^{-\alpha L})^2}{\alpha^2} \eta \quad (4.42)$$

In Equation (4.42), i, j and k represent the three input wavelengths at $Z = 0$, P_F is the power at wavelength F and at $Z = L$. A_{eff} is the effective area of the guide, α is the loss, and D_x is a degeneracy factor. The efficiency factor η is given by:

$$\eta = \frac{\alpha^2}{\alpha^2(\Delta\beta)^2}1 + \frac{4e^{-\alpha L}\sin^2(\Delta\beta L/2)}{(1-e^{-\alpha L})^2} \quad (4.43)$$

where
$$\Delta\beta = \beta(f_i) + \beta(f_j) + \beta(f_k) - \beta(f_F). \quad (4.44)$$

In Equation (4.44), $\Delta\beta$ is the difference in propagation constants, which depends on the dispersion of the fiber. Numerical and systems studies have shown the importance of using a finite amount of dispersion to avoid generating new wavelengths while depleting the signal wavelength. Finite dispersion is also important at single channel operation very long lengths because multiple wavelength mixing occurs through the wavelengths generated with amplifiers by amplified spontaneous emission. A major impact of four-wave mixing is that it forces designers to operate their signal wavelength away from a dispersion zero. Over long distances, this require chromatic or dispersion compensation.

Brillouin scattering is the interaction of light with acoustical vibrations in fiber. The signal or carrier wave is shifted to longer wavelengths (Stoke's shift) with the simultaneous emission of an acoustical phonon. The amount of power generated can be characterized with an exponential gain coefficient, g_B:

$$g_B = 4 \times 10^{-9} \text{ cm/W}. \quad (4.45)$$

The gain coefficient enables one to calculate the amount of stimulated Brillouin scattering with length as a function of incident light power.

The wavelengths generated are separated from the carrier by less than one thousandth of a nanometer. This slight shift in wavelength would not be expected to cause problems, except for the fact that the generated light is scattered backwards, depleting the carrier signal and, at times, affecting the transmission laser. It is estimated that Brillouin scattering becomes an issue when the power in an optical fiber is on the order of a milliwatt.

Raman scattering results from an interaction of the incident light with molecular vibrations in the fiber. As with Brillouin scattering, Raman scattering can be characterized with a gain coefficient:

$$g_B = 7 \times 10^{-12} \text{ cm/W}. \tag{4.46}$$

Comparison of Eqs (4.45) and (4.46) shows that the threshold for Raman scattering occurs at power levels three orders of magnitude higher than that for Brillouin scattering. Raman scattering can be a problem, however, because the wavelength or bandwidth of interaction is greater than a hundred nanometers. Raman scattering can cause serious cross-talk for multiple channel systems of significant power and will ultimately limit the information transmission capacity of optical fiber systems. It is estimated that Raman scattering becomes an issue when the total power in an optical fiber is on the order of 1 W.

Self- and cross-phase modulation refers to the fact that the index of glass is intensity-dependent:

$$n = n_0 + n_2 I \tag{4.47}$$

where n_0 represents the linear index and the second term includes the intensity-dependent refractive index n_2 and the optical intensity I.

In silica, n_2 is on the order of 3×10^{-16} cm^2/W and 6×10^{-16} cm^2/W for self- and cross-phase modulation, respectively. Both cross- and self-modulation affect the phase and, hence, the arrival time of a pulse. Changes in power and the modulation of power with other carriers limit the amount of power in fibers. It is estimated that self- and phase-modulation becomes an issue when the power in a fiber is on the order of 10 mW.

4.20 SOLITONS IN OPTICAL FIBER

Now, we will briefly introduce the concept of *soliton pulses in optical fiber*. This is a fascinating and current research topic, and it is expected that the use of soliton pulses in optical fibers will enable transmission over transoceanic distance at the highest of possible bit rates.

Soliton propagation in fiber is a nonlinear phenomenon, and such pulses are sensitive to the amount of optical power. Attenuation severely limits the distances over which a soliton pulse can travel without significant distortion. The development of the *erbium-doped optical amplifier* has spurred research activity toward commercializing transoceanic soliton systems. The coincidence of minimum fiber attenuation in the 1550 nm telecommunication window and the strong gain in this same window with the erbium amplifier is, indeed, encouraging. In fact, by amplifying the signal every 30 kms, soliton pulses can travel thousands of kilometers without distortion. One can also use multiple wavelengths *wavelength division multiplexing* (WDM) without the imposition of a number of the nonlinearities. For these reasons, it is expected that, ultimately, soliton trans-mission will be the technique of choice for *transoceanic communication*. However, more research is required before the potential of soliton communication can be fully realized.

Solitons occur in nature in many different ways, optical pulses being one of them. Scott Russel is credited with first observing and recording solitary waves in a barge canal in Great Britain in 1938. He followed the "large solitary elevation" on horseback and noted that it did not change in form or speed for miles. He derived equations describing the velocity of the waves and reported on his work at the Liverpool meeting of the British Association for the Advancement of Science. In 1967, a group of mathematical physicists from Princeton University solved the so-called Kortewegde Vries equations describing the nonlinear hydrodynamic wave. The Russian pair, Zahkharov and Shabat in 1971 considered the nonlinear propagation of optical waves in a two-dimensional medium. They showed that the nonlinear Schroedinger equation could be solved using the inverse scattering theorem and pointed out the relationship between spatial dispersion and optical intensity.

Hasegawa and Tappert in 1973 showed, theoretically, that temporal solitons should exist in fiber and pointed out the possibility of using them for optical communication. Mollenauer and collaborators in 1980 observed solitons experimentally in fiber. They built a "color center" laser that enabled them to generate and launch narrow temporal pulses of power levels significant enough to develop into soliton pulses. This experimental observation can be considered to mark the beginning of a new technology aimed at enabling dispersionless transmission over transoceanic distances. There are, however, many obstacles yet to be overcome before this technology can be fully utilized.

Nonlinear pulse propagation in fiber is described with the so-called nonlinear Schroedinger equation:

$$-i\frac{\partial u}{\partial z} = \frac{\lambda^2 D}{4\pi c}\frac{\partial^2 u}{\partial t^2} + \frac{2\pi n_2}{\lambda A_{\text{eff}}}|u|^2 u. \tag{4.48}$$

Here, D is the dispersion and u is the pulse amplitude which varies both spatially and temporally. The soliton pulse has both a temporal and spectral width, both of which are described with hyperbolic secant functions:

$$u(t) = (\text{sech}\, t)\, e^{iz/2} \tag{4.49}$$

and
$$\bar{u}(w) = \frac{1}{2} sech\left[(w-w_0)\frac{\pi t_c}{2}\right].$$ (4.50)

In Equation (4.48), t_c is the temporal width and w_0 the center frequency. An important parameter is the power P_c at which the nonlinearity and dispersion balance:

$$P_c = \frac{\lambda A_{eff}}{2\pi n_2 Z_c}$$ (4.51)

where
$$Z_c = \frac{2\pi c}{\lambda^2 D} t_c^2.$$ (4.52)

The parameter Z_c characterizes the distance at which the pulse begins to spread. Another important parameter is the soliton period, characterized by Z_c times p/2.

The possibility of switching light at ultrafast speeds using solitons is also a topic for current research. Soliton pulses have properties that make them attractive in this regard.

4.21 APPLICATIONS OF FIBER OPTIC CABLES

Since the discovery of the laser, fiber optic cables and optical fiber devices have seen increased applications in every sector of industry. Light is a very important element in our lives, controlling and operating many types of devices, instruments, and systems. Fiber optics has emerged as a practical technology that is easy to work with. Fiber optic cables with other optic components are used in building optical fiber devices and systems. One of the large-scale applications of fiber optics is its use in communication systems. The small sizes and wide bandwidths, as well as their capacity to carry large amounts of information, make optical fibers very attractive for use in these systems. Later chapters in this book will explain in more detail communication systems that use optical fiber technology. Video, including broadcast television, is one of the main telecommunication applications. Other applications include cable television, is one of the main telecommunication applications. Other applications include cable television, high-speed internet, wireless transition, remote monitoring, and surveillance. Fiber optic video transmission is successfully used around the world in surveillance and remote monitoring systems with many applications. Fiber optics applications in the military include communications, command-and-control links on ships and aircrafts, data links for satellite earth stations, and transmission lines for tactical command-post communications.

Fiber cables can be used throughout the communication network, including in the final link into the subscriber's home to wall outlets. This field has continued to develop since the discovery of optical amplifiers, dense wavelength division multiplexes (DWDM), fiber Bragg grating, and photonics crystal fibers.

One particular advantage of fiber optic cables is that they are immune to electromagnetic interference (EMI) from electricity. Therefore, optic cables can be placed near high-voltage power cables without any effect on data transmission. Similarly, the cables also can be laid along railway lines without suffering from EMI.

Optical fibers are applied in building *night-vision viewing devices, scanning and sensing instruments*, and *vibration sensors*, which are extensively used in military, medical, and other applications. Imaging techniques have been rapidly developed for a variety of medical applications, such as viewing inside human tissues and scanning microscopic particles.

Fibers are also used in monitoring and sensing technology. They are used as sensors to monitor the vibration in the structures of bridges and high buildings. They are also used as gas and DNA sensors.

The use of fiber optic cables in lighting systems can reduce energy consumption. Fiber optic lighting systems are developing quickly, with wide applications. Fiber optic lighting systems can be applied to the interior and exterior of commercial, retail, and residential buildings. New applications are being explored in landscaping, waterscaping, medical lighting instruments, and theme environments.

Fiber optic cables can also be coated for special handling requirements and resistance to temperature, chemicals, or radiation. Radiation-resistant fiber is suitable for use in environments where electronics-based optical solutions are not viable, such as monitoring nuclear waste disposal in storage facilities. To make the fibers heat resistant, a chemical-resistant polyimide coating that can withstand temperatures of up to 300°C is applied. This is especially useful for manufacturers designing medical equipment for applications in which autoclave sterilization is necessary.

4.21.1 Applications of Advanced Fiber Cables

Fiber-optic cables are used in many applications. One of many applications for different fiber cable types is fiber-optic sensing. Fiber-optic sensing can be classed as intrinsic and extrinsic. In intrinsic sensors, the fiber simply conducts light from a sensing head to a detector; the interaction between light and the environment takes place outside the fiber cable. In intrinsic sensors, the interaction between the environment, the fiber, and the light itself generates information about a specific measurement. The key advantage of intrinsic fiber-optic sensors is the fundamental ability of a fiber to guide light around bends and over long distances. This enables the fiber to be confined within small physical volumes and magnifies the effects of very fine environment changes to a level that can be measured accurately and quantified.

Fiber Bragg gratings are commonly employed as passive temperature and pressure sensors. The ability to adjust these optical devices introduces a new dimension for the design of very fast, precise, and multifunctional fiber sensors without compromising their intrinsic advantages. As explained in the previous section, an FBG consists of a short length of fiber-optic having a periodic modulation of refractive index along it length. This periodic structure causes the FBG to act as a narrowline filter with a peak reflectivity at a specific wavelength, called Bragg wavelength (λ_B), which is determined by the period length of the FBG and the refractive index of the fiber. If the refractive index of the FBG is changed by a temperature shift, or the period is altered in some way (e.g. by compression or expansion), a shift in the peak reflected wavelength results. Detecting this shift is the basis of fiber sensing.

Fiber Bragg gratings are also used in many other applications, such as tunable FBGs. Tuning can be accomplished by several mechanisms, such as electric heating, the piezoelectric effect, mechanical stretching and bending, and acoustic modulation. These sensors are available in the market in different types, sizes, and wavelength ranges.

Photonics bandgap and photonics crystal fibers are both members of the family of MOFs, called *holey fibers*. Both types can be used in gap sensors. The sensors are made of a holey fiber for the 1550 nm spectral range. One end of the fiber is spliced to conventional single mode fiber from an optical source, such as a laser or an LED, and placed on the outer end in a V-groove in a vacuum chamber, as shown in Fig. 4.35. A multi-mode fiber leading to a detector is installed 50 µm away from the end in the V-groove. The separation allows gas to flow into or out of the hollow core of the bandgap fiber, while still allowing for efficient optical coupling between the fibers. A 1-meter length of the fiber is filled with acetylene (C_2H_2) to a pressure of 10 m bar and illuminated with a tunable laser source. The expected spectral changes of acetylene are observed and displayed on a monitor.

Beyond pressure and temperature sensor technology, advanced fiber sensors types include chemical, strain, biomedical, electrical and magnetic, rotation, vibration, and displacement as major applications. One more application of the advanced fiber sensor technology is in the

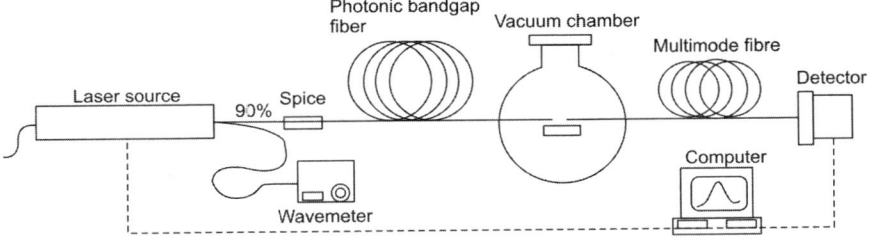

Fig. 4.35. Bandgap fiber used in gas sensor

170 Photonics | Optoelectronics

area of DNA analysis. Fiber-optic biosensors have the ability to detect the presence of short DNA sequences called *oligonucleotides*. This ability gives the diagnostic some excellent tools for early and accurate diagnosis. These fibers are efficiently constructed as waveguides with novel properties for communications and sensing applications.

Example 1

For a 3 µm diameter optical fiber with core and cladding indexes of refraction of 1.545 and 1.510 respectively, determine the cutoff wavelength.

Solution

$$\lambda_c = \frac{2\pi a n_1 \sqrt{2\Delta}}{2.405}$$

where, λ_c – cutoff wavelength (µm),
  n_1 – core index of refraction (unit less)
  n_2 – cladding index of refraction (unit less)
  a – core radius (µm)

$$\Delta = \frac{n_1 - n_2}{n_1}$$

$$\Delta = \frac{1.545 - 1.510}{1.545} = 0.023, \quad a = \frac{3}{2} = 1.5 \, \mu m$$

$$\lambda_c = \frac{(2\pi)(1.5)(1.545)\sqrt{2(0.023)}}{2.405} = 1.29 \, \mu m$$

Example 2

For a multimode index fiber $n_1 = 1.53$, $n_2 = 1.50$ and $\lambda = 1$ µm. If core radius is 50 µm, calculate the normalized frequency of the fiber (v) and the number of guided mode.

Solution

Normalized frequency,

$$v = \frac{2\pi a (\text{NA})}{\lambda}$$

$$= \frac{2 \times 3.14 \times 50 \times 10^{-6} \times \sqrt{(1.53)^2 - (1.50)^2}}{1 \times 10^{-6}} = 94.72$$

Total number of guided mode, $m_s = \frac{v^2}{2} = \frac{(94.72)^2}{2} = 4486$

Example 3

For a multimode stepped index fiber with a glass core ($n_1 = 1.5$) and fused quartz cladding ($n_2 = 1.46$), determined θ_c, θ_m and NA. The source to fiber medium is air.

Solution

Critical angle, $\theta_c = \sin^{-1} \dfrac{n_2}{n_1} = \sin^{-1} \dfrac{1.46}{1.5} = 76.7°$

$$\theta_m = \sin^{-1} \dfrac{\sqrt{n_1^2 - n_2^2}}{n_0}$$

$\theta_m \text{ (max)} = \sin^{-1} \sqrt{n_1^2 - n_2^2} = \sin^{-1} \sqrt{(1.5)^2 - (1.46)^2} = 20.2°$

$\text{NA} = \sin \theta_m = \sin 20.2 = 0.344$

The maximum diameter a single-mode optical fiber can have is proportional to the wavelength of light ray entering the cable and numerical aperture of the fiber. Mathematically, the maximum radius of core of a single-mode fiber is:

$$r_{max} = \dfrac{0.383 \text{ Å}}{\text{NA}} = \dfrac{0.383 \text{ Å}}{0.344} = 1.113 \text{Å}$$

where r_{max} is maximum core radius, NA → Numerical Aperture (unitless), λ → light ray wavelength.

Example 4

The numerical aperture of an optical fiber is 0.5 and core refractive index 1.54. Determine (i) Refractive Index (RI) of cladding, (ii) change in core cladding refractive index per unit RI of the core.

Solution

(i) $\text{NA} = \sin \theta_m = \sqrt{n_1^2 - n_2^2}$

$\therefore (0.5)^2 = (1.54)^2 - n_2^2$ or $n_2 = 1.456$

(ii) RI of the core $= \dfrac{n_1 - n_2}{n_1} = \dfrac{1.54 - 1.456}{1.54} = 0.0542$

Example 5

Determine the numerical aperture of a step index fiber when core refractivve index $n_1 = 1.5$ and cladding refractive index $n_2 = 1.48$. Find the maximum angle for entrance of light if the fiber is placed in air.

Solution

(i) $NA = \sqrt{(n_1^2 - n_2^2)} = \sqrt{(1.5)^2 - (1.48)^2} = \sqrt{(2.98 \times 0.02)} = 0.24413$

(ii) The maximum entrance angle i_0 can be found from:

$$i_0 = \sin^{-1}\left(\frac{NA}{n}\right) = \sin^{-1}\frac{0.24413}{1} = 14.13°$$

Example 6

An optical fiber has NA of 0.20 and cladding refractive index of 1.59. Determine acceptance angle for the fiber in water with refractive index 1.33.

Solution

$$NA = \frac{\sqrt{(n_1^2 - n_2^2)}}{n_0}$$

When the fiber is in air, $n_0 = 1$, then:

$$NA = \sqrt{(n_1^2 - n_2^2)} = 0.20$$

$$\therefore \quad n_0 = \sqrt{(NA)^2 + n_2^2} = \sqrt{(0.20)^2 + (1.59)^2} = 1.6025$$

When the fiber is in water, $n_0 = 1.33$.

Now, $\quad NA = \frac{\sqrt{(n_1^2 - n_2^2)}}{n_0} = \frac{\sqrt{(1.6025)^2 - (1.59)^2}}{1.33} = 0.15$

$\therefore \quad i_0 = \sin^{-1}(NA) = \sin^{-1}(0.15) = 8.6°$.

Example 7

Calculate the refractive indices of core and cladding material of a fiber from the following data: NA = 0.22 and $\Delta = 0.012$.

Solution

$$NA = n_1\sqrt{2\Delta}$$

$\therefore \quad n_1 = \frac{NA}{\sqrt{2\Delta}} = \frac{0.22}{\sqrt{2 \times 0.012}} = 1.42$

$\therefore \quad \Delta = \frac{n_1 - n_2}{n_1}$

$\therefore \quad 0.012 = \frac{1.42 - n_2}{1.42} \quad \text{or} \quad n_2 = 1.40$

Example 8

Calculate numerical aperture, acceptance angle and critical angle of an optical fiber, having refractive index of core 1.52 and refractive index of cladding 1.46.

Solution

We have, $\Delta = \dfrac{n_1 - n_2}{n_1} = \dfrac{1.52 - 1.46}{1.52} = 0.0395$

Now, $NA = n_1\sqrt{(2\Delta)} = 1.52 \times \sqrt{(2 \times 0.0395)} = 1.52 \times 0.281 = 0.427$

Acceptance angle, $i_0 = \sin^{-1}(NA) = \sin^{-1}(0.427)$

Critical angle, $\theta_c = \sin^{-1}(n_2/n_1) = \sin^{-1}(1.46/1.52) = \sin^{-1}(0.96)$

Example 9

A step index in air has NA of 0.16, core refractive index of 1.45 and core diameter of 60 cm. Determine the normalized frequency for the fiber when light at a wavelength of 0.9 μm is transmitted.

Solution

$$V = \dfrac{\pi d}{\lambda}\sqrt{(n_1^2 - n_2^2)} = \dfrac{\pi d}{\lambda}(NA)$$

Substituting the given values, one obtains:

$$V = \dfrac{3.14 \times 0.60 \text{ m}}{9 \times 10^{-7} \text{ m}} \times 0.16 = 335103 = 3.35 \times 10^5$$

Example 10

Calculate the numerical aperture, acceptance angle and critical angle of an optical fiber, having refractive index of the core is 1.5 and refractive index of the cladding is 1.45.

Solution

Numerical aperture $NA = n_1\sqrt{2\Delta}$

where $\Delta = \dfrac{n_1 - n_2}{n_1} = \dfrac{1.5 - 1.45}{15} = 0.0333$

∴ $NA = n_1\sqrt{2\Delta} = 1.5\sqrt{2 \times 0.0333} = 0.387$

Maximum acceptance angle:

$\sin \alpha_m = 0.387$

$\alpha_m = \sin^{-1}(0.387) = 22.8°$

Let θ_c be the critical angle:

$\alpha_m = \sin^{-1}(0.387) = 22.8°$

Example 11

A glass fiber is made with core material whose refractive index is 1.5 and the cladding is doped to give a fractional index difference of 0.0005. Calculate the refractive index of the cladding, the numerical aperture and critical internal angle.

Solution

The fractional difference of refractive index:

$$\Delta = \frac{n_1 - n_2}{n_1}$$

$$0.0005 = \frac{1.5 - n_2}{1.5}$$

$$n_2 = 1.49925$$

Numerical aperture $\quad NA = n_1\sqrt{2\Delta} = 1.5\sqrt{2 \times 0.0005} = 0.0474$

Let θ_c be the critical angle:

$$\therefore \quad \sin \theta_c = \frac{n_1}{n_2} = \frac{1.49925}{1.5} - 0.9995$$

$$\theta_c = \sin^{-1}(0.9995) = 88.2°$$

Thus, numerical aperture (NA) is 0.0474 and critical angle (θ_c) is 88.2°.

Example 12

The angle of incidence for a ray in a step-index fiber for which $n_1 = 1.5$ is 11°.

Compute the refractive index of the cladding material.

Solution

$$i + \theta_c = 90°$$
$$\theta_c = 90 - 11 = 79°$$

Now:

or
$$79° = \sin^{-1}\left[\frac{n_2}{n_1}\right]$$

$$\left(\frac{n_2}{n_1}\right) = \sin 79° = 0.982$$

$$n_2 = n_1 \times 0.982 = 1.5 \times 0.982 = 1.472$$

Thus, the refractive index of cladding (n_2) is 1.472.

Example 13
A fiber has diameter of 5 μm and its core refractive index is 1.45 and cladding refractive index is 1.447. If $\lambda = 1$ μm, how many modes can be propagated inside the fiber?

Solution
The number of modes in Si fiber is given by:

$$N = 4.9 \left(\frac{d \times \text{NA}}{\lambda} \right)^2 = 4.9 \left[\frac{5 \times 10^{-6} \times 0.092}{1 \times 10^{-6}} \right] = 1.04$$

where $\quad \text{NA} = \sqrt{n_1^2 - n_2^2} = \sqrt{1.45^2 - 1.447^2} = \sqrt{0.0085} = 0.092$

Therefore, there is a single mode propagation through this fiber.

Example 14
Calculate the V number and number of modes propagating through the fiber of diameter 100 μm having core and cladding refractive indices of 1.53 and 1.50 respectively? The propagating wave has a wavelength of 1 μm.

Solution

The V number $= \dfrac{2\pi a}{\lambda} (\text{NA})$

$$\text{NA} = \sqrt{(n_1^2 - n_2^2)} = \sqrt{(1.53)^2 - 1.50^2} = 0.301$$

$\therefore \quad$ V number $= \dfrac{2\pi \times 50 \times 10^{-6} \, (0.301)}{1 \times 10^{-6}} = 94.6$

The maximum number of modes propagating through the step index fiber is:

$$N = \frac{V^2}{2} = \frac{(94.6)^2}{2} = 4474$$

V number is 94.6; number of modes propagating through the step index fiber is 4474.

Example 15
Calculate the fractional index change for a given optical fiber if the refractive indices of the core and the cladding are 1.563 and 1.498 respectively.

Solution
Fractional index change:

$$\Delta n = \frac{n_1 - n_2}{n_1} = \frac{1.563 - 1.498}{1.563} = 0.0416$$

Fractional index change is 0.0416.

Problem 16

A step-index has a diameter for 200 mm and numerical aperture 0.3. Calculate the number of propagating modes at wavelength of 0.85 mm.

Solution

The number of modes:

$$N = \frac{V^2}{2} = \frac{1}{2}\left[\frac{2\pi a}{\lambda}(NA)\right]^2$$

$$= \frac{1}{4}\frac{4\pi^2 a^2}{\lambda^2}(NA)^2 = \frac{2\pi^2 a^2}{\lambda^2}(NA)^2$$

$$= \frac{2 \times (3.14)^2 \times (100 \times 10^{-6})^2 \times (0.3)^2}{(0.85 \times 10^{-6})^2}$$

$$= 24{,}560$$

Taking into account the two possible polarization, the number of modes = $2 \times 24{,}560 = 49{,}120$.

Example 17

Calculate the total number of guided modes propagating in the multi-mode step-index fiber having core diameter of 50 μm and numerical aperture of 0.2 and operating at wavelength of 1 μm and also calculate the number of modes of graded index fiber having the same parameter.

Solution

Let N be the number of modes in a step index:

$$N_{SI} = 4.9\left[\frac{d \times NA}{\lambda}\right]^2$$

where d = core dia = 50×10^{-6} meter
λ = wavelength = 1×10^{-6} meter

$$\therefore \quad N_{SI} = 4.9\left[\frac{50 \times 10^{-6} \times 0.2}{1 \times 10^{-6}}\right]^2 = 4.9[10]^2 = 490$$

Thus, the number of modes propagated in a step index is 490. Number of modes propagated in a graded index fiber is 245.

Example 18

If the core and cladding, refractive indices for a step index fiber are 1.47 and 1.46 respectively, what will be the broadening of a pulse after a distance of 5 km?

Solution
Given:
$$L = 5 \times 10^3 \text{ m}$$
$$n_1 = 1.47$$
$$n_2 = 1.46$$
$$c = 3 \times 10^8 \text{ m/s}$$

We know that the pulse broadening is given by:
$$\Delta t = \frac{n_1 L}{c n_2}(n_1 - n_2)$$
$$= \frac{1.47 \times 5 \times 10^3}{3 \times 10^8 \times 1.46}(1.47 - 1.46)$$
$$= \frac{7.35 \times 10^3 \times 0.01}{4.38 \times 10^8}$$
$$\simeq 0.17 \text{ ms}$$

Example 19
Calculate the index difference between the core and the cladding of a fiber with an NA of 0.1. We will solve this in two ways. First we will make as estimate:

$$\sin(\theta_a) = \sqrt{n_1^2 - n_2^2} = \text{NA} = 0.1$$
$$n_1^2 - n_2^2 = 0.01$$
$$(n_1 - n_2)(n_1 + n_2) = 0.01$$
$$\Delta(n_1 + n_2) = 0.01$$
$$\Delta = \frac{0.01}{n_1 + n_2} \cong \frac{0.01}{2.90} = 0.003448$$

In this case, we aqssume that $n_1 \cong n_2 = 1.45$.

In the second case, we will evaluate the approximation made above, taking the index of the cladding to be 1.45:
$$\Delta(n_1 + n_2) = 0.01$$
$$\Delta(n_1 - 2n_2 + 2n_2 + n_2) = \Delta^2 + 2\Delta n_2 = 0.01$$
$$\Delta = \frac{-2n_2 \pm \sqrt{4n_2^2 + 0.04}}{2} = -1.45 + 1.453445 = 0.003444$$

The index difference is less, much less the 1%. The accuracy of the approximation is better than three significant figures.

Example 20
It is given that for a GaAs LED, the relative spectral width γ at $\lambda = 0.85$ μm is 0.035. This source is coupled to a pure silica fiber

(with $|\lambda^2(d^2n/d\lambda^2)| = 0.021$ for $\lambda = 0.85$ mm). Calculate the pulse broadening per kilometer due to material dispersion.

Solution

The pulse broadening per unit length due to material dispersion is given by:

$$\frac{\tau}{L} = \frac{\gamma}{c}\left|\lambda^2 \frac{d^2n}{d\lambda^2}\right| = \frac{0.035}{3\times 10^8 \text{ ms}^{-1}} \times 0.021 = 2.45\times 10^{-12} \text{ s m}^{-1}$$

or $\quad \dfrac{\tau}{L} = 2.45 \text{ ns}\cdot\text{km}^{-1}$

Example 21

Calculate the total pulse broadening due to material dispersion for a graded-index fiber of total length 80 km when a LED emitting at (a) $\lambda = 850$ nm and (b) $\lambda = 1300$ nm is coupled to the fiber. In both the cases, assume that $\Delta\lambda = 30$ nm. The material dispersion parameters of the fiber for the two wavelengths are 105.5 ps nm$^{-1}\cdot$km^{-1} and 2.8 ps nm^{-1}, respectively:

Solution

(a) We have:
$\tau = D_m L \Delta\lambda = 105.5 \times 80 \times 30 = 253{,}200$ ps $= 253.2$ ns
(b) $\tau = 2.8 \times 80 \times 30 = 6720$ ps $= 6.72$ ns

This example shows that proper selection of wavelength can reduce pulse broadening considerably.

Example 22

The core diameter of multimode step-index fiber is 46 μm and numerical aperture is 0.3. Determine number of guided modes at operating wavelength of 0.8 μm.

Solution

$$f = \frac{2\pi}{\lambda}a\,(\text{NA}) = \frac{2\pi}{0.8\times 10^{-6}} \times 23 \times 10^{-6} \times 0.3 \approx 54.16$$

$$N_g = \frac{f^2}{2} \approx 1466$$

Example 23

The mean optical power launched into an optical fiber 150 μW. The mean optical power at the fiber output is 10 μW. The length of the fiber is 5 km. Calculate the signal attenuation per kilometer for the fiber.

Solution

$$\alpha_{dB} = 10 \log_{10} \frac{P_i}{P_o} = 10 \log_{10} \frac{150 \times 10^{-6}}{10 \times 10^{-6}} = 11.76 \text{ dB}$$

$$\alpha_{dB} \div L = \frac{11.76}{5} = 2.352 \text{ dB/km}$$

Example 24

For a step index fiber, the normalized frequency (V parameter) is 26.6 at a wavelength of 1300 nm. Determine the numerical aperture (NA), if the core radius is 25 µm.

Solution

Given:
$$V = 26.6$$
$$\lambda = 1300 \text{ nm} = 1.3 \text{ mm}$$
$$r = 25 \text{ µm}$$

The normalized frequency for a step index fiber is given by:

$$V = \frac{2\pi r}{\lambda}(NA)$$

$$NA = \frac{V\lambda}{2\pi r}$$

$$= \frac{26.6 \times 1.3}{2 \times 3.14 \times 25}$$

$$= \frac{24.58}{157} = 0.22$$

Example 25

An optical fiber has an attenuation 3.5 dB/km. If 0.5 mW of optical power is mainly launched into the fiber, what is the power level in µW after 4 km?

Solution

Given:
$$\alpha = 3.5 \text{ dB/km}$$
$$P_i = 0.5 \text{ mW}$$
$$L = 4 \text{ km}$$

We know that the attenuation of an optical fiber is given by:

$$\alpha = \frac{10}{L} \log \frac{P_i}{P_o}$$

$$3.5 = \frac{10}{4} \log \left(\frac{0.5}{P_o}\right)$$

or
$$\frac{3.5 \times 4}{10} = \log \frac{0.5}{P_o}$$

or
$$1.4 = \log \frac{0.5}{P_o}$$

or
$$10^{1.4} = \frac{0.5}{P_o}$$

$$P_o = \log \frac{0.5}{10^{1.4}}$$

$$= \frac{0.5}{25.11}$$

$$= 19.9 \text{ mW}$$

Example 26
If the core and cladding, refractive indices for a step index fiber is 1.47 and 1.46 respectively, what will be the broadening of a pulse after a distance of 5 km?

Solution
Given:
$$L = 5 \times 10^3$$
$$n_1 = 1.47$$
$$n_2 = 1.46$$
$$c = 3 \times 10^8 \text{ m/s}$$

We know that the pulse broadening is given by:
$$\Delta t = \frac{n_1 L}{c n_2}(n_1 - n_2)$$

$$= \frac{1.47 \times 5 \times 10^3}{3 \times 10^8 \times 1.46}(1.47 - 1.46)$$

$$= \frac{7.35 \times 10^3 \times 0.01}{4.38 \times 10^8}$$

$$= 0.17 \text{ μs}$$

SUGGESTED READINGS
1. Senior JM, 'Optical Fiber Communications', Pearson (2009).
2. Azzawi AA, 'Fiber Optics', CRC Press (2007).
3. Agrawal GP, 'Fiber-Optic Communication Systems', 2nd ed., Wiley, New York (1997).

4. Goff DR and Hansen K, 'Fiber-Optic Reference Guide: A Practical Guide to the Technology', 2nd ed., Butterworth-Heinemann, London (1999).
5. Ungar S, 'Fiber Optics: Theory and Applications', Wiley, New York (1991).
6. Yeh C, 'Handbook of Fiber Optics: Theory and Applications', Academic Press (1990).
7. Yeh C, Applied Photonics, Academic Press (1994).
8. Nolan DA, Optical Fibers in 'The Handbook of Photonics', pp 12.1–12.21, CRC Press, 2nd ed. (2007).
9. Degiorgio V and Cristiani I, 'Photonics: A short course', Springer (2014).
10. Kumar S and Jamal Deen M, 'Fiber Optic Communications', Wiley (2014).
11. Kakani SL and Shubhra Kakani, Engineering Physics, 3rd ed. (2015), CBS Publishers & Distributors, New Delhi-1.

GLIMPSES

- The refractive index (n) of a transparent medium is the ratio of the speed of light in vacuum (c) to the speed of light in the medium (v), thus:

$$n = \frac{c}{v}$$

- The phenomenon of refraction of light is governed by *Snell's law*:

$$n_1 \sin \theta_1 = n_2 \sin \theta_2$$

- An optical fiber is a thin dielectric waveguide made up of a solid cylindrical core of glass (silica) or plastic of refractive index (RI), n_1 surrounded by a coaxial cylindrical cladding of RI, n_2. When n_1 and n_2, both are constant, the fiber is said to be a step-index fiber.
- Within an optical fiber, light is guided by *total internal reflection* (TIR). For light guidance, two conditions have to be satisfied: (i) n_1 must be greater than n_2 ($n_1 > n_2$), and (ii) at the core-cladding interface, the light ray must strike at an angle greater than the critical angle θ_c, where $\theta_c = \sin^{-1}(n_2/n_1)$.

This requires that the ray of light must enter the core of the fiber at an angle less than the acceptance angle θ_a:

$$n_a \sin \theta_a = n_1 \sin \theta_c$$

- The light gathering capacity of a fiber is expressed in terms of the *numerical aperture* (NA), which is expressed as:

$$\text{NA} = n_a \sin \theta_a = \sqrt{(n_1^2 - n_2^2)} = n_1 \sqrt{2\Delta}$$

- When a pulse of light propagates through the fiber, it gets broadened. *Pulse broadening* is caused due to following two mechanisms:
  (i) *Multipath time dispersion*, given by:
  $$\frac{\Delta T}{L} = \frac{n_1}{n_2}\left(\frac{n_1 - n_2}{c}\right)$$
  To some extent, one can reduce this type of dispersion by *index grading*, i.e., by varying the RI of the core in a specific manner. It is reported that the *parabolic profile* gives best results.
  (ii) *Material dispersion*, caused by change in the RI of the fiber material with wavelength. This is given by:
  $$\frac{\tau}{T} = \frac{r}{c}\left|\lambda^2 \frac{d^2 n}{d\lambda^2}\right|$$
- Fiber optics is a technology related to the transportation of optical energy (light energy) in guiding media, specifically glass fiber.
- In case of optical fibers, there must be very little absorption of light as it travels through a long distance inside the fiber. One can achieve this by purification and special preparation of the material used.
- Fibers that are used in optical communication are wave guides made of transparent dielectrics. Its function is to guide visible and infrared light over long distances.
- Optical fibers are classified into two categories based on: (i) number of modes, and (ii) refractive index.
- On the basis of number of modes of propagation, optical fibers are classified into two types: (i) single mode fiber (SMF) and (ii) multi-mode fiber (MMF).
- On the basis of refractive index, fibers can be classified into two types: (i) step-index optical fiber, and (ii) graded-index optical fiber.
- The condition for propagation of light within the optical fiber is $\sin i < NA$.
- V-number is an important parameter for optical fiber and is generally called normalized frequency of the fiber.
$$V = \frac{2\pi a}{\lambda}(NA) = \frac{2\pi a}{\lambda} n_1 \sqrt{2\Delta},$$
where $\Delta = \dfrac{n_1 - n_2}{n_1}$ is fractional refractive index change.

The wavelength corresponding to the value of $V = 2.405$ is known as cutoff wavelength.
- Losses occurring in optical fibers may be mainly due to the following mechanisms: (i) absorption, (ii) scattering, and (iii) fiber bends.

- The index of refraction in graded index fibers varies gradually from the axis of fiber.
- Mode of fiber refers to the number of paths the light ray can travel down the optical fiber.
- The single mode propagation of LP_{01} mode in step index optical fiber is possible over the region $0 < V < 2.405$.
- *Step index fibers* are those fibers where the refractive index (RI) of the core is constant throughout and steps down to the refractive index of the cladding. The light in these fibers travels down the fibers by bouncing off the core-cladding interface.
- *Graded index fibers* have a core with a RI which varies quadratically according to the distance out from the center of the fiber, decreasing as you move further out. Ligfht travelling in these fibers actually follows a curved path due to the continuously varying RI. In these fibers, ray of light follows sinusoidal paths.

REVIEW QUESTIONS

1. Define an optical fiber system. What is numerical aperture of an optical fiber? What is its significance?
2. Obtain the expression for NA in terms of RI of the core and cladding of an optical fiber.
3. Outline the primary building blocks of a fiber optic system.
4. Draw a block diagram for optical communication through fiber optic cables. Mention the advantages of optical communication.
5. Explain the working of monomode and multimode optical fibers. For long distance communication which optical fiber is preferred and why?
6. Contrast glass and plastic fiber cables.
7. Explain the principle and working of fiber optic sensors.
8. Write short notes on:
    (i) Fiber optic cable
    (ii) Step index fiber
    (iii) Graded index fiber
    (iv) Fiber optic communication system
9. What are the functions of the core and cladding in an optical fiber? Why should their refractive indices be different? Can it be possible for the light to be guided without cladding?
10. Explain total internal reflection? Why it is necessary to meet the condition of total internal reflection at the core-cladding interface.
11. Define NA of a fiber. On what factors it depend?
12. What is acceptance angle? What is the condition for propagation of light through an optical fiber?
13. Explain how right propagates through a graded index fiber.
14. Differentiate between single mode and multimode fibers.
15. Differentiate between step index fiber and graded index fiber.

184 Photonics | Optoelectronics

16. Explain multipath time dispersion and material dispersion. How can these be minimized?
17. Distinguish between step index multimode fiber and graded index multimode fiber.

PROBLEMS

1. A multimode step index fiber has $n_1 = 1.53$, $n_2 = 1.50$ and $\lambda = 1$ μm. In order to make this fiber single mode, what will be core radius?

   [Hint: $a \leq \dfrac{2.405\lambda}{2\pi(NA)}$, $NA = \sqrt{(1.53)^2 - (1.50)^2}$

   $a \leq \dfrac{2.405 \times 1 \times 10^{-6}}{2 \times 3.14 \times \sqrt{(1.53)^2 - (1.50)^2}}$, $\therefore < 1.27$ μm]

2. Numerical aperture of an optical fiber is 0.5 and core refractive index is 1.48. Show that cladding refractive index is 1.393 and acceptance angle is 30°.

   [Hint: $NA = \sqrt{n_1^2 - n_2^2} = 0.5$, RI of core $= n_1 = 1.48$; that is $n_2 = 1.393$; RI of cladding $= n_2 = 1.393$; Acceptance angle $\theta = \sin^{-1}(NA) = 30°$]

3. A step index fiber in air has NA of 0.16, core refractive index of 1.45 and core diameter of 60 cm. Show that the normalized frequency for the fiber when light at a wavelength of 0.9 mm transmitted is 3.35×10^5.

   $\left[\text{Hint}: v = \dfrac{\pi d}{\lambda_0}\sqrt{n_1^2 - n_2^2} = \dfrac{3.143 \times 0.6 \text{ m}}{9 \times 10^{-7} \text{ m}} \times 0.16 = 3.35 \times 10^5 \right]$

4. An optical fiber has NA of 0.20 and cladding refractive index of 1.59. Show that the acceptance angle for the fiber in water which has a refractive index of 1.33 is 8.6°.

   [Hind: $NA = \dfrac{\sqrt{(n_1^2 - n_2^2)}}{n_0}$

   When the fiber is in air, $n_0 = 1$ and $NA = \sqrt{(n_1^2 - n_2^2)} = 0.20$

   $\therefore n_1 = \sqrt{(NA)^2 + n_2^2} = \sqrt{(0.20)^2 + (1.59)^2} = 1.6026$

   When the fiber is in water, $n_0 = 1.33$

   $NA = \dfrac{\sqrt{(n_1^2 - n_2^2)}}{n_0} = \dfrac{\sqrt{(1.6025)^2 - (1.59)^2}}{1.33} = 0.15$

   $\theta_0 (\max) = \sin^{-1}(NA) = \sin^{-1}(0.15) = 8.6°$]

5. Show that the core radius necessary for single mode operation at 850 nm Si fiber with $n_1 = 1.480$ and $n_2 = 1.47$ is 1.89 μm. Also show that NA and maximum acceptance angle of this fiber are 0.1717 and 9° 53′ 12″ respectively.

[Hint:

(i) $v = \dfrac{\pi d}{\lambda_0}\sqrt{n_1^2 - n_2^2}$

$2.405 = \dfrac{\pi d}{450 \times 10^{-9}\text{ m}} \times 0.1717$

or $d = 3.79$ mm, $\therefore r = 1.89$ mm

(ii) $NA = \sqrt{n_1^2 - n_2^2} = \sqrt{(1.48)^2 - (1.47)^2} = 0.1717$

(iii) $\sin\theta_0(\max) = NA$

$\therefore \theta_0(\max) = \sin^{-1}(NA) = (n_1^2 - n_2^2)^{1/2}$

$= \sin^{-1}(0.1717) = 9°53'12''$]

6. A fiber cable has index of refraction core 1.52 and cladding 1.31. Show that numerical aperture is 0.77.

7. For a multimode step index fiber with glass core ($n_1 = 1.5$), acceptance angle is required to be 20.2°. Find numerical aperture and refractive index of cladding (n_2).

[**Hint:** $NA = \sin(\theta_{in})$, Here $\theta_{in} = 20.2°$, hence $NA = \sin 20.2° = 0.344$,

$NA = \sqrt{n_1^2 - n_2^2}$

Hence, $0.344 = \sqrt{1.5^2 - n_2^2} = \sqrt{2.25 - n^2}$. By solving, we get $n_2 = 1.45$]

8. A light ray travels from a medium of refractive index $n_1 = 1.5$ to another with refractive index $n_2 = 1.46$. Calculate the critical angle of incidence measured from normal at the boundary. What will be this angle if the second medium is air?

[**Hint:** For $n_1 = 1.5$ and $n_2 = 1.46$:

Critical angle,

$\theta_c = \text{arc sin}\left(\dfrac{n_2}{n_1}\right) = \text{arc sin}\left(\dfrac{1.46}{1.5}\right) = \text{arc sin}(0.9733) = 76.7°$

When second medium is air:

$n_2 = 1.0003$, $\theta_c = \text{arc sin}\left(\dfrac{1.0003}{1.5}\right) = 41.8°$]

9. A step index fiber has an acceptance angle (θ_a) of 20° in air and relative refractive index difference of 3%. Calculate the NA and the critical angle at the core-cladding interface. [**Ans.** 0.34, 76°]

10. The speed of light in vacuum and in the core of a Si fiber is 3×10^8 ms^{-1} and 2×10^8 ms^{-1} respectively. When the fiber is placed in air, the critical angle (θ_c) at the core-cladding interface is 75°. Calculate the (i) NA of the fiber, and (ii) multipath time dispersion per unit length. [**Ans.** (i) 0.388, (ii) 1.7×10^{-10} sm^{-1}]

11. A Si fiber has NA = 0.17 and a cladding refractive index of 1.46. Determine the (a) the acceptance angle of the fiber when it is placed in water (the refractive index of water = 1.33), and (b) the critical angle (θ_c) at the core-cladding interface. [**Ans.** (a) 7.34°, (b) 83.35°]

SHORT ANSWER QUESTIONS

1. To prevent energy losses via absorption and scattering, what should be the order of cladding in an optical fiber?

Ans. At least a few wavelengths thick.

2. How numerical aperture (NA) of an optical fiber is defined?

Ans. NA = $\sqrt{n_1^2 - n_2^2}$, where n_1 is refractive index of the core of a fiber and n_2 is that of cladding ($n_1 > n_2$): One can define the numerical aperture of a fiber as the physical area that light must enter in order to propagate through the fiber. The numerical aperture for light sources can be calculated in terms of angle involved with the output.

3. How will you define acceptance angle (θ_a) for an optical fiber?

Ans. $\theta_a = \sin^{-1}\sqrt{n_1^2 - n_2^2}$.

4. What is major economic benefit offered by fiber optics in information technology?

Ans. Very high information transmission rate at low cost per circuit-km.

5. What type of reflection is expected from an optical fiber?

Ans. Total internal reflection.

6. How is an optical fiber fabricated?

Ans. In the fabrication of optical fibers one can use either different glasses with different refractive indices for core and cladding or use fused silica (quartz, glass) whose refractive index is modified by doping. The type of glass used will also vary with application, but purity (lack of foreign materials as well as imperfections such as cracks or bubbles) must also be kept very high.

7. What is kevlar?

Ans. It is a yarn type material having high tensile strength. It is used in fiber construction to provide additional strength to the cable.

8. Explain the relationship between information capacity and bandwidth.

Ans. The information carrying capacity of a communication system is directly proportional to its bandwidth, i.e. higher the bandwidth, the greater its information carrying capacity.

9. Explain the purpose of cladding in optical fiber.

Ans. Due to its lower refractive index than core, cladding helps in retaining light with the core. Moreover, it provides mechanical strength, reduces scattering loss and also protects the core from absorbing surface contaminates with which is could come in contact.

10. Mention the advantages of optical fiber communication over metallic cable communication.

Ans. (i) Extremely wide bandwidth
(ii) Light in weight and small in size
(iii) Immune to electrostatic interference
(iv) More secured, greater safety and no crosstalk.

11. Can light propagate through an optical fiber cable with the angle of incidence at entrance end greater than acceptance angle?

Ans. No. Only when light enters at an angle less than the acceptance angle, light can propagate through the cable by undergoing total internal reflection.

12. What is an optical fiber sensor?

Ans. It is a transducer which can convert various input variables (physical quantities) into an electrical signal in measurable form.

13. What are the advantages of optical sensors in sensing applications?

Ans. (i) They are electrically passive
(ii) They have good geometrical flexibility
(iii) They are light in weight and small in size
(iv) Chemical and environmental ruggedness is high

14. What are the characteristics of wave that may get modulated in fiber optic sensors?

Ans. Amplitude or intensity, phase, polarisation, frequency and direction of propagation may get modulated.

15. Which type of optical fiber (single mode or multimode) tends to have lower loss and produces less signal distortion?

Ans. Single mode fiber.

16. Which properties of optical fibre reduce connection loss in short distance systems?

Ans. (i) Higher fiber numerical aperture, and (ii) Large fiber core.

17. Which basic optical material property is considered relevant for optical fiber light communication?

Ans. Index of refraction.

18. Out of core or cladding, which optical fiber has a higher index of refraction?

Ans. Core.

19. Multimode step index fibers have a core and cladding of constant refractive index n_1 and n_2 respectively. Out of these, which refractive index core or cladding is lower?

Ans. Cladding.

20. What is the benefit of having large core diameters and large NA for multimode step index fibres?
Ans. This makes it easier to couple light from an LED into the optical fiber.
21. Meridional rays are classified as either bound or unbound rays. While propagating through the fiber, the bound rays follow what property of light?
Ans. Total internal reflection.
22. What are the advantages of optical fiber over copper wire conductors?
Ans. Optical fiber transmission has following advantages over convention copper wire system:
    (i) More data can be sent over the long distances through optical fiber due to its lower transmission losses
    (ii) The light weight and the small thin hair-szied dimension of the optical fibers are quite advantageous, e.g. in aircraft
    (iii) Dielectric nature of optical fiber provides optical immunity to the electromagnetic interferences
    (iv) The principal material for the optical fiber, i.e. silica, whose source is sand, is abundantly available
23. What are the advantages of optical fibers?
Ans. See text.
24. What are fiber lasers?
Ans. Fiber lasers also called as fiber sources are light sources made from fiber. They are built around fiber cores, which are doped with materials that can be stimulated to emit light.
25. What are the advantages of Fiber optic systems?
Ans. Fiber optic systems offer several advantages. Few of them are as follows:
    (i) Optical fibers are light in weight and occupy less space than coaxial (bundles of twisted pair) cables.
    (ii) Optical fibers are composed of dielectric materials, and therefore, these are totally immune to extraneous interfering electromagnetic signals.
    (iii) There is virtually no signal leakage from optical fibers and hence cross-talk is not possible.
    (iv) Attenuation of signals in optical fibers is much smaller than in coaxial cables or twisted pair.
    (v) Optical fibers are immune to electromagnetic interference and do not pick up line currents. Obviously, these can be safely used in high voltage environments, e.g. power station, and can be laid alongside metallic power cables.
    (vi) The basic raw material used in the fabrication of low loss fibers is silica, which is abundantly available in nature

(although in impure form), whereas copper constitutes the basic raw material in coaxial cables.

(vii) Optical fibers can withstand environmental hazards in a better manner and also have longer life compared to copper cables.

(viii) Optical fibers are non-conductive and non-inductive. This reveals that there is no interference with other communication systems.

(ix) The speed of transmission through optical fibers is very high because the signals are carried by light.

(x) The channel capacity of optical fibers is very large as compared to the audio band (5 to 10 Hz); the frequency of carrier light wave is very high ($\sim 10^{14}$ Hz). Obviously, one can accommodate a large number of speed signal channels in fibers whose bandwidth is larger than that of a wire transmission line.

26. What are the disadvantages of optical fiber systems?

Ans. The major disadvantages of optical fiber systems are as follows:

(i) Optical fiber systems are virtually of less use unless they are connected to standard electronic facilities which often require expensive interfaces.

(ii) Optical fibers have significantly lower tensile strength than coaxial cables. This can be improved by coating the fiber with standard Kevlar and a protective jacket of PVC.

(iii) Occasionally, it is necessary to provide electrical power to remote interface or regenerating equipment. Thus, additional metallic cables have to be included in the cable assembly.

(iv) Optical fibers require special tools to splice and repair cables and special test equipments to make routine measurements. Repairing fiber cables is also difficult and expensive and working on optical fiber cables also requires special skills and training

MULTIPLE CHOICE QUESTIONS

1. In fiber-optic communication, the propagation of light through the fiber is based on the principle of:
   (a) diffraction of light   (b) interference of light
   (c) total internal reflection   (d) polarization of light

2. The refractive index in a graded index fiber:
   (a) increases throughout the core
   (b) remains constant throughout the core
   (c) decreases parabolically throughout the core
   (d) None of the above

3. The acceptance angle or acceptance cone half angle for the optical fiber is:
   (a) $\theta_m = \sin^{-1}(n_1 - n_2)$
   (b) $\theta_m = \sin^{-1}\sqrt{n_1^2 - n_2^2}$
   (c) $\theta_m = \sqrt{n_1^2 - n_2^2}$
   (d) $\theta_m = n_1^2 / n_2^2$

4. Numerical aperture of step index optical fiber is:
   (a) $NA = n_1 \sqrt{2\Delta}$
   (b) $NA = n_1^2 \sqrt{2\Delta}$
   (c) $NA = n_1^2 - n_2^2$
   (d) $NA = n_1 2\Delta$

5. The jacket of an optical fiber enables:
   (a) to prevent from moisture trapping
   (b) to prevent from mechanical abrasions
   (c) to prevent from interaction with internal atmosphere
   (d) All of the above

6. When V parameter is less than 2.405, then optical fiber will support:
   (a) three modes
   (b) two modes
   (c) one mode
   (d) none of the above

7. The optical fibers are made of:
   (a) dielectrical material
   (b) metallic material
   (c) magnetic material
   (d) plastic doped with metallic impurities

8. Pulse broadening can be minimised by making use of:
   (a) multimode graded index fiber
   (b) single mode index fiber
   (c) multimode step index fiber
   (d) none of the above

9. The total number of modes in a multi-mode fiber is given by:
   (a) $M = \dfrac{1}{2}\left[\dfrac{\pi d}{\lambda} NA\right]^2$
   (b) $M = \left[\dfrac{2\pi d}{\lambda} NA\right]$
   (c) $M = \dfrac{1}{2}\left[\dfrac{\pi d}{\lambda} NA\right]^{1/2}$
   (d) $M = \dfrac{1}{2}\left[\dfrac{\pi d}{\lambda} NA\right]^3$

10. Attenuation in an optical fiber is measured by:
    (a) $loss = 10 \log_{10}\left(\dfrac{P_i}{P_o^2}\right)$
    (b) $loss = -10 \log_{10}\left(\dfrac{P_o}{P_i}\right)$
    (c) $loss = 10 \log_{10}\left(\dfrac{P_i^2}{P_o}\right)$
    (d) $loss = -10 \log_{10}\left(\dfrac{P_i}{P_o}\right)$

11. Information signals are transmitted from optical fiber in the form of:
    (a) analogue signals
    (b) digital signals
    (c) binary and hexadecimal bits
    (d) None of the above
12. Dispersion in optical fibers is due to:
    (a) chromatic dispersion   (b) material dispersion
    (c) intermodal dispersion   (d) All of the above
13. The propagation of light along the wavelength is decided by the:
    (a) interference
    (b) dispersion
    (c) modes of the waveguides
    (d) None of the above
14. In multimode propagation the light propagates the fiber in:
    (a) straight line   (b) zig zag fashion
    (c) circular path   (d) elliptical path
15. In order that mode remain guided, the propagation factor β must safisfy:
    (a) $n_2 < \beta < n_1$
    (b) $n_2 \frac{2\pi}{\lambda} < \beta < n_1 \frac{2\pi}{\lambda}$
    (c) $n_2 < \beta = n_1$
    (d) $n_2 = \beta < n_1$
16. A ray of light is passing from a silica glass of refractive index 1.48 to another silica glass of refractive index 1.46. What is the range of angles (measured with respect to the normal to the interface) for which this ray will undergo total internal reflection?
    [B.Tech.]
    (a) 0°–80°   (b) 81°–90°
    (c) 90°–180°   (d) 180°–360°
17. Light is guided within the core of a step-index fiber by:
    (a) refraction at the core-air-interface
    (b) total internal reflection at the core-cladding interface
    (c) total internal reflection at the outer surface of the cladding
    (d) change in the speed of light within the core
18. A step-index fiber has a core with a refractive index of 1.50 and a cladding with a refractive index of 1.46. Its numerical aperture is: [B.Tech.]
    (a) 0.156   (b) 0.244
    (c) 0.344   (d) 0.486
19. The optical fiber of Question 18 is placed in water (refractive index 1.33). The acceptance angle of the fiber will be approximately:

(a) 10° (b) 15°
(c) 20° (d) 25°

20. The axial refractive index of the core, n_0 of a graded-index fiber is 1.50 and the maximum relative refractive index difference Δ is 1%. What is the refractive index of the cladding? [B.Tech.]
    (a) 1.485 (b) 1.50
    (c) It will depend on the profile parameter
    (d) It will depend on the radius of the core

21. An impulse is launched into one end of a 30-km optical fiber with a rated total dispersion of 20 ns/km. What will be the width of the pulse at the other end? [B.Tech.]
    (a) 20 ns (b) 100 ns
    (c) 300 ns (d) 600 ns

22. For a typical LED, the relative spectral width γ is 0.030. This source is coupled to a pure silica fiber with $|\lambda^2(d^2n/d\lambda^2)| = 0.020$ at the operating wavelength. What is the pulse broadening per km due to material dispersion? [B.Tech.]
    (a) 2 ps·km^{-1} (b) 2 ns·km^{-1}
    (c) 2 ms·km^{-1} (d) 2 ms·km^{-1}

23. A pulse of 100 ns half-width is transmitted through an optical fiber of length 20 km. The fiber has a rated multipath time dispersion of 10 ns·km^{-1} and a material dispersion of 2 ns·km^{-1}. What will be the half-width of the received pulse? [B.Tech.]
    (a) 100 ns (b) 227 ns
    (c) 240 ns (d) 340 ns

24. A LED is emitting a mean wavelength of $\lambda = 0.90$ μm and its spectral half-width $\Delta\lambda = 18$ nm. What will be the half-width of the received pulse? [B.Tech.]
    (a) 0.02 (b) 0.05
    (c) 0.90 (d) 18

25. A laser diode has a relative spectral width of 2×10^{-3} and is emitting a mean wavelength of 1 μm. What is its spectral half-width? [B.Tech.]
    (a) 1 μm (b) 0.2 μm
    (c) 20 nm (d) 2 nm

ANSWERS

1. (c)   2. (c)   3. (b)   4. (b)   5. (d)   6. (c)   7. (a)
8. (a)   9. (a)   10. (b) 11. (b) 12. (d) 13. (c) 14. (b)
15. (b) 16. (b) 17. (b) 18. (c) 19. (b) 20. (a) 21. (d)
22. (b) 23. (b) 24. (a) 25. (d)

5

Fiber Optic Cables and Connectors

5.1 INTRODUCTION

The production, application and installation of optical fibers within a line transmission system are of paramount importance if optical fiber communication systems are to be considered as viable replacements for conventional metallic line communication systems. It is, therefore, essential that:

(i) Optical fibers may be produced with good stable transmission characteristics in long lengths at a minimum cost with maximum reproducibility.

(ii) A range of optical fiber types with regard to size, refractive indices and index profiles, operating wavelengths, materials etc. be available in order to fulfill many different system applications.

(iii) The fibers may be converted into practical cables which can be handled in a similar manner to conventional electrical transmission cables without problems associated with the degradation of their characteristics or damage.

(iv) The fibers and fiber cables may be terminated and connected together (jointed) with excessive practical difficulties and in ways which limit the effect of this process on the fiber transmission characteristics to keep them within acceptable operating levels. It is important that these joining techniques may be applied with ease in the field locations where cable connection take place.

(v) The fiber optic cables are to be used under a variety of situations such as underground, outdoor poles or submerged under water. The structure of cable depends on the situation where it is to be used, but the basic cable design principles remains same.

(vi) Mechanical property of cable is one of the important factor for using any specific cable. Maximum allowable axial load on cable decides the length of the cable can be reliably installed.

(vii) Also the fiber cables must be able to absorb energy from impact loads. The outer sheath must be designed to protect glass fibers from impact loads and from corrosive environmental elements.

5.2 FIBER MATERIALS

Light guidance through a step-index optical fiber requires the refractive indices of the core and cladding to be different. Hence, two compatible materials that are transparent in the operating wavelength range are required. For graded-index fibers, in order to produce a particular index profile, a variation of the refractive index within the core is also required. This is possible only by varying the dopant concentration. Here again two compatible materials, which are mutually soluble and have similar transmission characteristics, will be required. Further, these materials should be such that long, thin, flexible fibers can be drawn. We summarize these requirements of fiber optic material as follows:

(i) The material must be transparent for efficient transmission of light.
(ii) It must be possible to draw long thin fibers from the material.
(iii) Fiber material must be compatible with the cladding material.

Glass and plastics fulfills the above mentioned requirements of fiber materials.

Most fiber consists of silica (SiO_2) or silicate. Various types of high loss and low loss glass fibers are available to suit the requirements. Plastic fibers are not popular because of high attenuation but they have better mechanical strength.

Glass Fibers

Glass is made by fusing mixtures of metal oxides having refractive index of 1.458 at 850 nm. For changing the refractive index different oxides such as B_2O_3, GeO_2 and P_2O_5 are added as dopants. Fig. 5.1 shows variation of refractive index with doping concentration.

Figure 5.1 shows that addition of dopants GeO_2 and P_2O_5 increases refractive index, while dopants fluorine (F) and B_2O_3 decreases refractive index. One important criteria is that the refractive index of core is greater than that of the cladding, hence some important compositions are used as:

Composition	Core	Cladding
1	GeO_2–SiO_2	SiO_2
2	P_2O_5–SiO_2	SiO_2
3	SiO_2	B_2O_3–SiO_2
4	GeO_2–B_2O_3–SiO_2	B_2O_3–SiO_2

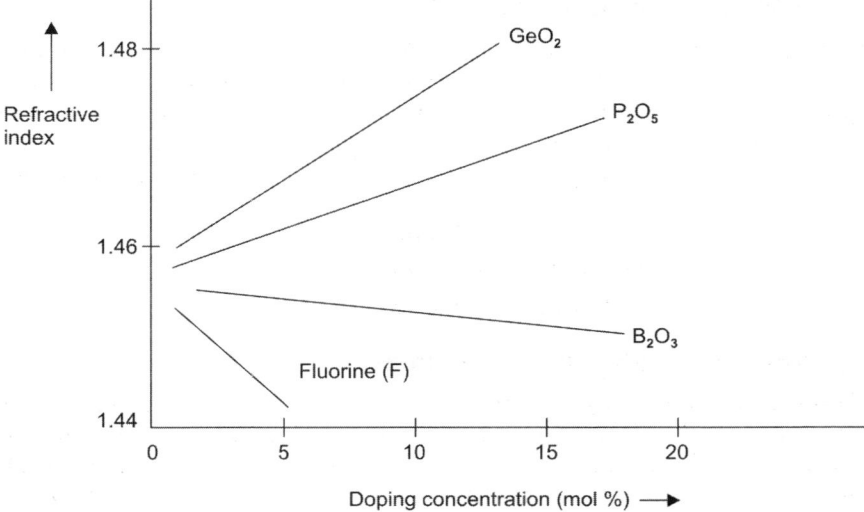

Fig. 5.1. Variation of refractive index of silica glass with doping concentration

The principal raw materials for silica is sand and glass. The fiber composed of pure silica is called as *silica glass*. The desirable properties of silica glass are:
- Resistance to deformation even at high temperature.
- Resistance to breakage from thermal shocks (low thermal expansion).
- Good chemical durability.
- Better transparency.

Other types of glass fibers are:
(i) Halide glass fibers.
(ii) Active glass fibers.
(iii) Chalcogenide glass fibers.
(iv) Plastic optical fibers.

A relatively newer class of fibers is being developed using fluoride glasses, which have extremely low transmission losses at mid-infrared wavelengths (0.2–8 μm), with least loss at ~2.55 μm. A typical core glass consists of ZBLAN glass (named after its constituents ZrF_4, BaF_2, LaF_3, AlF_3, and NaF) and ZHBLAN cladding glass (H standing for HaF_4).

Another class of fibers, called *active fibers*, incorporates some rare-earth elements into the matrix of passive glass. These dopant ions absorb light from the optical source, get excited, and emit fluorescence. Erbium and neodymium have been widely used. Thus, it is possible to fabricate fiber amplifiers, using selective doping.

5.3 FIBER FABRICATION METHODS

Fabrication of all glass-fibers is a two-stage process in which initially the pure glass is produced and converted in a form (rod or preform) suitable for making the fiber. A drawing or pulling technique is then employed to acquire the end product. The methods of preparing extremely pure optical glasses may be placed into following two major categories:

(i) Liquid-phase (melting) techniques, and
(ii) Vapour-phase deposition techniques.

Now, we briefly describe these techniques.

(i) Liquid-Phase (Melting) Techniques

These techniques employ conventional glass refining techniques for ultra pure material powders which are usually oxides or carbonates of required constituents. These include oxides such as SiO_2, GeO_2, B_2O_2 and Al_2O_3, and carbonates such as Na_2CO_3, K_2CO_3, $CaCO_3$, and $BaCO_3$ which will decompose into oxides during the glass melting. An appropriate mixture of these materials is then melted at temperatures varying from 900°C to 1300°C in silica or platinum crucibles. After the melt has been processed suitably, it is cooled and drawn into rods or tubes (or about 1 m in length) of multicomponent glass. The rod of core glass is then inserted into a tube of cladding glass to make a perform.

Fig. 5.2 shows the apparatus used for drawing fiber from this preform. Here, the preform is precision-fed into a cylindrical furnace capable of maintaining high temperature, normally called a drawing furnace. During its passage through the hot zone, its end is softened to the extent that a vary thin fiber can be drawn from it. The outer diameter

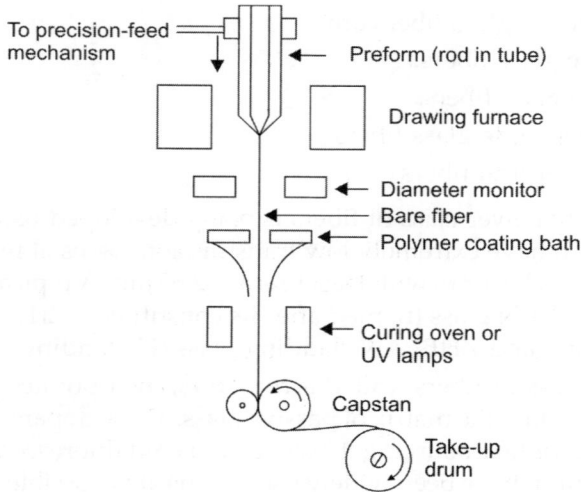

Fig. 5.2. Schematic of a fiber-drawing apparatus

of the fiber is monitored through a feedback mechanism, which controls the feed rate of the preform and also the winding rate of the fiber. The bare fiber is then given a primary protective coating of polymer by passing it through the coating bath. This coating is cured either by UV lamps or thermally. The furnished fiber is then wound on a take-up drum.

A fiber about 20–30 km long can be drawn from a preform of about 1 m in 2–3 h. Higher pulling rates are limited by the pulling process as well as the subsequent primary coating operation. This method of preparing fibers tends to be a bach process and hence continuous production is not possible. Continuous manufacture is possible using another technique, which is called the *double crucible method*. Fig. 5.3 shows the double crucible apparatus.

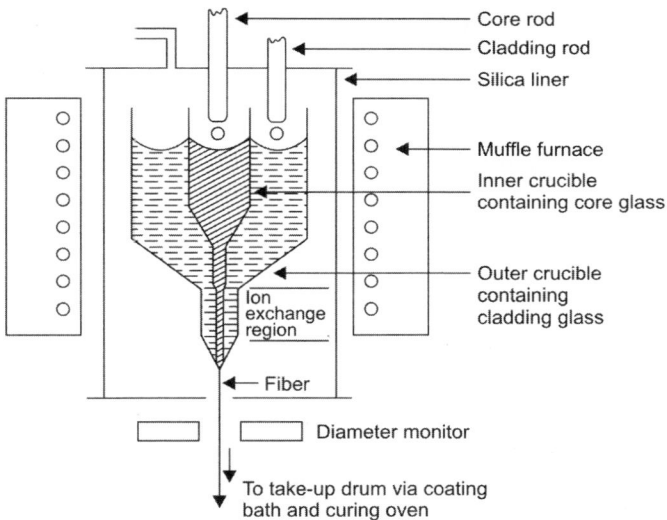

Fig. 5.3. Double crucible apparatus for continuous production of fibers

Double crucible consists of two concentric platinum crucibles mounted inside a vertical cylindrical muffle furnace whose temperature may be varied from 800°C to 1200°C. The starting material—core and cladding glasses, either directly in the powdered form or in the form of preformed rods—is fed into the two crucibles separately. Both the crucibles have nozzles at their bases from which a clad fiber may be drawn from the melt in a manner similar to that shown in the Fig. 5.2. Index grading may be achieved by diffusion of dopant ions across the core-cladding interface, within the melt. Relatively inexpensive fibers of large core diameters and, therefore, large numerical apertures may be produced continuously by this method. An attenuation level of the order of 3 dB/km for sodium borosilicate glass fiber, which is prepared using this technique, has been reported in the literature.

(ii) Vapour-Phase Deposition Techniques

These techniques are used to produce silica rich glasses of highest transparency and with the optimal optical properties. The melting temperature of silica-rich glasses are too high for liquid-phase melting techniques; therefore, vapour-phase deposition methods are used. Herein, the starting materials are halides of silica (e.g. $SiCl_4$) and of the dopants, e.g., $GeCl_4$, $TiCl_4$, BBr_3, etc., which are purified to reduce the concentration of transition-metal impurities to below 10 ppb. Gaseous mixtures of halides of silica and the dopants are combined in vapour-phase oxidation through either flame hydrolysis or chemical vapour deposition methods. Some typical reactions are as follows:

$$SiCl_4\text{(vapour)} + 2H_2O\text{(vapour)} \xrightarrow{heat} SiO_2\text{(solid)} + 4HCl\text{(gas)} \quad (5.1)$$

$$GeCl_4\text{(vapour)} + 2H_2O\text{(vapour)} \xrightarrow{heat} GeO_2\text{(solid)} + 4HCl\text{(gas)} \quad (5.2)$$

$$2BBr_3\text{(vapour)} + 3H_2O\text{(vapour)} \xrightarrow{heat} B_2O_3\text{(solid)} + 4HBr\text{(gas)} \quad (5.3)$$

$$SiCl_4\text{(vapour)} + O_2\text{(gas)} \xrightarrow{heat} SiO_2\text{(solid)} + 2Cl_2\text{(gas)} \quad (5.4)$$

$$GeCl_4\text{(vapour)} + O_2\text{(gas)} \xrightarrow{heat} GeO_2\text{(solid)} + 2Cl_2\text{(gas)} \quad (5.5)$$

$$TiCl_4\text{(vapour)} + O_2\text{(gas)} \xrightarrow{heat} TiO_2\text{(solid)} + 2Cl_2\text{(gas)} \quad (5.6)$$

The oxides resulting from these reactions are normally deposited onto a substrate or within a hollow tube, which is built up as a stack of successive layers. Thus, the concentration of the dopant may be varied gradually to produce the desired index profile. This process results in either a solid rod or a hollow tube of glass, which should be collapsed to produce a solid perform. Using the apparatus shown in Fig. 5.2, fibers may be drawn from this perform.

Based on above principle, various techniques have been developed. We will discuss few of them.

(a) Outside-Vapour-Phase Oxidation (OVPO) Process

This process which uses flame hydrolysis stems from work on 'soot' processes originally developed by Hyde in 1942. Fig. 5.4 shows the flame hydrolysis to deposit the required glass composition onto a rotating mandrel (an aluminium rod). The mixtures of vapours of the starting materials, e.g., $SiCl_4$, $GeCl_4$, BBr_3, etc., is blown through the oxygen-hydrogen flame. The soot produced by the oxidation of halide vapours is deposited on a cool mandrel. The flame is moved back and forth

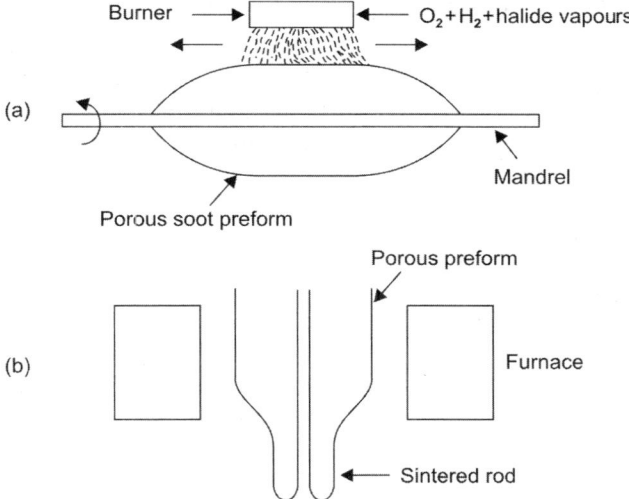

Fig. 5.4. Schematic of glass deposition by the OVPO method: (a) soot deposition and (b) preform sintering

over the length of the mandrel so that a sufficient number of layers is deposited on it. The concentration of the dopant halides is either varied gradually, if index grading is required, or maintained constant, if step-index is required.

After the deposition of the core and cladding layers is complete, the mandrel is removed. The hollow and porous preform left behind is then sintered in a furnace to form a solid transparent glass rod. This is then drawn into a fiber by the apparatus discussed earlier. With a single preform, 30–40 km of fiber can be easily prepared. Nevertheless, a number of proprietary approaches to scaling up the process have provided preforms capable of producing 250 km of fiber. The index profile can be controlled well using this method, as the flow rate of vapours can be adjusted after the deposition of each layer. An attenuation of less than 1 dB/km at an operating wavelength of 1.3 µm has been reported, but the fibers exhibit an axial dip in the refractive index.

The index profile can be controlled well using this method, as the flow rate of vapours can be adjusted after the deposition of each layer. An attenuation of less than 1 dB/km at an operating wavelength of 1.3 µm has been reported, but the fibers exhibit an axial dip in the refractive index.

The purity of the glass fiber depends on the purity of the feeding materials and also upon the amount of OH impurity from the exposure of the silica to the water vapour in the flame following the reactions described earlier.

(b) Vapour Axial Deposition (VAD) Process

This process was developed by Zaw in 1977. The VAD technique uses an end-on deposition onto a rotating fused silica target. In this process, core and cladding glasses are simultaneously deposited onto the end of a speed rod, which is rotated to maintain azimuthal homogeneity and also pulled up, as illustrated in Fig. 5.5. The porous preform so deposited, while the seed rod is being pulled up, is heated to about 1100°C in an electric furnace in an atmosphere of O_2 and thionyl chloride. Any water vapour in the preform is removed through the following reaction:

$$SOCl_2 + H_2O \rightarrow SO_2 + 2HCl \tag{5.7}$$

The porous preform is then heated to about 1500°C in a carbon furnace, where it is sintered into a transparent solid glass rod.

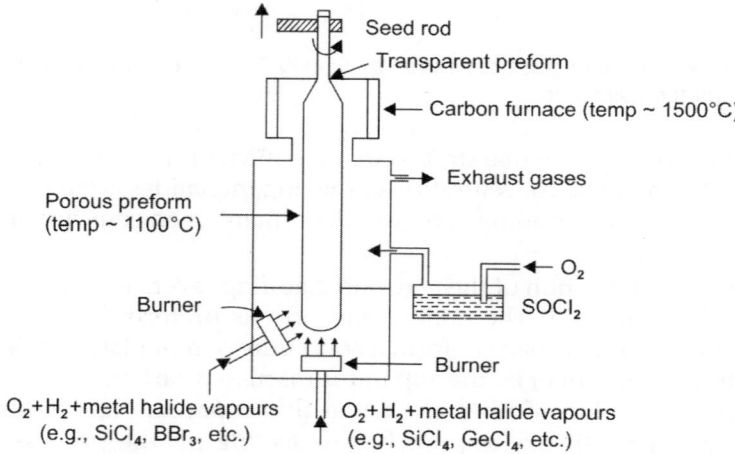

Fig. 5.5. Schematic of the apparatus for the VAD technique

A good control over the index profile may be achieved with germania-doped cores. Graded-index with an attenuation level less than 0.5 dB/km at 1.3 µm has been produced by this method. In principle, this process may be adopted to draw fiber continuously, although at present, it tends to be operated as a batch process partly because the resultant performs can yield more than 100 km of fiber.

Modified chemical vapour deposition (MCVD method): Chemical vapour deposition processes are generally used at very low deposition rates in the semiconductor industry to produce protective SiO_2 films on silicon semiconductor devices. Basically MCVD is a vapour-phase oxidation process taking place inside a hollow silica tube as shown in Fig. 5.6. The tube has a length of about 1 m and a diameter of about 15 mm. It is horizontally mounted and rotated on a glass-working lathe, with an arrangement (normally an oxygen-hydrogen flame) for heating the outer surface of the tube to about 1500°C. The reactants in the form of halide vapours and oxygen are passed at a controlled rate through the tube. The halides are oxidized in the hot zone of the tube and the generated soot (glass particles) is deposited on the inner wall. The hot zone (i.e., the flame) is moved back and forth allowing the layer-by-layer deposition of the soot. Index grading may be achieved by varying the concentration of the dopants layer by layer. The tube can form a cladding material or serve as a support structure only for the porous preform. After the deposition is complete, the tube or the porous preform is sintered at a higher temperature (1700–1900°C) to form a transparent glass rod. Fiber is then drawn from this rod in the usual manner.

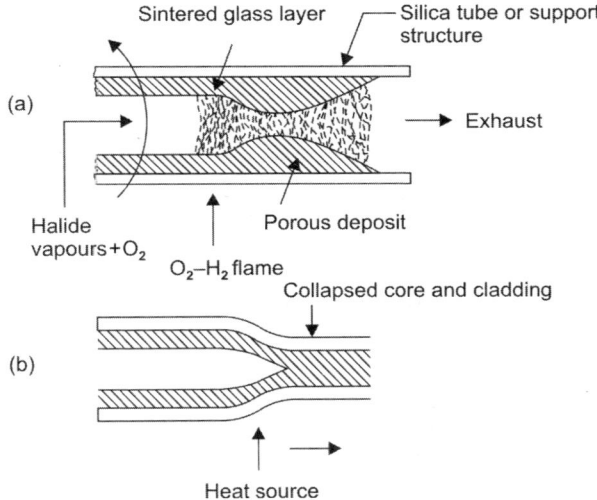

Fig. 5.6. Schematic diagram: (a) Apparatus for the MCD process, (b) Preform sintering

Presently, this is a widely used method for fabricating fibers, as it permits the deposition to occur in a clean environment, with reduced OH impurity. This method is suitable for preparing a variety of glass compositions for multimode or single-mode step-index (SI) or graded-index fibers. Typically, attenuation of the order of 0.2 dB/km at a wavelength of 1.55 μm has been reported for single-mode germania-doped silica fibers prepared by this method. Further, this technique is also suitable for preparing polarization-maintaining single-mode fibers.

Although, MCVD is not a continuous process, but this technique has proved suitable for the widespread mass production of high-performance of optical fibers. Further, it can be scaled up to produce preforms which provide 100 km to 200 km of fiber.

(d) Plasma-Activated Chemical Vapour Deposition (PCVD) Method

A variation on the MCVD technique is the use of various types of plasma to supply energy for the vapour-phase oxidation of halides. This method involves plasma-induced chemical vapour deposition inside a silica. The deposition rates of the MCVD process may be increased if microwave-frequency plasma is created in the reaction zone. This process is illustrated in Fig. 5.7. Herein, a microwave cavity (operating at 2.45 GHz) surrounds the substrate tube.

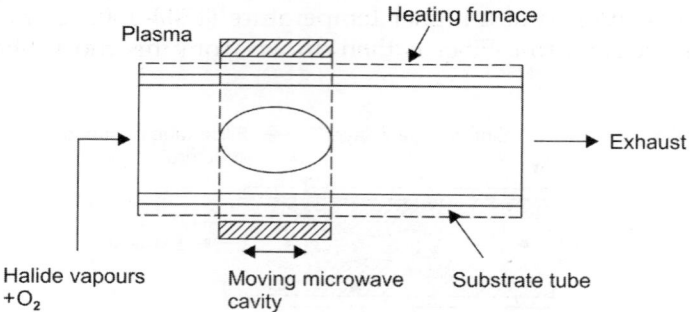

Fig. 5.7. Schematic of the apparatus used for the PCVD process

The halide vapours of the silica-based compound or the dopants along with oxygen are introduced into the tube where they react in the microwave-excited plasma zone. The tube temperature is maintained at about 1900°C using a stationary furnace. The reaction zone is moved back and forth along the tube enabling circularly symmetrical deposition of glass layers onto the inner wall of the substrate tube. High deposition efficiency and an excellent control of index grading is possible with this method. Attenuation of the order of 0.3 dB/km at $\lambda = 1.55$ μm has been obtained for dispersion flattened single-mode fibers.

The PCVD method also lends itself to large-scale production of optical fibers with preform sizes that would allow the preparation of over 200 km of fiber.

Salient features of the four vapour-phase deposition (VPD) techniques discussed above are summarized in Table 5.1.

5.4 OPTICAL FIBER CABLES

Fiber optic cables transmit data through very small cores at the speed of light. Significantly different from copper cables, fiber optic cables offer *high band width* and *low losses* which allow high data-transmission rates over long distances. Light propages throughout the fiber cables according to the principle of total internal reflection (TIR).

There are three common types of fiber optic cables: single mode, multimode, and graded index. Each has its advantages and dis-

Table 5.1. Salient features of VPD techniques used in the preparation of low-loss optical fibers

• Reaction type	OVPO, VAD
• Flame hydrolysis	MCVD
• High-temperature oxidation	PCVD
• Low-temperature oxidation	
Depositional direction	
• Outside layer deposition	OVPO
• Inside layer deposition	MCVD, PCVD
• Axial layer deposition	VAD
Refractive index profile formation	
• Layer approximation	OVPO, MCVD, PCVD
• Simultaneous formation	VAD
Process	
• Batch	OVPO, MCVD, PCVD
• Continuous	VAD

advantages. There are also several different designs of fiber optic cables, each made for different applications. In addition, new fiber optic cables with different core and cladding designs have been emerging; these are faster and can carry more modes. While fiber optic cables are used mostly in communication systems, they also have established medical, military, scanning, imaging and sensing applications. They are also used in optical fiber devices and fiber opting lighting.

The scaling down of light sources brings the benefits of fiber technology to a wide range of applications. These developments have led to a surge of interest in advanced fiber cables for both industrial and military applications. Advanced fiber cables are also used for transmitting high volumes of data in communication systems over long distances for getting clear images, and in building many sophisticated instruments for a variety of applications. By creating new core designs, adding dopants to the fiber core and cladding, and developing manufacturing processes, engineers achieve advanced fiber-optic cable technology, e.g., the core of *holey fibers* consists of many air holes; each hole acts as a single fiber. This fiber enables a high data transmission rate and capacity, and consequently, reduces the cost of the network.

Optical glass fibers are brittle and have very small cross-sectional areas (typical outer diameters range from 100 to 250 µm). They are quite susceptible to damage during normal handling and use. Thus, it is necessary to improve their tensile strength and protect them from external influences. Therefore, it is necessary to encase them in cables.

The exact design of the optical-fiber cable may vary depending on the application; i.e., it will depend on whether the cable is required to be used in underground ducts, buried directly, hung from the poles, or submerged underwater, etc. Thus, one can develop a general criterion to be applied to all the designs of cables to be formulated. A cable design should be such that it (i) protect the optical fiber from damage and breakage, (ii) does not degrade the transmission characteristics of the optical fiber, (iii) prevents the fiber from being subjected to excessive strain and limits the bending radius, (iv) provides a strength member which can improve its mechanical strength, and (v) provides for (in the case of multifiber cables) the identification and jointing of optical fibers within the cable.

In the light of above factors, the primary coated fibers are given a secondary or buffer coating for protection against external influences. It is also possible to place the fibers in an oversized extruded tube normally called a *loose buffer jacket*. This structure isolates the fiber mechanically from external forces as well as microbending losses. The empty space in the loose tube may be filled with soft, self-healing material that remains stable over a wide range of temperatures. The buffered fibers are then either stranded helically around a central strength member of placed in the slots of a structural member. The structural member may also serve as a strength member if made of load-bearing material.

The common desirable features of strength and structural members are *high Young's modulus, high tolerance to strain, flexibility,* and *low weight per unit length*. An additional requirement of the strength member is that it should have *high tensile strength*. In order to provide cushion to the entire assembly consisting of buffered fiber, and structural and strength members, a coating of extruded plastic is applied or a tape is helically wound. A further thick outer sheath of plastic is necessary to provide the cable with extra protection against mechanical forces such as crushing. Some designs include copper wires in the cable. These wires are used to feed electrical power to the remote online repeaters and also to serve as voice channels during installation and repair. Some designs of cable are shown in Figs 5.8–5.11.

Some important points about the fiber optic cables are as follows:
(i) The fiber optic cables are to be used under variety of situations such as underground, outdoor poles or submerged under water. The structure of cable depends on the situation where it is to be used, but the basic cable design principles remains same.
(ii) *Mechanical property* of the cable is one of the important factor for using any specific cable. Maximum permissible axial load on cable decides the length of the cable that can be reliably installed.
(iii) Fiber optic cables must be able to absorb energy from impact loads and from corrosive environmental elements.

Fiber Optic Cables and Connectors   205

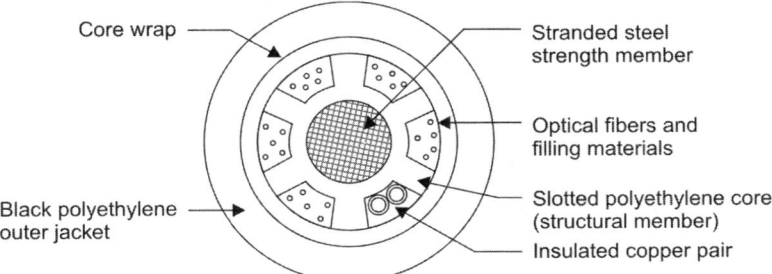

Fig. 5.8. Schematic of a slotted core cable. Here, a slotted polyethylene core is extruded over the stranded steel strength member. The buffered fibers are placed in the slots. The design is quite easy to fabricate and may be adopted for a variety of applications

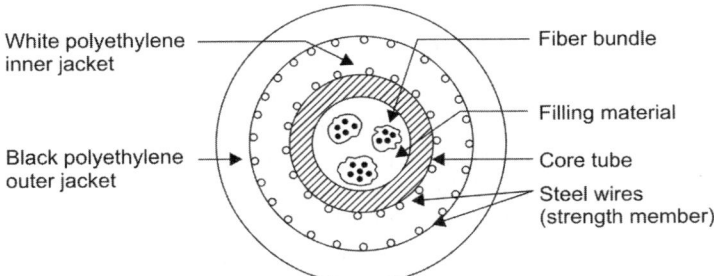

Fig. 5.9. Schematic of a loose bundle cable. Here, a tube is extruded over the fiber bundles, each bundle containing several fibers. The steel wires surrounding this tube serve as strength members. This permits a large number of fibers to be accommodated in a compact design

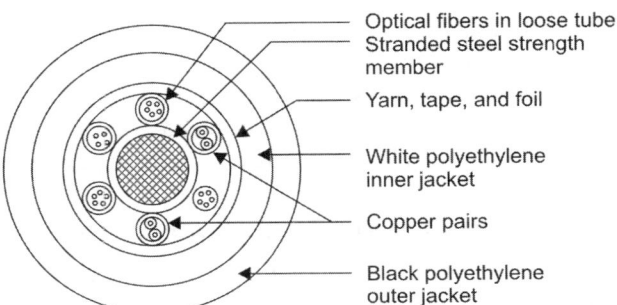

Fig. 5.10. Schematic of a multifiber cable. Here, buffered fibers are placed in loose tubes. Out of the six tubes shown, four contain optical fibers and two contain insulated copper pairs. A central steel strength member has been provided. This cable is suitable for underground ducts

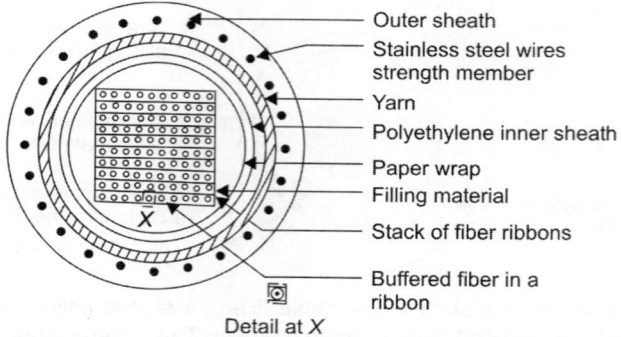

Fig. 5.11. Schematic of a multifiber ribbon cable. This design allows a large number of fibers to be placed in a single cable. The cable design by AJ and T can accommodate 144 fibers in the form of a stack of 12 ribbons, each ribbon containing 12 optical fibers. Ribbon cables are being developed, which can accommodate several hundred fibers

5.5 FIBER-OPTIC CONNECTIONS AND RELATED LOSSES

There are three factors which determine the continuous length of an optical fiber along a communication link. These are: (i) the continuous length of optical fiber that can be produced by prevalent manufacturing methods, (ii) the length of the cable that can be produced and *installed as a continuous section along the link,* and (iii) *the cable length between the repeaters.* This uninterrupted length of optical fiber along a link, therefore, is not more than 10 km. Obviously, for establishing long-haul transmission links, optical fibers are required to be connected. The fiber-to-fiber connection may be achieved in two ways: (i) *splices,* which are permanent joints between two fibers (splicing is analogous to the electrical soldering of two metallic wires), and (ii) *connectors,* which are demountable joints (analogous to a plug-in-socket arrangement). The major consideration in making these connections is the optical loss associated with them. Thus prior to discussion of technique used for connecting optical fibers through splices or connectors, let us briefly review the loss mechanisms associated with fiber-to-fiber connections. Connection losses may be grouped into two categories: (i) losses due to extrinsic parameters, and (ii) losses due to intrinsic parameters.

(i) Connection Losses Due to Extrinsic Parameters

There are some factors extrinsic to the fibers that contribute to coupling losses. The important ones among these are: (i) Fresnel reflection (e.g., at glass-air-glass interfaces), (ii) longitudinal, lateral, and angular misalignment of fibers, and (iii) lack of parallelism and flatness in the end faces. We now determine the order of magnitude of joint losses due to these parameters.

Fiber Optic Cables and Connectors 207

Loss due to Fresnel reflection: When light passes from one medium to another, a part of it is reflected back into the first medium. This phenomenon is called Fresnel reflection. Therefore, even if the end faces of the fibers (to be connected) are perfectly flat and the axes of the fibers are perfectly aligned; there will be some loss at the joint due to Fresnel reflection. The magnitude of this loss be determined as follows:

Let us assume that the two fibers are identical and have a core index n, then the fraction of optical powe, R, that is reflected at the core-medium interface (for normal incidence) is given by:

$$R = \left(\frac{n_1 - n}{n_1 + n}\right)^2 \tag{5.8}$$

Therefore, the fraction of power that is transmitted by the interface will be given by:

$$T = 1 - R = 1 - \left(\frac{n_1 - n}{n_1 + n}\right)^2 = \frac{4k}{(k+1)^2} \tag{5.9}$$

where $k = n_1/n$. As there are two interfaces (glass-medium-glass) at a joint, the coupling efficiency η_F in the presence of Fresnel reflection for two compatible fibers will be given by:

$$\eta_F = \frac{4k}{(k+1)^2} \frac{4k}{(k+1)^2} = \frac{16k^2}{(k+1)^4} \tag{5.10}$$

However, if the cores of the two fibers have the same size (i.e., same core diameter) but different refractive indices, say, n_1 and n_1' then the coupling efficiency η_F' in this case will be given by:

$$\eta_F' = \frac{4k}{(k+1)^2} \frac{4k'}{(k'+1)^2} \tag{5.11}$$

where $k = n_1/n$ and $k' = n_1'/n$.

Thus the loss, in decibels (dB), at a joint due to Fresnel reflection will be given by:

$$L_F = -10 \log_{10}(\eta_F) \tag{5.12a}$$

or
$$L_F' = -10 \log_{10}(\eta_F') \tag{5.12b}$$

Typically, if $n_1 = 1.5$ and $n = 1$ (for air), $L_F = 0.36$ dB. Normally, in order to minimize such losses, index-matching fluid is used in the gap between fiber ends.

Fiber-to-fiber misalignment losses: In an optical fiber connection, the alignment of the two fibers to be connected is very important. The three types of misalignment that may occur are shown in Fig. 5.12. It is possible that there is (i) separation between the fiber ends along the common axis (end separation or longitudinal misalignment), (ii) a lateral offset between the axes of the two fibers (lateral misalignment), and (iii) an angle between the axes of the two fibers (angular misalignment). In each of these cases, the loss at a joint is determined by the optical coupling efficiency between the two fibers. It has been shown that for the three types of misalignment shown in parts (a), (b), and (c) of Fig. 5.12, the coupling efficiencies for two compatible multimode step-index fibers are given by Eqs (5.13), (5.14), and (5.15), respectively. The coupling efficiency η_{long} for a longitudinal misalignment Δx between the two fibers is given by:

$$\eta_{long} = \left[1 - \frac{\Delta x \text{NA}}{4an}\right] \quad (5.13)$$

where a is the core radius and NA is the numerical aperture of both fibers. The coupling efficiency η_{lat} for a lateral offset Δy between the axes of the two fibers is expressed as:

$$\eta_{lat} \approx \frac{2}{\pi}\left[\cos^{-1}\left(\frac{\Delta y}{2a}\right) - \left(\frac{\Delta y}{2a}\right)\left\{1 - \left(\frac{\Delta y}{2a}\right)^2\right\}^{1/2}\right] \quad (5.14)$$

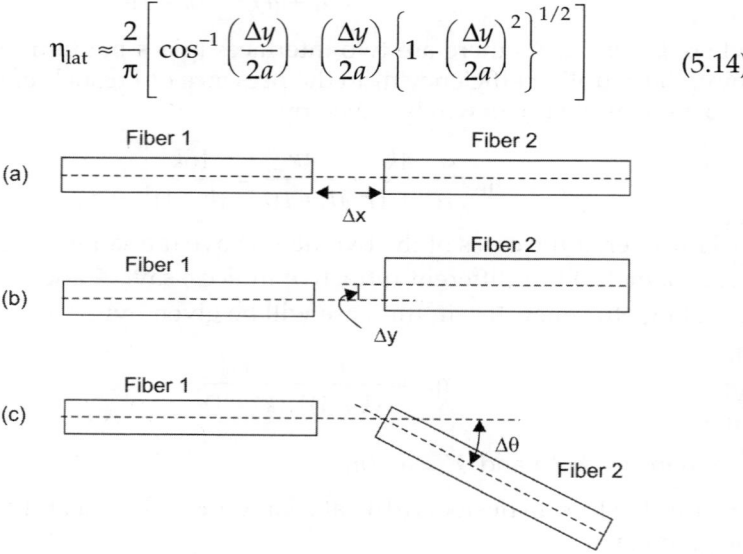

Fig. 5.12. Schematic of (a) longitudinal, (b) lateral, and (c) angular misalignment of fiber 2 with respect to fiber 1

Here $\cos^{-1}(\Delta y/2a)$ is expressed in radians. Finally, the coupling efficiency η_{ang} for an angular misalignment $\Delta\theta$ between the axes of the two fibers is given by:

$$\eta_{ang} \approx \left[1 - \frac{n\Delta\theta}{\pi \text{NA}}\right] \quad (5.15)$$

The main assumptions in deriving these formulae are as follows: (i) All the modes are uniformly excited. (ii) In Eq. (5.13), the optical wave propagating through the fiber is assumed to be expressed by a meridional ray and the gap (Δx) between the two fibers is assumed to be less than $a/\tan\phi_0$, where $\phi_0 = NA/n$. (iii) In Eq. (5.14), it has been assumed that the overlapped area between both the cores gives the coupling efficiency η_{lat} approximately, and the change in the optical ray angular component is small. This relation is valid for $0 \leq \Delta y \leq 2a$. (iv) In Eq. 5.15, the propagation angle (in radians) in the second fiber is restricted to $\Psi \leq (2\Delta)^{1/2}$, where Δ is the relative refractive index difference of the fiber.

The loss, in decibel, due to the three types of misalignment described above may be determined using the following relations. The loss due to longitudinal misalignment is given by:

$$L_{long} = -10 \log_{10}\eta_{long} \quad (5.16)$$

the loss due to lateral misalignment is given by:

$$L_{lat} = -10 \log_{10}\eta_{lat} \quad (5.17)$$

and the loss due to angular misalignment is given by:

$$L_{ang} = -10 \log_{10}\eta_{ang} \quad (5.18)$$

Losses due to other factors: The other extrinsic factors that may cause losses at a joint are related to the state of the end faces of the two fibers. For example, the end faces of the fibers may not be orthogonal with respect to the fiber axes. Further, they may not be flat. The mathematical expressions for such losses are rather difficult to arrive at. Nevertheless, such factors are taken care of during the cleaving and polishing of the fiber end faces before making a connection.

(ii) Connection Losses Due to Intrinsic Parameters

When the fibers to be connected are not compatible, i.e., they have different geometrical and/or optical parameters, then power may be lost at the joint. In this context, the following are quite important:
  (i) the core diameter,
  (ii) the numerical aperture or the relative refractive index difference, and
  (iii) the refractive index profile.

The mismatch of these parameters is illustrated in Fig. 5.13. In a multimode step-index or graded-index fiber, if we assume that all the modes are uniformly excited and that all the parameters of the two fibers are same except their diameters, then the coupling efficiency η_{cd} may be estimated by the ratio of the core areas. Thus, one finds:

$$\eta_{cd} = \begin{cases} \dfrac{\pi d_2^2/4}{\pi d_1^2/4} = \left(\dfrac{d_2}{d_1}\right)^2 & \text{for } d_2 < d_1 \\ 1 \text{ for } d_2 \geq d_1 \end{cases} \quad (5.19\text{a and b})$$

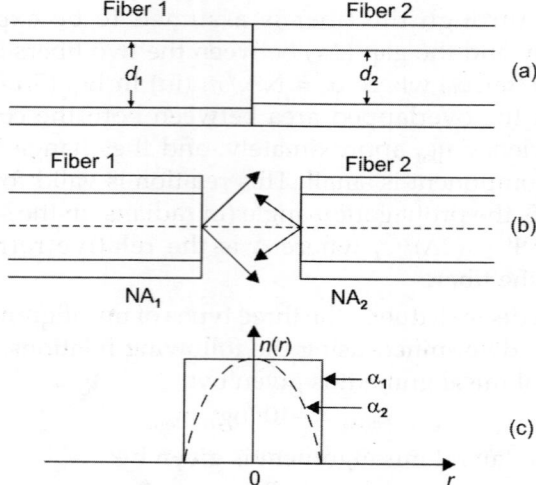

Fig. 5.13. Schematic of mismatch of intrinsic parameters: (a) core diameter, (b) numerical aperture (NA), and (c) refractive index profile

where d_1 and d_2 are the core diameters of the transmitting and receiving fibers, respectively. The corresponding loss (in dB) is obtained as:

$$L_{cd} = -10 \log_{10} \eta_{cd} \qquad (5.20)$$

Diameter discrepancies of the order of 5% can result in a loss of the order of 0.42 dB. If there is a mismatch in the numerical aperture of the two fibers, the light cone transmitted by one fiber either overfills or underfills the receiving fiber. If we assume that NA_1 and NA_2 are respectively, the numerical apertures of the transmitting and receiving fibers and all their other parameters are same, then the coupling efficiency [Fig. 5.13(b)] is given by:

$$\eta_{NA} = \begin{cases} \left(\dfrac{NA_2}{NA_1}\right)^2 & \text{for } NA_1 > NA_2 \\ 1 & \text{for } NA_2 \le d_1 \end{cases} \qquad (5.21\text{a and b})$$

The corresponding loss (in dB) is given by:

$$L_{NA} = -10 \log_{10} \eta_{NA} \qquad (5.22)$$

Further, the coupling efficiency due to a mismatch of the refractive index profiles [Fig. 5.13(c)] is given by:

$$\eta_\alpha = \begin{cases} \left(\dfrac{1 + 2/\alpha_1}{1 + 2/\alpha_2}\right)^2 & \text{for } \alpha_1 > \alpha_2 \\ 1 & \text{for } \alpha_1 \le \alpha_2 \end{cases} \qquad (5.23\text{a and b})$$

and the corresponding loss is given by:
$$L_\alpha = -10 \log_{10} \eta_\alpha \qquad (5.24)$$

where α_1 and α_2 are the profile parameters for the transmitting and receiving fibers, respectively. Thus, if the transmitting fiber has a step-index profile ($\alpha = \infty$) and the receiving fiber has a parabolic profile, and if both fibers have the same core diameter and axial NA, then, for an index-matched joint ($k = 1$), there is a loss of 3 dB. However, if the direction of light propagation is reversed, there will be no loss at this joint.

5.6 CONNECTORS AND SPLICES

The interconnection of optical components is a vital part of an optical system, having a major effect on performance. Interconnection between two fiber-optic cables is achieved by either *connectors* or *splices* which link the ends of the fiber cables optically and mechanically.

Connectors are devices used to connect a fiber-optic cable to an optical fiber device, such as a detector, optical amplifier, optical light meter, or link to another fiber cable. They are designed to be easily and reliably connected and disconnected. The connectors create an intimate contact between the mated halves to minimize the power loss across the junction. They are appropriate for indoor applications. Splices are used to permanently connect one fiber-optic cable to another. Splices are suitable for outdoor and indoor applications. Some types of splices are used to temporarily connect for quick testing purposes.

The key to a fiber-optic interconnection is precise alignment of the mated fiber cable cores so that the light couples from one fiber, across the junction, into the other fiber. This precise alignment creates a challenge for designers.

There is a difference between a connection of two fiber cables and a coupling of a light source into a fiber cable. This chapter presents the operating principles of the connectors and splices, and describes their types, properties, and operations.

5.7 APPLICATIONS OF CONNECTORS AND SPLICES

Connectors and splices make optical and mechanical connections between two fiber cables. It is easy to connect and disconnect a cable with a connector from another cable or a device. There are many applications for fiber connectors and splices in fiber systems, such as:
- Connecting between a pair of fiber cables, using connectors or a splice, is an essential part of any fiber system.
- Interfacing devices to local area networks.

- Connecting and disconnecting fiber cables to patch panels where signals can be checked and routed in a fiber system.
- Connecting and splicing may be required on short fiber cables for wiring, testing devices, connecting instruments and devices, and at other intermediate points between transmitters and receivers.
- Dividing a fiber system into subsystems, which simplifies the selection, installation, testing, and maintenance of fiber systems.
- Temporarily connecting remote mobile systems and recording equipment in many fiber systems.

5.8 REQUIREMENTS OF CONNECTORS AND SPLICES

It is very difficult to design a connector or a splice that meets all the requirements. A low-loss connector may be more expensive than a high-loss connector, or it may require relatively expensive application tooling. The lowest losses are desirable, but the other factors clearly influence the selection of the connector or splice as well.

The most desirable features for fiber connectors or splices required by customers and industry are as follows:

(i) *Low loss* (insertion and return): The connector or splice causes low loss of optical power across the junction between a pair of fiber cables.

(ii) *Easy installation and use*: The connector or splice should be easily and rapidly installed without the need for special tools or extensive training.

(iii) *Repeatability*: There should be no variation in power loss. Loss should be consistent whenever a connector is connected, disconnected and reconnected again, as many times as required.

(iv) *Economical*: The connector, splice, and special application tooling should be inexpensive.

(v) *Compatibility with the environment*: The connector or splice should be waterproof and not affected by temperature variations.

(vi) *Mechanical properties*: The connector or splice should have high mechanical strength and durability to withstand the application and tension forces.

(vii) *Long life*: The connector or splice should be built with a material that has a long life in various applications.

5.9 FIBER CONNECTORS

Fiber connectors are designed to be easily connected and disconnected. Fiber-optic cable can be easily connected to a transmitter, receiver, power, meters, or another fiber cable. The key optical parameter for a fiber-optic connector is its attenuation. Signal attenuation in connectors

Fiber Optic Cables and Connectors 213

is the sum of losses caused by several factors. The major factors are as follows:
(i) Overlap of fiber cable cores (also called lateral displacement)
(ii) Alignment of fiber axes
(iii) Fiber cable numerical aperture
(iv) Reflection at the fiber cable junction/interface
(v) Connector end polishing
(vi) Fiber cable spacing
(vii) Connector end face profiles
(viii) Insertion loss

When the diameter of the transmitting fiber cable is greater than that of the receiving fiber cable, as shown in Fig. 5.14, the diameter–mismatch loss ($Loss_{dia}$) is given by:

$$Loss_{dia}\ 10\ \log_{10}\ \frac{(dia_t^2 - dia_r^2)}{dia_t^2} \tag{5.25}$$

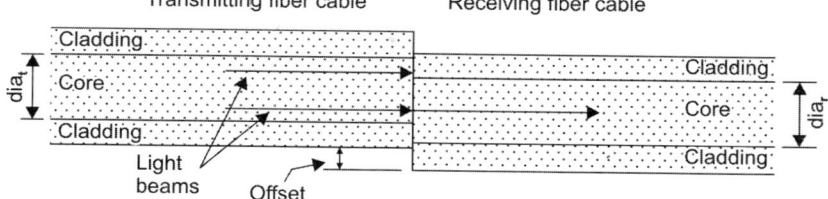

Fig. 5.14. Overlap of fiber cable cores

When the numerical aperture NA of the transmitting fiber cable is greater than that of the receiving fiber cable, as shown in Fig. 5.15, the NA–mismatch loss ($Loss_{NA}$) is given by:

$$Loss_{NA}\ 10\ \log_{10}\left(\frac{NA_r}{NA_t}\right)^2 \tag{5.26}$$

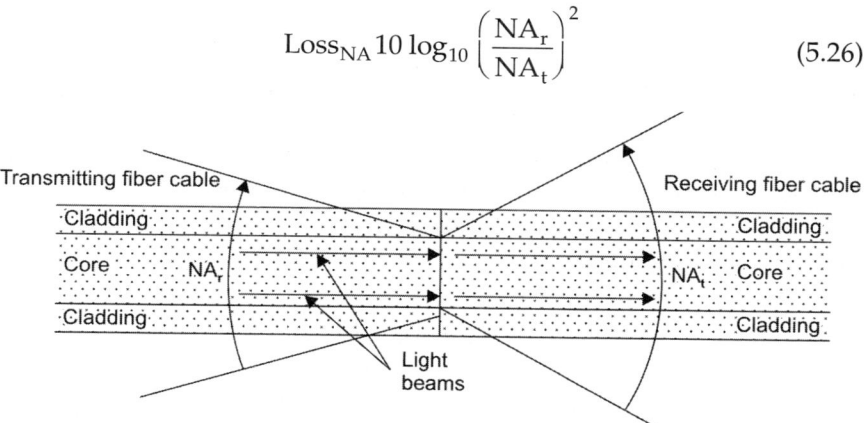

Fig. 5.15. Numerical aperture (NA)–mismatch loss

214 Photonics | Optoelectronics

The formula for the loss due to end separation ($Loss_{Seperation}$) between two fiber-optic cables (at separation distance) is rather involved. Assume that the transmitting and receiving fibers are identical. Figure 5.16 illustrates the separation (some times called air gap) between a pair of fiber cables. The formula for end separation loss ($Loss_{Seperation}$) is given by:

$$Loss_{Separation} = 10 \log_{10} \left(\frac{\frac{d}{2}}{\frac{d}{2} + S \tan \left(\arcsin \left(\frac{NA}{n_0} \right) \right)} \right) \quad (5.27)$$

where, d is core diameter, S is the fiber spacing, NA is the numerical aperture, n_0 is the refractive index of the material between the two fiber cables.

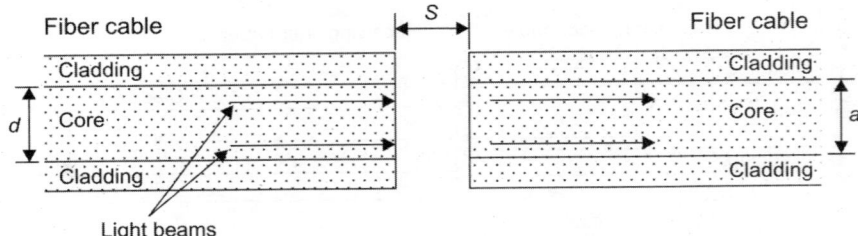

Fig. 5.16. End separation loss

A material known as index-matching fluid or gel applied between the two fiber cables reduces the reflection loss between the surfaces of the fiber cable ends. This loss, called *Fresnel reflection loss*, generally occurs between parallel optical surfaces. Most mechanical splices also use an index-matching gel to fill the gap between the connected fiber cable ends. An antireflection coating can also be applied to reduce this loss.

Additional losses may be experienced when two different types of fiber cable connectors are connected using an adapter.

Insertion loss is a measure of the performance of a connector or splice. Insertion loss is calculated by:

$$\text{Loss (dB)} = 10 \log_{10} \left(\frac{P_2}{P_1} \right) \quad (5.28)$$

where P_1 is the initial power measured and P_2 is the power measured after the connector has been mated.

5.10 MECHANICAL CONSIDERATIONS

The optical characteristics of the fiber connectors are significant. However, the mechanical characteristics are also important, and in some cases critical. Virtually all fiber connectors are designed to remain in place under working conditions. Connectors must withstand physical stresses, such as forces encountered during mating and unmating, and sudden stress induced by bending and tension. Connectors must also prevent contamination caused by dirt and moisture in the fiber cable ends.

5.10.1 Durability

Durability is a concern with any type of connector. Repeated mating and unmating of the fiber connectors can cause wear in the mechanical components. Allowing dirt into the optics and straining the fiber cable will damage the exposed fiber cable ends.

5.10.2 Environmental Consideration

Fiber connectors designed for indoor applications must be protected from environmental extremes to avoid excessive connector loss and poor system performance. Special *hermetically-sealed* connectors are required for outdoor use.

5.10.3 Compatibility

Compatibility refers to the need for the connector to be compatible with other connectors or with specifications. Specifications describe the type of connector to be used in specific applications. Compatibility exists on several levels the most basic level being physical compatibility. The connector must meet certain dimensional requirements to allow it to mate with other connectors of the same style. The second level of compatibility involves connector performance, such as insertion loss, durability, operating temperature range, and other requirements specified by the customers.

5.11 FIBER-OPTIC CONNECTOR TYPES

Figure 5.17 shows the most common types of fiber-optic connectors. Fiber connectors are unique in that they must make both optical and mechanical connections. They must also allow the fiber cables to be precisely aligned to ensure that a connection is robust. Fiber connectors use various methods to achieve solid connections. Some of the types of fiber-optic connectors currently in use are listed below:
- Subscriber Connectors (SC)
- SC/APC Connectors
- FC/PC Connectors
- FC/APC Connectors

216 Photonics | Optoelectronics

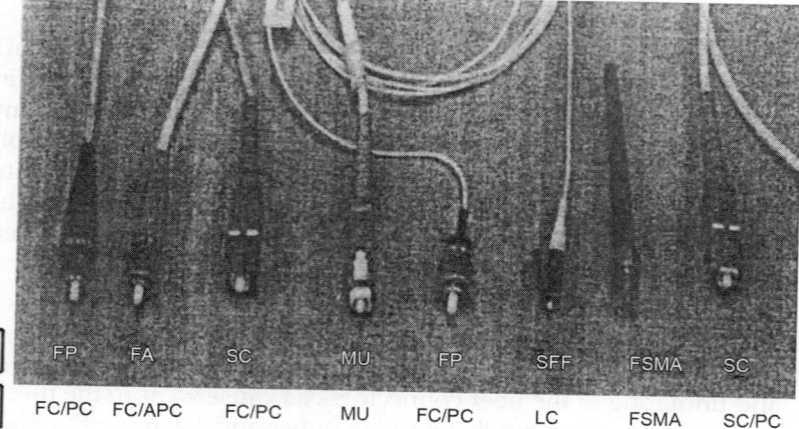

Fig. 5.17. Fiber-optic connector types

- LC Connectors
- MU Connectors
- Straight-tip Connectors (ST)
- 5685C Connectors (duplex SC)
- FDDI Connectors (MIC)
- Biconic Connectors
- SMA Connectors
- Enterprise System Connection (ESCON)
- Duplex Connectors (ST)
- Polarizing Connectors
- MT Multifiber Connectors
- MT-RJ Connectors
- D4-style Connectors
- Biconic Connectors
- MFS/MPO Connectors
- Plastic-Fiber Connectors
- E-2000 Diamond
- Fiber-Optic Connectors Self-Latch in Push/Pull System
- Special Connectors

5.12 ADAPTERS FOR DIFFERENT FIBER-OPTIC CONNECTOR TYPES

An adapter is a passive device used to join two different types of connectors together. The type of adapter is identified by a nomenclature, such as SC, FC, ST, or 568SC. Hybrid adapters join dissimilar connectors together, such as SC to FA. Figure 5.18 shows examples of some adapters.

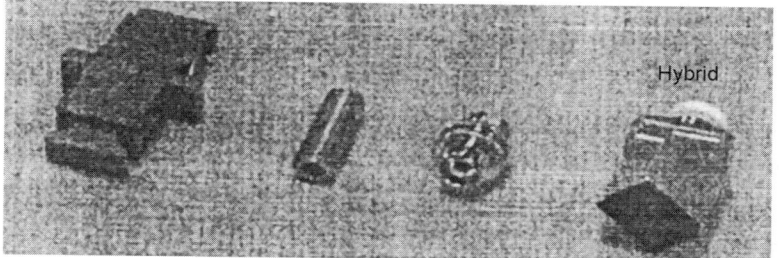

Fig. 5.18. Fiber-optic adapter types

5.13 FIBER-OPTIC CONNECTOR STRUCTURES

Most fiber-optic connectors are built from a ferrule, a connector body, an epoxy material, and a strain relief boot. Most connectors use a ferrule to hold the fiber and provide alignment. The most popular ferrule size is a 2.5 mm diameter, which is standard. Manufacturers offer a few types of ferrules made from different materials, such as ceramic, plastic, and stainless steel.

Connectors may be attached to a device-outlet box or adapter—by direct connection, by coupling a threaded nut, or by twisting a spring-loaded bayonet socket. The connector body is made from steel, ceramic, or plastic. Epoxy is usually applied to secure the fiber cable end in the connector body. A strain relief boot made from plastic or rubber is used at the junction between the connector body and the fiber cable.

5.14 FIBER-OPTIC CONNECTOR ASSEMBLY TECHNIQUES

We now present common assembly techniques that are used in building fiber-optic connectors.

5.14.1 Common Fiber Connector Assembly

The most common fiber connector assembly techniques use a fiber cable and a suitable connector. The fiber cable is most often epoxied into the connector. Epoxy provides good tensile strength to the connector to prevent the fiber cable from moving within the connector body, maintaining a good alignment. After the epoxy cures, the ferrule end is polished to a smooth finish by one of the many available procedures. Then the connector undergoes many inspections and test procedures to issue a data sheet for the customer.

5.14.2 Hot-Melt Connector

The hot-melt connectors use preloaded epoxy so that external mixing is not required. The prepared end of the fiber-optic cable is inserted into the connector ferrule. The cable (with the connector inserted) is loaded onto the connector holder and placed in an oven for a few minutes, which softens the epoxy around the fiber cable and cures the epoxy at the same time. The curing time is dependent on the type of epoxy. The end of the connector is then polished to a smooth finish. The polishing can be done by hand or by an industrial polishing machine. When such connectors are assembled in the field, a portable hand polisher is used.

5.14.3 Epoxyless Connector

Epoxyless connectors, also called crimp connectors, have been widely used for quick cable connections in telecommunication systems. When the connector is crimped, an insert compresses around the fiber cable. A front clamp on the bare fiber cable and a rear compression clamp add a higher clamping force on the fiber able buffer coating to provide the necessary tensile strength. Special gripping tools are used in the assembly of the epoxyless connectors. The end of the fiber connector is polished to a smooth finish using a portable hand polisher before the connector is assembled in the field. The main advantage of an epoxyless connector is the speed of assembly. Some customers will tolerate a slightly higher loss to achieve a fast, easy termination. The epoxyless approach is a technology that is not limited to one connector type.

5.14.4 Butt-Jointed Connectors

Butt-jointed connectors are based on the principle of aligning the two fiber ends and keeping them in close proximity (i.e., butted to each other). For this purpose, the plug-in-socket configuration shown in Fig. 5.19 is normally employed.

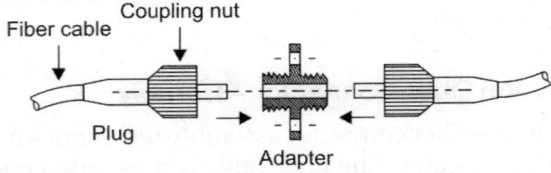

Fig. 5.19. A plug-adapter-plug configuration

The mechanical connection between the plug and the adapter on both the ends is made with the help of either threaded nuts or bayonet locks. Some connectors employ standard BNC or SMA configurations. The design of connectors differs mainly in the technique of aligning fiber ends. The simplest connector design is shown in Fig. 5.20.

It consists of metal plug (normally called ferrules), which are precision-drilled along the central axis. The prepared fiber ends (to be connected) are placed in these holes. They are then permanently bonded to the ferrules by an epoxy resin. A spring retains the ferrule in its

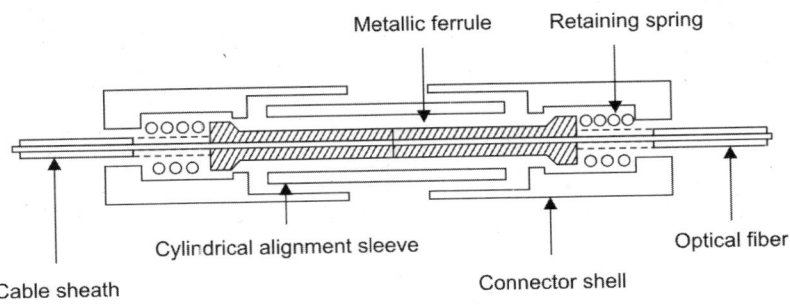

Fig. 5.20. The basic ferrule connector

position. The two opposite ferrules are aligned by a coaxial cylindrical alignment sleeve.

Another plug-adapter-plug design is shown in Fig. 5.21. Instead of metal ferrules, it employs ceramic capillary ferrules. Ceramic has better thermal, mechanical, and chemical resistance than metal or plastic.

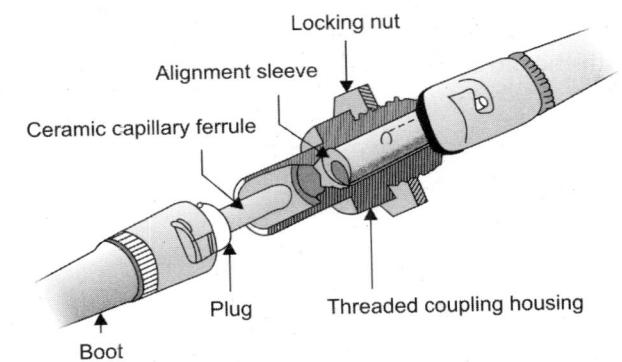

Fig. 5.21. Typical connector design employing ceramic ferrules

5.14.5 Expanded-Beam Connectors

An alternative design of connectors is based on expanded-beam coupling, illustrated in Fig. 5.22.

This technique uses two microlenses for collimating and refocusing light from one fiber end to another. As the beam diameter is expanded, the requirement of lateral alignment of the two plugs in an adapter becomes less critical as compared to butt-jointed connectors. Fresnel reflection losses may increase in this case but are normally reduced with the help of antireflection coating on the lenses.

5.14.6 Multifiber Connectors

In order to couple a number of fibers from two multifiber cables, multiple connectors are normally used. High-precision grooved silicon chips are employed to position fiber arrays. One chip can accommodate

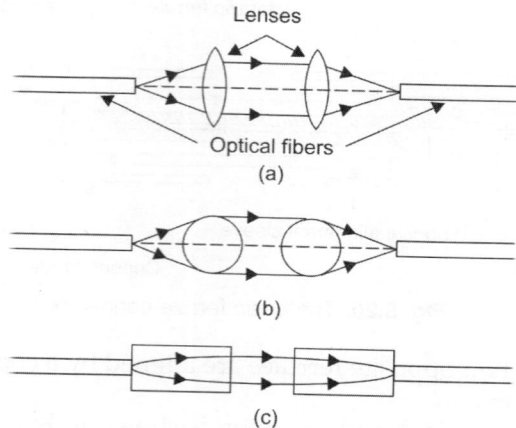

Fig. 5.22. Expanded-beam coupling using (a) a convex microlens, (b) a spherical microlens, and (c) GRIN rod lenses

12 fibers, and it is possible to stack many such chips. This structure is secured with the aid of spring clips and metal-backed chips as shown in Fig. 5.23.

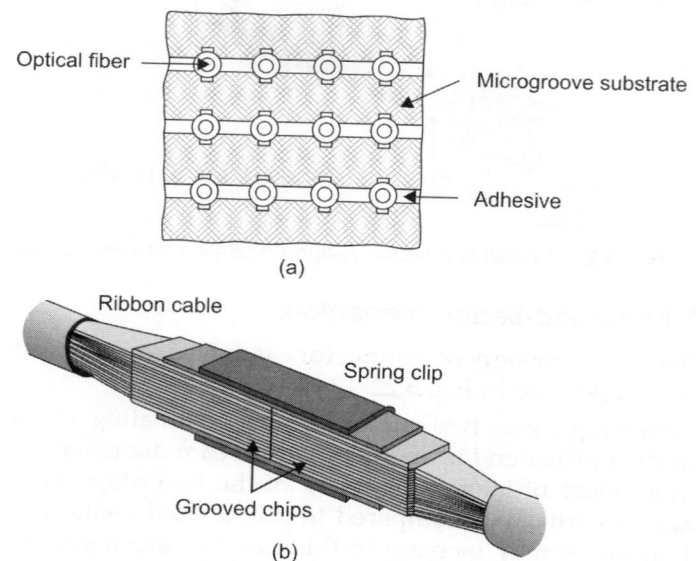

Fig. 5.23. Multiple connector: (a) cross-section of a grooved chip connector, (b) grooved chip assembly

5.14.7 Automated Polishing

All fiber-optic connector-polishing machines are designed for accuracy, easy setup, and production efficiency. The polishing pressure, speed, and duration can be adjusted to meet exact requirements. These machines precisely polish the ends of fiber-optic connectors in a repeatable and reliable manner. Polishing machines are available for dry or wet polishing process.

5.14.8 Fluid Jet Polishing

Fluid jet polishing (FJP) is another technique for shaping and polishing small surface areas of complicated optics made of brittle materials. This technique uses a fluid jet system to guide pre-mixed slurry, at low pressure, onto the optical surface being machined. The surface is altered by the erosive effect of the abrasive particles in the stream.

5.14.9 Fiber-Optic Connector Cleaning

Contamination of connector ends can occur from something as simple as dust particles or fingerprints that can reduce light propagation through the fiber cable. This will degrade device performance, causing data error and loss. To avoid this, it is common practice to clean fiber connectors prior to assembly and testing.

There are three major components of the fiber-optic connector system that users must consider when cleaning, mating, and testing fiber-optic connectors: the adapter split sleeve, the outer diameter of the ferrule, and the tip of the ferrule. There are many techniques for cleaning connectors, either wet or dry, by hand with recommended cleaning chemicals, or with automated machines. Follow the cleaning procedure for each fiber connector type. Do not use the same procedure for other types of connectors. Cleaning standards for fiber-optic connectors promise savings in time and cost.

5.14.10 Connector Testing

There are many testing instruments available for testing connectors. Testing instruments range from a simple view scope to a sophisticated system. The condition of the end of the ferrule after the polishing process is usually inspected using simple instruments, as shown in Fig. 5.24. This procedure is adequate for inspecting a connector build and polished in the field.

Handheld devices can measure the losses, optical powers, light sources, etc. The basic test measures the attenuation of the fiber cable with connectors by comparing the power through the fiber cable to that of a known reference fiber cable. The power through the fiber cable under test is measured in absolute units. The power through the reference fiber cable is also measured. Fig. 5.25 illustrates a connector test set-up.

Using sophisticated systems for testing connectors saved time and cost in industrial production. These systems are very accurate. Each connector type refers to a standard test.

222 Photonics | Optoelectronics

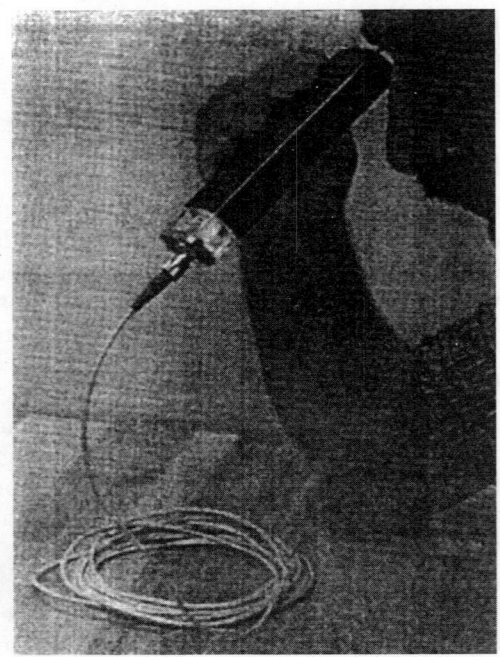

Fig. 5.24. Connector end inspection

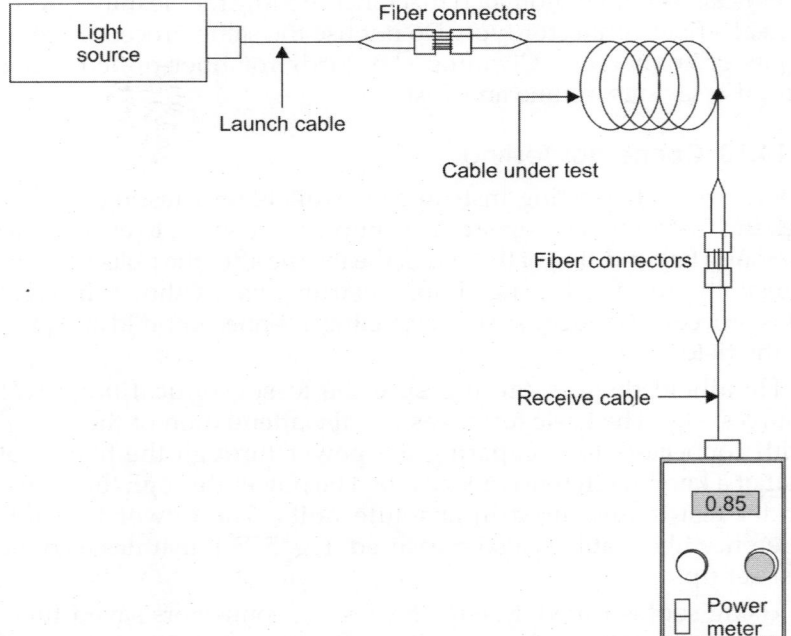

Fig. 5.25. Connector testing set-up

5.15 FIBER SPLICING

The splicing process joins fibre-optic cable permanently. In general, a splice has a lower loss than a connector. Splices are typically used to join lengths of cable for outside applications. Splices may be incorporated into lengths of fiber-optic cable or housed in indoor/outdoor splice boxes, whereas connectors are typically found in patch panels or attached to equipment at fiber cable interfaces. Splicing is required (i) when the length of the system span is more than the manufactured cable length, and (ii) when the cable is broken and needs to be repaired. The primary objective of splicing is to establish transmission continuity in the fiber-optic link. This can be done in two ways, namely, through (i) *fusion splices*, or (ii) *mechanical splices*.

In order to achieve a low-loss splice, it is essential for the fiber ends (to be joined) to be smooth, flat, and perpendicular to the core axes. This is normally achieved using a cleaving tool (a blade of hard metal or diamond). The technique is called '*scribe and break*' or '*score and break*'. It involves scoring the fiber under tension with a cleaving tool, as shown in Fig. 5.26. This generates a crack in the fiber surface that propagates in the transverse direction and a flat end is produced.

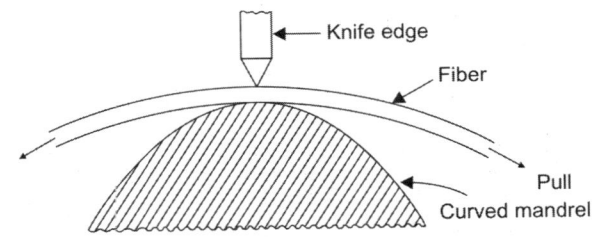

Fig. 5.26. 'Score and break' technique of cleaving optical fibers

5.15.1 FUSION SPLICES

Fusion splice may be made by thermally bonding together prepared fiber ends as shown in Fig. 5.27.

Herein, the prepared fiber ends are placed in a precision alignment jig. The alignment is done with the help of an inspection microscope (not shown). After the initial setting, a short arc discharge is applied to 'fire polish' the fiber ends. This removes any defects due to imperfect cleaving. In the final step, the two ends are pressed together and fused with a stronger arc, thus producing a fusion splice. A possible drawback of such a splicing mechanism is that the heat produced by the welding arc may weaken the fiber in the vicinity of the splice.

Summarising, we have:
(i) Fusion splicing involves butting two cleaned fiber end faces and heating them until they melt together or fuse.

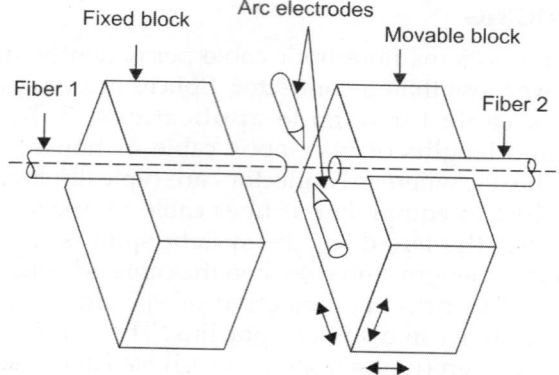

Fig. 5.27. Schematic of fusion splicing apparatus

(ii) Fusion splicing is normally done with a fusion splice that controls the alignment of the two fibers to keep losses as low as 0.05 dB.

(iii) Fiber ends are first prealigned and butted together under a microscope with micromainpulators. The butted joint is heated with electric arc or laser pulse or melt the fiber ends so can be bonded together.

5.15.2 Mechanical Splices/V-Groove

Mechanical splices join two fiber cable ends together, both optically and mechanically by clamping them within a common structure. In general, mechanical splicing requires less expensive equipment; however, higher consumable costs are experienced. A few important types of mechanical splices are listed below:

- Table-type splices
- Key lock splices
- Fiber lock splices
- Twist lock splices
- Fastomeric splices
- Capillary splices
- Rotary or polished-ferrule splices
- V-groove splices
- Elastomeric splices
- Finger splices
- Inner lock splices

Many other types are available. These normally use appropriate fixtures for aligning the fibers and holding them together. A popular technique, known as the snug tube splice, uses a glass or ceramic capillary with an inner diameter just large enough to accommodate the optical fibers (Fig. 5.28). The prepared fiber ends are gently inserted into the capillary and a transparent adhesive (e.g., epoxy resin) is injected through a transverse hole. The adhesive ensures both mechanical bonding and index-matching. A stable low-loss splice may be obtained in this way but it poses stringent limits on the capillary diameters.

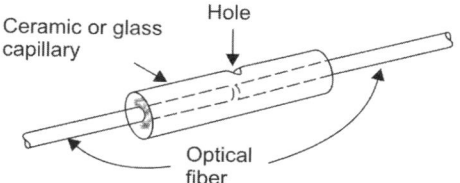

Fig. 5.28. Capillary splicing technique

A slightly different technique uses an oversized metallic capillary of square cross-section (Fig. 5.29). The capillary is first filled with the transparent adhesive, after which the prepared fiber ends are inserted into it. The two fiber ends are forced against one of the four inner corners of the capillary.

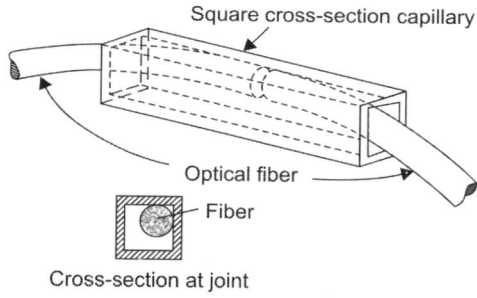

Fig. 5.29. Loose tube splicing technique

Other techniques of mechanical splicing normally employ V-grooves for securing optical fibers. The simplest technique uses an open V-goove, into which the prepared fiber ends are placed as shown in Fig. 5.30. The splice is accomplished with the aid of epoxy resin.

It is also possible to obtain a suitable groove by placing two precision pins (of appropriate diameter) close to each other. The fibers may then

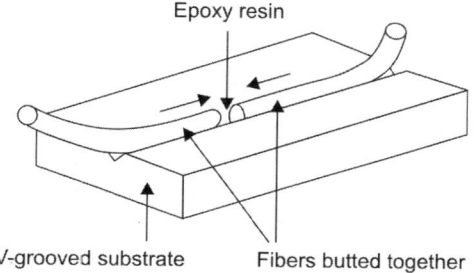

Fig. 5.30. V-groove splicing technique

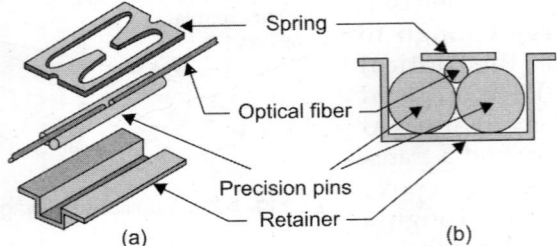

Fig. 5.31. Spring-groove splicing technique: (a) exploded view illustrating the spring, fibers on pins, and retainer; (b) cross-sectional view

be placed in the cusp as shown in Fig. 5.31. A transparent adhesive ensures bonding as well as index-matching, and a flat spring on the top applies pressure ensuring that fibers remain in their positions. Such a groove is called a *spring groove*.

There is yet another technique that utilizes the V-groove principle to realize what is known as an elastomeric splice (Fig. 5.32). In this method, the prepared fiber ends are sandwiched between two elastomeric internal parts, one of which contains a V-groove. An outer sleeve holds these two parts compressed so that the fibers are held tightly in alignment. Index-matching gel is employed to improve its performance. Originally, the technique was developed for coupling multimode fibers, but it can also be used for single-mode fibers as well as fibers with different core diameters.

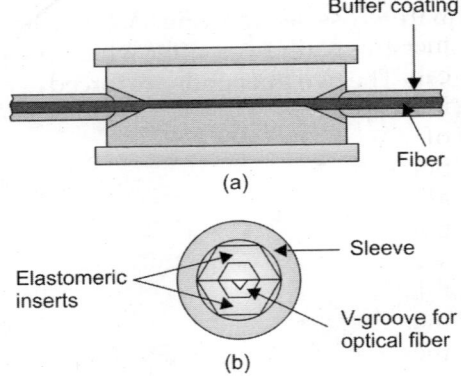

Fig. 5.32. An elastomeric splice: (a) longitudinal section, (b) cross-section

Splicing with most of these techniques, if properly carried out, results in splice loss of about 0.1 dB for multimode fibers. Some of these can also be used for splicing single-mode fibers.

Key-Lock Mechanical Fiber-Optic Splices

Key-lock mechanical fiber optic splices are commonly used to quickly mate and unmate fiber optic cables. It is made for a U-shaped metal part covered by a transparent plastic body with two holes on each end. The prepared ends of the fiber cables are made longer than half of the length of the metal part. The fiber cable is inserted in the center hole. When the key is inserted in the second hole towards the edge of the splice and turned by 90°, the metal part opens and one fiber cable end can be easily inserted. This operation can be repeated on the other side to insert the second fiber cable. This type of splice provides a quick and easy way of joining two fiber cables with low signal loss. It may be used to temporarily or permanently connect fiber cables, wavelength division multiplexing components, and other fiber-optic elements.

Table-Type Mechanical Fiber-Optic Splices

A custom-made mechanical splice, used for quick mating and unmating of connections. This splice works like any other mechanical splice. The fiber cable ends are prepared and inserted into the mid-point of the block assembly. Screws are tightened to align the fiber cables on

Fig. 5.33a. Table-type mechanical fiber-optic splice

both sides. L-clamps and K-clamps (Fig. 5.33a) are placed in position to secure the fiber cables on both sides. Most fiber-optic companies use this kind of mechanical splice for quickly mating and unmating during manu-facturing and testing processes. The splice loss associated with these instruments is acceptable by industry standards.

5.15.3 Multiple Splices

For ribbon cables containing linear arrays of fibers, the following technique has been used. In this method (Fig. 5.33b), the fiber ends are individually prepared, and then placed in a grooved substrate. Adhesive is then used for bonding and

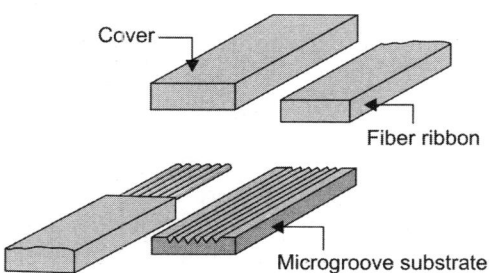

Fig. 5.33b. Multiple splicing technique

index-matching. A cover plate retains the fibers in their position and also maintains mechanical stability.

5.16 CONNECTORS VERSES SPLICES

There are definite differences between connectors and splices. Most companies make connectors and splices to satisfy customer requirements for smaller size and lower loss. As mentioned in this chapter, fiber-optic technology is moving forward to create high-durability connectors and splices with small sizes and low cost in production and installation. Table 5.2 compares general important factors between the fiber connectors and splices.

Table 5.2. Connectors versus splices

Connectors	Splices
Provide temporary connections	Provide permanent connections
Higher loss	Lowest loss
Larger sizes	Smaller sizes
Immune, or not immune, to environmental effects (depends on the connector type)	Immune to environmental effects
It takes a long time to build a connector	It takes a very short time to build a splice
Diverse applications	Connection between a pair of fiber cables
Many types	Few types
New technology reduces installation time	Conventional technology keeps the same installation time
Building reasonable mechanical stability at the connection points	Building better mechanical stability at the connection points

5.17 CHARACTERIZATION OF OPTICAL FIBERS

There are various parameters of the optical fibers, which have to study for the evaluation of the performance of an optical fiber, e.g. optical attenuation, dispersion, numerical aperture, and refractive index profile. There are number of methods for measuring each of these parameters. We will discuss only a few of them.

5.17.1 Measurement of Optical Attenuation

Attenuation of a light signal as it propagates along a fiber is an important consideration in the design of an optical communication system, since it plays a major role in determining the maximum transmission distance between transmitter and a receiver or an in-line amplifier. The basic attenuation mechanisms in a fiber are absorption, scattering and radiative losses. There are three types of techniques developed to measure (i) total attenuation, (ii) absorption loss, and (iii) scattering loss. The total attenuation is of interest to the system designer, whereas the contribution to this total attenuation by the absorption loss and scattering loss mechanisms is of interest in the development of low-loss optical fibers.

Cutback or differential technique method is used for measuring total fiber attenuation. This method is based on the following principle. Power P_0 is launched at one end of a long length L_1 of the test fiber; the power P_1 is received at the other end and is measured. Then, the fiber is cut back to a smaller length L_2 and the power P_2 received at the other end is measured again. Assuming wavelength λ remains same, the optical attenuation per unit length α (dB/km) may be expressed by the following relation:

$$\alpha = \frac{10}{L_1 - L_2} \log_{10} \left(\frac{P_2}{P_1} \right) \qquad (5.29)$$

We now discuss the criterion for designing the equipment for studying this parameter by the cutback method.

First, a polychromatic continuous source of radiation (containing sufficient power at all the wavelengths of interest) is required. As the attenuation is to be studied for all wavelengths, a wavelength-isolation device (e.g. a monochromator) is required to follow the source. Then suitable optics has to be designed for launching the optical power at one end of the test fiber. At the other end of it, again suitable optics is required so that most of the power transmitted by the fiber is received by the detector. The detector signal should be processed and then output read on a meter or recorded on a chart recorder. Accordingly, the modules may be arranged as shown in Fig. 5.34.

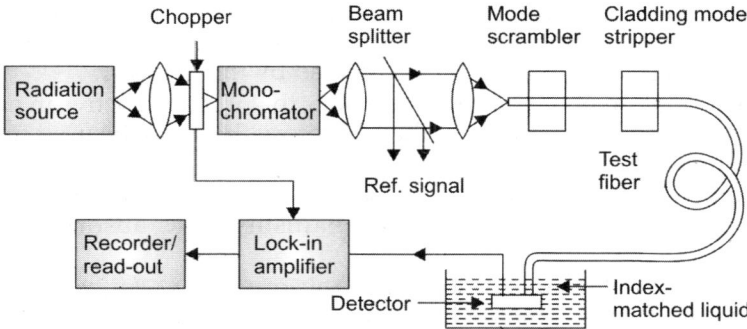

Fig. 5.34. Experimental arrangement for the measurement of total attenuation

One can understand the importance of each of these modules by investigating the dependence of the signal S developed by the detector on the pertinent variables:

$$S = P(\lambda)\, M(\theta, \lambda)\, (n_a \sin \alpha_m)\, T(\alpha_{ab}, \alpha_{sc}, L)\, D(\lambda) \qquad (5.30)$$

where $P(\lambda)$ is the power furnished by the source at a specific wavelength λ; $M(\theta, \lambda)$ is a function governing the solid angle θ seen by the monochromator and its transmittance with wavelength λ; $n_a \sin \alpha_m$ gives the numerical aperture of the fiber, n_a is the refractive index of the medium surrounding the launching end of the fiber; $T(\alpha_{ab}, \alpha_{sc}, L)$ is a function determining the transmittance of the fiber. This is dependent on the absorption loss per unit length α_{ab}, the scattering loss per unit length λ_{sc}, the total length L of the fiber, and $D(\lambda)$ is the responsivity of the detector as a function of wavelength λ.

Clearly, the source should have high radiance and be continuous. A black body radiator, e.g., a tungsten halogen lamp or a high-pressure discharge lamp, e.g., a xenon arc lamp may be used. A monochromator should collect as much as possible. The components in the monochromator should have high transmittance in the regions of investigation. If a grating is used as a monochromator, overlapping orders may cause problems and hence, an order sorting filter may also be required at the exit slit of the monochromator. In order to improve the S/N ratio, the signal from the source is generally chopped at a low frequency and at the receiver end a lock-in amplifier is used to perform phase-sensitive detection.

A beam splitter is placed as shown in Fig. 5.34 for obtaining a reference signal as well as for viewing the optics. If viewing the optics is not required, a rotating sector mirror may be used in place of a beam splitter and the chopper in between the source and the monochromator may be omitted. This will provide a greater energy throughput to the optical fiber as well as a reference signal for comparison. A mode scrambler has been used to obtain equilibrium mode distribution. The fiber is also put through a cladding mode stripper, which is a device for removing the light launched into the fiber cladding. At the receiver end, the optical power is detected using either a *p-i-n* or an avalanche photodiode. The other end of the fiber terminates in an index-matched liquid so that most of the light is received by the detector.

The limitation of the cutback method is that it is destructive in nature and hence, can be used only in the laboratory. It cannot be used in field measurements.

The question arises, how does one isolate the contribution to total attenuation by the major loss mechanisms (e.g., absorption and scattering)?

In order to determine the loss due to absorption, a *calorimetric method* may be used. In this method, two similar fibers are taken and light is launched through one of them (Fig. 5.35). Absorption of light (of specific wavelength) by the bulk material of the test fiber raises the temperature of the fiber, which can be measured using a thermocouple. The rise in temperature may then be related to the absorption loss.

The power loss due to scattering alone may be measured with the help of cell (Fig. 5.36). Light from a powerful source is launched into

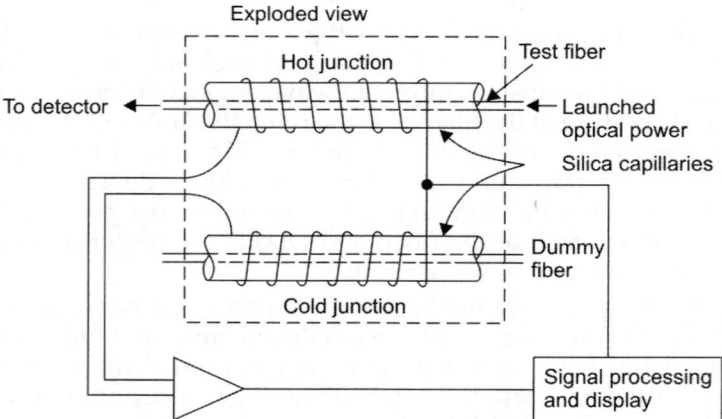

Fig. 5.35. Schematic of measurement of the temperature of an optical fiber using a thermocouple

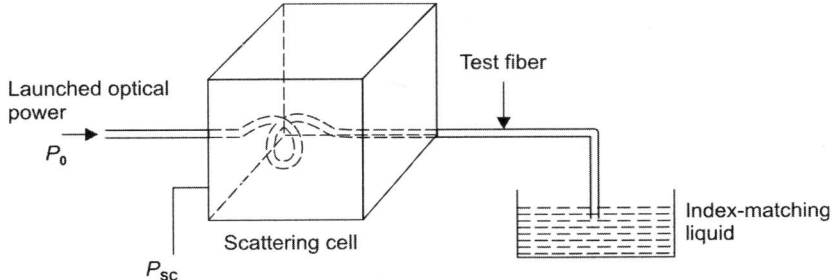

Fig. 5.36. Schematic of experimental set-up for measurement of scattering loss

the optical fiber through appropriate launch optics. A certain length (say, L) of the fiber is enclosed inside the scattering cell. All the six inner surfaces of the cell are fitted with six photovoltaic detectors. These detectors measure the optical power (P_{sc}) scattered by the enclosed length L of the fiber. The scattering loss α_{sc} (dB/km) may be expressed as:

$$\alpha_{sc} = \frac{10}{L \text{ (km)}} \log_{10}\left(\frac{P_0}{P_0 - P_{sc}}\right) \text{ dB/km} \qquad (5.31)$$

where P_0 is the power launched.

5.17.2 Measurement of Dispersion

Distortion of optical signals propagating down an optical fiber is caused due to two mechanisms and this causes to limit the information-carrying capacity of the fiber. These are *intermodal* and *intramodal* dispersions.

One can study the dispersion effects by measuring the impulse response of the fiber in the time domain or by measuring the baseband frequency response in the frequency domain. A general method used for measuring the pulse distortion in optical fibers in the time domain is shown in Fig. 5.37. Short-duration pulses (of the order of a few hundred picoseconds) are launched at the one end of the optical fiber from a pulsed laser. As the pulses propagate through the fiber, they get broadened due to the various dispersion mechanisms. At the other end, these pulses are received by a high-speed photodetector [e.g., an avalanche photodiode (ADP)], and the detector signal is displayed on the cathode ray oscilloscope (CRO). A reference signal is utilized for triggering the CRO and also for measuring the input pulse. If τ_i and τ_o are the half-widths of the input and output pulses respectively and if the shape of the pulses is assumed to be Gaussian, then the impulse response of the fiber is given by:

$$\tau = \frac{(\tau_o^2 - \tau_i^2)^{1/2}}{L} \text{ (say, ns/km)} \qquad (5.32)$$

where L is the length of the optical fiber.

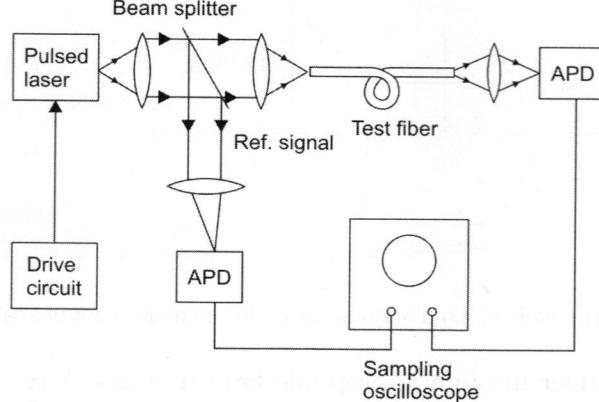

Fig. 5.37. Schematic of experimental set-up for the measurement of intermodal dispersion

In order to evaluate the the bandwidth of the fiber, measurements in the frequency domain are required. In this case, the apparatus is almost the same (Fig. 5.37), except that a sampling oscilloscope is replaced by a spectrum analyser. The latter takes the Fourier transform of the output pulse in the time domain and displays its constituent frequency components. For the measurement of the intramodal or chromatic dispersion, a polychromatic source is required in place of a laser.

5.17.3 Measurement of Numerical Aperture (NA)

NA is directly related to the light-gathering capacity of the fiber and is an important characterizing parameter. NA decides the number of modes propagating through the multimode fiber. For a step index fiber, NA is given by:

$$\text{NA} = n_a \sin \alpha_m = (n_1^2 - n_2^2)^{1/2} \qquad (5.33)$$

where α_m is the angle of acceptance, n_a is the refractive index of the medium in which the fiber is placed, and n_1 and n_2 are the refractive indices of the core and cladding, respectively. For a graded-index fiber, NA is not constant but varies with the distance r from the core axis. The local NA at a radial distance r is given by:

$$\text{NA}(r) = n_a \sin \alpha_m(r) = [n_1^2(r) - n_2^2]^{1/2} \qquad (5.34)$$

From Eqs (5.33) and (5.34), it becomes clear that the NA can be calculated if the refractive index profile of the fiber is known. However, this method is seldom used.

A general method, shown in Fig. 5.38 involves the measurement of the far-field pattern of the fiber. Light from a powerful source such as a laser is launched at one end of the fiber. The other end is held in the chuck of the fiber holder on a rotating stage. As the tip of the fiber is

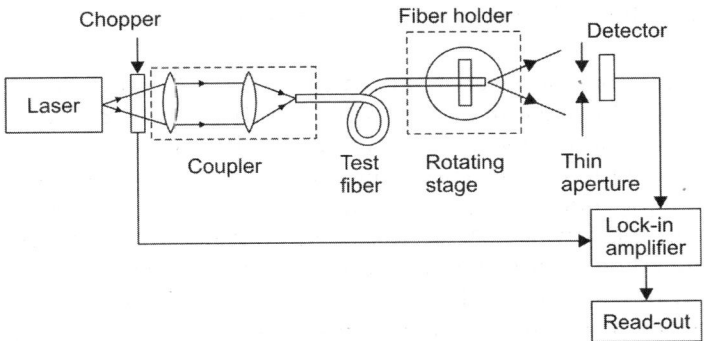

Fig. 5.38. Schematic of experimental set-up for evaluating the NA of an optical fiber

rotated, the intensity of light reaching the detector falls off on either side and an approximately Gaussian curve results. The angle at which the intensity falls to 5% of its maximum gives the value of α_m.

Alternatively, the light from the laser source may be made to fall a different angles on one end of the fiber and the output at the other end may be measured with the help of a detector. Again, an approximately Gaussian curve results when the output power is plotted as a function of the angle of rotation. The angle for which the power falls to 5% of its maximum gives us the value of α_m.

5.17.4 Measurement of Refractive Index (RI) Profile

RI profile of an optical fiber plays a significant role in its characterization. One can determine the NA of the fiber and the number of guided modes propagating within the fiber with the knowledge of RI profile. RI profile of the fiber also enable one to predict the response of the impulse and hence the information-carrying capacity of the optical fiber.

There are several methods to measure the RI profile. We discuss the *end-reflection method*. The method is based on the principle that when a focused beam of light is incident normally on the flat end face of a fiber, a part of the light is reflected back. The fraction R of the light reflected at the fiber-medium interface is given by the Fresnel reflection coefficient given by:

$$R = P_r / P_i = \left(\frac{n_1 - n}{n_1 + n}\right)^2 \quad (5.35)$$

where P_r and P_i are the reflected and incident powers respectively, n_1 is the RI at the striking point of the fiber, and n is the RI of the medium surrounding the fiber. For a small variation in the value of n_1, we have:

$$\Delta R = 4n \left\{\frac{n_1 - n}{(n_1 + n)^3}\right\} \Delta n_1 \quad (5.36)$$

Clearly, the variation in the reflected light intensity can be used to calculate the RI.

The experimental set-up is shown in Fig. 5.39. A highly focused laser beam is used to measure the RI profile. The beam is first modulated by the chopper and purified by passing through a polarizer and quarter-wave plate combination. This combination also decouples the incident light from the reflected light. This light is then focused on the polished flat end of the test fiber. The other end of the fiber is dipped into an index-matching liquid so that light is not reflected back from this end. The light reflected from the flat end of the fiber is directed onto the detector via the beam splitter. The modulated output of the detector is amplified and recorded on the recorder.

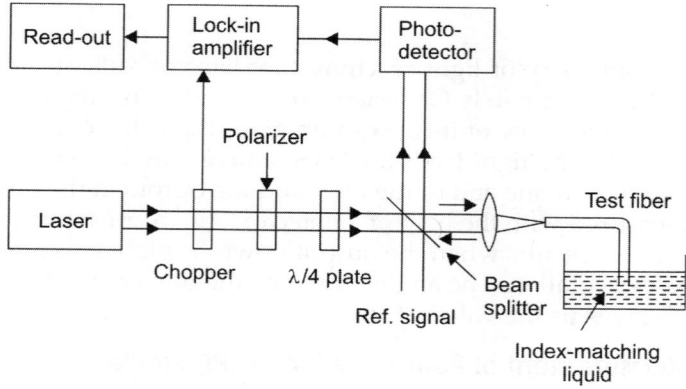

Fig. 5.39. Schematic of experimental set-up for studying the RI profile of an optical fiber

5.17.5 Field Measurements: Optical Time Domain Reflectometry (OTDR)

The method discussed above are primarily suited to the laboratory environment. However, there is a technique that can measure attenuation, connector and splicing losses, and at the same time, can also locate faults in optical fiber links in the field required. A method that finds wide applications in this field is called OTDR or the backscatter technique.

Figure 5.40 shows a schematic diagram of the OTDR apparatus. A powerful beam of light is launched through a bidirectional coupler into one end of the fiber and the backscattered light is detected with the help of an APD receiver. The received signal is integrated and amplified, and the averaged signals for successive points within the fiber are presented on the recorder or the cathode ray tube (CRT). The information displayed on the chart of the recorder or the screen of the CRT is the signal strength along the y-axis and time along the x-axis.

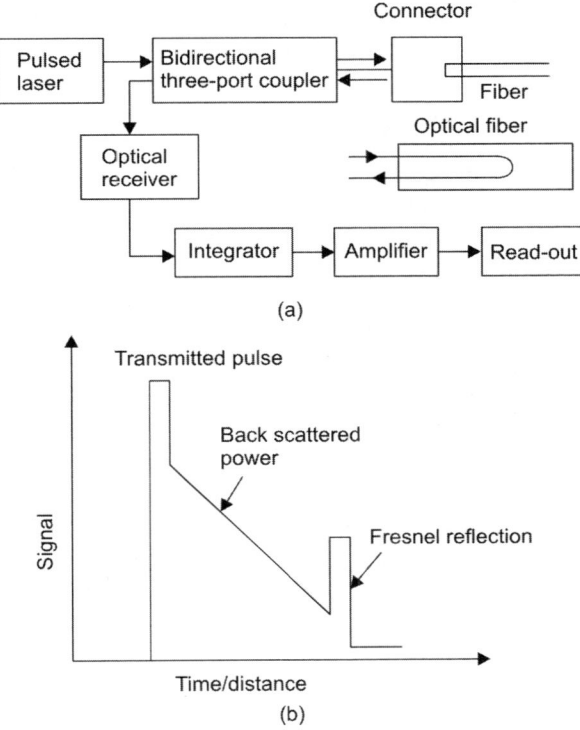

Fig. 5.40. Schematic of (a) OTDR apparatus, (b) Backscatter plot for an ideal fiber

The time is usually multiplied by the velocity of propagation to give an indication of the distance. The display expected for an ideal optical fiber of finite length is shown in Fig. 5.40(b). Any deviation from this is due to some kind of fault.

OTDR provides information about the location dependence of the attenuation. The slope of the plot shown in Fig. 5.40(b) simply provides the attenuation per unit length for the fiber. Thus, it is superior to other methods of measuring attenuation, which provide the average loss over the whole length. Moreover, it gives information about the splice or connector losses and the location of any faults on the link. Finally, the overall link length can be calculated from the time difference between the Fresnel reflections from the two ends of the fiber. Clearly, it requires access to only one end of the fiber for performing measurements.

5.18 APPLICATIONS OF THE FIBER OPTIC CABLES

Fiber optic cables with other components are used in building optical fiber devices and systems. One of large-scale applications is its use in communication systems. Video, including broadcast television, is one of the main telecommunication applications. Other applications include cable television, high-speed internet, wireless transition, remote monitoring, and surveillance. Fiber optic video transmission is successfully used around the world in surveillance and remote monitoring systems with many applications. Fiber optics applications in the military include communications, command-and-control links on ships and aircrafts, data link for satellite earth stations, and transmission lines for tactical command post communications.

Fiber cables can be used throughout the communication network, including in the final link into the subscriber's home to wall outlets. This field has continued to develop since the discovery of optical amplifiers, dense wavelength division multiplexers (DWDM), fiber Bragg grating, and photonics crystal fibers.

One particular advantage of fiber optic cables is that they are immune to electromagnetic interference (EMI) from electricity. Therefore, optic cables can be placed near high-voltage power cables without any effect on data transmission. Similarly, the cables also can be laid along railway lines without suffering from EMI.

Optical fibers are applied in building night-vision viewing devices, scanning and sensing instruments, and vibration sensors, which are extensively used in military, medical, and other applications. Imaging techniques have been rapidly developed for a variety of medical applications, such as viewing inside human tissues and scanning microscopic particles.

Fibers are also used in monitoring and sensing technology. They are used as sensors to monitor the vibration in the structures of bridges and high buildings. They are also used as gas and DNA sensors.

The use of fiber optic cables in lighting systems can reduce energy consumption. Fiber optic lighting systems are developing quickly, with wide applications. Fiber optic light systems can be applied to the interior and exterior of commercial, retail, and residential buildings. New applications are being explored in landscaping, waterscraping, medical lighting instruments, and theme environments.

Fiber optic cables can also be coated for special handling requirements and resistance to temperature, chemicals, or radiation. Radiation-resistant fiber is suitable for use in environments where electronics-based optical solutions are not viable, such as monitoring nuclear waste disposal in storage facilities. To make the fibers heat resistant, a chemical-resistant polyimide-coating that can withstand temperatures upto 300°C is applied. This is especially useful for manufacturers designing medical equipment for applications in which autoclave sterilization is necessary.

One of many applications of different fiber cable types is fiber-optic *sensing*. Fiber optic sensing can be classed as intrinsic and extrinsic. In intrinsic sensors, the fiber simply conducts light from a sensing head to a detector, the interaction between light and the environment take place outside the fiber cable. In intrinsic sensors, the interaction between the environment, the fiber, and the light itself generates information about a specific measurement. The key advantage of intrinsic fiber-optic sensors is the fundamental ability of the fiber to guide light around bends and over long distances. This enables the fiber to be confined within small physical volumes and magnifies the effects of very fine environment changes to a level that can be measured accurately.

Fiber Bragg gratings (FBGS) are employed as passive temperature and pressure sensors. The ability to adjust these optical devices introduces a new dimension for the design of very fast, precise and multifunctional fiber sensors without compromising their intrinsic advantages. Fiber Bragg gratings are also used in many other applications. Tuning can be accomplished by several mechanisms, such as electric heating, the piezoelectric effect, mechanical stretching and bending, and acoustic modulation. These sensors are available in the market in different types, sizes and wavelength ranges.

Photonic bandgap and photonic crystal fibers are both members of the family of grape fruit microstructure fibers. Both types can be used in gas sensors.

Beyond pressure and temperature sensor technology, advanced fiber sensors types include chemical strain, biomedical, electrical and magnetic, rotation, vibration, and displacement as major applications. One more application of the advanced fiber sensor technology is in the area of DNA analysis. This ability gives the diagnostic some excellent fibers are efficiently constructed as waveguides with novel properties for communications and sensor applications.

Example 1

The Si–O bond has a theoretical cohesive strength of 2.6×10^6 psi which corresponds to a bond distance of 0.16 nm. A silica optical fiber has an elliptical crack of depth 10 nm at a point along its length. Calculate (a) the fracture stress in psi for the fiber if it is dependent upon this crack, and (b) the percentage strain in the break. Given Young's modules for silica = 9×10^{10} N·m^{-2} and 1 psi = 6894.76 N·m^{-2}.

Solution

The theoretical cohesive strength for the Si–O bond is given by:

$$S_t = \left(\frac{\gamma_p E}{4 l_a}\right)^{1/2} \text{ or } \gamma_p = \frac{4 l_a S_t^2}{E}$$

where γ_p is surface energy of the material, E is Young's modulus for the material and l_a is the atomic spacing or bond distance. Substituting the values, one obtains:

$$\gamma_p = \frac{4 \times 0.16 \times 10^{-9} \, (2.6 \times 10^6 \times 6894.76)^2}{9 \times 10^{10}}$$

$$= 2.29 \text{ J}$$

The fracture stress for silica fiber may be obtained from:

$$S_f = \left(\frac{2E\gamma_p}{Y^2 C}\right)^{1/2}$$

where Y is a constant dictated by the shape of the crack (e.g. $Y = \pi^{1/2}$ for an elliptical crack) and C is the depth of the crack. For an elliptical crack:

$$S_f = \frac{2E\gamma_p}{Y^2 C}$$

$$= \left(\frac{2 \times 9 \times 10^{10} \times 2.29}{3.14 \times 10^{-8}}\right)^{1/2}$$

$$= 3.62 \times 10^9 \text{ N·m}^{-1}$$

$$= 5.25 \times 10^5 \text{ psi}$$

We may note that the fracture stress is reduced from the theoretical value for flawless silica of 2.6×10^6 psi by a factor of approximately 5.

$$\text{Young's modules, } E = \frac{\text{Stress}}{\text{Strain}}$$

$$\therefore \quad \text{Strain} = \frac{\text{Stress}}{E} = \frac{3.62 \times 10^9}{9 \times 10^{10}}$$

$$= 0.04$$

Obviously, the strain at break is 4%, which corresponds to the change in length over the original length for the fiber.

Example 2

An optical fiber has a core refractive index of 1.5. Two lengths of the fiber with smooth and perpendicular (to the core axes) end faces are butted together. Assuming the fiber axes are perfectly aligned, calculate the optical loss in decibels at the joint (due to Fresnel reflection) when there is a small air gap between the fiber end faces.

Solution

The magnitude of the Fresnel reflection at the fiber-air interface is given by:

$$\gamma = \left(\frac{n_1 - n}{n_1 + n}\right)^2$$

where n_1 is the refractive index of the fiber core and n is the refractive index of the medium between the two jointed fibers (i.e. for $n = 1$).

$$\therefore \quad \gamma = \left(\frac{1.5-1}{1.5+1}\right)^2 = \left(\frac{0.5}{2.5}\right)^2 = 0.04$$

The value of γ obtained corresponds to a reflection of 4% of the transmitted light at the single interface. One can obtain the optical loss in decibels at the single interface from:

$$\begin{aligned} \text{Loss}_{\text{Fres}} &= -10 \log_{10}(1-\gamma) \\ &= -10 \log_{10}(1-0.04) \\ &= -10 \log_{10} 0.96 \\ &= -0.18 \text{ dB} \end{aligned}$$

Similarly, we obtain the loss at the second interface also 0.18 dB.

Hence, the total loss due to Fresnel reflection at the fiber joint is approximately = 0.36 dB.

Example 3

Two compatible multimode SI fibers are jointed with a small air gap. The fiber axes and end faces are perfectly aligned. Determine the refractive index of the fiber core if the joint is showing a loss of 0.47 dB.
[B.Tech.]

Solution

We have:

$$L_F = 10 \log_{10}(\eta_F) = 0.47 \text{ dB}$$

This gives

$$\eta_F = 0.897 = \frac{16k^2}{(k+1)^4}$$

For air, $n = 1$,

$$\therefore \quad k = \frac{n_1}{n} = n_1$$

Thus $\quad n_1^2 - 2.22 n_1 + 1 = 0 \quad$ or $\quad n_1 = 1.59$ and $n_1 = -1$

Taking the positive root, $n_1 = 1.59$. The negative root gives n_1 less than 1, which is not possible.

Example 4

Two multimode step index fibers have numerical aperture of 0.2 and 0.4 respectively, and both have the same core refractive index of 1.48. Estimate the insertion loss at a joint in each fiber caused by a 5° angular misalignment of the fiber core axes.

Solution

The angular coupling efficiency is given by:

$$\eta_{ang} = \frac{16\left(\frac{n_1}{n}\right)^2}{\left[1+\left(\frac{n_1}{n}\right)\right]^4}\left[1 - \frac{n\theta}{\pi n_1 (2\Delta)^{1/2}}\right]$$

where $NA = n_1(2\Delta)^{1/2}$

$$\therefore \quad \eta_{ang} \simeq \frac{16\left(\frac{n_1}{n}\right)^2}{\left[1+\left(\frac{n_1}{n}\right)\right]^4}\left[1 - \frac{n\theta}{\pi \times 0.2}\right]$$

For $NA = 0.2$ fiber

$$\eta_{ang} \simeq \frac{16\,(1.48)^2}{[1+1.48]^4}\left[1 - \frac{\frac{5\pi}{180}}{\pi \times 0.2}\right]$$

$$= 0.797$$

The insertion loss due to misalignment may be obtained from:

$Loss_{ang} = -10\log_{10} \eta_{ang}$
$\qquad = -10\log_{10}(0.797) = 0.90$ dB

For $NA = 0.4$ fiber

$$\eta_{ang} \simeq 0.926\left[1 - \frac{\frac{5\pi}{180}}{\pi \times 0.4}\right]$$

$$= 0.862$$

The insertion loss due to the angular misalignment:

$Loss_{ang} = -10\log_{10}(0.862)$
$\qquad = 0.93$ dB

Example 5

Two compatible multimode SI fibers are jointed with a lateral offset of 3 μm, an angular misalignment of the core axes of 3°, and a small air gap (but negligible end separation). If the core of each fiber has a refractive index of 1.48, a relative refractive index difference of 2%, and a diameter of 100 μm, calculate the total insertion loss at the joint, which may be assumed to complete the sum of all the misalignment losses.

[B.Tech.]

Solution

Given: $\Delta x \approx 0$ (negligible), $\Delta y = 3$ μm, $\Delta \theta = 3° = 0.052$ rad, $n_1 = 1.48$, $\Delta = 2\% = 0.02$, $2a = 100$ μm, $n = 1$ (for air), and therefore, $k = n_1/n = n_1 = 1.48$.

We have
$$\eta_F = \frac{4k}{(k+1)^2} \frac{4k}{(k+1)^2} = \frac{16k^2}{(k+1)^4}$$

or
$$\eta_F = \frac{16 \times (1.48)^2}{(1.48+1)^4} = 0.9264$$

Using
$$\eta_{\text{lat}} \approx \frac{2}{\pi}\left[\cos^{-1}\left(\frac{\Delta y}{2a}\right) - \left(\frac{\Delta y}{2a}\right)\left\{1 - \left(\frac{\Delta y}{2a}\right)^2\right\}^{1/2}\right]$$

We obtain:
$$\eta_{\text{lat}} = \frac{2}{\pi}\left[\cos^{-1}\left(\frac{3}{100}\right) - \frac{3}{100}\left\{1 - \left(\frac{3}{100}\right)^2\right\}^{1/2}\right] = 0.962$$

and using:
$$\eta_{\text{ang}} \approx \left[1 - \frac{n\Delta\theta}{\pi\,\text{NA}}\right]$$

One obtains:
$$\eta_{\text{ang}} = \left[1 - \frac{1 \times 0.052}{\pi \times 0.296}\right]$$

as
$$\text{NA} = n_1\sqrt{2D} = 1.48\sqrt{2 \times 0.02} = 0.296$$

$$\eta_{\text{ang}} = 0.944$$

Therefore the total coupling efficiency η_τ will be given by:
$$\eta_T = \eta_F \eta_{\text{lat}} \eta_{\text{ang}} = 0.9264 \times 0.962 \times 0.944 = 0.8412$$

Thus the total loss:
$$L_T = -10\log_{10}\eta_T = 0.775 \text{ dB}$$

Example 6

A single mode fiber has the following parameters: (i) normalized frequency $(V) = 2.40$, (ii) core refractive index $(n_1) = 1.46$, (iii) core diameter $(2a) = 8$ μm, and (iv) numerical aperture (NA) = 0.1

Estimate the total insertion loss of a fiber joint with a lateral misalignment of 1 mm and an angular misalignment of 1°.

Solution

Initially, we have to determine the normalized spot size in the fiber. We have:
$$\omega = a\frac{(0.65 + 1.62\,V^{-3/2} + 2.88\,V^{-6})}{2^{1/2}}$$

$$= 4\frac{[0.65 + 1.62\,(2.4)^{-1.5} + 2.88\,(2.44)^{-6}]}{\sqrt{2}}$$

$$= 3.12\ \mu m$$

The loss due to lateral offset can be obtained from the relation:

$$T_1 = 2.17\left(\frac{1}{3.12}\right)^2$$

or $\qquad T_1 = 0.22$ dB

The loss due to angular misalignment may be obtained from the relation:

$$T_a = 2.17\left(\frac{\theta\omega\,n_1 V}{a\,NA}\right)^2$$

$$= 2.17\left(\frac{\left(\frac{\pi}{180}\right) \times 3.13 \times 1.46 \times 24}{4 \times 0.1}\right)$$

$$= 0.49\ dB$$

∴ Total insertion loss is:

$$T_T = T_1 + T_a$$
$$= 0.22 + 0.49 = 0.71\ dB$$

Example 7

A 60/120 μm graded-index fiber with a numerical aperture of 0.25 and a profile parameter of 1.9 is jointed with a 50/120 μm graded-index fiber with a numerical aperture of 0.20 and a profile parameter of 2.1. If the fiber axes are perfectly aligned and there is no air gap, calculate the insertion loss at the joint in the forward and backward directions.

[B.Tech.]

Solution

In the forward direction:

$$\eta_{cd} = \left(\frac{50}{60}\right)^2 = 0.6944$$

$$\eta_{NA} = \left(\frac{0.20}{0.25}\right)^2 = 0.64$$

and $\qquad \eta_a = 1$

∴ $\qquad \eta_T = 0.6944 \times 0.64 \times 1 = 0.444$

and $\qquad L_T = -10\log_{10}\eta_T = 3.52$ dB

In the backward direction, $\eta_{cd} = 1$ and $\eta_{NA} = 1$.

But
$$\eta_a = \left(\frac{1+2/2.1}{1+2/1.9}\right) = 0.95$$

$$\eta_T = 1 \times 1 \times 0.95 = 0.95$$

and
$$L_T = -10\log_{10}(0.95) = 0.218 \text{ dB}$$

Example 8

Two single mode fibers with mode field diameters of 9.2 μm and 8.4 μm are to be connected together. Assuming no extrinsic losses, determine the loss at the connection due to the mode-field diameter mismatch.

Solution

We have the expression for intrinsic loss as:
$$\text{Loss}_{int} = -10\log_{10}\left[4\left(\frac{\omega_{02}}{\omega_{01}} + \frac{\omega_{01}}{\omega_{02}}\right)^{-2}\right]$$

where ω_{01} and ω_{02} are the spot sizes of the transmitting and receiving fibers respectively. Substituting the values in the above relation, we obtain: Here $\omega_{01} = \frac{9.2}{2} = 5.6$ μm, $\omega_{02} = \frac{8.4}{2} = 4.2$ μm

$$\text{Loss}_{int} = -10\log_{10}\left[4\left(\frac{4.2}{5.6} + \frac{5.6}{4.2}\right)^{-2}\right]$$

$$= -10\log_{10} 0.922$$
$$= 0.35 \text{ dB}$$

Example 9

A four-port multimode fiber FBT coupler has 60 μW optical power launched into port 1. The measured output powers at ports 2, 3 and 4 are 0.004, 26.0 and 27.5 μW respectively. Determine the excess loss, the insertion losses between the input and output ports, the crosstalk and the split ratio for the device.

Solution

We have the expression for excess loss (four-port coupler) as:

$$\text{Excess loss (four-port coupler)} = 10\log_{10}\left(\frac{P_1}{P_3 + P_4}\right) \text{ (dB)}$$

$$= 10\log_{10}\frac{60}{53.5}$$

$$= 0.5 \text{ dB}$$

The insertion loss (parts 1 to 3) $= 10 \log \dfrac{P_1}{P_3}$ (dB)

$= 10 \log \left(\dfrac{60}{26} \right)$

$= 3.63$ dB

Insertion loss (parts 1 to 4) $= 10 \log_{10} \left(\dfrac{60}{27.5} \right)$

$= 3.39$ dB

Cross talk (four-port coupler) $= 10 \log_{10} \left(\dfrac{P_2}{P_1} \right)$ dB

$= 10 \log_{10} \left(\dfrac{0.004}{60} \right)$

$= -41.8$ dB

Split ratio $= \left[\dfrac{P_3}{(P_3 + P_4)} \right] \times 100\%$

$= \dfrac{26}{53.5} \times 100$

$= 48.6\%$

GLIMPSES

- Fiber optic cables are significantly different from copper cables. Fiber-optic cables transmit data through very small cores over long distances at the speed of light. These cables come in a wide variety of configurations. Important considerations in any cable installation and operation are the bending radius, tensile strength, ruggedness, durability, flexibility, environmental conditions, e.g. temperature extremes and even appearance.
- From the considerations of optical waveguiding, it is clear that variation of refractive index inside the optical fiber (i.e. between the core and the cladding) is a fundamental necessity in the fabrication of fibers for light transmission. This means at least two different compatible materials which are transparent to light over the major operating wavelength range (0.8 to 1.7 µm) are required. For most applications, silica based glass is the ultimate choice for producing optical fibers.
- Most fibers consists of silica (SiO_2) or silicate. Glass is made by fusing mixtures of metal oxides having refractive index of 1.458 at 850 nm. For changing the refractive index different oxides such as B_2O_3, GeO_2 and P_2O_5 are added.

- The fabrication of all glass fibers is a two-stage process. In the first stage, pure glass is produced and transforms it into a rod or perform. The second stage employs a pulling technique to draw fibers of required diameters.
- Vapour-phase methods are employed to produce silica-rich glass fibers whereas liquid-phase methods are used for manufacturing multicomponent glass fibers.
- The cabling of fibers requires that (i) the fiber be given primary and buffer coatings to protect against external influences, (ii) a strength member be provided to improve mechanical strength, and (iii) a structural member be provided to place them in multifiber cables. Several designs are available, e.g. two fiber cable, multiple fiber cables, etc.
- The basic fiber building blocks are used to form large cable. These units are bound on a buffer material which acts as strength element along with insulated copper conductor. The fiber building blocks are surrounded by paper tape, PVC jacket, yarn and outer sheath.
- Optical fiber links, in common with any communication system, have a requirement for both joining and termination of the transmission medium. The number of intermediate fiber connections or joints is dependent upon the link length (between repeaters), the continuous length of fiber cable that may be produced by the preparation methods, and the length of the fiber cable that may be practically or conveniently installed as a continuous section on the link.
- Fiber-to-fiber connection may be achieved through (i) splices (permanent joints) or connectors (demountable joints). It must be ensured that there are no misalignments in this jointing process. Connection losses may occur due to extrinsic parameters, e.g., Fresnel reflection, end separation, lateral offset, or angular misalignment of the two fiber ends. Losses can also occur due to intrinsic parameters, e.g., mismatches of core diameters, RI profiles, or numerical apertures.
- In order to evaluate the performance of an optical fiber, it is essential to measure its properties. Important among these are optical attenuation, pulse dispersion, numerical aperture, RI profile, etc. In the field, however, optical time domain reflectometry (OTDR) is an essential tool.

SUGGESTED READINGS

1. Buck JA, 'Fundamentals of Optical Fibers', Wiley (2004).
2. Doremus DH, 'Glass Science', Wiley, 2nd ed. (1994).
3. Mynbaev DK and Scheiner LL, 'Fibre-Optic Communication Technology – Fiber Fabrication', Prentice Hall (2005).

4. Kolimbiris H, 'Fiber Optics Communications', Prentice Hall (2004).
5. Yeh C, 'Handbook of Fiber Optics: Theory and Applications', Academic Press (1990).
6. Kumar S, Jamal Deen M, Fiber Optic Communications, Wiley (2014).

REVIEW QUESTIONS

1. Describe in general terms liquid-phase techniques for the preparation of multicomponent glasses for optical fibers. Discuss with the help of a suitable diagram one melting method for the preparation of multicomponent glass.
2. Indicate the major advantages of vapour-phase deposition in the preparation of glasses for optical fibers. Briefly describe the various vapour-phase techniques currently in use.
3. Compare and contrast, using suitable diagrams, the outside vapour-phase oxidation (OVPO) and the modified chemical vapour deposition (MCVD) technique for the preparation of low-loss optical fibers.
4. Describe the double crucible method for producing optical fibers. Mention the limitations of this method.
5. Distinguish between outside vapour-phase oxidation (OVPO) method and inside vapour-phase oxidation (IVPO) method for manufacturing optical fibers. Compare the salient features of these two methods.
6. Discuss the design of optical fibers from prepared glasses with regard to (a) multicomponent glass fibers, (b) silica rich fibers.
7. List the silica based various optical fiber types currently on the market indicating the important features. Hence, briefly describe the general areas of applications for each type.
8. Describe the effects of stress corrosion on optical fiber strength and durability.
9. Discuss optical fiber cable design with regard to:
    (a) fiber buffering
    (b) cable strength and structural members
    (c) layered cable construction
    (d) cable sheath and water barrier.
10. State the two major categories of fiber-fiber joint, indicating the difference between them.
11. Describe the three types of fiber misalignment which may contribute to insertion loss at an optical fiber joint.
12. Briefly outline the factors which cause intrinsic losses of fiber-fiber joints.
13. Describe with aid of suitable diagrams, three common techniques used for the mechanical splicing of optical fibers.

14. Describe with aid of suitable diagrams, the following techniques of mechanical splicing of optical fibers: (a) snug tube splice, (b) spring-groove splice, and elastomeric splice.
15. Describe with the aid of suitable diagrams, the design of the following connectors: (a) ferrule connector, and (b) expanded-beam connector.
16. The fraction of reflected light (R) at an air-fiber interface is given by:

$$R = \frac{P_r}{P_i} = \left(\frac{n_1 - n}{n_1 + n}\right)^2$$

where P_r and P_i are the reflected and incident powers, n_1 is the RI at the striking point of the fiber, and n is the RI of the medium surrounding the fiber. Show that for a small variation in the value of core index n_1, the change in R is given by:

$$\Delta R = 4n\left[\frac{n_1 - n}{(n_1 + n)^2}\right]\Delta n_1$$

17. With the aid of simple sketches, outline the major categories of multiport optical fiber coupler.

PROBLEMS

1. Silica has a Young's modulus of 9×10^{10} N·m^2 and a surface energy of 2.29 J. Estimate the fracture stress in psi for a silica optical fiber with a dominant elliptical crack of depth 0.5 µm. Also, calculate the strain at the break for the fiber. Given 1 psi = 6.894.76 N·m^{-2}.
   [**Ans.** 7.43×10^4 psi, 0.6%]
2. A fusion splice is made for a multimode step index fiber. The splice exhibit a loss of 0.36 dB, which seems to be mainly due to an air gap. Show that the refractive index of fiber core is 1.5
3. Another length of the optical fiber described in problem 1 is found to break at 1% strain. The failure is due to dominant elliptical crack. Estimate the depth of this crack. [**Ans.** 0.2 µm]
4. It is reported that a 20 m length of fused silica optical fiber may be extended to 24 m at liquid nitrogen temperatures (i.e. little stress corrosion) before failure occurs. Estimate the failure stress under these conditions. Young's modulus for silica = 9×10^{10} N·m^2 and 1 psi = 6894.76 N·m^{-2}. [**Ans.** 2.61×10^6 psi]
5. The Fresnel reflection at a butt joint with an air gap in multimode step index fiber is 0.46 dB. Determine the refractive index of the fiber core. [**Ans.** 1.59]
6. A graded index fiber with a 50 µm core diameter has a characteristic refractive index profile (α) of 2.25. The fiber is joined with index

matching and the connection exhibits an optical loss of 0.62 dB. This is found to be solely due to a lateral offset of the fiber ends. Calculate the magnitude of the lateral offset assuming the uniform illumination of all guided modes in the fiber core. [**Ans.** 4.0 µm]

7. A fiber Bragg grating assisted coupler is designed to block an incoming optical signal present at the input part of the device. When the fiber core refractive index is 1.6 and the grating period is 0.42 µm, calculate the wavelength of the blocked signal.

[**Ans.** 1.34 µm]

8. A single mode step index fiber of 5 µm core diameter has a normalized frequency of 1.7, a core refractive index of 1.48 and a numerical aperture of 0.14. The loss in decibels due to angular misalignment at a fusion splice with a lateral offset of 0.4 µm is twice that due to the lateral offset. Calculate the degrees of the angular misalignment. [**Ans.** 0.65°]

9. Two compatible multimode SI fibers are jointed with a lateral offset of 10% of the core radius. The refractive index of the core of each fiber is 1.50. Estimate the insertion loss at the joint when (a) there is small air gap, and (b) an index-matching fluid is inserted between the fiber ends. [B.Tech.]

[**Hint:** We have $\Delta y/a = 10\% = 0.1$ and $n_1 = 1.50$. With the air gap ($n = 1$), $k = n_1/n = 1.5$, and with index-matching fluid ($n = n_1$), $k = 1$.

(a) We have $\eta_F = \dfrac{16k^2}{(k+1)^4}$

∴ $\eta_F = \dfrac{16(1.5)^2}{(1.5+1)^4} = 0.9216$

Also, $\eta_{lat} \approx \dfrac{2}{\pi}\left[\cos^{-1}\left(\dfrac{\Delta y}{2a}\right) - \dfrac{\Delta y}{2a}\left\{1 - \left(\dfrac{\Delta y}{2a}\right)^2\right\}^{1/2}\right]$

∴ $\eta_{lat} = \dfrac{2}{\pi}\left[\cos^{-1}\left(\dfrac{0.1}{2}\right) - \left(\dfrac{0.1}{2}\right)\left\{1 - \left(\dfrac{0.1}{2}\right)^2\right\}^{1/2}\right]$

Thus, the total coupling efficiency, η_T is obtained as:

$\eta_T = \eta_F \eta_{lat} = 0.9216 \times 0.936 = 0.8629$.

Thus the total loss $L_T = -\log_{10}\eta_T = (-10 \log_{10}(0.8629) = 0.64$ dB.

(b) Here, $\eta_F = 1$ and $\eta_{lat} = 0.936$, and hence:

$\eta_T = \eta_F \eta_{lat} = 1 \times 0.936 = 0.936$

and the total loss:

$L_\tau = -\log_{10}(0.936) = 0.287$ dB.]

Fiber Optic Cables and Connectors 249

10. A 80/125 µm graded-index (GI) fiber with a NA of 0.25 and α of 2.0 is jointed with a 60/125 mm GI fiber with an NA of 0.21 and α of 1.9. The fiber axes are perfectly aligned and there is no air gap. Calculate the insertion loss at a joint for the signal transmission in the forward and backward directions.

[Hint:
For **Fiber 1**: Core diameter: $d_1 = 80$ µm, $NA_1 = 0.25$, $a_1 = 2.0$.
For **Fiber 2**: Core diameter: $d_2 = 60$ µm, $NA_2 = 0.21$, $a_2 = 1.9$.
In the *forward direction*, i.e. when the signal is propagating from fiber 1 to fiber 2, we have

$$\eta_{cd} = \left(\frac{d_2}{d_1}\right)^2 = \left(\frac{60}{80}\right)^2 = 0.5625$$

$$\eta_{NA} = \left(\frac{NA_2}{NA_1}\right)^2 = \left(\frac{0.21}{0.25}\right)^2 = 0.7056$$

$$\eta_\alpha = \left(\frac{1+2/\alpha_1}{1+2/\alpha_2}\right) = \left(\frac{1+2.0/2.0}{1+2.0/1.9}\right) = 0.9743$$

Thus, the total coupling efficiency at the joint, $\eta_T = \eta_{cd}\eta_{NA}\eta_\alpha$. Substituting the values of η_{cd}, η_{NA} and η_α, one obtains

$$\eta_T = 0.5625 \times 0.7056 \times 0.9743 = 0.3867$$

Therefore, the total loss at the joint will be

$$L_T = -\log_{10}\eta_T = -\log_{10}(0.3867) = 4.1259 \text{ dB}$$

In the *backward direction*, η_{cd}, η_{NA} and η_a are all unity and hence there will be no loss.]

SHORT ANSWER QUESTIONS

1. What optical fiber properties reduce connection loss in short distance systems?

Ans. (i) Larger fiber core, and
(ii) Higher fiber numerical

2. What trade offs are considered by designers in fiber optic systems?

Ans. (i) Types of connections
(ii) Optical sources, and
(iii) detector types in military and subscriber loop applications.

3. Which fiber part, core or cladding has a higher index of refraction?

Ans. Core.

4. In multimode step index fibers, the majority of light propagates in the fiber core, why?

Ans. Most modes in multimode step index fibers propagates far from cutoff.

5. What is basic requirement for light guidance through an optical fiber?

Ans. Refractive indices of the core and cladding should be different. This is why, two compatible materials that are transparent in the operating wavelength range are required.

6. Why multimode step index fibers have relatively large core diameters and large NA?

Ans. To make it easier to couple light from a LED into the fiber.

7. For most applications, which material is the ultimate choice for producing optical fibers?

Ans. Silica based glass.

8. How are the source to fiber coupling and micro-bending and macro-bending losses affected by core diameter and Δ?

Ans. Coupling efficiency increases with core diameter and Δ, whereas bending losses increases with core diameter and inversely with Δ.

9. How glass fibers are fabricated?

Ans. The fabrication of all-glass fibers is a two-stage process. In the first stage pure glass is produced and transforms it into a rod or perform. The second stage employs a pulling technique to draw fibers of required diameters.

10. Which fiber optic component (splice, connector or coupler) makes a permanent connection in a distributed system?

Ans. Splice.

11. Which methods produce multicomponent glass fibers?

Ans. Liquid phase methods.

12. Which methods are employed to produce silica-rich glass fibers?

Ans. Vapour phase methods.

13. Mention the three basic errors that occur during fiber alignment.

Ans. (i) Longitudinal misalignment
(ii) Lateral misalignment, and
(iii) Angular misalignment.

14. Mention the requirements of cabling of fibers.

Ans. (i) The fiber should be given primary and buffer coating to protect against external influences
(ii) A strength member to be provided to improve mechanical strength, and

(iii) A structural member be provided to place them in multifiber cable

15. When the axes of two connected fibers are no longer in parallel, what this misalignment termed?

Ans. Angular misalignment.

16. How fiber-to-fiber connections may be achieved?

Ans. Through splices (permanent joints) or connectors (demountable joints). One will have to ensure that there are no misalignments in this joining process. One will have to take care of connection losses. Connection losses may occur due to extrinsic parameters, e.g. Fresnel's reflection, end separation, lateral offset, or angular misalignment of the two fiber ends.

Losses can also occur due to intrinsic parameters, e.g. mismatches of core diameters, RI profiles or aperatures.

17. Which single mode or multimode fibers are more sensitive to alignment errors?

Ans. Single mode fibers.

18. For reliable operation quality end preparation is needed. What properties must an optical fiber end face should have to ensure proper fiber connections?

Ans. Flat, smooth and perpendicular to the fiber axis.

19. What are the important properties of an optical fiber, which are to be measured essentially for the evaluation of its performance?

Ans. Optical attenuation, pulse dispersion, numerical aperture (NA), RI profile, etc.

20. What is a fiber optic splice?

Ans. A fiber optic splice is a permanent fiber joint whose purpose is to establish an optical connection between two individual optical fibers.

21. Mention the techniques used for fiber splicing.

Ans. (i) Mechanical splicing
(ii) Fusion splicing

22. What type of faults are produced by (a) Fiber breaks, (b) Fiber cracks, and Fiber microbends?

Ans. (a) Reflective
(b) Non-reflective

MULTIPLE CHOICE QUESTIONS

1. Most low-loss optical fibers are made of oxide glasses, the most widely used material is:
   (a) silica (SiO_2) (b) germanium
   (c) arsenic (d) None of the above

2. Optical glass fibers are brittle and have very small cross-sectional area (~100 to 250 μm). They are, therefore, highly susceptible to damage during handling and use. In order to improve their tensile strength and protect them from external influences, it is necessary to:
   (a) polish them
   (b) encase them in cables
   (c) wind up them
   (d) None of the above
3. In a multifiber cable, the strength member: [B.Tech.]
   (a) must be placed along the central axis of the cable
   (b) must be placed in a coaxial cylindrical configuration
   (c) can be placed anywhere within the cable
   (d) is not required at all
4. An air gap is introduced while splicing two compatible fibers with core indices of 1.46. What is the loss due to Fresnel reflection at the joint? [B.Tech.]
   (a) Zero
   (b) 0.154 dB
   (c) 0.309 dB
   (d) 0.36 dB
5. Two optical fibers with numerical apertures 0.17 and 0.20 are to be spliced. What will be the loss at the joint in the forward direction? [B.Tech.]
   (a) Zero
   (b) 1.41 dB
   (c) 1.82 dB
   (d) 2.50 dB
6. Increase in the concentration of GeO_2 in SiO_2 will: [B.Tech.]
   (a) decrease the RI
   (b) increase the RI
   (c) change RI randomly
   (d) not change RI at all
7. What type of optical fibers can be drawn from a solid perform (formed by collapsing a solid rod or hollow tube deposited by the vapour-phase oxidation method)? [B.Tech.]
   (a) Multimode SI fibers
   (b) Multimode GI fiber
   (c) Single-mode fibers
   (d) All of these
8. Increase in the concentration of B_2O_3 in SiO_2 will:
   (a) increase the RI
   (b) decrease the RI
   (c) change RI randomly
   (d) RI remains unaffected
9. For the optical fibers of Question 5, what will be the joint loss in the backward direction? [B.Tech.]
   (a) Zero
   (b) 1.41 dB
   (c) 1.82 dB
   (d) 2.50 dB
10. A 62.5/125 μm SI fiber is to be spliced to a 50/125 μm SI fiber. Both the fibers have a core index of 1.50. What will be the joint loss in the forward direction? [B.Tech.]

(a) Zero (b) 0.97 dB
(c) 1.94 dB (d) 2.45 dB

11. A multimode SI fiber with a core RI of 1.50 is spliced with an identical fiber. What is the NA of the fiber if the splice loss is 0.7 dB, which is mainly due to a 5° angular misalignment of the fiber core axes? [B.Tech.]

(a) 0.17 (b) 0.21
(c) 0.28 (d) 0.30

12. If two optical fibers with different diameters are to be spliced, which of the following mechanical splices will be most suitable? [B.Tech.]

(a) Snug tube splice (b) Loose tube splice
(c) Spring-groove splice (d) V-groove splice

13. With an OTDR, it is possible to know:
(a) the location dependence of attenuation
(b) the overall link length
(c) splice and connector losses
(d) All of the above

14. If τ_i and τ_o are the half width of the input and output pulses and if, the shape of the pulses assumed to be Gaussian, then the impulse response of the fiber in ns/km is given by: [B.Tech.]

(a) $\tau = \dfrac{\tau_o - \tau_i}{L}$

(b) $\tau = \dfrac{\left(\tau_o^2 - \tau_i^2\right)^{1/2}}{L}$

(c) $\tau^2 = \sqrt{\dfrac{\left(\tau_o^2 - \tau_i^2\right)}{L}}$

(d) $\tau^2 = \dfrac{L}{\tau_o^2 - \tau_i^2}$

15. The refractive index profile of an optical fiber helps in:
(a) its characterization
(b) determination of the NA
(c) determination of guided modes propagating within the fiber
(d) All of the above

ANSWERS

1. (a) 2. (b) 3. (c) 4. (c) 5. (a) 6. (b) 7. (d)
8. (b) 9. (b) 10. (c) 11. (c) 12. (d) 13. (d) 14. (b)
15. (d)

6

Optical Sources

6.1 INTRODUCTION

LASER is the acronym of Light Amplication by Stimulated Emission of Radiation. This is a device which produces light beam with some extraordinary properties. It is used in a large variety of instruments from household to industry. Theoretical idea of stimulated emission or laser was put forward by Einstein in 1917, but it took a couple of decades to become a reality. CH Townes demonstrated the operation of **MASER** (Microwave Amplication by Stimulated Emission of Radiation). It was a *pulse laser*, in the same year *continuous wave laser* was developed and very soon, in 1962, many laboratories of the world developed *semiconductor lasers*.

In this chapter, we study fundamental concepts and formulas related to *miniature* solid-state lasers. Diode-pumped, miniature, monolithic, solid state lasers offer an efficient, compact, and robust means of generating diffraction-limited, single frequency radiation. In addition, their diminutive size results in high-speed tuning capabilities and short-pulsed operation unmatched by larger devices. Application areas are as diverse as communications, spectroscopy, remote sensing, nonlinear optics, projection displays, and micromatching.

6.1.1 Semiconductors

According to bond theory of solids, materials may be classified into three categories from the point of view of electric conduction: (i) conductors, (ii) insulators, and (iii) semiconductors. The sensitivity of these materials lies in the following range of values:

Conductor: 10^{-6}–10^{-4} Ω cm
Insulator: 10^{10}–10^{20} Ω cm
Semiconductor: 10^{-2}–10^{8} Ω cm

6.1.2 Energy Bands

Semiconductor materials have conduction properties that lie somewhat between those of metals and insulators. As an example, we consider silicon (Si), which is located in the fourth column (group IV) of the periodic table of elements. A Si atom has four electrons in its outer shell, by which it makes covalent bonds with its neighbouring atoms in a crystal. Such outer-shell electrons are called *valence electrons*.

The conduction properties of a semiconductor can be interpreted with the aid of the *energy-band diagrams* shown in Fig. 6.1(a). In a semiconductor the valence electrons occupy a band of energy levels called the *valence band*. This is the lowest band of allowed states. The next higher band of allowed energy levels for the electrons is called the *conduction band*. In a pure crystal at low temperatures, the *conduction band* is completely empty of electrons and the *valence band* is completely full. These two bands are separated by an *energy gap*, or *band gap*, in which no energy levels exist. As the temperature is raised, some electrons are thermally excited across the band gap. For Si, this excitation energy must be greater than 1.1 eV, which is the band-gap energy. This gives rise to a concentration n of free electrons in the conduction band, which leaves behind an equal concentration p of vacancies, or *holes*, in the valence band, as is shown schematically in Fig. 6.1(b). Both the free electrons and the holes are mobile within the material, so that both can contribute to electrical conductivity; that is, an electron in the valence band can move into a vacant hole. This action makes the hole move in the opposite direction to the electron flow, as is shown in Fig. 6.1(a).

When an electron propagates in a semiconductor, it interacts with the periodically arranged constituent atoms of the material and thus,

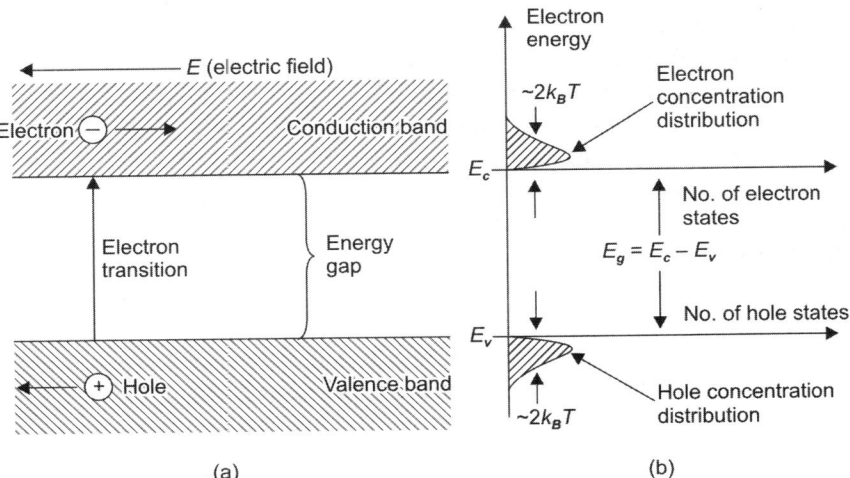

Fig. 6.1. (a) Energy level diagrams exhibiting the excitation of an electron from the valence band to the conduction band. The resultant free electrons and holes move under the influence of an external electric field E; (b) equal electron and hole concentrations in an intrinsic semiconductor created by the thermal excitation of electrons across the band gap

experiences *external forces*. As a result, to describe its acceleration a_{crys} in a semiconductor crystal under an external force F_{ext}, its mass needs to be described by a quantum mechanical quantity m_e called the *effective mass*. That is, when using the relationship $F_{ext} = m_e a_{crys}$ (force equals mass times acceleration), the effects of all the forces exerted on the electron within the materials are incorporated into m_e.

The concentration of electrons and holes is known as the *intrinsic carrier concentration* n_i, and for a perfect material with no imperfections or impurities it is given by:

$$n = p = n_i = K \exp\left(-\frac{E_g}{2k_B T}\right) \tag{6.1}$$

where

$$K = 2(2\pi k_B T / h^2)^{3/2} (m_e m_h)^{3/4}$$

is a constant that is characteristic of the material. Here, T is the temperature in degree kelvin, k_B is Boltzmann's constant, h is Planck's constant, and m_e and m_h are the effective masses of the electrons and holes, respectively, which can be smaller by a factor of 10 or more than the free-space electron rest mass of 9.11×10^{-31} kg.

The conduction can be greatly increased by adding traces of impurities from the group V elements (e.g. P, As, Sb). This process is called *doping* and the doped semiconductor is called an *extrinsic material*. These elements have five electrons in the outer shell. When they replace a Si atom, four electrons are used for covalent bonding, and the fifth, loosely bound electron is available for conduction. As shown in Fig. 6.1(c), this gives rise to an occupied level, just below the conduction band, called the *donor level*. The impurities are called *donors* because they can give up an electron to the conduction band. This is reflected by the increase in the free-electron concentration in the conduction band, as shown in Fig. 6.1(d). Since in this type of material, the current is carried by (negative) electrons (because the electron concentration is much higher than that of holes), it is called *n*-type material.

The conduction can also be increased by adding group III elements, which have three electrons in the outer shell. In this case, three electrons make covalent bonds, and a hole with properties identical to that of the donor electron is created. As shown in Fig. 6.1(e), this gives rise to an unoccupied level just above the valence band. Conduction occurs when electrons are excited from the valence band to this *acceptor level* (so called because the impurity atoms have accepted electrons from the valence band). Correspondingly, the free-hole concentration increases in the valence band, as shown in Fig. 6.1(f). This is called *p*-type material because the conduction is a result of (positive) hole flow.

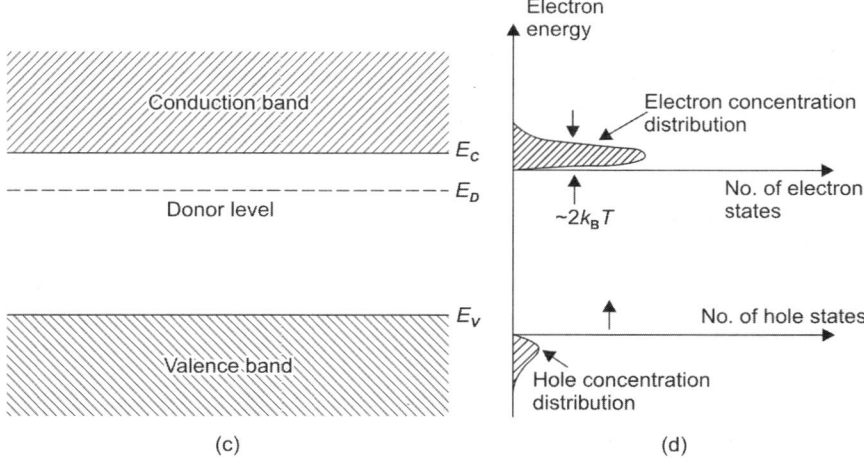

Fig. 6.1. (c) Donor level in an *n*-type material; (d) the ionization of donor-impurities increases the electron concentration distribution in the conduction band

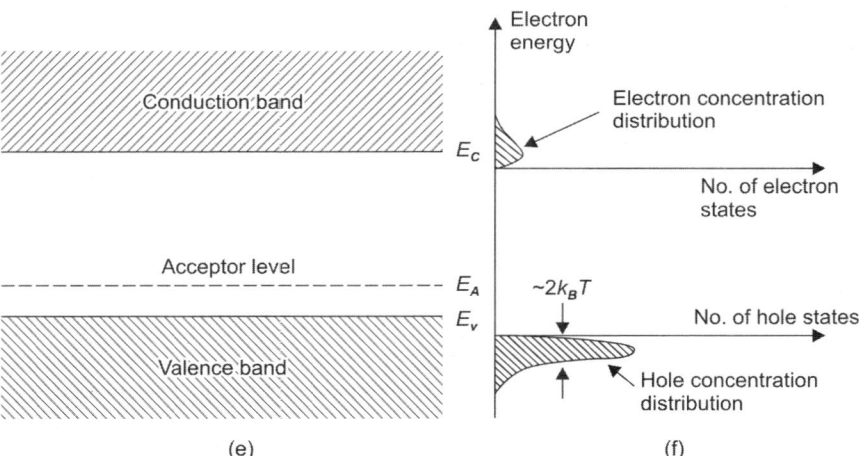

Fig. 6.1. (e) Acceptor level in a *p*-type material; (f) the ionization of acceptor impurities increases the hole concentration distribution in the valence band

6.1.3 Intrinsic and Extrinsic Materials

A perfect material containing no impurities is called an *intrinsic material*. Because of thermal vibrations of the crystal atoms, some electrons in the valence band gain, enough energy to be excited to the conduction band. This *thermal generation process* produces free electron-hole pairs, since every electron that moves to be conduction band, leaves behind a hole. Thus, for an intrinsic material the number of electrons and holes are both equal to the intrinsic carrier density, as denoted by Eq. (6.1). In the opposite *recombination process*, a free electron releases its energy and drops into a free hole in the valence band. For an extrinsic semiconductor, the increase of one type of carrier reduces the number of the other type. In this case, the product of the two types of carriers remains constant at a given temperature. This gives rise to the *mass-action law*:

$$pn = n_i^2 \tag{6.2}$$

which is valid for both intrinsic and extrinsic materials under thermal equilibrium.

Since the electrical conductivity is proportional to the carrier concentration, two types of charge carriers are defined for this material:
  (i) *Majority carriers* refer either to electrons in n-type material or to holes in p-type material.
  (ii) *Minority carriers* refer either to hole in n-type material or to electrons in p-type material.

The operation of semiconductor devices is essentially based on the *injection* and *extraction* of minority carriers.

6.1.4 The *pn* Junction

Doped n- and p-type semiconductor material by itself, serves only as a conductor. To make devices out of these semiconductors, it is necessary to use both types of materials (in a single, continuous crystal structure). The junction between the two material regions, which is known as the *pn junction*, is responsible for the useful electrical characteristics of a semiconductor device.

When a *pn* junction is created, the majority carriers diffuse across it. This causes electrons to fill holes in the p side of the junction and cause holes to appear on the n side. As a result, an electric field (or *barrier potential*) appears across the junction, as is shown in Fig. 6.1(g). This field prevents further net movements of charges once equilibrium has been established. The junction area now has no mobile carriers, since its electrons and holes are locked into a covalent bond structure. This region is called either the *depletion region* or the *space charge region*.

When an external battery is connected to the *pn* junction with its positive terminal to the n-type material and its negative terminal to the p-type material, the junction is said to be *reverse-biased*. This is shown in Fig. 6.1(h). As a result of the reverse bias, the width of the depletion region will increase on both the n side and the p side. This effectively increases the barrier potential and prevents any majority carriers from flowing across the junction. However, minority carriers can move with the field across the junction. The minority carrier flow is small at normal temperatures and operating voltages, but it can be significant when excess carriers are created as, for example, in an illuminated photodiode.

When the *pn* junction is *forward-biased*, as shown in Fig. 6.1(i), the magnitude of the barrier potential is reduced. Conduction-band

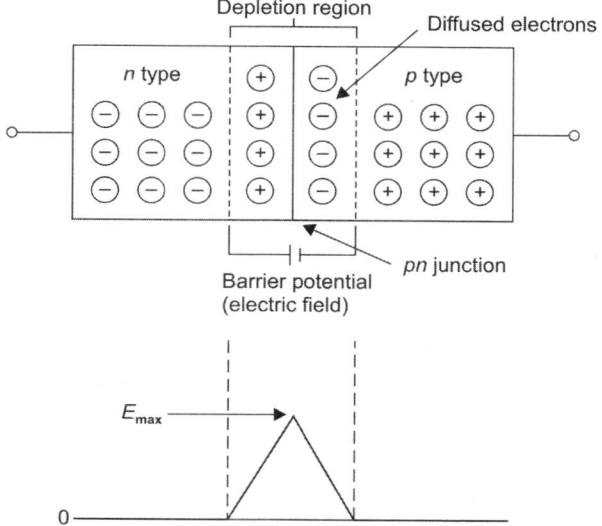

Fig. 6.1. (g) Electron diffusion across a *pn* junction creates a barrier potential (electric field) in the depletion region

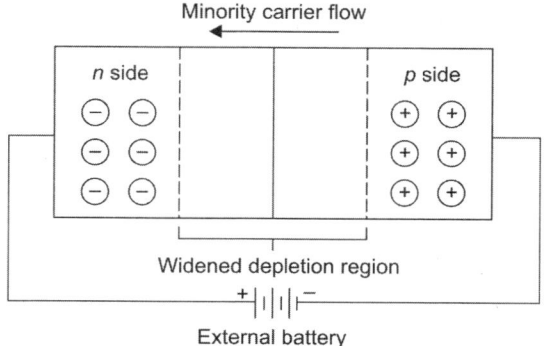

Fig. 6.1. (h) A reverse bias widens the depletion region, but allows minority carriers to move freely with the applied field

electrons on the *n* side and valence-band holes on the *p* side are, thereby, allowed to diffuse across the junction. Once across, they significantly increase the minority carrier concentrations, and the excess carriers recombine with the oppositely charged majority carriers. The recombination of excess minority carriers is the mechanisms by which optical radiation is generated.

6.1.5 Direct and Indirect Band Gaps

In order for electron transitions to take place to or from the conduction band with the absorption or emission of a photon, respectively, both

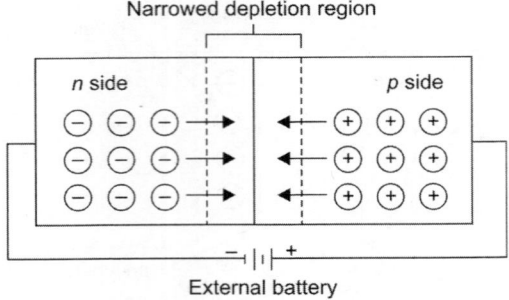

Fig. 6.1. (i) Lowering the barrier potential with a forward bias allows majority carriers to diffuse across the junction

energy and momentum must be conserved. Although a photon can have considerable energy, its momentum $h\nu/c$ is very small.

Semiconductors are classified as either *direct-band-gap* or *indirect-band-gap* materials depending on the shape of the band gap as a function of the momentum k, as shown in Fig. 6.1(j). Let us consider recombination of an electron and a hole, accompanied by the emission of a photon. The simplest and most probable recombination process will be that where the electron and hole have the same momentum value (see Fig. 6.1(j)). This is a direct-band-gap material.

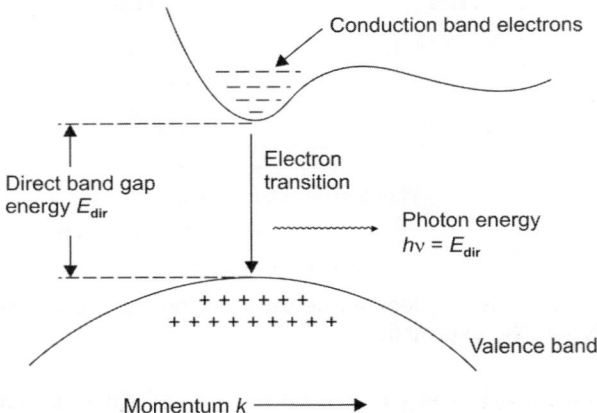

Fig. 6.1. (j) Electron recombination and the associated photon emission for a direct-band-gap material

For indirect-band-gap material, the conduction-band minimum and the valence-band maximum energy levels occur at different values of momentum, as shown in Fig. 6.1(k). Here, band-to-band recombination must involve a third particle to conserve momentum, since the photon momentum is very small. *Photons* (i.e., crystal lattice vibrations) serve this purpose.

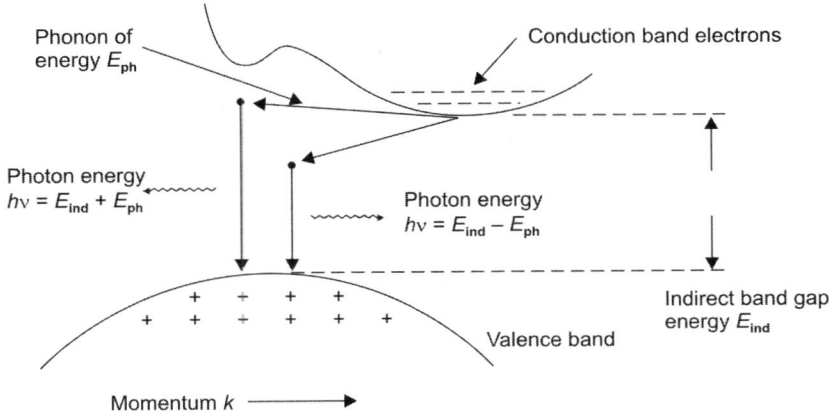

Fig. 6.1. (k) electron recombination for indirect-band-gap materials requires a photon of energy E_{ph} and momentum k_{ph}

6.2 SPONTANEOUS AND STIMULATED EMISSION

A material medium is composed of identical atoms or molecules, each of which is characterized by a set of discrete allowed energy states. An atom can move from one energy state to another when it receives or releases an amount of energy equal to the difference of energy between these two states. This is called *quantum transition*.

Let there are two states E_1 and E_2 of an atom. E_1 is lower energy state and E_2 is the excited state. As the constituent atoms of the medium are identical, the energy states E_1 and E_2 will be common for all atoms in the medium.

Let a monochromatic radiation of frequency (ν) be incident on the medium. In accordance with Planck's quantum theory of radiations, the radiation may be viewed as a stream of photons, each photon carrying an energy $h\nu$.

Now, if $E_2 - E_1 = h\nu$, the interaction of radiation with atom leads to the following three distinct processes in the medium: (i) *Stimulated absorption*, (ii) *Spontaneous emission* and (iii) *Stimulated emission*.

If the energy of incident photon is not equal to the energy difference between two permitted atomic levels, it passes without interacting with atoms. If it is equal to the energy difference, it is either absorbed or emitted. When more atoms are in lower energy level, it is absorbed and when there exists population inversion it is emitted with additional radiation. When atom absorbs energy, it goes in the excited state. It is called *stimulated absorption*. The excited state being unstable, atom immediately returns to lower state by radiating the energy difference. The frequency of radiation is $\nu = (E_2 - E_1)/h$, where $(E_2 - E_1)$ is the energy difference between the levels and h is Planck's constant. This process is called *spontaneous emission*. The radiation is random and incoherent. But, if there exists population inversion and a photon of correct energy is incident on excited atoms, all atoms return to lower energy level by emitting photons which are in phase with triggering photon. This process is called *stimulated emission*. Absorption, spontaneous processes and stimulated emission processes are illustrated in Fig. 6.2. The radiated (photon or) wave by this process is coherent with (triggering or) stimulating (photon or) wave. It has same frequency, direction of propagation, polarization etc. to that of triggering wave. Thus output is the amplification of input (triggering) wave. The wave needed for amplification is supplied by pumping process.

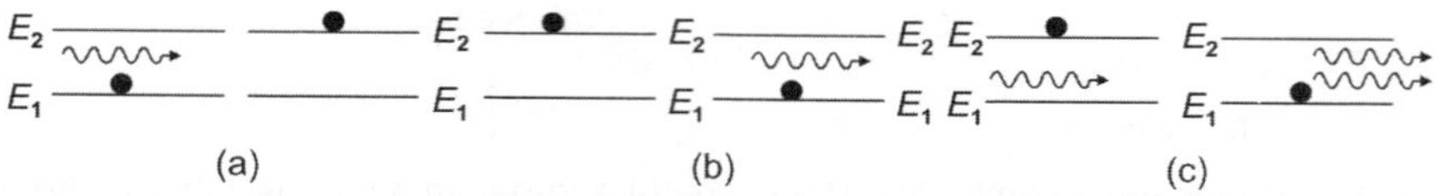

Fig. 6.2. (a) Absorption, (b) spontaneous emission, (c) stimulated emission

As shown in Fig. 6.2(c) that one original photon of energy $h\nu$ and other emitted photon on application of external electromagnetic radiation energy move together. The direction of propagation, phase energy and state of polarization of emitted photon is quite same as that of stimulating photon. This results in an enhancement of *coherent light*.

The probable rate of stimulated transition from excited state to lower energy state is proportional to the energy density $u(\nu)$ of the stimulating photon and is expressed as:

$$(P_{21})_{\text{stimul.}} = B_{21}\, u(\nu) \tag{6.3}$$

where B_{21} is the Einstein's constant of stimulated emission of radiation.

Now, the total probability (probable rate) of emission transition from excited state to lower energy state is summation of probable rates of spontaneous and stimulated emission:

$$P_{21} = A_{21} + B_{21}\, u(\nu) \tag{6.4}$$

where A_{21} is the Einstein's constant of stontaneous emission of radiation $((P_{21})_{\text{spont.}} = A_{21})$.

6.3 RELATION BETWEEN EINSTEIN'S COEFFICIENTS

Let us consider an assembly of atoms in thermal equilibrium at a particular temperature T with radiation of frequency ν and energy density $u(\nu)$ in a two level system (TLS). Let at any instant N_1 and N_2 are the number of atoms in lower energy state E_1 and excited energy state E_2 respectively.

In state E_1, the number of atoms that absorb a quantum (photon) and rise to excited state E_2 per unit time is:

$$N_1 P_{12} = N_1 B_{12}\, u(\nu) \tag{6.5}$$

Optical Sources **263**

The number of atoms in excited state E_2 that jump to lower energy state E_1 either spontaneously or under stimulating emitting a photon per unit time is:

$$N_2 P_{21} = N_2 [A_{21} + B_{21} u(\nu)] \tag{6.6}$$

At thermal equilibrium between atomic system and the radiation field, the number of upward transitions should be equal to the number of downward transitions. Thus, at thermal equilibrium:

$$N_1 P_{12} = n_2 P_{21}$$

or
$$N_1 B_{12} u(\nu) = N_2 [A_{21} + B_{21} u(\nu)]$$
$$= N_2 A_{21} + N_2 B_{21} u(\nu) B_{21} u(\nu) \tag{6.6a}$$

or $\quad u(\nu) N_1 B_{21} - N_2 B_{21} u(\nu) = N_2 A_{21}$

or
$$u(\nu) = \frac{N_2 A_{21}}{N_1 B_{12} - N_2 B_{21}}$$

or
$$u(\nu) = \frac{A_{21}}{B_{21}} \frac{1}{\left[\frac{N_1}{N_2} \left(\frac{B_{12}}{B_{21}}\right) - 1\right]} \tag{6.7}$$

However, from *Maxwell's distribution law*, the number of atoms N_1 and N_2 in energy states E_1 and E_2 in thermal equilibrium at temperature T are given by:

$$N_1 = N_0 \exp(-E_1/k_B T)$$

and
$$N_2 = N_0 \exp(-E_2/k_B T)$$

where N_0 is the number of atoms present in the ground state, and k_B is Boltzmann's constant.

Thus, we have:

$$\frac{N_2}{N_1} = \frac{e^{-E_2/k_B T}}{e^{-E_1/k_B T}} = e^{-(E_2/E_1)/k_B T}$$

But, $E_2 - E_1 = h\nu$, i.e. energy of emitted or absorbed photon. Then, we have:

$$\frac{N_2}{N_1} = e^{-h\nu/k_B T}$$

or
$$\frac{N_1}{N_2} = e^{-h\nu/k_B T} \tag{6.8}$$

Using Eq. (6.8), Eq. (6.7) takes the form:

$$u(\nu) = \frac{A_{21}}{B_{21}} \frac{1}{\left[e^{h\nu/k_B T} \left(\frac{B_{12}}{B_{21}}\right) - 1\right]} \tag{6.9}$$

The energy density $u(\nu)$ at a particular frequency ν is given by Planck's radiation formula is:

$$u(\nu) = \frac{8\pi h \nu^3}{c^3} \frac{1}{\left[e^{h\nu/k_B T} - 1\right]} \qquad (6.9a)$$

Assuming $B_{12} = B_{21} = B$ (as per assumption of Einstein) and comparing Eq. 6.9(a) and (6.9), one obtains:

$$\frac{A_{21}}{B_{21}} = \frac{8\pi h \nu^3}{c^3} \qquad (6.9b)$$

Equation 6.9(b) is known as *Einstein's relation* for his coefficients A and B.

We may note that the ratio of spontaneous emission and stimulated emission is proportional to ν^3. This means that the probable rate of spontaneous emission primes over stimulated emission more and more as the energy difference between the two states increases.

6.4 LASER COMPONENTS

A laser generally consists of three components: an *active medium* with energy levels that can be selectively populated, a *pump* to produce *population inversion* between some of these levels, and a *resonant electromagnetic cavity* that contains the active medium and provides feedback to maintain the coherence of electromagnetic field. In a continuously operating laser, coherent radiation will built up in the cavity to the level required to balance stimulated emission and cavity losses. The system is then said to be *lasing*, and the radiation is emitted in a direction defined by the cavity. Figure 6.3a shows a schematic block diagram of laser.

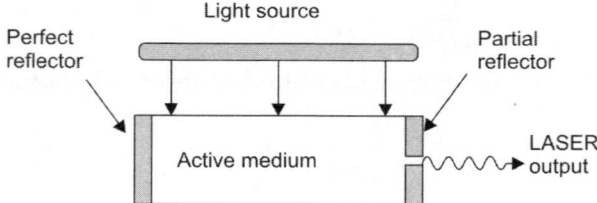

Fig. 6.3a. Schematic block diagram of laser

To achieve the population inversion a right group of atoms or molecules, which is called *active medium*, is needed. This active medium is gas or mixture of gas or crystal or semiconductor, etc. The population inversion is created by pumping process. Various techniques are used for pumping, for example, optical, pressure-temperature cycle, physical separation etc. The optical cavity is used to generate feedback. It is formed by dielectric mirrors or polished ends of crystal rod which act as reflectors. The pumping process creates population inversion. If that particular energy state is short lived or unstable, atoms go to another lower state by way of fast decay. Generally, this state (or level) is metastable. The *metastable state* means long life state compared to other states. In general, life time of an energy state is 10^{-8} s, but for metastable (temporarily stable) state, it is 10^{-3} s or more. This is the key to the laser. In many atoms, one or more metastable states are available.

After the population inversion, stimulated emission is triggered by a photon of appropriate frequency. In some cases, the first atom, making the transition from metastable state to lower state generates triggering photon. Thus, the stimulated emission produces more photons than the photons entering the system.

A material system is excited and displaced from normal thermal equilibrium by external processes such as chemical reaction, electron beam, optical field, etc. It selectively excites energy level of material, resulting in population inversion and finally in laser operation. The population inversion and its duration depends upon the relaxation rates of different energy levels, the degrees of freedom of the system and rate of stimulated emission. Consider a three-level and four-level laser system (see Figs 6.3b and c).

The pumping process excites the system from ground state to excited state. The excited state relaxes to the upper laser level. The stimulated emission (lasing action) occurs between upper and lower laser level. Finally lower laser level can either relax to the ground state or absorb the laser radiation and repopulate the upper laser level.

For ideal operation of laser, some conditions should be satisfied by the system. The relaxation rates from excited state to upper laser level and from lower laser level to ground state should be high to maintain maximum population inversion between upper to lower laser levels. The thermal equilibrium population of lower laser level should be as small as possible. The decay of excited state to any level other than upper laser level should be as slow as possible. The non-radiative decay of upper laser level should be slow.

The light emitted by laser is significantly sharper in wavelength than ordinary light because it is reflected back and forth in the cavity to form an intense pattern. We know the theory of stationary waves. The resonance occurs only for certain definite wavelengths, depending upon the length of resonator. The intense pattern is built up between two mirrors only if an exact number of waves fit between the mirrors. If there is only one wavelength that resonates, the laser gives out a single sharp line; if two or three wavelengths resonate, laser emits two or three lines.

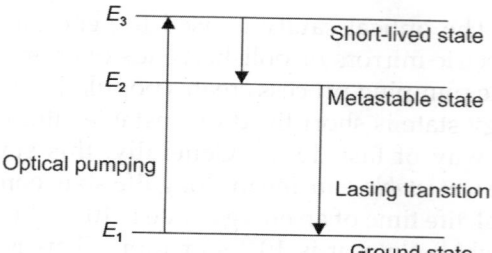

Fig. 6.3b. Transition in a three level laser. The lasing transition takes the system from the metastable state to the ground state

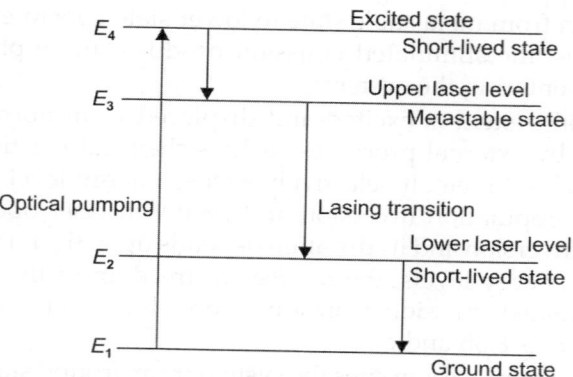

Fig. 6.3c. Four level system. The lasing transition takes the system from the metastable state (E_3) to another short-lived (lower laser level) state (E_2). The system then returns quickly to the ground state (E_1), so the photons cannot be reabsorbed in returning the system from E_2 to E_3

6.5 POPULATION INVERSION

The number of active atoms occupying an energy state is called *population* of that state. Hence, N_1 and N_2 are population of E_1 (lower) and E_2 (excited) levels or states respectively.

Population inversion is the state of a system at which the population of a particular higher energy state is more than that of a specified lower energy state.

Let us consider a two level system consisting of lower energy level E_1 and higher (excited) energy level E_2. At normal condition, the number of atoms in lower level, i.e. N_1 is more (greater) than that of the atoms in excited states, i.e. N_2.

In order to produce emission from higher energy level, the population of higher (excited) level must be as high as possible. If the number of atoms (population) of higher energy level is more than that of the low energy level, then we say, it is *population inversion*.

As stated earlier, the laser beam emission takes place only when the rate of emission is greater than the rate of absorption, i.e.

Rate of (stimulated + spontaneous) emission > Rate of absorption

or
$$A_{21}N_2 + B_{21}\, u(v)\, N_2 > B_{12}\, u(v)\, N_1$$

For lasing action to be dominant, only stimulated emission is to be considered here. Thus, deleting spontaneous emission, one obtains:
$$B_{21}\, u(v)\, N_2 > B_{12}\, u(v)\, N_1$$
Taking $B_{12} = B_{21} = B$ (say), one obtains:
$$N_2 > N_1$$

Gain: The probability that a randomly chosen photon in an optical field of cross-section A will stimulate a given inverted site with a radiative cross-section σ_r as it passes through a material is σ_r/A. The radiative cross-section of the transition is proportional to the *dipole strength* of the transition and is the same for absorption and emission. If the low-energy state is degenerate, with a degeneracy of g_1, the probability of stimulated emission becomes $g_1\sigma_r/A$. The product g_1, σ_r is known as the *emission or gain cross-section* σ_g. When all of the photons in the optical field are accounted for an absorption and stimulated emission are included, then we obtain:

$$I(l) = I(o)\, \exp\,[\sigma_g \rho_u - \sigma_a \rho_l)\, l] \qquad (6.9c)$$

[When all the photons in an optical field of intensity I and all of the absorption sites in a material of length dl are accounted for, the intensity of an optical field passing through the material changes by:

$$dI = -I\rho_l \sigma_a\, dl$$

where $\sigma_a = g_u \sigma_r$ is known as *absorption coefficient* and g_u are identical (degenerate) high energy states. P_1 is the density of absorption sites in the material. The above equation has the solution:

$$I(l) = I(o)\, \exp\,(-\alpha l)$$

where $I(o)$ is the intensity of the optical field as it enters the material at position $Z = o$ and $\alpha = \rho_1 \sigma_a$ is the absorption coeffient of the material. When all the photons in the optical field are accounted for and absorption and stimulated emission are included, this equation becomes 6.9(c)].

where ρ_u is the density of inverted sites. If the material is forced out of thermal equilibrium (pumped) to a sufficient degree, so that $\sigma_g \rho_u > \sigma_a \rho_l$, stimulated emission occurs at a higher rate than absorption. The material is now said to have *gain*, with a gain coefficient:

$$g = \left[\rho_u - \left(\frac{g_u}{g_1}\right)\rho_1\right]\sigma_g \qquad (6.9d)$$

The term $\rho_{\text{eff}} = \left[\rho_u - \left(\dfrac{g_u}{g_1}\right)\rho_1\right]$ is referred to as the effective inversion density, and $g = \rho_{eff}\sigma_g$. When $\rho_g\rho_u = \sigma_a\rho_1$, there is no change in the intensity of an optical field passing through the material and the material is said to be in a state of *transparency*.

6.6 CHARACTERISTICS OF LASER RADIATION

Laser radiation is available both in *continuous wave mode* and in pulses. Like all other radiations, laser radiation is electromagnetic in nature. Important characteristics of laser radiations are as follows:

(i) *Coherence*: The laser radiation is highly coherent both *temporarily* and *spatially*. Coherence means that two or more waves in a radiation field bear the *same phase* relationship to each other at all times. Coherence is of two types: *spatial coherence* and *temporal coherence* or *longitudinal coherence* and *transverse coherence*.

Spatial coherence or transverse coherence means the phase relationship between waves travelling in a plane perpendicular to the direction of propagation, i.e., between waves travelling side by side. Spatial coherence requires that the waves not only are of same frequency, but they should be in phase in space. *Temporal* coherence or *longitudinal* spatial coherence applies to waves travelling the same path.

(ii) *Monochromaticity*: The laser light is nearly monochromatic. Truly speaking, no light is perfectly monochromatic, i.e., it is not characterized by spread in frequency Δv about the central frequency (or $\Delta \lambda$ in case of wavelength λ). One can define the monochromaticity of light by $\Delta v/v$. For perfect monochromaticity $\Delta v = 0$, which is not possible to attain in practice. However, the value of Δv is much smaller for laser light compared to ordinary light. For an ordinary light $\Delta v \approx 10^{10}$ Hz whereas for a laser $\Delta v \approx 500$ Hz. Thus, for ordinary light of wave length $\lambda = 6000$Å (or frequency $v = 5 \times 10^{14}$ Hz) monochromaticity is:

$$\frac{\Delta v}{v} = \frac{10^{10} \text{ Hz}}{5 \times 10^{14} \text{ Hz}} = 2 \times 10^{-5}$$

Monochromaticity of laser light:

$$\frac{\Delta v}{v} = \frac{500 \text{ Hz}}{5 \times 10^{14} \text{ Hz}} = 10^{-12}$$

Obviously, laser light is highly monochromatic compared to ordinary light.

(iii) *Directionality*: The output beam of a laser light has a well-defined wavefront and therefore, it is highly directional except for the divergence caused due to diffraction effects. The high directionality of laser light permit us to focus it into a point by passing the beam through a suitable convex lens. For example, if the focal length of the lens is 5×10^{-2} m, $\lambda = 6000$ Å, and the radius of the beam, $r = 2$ mm, then the area of the spot at the focal plane is:

$$\frac{\pi\lambda^2 f^2}{a^2} = \frac{\pi(6\times 10^{-7}\text{ m})^2 \times (5\times 10^{-2}\text{ m})^2}{(2\times 10^{-3}\text{ m})^2}$$
$$= 7.1\times 10^{-10}\text{ m}^2$$

Obviously, this is very small.

(iv) *Intensity*: The laser light beam is highly intense compared to ordinary light. Since the power of a laser is concentrated in a beam of very small diameter (≈ few mm), even a small laser can deliver extremely high intensity at the focal plane of the lens. For example, if P (power) of a laser beam is 1 W, the intensity at the point is obtained as:

$$I = \frac{P}{A\text{ (area)}} = \frac{Pa^2}{\pi\lambda^2 f^2}$$
$$= \frac{1\text{ W}}{7.0\times 10^{-10}\text{ m}^2} = 1.4\times 10^9 \text{ Wm}^{-2}$$

This means that even a small power of 1 W can give an intensity of 10^9 Wm^{-2}, which is quite large.

6.7 TYPES OF LASERS

Broadly, there are six different types of lasers: solid state laser, gas laser, dye laser, semiconductor laser, UV and X-ray laser and free-electron laser. In solid-state lasers, active medium is an insulating dielectric solid. These solids can be crystalline or amorphous. These lasers are used to generate a wide range of wavelength from the vaccuum ultraviolet to mid-infrared. They are capable of generating high peak powers (~10^{-14} W) because of long life metastable states which allow higher energy storage compared to other active media. The common examples of solid-state lasers are ruby, which is a chromium-doped Al_2O_3, titanium-doped Al_2O_3 (Ti:Al_2O_3), neodymium-doped yttrium aluminium garnet (Nd:YAG) and neodymium-doped glasses (Nd:glass).

In gas lasers, gas or gas mixture is capable of withstanding a large amount of power. The pointing stability and high optical quality beams are their greatest assets. Common examples are copper-vapour laser, helium-neon laser, helium-cadmium laser and carbon-dioxide laser.

The free electron lasers (FEL) are operated in far infrared and ultraviolet (below 100 nm) where atomic or molecular lasers are not readily available; and for large average power and high efficiency systems. The disadvantages of FEL are greater complexity and the cost of a particle accelerator.

The important characteristics of laser dye media are their broad wavelength tunability, wide spectral coverage and practical simplicity. Hundreds of dyes are reported to have lasing action. The examples of various dye classes are Oligophenylenes, Coumarins, Xanthenes,

Merocyanines and Cyanines with spectral emission from 300 nm to more than 1100 nm.

The semiconductor lasers have some special features such as compactness, high efficiency, capability for high-speed direct modulation, wide emission spectrum and high reliability. But disadvantage is that they are sensitive to temperature.

A significant percentage of today's lasers are fabricated using semiconductor technology. Those devices are known as semiconductor lasers. In the present chapter we will restrict to the study of semiconductor lasers only.

6.8 LASER OSCILLATIONS AND RESONANT MODES

Light propagation with amplification is illustrated in Fig. 6.4. Mathematically, it is described by assuming that there is no phase change on reflection at either end (left and right). Left end is defined as $z = 0$) and right end as $z = L$. At the right facet, the forward optical wave has a fraction r_R reflected (amplitude reflection) and after reflection that fraction travels back (from right to left).

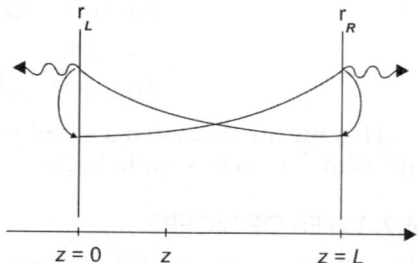

Fig. 6.4. Schematic illustration of the amplification in a Fabry-Perot (FP) semiconductor laser with homogeneously distributed gain

In order to form a stable resonance, the amplitude and phase of the wave after a single round trip must match the amplitude and phase of the starting wave. At arbitrary point z inside the cavity (Fig. 6.4), the forward wave is:

$$E_0 e^{gz} e^{-j\beta z} \qquad (6.10)$$

where we have dropped $e^{i\omega t}$ term which is common and defined $g = g_m - \alpha_m$, where g_m describes gain (amplification) of the wave and α_m its losses. Also r_R and r_L are, respectively, right and left reflectivities, L length of the cavity and β propagation constant.

The wave travelling one full round will be:

$$\{E_0 e^{gz} e^{-j\beta z}\} \{e^{g(L-z)} e^{-j\beta(L-z)}\} \{r_R e^{gL} e^{-j\beta L}\} \{r_L e^{gz} e^{-j\beta z}\} \qquad (6.11)$$

The above terms are interpreted as follows. In the first bracket, there is an original forward propagating wave which started at z, in the second bracket, there is a wave travelling from z to L, the third bracket describes a wave propagating from $z = L$ to $z = 0$, and the last one contains a wave travelling from $z = 0$ to the starting point z. At that point, the wave must match the original wave as given by Eq. (6.10). From the above, one obtains a condition for stable oscillations:

$$r_R r_L e^{2gL} e^{-2j\beta L} = 1 \qquad (6.12)$$

Optical Sources 271

This condition can be split into an amplitude condition:
$$r_R r_L e^{2(g_m - \alpha_m)L} = 1 \qquad (6.13)$$

and phase condition:
$$e^{-2j\beta L} = 1 \qquad (6.14)$$

From the amplitude condition one obtains:
$$g_m = \alpha_m + \frac{1}{2L} \ln \frac{1}{r_R r_L} \qquad (6.15)$$

From the phase condition, it follows:
$$2\beta L = 2\pi n \qquad (6.16)$$

where n is an integer. The last equation determines wavelengths of oscillations since:
$$\beta = \frac{2\pi}{\lambda_n} = \frac{\omega_n}{c} \qquad (6.17)$$

with λ_n being the wavelength. Typical gain spectrum and location of resonator modes are shown in Fig. 6.5(a). Longitudinal modes with angular frequencies ω_{n-1}, ω_n and ω_{n+1} are shown. In time, the mode which has the largest gain will survive; the other modes will diminish (Fig. 6.5(b)).

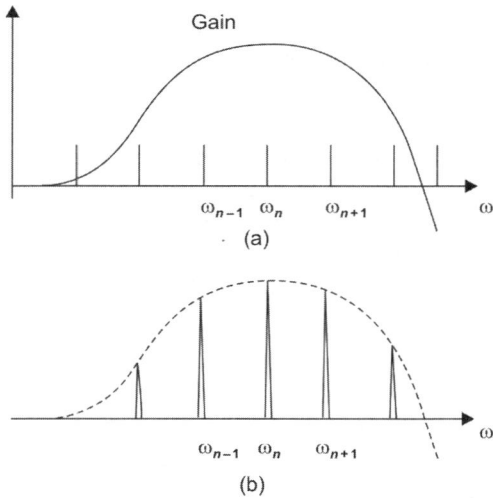

Fig. 6.5. Gain spectrum of semiconductor laser and location of longitudinal modes. ω_n are FP resonances determined from phase condition

6.9 SEMICONDUCTOR LASERS

Coherent laser radiation ranging from the infrared, through the ultraviolet regions of the spectrum has been obtained from semiconductors and semiconductor junction diodes.

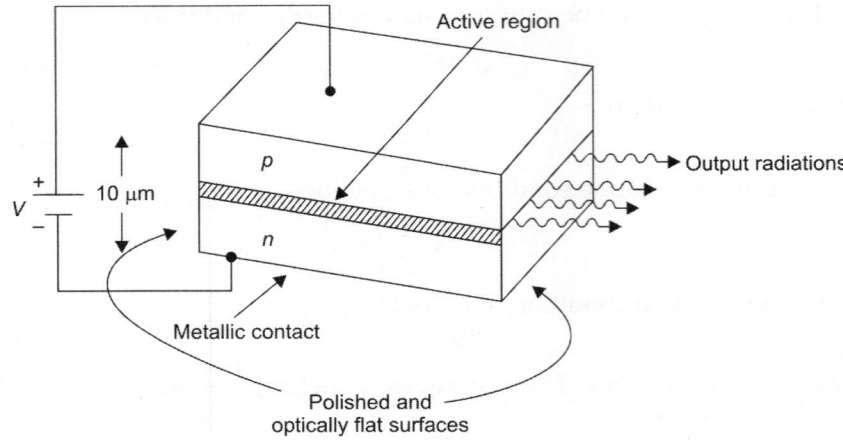

Fig. 6.6. (a) Semiconductor laser

Laser radiation can be obtained by pumping a solid intrinsic semiconductor or a *pn* junction diode. The method of pumping varies with the type of semiconductor used (Fig. 6.6(a)). The most common type of semi-conductor used is a III-V compound. Semiconductor lasers are different from other lasers in the following aspects:

(i) Transition occurs between energy bands rather than between discrete energy levels, the emission being a result of electron transitions from the conduction to the valence bands.

(ii) The laser is very small in size, typically 50 μm × 250 μm × 50 μm.

(iii) The characteristics of the laser beam are strongly influenced by the properties of the junction material.

(iv) In *pn* junction lasers, population inversion occurs in the very narrow region about the junction. The pumping of the *pn* junction laser is accomplished by the application of a forward bias to the diode.

It is a property of semiconductors that the downward transitions of electrons recombining with holes may or may not result in the emission of option of optical energy. Based on this property, semiconductors can be classified as *indirect and direct*. This classification is related to the energy band structure, silicon and germanium are indirect types, where as most of the III-V compounds are direct. In direct types, radiative transitions take place faster than non-radiative and impurity transitions. In direct semiconductors, radiative recombination's occur faster than radiative recombinations, so that in essence photon emission is suppressed. In fact, some radiation is emitted by almost all semiconductor *pn* junctions, but in junctions using the indirect type of semiconductors, this radiation is very inefficient.

6.9.1 Types of Semiconductor Laser

In a simple homojunction semiconductor laser, one can achieve relatively high threshold current densities because both carriers and optical confinement within the active region of device are minimum. Carriers injected into active region of device pass through it without any recombination and do not contribute to laser action. It is also seen that the optical fields leak into the surrounding inactive layers. Due to these two reasons, the threshold current density increases which reduces the efficiency of laser action.

Heterostructure provides a mean of confining both the optical fields and carriers within the active region and reducing the threshold current density. For heterostructure, it is required that the lattice constants of two materials should match, e.g., GaAs and AlGaAs. The structure of a simple GaAs-AlGa is shown in Fig. 6.6(b) which is called a **double heterostructure laser** because it contains two GaAs-AlGaAs interface. Since the energy bandgap of AlGaAs is greater than GaAs, therefore, there is an energy gap discontinuity between two materials. GaAs and AlGaAs are said to form a type I heterojunction in which the energy gap discontinuity is equal to the sum of the conduction and valence band edge discontinuities.

The presence of conduction and valence band energy discontinuities in the two heterostructures act to greatly confine the injected carriers within the narrow gap GaAs layer as shown in Fig. 6.6(c) which ultimately can result in a radiative recombination event with the subsequent emission of a photon. Heterostructure has lower threshold current density than the homojunction laser. The double heterostructure

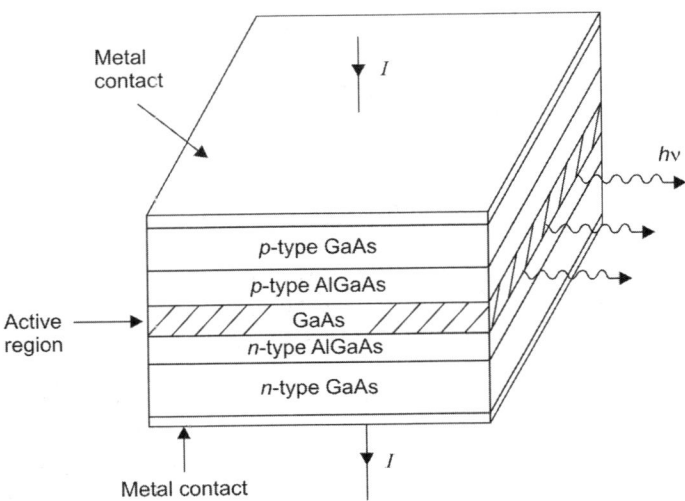

Fig. 6.6. (b) Structure of GaAs-AlGa laser

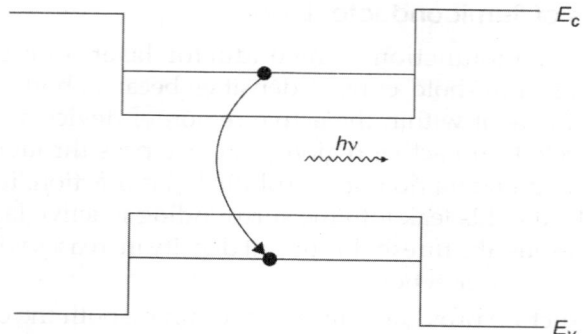

Fig. 6.6. (c) Radiative event in GaAs layer

also acts to confine the optical fields within the GaAs layer due to different refractive indices of the used semiconductor materials.

In the laser action photon is emitted by stimulated emission. The light bounces back and forth due to reflection from the boundaries of the device. The reflecting surfaces are produced by cleaving the crystal such that the cleaved surface provides for nearly complete reflection. Stimulated emission with the GaAs layer is triggered by reflected light. The emitted photon has the same phase as that of the stimulating light, i.e., coherence is maintained. In double heterostructure laser, the emitted light has an energy close to that of the energy gap of narrow gap material. The energy of the emitted photon thus depends on the intrinsic properties of the active medium.

6.9.2 Quantum Well Lasers

Quantum well lasers offer an alternative to double heterostructures. In *quantum well lasers*, the energy of emitted photon can be made subsequently higher than that of the bandgap. By proper choice of well width, the energy of the emitted photon can be adjusted so that lasing can be achieved at various wavelengths using the same material. A single quantum well laser is similar in design to a double heterostructure laser except that the narrow gap. This means active region is intentionally made thin so that the spatial quantization effects occur. As we know that when the dimension of the confining region is equal to that of electron de Broglie wavelength, quantum mechanical effects occur. According to quantum mechanical theory, only certain discrete energy levels are allowed in the atom and electrons/holes can occupy only these allowed energy levels. Spatial quantization results in the production of energy levels above the conduction band minimum in a quantum well. The minimum electron and hole energies within the well are both above the conduction band and valence band minima respectively. If electron and hole recombine, the energy of the emitted photon is greater than that of the energy gap. It is found that the energy of the emitted photon is a strong function of the well width. The narrower the width, the greater the quantum state energy. This means in a very narrow width device, the energy of emitted photon is greater than that of a wider well width device. By adjusting the well width, the energy of the emitted photon can be tuned.

Limitation of quantum well lasers: The cross-section for the capture of carriers in the well is relatively small. In order to obtain a radiative recombination event, it is necessary for the injected electrons and holes to be captured within the well. To confine the carriers within the well, it is required that they must undergo an inelastic scattering event in order to lose sufficient energy. If there is no scattering events within the well width, the injected carriers will traverse the well without being captured. In other words, we can say that the injected carriers can pass directly over the well without becoming trapped and do not contribute to the optical output. As a result, threshold current density can be relatively high.

Multiple quantum well lasers: The capture probability of a single quantum well laser can be improved by using multiple quantum wells as shown in Fig. 6.7(a), the single quantum well within the active region is replaced by a series of quantum wells. Since, more quantum wells are present in the structure, therefore, there is a higher probability that the injected carrier will suffer an inelastic scattering event and be captured within a well before it can completely traverse the device. Main drawback of multiple quantum well device is that the injected electrons and holes enter the active region from opposite ends. Therefore, the electrons and holes are not necessarily captured within the same well which acts to reduce the radiative recombination rate. For this reason the multiple quantum well devices do not have the lowest threshold current density of semiconductor laser.

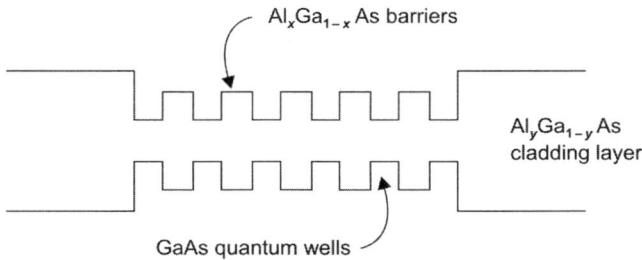

Fig. 6.7. (a) Structure of multiple quantum well laser

GRINSCH laser: The acronym stands for *graded index separate confinement heterostructures*. This has the lowest threshold current density of existing semiconductor lasers. The structure of this laser is shown in Fig. 6.7(b).

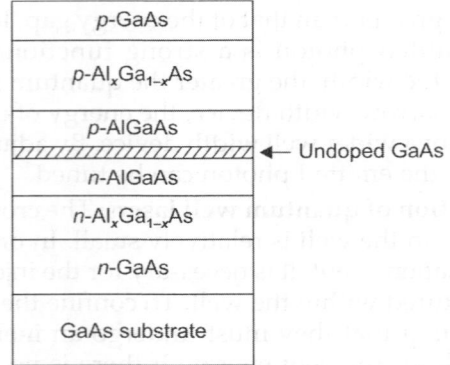

Fig. 6.7. (b) Structure of GRINSCH laser

In the structure as shown in Fig. 6.7(b), a single quantum well is embedded within a graded funlike region. The graded region surrounding the quantum well provide a means by which the injected carriers can be funneled into the well. The bandgap grading induces an electric field that directs the electrons and holes into the same quantum well where they can recombine radiatively. As a result, fewer electrons and holes can traverse the quantum well without being trapped. Thus, the threshold current density of GRINSCH laser is relatively lower than any other lasers.

6.9.3 Quantum-Dot Lasers

More recently, quantum-well lasers have been developed in which the device contains a single discrete atomic structure or so-called *quantum dot* (QD). Quantum dots are small elements that contain a tiny droplet of free electrons forming a quantum-well structure. Hence a QD laser is also referred to as a dot-in-a-well device. They are fabricated using semiconductor crystalline materials and have typical dimensions between nanometers and a few microns. The size and shape of these structures, and therefore, the number of electrons they contain may be precisely controlled such that a QD can have anything from a single electron to several thousand electrons. Theoretical treatment of QDs indicates that they do not suffer from thermal broadening and their threshold current is also temperature insensitive. If the conventional injection laser diode is regarded as three dimensional and a quantum well (i.e. an SQW where an array of SQWs forms an MQW structure) is confined to two dimensions, then the QD structure can be considered to be zero dimensional. It should be noted, however, that the single dimensional structure forms a quantum wire or dash.

The above hierarchy is illustrated in Fig. 6.8 which identifies four different possible structure for the semiconductor laser with their corresponding energy responses with respect to carrier densities shown underneath. The three-dimensional structure of the conventional injection laser diode on the left display an exponential variation in the density of states for the charge carriers. As SQW structure exhibits two dimensions (i.e. length and height) where the corresponding energy representation is shown in Fig. 6.8 by a staircase response in the carrier density of states. However, when this structure is reduced to one dimension (i.e. length only), it displays a sharp rise and an exponential fall in the carrier density variation. Since this one-dimensional quantum-well structure is confined to only the device length, then, in general, it appears as a long wire and hence, it is known as a quantum wire. The zero-dimensional (i.e. single-point) structure shown on the right of Fig. 6.8, however, corresponds to a single QD which results in an impulse response for the variation in the charge carrier density with increasing number of carriers.

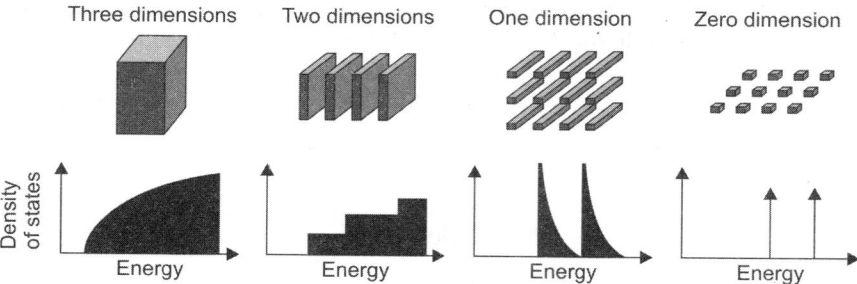

Fig. 6.8. Schematic illustration and density states for semiconductor lasers. From left to right: conventional injection laser diode; multiple quantum wells; array of quantum wires; an array of quantum dots. Shown underneath each of these illustrations is the corresponding density of states for each type of laser structure

The size and shape of the structure for a QD laser can be altered as required during the fabrication process. For example, in fabrication arrays of QDs can be formed on a GaAs substrate with different shapes being produced. Shapes such as the cube, circular disk, cylinder, pyramid or truncated pyramid can be created from self-organized crystalline growth of InGaAs material on the GaAs substrate. Each of these crystal shapes possesses different material characteristics (i.e. elasticity, stress, strain distribution, etc.) and therefore, their different shapes and sizes produce a varying impact on the operation of the QD laser (i.e. emission wavelength, polarization and operating temperature). By contrast, regularity of size and shape in an array improves the control of the QD device lasing frequency and intensity.

One of the important features of the QD laser is its *very low-threshold current density*. For example, low-threshold current densities between 6 and 20 $A \cdot cm^{-2}$ have been obtained with InAs/InGaAs QD lasers emitting at the wavelength of 1.3 μm and 1.5 μm. These low values of threshold current density make it possible to create stacked or cascaded QD structures thus providing high optical gain suitable for the short-cavity transmitters and vertical cavity surface-emitting lasers. Despite the potential benefits of QD technology, issues remain in relation to materials technology and in the design and fabrication techniques to facilitate the large-scale production of QD devices.

6.9.4 Vertical Cavity Surface Emitting Lasers (VCSELs)

In this structure, the light is emitted normal to the surface as shown in Fig. 6.9. The primary advantages of VCSELs is that the single mode operation can be achieved due to the short cavity length of the device. These devices are better suited than the edge emitting lasers to forming a two-dimensional away. Single mode operation is possible with VCSELs since the mode spacing is inversely proportional to the cavity length. Thus, the smaller the cavity length, the greater the mode spacing which leads to single mode operation. The output wavelength l_0 is given by:

$$\lambda_0 = \lambda n' \qquad (6.18)$$

where λ is wavenelgth of incident light and n' is the refractive index within the laser:

$$\text{Mode index } m = \frac{2Ln'}{\lambda_0}$$

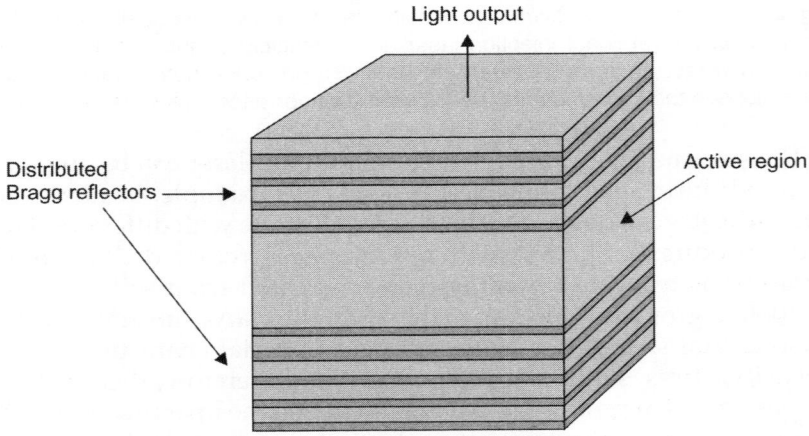

Fig. 6.9. Structure of VCSELs laser

Hence:

$$\frac{dm}{d\lambda_0} = \frac{-2Ln'}{\lambda_0^2} + \frac{2L}{\lambda_0}\frac{dn'}{d\lambda_0} \qquad (6.19)$$

where $dm/d\lambda_0$ defines the mode spacing with wavelength. Solving Eq. (6.19), we have:

$$d\lambda_0 = \frac{\left(\dfrac{\lambda_0^2}{2Ln'}\right) dm}{\left[1 - \dfrac{\lambda_0^2}{n'}\dfrac{dn'}{d\lambda_0}\right]} \tag{6.20}$$

The value of dm is 1 for adjacent modes.

From Eq. (6.20), it is clear that the wavelength separation between adjacent modes is inversely proportional to the cavity length.

The major disadvantage of VCSELs is that the cavity length is short as the round trip gain of the device is low. The threshold current density of VCSELs is significantly higher than that of an edge emitting laser.

The operation of semiconductor lasers as sources of electromagnetic radiation is based on the interaction between EM radiation and the electrons and holes in semiconductors (structure of basic *pn* junction semiconductor laser is shown in Fig. 6.10). These are: *vertical cavity surface emitting lasers* (VCSELs), where light propagates perpendicularly to the main plane and in-plane laser where light propagates in the main plane (Fig. 6.11). The largest dimensions of in-plane structures is typically in the range of 250 μm (longitudnial direction), whereas the typical diameter of VCSEL cylinder is about 10 μm.

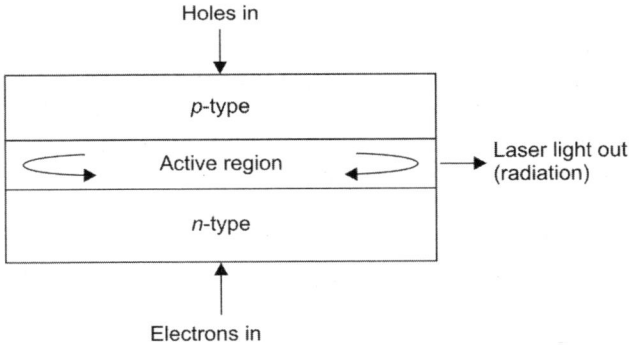

Fig. 6.10. The basic *p-n* junction laser

The basic semiconductor laser is just a *p-n* junction (Fig. 6.10), where cross-section along lateral-transversal directions is shown. Current flows (holes on *p*-side and electrons on *n*-side) along the transversal direction, whereas light travels along the longitudinal direction and leaves device at one or both sides.

In VCSEL, the cavity is formed by the so-called *Bragg mirrors* and an active region typically consists of several *quantum well layers* separated by barrier layers (Fig. 6.11). Bragg mirrors consists of several layers of different semiconductors which have different values of refractive index. Due to the Bragg reflection such structure shows a very large reflectivity

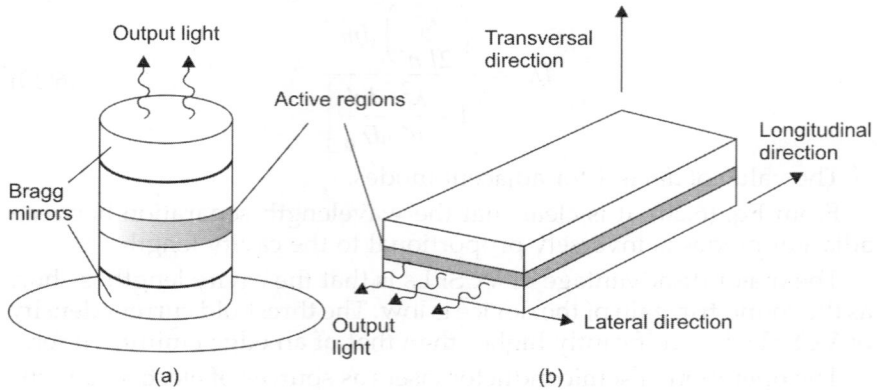

Fig. 6.11. VCSEL (left) and in-plane laser (right)

(around 99.9%). Such large values are needed because a very short distance of propagation of light does not allow to build enough amplification when propagating between distributed mirrors.

A *three-dimensional prospective view* of some generic semiconductor lasers is shown in Fig. 6.11. The structure consists of many layers of various materials, each engaged in a different role. Those layers are responsible for the efficient transport of electrons and holes from electrodes into an active region and for confinement of carriers and photons, so they can strongly interact. Modern structures contain so-called quantum wells which form an active region and where conduction-valence band transitions are taking place. It is possible to have different types of mirrors as they also provide mode selectivity. Two basic types are illustrated in Figs 6.12 and 6.13. They are known as *distributed feedback* (DFB) and *distributed Bragg* (DBR) structures.

In DFB lasers, grating (corrugation) is produced in one of the cladding layers, thus creating Bragg reflections at such periodic structure. The structure causes a wavelength sensitive feedback. It should be emphasized that the grating extends over the entire laser structure. When one restricts corrugation to the mirror regions only and leaves a flat active region in the middle, then the so-called DBR structure

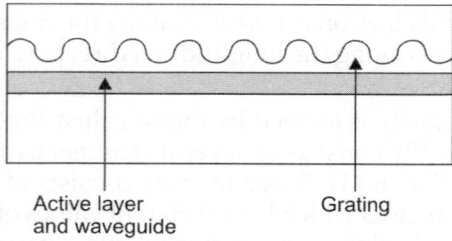

Fig. 6.12. Basic DFB laser structure

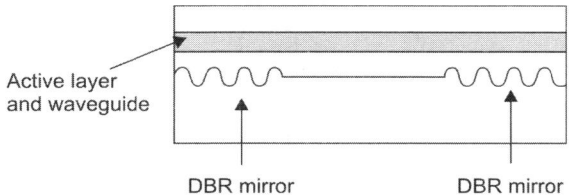

Fig. 6.13. Basic DBR laser structure.

is created (Fig. 6.13), which also provides wavelength sensitive feedback

6.9.5 Electron Transitions in Semiconductors

We now extend the previous discussion of TLS to describe the transitions between bands in semiconductors (Fig. 6.14). We introduce:

- f_v the probability of the state of energy E_v in the valence band being filled.
- f_c the probability of the state of energy E_c in the conduction band being filled.

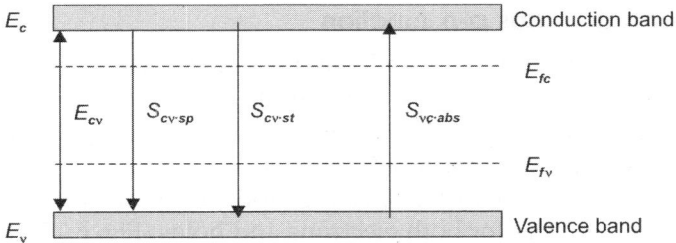

Fig. 6.14. Electron transitions between conduction and valence bands

Rates of transitions can now be determined, similarly to the TLS description. Here, we use index rotation appropriate to semiconductors, i.e. c, v instead of 1, 2. The rates are:

$$S_{spon} = S_{cv \cdot sp} = A_{21} f_c (1 - f_v) \qquad (6.21)$$

$$S_{stim} = S_{cv \cdot st} = B_{21} f_c (1 - f_v)(E_{cv}) \qquad (6.22)$$

$$S_{abs} = S_{cv \cdot abs} = B_{12} f_v (1 - f_c)(E_{cv}) \qquad (6.23)$$

Fermi-Dirac statistics are assumed for the appropriate probabilities as:

$$f_v = \frac{1}{\exp\left(\dfrac{E_v - E_{Fc}}{kT} + 1\right)} \qquad (6.24)$$

$$f_c = \frac{1}{\exp\left(\dfrac{E_c - E_{Fc}}{kT} + 1\right)} \quad (6.25)$$

Here E_{Fc} and E_{Fv} are the quasi-Fermi levels for electrons and holes.

The condition for stimulated emission to exceed absorption is:

$$S_{stim} > S_{abs}$$

which gives:

$$B_{21}f_c = (1 - f_v) > B_{12}f_v (1 - f_c)$$

Since the coefficnents B_{21} and B_{12} are equal, after substituting expressions (6.24) and (6.25), one obtains:

$$\exp\left(\frac{E_v - E_{Fv}}{kT}\right) > \exp\left(\frac{E_c - E_{Fc}}{kT}\right)$$

which can be written in the form:

$$E_{Fc} - E_{Fv} > E_c - E_v = h\nu \quad (6.26)$$

The above inequality is known as *Bernard-Duraffourg condition*.

6.9.6 Homogeneous *p-n* Junction

Figure 6.15 shows energy band diagram of *p-n* junction in thermal equilibrium and under *forward bias* condition.

Electrons at the *n*-type side do not have enough energy to climb over the potential barriers to the left. Similar situation exists for holes in the *p*-type region (Fig. 6.15(a)). One needs to apply forward voltage to lower the potential barriers for both electrons and holes (Fig. 6.15(b)).

An application of the forward voltage also separates Fermi levels. Thus, two different so-called quasi-Fermi levels E_{Fc} and E_{Fv} are created which are connected with an external bias voltage as:

$$E_{Fc} - E_{Fv} = eV_{bias} \quad (6.27)$$

With the lowered potential barriers, electrons and holes can penetrate central region where they can recombine and produce photons. However, the confinement of both electrons and holes into the central region is very poor (there is no mechanism to confine those carriers). Also, there is no confinement of photons (light) into the region where electrons and holes recombine (the region is known as an active region). Therefore, the interaction between carriers and photons is weak, which makes homojunctions a very poor light source. One must, therefore, provide some mechanism which will confine both carriers and photons into the same physical region where they will strongly interact. Such a concept is possible with the invention of heterostructures.

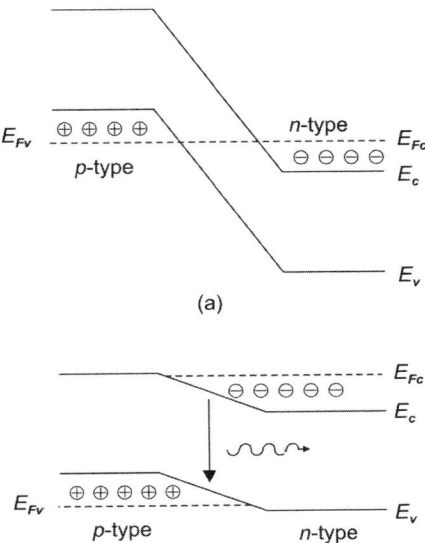

Fig. 6.15. Energy-band diagram of a *p-n* junction. (a) thermal equilibrium, (b) forward bias

6.9.7 Heterostructures

A heterojunction is formed by joining dissimilar semiconductors. The basic type is formed by two heterojunctions and it is known as a *double-heterojunction* (more popular name is *double-heterostructure*).

The materials forming double heterostructures have different bandgap energies and different refractive indices. Therefore, in a natural way potential wells for both electrons and holes are created. Schematic energy-band diagram of a double-heterostructure *p-n* junction in thermal equilibrium and under forward bias conditions are shown in Fig. 6.16. The bandgap and refractive index for *InGaAsP* material will now be summarized.

Band Gap

For the $In_{1-x}Ga_xAs_yP_{1-y}$ system, the relation between compositions x and y which results in the lattice match to *InP* is:

$$y = \frac{0.1894y}{0.4184 - 0.013y} \quad (6.28)$$

In such case the bandgap is:

$$E_{gap}[eV] = 1.35 - 0.72y + 0.12y^2 \quad (6.29)$$

with the extra relations $y \approx 2.20x$ and $0 \leq x \leq 0.47$. The material for such compositions thus covers bandgaps in the range of wavelengths

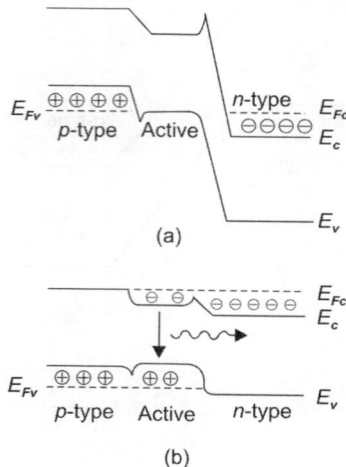

Fig. 6.16. Energy-band diagram of a double-heterostructure *p-n* junction. (a) Thermal equilibrium, (b) forward bias

0.92 µm to 1.65 µm. For example, the material $In_{0.74}Ga_{0.26}As_{0.57}P_{0.43}$ (i.e. $x = 0.26$, $y = 0.57$) has bandgap $E_g = 0.97$ eV, which corresponds to the wavelength $\lambda = 1.27$ µm.

6.9.8 Refractive Index

As explained earlier, the dependence of refractive index as a function of wavelength is described by *Sellmeier equation*. For important telecommunication wavelengths, namely 1.3 µm and 1.55 µm, the y dependence of refractive index $In_{1-x}Ga_xAs_yP_{1-y}$ that is lattice matched to *InP* is shown in Table 6.1.

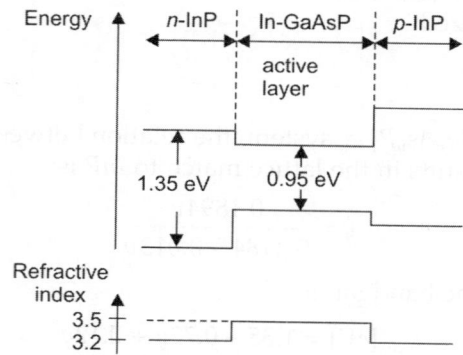

Fig. 6.17. Band structure and refractive index for *InGaAsP/InP* heterostructure

Table 6.1. Typical values of refractive indices for various compositions

Wavelength (λ)	Range of y compositions	Refractive index (n)
$\lambda = 1.3$ mm	$0 \leq y \leq 0.6$	$n(y) = 3.205 + 0.34y + 0.21y^2$
$\lambda = 1.55$ mm	$0 \leq y \leq 0.9$	$n(y) = 3.166 + 0.26y + 0.09y^2$

6.9.9 Optical Gains

For efficient and reliable numerical simulations, an exact mathematical expression of the material gain is critical. Such an expression should also be subject to an experimental verification. This problem has been investigated since the early developments of semiconductor lasers. The first step is usually the determination of gain spectra and its comparison with experiment. The simplest approach will be provided in the next section. Typical curves of gain spectra for a *four quantum well system* & comparison with experimental measurements are shown in Fig. 6.18. Mathematics involved in determining optical gain of semiconductor quantum-well structures is complex. As a result, analytic approximations of the optical gain which can be used in fast calculations were determined. From an extensive discussion, it was determined that the peak material gain for bulk materials varies linearly with carrier density.

$$g(N, \lambda) = a(\lambda) \min(N, P) - b(\lambda) \quad (6.30)$$

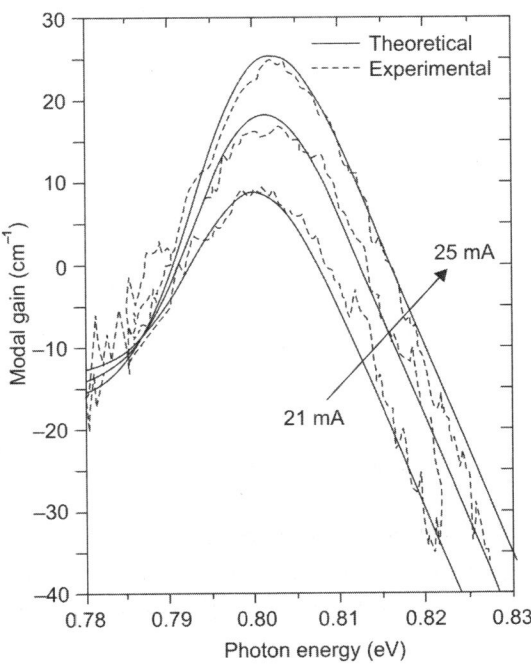

Fig. 6.18. Fitted theoretical and experimental net modal gain versus photon energy including leakage terms for three values of injected current

The parameter $a(\lambda)$ is commonly called the *differential gain*.

On the basis of experimental observations, Westbrook in 1986, 87 extended the linear gain peak model to allow for wavelength dependence. In its simplest form, it can be written as:

$$g(N, \lambda) = a(\lambda_p) N - b(\lambda_p) - b_a (\lambda - \lambda_p)^2 \quad (6.31)$$

where λ_p is the wavelength of the peak gain and b_a governs the base width of the gain spectrum. Wavelength peak λ_p can be also carrier density dependent.

Gain (absorption) in a semiconductor is a function of carrier density n [cm^{-3}] (Fig. 6.19), which shows gain spectrum for several values of carrier concentration. When n is below *transparency density* n_{tr}, the medium absorbs optical signal. For $n > n_{tr}$, optical gain in the material exceeds loss. The dependence of the so-called *gain peak* on the carrier density for quantum well systems is logarithmic (Fig. 6.20). It is described by the expression:

$$g = g_0 \ln \frac{N}{N_{tr}} \quad (6.32)$$

For modelling purposes, a linear approximation is often employed. In the above formulas g_0 is the differential gain, N_{tr} transparency carrier density and N_{th} is the carrier density at threshold.

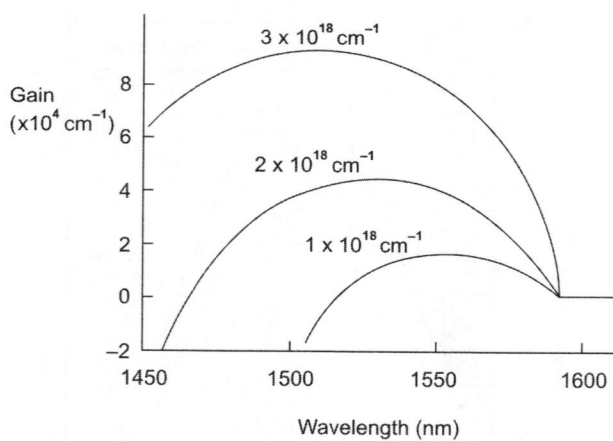

Fig. 6.19. Schematic of optical gain spectra for various carrier densities

6.9.10 Determination of Optical Gain

The simplest approach to determine *optical gain* is based on *Fermi golden rule*. Here we derive it for $T = 0$. Transition rate between two levels a and b is:

$$W_{ab} = \frac{2\pi}{\hbar} |H'_{ab}|^2 \delta(E_b - E_a - \hbar\omega) \quad (6.33)$$

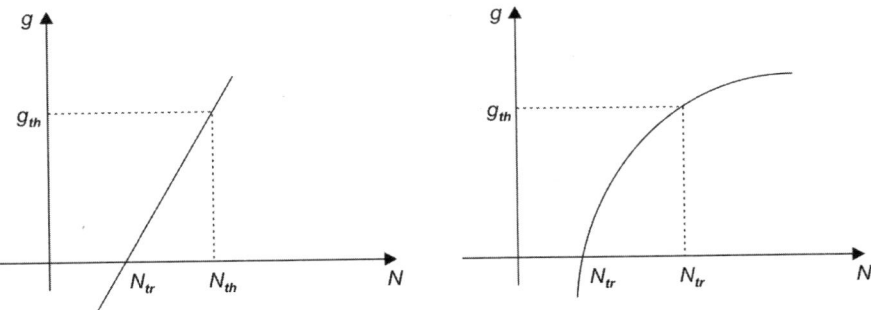

Fig. 6.20. Illustration of threshold and transparency densities for linear (left) and logarithmic (right) gain models

where H'_{ab} is the matrix element describing transitions between a and b. Assume parabolic bands for both electrons and holes with the effective mass m^* taking the value of m^*_c for conduction band and m^*_v for a valence band as follows:

$$E(k) = \frac{\hbar^2 k^2}{2m^*} \tag{6.34}$$

From the above:

$$E_b - E_a = \frac{\hbar^2 k^2}{2}\left(\frac{1}{m_c} + \frac{1}{m_v}\right) + E_g \tag{6.35}$$

Due to conservation of the crystal momentum in the transition, one has $k_{In} = k_{fi}$. For a single transition at specific value of k, one therefore has:

$$W(k) = \frac{2\pi}{\hbar}|H'_{vc}(k)|^2 \, \delta\left(\frac{\hbar^2 k^2}{2m_r} + E_g - \hbar\omega\right) \tag{6.36}$$

where $m_r = \dfrac{m_v m_c}{m_v + m_c}$ is reduced effective mass. If N is a total number of transitions per second in crystal volume V and $g(k) = \dfrac{k^2 V}{\pi^2}$ is number of states per unit k in volume V, the total number of transitions per second is:

$$N = \int_0^\infty W(k) \cdot g(k)\, dk$$

$$= \frac{2V}{\pi \hbar} \int_0^\infty |H'_{vc}(k)|^2 \, \delta\left(\frac{\hbar^2 k^2}{2m_r} + E_g - \hbar\omega\right) k^2 \, dk \tag{6.37}$$

To evaluate integral, let us introduce new variable:
$$X \equiv \frac{\hbar^2 k^2}{2m_r} + E_g - \hbar\omega$$
Integral takes the form:
$$\begin{aligned}N &= \frac{2v}{\pi\hbar} \int |H'_{vc}(k)|^2 \frac{m_r}{\hbar} \delta(X) \sqrt{\frac{2m_v}{\hbar^2}} (X + \hbar\omega - E_g) dX \\ &= \frac{V}{H} |H'_{vc}(k)|^2 \frac{(2m_c)^{3/2}}{\hbar^4} (\hbar\omega - E_g)^{1/2} \end{aligned} \quad (6.38)$$

where $\frac{\hbar^2 k^2}{2m_r} + E_g = \hbar\omega$. We are in a position now to determine absorption coefficient α. It is defined as:
$$\alpha(\omega) \equiv \frac{\text{power absorbed per unit volume}}{\text{power crossing a unit area}} = \frac{N \cdot (\hbar\omega)/V}{\varepsilon_0 n E_0^2 c / 2} \quad (6.39)$$
where n is refractive index, c is the velocity of light in a vacuum and E_0 is the amplitude of electric field. Use:
$$H'_{vc}(k) = \frac{eE_0 \chi_{vc}}{2}, \quad \chi_{vc} = \langle u_{vk'} | x | u_{ck} \rangle \quad (6.40)$$
Using the above and expression for N, one finds:
$$a_0(\omega) = \frac{\omega e^2 \chi_{vc}^2 (2m_r)^{3/2}}{2\pi\varepsilon_0 n c \hbar^3} (\hbar\omega - E_g)^{1/2} = K(\hbar\omega - E_g)^{1/2} \quad (6.41)$$

Example: Estimates of K

Data for GaAs. $\hbar\omega = 1.5$ eV, $m_v = 0.46 m_0$, $m_c = 0.067 m_0$, $\chi_{vc} = 3.2$Å, $n = 3.64$. Those data give $K \approx 11700$ cm^{-1} (eV)$^{-1/2}$, and $\alpha_0(\omega) = 1170$ cm^{-1}.

At T = 0 depending on the value of $\hbar\omega$, one has the following possibilities summarized next:

$\hbar\omega < E_g$ $\qquad \alpha(\omega) = 0$
$E_g < \hbar\omega < E_{Fc} - E_{Fv}$ $\quad \alpha(\omega) = -\alpha_0(\omega) = -K(\hbar\omega - E_g)^{1/2}$ amplification (6.42)
$E_{Fc} - E_{Fv} < \hbar\omega$ $\qquad \alpha(\omega) = -\alpha_0(\omega) = -K(\hbar\omega - E_g)^{1/2}$ absorption

6.10 RATE EQUATIONS

With all basic elements in place, we are now in a position to provide the simplest (phenomenological) description of semiconductor lasers based on rate equations. The main role in those devices is played by two subsystems: carriers (electrons and holes) and photons. They interact in the so-called **active region**, defined as part of the structure where recombining carriers contribute to useful gain and photon emission. We describe both subsystems separately, starting with carriers.

In Fig. 6.21, we schematically showed model of a laser operating below threshold. It resembles a tank partially filled with water continuously flowing in and at the same time water leaves the tank. The water models carriers which are continuously provided by current flowing in. Not all the current at electrodes actually reaches the device (tank). Some of it is lost as so-called leakage current. Below threshold, carriers disappears through losses in the device (R_{loss}), non-radiative recombination (R_{nr}) and spontaneous emission (R_{sp}). Above threshold (Fig. 6.22), the tank is completely filled with water (in laser the situation corresponds to the so-called threshold with density (N_{th}). Its operation is mostly dominated by yet another process, namely stimulated emission (here characterized by coefficient R_{st}).

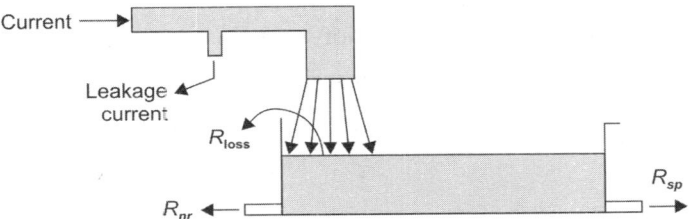

Fig. 6.21. Model of a laser operating below threshold

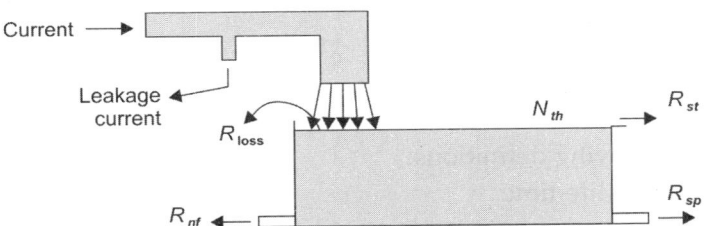

Fig. 6.22. Model of a laser operating above threshold

Now, based on the above picture, we will formulate equations describing dynamics of carriers and photons.

6.10.1 Carriers

Inside the laser, there exist two types of carriers: electrons and holes. Both are described in a similar way; however, parameters used will be different for both types.

The rate of change of carrier density is governed by:

$$\frac{dN}{dt} = G_{gener} - R_{recom} \qquad (6.43)$$

where the term on right are responsible for generation and recommendation of carriers and are given by:

$$G_{gener} = \eta_i \frac{1}{q \cdot V}$$

with V being the volume of the active region, q is the electron's charge and I is the electric current. The internal efficiency n_i describes the fraction of terminal current that generates carriers in the active region. The recombination term consists of several contributions:

$$R_{recom} = R_{sp} + R_{nr} + R_l + R_{st}$$

The meaning of terms is as rollows:
R_{sp} – spontaneous recombination rate
R_{nr} – nonradiative recombination rate
R_l – carrier leakage rate
R_{st} – net stimulated recombination rate

Recombination processes are described phenomenologically as:

$$R_{recom} = \frac{N}{\tau} + v_g g(N) S \qquad (6.44)$$

6.10.2 Photons

Let S be the photon density. We postulate the following rate of change of photon density:

$$\frac{dS}{dt} = \Gamma R_{st} - \frac{S}{\tau_p} + \Gamma \beta_{sp} R_{sp}$$

with the following definitions:
τ_p – photon life-time
β_{sp} – spontaneous emission factor (reciprocal of the number of optical modes)
R_{st} – stimulated recombination
Γ – confinement factor which is the ratio of the active layer volume to the volume of the optical mode.

Consider growth of a photon's density over the active region, (assume $\Gamma = 1$):

$$S + \Delta S = S e^{g \cdot \Delta z}$$

where g is gain.
if $\Delta z \ll 1$ then $\exp(g \cdot \Delta z) \approx 1 + g \Delta z$

Using the relation $\Delta z = v_g \Delta t$ (v_g – group velocity), one finds $S = S g \cdot v_g \cdot \Delta t$. Thus, the generation term can be written as:

$$\left(\frac{dS}{dt}\right)_{gen} = R_{st} = \frac{\Delta S}{\Delta t} + v_g g S$$

Finally, the rate equations used in this section are:

$$\frac{dN}{dt} = \eta_i \frac{I}{qV} - \frac{N}{\tau} - v_g g(N)S \qquad (6.45)$$

$$\frac{dS}{dt} = \Gamma v_g g(N)S - \frac{S}{\tau_p} + \Gamma \beta sp R_{sp} \qquad (6.46)$$

We have explicitly indicated that gain g depends on a carrier's concentration.

6.10.3 Rate Equation Parameters

Other important parameters which in the simple model can be taken as constants due to complicated dependencies. Those parameters are:

(i) the carrier's lifetime τ which strongly depends on carrier's density. Typical dependence is shown below:

$$\frac{1}{\tau} = A + BN + CN^2$$

where coefficient A describes non-radiative processes, B is responsible for spontaneous recombination and C describes *nonradiative Auger recombinations*.

(ii) photon lifetime τ_p which is:

$$\frac{1}{v_g \tau_p} = \alpha_i + \alpha_m = \Gamma g_{th}$$

Here α_m describes mirror reflectivity:

$$\alpha_m = \frac{1}{L} \ln \frac{1}{R}$$

and α_i account for all losses.

The meaning of all parameters appearing in rate equations and gain models is summarized in Table 6.2 along with typical values for those symbols.

In the following, rate equations will provide a starting point for analysis of dynamical properties of semiconductor lasers. Before we start detailed analysis based on rate equations, we will establish a rate equation for electric field.

6.10.4 Derivation of Rate Equation for Electric Field

The relevant equation will be derived starting from the wave equation.

$$\nabla^2 E(r,t) - \frac{1}{c^2} \frac{\partial^2}{\partial t^2} E(r,t) = \mu_0 \frac{\partial^2}{\partial t^2} P(r,t) \qquad (6.47)$$

Assume the following decomposition:

$$E(r, t) = \hat{a} E_t(x, y) \sin(\beta_z z) E(t) e^{j\omega t} \qquad (6.48)$$

where $\hat{a}$ is a unit vector describing polarization, β_z propagation constant in the z-direction. The transversal field $E_t(x, y)$ obeys the following equation:

$$\left(\frac{\partial^2}{\partial x^2} + \frac{\partial^2}{\partial y^2} \right) E_t(x, y) = - k_t^2 E_t(x, y) \qquad (6.49)$$

It is further assumed that $E(t)$ is slowly varying compared to $e^{j\omega t}$. Substituting solution (6.48) into (6.47), neglecting fast-varying terms in time and using (4.49), one obtains:

$$\left\{ k_t^2 - \beta_z^2 - \frac{1}{c^2} \left[2j\omega \frac{\partial E(t)}{\partial t} - \omega^2 E(t) \right] \right\} a E_t(x, y) \sin(\beta_z z) e^{j\omega t} = \mu_0 \frac{\partial^2}{\partial t^2} P(r, t)$$

Table 6.2. Basic parameters appearing in rate equation approach and their typical values

Symbol	Description	Value and unit
N	carrier density	cm^{-3}
S	photon density	cm^{-3}
I	current	mA
q	elementary charge	1.602×10^{-19} C
L	cavity length	250 μm
w	width of active region	2 μm
d	thickness of active region	80 Å
h_i	fraction of the injected current I into active region	0.8
V_{active}	volume of active region	$L \cdot w \cdot d$
τ	carrier lifetime	2.71 ns
v_g	group velocity	c/n_{ref}
n_{ref}	refractive index	3.4
τ_p	photon lifetime	2.77 ps
Γ	confinement factor	0.01
β_{sp}	spontaneous emission factor	10^{-4}
a	differential gain (linear model)	5.34×10^{-16} cm^2
N_{tr}	carrier density at transparency	3.77×10^{18} cm^{-3}
I_{th}	current at threshold	1.11 mA
α_m	facet loss	45 cm^{-1}
λ_{ph}	laser wavelength	1.3 μm

To evaluate terms on the right hand side, we need to account for a nonlocal relation between polarization and susceptibility:

$$P(r, t) = \varepsilon_0 \int \chi(r, t') E(r, t - t') dt' \qquad (6.50)$$

In the above $E(r, t - t')$ is given by (6.48). As $E(t)$ is slowly varying in time, we can expand it into Taylor series around t. The general formula

Optical Sources 293

for Taylor expansion of the function $f(x)$ around a is $f(x) + \dfrac{df}{dx}(x-a)$. In our case, substituting $x = t - t'$ and $a = t$, we obtain:

$$E(t-t') \approx E(t) + \frac{dE(t)}{dt}(-t')$$

Full electric field is therefore:

$$E(\mathbf{r}, t-t') = E_t(x,y)\sin(\beta_z z)\, E(t-t')\, e^{j\omega(t-t')}$$

Substituting the last result and Taylor expansion for $E(t-t')$ into expression for polarization (6.50), one finally obtains the wave equation for $E(t)$:

$$\left[-\beta_0^2 + \frac{\omega^2}{c^2}\varepsilon_r(\omega)\right]E(t) - \frac{2j\omega}{c^2}\left[\varepsilon_r(\omega) + \frac{1}{2}\omega\frac{d\varepsilon_r(\omega)}{d\omega}\right]\frac{dE(t)}{dt} = 0 \quad (6.51)$$

Here $\varepsilon_r(\mathbf{r},\omega) = 1 + \chi(\mathbf{r},\omega)$ and Fourier transform of susceptibility is (we have dropped dependence):

$$\chi(\omega) = \int dt'\, \chi(t')\, e^{-j\omega t'}$$

Susceptibility is further separated into terms which account for various physical effects as follows:

$$\chi(\omega) = 1 + \chi_b + \chi_p - j\chi_{loss}$$

where $\chi_b = \chi_b' = j\chi_b''$ is due to background $\chi_p = \chi_p' = j\chi_p''$ is induced by pump and χ_{loss} accounts for losses. The relative dielectric constant is written as (dropping argument dependencies) $\varepsilon_r = \varepsilon_r' + j\varepsilon_r''$. The real part of ε_r is approximated in terms of refractive index:

$$\varepsilon_r = (n_0 + \Delta n_p)^2 \approx n_0^2 + 2n_0\Delta n_p,$$

where n_0 is the background refractive index and Δn_p is the change induced by pump. Using the above results, ε_r takes the form:

$$\begin{aligned}\varepsilon_r &= \varepsilon_r' + j\varepsilon_r'' = n_0^2 + 2n_0\Delta n_p + j\varepsilon_r'' \\ &= 1 + \chi_b' + \chi_p' + j(\chi_b'' + \chi_p'' - \chi_{loss})\end{aligned} \quad (6.52)$$

Explicitly, one has the following identifications:

$$\begin{aligned}1 + \chi_b' &= n_0^2 \\ \chi_p' &= 2n_0\Delta n_p \\ \chi_b'' + \chi_p'' - \chi_{loss} &= \varepsilon_r''\end{aligned}$$

Using the above relations, the term in the equation for slowly varying amplitude $E(t)$, Eq. (6.51) is:

$$\frac{\omega^2}{c^2}\varepsilon_r(\omega) - \beta_0^2 = \frac{\omega^2 - \omega_0^2}{c^2} n_0^2 + \frac{\omega^2}{c^2}\chi_p' + j\frac{\omega^2}{c^2}(\chi_b'' + \chi_p'' - \chi_{loss}) \quad (6.53)$$

where we have used the approximate relation:

$$\beta_0 \approx \frac{\omega_0}{c} n_0$$

Imaginary parts of background and pump susceptibilities are related to gain via an experimental relation:

$$\frac{\omega}{cn_0}(\chi_b'' + \chi_p'') = \Gamma g(N)$$

where $g(N)$ is gain. Assume also:

$$\chi_{loss} = \frac{\omega}{cn_0}\alpha_{loss}$$

With the last two relations, term given by Eq. (6.53) can be expressed as:

$$\frac{\omega^2}{c^2}\varepsilon_r(\omega) - \beta_0^2 = \frac{\omega^2 - \omega_0^2}{c^2}n_0^2 + \frac{\omega^2}{c^2}2n_0\Delta n_p + j\frac{\omega n_0}{c}[\Gamma g(N) - \alpha_{loss}]$$

Using $n^2(\omega) = \varepsilon_r(\omega)$, the dispersion term in (6.51) is evaluated as:

$$\varepsilon_r(\omega) + \frac{1}{2}\omega\frac{d\varepsilon_r(\omega)}{d\omega} = n^2(\omega) + \frac{1}{2}\omega n(\omega)\frac{dn(\omega)}{d\omega} \quad (6.54)$$

$$= n(\omega) + \omega n_g(\omega)$$

where we have defined group index $n_g(\omega)$ as:

$$n_g(\omega) = n(\omega) + \omega\frac{dn(\omega)}{d\omega} \quad (6.55)$$

Using the results (6.53) and (6.54), Eq. (6.53) takes the form:

$$\frac{dE(t)}{dt} = \left\{-j(\omega - \omega_0)\frac{n_0}{n_g} - j\frac{\omega}{n_g}\Delta n_p + \frac{1}{2}v_g[\Gamma g(N) - \alpha_{loss}]\right\}E(t) \quad (6.56)$$

where we approximated $\omega^2 - \omega_0^2 = (\omega - \omega_0)(\omega - \omega_0) \approx 2\omega(\omega - \omega_0)$ and also $\frac{n_0(\omega)}{n(\omega)} \approx 1$ and used the relation $\frac{c}{n_g} = v_g$.

6.11 ANALYSIS BASED ON RATE EQUATIONS

Rate equation just introduced will now be used to study some *dynamical properties* of semiconductor lasers. Let us start with steady-state.

6.11.1 Steady-State Analysis

In this situation there is no change in time; i.e.:

$$\frac{dN}{dt} = \frac{dS}{dt} = 0$$

Rate equations take the form:

$$\eta_i \frac{I_0}{qV} - \frac{N_0}{\tau} - v_g g(N_0) S_0 = 0 \tag{6.57}$$

and

$$\Gamma v_g g(N_0) S_0 - \frac{S_0}{\tau_p} + \Gamma \beta_{sp} R_{sp} = 0 \tag{6.58}$$

If in the last equation, we neglect the small spontaneous emission term, we obtain an expression for a photon life-time:

$$\frac{1}{\tau_p} = \Gamma v_g g(N_0) \tag{6.59}$$

6.11.2 Small-Signal Analysis with the Linear Gain Model

When neglecting spontaneous emission and using linear gain model where gain is given by $g = a(N - N_{tr})$, one obtains:

$$\frac{dN}{dt} = \eta_i \frac{I}{qV} - \frac{N}{\tau} - v_g a (N - N_{tr}) S \tag{6.60}$$

and

$$\frac{dS}{dt} = \Gamma v_g a (N - N_{tr}) S - \frac{S}{\tau_p} \tag{6.61}$$

Assume that all time-dependent quantities oscillate as:

$$I = I_0 + i(\omega)e^{j\omega t}$$
$$N = N_0 + n(\omega)e^{j\omega t}$$
$$S = S_0 + s(\omega)e^{j\omega t}$$

where ω is the angular frequency of an external (small) perturbation and I_0 is the bias value of current. Here N_0 and S_0 are the solutions in the steady-state. Substitute into the first rate equation and after multiplication and neglecting second-order term $(n(\omega)s(\omega))$, one obtains:

$$j\omega n(\omega)e^{j\omega t} = \frac{\eta_i}{qV}[I_0 + i(\omega)e^{j\omega t}] - \frac{N_0}{\tau} - \frac{1}{\tau}n(\omega)e^{j\omega t}$$
$$- v_g a\big((N_0 - N_{tr})S_0 + (N_0 - N_{tr})s(\omega)e^{j\omega t} + S_0 n(\omega)e^{j\omega t}\big) + O(n^2)$$

Using steady-state results, expression (6.59) for photon life-time and dropping $e^{j\omega t}$ dependence, one finally obtains:

$$j\omega n(\omega) = \frac{\eta_i}{qV} i(\omega) - \frac{n(\omega)}{\tau} - \frac{s(\omega)}{\Gamma \tau_p} - v_g a S_0 n(\omega) \tag{6.62}$$

The second equation for photons is obtained in a similar way. Substituting small signal expressions, neglecting the second-order term, using the photon life-time expression and finally dropping $e^{j\omega t}$ dependence, one obtains:

$$j\omega s(\omega) = \Gamma v_g a S_0 n(\omega) \tag{6.63}$$

From the above equations, we want to determine *modulation response* which is defined as:

$$M(\omega) = \frac{s(\omega)}{i(\omega)}$$

Expressing $n(\omega)$ from Eq. (6.63) and substituting into Eq. (6.62), one obtains:

$$\frac{s(\omega)}{i(\omega)} = \frac{\frac{n_i}{eV}\Gamma v_g a S_0}{D(\omega)}$$

where $D(\omega)$ is given by:

$$D(\omega) = -\omega^2 + j\omega\left(\frac{1}{\tau} + v_g a S_0\right) + \frac{v_g a S_0}{\tau_p}$$

We define response function as:

$$r(\omega) = \left|\frac{s(\omega)}{i(\omega)}\right| = \frac{\frac{n_i}{eV}\Gamma v_g a S_0}{|D(\omega)|}$$

If we write $D(\omega) = a + jb$, we have $|D(\omega)|^2 = a^2 + b^2$. Therefore:

$$|D(\omega)|^2 = \left(\omega^2 - \frac{v_g a S_0}{\tau_p}\right)^2 + \omega^2\left(\frac{1}{\tau} + v_g a S_0\right)^2$$

We will see that the response function has a peak at frequency f_R. To find that frequency, we evaluate first derivative $\dfrac{\partial |D(\omega)|^2}{\tau_p}$ and set it to zero. Explicity:

$$\left.\frac{\partial |D(\omega)|^2}{\partial \omega^2}\right|_{\omega = \omega_R} = -2\left(\frac{v_g a S_0}{\tau_p} - \omega_R^2\right) - \frac{1}{2}\left(\frac{1}{\tau} + v_g a S_0\right) = 0$$

From the above:

$$\omega_R^2 = \frac{v_g a S_0}{\tau_p} - \frac{1}{2}\left(\frac{1}{\tau} + v_g a S_0\right) \tag{6.64}$$

which is known as a *relaxation-oscillation frequency* and it describes the rate at which energy is exchanged between photon system and carriers. The second term in Eq. (6.64) is usually very small and can be neglected. Relaxation-oscillation frequency is thus approximated as:

$$\omega_r = \sqrt{\frac{1}{\tau_p}\frac{S_0 v_g a}{1 + \varepsilon S_0}} \tag{6.65}$$

To estimate the typical values of the relaxation-oscillation frequency, let us assume: $L = 300$ µm, $\tau_{ph} \approx 10^{-12}$ s and $\tau \sim 4 \times 10^{-9}$ s. One finds $AP_0 \sim 10^9$ s^{-1}. So, the zeroth-order expression is a very good one.

The modulation characteristics of semiconductor lasers is shown in Fig. 6.23.

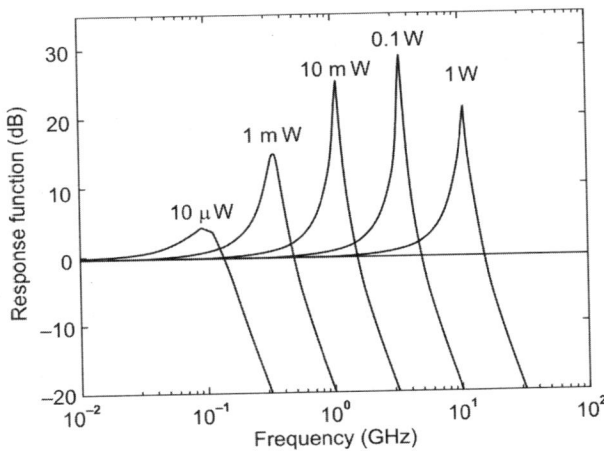

Fig. 6.23. Modulation characteristics of semiconductor lasers from small-signal analysis

6.11.3 Small-Signal Analysis with Gain Saturation

Let us remind ourselves of the rate equations which were established in the previous section:

$$\frac{dN}{dt} = n_i \frac{I}{qV} - \frac{N}{\tau} - v_g g(N) S \tag{6.66}$$

$$\frac{dS}{dt} = \Gamma v_g g(N) S - \frac{S}{\tau_p} + \beta_{sp} R_{sp} \tag{6.67}$$

Use small-signal assumptions:
$$I(t) = I_0 + i(t)$$
$$N(t) = N_0 + n(t)$$
$$S(t) = S_0 + s(t)$$

Also, generalize the gain model to include saturation as follows:

$$g(n) = \frac{g_0 + g'(N - N_0)}{1 + \epsilon S}$$

where $g' = \left.\frac{\partial g}{\partial N}\right|_{N=n_0}$ is differential gain evaluated at $N = N_0$ and $g_0 = g(N_0)$. The factor $1 + \epsilon S$ introduces nonlinear gain saturation. The gain compression parameter ϵ has a small value and the term ϵS is small compared with the one even at very high optical power. The effect of this term on dc properties is small and can be neglected. However, it significantly affects dynamics of a semiconductor laser. The reason is that the laser's dynamics depends on the small difference between gain and cavity loss.. This difference is only about a few percent. Therefore, even small gain compression due to ϵ produces significant effects.

One can evaluate the gain compression term as follows:

$$1+\epsilon S \quad 1+\epsilon (S_0 + s(t)) = (1+\epsilon S_0)\left(1+\frac{\epsilon s(t)}{1+\epsilon S_0}\right)$$

With the above result, gain can be approximated as:

$$g(n) = \frac{g_0}{1+\epsilon S_0} + \frac{g'n(t)}{1+\epsilon S_0} - \frac{g_0}{1+\epsilon S_0}\left(1+\frac{\epsilon s(t)}{1+\epsilon S_0}\right) \quad (6.68)$$

Using the above results, stimulated emission term takes the form:

$$g(n)S = \frac{g_0 S_0}{1+\epsilon S_0} + \frac{g' S_0}{1+\epsilon S_0}n(t) - \frac{g_0 S_0}{(1+\epsilon S_0)^2}\epsilon s(t)$$

$$+ \frac{g_0}{1+\epsilon S_0}s(t) + O(s^2) \quad (6.69)$$

The following analysis is separated into two parts: dc and ac analysis.

dc analysis: With the assumption that $\frac{d}{dt}=0$ from rate equations, one obtains:

$$\eta \frac{I_0}{qV} - \frac{N_0}{\tau} - v_g \frac{g_0 S_0}{1+\epsilon S_0} = 0$$

and

$$\Gamma v_g \frac{g_0 S_0}{1+\epsilon S_0} - \frac{S_0}{\tau_p} + \beta_{sp} R_{sp} = 0$$

Neglecting spontaneous emission ($\beta_{sp}=0$), from the second equation, one obtains an expression for the photon life-time τ_p:

$$\frac{1}{\tau_p} = \Gamma v_g \frac{g_0}{1+\epsilon S_0} \quad (6.70)$$

ac analysis: Substituting small-signal assumptions into rate equations, and using an approximation of stimulated emission Eq. (6.69) and eliminating dc terms, one obtains:

$$\frac{dn(t)}{dt} = \eta_i \frac{i(t)}{qV} - \frac{n(t)}{\tau} - \frac{v_g g' S_0}{1+\epsilon S_0}n(t) + \frac{1}{\tau_p \Gamma}\frac{S_0}{1+\epsilon S_0}\epsilon s(t) - \frac{1}{\tau_p \Gamma}s(t)$$

$$\frac{ds(t)}{dt} = \frac{\Gamma v_g g' S_0}{1+\epsilon S_0}n(t) - \frac{1}{\tau_p}\frac{S_0}{1+\epsilon S_0}\epsilon s(t)$$

In obtaining the above equations we used Eq. (6.70). Those equations can also be written in matrix form:

$$\frac{d}{dt}\begin{bmatrix} n(t) \\ s(t) \end{bmatrix} + \begin{bmatrix} A & B \\ -C & D \end{bmatrix}\begin{bmatrix} n(t) \\ s(t) \end{bmatrix} = \begin{bmatrix} \eta_i \dfrac{n(t)}{qV} \\ 0 \end{bmatrix}$$

where

$$A = \frac{1}{\tau} + \frac{v_g S_0 g'}{1 + \epsilon S_0}, \quad B = \frac{1}{\tau_p \Gamma} - \frac{1}{\tau_p \Gamma}\frac{S_0}{1+\epsilon S_0}\epsilon,$$

$$C = \frac{\Gamma v_g S_0 g'}{1+\epsilon S_0}, \quad D = \frac{1}{\tau_p}\frac{S_0}{1+\epsilon S_0}\epsilon$$

Assuming harmonic time dependence of the form $exp(j\omega t)$, the matrix equation is:

$$\begin{bmatrix} j\omega + A & B \\ -C & j\omega + D \end{bmatrix}\begin{bmatrix} n(t) \\ s(t) \end{bmatrix} = \begin{bmatrix} \eta_i \dfrac{i(t)}{qV} \\ 0 \end{bmatrix}$$

Solving the above system, one obtains modulation response which is expressed as:

$$\frac{s(t)}{i(t)} = \eta_i \frac{1}{qV}\frac{C}{H(\omega)}$$

where

$$H(\omega) = (j\omega + A)(j\omega + D) + CB$$

The function $H(\omega)$ can be written as:

$$H(\omega) = -\omega^2 + j\omega\gamma + \omega_R^2$$

Here ω_R is the relaxation-oscillation frequency and γ is the damping factor. One defines modulation response $r(\omega)$ as:

$$r(\omega) = \frac{H(\omega)}{H(0)}$$

The plot of response $r(\omega)$ is shown in Fig. 6.24 for three values of ϵ_r. The effect of gain compression parameter ϵ is clearly visible.

6.11.4 Large-Signal Analysis for QW Lasers

Main equations for large-signal analysis without spontaneous emission are:

$$\frac{dN}{dt} = \eta_i \frac{I}{eV} - \frac{N}{\tau} - v_g g(N) \cdot S \tag{6.71}$$

$$\frac{dS}{dt} = \Gamma \cdot v_g g(N) \cdot S - \frac{S}{\tau_p} \tag{6.72}$$

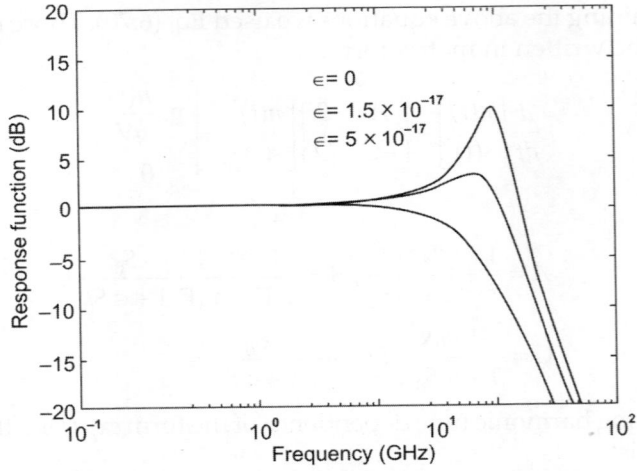

Fig. 6.24. The effect of gain compression on modulation response

All symbols have been explained previously and their values are summarized in Table 3.2.

Typical results for step current are shown in Fig. 6.25.

One can observe that when we neglect spontaneous emission term in the rate equation for photons, photon density tends to zero (steady-

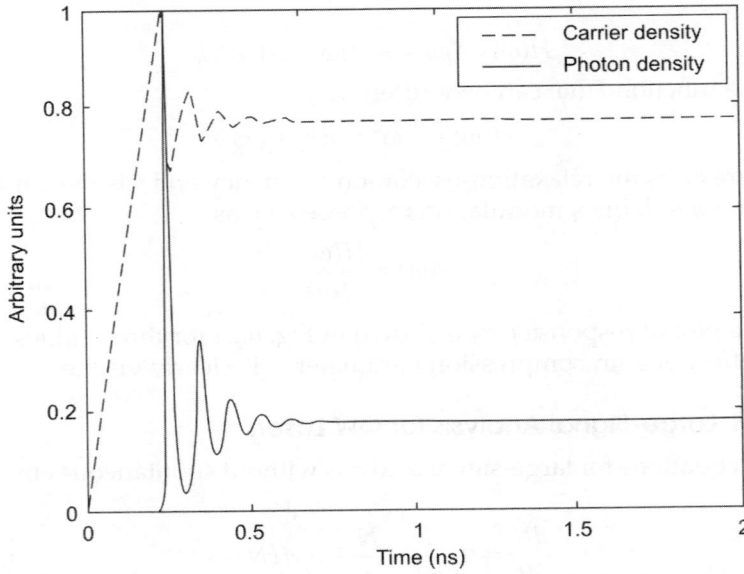

Fig. 6.25. Numerical solutions of large-signal rate equations after applying rectangular current pulse at $t = 0$. Normalized values of electron density $N(t)$ and photon density $S(t)$ are shown

state value) as $t \to \infty$, the result which is also evident from the steady-state analysis.

6.11.5 Frequency Chirping

Frequency chirping arises during direct modulation of semiconductor laser when carrier density undergoes abrupt change. As a result, material gain also changes. This change results in a variation of the refractive index, which in turn affects the phase of the electric field.

The rate of change of the phase is:

$$\frac{d\phi}{dt} = \frac{1}{2} v\alpha_H (\Gamma g - \alpha_{loss})$$

An associated change in frequency is created which is described by:

$$\Delta v(t) = \frac{1}{2\pi} \frac{d\phi}{dt} = \frac{1}{4\pi} v\alpha_{enh} (\Gamma g - \alpha_{loss})$$

This change in frequency is called 'chirp'. During direct modulation, it results in a shift in frequency of the order of 10–20 GHz.

To eliminate chirp effect, an external modulator should be used. In such cases a laser produces constant output power and a separate modulator provides modulation.

Another way to reduce chirp is to fabricate devices with small values of line-width enhancement factor α_H.

$$\alpha_H \equiv -\frac{\frac{dn_r}{dN}}{\frac{dn_i}{dv}}$$

where $n = n_r + jn_i$ (complex refractive index).

6.11.6 Equivalent Circuit Model for a Bulk Laser

Equivalent circuit models of semiconductor lasers provide further understanding of the laser properties. They are derived from rate equations. More recent works include additional effects like carrier transport or proper representation of optical gain. In the following, we outline the creation of equivalent circuit model for a bulk semiconductor laser.

For bulk lasers, rate equations are:

$$\frac{dS}{dt} = (G - \gamma)S + R_{sp} \tag{6.73}$$

$$\frac{dN}{dt} = \frac{1}{q} - \gamma_e N - G \cdot S \tag{6.74}$$

Here N, S are the total number of electrons inside the cavity and total number of photons is the lasing mode. Those are dimensionless quantities. We introduce small deviations from equilibrium as:

$$S(t) = S_0 + \delta S(t) \tag{6.75}$$
$$N(t) = N_0 + \delta N(t) \tag{6.76}$$

Optical gain is approximated as:

$$G = G_0 + \frac{\partial G}{\partial N}\delta N + \frac{\partial G}{\partial S}\delta S = G_0 + G_N \delta N + G_S \delta S \tag{6.77}$$

In the steady-state, one obtains:

$$0 = (G_0 - \gamma) S_0 + R_{sp,0}$$
$$0 = \frac{I_0}{q} - \gamma_e N_0 - G_0 \cdot S_0$$

Observing that spontaneous emission $R_{sp} = R_{sp}(N)$ depends on N, and performing an expansion:

$$R_{sp} = R_{sp \cdot 0} + \frac{\partial R_{sp}}{\partial N}\delta N \tag{6.78}$$

one can derive small-signal equations:

$$\frac{d\delta S}{dt} = -\Gamma_S \delta S + \sigma_N \delta N \tag{6.79}$$

$$\frac{d\delta N}{dt} = -\Gamma_N \delta N + \sigma_N \delta S \tag{6.80}$$

where

$$\Gamma_S = \frac{R_{sp}}{S_0} - S_0 G_S, \quad \Gamma_N = \gamma_e + N_0 \frac{\partial \gamma_e}{\partial N} + S_0 G_N \tag{6.81}$$

and

$$\sigma_N = S_0 G_N + \frac{\partial R_{sp}}{\partial N}, \quad \sigma_S = G_0 + S_0 G_S \tag{6.82}$$

Using these equations, one creates equivalent circuit model. It is based on an observation that the small-signal rate equations are similar to the voltage and current equations of the RLC circuit. We therefore assume that electrons in a laser cavity can be represented by the charge across a capacitor and the photons are represented by the magnetic flux of an inductor. Formally one introduces the following equivalence:

$$\delta S(t) = \frac{\phi(t)}{q \cdot unit}, \quad \delta N(t) = \frac{Q(t)}{q} \tag{6.83}$$

where q is an electron's charge, unit is a parameter in [H/s] to ensure proper units for the components of the electrical circuit and $\phi(t)$ is the magnetic flux in Wb. We also recall standard relations from electromagnetism:

$$Q = C \cdot V_C \tag{6.84}$$

$$i_C = \frac{dQ}{dt} \tag{6.85}$$

$$\phi(t) = L \cdot i_L \tag{6.86}$$

$$V_L = \frac{d\phi}{dt} \tag{6.87}$$

Now, we use (6.83) and relations (6.84)–(6.87) to convert small-signal equations (6.79) and (6.80) into voltage and current relations. After simple algebra, from (6.79), one obtains:

$$V_L(t) = \Gamma_S \cdot L \cdot i_L(t) + \sigma_N \cdot C \cdot unit \cdot V_C \tag{6.88}$$

This equation can be identified with loop equation for the RLC circuit shown in Fig. 6.26. A loop equation gives:

$$V_C = V_L + V_{Rp} \tag{6.89}$$

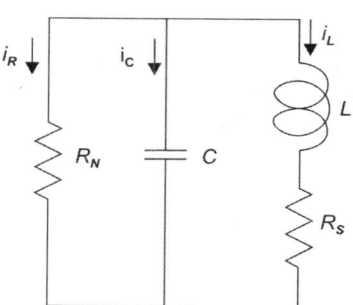

Fig. 6.26. Equivalent circuit model for a bulk laser

In order for the above equations to be identical, one makes the following identifications:

$$R_S = \Gamma_S \cdot L = \Gamma_S \cdot \frac{unit}{\sigma_S}, \quad C = \frac{1}{unit \cdot \sigma_N} \tag{6.90}$$

Similarly, from (6.80), one obtains:

$$i_C(t) = -\Gamma_N \cdot C \cdot V_C(t) - \frac{\sigma_S}{unit} \cdot L \cdot i_L(t) \tag{6.91}$$

That equation can be compared with the node equation from a circuit in Fig. 6.26:

$$i_C = -i_R - i_L \tag{6.92}$$

For those equations to be identical, one sets:

$$R_N = \frac{1}{\Gamma_N \cdot C}, \quad \text{and} \quad L = \frac{unit}{\sigma_S} \tag{6.93}$$

One also find the following equivalence:

$$\delta N = \frac{C \cdot V_C}{q}, \quad \text{and} \quad \delta S = \frac{L \cdot i_L}{q \cdot unit} \tag{6.94}$$

6.12 APPLICATIONS OF LASERS

Some of the important applications of lasers in brief are as follows:

Lasers are used for study of atoms and molecules by laser spectroscopy. They are used in probing the earth's upper atmosphere and in separation of atomic isotopes from one another. Like RADAR, LIDAR (Light Detection And Ranging) technique is used for accurate distance measurement. For example, the accurate distance and hence the variation in time of earth-moon separation is needed to evaluate, change in universal gravitational constant, drift of continents of earth, motion of the moon away from the earth, etc. Range finding by laser has many military applications, in target identification, imaging and tracking. The lidar technology is used in atmospheric pollution monitoring systm, detection of fog layers, surveying, altimeters in aircrafts etc.

The capability of coherent infrared, visible or ultraviolet radiation to focus to beam diameters below 100 µm is used by the electronics industry for drilling, trimming and marking of materials. For example, Xe laser which operates at 537 nm is used in resistor trimming, a CO_2 laser at 10.6 µm is used for marking. Other applications in electronics industry are microlithography, deposition and etching of thin films.

Other industrial uses consist of surface treatment (transformation hardening, melting, alloying, cladding), welding, drilling (piercing), cutting, marking, annealing, scribing, soldering, deposition, ablation, grain refining, curing, 3D laser machining, laser assisted machining etc.

Nd:YAG, along with CO_2 and argon ion are the most common lasers used in medicine. For example, photocoagulation of blood and tissue welding, precise cutting, ophthalmology, in medical research such as investigation of laser-tissue interaction etc.

In production and reading of holograms, laser is used. The holographic technique is used in testing of stresses and structural defects in materials. The holograms are produced with and without stress and then compared. In medical research three dimensional view of living cells is obtained using holographic principle.

The industries, where sheet materials are processed, like metal, plastic, photographic film, magnetic tapes, paper etc., the laser based system is used to detect small holes, flaws, folds, contaminants etc. This detection occurs at the speed of light. It reduces the cost and improves the quality.

Lasers are used in printing industry as well as in laser printers.

A small plastic disc, that is, compact disc (CD) can store large amount of data (MB to GB) in digital form. A laser is used in recording and reading this data, for example, audio and video CD, DVD etc. Laser is used in optical fibre communications.

Example 1

A laser beam has a power of 50 mW. It has an aperture of 5×10^{-3} m and it emits light of wavelength 7200 Å. The beam is focussed with a lens of focal length 0.1 m. Show that the areal spread and the intensity of the image are 2.074×10^{-10} m² and 2.4×10^8 Wm⁻² respectively.

Solution

$$d\theta = \frac{\lambda}{d} \qquad [\lambda = 7200 \times 10^{-10} \text{ m}]$$

or $\quad d\theta = \dfrac{7200 \times 10^{-10}}{5 \times 10^{-3}} = 1.44 \times 10^{-4}$ radian $\qquad [d = 5 \times 10^{-3} \text{ m}]$

(i) Areal spread = $(d\theta f)^2$, Here $f = 0.1$ m.

$\therefore$ Areal spread = $(1.44 \times 10^{-4} \times 0.1)^2 = 2.074 \times 10^{-10}$ m²

(ii) Intensity $= \dfrac{\text{Power}}{\text{Area}} = \dfrac{\text{Power}}{\text{Area spread}} = \dfrac{50 \times 10^{-3} \text{ W}}{2.074 \times 10^{-10} \text{ m}^2}$

$= 2.411 \times 10^8$ Wm⁻²

Example 2

A laser beam has a wavelength of 8×10^{-7} m and aperture 5×10^{-3} m. The laser beam is sent to moon. The distance of the moon is 4×10^5 km from the earth's surface. Determine (i) the angular spread of the beam, and (ii) the axial spread when it reaches the moon's surface.

Solution

We have $d\theta = \lambda/d$:

Angular spread, $d\theta = \dfrac{8 \times 10^{-7}}{5 \times 10^{-3}} = 1.6 \times 10^{-4}$ radiation,

$\lambda = 8 \times 10^{-7}$ m, $d = 5 \times 10^{-3}$ m

Areal spread = $(Dd\theta)^2$ where $D = 4 \times 10^5$ km = 4×10^8 m

$= (4 \times 10^8 \times 1.6 \times 10^{-4})^2 = 4.096 \times 10^9$ m²

Example 3

A laser beam of wavelength 6000 Å on earth is focussed by a lens (or mirror) of diameter 2 m on to a crater on the moon's surface. The distance of the moon from earth is 4×10^8 m. Neglecting the effect of earth's atmosphere, show that the angular spread of the spot on the moon will be 3×10^{-7} radian.

Solution

Angular spread:

$$d\theta = \frac{\lambda}{d} = \frac{6 \times 10^{-7}}{2} = 3 \times 10^{-7} \text{ radian} \qquad \lambda = 600 \times 10^{-10} \text{ m},$$
$$d = 2 \text{ m}, D = 4 \times 10^8 \text{ m}$$

Example 4

A laser beam has aperature $d = 1.8 \times 10^{-2}$ m and emits radiation of wavelength $\lambda = 5 \times 10^{-7}$ m. Calculate (i) the semi-angle of the cone of its beam, (ii) the solid angle of this cone.

Solution

(i) The semi-angle of the cone of the laser beam:

$$\theta = \frac{\lambda}{d} = \frac{5 \times 10^{-7} \text{ m}}{1.8 \times 10^{-2} \text{ m}} = 2.72 \times 10^{-5} \text{ radian}$$

(ii) Solid angle of this cone:

$$\phi = \pi \frac{D^2 \theta^2}{D^2} = \pi \theta^2 = \frac{22}{7} \times (2.72 \times 10^{-5})^2 = 2.5 \times 10^{-9} \text{ steradian}$$

Example 5

The coherence length for sodium light is 2.945×10^{-2} m. The wavelength of sodium light is 5890 Å. Determine (i) the number of oscillations corresponding to coherence length, and (ii) the coherence time.

Solution

(i) Number of oscillations in coherence length L:

$$n = \frac{L}{\lambda} = \frac{2.945 \times 10^{-2}}{5890 \times 10^{-10}} = 5 \times 10^4 \qquad \lambda = 5890 \times 10^{-10} \text{ m}$$
$$c = 3 \times 10^{10} \text{ m/s},$$
$$L = 2.945 \times 10^{-2} \text{ m}$$

(ii) Coherence time $= L/c = \dfrac{2.945 \times 10^{-2}}{3 \times 10^{10}} = 9.8 \times 10^{-11}$ s

Example 6

The coherence length for sodium D_1 line is 2.5 cm. Determine (i) the spectral width of the line $\Delta\lambda$, the purity factor, Q, and (iii) the coherence time τ. The wavelength of light is 6000 Å.

Solution

(i) $\Delta\lambda = \dfrac{\lambda}{2L} = \dfrac{36 \times 10^{-10}}{5} \text{ cm} = 7 \text{Å} \qquad L = 2.5 \text{ cm}$
$$c = 3 \times 10^{10} \text{ cm·s}^{-1}$$

(ii) $Q = \dfrac{\lambda}{\Delta\lambda} = \dfrac{6\times 10^{-5}}{7\times 10^{-8}} = 10^3$

(iii) $\tau = \dfrac{L}{c} = \dfrac{2.5}{3\times 10^{10}} \cong 0.8\times 10^{-10}$ s

Example 7
Light from a 2.5 mW water source of aperture diameter 1.8 cm and $l = 500$ nm is focused by a lens of focal length 20 cm. Compute: (a) the area, and (ii) the intensity of the image:

Solution
Given
$\lambda = 500$ nm $= 5\times 10^{-7}$ m
$2a = 1.8$ m
$\quad = 0.018$
$a = 0.009$ m
$f = 20$ cm $= 0.20$ m

(a) Area of the spot at focal plane $= \dfrac{\pi\lambda^2 f^2}{a^2}$

$= \dfrac{\pi\times(5\times 10^{-7})^2\times(0.20)^2}{(0.009)^2}$

$= 3.88\times 10^{-10}$ m².

(b) Intensity at the focus $I = \dfrac{Pa^2}{\pi^2\lambda^2 f^2}$

$= \dfrac{2.5\times 10^{-3}\text{ W}}{3.88\times 10^{-10}\text{ m}^2}$

$= 6.44\times 10^6$ W/m²

Example 8
Consider a He–Ne gas laser for which the laser radiation wavelength is 633 nm. (a) Calculate the n-value for a standing wave in a cylindrical cavity 80 cm long, (b) given that the spread in frequencies, Δv, of neon atoms at room temperature is 1.7×10^9 Hz, find the range of n values covered by Δv.

Solution
(a) We have $n = \dfrac{2L}{\lambda_n} = \dfrac{1.6\text{ m}}{633\times 10^{-9}\text{ m}} = 2.53\times 10^6$.

(b) $n\lambda_n = nc/v_n = 2L = $ a constant, one finds that $n = $ const. $\times f_n$. This in turn implies that $(\Delta_n)/n = (\Delta v_n)/v_n$. Thus, we have:

$$\Delta n = \frac{n}{v_n}\Delta v_n = \frac{2L}{c}\Delta v_n = \frac{1.6\,\text{m}}{3\times 10^8\,\text{m/s}}(1.7\times 10^9\,\text{Hz}) = 9$$

Example 9

A ruby rod contains typically a total of 3×10^9 Cr^{+3} ions. It lases at 694.3 nm. Determine the (i) energy of one photon, and (ii) total energy available per pulse assuming that there is total population inversion.

Solution

(i) Energy of Photon $E = \dfrac{1240}{\lambda\,(\text{nm})}\,\text{eV} = \dfrac{1240}{649.3} = 1.79\,\text{eV}$.

(ii) Energy available per pulse
  = (energy of one photon) (total no. of photons)
  = $1.79 \times 3 \times 10^{19} = 5.37 \times 10^{19}$ eV = 8.6 J.

Example 10

In a ruby laser, total number of Cr^{+++} ions are 2.8×10^{19}. If the laser emits radiation of wavelength 700 nm, what will be: (a) the energy of one emitted photon, and (ii) the total energy available per laser pulse?

Solution

Given:
  N = total number of photons = 2.8×10^{19}
  λ = 700 nm = 7.0×10^{-7} m

(a) The energy of photon:

$$E = \frac{12400}{\lambda(\text{Å})} = \frac{12400}{7000} = 1.77\,\text{eV}$$
  $= 1.77 \times 1.6 \times 10^{-19}$ J
  $= 2.832 \times 10^{-19}$ J

(b) Energy per pulse = energy of one photon × total no. of photons
    = $2.832 \times 10^{-19} \times 2.8 \times 10^{19}$ J
    = 7.93 J.

Example 11

A He–Ne laser operating at 632.8 nm has an output power of 1 mW with 1.0 mm beam diameter. The beam comes out through the mirror which has 1% transmittance at laser wavelength. Find the ratio of stimulated emission to spontaneous emission $B_{21}\,U(v)/A_{21}$. The line width of laser line is 1.5×10^8 Hz.

Solution

The frequency of laser is:

$$\nu = \frac{c}{\lambda} = \frac{3 \times 10^8}{632.8 \times 10^{-9}} = 4.74 \times 10^{14} \text{ Hz}$$

We know that
$$\frac{A_{21}}{B_{21}} = \frac{8\pi h \nu^3}{c^3}$$

Therefore,
$$\frac{A_{21}}{B_{21}} = \frac{c^3}{8\pi h \nu^3} = \frac{(3 \times 10^8)^3}{8 \times 3.14 \times 6.63 \times 10^{-34} \times (4.74 \times 10^{14})^3}$$
$$= 1.52 \times 10^{13} \text{ m}^3/\text{Js}$$

$U(\nu)$ is obtained from the fact that intensity is the product of energy density and speed of light. Intensity is the energy per unit area per second. The energy within resonator is 99 mW.

$$\therefore \quad U(\nu) = \frac{I}{c\, d\nu} = \frac{(99 \times 10^{-3})/(\pi \times 0.5 \times 10^{-3})^2}{3 \times 10^8 \times 1.5 \times 10^9 \times 1.5 \times 10^8}$$
$$= 2.80 \times 10^{-12} \text{ J/s/m}^3$$

The desired ratio:
$$\frac{B_{21} U(\nu)}{A_{21}} = \left(\frac{B_{21}}{A_{21}}\right) U(\nu)$$
$$= 1.52 \times 10^{13} \times 2.80 \times 10^{-12}$$
$$= 42.6$$

Example 12

The ruby laser has two states at 300 K and 500 K. If it emits light 700 nm, then compute relative population:

Solution

Given
  $T_1 = 300$ K
  $T_2 = 500$ K
  $\lambda = 700$ nm $= 7000$ Å
  We know that population ratio:
$$\frac{N_2}{N_1} = e^{-(E_2 - E_1)/k_B T}$$

$\therefore \quad E_2 - E_1 = h\nu = \dfrac{12400}{\lambda}$

$\therefore \quad E_2 - E_1 = \dfrac{12400}{7000} = 1.77$ eV

At $T_1 = 300$ K

$$\frac{N_2}{N_1} = \exp\left(\frac{-1.77}{8.62 \times 10^{-5} \times 300}\right)$$

$$= e^{-65}$$

At 500 K

$$\frac{N_2}{N_1} = \exp\left(\frac{-1.77}{8.62 \times 10^{-5} \times 500}\right)$$

$$= e^{-41.1}$$

Example 13

A ruby laser emits 0.1 J pulses of light of wavelength 720 nm. How many minimum number of Cr^{+++} ions are there in ruby?

Solution

Given

$E = 0.1$ J

$\lambda = 720$ nm $= 7.2 \times 10^{-7}$ m

The number of Cr^{+++} ions participating in laser action is given by:

$$N = \frac{E}{h\nu} = \frac{E\lambda}{hc}$$

$$= \frac{0.1 \times 7.2 \times 10^{-7}}{6.62 \times 10^{-34} \times 3 \times 10^8}$$

$$= 3.625 \times 10^{17}$$

Example 14

Consider an AlGaAs based heterostructure laser diode which has an optical cavity of length 200 microns. The peak radiation is at 870 nm and the refractive index of GaAs is about 3.7. What is the mode integer m of the peak radiation and the separation between the modes of the cavity? If the optical gain vs. wavelength characteristics has a FWHM wavelength width of about 6 mm, how many modes are there within the bandwidth? How many modes are there if the cavity length is 20 μm? [B.Tech.]

Solution

Figure 6.27 schematically illustrates the cavity modes, the optical gain characteristics and a typical output spectrum from a laser. The free-space wavelength λ of a cavity mode and length L are related by:

$$m\frac{\lambda}{2n} = L$$

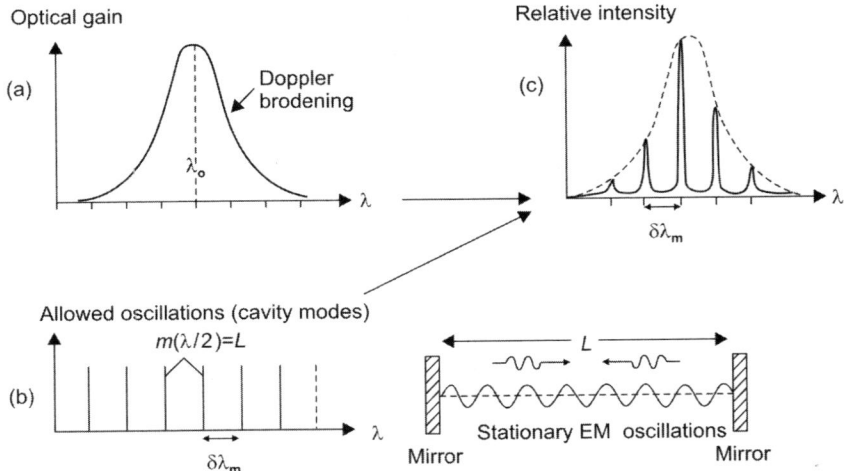

Fig. 6.27. (a) Optical gain vs. wavelength characteristics (called the optical gain curve) of the lasing medium. (b) Allowed modes and their wavelength due to stationary EM waves within the optical cavity. (c) The output spectrum (relative intensity vs. wavelength) is determined by satisfying (a) and (b) simultaneously, assuming no cavity losses

so that:

$$m = \frac{2nL}{\lambda} = \frac{2(3.7)(200 \times 10^{-6})}{(900 \times 10^{-9})} = 1644.4 \simeq 1644.$$

The wavelength separation $\Delta\lambda_m$ between the adjacent cavity modes m and $(m+1)$ in Fig. 6.27 is:

$$\delta\lambda_m = \frac{2nl}{m} - \frac{2nl}{m+1} \approx \frac{2nl}{m^2} = \frac{\lambda^2}{2nL}$$

Thus the separation between the modes for a given peak wavelength increases with decreasing L. When $L = 200$ μm:

$$\delta\lambda_m = \frac{(900 \times 10^{-9})^2}{2(3.7)(200 \times 10^{-6})} = 5.47 \times 10^{-10} \text{ m or } 0.5467 \text{ nm}$$

If the optical gain has a bandwidth of $\Delta\lambda_{1/2}$ as in Fig. 6.27, then there will be $\Delta\lambda_{1/2}/\Delta\lambda_m$ number of modes or (6 nm)/(0.547 nm), that is 10 modes.

When $L = 20$ μm, the separation between the modes becomes:

$$\delta\lambda_m = \frac{(900 \times 10^{-9})^2}{2(3.7)(200 \times 10^{-6})} = 5.47 \text{ nm}$$

Then $(\Delta\lambda_{1/2})/\Delta\lambda_m = 1.1$ and there will be one mode that corresponds to about 900 nm. In fact m must be an integer and when $m = 1644$,

$\lambda = 902.4$. It is apparent that reducing the cavity length suppresses higher modes. Note that the optical bandwidth depends on the diode current.

Example 15

Given that the refractive index n of GaAs has a temperature dependence $dn/dT \approx 1.5 \times 10^{-4}$ K^{-1} estimate the change in the emitted wavelength 870 nm per degree change in the temperature between mode hops.

[B.Tech.]

Solution

Consider a particular given mode with wavelength λ_m.

$$m\left(\frac{\lambda_m}{2n}\right) = L$$

Then,

$$\frac{d\lambda_m}{dT} \approx \frac{d}{dT}\left[\frac{2}{m}nL\right] \approx \frac{2L}{m}\frac{dn}{dT}$$

Substituting for L/m in terms of λ_m:

$$\frac{d\lambda_m}{dT} \approx \frac{\lambda_m}{n}\frac{dn}{dT} = \frac{870 \text{ nm}}{(3.7)}(1.5 \times 10^{-4} \text{ K}^{-1}) = 0.035 \text{ nm K}^{-1}.$$

Note that we have used n for a passive cavity whereas n above should be the effective refractive index of the *active* cavity which will also depend on the optical gain of the medium, and hence its temperature dependence is likely to be somewhat higher than the dn/dT value we used.

Example 16

Consider a GaAs quantum well (QW). Effective mass of a conduction electron in GaAs is 0.07 m_e, where m_e is the electron mass in vacuum. Calculate the first two electron energy levels for a quantum well of thickness 10 nm. What is the hole energy below E_c of GaAs if the effective mass of the hole is about 0.50 m_e?

What is the change in the emission wavelength with respect to bulk GaAs which has an energy bandgap of 1.42 eV?

[B.Tech.]

Solution

The lowest energy levels with respect to the CB edge E_c in GaAs are determined by the energy of an electron in a one-dimensional potential energy well.

$$\varepsilon_n = \frac{h^2 n^2}{8 m_h^* d^2}$$

where n is a quantum number 1, 2, ..., and ε_n is the electron energy with respect to E_c in GaAs or $\varepsilon_n = E_a - E_c$. Figure 6.28(b) using $d = 10 \times 10^{-9}$ m, $m_e^* = 0.07$ m, and $n = 1$ and 2, we find, $\varepsilon_1 = 0.0537$ eV and $\varepsilon_2 = 0.215$ eV respectively.

The hole energy levels below E_v in Fig. 6.28(b) are given by:

$$\varepsilon_n' = \frac{h^2 n^2}{8 m_h^* d^2}$$

Using $d = 10 \times 10^{-9}$ m, $m_h^* \approx 0.5 \, m_e$, and $n = 1$, we find, $\varepsilon_1' = 0.0075$ eV. The wavelength of emission from bulk GaAs with $E_g = 1.42$ eV is:

$$\lambda_g = \frac{hc}{E_g} = \frac{(6.626 \times 10^{-34})(3 \times 10^8)}{(1.42)(1.602 \times 10^{-19})} = 874 \times 10^{-9} \text{ m (874 nm)}$$

Whereas from the GaAs QW, the wavelength is:

$$\lambda_{QW} = \frac{hc}{E_g + \varepsilon_1 + \varepsilon_1'} = \frac{(6.626 \times 10^{-34})(3 \times 10^8)}{(1.42 + 0.0537 + 0.0075)/(1.602 \times 10^{-19})}$$

$$= 839 \times 10^{-9} \text{ m (839 nm)}$$

The difference is $\lambda_g - \lambda_{QW} = 35$ nm.

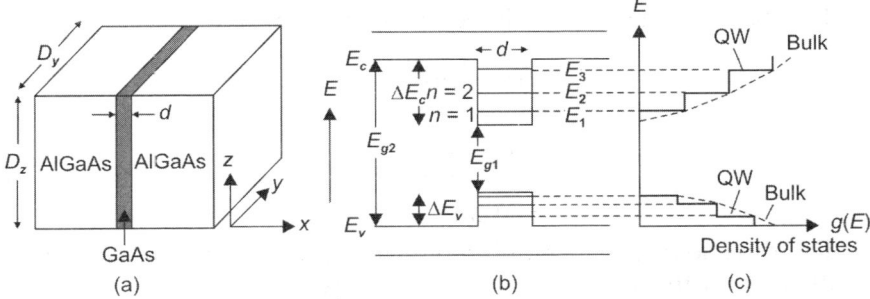

Fig. 6.28. A quantum well (QW) device. (a) Schematic illustration of a quantum well (QW) structure in which a thin layer of GaAs is sandwiched between two wider bandgap semiconductors (AlGaAs). (b) The conduction electron in the GaAs layer are confined by (ΔE_c) in the x-direction to a small length d so that their energy is quantized. (c) The density of states of a two-dimensional QW. The density of states is constant at each quantized energy level

Example 17

The reflection coefficient for a semiconductor laser are equal and have the value 0.5. The length of the cavity is 1 μm. Assuming that there are no additional loss terms, show that the gain of the laser is 6.93×10^3 cm^{-1}.

Solution

We have the gain for laser:

$$g = \alpha_i + \frac{1}{2L} \ln\left(\frac{1}{r_1 r_2}\right)$$

where α_i is the losses within the propagation medium, L is the length of the cavity and r_1 and r_2 are the reflection coefficients of the respective surface. Since there are no internal losses, means $\alpha_i = 0$. Thus, we have:

$$\begin{aligned} g &= \frac{1}{2L} \ln\left(\frac{1}{r_1 r_2}\right) \\ &= \frac{1}{2 \times 1} \ln\left(\frac{1}{0.5 \times 0.5}\right) \\ &= 6.93 \times 10^3 \text{ cm}^{-1}. \end{aligned}$$

Example 18

An injection laser has an active cavity with losses of 30 cm^{-1} and reflectivity of each cleaved laser facet is 30%. Show that the laser gain coefficient for the cavity when it has a length of 600 µm is 50 cm^{-1}.

Solution

Here, $r_1 = r_2 = r$.

The threshold gain per unit length is given by:

$$\begin{aligned} g &= \alpha + \frac{1}{L} \ln \frac{1}{r} \\ &= 30 + \frac{1}{0.06} \ln \frac{1}{0.3} \\ &= 50 \text{ cm}^{-1} \end{aligned}$$

The threshold gain per unit length is equivalent to the laser gain coefficient for the active cavity, which is 50 cm^{-1}

Example 19

A double-heterojunction InGaAsP LED emitting at a peak wavelength of 1310 nm has radiative and nonradiative recombination times of 30 and 100 ns, respectively. The drive current is 40 mA. The bulk recombination lifetime is obtained from:

$$\frac{1}{\tau} = \frac{1}{\tau_r} + \frac{1}{\tau_{nr}} \qquad (i)$$

where τ_r (= n/R_r) is radiative recombination lifetime and τ_{nr} is nonradiative recombination lifetime. Calculate τ, internal quantum efficiency (η_{int}) and internal power level.

Solution

$$\tau = \frac{\tau_r \tau_{nr}}{\tau_r + \tau_{nr}} = \frac{30 \times 100}{30 + 100} \text{ ns} = 23.1 \text{ ns}$$

The internal quantum efficiency using Eq. (i)

$$\eta_{int} = \frac{\tau}{\tau_r} = \frac{23.1}{30} = 0.77$$

This yields an internal power level of:

$$P_{int} = \eta_{int} \frac{hcI}{q\lambda}$$

$$= 0.77 \frac{(6.6256 \times 10^{-34} \text{ J} \cdot \text{s})(3 \times 10^8 \text{ m/s})(0.040\text{A})}{(1.602 \times 10^{-19} \text{ C})(1.31 \times 10^{-6} \text{ m})}$$

$$= 29.2 \text{ mW}$$

SUGGESTED READINGS

1. Davis CC, 'Lasers and Electro Optics: Fundamentals and Engineering', Cambridge University Press (2002).
2. Mroziewicz B et al., 'Physics of Semiconductor Lasers', Polish Scientific Publishers (1991).
3. Coldren LA and Crozine SW, 'Diode Lasers and Photonic Integrated Circuits', Wiley (1995).
4. Kapon E. (ed.), 'Semiconductor Lasers', Academic Press (1999).
5. Agrawal GP and Dutta NK, 'Semiconductor Lasers' (2nd ed.), Kulwar Academic Publishers (2000).
6. Chuang SL, 'Physics of Photonic Devices', Wiley (2009).
7. Kumar S and Jamal Deen, M, 'Fiber Optic Communications', Wiley (2014).
8. Kakani SL and Shubhra Kakani, Modern Physics, 2nd ed. (2014), Viva Books, New Delhi-2 (India)

GLIMPSES

- **LASER** (**L**ight **A**mplification by **S**timulated **E**mission of **R**adiation) is a light amplifier source of near monochromatic source of radiation usually used to produce monochromatic coherent radiation in the infrared, visible, and UV regions of the electromagnetic spectrum.

  The production of the laser beam depends on *stimulated emission*. The emission of photons that occur following *excitation* of electrons in a system is usually *spontaneous* and cannot be controlled. Stimulated emission is a process whereby an incoming photon of energy $h\nu$ can stimulate an electron in the high energy state E_2 to jump to a lower energy state E_1, where $E_2 - E_1 = h\nu$. The photon resulting from this process has the same frequency $v = (E_2 - E_1)/h$, as the stimulating photon and travels in the same direction. If there are sufficient electrons in the high-energy level, both stimulating and stimulated photons can cause further stimulated emission and a narrow beam of monochromatic radiation results, the intensity of which increases exponentially. The laser beam is *coherent* (i.e. spatially and temporarily in phase) and can have a very high energy density.

A laser beam is produced by stimulated emission but can only operate efficiently if a large number of electrons are in a particular high-energy level. This condition is called *population inversion*, is a non-equilibrium condition and power has to be fed into the system to maintain the population inversion.

- **Functioning of laser:** There are three basic requirements of laser: an *active medium, population inversion,* and some sort of feedback.
- To achieve *population inversion* a right group of atoms or molecules, which is called *active medium* is needed. This active medium is a gas or mixture of gas or crystal or semiconductor.
- **Metastatable state:** The metastable state means long life state compared to other states. In general lifetime of an energy state is 10^{-8} s or more. This is key to the laser. In many atoms one or more metastable states are available.
- **Operation of laser:** For an ideal operation of laser, some conditions have to be satisfied by the system. The relaxation rates from excited state to upper laser level and from lower laser level to ground state should be high to maintain population inversion between upper and lower laser levels.

  The *thermal equilibrium* population of laser level should be as small as possible.

  The *decay of excited state* to any level other than upper laser level should be as slow as possible.

  The *non-radiative decay* of upper laser level should be slow.
- **Properties of laser:**
  (i) *Monochromaticity*: The minimum frequency range for laser is 1 Hz.
  (ii) *Directionality*: Laser can travel large distance without deviation.
  (iii) *Coherence time*: $\Delta\tau = 1/\Delta v$. If $\Delta v = 1$ MHz, then $\Delta\tau = 1$ μs. This time is longer than atomic processes, which are of the order of ns. Sunlight has a bandwidth $\Delta v = 10^{14}$ Hz, hence the coherence time is very short, $\Delta\tau = 10^{-14}$ s.
  (iv) *Coherence length*: The distance $\Delta L = c\Delta\tau$ is called the coherence length of the beam. If $\Delta\tau = 1$ μs, the light travel $\Delta L = 3 \times 10^8$ ms^{-1} × 1 μs = 300 m.

(v) *Spectral brightness*: The power flow per unit area, per unit bandwidth and per unit solid angle is called brightness of the source:

$$B = \frac{P}{A\,\delta\nu\Delta\Omega}$$

where A is the surface area (source size), $\Delta\nu$ is the bandwidth and $\Delta\Omega$ is the solid angle (beam divergence). For sun $B = 10^{-12}$ W·cm^{-2} Hz sr^4.

- A typical *quantum well* device has an ultra thin, typically less than 50 nm, narrow bandgap semiconductor such as GaAs, sandwitched between two wider bandgap semiconductors, such as AlGaAs.
- The *gain coefficient g* of an active medium is defined as the fractional change in optical intensity per unit length as a light beam passes through. $g = N_{eff}\sigma_g/V_g$, where σ_g is the gain cross-section of a given inverted site, N_{eff} is effective population inversion and V_g is the volume of active medium.
- Every laser must fulfill three basic criteria:
  (i) Gain to provide stimulated emission
  (ii) Gain > loss, to sustain reflections
  (iii) Resonant cavity.
- The semiconductor lasers have some special features such as compactness, high efficiency, capability for high-speed direct modulation, wide emission spectrum and high reliability. However, they are sensitive to temperature.
- *Semiconductor laser* that have been well studied, such as III-V semiconductor lasers, involve strongly interacting states. In such cases, the large Bohr radii of ground state (1 S) *excitons* lead to strong overlap, as enough excitons are generated to reach the lasing threshold. Thus, the excitonic nature of the underlying recombination transition is replaced by a plasma of electrons and holes, and stimulated emission arises from recombining such electrons and holes in the plasma. This situation is naturally linked to the exciton population and to the size of excitons.
- **Exciton:** An electron in combination with a *hole* in a crystalline solid. The electron has gained sufficient energy to be in an excited state and is bound by electrostatic attraction to the positive hole. The exciton may migrate through the solid and eventually the hole and electron recombine with emission of a photon.

REVIEW QUESTIONS

1. Explain the terms: population inversion, spontaneous emission and stimulated emission.
2. Distinguish between spontaneous and stimulated emissions.

318 Photonics | Optoelectronics

3. What are Einstein's coefficients? Derive Einstein's relation. Explain the physical significance of Einstein's coefficients?
4. Derive relation between probabilities of spontaneous and stimulated emission in terms of Einstein's coefficients.
5. Differentiate between three level and four-level systems for lasing action.
6. Describe the principle and working of a semiconductor lasers. Mention its merits and demerits.
7. Write short notes on the following: (i) Metastable state, (ii) Population inversion, (iii) Pumping, (iv) Optical cavity, (v) Gain.
8. Explain laser oscillations and obtain the expression for wavelengths of oscillations.
9. What are homogeneous and heterostructure p-n junctions? Draw their energy-band diagrams.
10. What is optical gain? How it can be determined based on Fermi Golden rule.

PROBLEMS

1. The bandwidth of a He–Ne laser is 500 Hz. Obtain the coherence time and the longitudinal coherence length.
[**Ans.** (i) 2 ms, (ii) 600 km)
2. Calculate the gain constant β of a laser having the following parameters: wavelength = 650 nm, inversion density $(n_2 - n_1) = 5 \times 10^{22}/$m^3, life time for spontaneous emission 2×10^{-4} s and line width $\Delta\lambda = 15$ Å.   [**Ans.** 3.95]
3. Show that for a normal optical source with temperature about 10^3 K and wavelength 6000 Å, the emission is predominantly due to the spontaneous transitions.
4. A laser beam of wave length 6000 Å, power 10 mW, and angular spread 5×10^{-5} rad is focussed by a lens of focal length 10 cm. Calculate the radius and power density of the image. What is lateral coherence width?
[**Ans.** (i) radius = 2.5×10^{-4} cm, (ii) Power density = 5.1×10^4 W/cm^2, (iii) lateral coherence width = 1.2 cm]
5. For sodium line, the wave length is 5890 Å and coherence time is 10^{-10} s. Show that the monochromaticity of the source is (5890 ± 0.0578) Å.
6. A laser beam of wavelength 6328 Å from He-Ne laser can be focussed on an area equal to the square of its wavelength, i.e. λ^2. If the laser radiates energy at the rate of 10 MW, find the intensity of focussed beam.   [**Ans.** $I = 2.5 \times 10^{-16}$ W/m^2]

Optical Sources 319

7. A transition between the energy levels E_2 and E_1 produces a light of wavelength 632.8 nm. Calculate the energy of the emitted photons. **[Ans. 1.96 eV]**

8. Calculate the population of two states in He-Ne laser that produces light of wavelengths 700 nm at 27°C. $\left[\textbf{Ans. } \dfrac{N_2}{N_1} = 5.9 \times 10^{-29} \right]$

9. A semiconductor heterojunction laser is grown with GaAsP active region layer with lattice constant 0.56 nm. Assume that laser transition occurs from the bottom of conduction band involving band energy gap 1.85 eV. Determine wavelength of laser emitted. **[Ans. λ = 670 nm]**

10. The ratio of population inversion of two energy levels out of which upper one corresponds to a metastable state is 1.059×10^{-30}. Find the wavelength of light emitted at 330 K. **[Ans. λ = 632 nm]**

SHORT ANSWER QUESTIONS

1. What is a band width?

Ans. In any material system, the energy levels have a finite spectral (energy) width. This results in a bandwidth for the optical transitions of the system.

Bandwidth is the difference between the upper and lower frequency limits of a band, normally measured in Hz. The bandwidth may be determined by effects that are common to all sites within the system, resulting in a homogeneously broadened transition, or may be determined by local variations in material properties, leading to an inhomogeneously broadened transition.

2. What is a *p-n* junction?

Ans. The region at which two semiconductors of opposite polarity (*p*-type and *n*-type) meet. A *p-n* junction can perform various functions depending on the geometry, the bias conditions, and the doping level in each semiconductor region. Most diodes transistors, etc. utilize the properties of one or more *p-n* junctions. If the materials are dissimilar, e.g. silicon and germanium, the junction is a *heterojunction*. Normally, the same material is used but doped so as to produce two different conductivity types, in this case the junction is a simple homojunction.

Under reverse bias conditions (i.e. negative bias applied to the *p*-type semiconductor) very little current flows, until *breakdown* occurs. Under forward bias conditions, carriers are attracted across the junction into the region of opposite type (where they become *minority carriers*, and a current flows in the external circuit. The forward in a homojunction increases exponentially with the voltage, i.e.:

$$I = I_o (e^{-eV/k_BT} - 1)$$

where I_o is reverse saturation current, e is electronic charge and V is applied voltage. Resistance in the material reduces the rate of rise of current through the device after a few tenths of a volt.

3. What are laser diodes?

Ans. Laser diodes are light-emitting devices, capable of emitting laser beams. One of the commonly used laser diode is the *injection laser diode* (ILD). The laser diode works exactly like an ordinary diode. The *p*-type and *n*-type materials are made of AlGaAs (aluminium gallium arsenide).

When forward biased, light is emitted at the *pn* junction. However, unlike the LED the emitted light is coherent and monochromatic. In the LED, the light emitted is scattered in all directions. In the laser diode, however, the light emitted can escape only from the end faces because the edges are roughened. In most laser diodes, one of the end faces is coated with a reflective material so that the radiation of the laser is emitted only in one direction.

The laser diode emits light when forward biased and can withstand only relatively small reverse-biased voltages. A high reverse voltage can damage or destroy the laser diode.

Laser diodes are used primarily in fiber-optic communications. It is the only device capable of producing optical energy high enough in concentration to pass through lengthy fiber-optic cables.

4. What is a light emitting diode (LED)?

Ans. A LED is a semiconductor device that generates light. Light generation begins at the atomic level.

A LED is a *pn* junction diode of special construction. Like all semiconductor diodes, its behaviour is influenced by biasing conditions. Under forward bias, the electrons from the *n*-type region and the holes in the *p*-type material move toward the *pn* junction.

At the junction, the holes and electrons combine. Electrons are at higher energy level than the holes. The electrons literally fall into the holes as they combine, and in doing so release the excess amount of energy in the form of light. There is a current through the LED in the forward direction, just as in the case of an ordinary diode.

When the *pn* junction is reverse biased the LED does not conduct and there is no release of light. Hence LED emits light when

Optical Sources **321**

forward biased and does not emit when reverse biased. The LED operates at forward voltages typically ranging from 1.2 V to 4.34 V. The reverse breakdown voltage is in the range of –3 V to –10V.

Multicolour LEDs are available that will:
 (i) emit one colour light when the supply voltage is one polarity.
 (ii) emit a second colour when the polarity is reversed.
 (iii) emit a third colour when the bias polarity is rapidly switched.

Multicolour LEDs are usually two LEDs connected in antiparallel. That is, the anode of each diode is connected to the cathode of the other. Each LED emit light when forward biased. Thus, when voltage of either polarity is applied one LED is forward biased and emits its native colour.

The LEDs most commonly used are red and green in colour. The green LED is normally used to indicate whether something is functioning properly, and the red LED is used to indicate that there is a problem. If the multicolour LED is rapidly switched between the two polarities, the red/green LED appears to produce a third colour (yellow).

Infrared-emitting diode (IREDs) are used for home electronics such as remote controls, fiber optic communications, discriminating organic solvents in the field of medicine, and other such applications. Diodes made of *gallium aresnide* release energy by way of heat and infrared light. Such a diode is called a IRED.

LEDs are very widely used in *multisegment* displays. By lighting a combination of different LEDs, any number from 0 to 9 can be displayed.

5. In how many modes photodiodes can operate?

Ans. Two modes:
 (i) *Photovoltaic mode*: When operating in the photovoltaic mode, the photodiode generates a voltage in response to light. The incidence of light on the photodiode creates electron-hole pairs. The electrons generated in the depletion region are attracted to the positively charged ions in the *n*-type material, and the holes are attracted to the negatively charged ions in the *p*-type material. This creates a separation of charges, and a small voltage drop of about 0.45 V is developed across the diode. In this mode of operation, the photodiode is connected across the voltage source, a small current will flow from the cathode to the anode.

(ii) *Photoconductive mode*: In this mode, the conductance of the diode changes when light is applied. In this mode, the photodiode is reverse biased. The depletion region of the reverse biased photodiode is very wide and the resistance of the diode is high, and hence there will be only a reverse current through it. This reverse current that flows through the diode when there is no light being applied is called *dark current* (I_D).

When light is applied, electron-hole pairs are generated. The electrons are attracted to the positive bias voltage, and the holes are attracted to the negative bias voltage. This movement of electrons and holes causes a considerable *reverse current* to flow through the photodiode. The resistance of photodiode is very low when light is applied. If the intensity of light is increased, the resistance decreases and therefore, the reverse current increases. The current that passes through the photodiode when light is being applied is called the light current (I_L).

The conductivity of photodiode is low when there is no light applied, and the conductivity increases as the intensity of light increases. Consequently, the magnitude of the dark current is very much smaller than that of the light current.

6. Write the advantages of LEDs.

Ans. (i) LEDs have higher efficiency than incandescent lighting and can ultimately or possibly exceed fluorescence in efficiency.

(ii) LEDs are highly durable and reliable.

(iii) LED white light sources can provide significant energy saving compared with the conventional lighting.

7. Why heterojunction LEDs are used?

Ans. Since, a photon emitted due to radiative recombination process of direct bandgap material in one part of the device and reabsorbed before it reached the surface, therefore, one obvious solution is to put the junction close to the surface to avoid reabsorption. However, the presence of surface states, producing non-radiative recombination channels, can result in a reduction in the radiative efficiency the device. Obviously, the solution of this problem is to use heterojunction instead of avoiding absorption. A commonly used heterojunction is that of GaAs and AlGaAs. The heterojunction LED is designed in such a manner that the recombination occurs on the narrow gap semiconductor. The light is subsequently emitted through an AlGaAs window into the external environment. The band of AlGaAs is greater than that of GaAs and therefore, most of the light will transmit through the AlGaAs being absorbed.

8. What are quantum well lasers?

Ans. These lasers offer an alternative to double heterostructures. In these lasers, the energy of the emitted photon can be made subsequently higher than that of the bandgap. By proper choice of well width, the energy of the emitted photon can be adjusted so that the lasing can be achieved at various wavelength using the same material. A single quantum well laser is similar in design to a double heterostructure laser except that the narrow gap. This means that the active region is intensionally made so that the spatial quantization effects occur. Spatial quantization results in the production of energy levels above the conduction band minimum in a quantum well. The minimum electron and hole energies within the well are both above the conduction band and valence band minimum respectively. If electron and hole recombine, the energy of the emitted photon is a strong function of the well width. The narrower the width, the greater the quantum state energy. The energy of the emitted photon can be tuned by adjusting the well width.

Limitations of quantum well lasers: The cross-section for the capture of carriers in the well is relatively small. In order to obtain a radiative recombination event, there is necessity for the injected electrons and holes to be captured within the well. To confine the carriers within the well, it is desired that they should undergo an inelastic scattering event in order to lose sufficient energy. If there is no scattering events within the well, the injected carriers can pass directly over the well without getting trapped and do not contribute to the optical output. This means that the threshold current density can be relatively high.

9. How one can improve the extraction efficiency of LEDs?

Ans. One can improve the extraction efficiency of LEDs by choosing a transparent substance. In such a device the light that is internally reflected, but is reabsorbed, will eventually escape after being randomly scattered by rough spots on the surface of chip. This means that the extraction efficiency can be quite high.

10. Explain the importance of infrared LEDs?

Ans. Infrared LEDs include GaAs LEDs which can emit light near 0.9 µm and several iii-iv compounds, e.g. quantenary $Ga_xIn_{1-x}As_yP_{1-x}$ LEDs, which emit light in the range 1.1 to 1.6 µm.

Infrared LEDs find application in *optoisolators* where an input or control signal is decoupled from the output. Infrared LEDs also find application for transmission of an optical signal through an optical fiber.

324 Photonics | Optoelectronics

11. What is major difference between LEDs and laser diodes?

Ans. The optical output from LED is *incoherent*, whereas from laser diode is coherent.

12. Draw energy level diagrams showing various types of quantum well structures.

Ans. See Fig. 6.29.

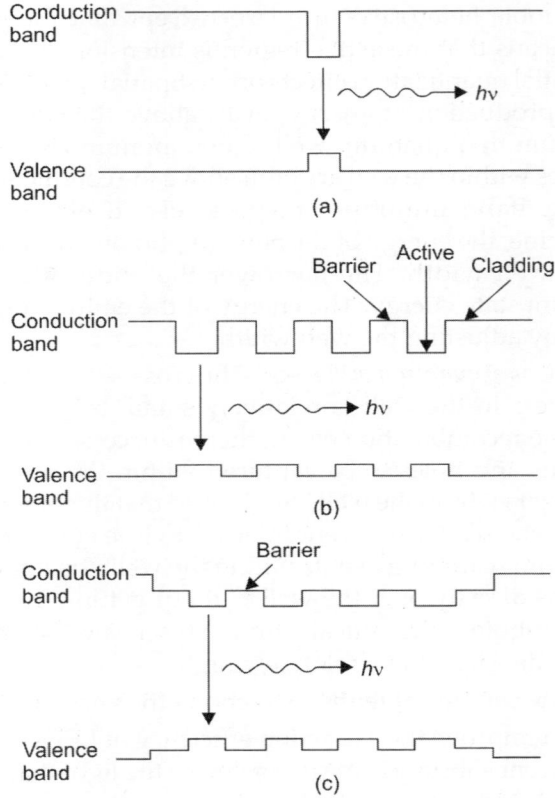

Fig. 6.29. Energy band diagram showing various types of quantum-well structure: (a) single quantum well; (b) multiquantum well; (c) modified multiquantum well.

MULTIPLE CHOICE QUESTIONS

1. Laser light is considered to be coherent because it consists of:
    (a) many wavelengths (b) uncoordinated wavelengths
    (c) coordinated wavelengths (d) divergent waves
2. The relation between Einstein's A_{21} and B_{21} coefficients is:
    (a) $\dfrac{A_{21}}{B_{21}} = 1:2$ (b) $\dfrac{A_{21}}{B_{21}} = \dfrac{8\pi h v^3}{c^3}$

(c) $\dfrac{A_{21}}{B_{21}} = 1:1$ (d) $\dfrac{A_{21}}{B_{21}} = \dfrac{v}{c}$

3. The ratio of the number of spontaneous to stimulated transition is:
   (a) $\dfrac{A_{21}}{B_{21}uv} = 1:1$
   (b) $\dfrac{A_{21}}{B_{21}uv} = 1:2$
   (c) $\dfrac{A_{21}}{B_{21}uv} = \exp\left(\dfrac{hv}{k_BT}\right) - 1$
   (d) $\dfrac{A_{21}}{B_{21}uv} = 8\pi\dfrac{hv^3}{c^3}$

4. When the atoms in the source go from an excited state (E_2) to a lower energy state (E_1), spontaneous emission of light takes place, the energy of the emitted photon being:
   (a) $hv_{12} = E_2 - E_1$
   (b) $hv_{12} = E_2/E_1$
   (c) $hv_{12} = E_1/E_2$
   (d) $hv_{12} = 0$

5. The order of the pumping power necessary to achieve the population inversion in a ruby is:
   (a) 10^7 W/m^2
   (b) 10^2 W/m^2
   (c) 10^{10} W/m^2
   (d) 10^{16} W/m^2

6. The most important characteristic of a laser is:
   (a) polarization
   (b) coherence
   (c) high intensity
   (d) directionality

7. The intensity of a laser beam does not decrease with distance in accordance with the inverse square law because:
   (a) the laser light is monochromatic
   (b) the laser light is very intense
   (c) the laser light is very directional
   (d) the laser light obeys Planck's law

8. For a laser beam λ = 4400Å and coherence time = 4×10^{-5} s, the coherence length will be:
   (a) 12 km
   (b) 1.2 km
   (c) 0.12 km
   (d) 0.012 km

9. For a laser beam minimum angular divergence depends on:
   (a) wavelength λ only
   (b) diameter of mirror D only
   (c) both λ and D
   (d) alignment of mirrors only

10. The wavelengths produced by a He-Ne laser correspond to transition in:
    (a) both helium and neon
    (b) helium
    (c) neon
    (d) neither helium nor neon

11. The term 'population inversion' means:
    (a) population of the ionised state is maximum
    (b) population of the lowest state is maximum

(c) population of the lower level is more than that of the higher level
(d) population of the higher level is more than that of the lower level

12. A laser operates at a frequency of 3×10^{14} Hz and has an aperture of 10^{-2} m. The angular spread will be:
    (a) 10^{-2} rad
    (b) 10^{-3} rad
    (c) 10^{-4} rad
    (d) 10^{-5} rad

13. The function of He atoms in the He-Ne laser is:
    (a) to quench the neon atoms
    (b) to provide energy to the neon atoms
    (c) to make neon atoms inactive
    (d) None of the above

14. Which of the following characteristics is not associated with a laser light?
    (a) Coherence
    (b) Brightness
    (c) Polarization
    (d) Birefringes

15. In ruby laser, the rod is surrounded by a helical photographic flash lamp filled with:
    (a) cromium
    (b) aluminium
    (c) xenon
    (d) neon

16. Who invented semiconductor laser?
    (a) Albert Einstein
    (b) T.H. Maiman
    (c) Ali Javan
    (d) Robert Hall

17. Which of the following modes of operations should be used to achieve laser pulses of very high power?
    (a) Q-switched operation
    (b) Pulsed operation with free oscillation
    (c) Continuous-wave operation
    (d) All of the above

18. According to the mode of oscillation, lasers may be categorized in the following group/groups.
    (a) Continuous wave (CW) lasers
    (b) Pulsed lasers with free oscillations
    (c) Q-switched lasers
    (d) All the above

19. The temporal width of a mode-locked rhodamine 6G dye laser emitting at 0.6 μm if $\Delta v = 10^{13}$ Hz is:
    (a) 10^{-13} s
    (b) 8.4×10^{-11} s
    (c) 3.3×10^{-13} s
    (d) 6.6×10^{-10} s

20. The spatial length of a mode-locked Nd^{3+} glass laser if $\Delta v = 3 \times 10^{12}$ Hz is:
    (a) 1 m
    (b) 3.3 m
    (c) 5 m
    (d) 0.1 mm

ANSWERS

1. (c)   2. (b)   3. (c)   4. (a)   5. (a)   6. (b)   7. (c)
8. (a)   9. (c)   10. (c) 11. (d) 12. (c) 13. (b) 14. (d)
15. (c) 16. (d) 17. (a) 18. (d) 19. (a) 20. (a)

7

Optoelectronic Devices

7.1 INTRODUCTION

The advent of low loss fibers has been a key to the recent emergence of optical methods as a backbone of long distance communications. The enormous scientific and economic potential is largely responsible for the dynamic pace of research and development activities in the area of *optoelectronic devices* and *optoelectronic integrated circuits* (OEICs). These devices and circuits alongwith optical fibers constitute the basic components for the physical network needed to transmit information. This network has been dubbed the information superhighway because of dense amount of traffic, it is expected to support.

Optoelectronic devices and related emerging technologies have a profound effect on mankind. One can feel the presence of these technologies in areas of computers, home entertainment systems, such as compact disks, medical science, telecommunications, etc.

The basic optoelectronic devices which are usually made with semiconductor materials are heterojunction lasers, light emitting diodes (LEDs), and photodetectors. The ingeration of lasers or photodetectors with electronic devices, such as field effect transistors and/or bipolar transistors constitutes OEICs. This integration can be of the hybrid or the monolithic form. Although, silicon is the predominant material for microelectronics, it is not suitable for optoelectronic devices in which energy conversion from electrical to optical and vice versa is the fundamental mechanism of the operation. Optoelectronic devices are made from *compound semiconductors* which have direct energy gaps. Energy conversion is more efficient in these materials. III-V compound semiconductors have been the traditional materials used for optoelectronic devices. However, the II-VI compounds have recently demonstrated the potential for optoelectronic devices operating at short wavelengths. The wide *bandgap nitrides*, which also have direct bandgaps, have been found quite impressive for lasers and photodetectors at short wavelengths.

Basically, optoelectronic devices can be placed under two groups (Fig. 7.1). Devices that fall under the first type are **light emitting devices**, e.g. light emitting diodes and laser diodes. Devices that fall under the second category work as **light detectors**, e.g. phototransistors, photoresistors, photovoltaic cells and liquid crystal displays (LCDs). There is another device called **optocoupler** or **optoisolator**, which is a combination of two types of devices in a single unit.

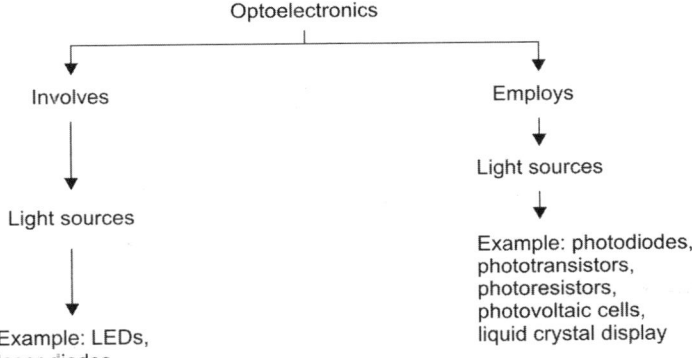

Fig. 7.1. Optoelectronic devices

There are several semiconductor devices for high frequency applications, which take the advantage of instabilities that occur in the semiconductor. The important type of instability is negative conductance. Devices that work with negative conductances are microwave devices, tunnel diode, IMPATT diode and Gunn diode. These are mainly two-terminal semiconductor devices that can be operated in negative conductance mode to provide amplification or oscillation at microwave frequencies in a proper circuit.

The intent of this chapter is to discuss the general principles and characteristics of semiconductor lasers and photodetectors. Although, the types of lasers and photodetectors available are diverse, there is an underlying unity in the basic principles and characteristics so that figures of merit can be used to compare the various devices. Quantum efficiency, responsivity, bandwidth, and noise equivalent power are the important parameters used to characterize photodetectors. The different types of photodetectors—photoconductors, p-i-n photodiodes, and avalanche photodiodes, commonly used along with the relatively new metal-semiconductors—metal photodiodes are discussed in this chapter.

7.2 PHOTOELECTRIC EFFECT

Photoelectric effect (photoeffect) is the process of emission of electrons from a metal surface when it is illuminated by high-frequency electro-

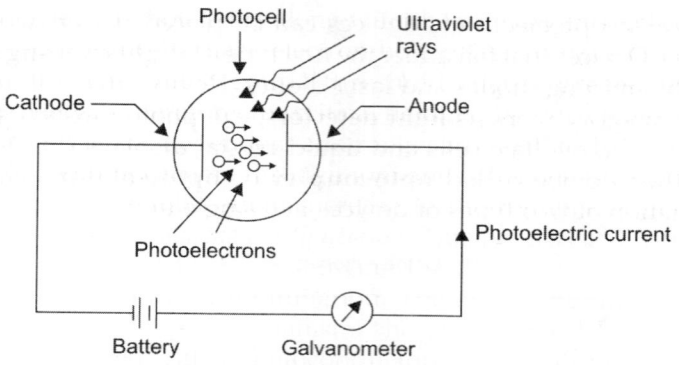

Fig. 7.2. Photoelectric effect

magnetic radiation (Fig. 7.2). The photoelectric effect is of different forms:

7.2.1 Photoemissive Effect

When a radiation (photon) of wavelength less than a critical value is incident upon a metal surface, electrons are found to be emitted; this is called photoemissive or photoelectric effect, e.g., vacuum and gas phototubes. Kinetic energy E of emitted electrons is given by:

$$E = h\nu - e\Phi \qquad (7.1)$$

The $e\Phi$ is the surface work function. No electrons will be emitted when $h\nu < e\Phi$ (or, in terms of wavelength, $\lambda > hc/e\Phi$). The ratio of the number of emitted electrons to the number of absorbed photons is called *quantum yield* or *quantum efficiency*. The minimum frequency that can cause a photoelectric effect is called threshold frequency, and the corresponding wavelength of light beyond which the photoelectric effect ceases is called *threshold wavelength* or *cut-off wavelength*.

7.2.2 Photoconductive Effect

An electron may be raised from the valence band to the conduction band in a semiconductor by the absorption of a photon of frequency ν provided that:

$$h\nu \geq E_g \qquad (7.2)$$

where energy gap is E_g, or in terms of wavelength:

$$\lambda \leq \frac{hc}{E_g} \qquad (7.3)$$

We define bandgap wavelength λ_g as the largest value of wavelength that can cause this transition, i.e.:

$$\lambda_g = \frac{hc}{E_g} \qquad (7.4)$$

Optoelectronic Devices **331**

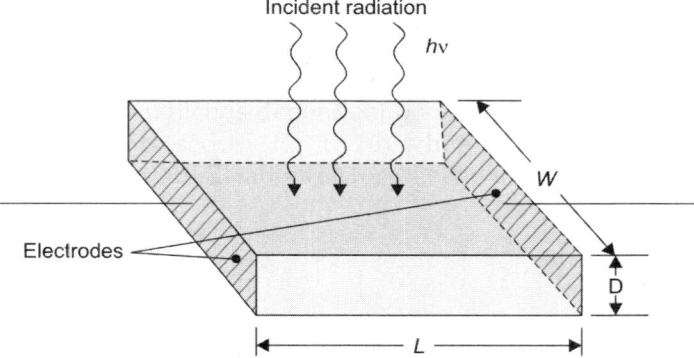

Fig. 7.3. Illustration of photoconductive effect

As long as the electron remains in the conduction band, it will cause an increase in the conductivity of the semiconductor. This is the phenomenon of **photoconductivity**. This is the basic mechanism in the working of photoconductive detectors. Figure 7.3 illustrates the photoconductive effect and a photoconductor bias circuit is shown in Fig. 7.4.

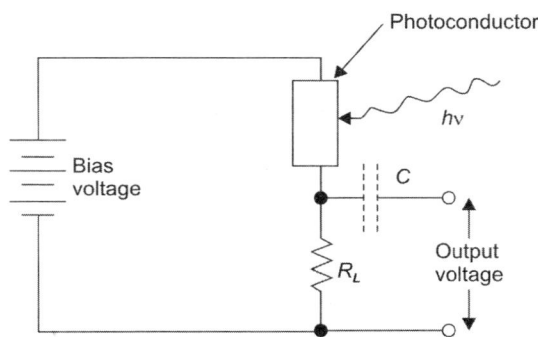

Fig. 7.4. Photoconductor bias circuit

Photoconductive mode: In this mode, the conductance of the diode changes when light is applied. In this mode, the photodiode is reverse-biased. Figure 7.5 shows a reverse-biased photodiode in the photoconductive mode. The depletion region of the reverse-biased photodiode is very wide and the resistance of the diode is high, and hence there will be only a small reverse current through it. This reverse current that flows through the diode when there is no light being applied is called dark current (I_o).

When light is incident, electron-hole pairs are generated. Electrons are attracted to the positive bias voltage, and holes are attracted to the negative bias voltage. This movement of electrons and holes causes a

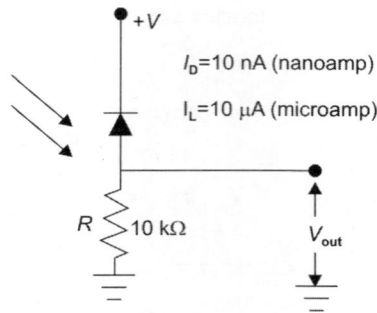

Fig. 7.5. Photoconductive mode of operation

considerable reverse current to flow through the photodiode. The resistance of the photodiode is extremely low when light is incident. If the intensity of light is increased, resistance decreases and therefore, reverse current increases. Current that is passed through the photodiode when light is being applied is called **light current** (I_L).

The conductivity of the photodiode is low when there is no light incident, and conductivity increases as the intensity of light increases. Consequently, the magnitude of dark current is very much smaller than that of light current. Consider the example shown in Fig. 7.5. Assume that dark current (I_D) is dependent on the amount of current flowing in the circuit. With no light present:

$$V_{out} = I_p \times R = 10 \text{ nA} \times 10 \text{ k}\Omega = 100 \text{ μV}$$

With light present:

$$V_{out} = I_L \times R = 10 \text{ μV} \times 10 \text{ k}\Omega = 1 \text{ V}$$

7.2.3 Photovoltaic Effect

In the photovoltaic effect, an e.m.f. is generated in a pn junction (when reverse-biased) under the influence of incident light.

Photovoltaic mode: While in operation in the photovoltaic mode, the photodiode generates a voltage in response to light. The incidence of light on the photodiode creates electron-hole pairs. Electrons generated in the depletion region are attracted to the positively charged ions in the *n*-type material, and holes are attracted to the negatively charged ions in the *p*-type material. This creates a separation of charges, and a small voltage drop of about 0.45 V is developed across the diode.

Figure 7.6 shows the photovoltaic mode of operation. In this mode, the photodiode acts as a **solar cell**. If a load resistor is connected across the voltage source, a small current will flow from the cathode to the anode.

Table 7.1 provides a brief summary of optoelectronic devices and related effect on which they are based.

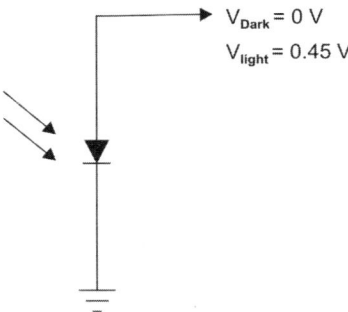

Fig. 7.6. Photovoltaic mode of operation

Table 7.1. Optoelectronic devices	
S. No.	Optoelectronic devices	Related photoelectric effect
1.	Photoresistor	Photoconductive
2.	Photodiode	Photovoltaic
3.	Phototransistor	Photovoltaic
4.	Photo SCR	Photovoltaic
5.	Solar cell	Photovoltaic
6.	Photomultiplier tube	Photovoltaic
7.	Photo FET	Photoemission

The details of a few materials generally used in optoelectronic devices are given in Table 7.2.

Table 7.2. Materials used in optoelectronic devices		
S. No.	Materials/ compounds	Forbidden energy gap (E_g) (eV)	Emission wavelength (λ) (nm)
1.	Si	1.07	1142
2.	Ge	0.65	1882
3.	GaAs	1.42	910
4.	GaP	2.26	561
5.	SiC	3.2	560
6.	AlSb	1.62	77.7
7.	InSb	0.20	6910

7.3 PHOTODETECTORS

Principle

A photodetector is a device that *converts optical energy to electrical energy*. The principal mechanism responsible for this transformation is photoconductivity. This property is exhibited by all semiconductors and it is the increase in conductivity brought about by the absorption of photons.

The absorption of a photon results in the generation of an electron-hole pair. The electrons and holes separate to become mobile carriers which are transported through the semiconductor under the influence of an externally applied electric field. The transport of these carriers enhances the conductivity of the semiconductor. In concise terms, the three basic processes involved in this conversion are:
1. The absorption of photons and the resulting generation of carriers.
2. The transport of the generated carriers across the absorption or drift region under the influence of an applied field. Internal amplification of carriers can occur at this stage via various mechanisms. An example of a mechanism is impact ionization which occurs with the application of large electric fields.
3. Collection of carriers constituting a photocurrent which flows through an external circuitry.

Perhaps, the simplest type of photodetector is the photoconductor which is simply a slab of an intrinsic semiconductor with two contacts. The electron-hole pairs, generated by absorption of photons in the material, are collected by oppositely biased contacts to constitute a photocurrent. Other types of photodetectors are photodiodes which are based on either the *p-n* junction or the metal-semiconductor junction (also called a Schottky-barrier). A *p-i-n* (or PIN) photodiode is a reverse-biased *p-n* junction with an intrinsic layer interposed between the *p* and *n* layers. Because the depletion area is the only region supporting an electric field in a *p-n* junction, the intrinsic layer serves to increase the depletion layer width, and therefore, the photon absorption region of the device. The PIN photodiode is normally operated in a bias mode in which the device does not exhibit gain. Another type of photodetector based on a *p-n* junction is the *avalanche photodiode* (APD). APDs are operated at electric fields which are highly enough to cause impact ionization and, thereby, generate more carriers, leading to the avalanche effect. The net effect of the avalanche process is a multiplication of carriers, resulting in gain for the output photocurrent of the device. As mentioned above, another important class of photodetectors, which have recently gained prominence, are the *metal-semiconductor photodiodes* which are made from metal-semiconductor-metal (MSM), photodiodes attractive for some applications requiring monolithic integration.

As described above, there are some basic properties exhibited by all semiconductor photodetectors. These general properties can be quantified, to a certain extent, and have become figures of merit used in comparing photodetectors. The properties can be quantified in terms of quantum efficiency, responsivity, bandwidth, and noise equivalent power (NEP).

The operation of photodetector involves three steps:
- Carrier generation, i.e., the absorption of photon reate electron-hole pairs

- Carrier transport
- Interaction of current with external circuit to provide output signal. Requirements for photodetector are high sensitivity at the operating wavelengths, high response speed and low noise. It should be compact, use low biasing voltage or currents and should be reliable.

7.3.1 Photoconductor

Figure 7.7 shows a bar of semiconductor material with ohmic contacts at each end of the bar. When the incident light falls on the surface of the semiconductor bar (which is photoconductor) electron-hole pairs (E_iHP) are generated by band-to-band transition or by extrinsic transitions. This extra electron-hole pairs increase the conductivity of the material. The initial thermal-equilibrium conductivity is (before light falls):

$$\sigma_0 = q(n_0\mu_n + p_0\mu_p) \quad (7.5)$$

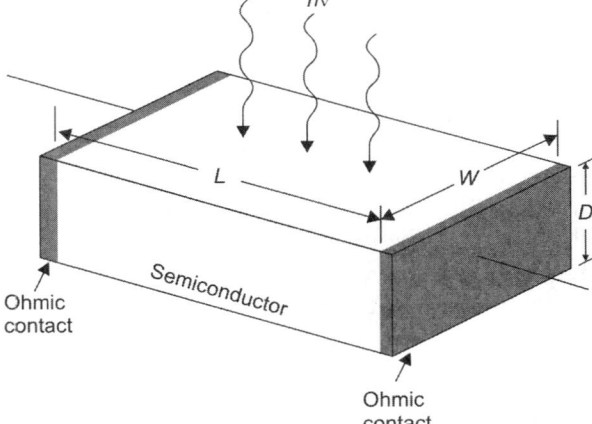

Fig. 7.7. A semiconductor bar exposed to light

The increase in conductivity under illumination is mainly due to the increase in the number of carriers, and is given as:

$$\Delta\sigma = q\delta p(\mu_n + \mu_p) \quad (7.6)$$

This change in conductivity due to optical excitation is known as **photoconductivity**.

When a voltage is applied across the bar of length L, then the electric field produces current in the bar. The current density is written as:

$$J = J_0 + J_L = (\sigma_0 + \Delta\sigma)E \quad (7.7)$$

where J_0 is the current density (when no optical excitation, i.e., dark condition) and J_L is photocurrent density.

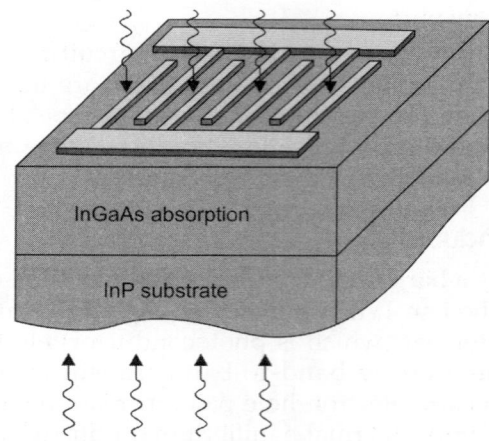

Fig. 7.8. Schematic of an InGaAs photoconductor which is illuminated on the front and back sides

Let us assume electrons and holes are generated uniformly throughout the bar then the current due to the optical excitation is:

$$\begin{aligned} I_L &= J_L A = \Delta\sigma A E \\ &= \delta_p (\mu_n + \mu_p) \, A E q \\ &= G_L \tau_p (\mu_n + \mu_p) \, A E q \end{aligned} \quad (7.8)$$

where A is the cross-sectional area of the device. Equation (7.8) shows that photocurrent is directly proportional to the area of the device and the rate at which excess-carriers are generate.

The time required by electron to flow through the bar of the photoconductor is known as **transit time**, and is given as:

$$t_n = \frac{L}{\mu_n E} \quad (7.8a)$$

Substituting this value in Eq. (7.8), one finds:

$$\begin{aligned} I_L &= G_L \tau_p (\mu_n + \mu_p) \, A q \frac{L}{t_n} \\ &= q A G_L \left(\frac{\tau_p}{t_n}\right) (\mu_n + \mu_p) \frac{L}{\mu_n} \\ &= q A G_L L \left(\frac{\tau_p}{t_n}\right) \left(1 + \frac{\mu_p}{\mu_n}\right) \end{aligned} \quad (7.9)$$

The *photocurrent gain is defined as the ratio of the rate at which charges are collected by the contacts to the rate at which charges are generated within the photoconductor*, i.e.:

$$\text{Gain} = \frac{I_L}{q G_L A_L} \quad (7.10)$$

Substituting Eq. (7.9) into Eq. (7.10), one obtains:

$$\text{Gain } (g) = \frac{\tau_p}{t_n}\left(1 + \frac{\mu_p}{\mu_n}\right) \qquad (7.11)$$

If $m_p/m_n \ll 1$:

$$\text{Gain } (g) = \frac{\tau_p}{t_n} \qquad (7.12)$$

Figure 7.8 shows an InGaAs photoconductor.

7.3.2 Quantum Efficiency (η)

The quantum efficiency η of a photodetector is the number of electron-hole pairs generated per incident photon collected at the contacts. Quantum efficiency is determined by many factors. These include the fact that not all photons incident on the semiconductor will produce electron-hole pairs and that some of the photons may be reflected at the surface of the semiconductor. All these factors combine to reduce η. Quantum efficiency can therefore be given by:

$$\eta = (1-r)\,\zeta[1 - \exp(-\alpha L)] \qquad (7.13)$$

where r is the optical power reflectance at the surface of the detector, ζ is the fraction of electron-hole pairs that actually contribute to the photocurrent, α is the absorption coefficient of the detector material per centimeter, and L is the width of the detector's absorption region. By applying an antireflection coating on the detector's surface for the wavelength of operation, reflection can be reduced and, thereby, the factor $(1-r)$ can be maximized. It is difficult to estimate the factor ζ because it depends on the quality of the material. Carriers can be lost through recombination at the surface or in the bulk of the material which reduces the photocurrent. Modern epitaxial growth methods are now capable of producing high quality materials, and therefore, for a practical estimation of quantum efficiency, ζ can be assumed as unity. The last factor $[1 - \exp(-\alpha L)]$ denotes the *fraction of the incident optical power* absorbed in bulk of the detector.

In terms of the quantities easily measured in the laboratory, quantum efficiency is given by:

$$\eta = \frac{I_p/q}{p_i/h\nu} = \frac{h\nu}{q}\cdot\frac{I_p}{p_i} \qquad (7.14)$$

where I_p is the detector photocurrent, p_i is the incident optical power, and $h\nu$ is the photon energy. The quantity η given above is known as the external quantum efficiency η_{ext}. The quantum efficiency is dependent on the absorption coefficient α which is a function of wavelength λ. Figure 7.9 shows the dependence of α on λ for some detector materials. Because only photons with energy greater than or equal to the bandgap energy E_g can be absorbed (i.e. $hc/\lambda \geq E_g$), the long-wavelength limit for a practical detector is the bandgap wavelength.

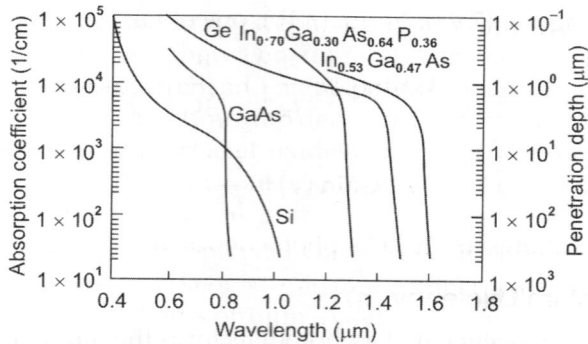

Fig. 7.9. Wavelength dependence of the optical absorption coefficients of several semiconductor materials

Bandgap energies at 300 K for representative photodetector materials are displayed in Table 7.3. There is also a short-wavelength limit because α is very large at short wavelengths for most semiconductors, and consequently, all of the incident photons are absorbed near the surface of the detector.

Table 7.3. Bandgap energies (in eV) at 300 K for some photodiode materials

Material	Bandgap energy (E_g)
GaAs	1.42
GaSb	0.73
GaAs$_{0.88}$Sb$_{0.12}$	1.15
Ge	0.67
InAs	0.35
In$_{0.53}$Ga$_{0.47}$As	0.75
InP	1.35
Si	1.14

Responsivity (R)

The responsivity, R of a detector is the photocurrent in the device divided by the input optical power and is given by:

$$R = \frac{I_p}{P_i} = \frac{nq}{h\nu} = \eta \frac{\lambda(\mu m)}{1.24} \qquad (7.15)$$

The unit of responsivity in A/W. It is seen from this expression that, for a constant η, R should increase with λ. This is illustrated in Fig. 7.10. However, because α depends on λ, there is a region between the short- and long-wavelength limits over which R increases. For photodetectors which exhibit gain, the gain factor G can be accommodated in a more general equation for responsivity given by:

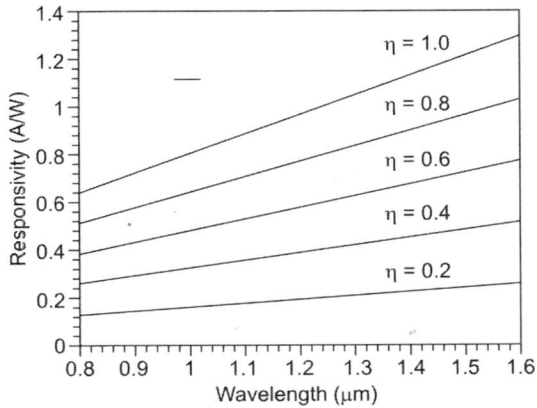

Fig. 7.10. Responsivity (R) vs wavelength (λ) for various external quantum efficiencies

$$R = G\eta \frac{\lambda(\mu m)}{1.24} \quad (7.16)$$

It is possible to degrade the responsivity of a detector by applying excessive incident optical power. The detector becomes saturated, thus, limiting its linear dynamic range, which is the range over which the relationship between the detector's output and the incident optical power in linear.

7.3.3 Bandwidth

The bandwidth B of a photodetector measures the shortest response time of the device. This property becomes very important when a photodetector is used in a data transmission circuit. The faster a detector can respond to a stream of optical pulses, the higher the density of the transmitted data can be. The response time of a photodetector is determined by three factors—*transit time, diffusion time*, and the *device RC time constant*.

Electron-hole pairs created by photons in the active region of a photodetector move in directions opposite to the contacts for collection under an applied electric field. The carriers move by drift and diffusion. If the electric field is sufficiently large, most of the carriers travel by drift, and they reach their scattering-limited or saturation velocity in the material. The velocity of holes is usually smaller than that of electrons, therefore, the time (i.e., transit time), it takes holes to drift across the active region of the detector limits the response time of the device. If electron-hole pairs are generated uniformly throughout the material, then, a severe transit time spread between electrons and holes can occur. Diffusion time limitations can occur only at low bias where the drift field is low. Because the diffusion process is slow, it can be a severe problem even though only a small number of carriers may be involved.

A judicious design of the active area of the detectors and the application of an appropriate bias can make this limitation insignificant. The last factor is the resistance R and the capacitance C of the device and its associated circuitry. This composite RC network integrates the output current of the detector and, therefore, increase the response time. Different types of photodetectors are influenced by different combinations of these limitations which set their bandwidths. However, photodetectors of a given design and material do exhibit a constant gain-bandwidth product.

7.3.4 Noise Equivalent Power (NEP)

Photodetectors are subject to several sources of noise that degrades their performance. The inherent randomness in the arrival of photons and the absorption of photons in the device save as sources of noise. Various sources of current generation exist in all photodetectors. Some of these include current due to the incoming optical signal, current due to background radiation, and the dark current that is due to surface leakage, tunneling, and thermal generation of electron-hole pairs in or around the active region. All of these currents are generated randomly and contribute to *shot noise*. The amplification process that produces gain in some detectors is the *avalanche mechanism*. This is a random effect, and, therefore, there is a gain noise associated with such detectors. Another source of noise involves the *random motion of carriers* in resistive electrical materials at finite temperatures. There are parasitic resistances intrinsic to photodetectors and also resistances in circuits in which photodetectors are utilized. An example is a receiver circuit in which a detector serves as a source of input current to a preamplifier. The noise generated by these resistive elements is called *thermal*, or *Johnson*, or *Nyquist noise*. This noise is given by:

$$\langle i_j^2 \rangle = \frac{4kT_{\text{eff}}B}{R_{\text{eff}}} \tag{7.17}$$

where R_{eff} is the parallel combination of the detector and the preamplifier input resistances, B is the bandwidth, and T_{eff} is the effective temperature which is related to the noise figure NF of the amplifier:

$$T_{\text{eff}} = T\,(10^{\text{NF}/10} - 1) \tag{7.18}$$

where T is the ambient temperature. It is, therefore, evident that, in the operation of a detector, the output signal must be above the noise level. The signal-to-noise ratio (SNR) is, therefore, an important characteristic in photodetectors, and it is related to sensitivity. The sensitivity of a photodetector is the minimum optical input power needed to achieve a given value of SNR. A measure of sensitivity is called *noise equivalent power* (NEP). NEP is the optical power (or photocurrent) required for the SNR to be unity over a 1-Hz bandwidth. Essentially, this measures when the photocurrent is exactly equal to the noise current. Thus, NEP measures the minimum detectable power in a photodetector. NEP depends on bandwidth, and, to find the optical power required to produce a SNR of unity for an entire measurement bandwidth, we have:

$$-P_i = \text{NEP}\sqrt{B}. \qquad (7.19)$$

Another figure of merit also useful for determining the ultimate detection limit is *detectivity* D^* given by:

$$D^* = \frac{\sqrt{AB}}{\text{NEP}} \ (\text{cm Hz}^{1/2} \cdot \text{W}^{-1}) \qquad (7.20)$$

where A is the area of the photodetector on which light is incident. As with NEP, the reference bandwidth is taken as 1 Hz. D^* is usually expressed as $D^*(\lambda, f, 1)$ where λ is the wavelength and f is the modulation frequency of the input optical signal. It must be noted that NEP and D^* are not equal to system sensitivity in actual applications because other noise sources, such as preamplifier noise, may dominate, especially in high speed (GHz) systems.

Note

The quantum efficiency of a photoconductor is given by Eq. (7.13) and the responsivity is given by Eq. (7.16), where G is the internal photocurrent gain of the device. The gain is brought about by the fact that photogenerated carriers contribute to current until they recombine. The gain is given by:

$$G = \frac{\tau}{t_{tr}} \qquad (7.21)$$

where τ is the excess-carrier recombination life-time and τ_{tr} is the transit time of the majority carrier. The transit time is given by:

$$\tau_{tr} = \frac{L}{v} \qquad (7.22)$$

where L is the channel length (distance between contacts) and v is the carrier velocity. It is seen from these equations that, if the recombination lifetime is greater than the majority-carrier transit time, then many carriers will travel between the contacts before recombination takes place. This is photoconductor gain, and it can be below unity or well above unity depending on various factors including semiconductor material, size of the device, and the magnitude of the applied voltage.

The bandwidth of a photoconductor is given by:

$$B = \frac{1}{2\pi\tau} \qquad (7.23)$$

It is seen that, whereas a long recombination time makes for a high gain, it also reduces bandwidth. So, a trade-off between gain and bandwidth exists for photoconductors. The gain-bandwidth product of a photoconductive detector is given by:

$$GB = \frac{1}{2\pi t_{tr}} \quad (7.24)$$

where t_{tr} is a *constant for a given material and detector configuration*.

The primary contributions to noise in photoconductive detectors are made by thermal or Johnson noise and the generation-recombination noise. The thermal noise given by Eq. (7.17) results from the random motion of carriers with average energy of kT contributing to the dark current of the device. The generation-recombination noise is due to fluctuation in the generation and recombination of carriers which, in turn, leads to fluctuations in the conductivity of the device. The generation-recombination noise is given by:

$$\langle i^2_{G-R} \rangle = \frac{4qI_0 GB}{1 + \omega^2 \tau^2} \quad (7.25)$$

where I_o is steady-state output photocurrent and ω is the angular modulation frequency of the input optical signal.

To describe the overall noise performance of a photoconductor, the NEP is given by:

$$NEP = \frac{8h\nu}{\eta} \left[1 + \frac{kT}{qG}(1 + \omega^2 r^2)\frac{G_c}{I_0} \right] \quad (7.26)$$

where G_c is the conductance of the channel. The dominant noise mechanism in a photoconductive detector is the thermal noise of the conducting channel. An increase in channel resistance is necessary to reduce thermal noise. If the thickness of the channel is reduced to increase the resistance, then quantum efficiency is reduced. To obtain the highest resistance and, hence, the lowest thermal noise achievable while maintaining high gain and quantum efficiency, it is necessary to utilize materials with the lowest carrier concentration. Thermal noise, generation-recombination noise, and dark current are high in semiconductors with smaller bandgaps. Although, photoconductors can have large gains, the gain may not be enough to surmount the inherent noise limitations to make them useful in many applications.

7.4 PHOTODIODE

A *photodiode is basically p-n junction* or *metal semiconductor* contact operated under reverse bias. A cross-section of a typical silicon photodiode is shown in Fig. 7.11. The starting material is *n*-type silicon. A thin *p* layer is formed on the front surface of the device by thermal diffusion or ion implantation of the appropriate doping. Small metal contacts are applied to the front surface of the device and entire back is coated with a contact metal. The back contact is cathode and front

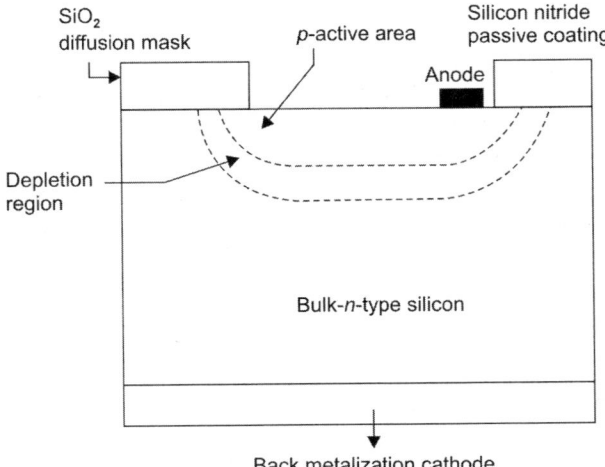

Fig. 7.11. Cross-section of silicon photodiode

contact is anode. The active area is coated either with silicon nitride, silicon monoxide or silicon oxide for protection and to serve as the anti-reflection coating. The thickness of the top p layer is determined by the wavelength of the radiation to be detected. The depth of the depletion region can be varied by applying a reverse bias voltage across the junction. When the depletion region reaches the back of the photodiode, then it is said to be *fully depleted*. The depletion region is important to photodiode performance since most of the sensitivity to radiation originates these.

When a light is absorbed in the active region, an electron-hole pair is created. The electrons are collected in the n-region and holes in the p-region. The migration of electrons and holes to their respective region is called *photovoltaic effect*. Photodiodes are most useful as current generators although a voltage is also generated by illumination. Also, the thermally generated minority carriers within the diffusion length of each side of the junction diffuse to the depletion region, and are swept to the other side by the electric field. Following points are important as regards photodiode:

(i) When a reverse bias is applied, some current will flow without illumination. The dark current is specified for every device. In cases, where a very low bias voltage is applied, shunt resistance is specified. This is determined by measuring dark current with +/− 0.010 V applied bias.

(ii) For high-frequency operation, the depletion region must be kept thin to reduce the transit time.

Let us assume that the junction is uniformly illuminated by photons of energy $h\nu > E_g$ and due to that FHPs are created at the rate of g_{op}

(EPH/cm^{-3} sec) which take part in the conductivity. The number of holes created per second within the diffusion length of transition region on n side is $AL_p g_{op}$, where A is the cross-sectional area. Similarly, $AL_n g_{op}$ represents the number of electrons created per second within the diffusion length of the transition region on p-side. $AW g_{op}$ is the number of generated carriers within the transition region of width W. These all optically generated carriers take part in conduction and hence resulting current is:

$$I_{op} = qAg_{op}(L_n + L_p + W) \tag{7.27}$$

Adding

$$I = I_{Th}(e^{qV/kT} - 1) - I_{op}$$

$$= qA\left(\frac{L_p}{\tau_p} p_n + \frac{L_n}{\tau_n} n_p\right)(e^{qV/kT} - 1) - qAg_{op}(L_n + L_p + W) \tag{7.28}$$

Equation (7.28) clearly shows that the total diode current reduces due to the generation rate. The first term of the RHS of Eq. (7.27) represents the usual diode current equation and the second term gives current due to optical generation. The I–V characteristics in the presence of optical generation is shown in Fig. 7.12.

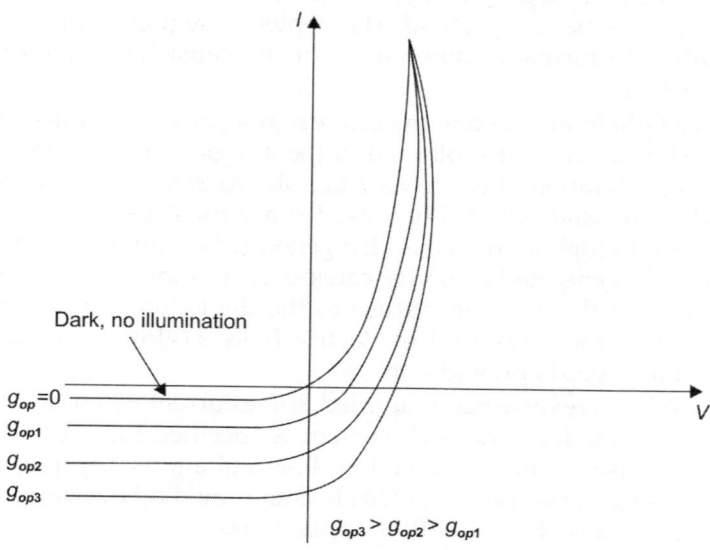

Fig. 7.12. I–V characteristic of photodiode

For $V = 0$, means device is short circuited then Eq. (7.28) reduces to:

$$I = -I_{op}$$

This shows that the short circuit current from n to p is equal to optical generation current in magnitude, i.e.:

$$I = -qAg_{op}(L_n + L_p + W)$$

or for constant values of A, L_n, L_p and W:

$$I \propto g_{op}$$

Therefore, the short circuit current of the device is proportional to the optical generation rate.

Let us assume V_{OC} represents the open circuit voltage for $I = 0$ and substituting $I = 0$ in Eq. (7.28), we have:

$$qA\left(\frac{L_p}{\tau_p}p_n + \frac{L_n}{\tau_n}n_p\right)(e^{qV_{oc}/kT} - 1) = qAg_{op}(L_n + L_p + W) \quad (7.29)$$

or

$$(e^{qV_{oc}/kT} - 1) = g_{op}\left(\frac{L_n + L_p + W}{(L_p/i_p)p_n + (L_n/i_n)n_p}\right)$$

or

$$V_{OC} = \frac{kT}{q}\ln\left(\frac{L_n + L_p + W}{(L_p/i_p)p_n + (L_n/\tau_n)n_p} \cdot g_{op} - 1\right) \quad (7.30)$$

Consider a special case, when junction is symmetrical:

$$\tau_p = \tau_n \text{ and } n_p = p_n$$

Equation (7.30) can be further expressed as:

$$V_{OC} = \frac{kT}{q}\ln\left(\frac{L_n + L_p + W}{(L_p + L_n)p_n/\tau_n} \cdot g_{op} - 1\right) \quad (7.31)$$

Defining $p_n/\tau_n = g_{Th}$ = Thermal generation rate, Eq. (7.31) can be further simplified and one finds:

$$V_{OC} = \frac{kT}{q}\ln\left(\frac{L_n + L_p + W}{L_p + L_n} \cdot \frac{g_{op}}{g_{Th}} - 1\right) \quad (7.32)$$

Assume $L_n + L_p \gg W$ and $(g_{op}/g_{Th}) \gg 1$. Then Eq. (7.33) reduces to:

$$V_{OC} = \frac{kT}{q}\ln\left(\frac{g_{op}}{g_{Th}}\right) \quad (7.33)$$

We may note that V_{OC} cannot increase indefinitely with increased g_{op} and limit on V_{OC} is that it can take maximum value of equilibrium contact potential.

Since the minority-carrier concentration increases due to optical generation rate, the lifetime τ_n becomes small and hence the term (p_n/τ_n) becomes larger which results in increase of g_{Th}. And hence, the ratio (g_{op}/g_{Th}) cannot increase for indefinite period. We may note:

(i) The magnitude of the photocurrent generated by a photodiode is dependent upon the wavelength of the incident light. Silicon photodiodes exhibit a response from the UV through the visible and into the near infrared part of the spectrum.

(ii) Radiometric sensitivity of a photodiode is defined as the ratio of short-circuit photocurrent generated by photodiodes divided by the energy of incident light.

Responsivity (R_e) is a measure of sensitivity which takes into account the active area of the photodiode chip. The parameter is obtained by dividing the short-circuit current by the energy of light per unit area.

Photodiodes have unity internal gain. In order to increase the sensitivity of photodiodes to light, one can either increase the active area of the photodiode chip itself or use lenses to increase the effective active area. There is a linear relationship between sensitivity and active area.

We may note that depending upon the required applications of the photodiode, it can be operated in either quadrants of its I–V characteristics. Since, the photodiode is reverse biased, so the preferable quadrants are either third or fourth as shown in Fig. 7.13.

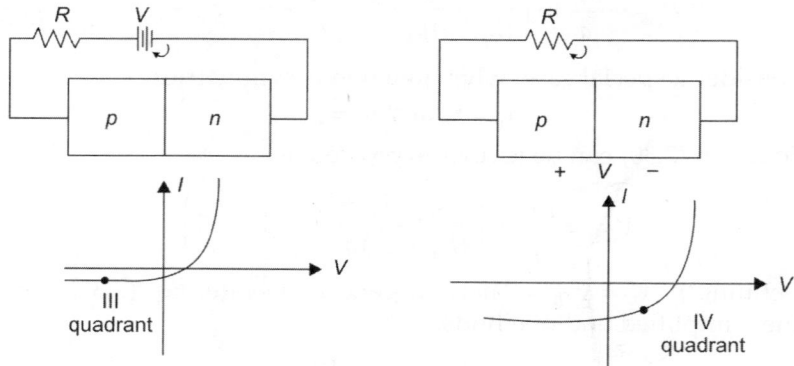

Fig. 7.13. Schematic representation of operating quadrant in photodiode

In third quadrant, both the current and junction voltage are either positive (or negative). In this quadrant power is delivered to the device from the external circuit because $P = VI = (-V)(-I)$. Power quantity is always positive.

In fourth quadrant, junction voltage is positive and current is negative and hence, power is delivered from the junction to the external circuit.

If one is interested to extract the power from the device then fourth quadrant is used. But usually in photodetector, we are using reverse bias and hence third quadrant will be chosen for operation.

Quantum Efficiency (η)

The quantum efficiency relates, as a percentage, the energy per photon and the quantum yields (electrons per photon). The mathematical expression for quantum efficiency is given as:

$$\eta = \left(\frac{I_{op}}{q}\right)\left(\frac{P_{op}}{h\nu}\right)^{-1} \qquad (7.34)$$

where I_{op} is optically generated current and P_{op} is the optical power. It can also be expressed as:

$$\eta = \left(\frac{(124)\, S_R}{\lambda}\right) \qquad (7.35)$$

where S_R is radiometric sensitivity and λ is the wavelength of incident light.

Increasing the operating temperature of a photodiode device results in two distinct changes in operating characteristics. The first change is a shift in the quantum efficiency due to changes in the radiation absorption of the device. QE value shifts lower in the ultraviolet (UV) region and higher in the IR region due to negative temperature coefficient at shorter wavelength of light and positive temperature coefficient in the infrared (IR) part of the spectrum.

The second change is caused by the exponential increase in the thermally excited electron-hole pairs resulting in increasing *dark current*.

In ultraviolet and visible region, metal-semiconductor photodiodes show good quantum efficiencies. In the near-infrared region, silicon photodiodes can reach 100% quantum efficiency near the 0.8 and 0.9 µm region, germanium photodiodes and group III–V photodiodes have shown high quantum efficiencies.

Dark Current

The dark current is the leakage current that flows when the photodiode is in dark, i.e., no illumination of light and a reverse bias voltage is applied across the junction. The dark current may vary from pA to µA depending upon the junction area and process used. Dark current is always specified at a particular value of reverse applied voltage. The dark current is temperature dependent and becomes double for every 10°C increase in ambient temperature.

Other Parameters

Shunt resistance (R_{SH}): The shunt resistance or dynamic junction resistance at zero voltage is determined by applying a small voltage to the photodiode (~mV) and measuring the resulting current. Values of shunt resistance may vary from 100 kΩ to 100 GΩ. Shunt resistance is voltage dependent and also depends on the active area of the diode chip and on the type of processing used. Shunt resistance decreases with increasing temperature.

Junction capacitance (C_j): This is the capacitance associated with the depletion region which exists at the *p-n* junction. The response time of a photodiode is dependent upon the product of junction capacitance and the external load resistor. Junction capacitance increases with increasing junction area and decreases with reverse applied voltage.

Reverse breakdown voltage (V_{BR}): This is the maximum reverse voltage that can safely be applied across the photodiode before breakdown occurs at the junction. The breakdown voltage is determined by process. Typical values for V_{BR} range from 5 V to over 100 V.

Open circuit voltage: V_{OC} is that voltage which is generated by photodiode when photocurrent is zero. V_{OC} varies logarithmically with light. Typical range of V_{OC} is 300 MV to 450 MV. Due to large temperature coefficient V_{OC} is not recommended as an accurate measure of light level.

Response time: A photodiode takes a certain amount of time to respond to a sudden change in light levels. It is expressed in terms of rise time t_r and fall time t_f.

Rise time t_r is the time required for the output to rise from 10% to 90% of its final value. t_f is the fall time, defined as the time required for the output to fall from 90% to 10% of its onstate value.

The response time of a photodiode depends on many factors, like wavelength of incident light, the value of applied voltage across the diode (affects the junction capacitor) and the load resistance.

Noise current: A photodiode will act as a source for electrical noise and generate a noise current J_N. The noise current will limit the usefulness of the photodiode at very low light levels where the magnitude of noise approached that of the signal photocurrent. The amount of noise generated depends on the characteristics of the photodiode and operating conditions.

There are three main components which contribute on the total noise of photodiode, namely, thermal noise, short noise and flicker noise. Thermal or Johnson noise is inversely related to the shunt resistance of the photodiode. Thermal noise tends to be dominant noise component when the diode is operated under zero applied reverse bias. Short noise is dependent on the leakage or dark current of the photodiode. It tends to dominate when the photodiode is used in the photoconductive mode. Flicker noise is unlike thermal or shot noise. It possesses a ($1/f$) spectral density. Flick noise may dominate when the bandwidth of the interest contains frequencies less than 1 kHz.

Noise equivalent power (NEP): In many design applications, the designer needs to know the minimum detectable light of the photodiode. The minimum incident power required on a photodiode to generate a photocurrent equal to the total photodiode noise current is defined as the noise equivalent power (NEP), i.e.:

$$\text{NEP} = \frac{\text{Noise current}}{\text{Responsivity}} \quad (7.36)$$

The NEP is dependent on the bandwidth of the measuring system. The NEP is always quoted at a particular wavelength and is nonlinear over the wavelength range.

We may note that the response speed of photodiode is limited by following three factors:

(i) Diffusion of carriers
(ii) Drift time in the depletion region
(iii) Capacitance of the depletion region.

Equivalent Operating Circuits

The equivalent circuit and symbol of a photodiode is shown in Fig. 7.14, where I_s is signal current, I_L is leakage current, I_n is noise current, C_d is diode junction capacitance, R_s is diode series resistance, R_{SH} is diode shunt resistance and R_L is load resistance.

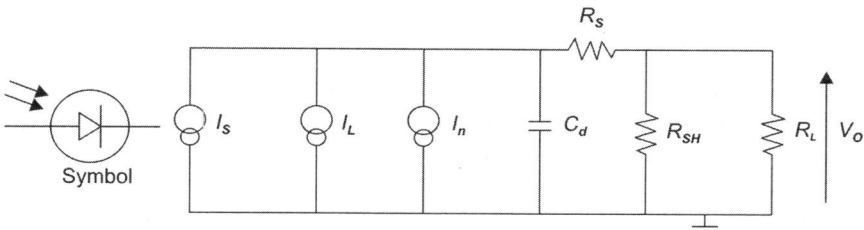

Fig. 7.14. Schematic of equivalent operating circuit model of photodiode

The output voltage V_o is given as:

$$V_o = (I_s + I_L + I_n)\left(\frac{R_L R_{SH}}{R_L + R_{SH} + R_s}\right) \quad (7.37)$$

Fundamentally a photodiode is a current generator. The junction capacitance of the photodiode depends on the depletion layer width and hence on bias voltage. Generally R_{SH} is very high of order MΩ and R_s is very low. We may note that:

(i) Photodiodes are attractive for light wave communication system because they provide high quantum efficiency and very high bandwidth.

(ii) In order to fully absorb the incident radiation in the photodiode, it is necessary to utilize a wide depletion region. This is the reason why *p-i-n* junction diode is often used. However, as the width of the depletion region increases the transit time across the device increases and hence the bandwidth reduces. This is a *tradeoff* in photodiode.

7.4.1 Basic Operation and Construction of Photodiode

A photodiode is a light receiving device that contains a semiconductor *pn* junction. Fig. 7.15 shows a typical photodiode package. A glass window or convex lens allows light to enter the case and strike the photodiode mounted within the glass case.

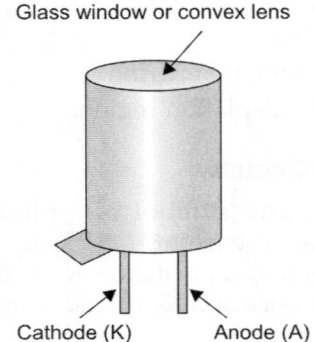

Fig. 7.15. Typical package of a photodiode

Fig. 7.16 shows the construction of a photodiode. It is constructed basically with a *p* type region diffused into an *n* type region. The metal base makes a connection between the cathode terminal and *n* type region, and the metal ring makes a contact between the anode and *p* type region. Light enters the photodiode through a hole in the metal ring, which accommodates the glass or convex lens. Figure 7.17 shows the most commonly used symbol of a photodiode. Note that the two arrows point towards the photodiode, indicating that it responds to light. Photodiodes can operate in two modes:

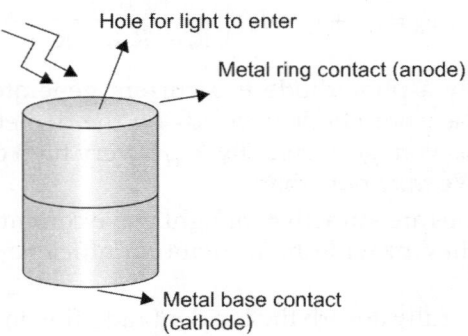

Fig. 7.16. Construction of a *pn* photodiode

(i) Photovoltaic mode
(ii) Photoconductive mode.

When a pn junction is formed from the semiconductor material, a region depleted of mobile charge carriers is formed with high internal electric field across it and called as depletion region. If an electron-hole pair is generated by photon absorption within this region, the internal field will cause the electron and hole to separate (Fig. 7.18).

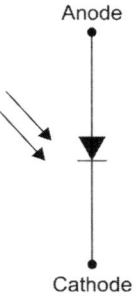

Fig. 7.17. Symbol of a photodiode

One may detect this charge separation in two ways: (i) when the device is left on open circuit, an externally measurable potential appear between p and n regions, and this is termed **photovoltaic mode of operation**. (ii) When the device is shortcircuited externally or operating under reverse-bias, then external current flow between p and n regions. This is termed as **photoconductive mode of operations**.

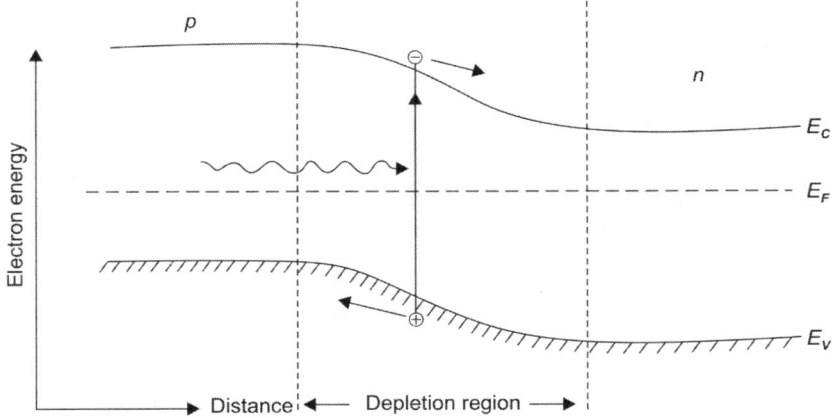

Fig. 7.18. Electron energy level diagram illustrating the generation and subsequent separation of an electron-hole pair by photon absorption within the depletion region of a *pn* junction

The *pn* junction will also respond to electron-hole pairs generated away from the depletion region provided they are able to diffuse to the edge of the depletion region prior the recombination takes place. Obviously, only carriers generated within a minority carrier diffusion length or so of the edge of the depletion region are likely to be able to do this; however, this does increase the sensitive volume of the device.

In operation, one may represent the photodiode by a constant current generator (current flow i_λ being generated by light absorption) with an ideal diode across it (to stimulate the effect of pn junction) (Fig. 7.19). The internal characteristics of the cell may be better modelled by the

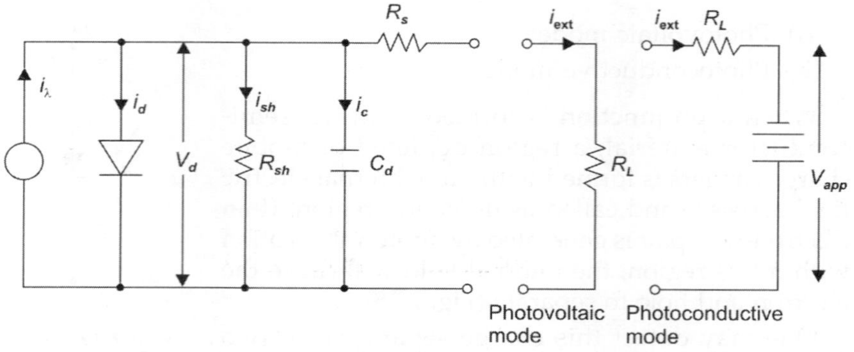

352 Photonics | Optoelectronics

Fig. 7.19. Photodiode equivalent circuit

introduction of a shunt resistor (R_{sh}), a shunt capacitor (C_d) and a series resistor (R_s). If we assume quantum efficiency η for the photon absorption process and also that all the incident radiation is absorbed within the cell, then one may write:

$$i_\lambda = \frac{\eta I_o A e \lambda}{hc} \qquad (7.38)$$

where I_o is light irradiance falling on a cell of area A.

One can develop expressions for sensitivities of the photodiode when operated in two different modes. For simplicity, we may assume that the operation is at fairly low optical modulation frequencies, so that the effects of shunt capacitance may be neglected (i.e. $i_c = 0$).

Considering current to flow as shown in Fig. 7.19, one finds:

$$i_\lambda = i_d + i_{sh} + i_{ext} \qquad (7.39)$$

also
$$V_{ext} = V_d - i_{ext} R_s \qquad (7.40)$$

and
$$V_d = i_{sh} R_{sh}$$

We now take current-voltage behaviour of the diode to be given by:

$$i_d = i_o \left[\exp\left(\frac{eV_d}{kT}\right) - 1 \right]$$

where i_o is diode reverse bias leakage current.

In photovoltaic mode, external current flow is very small; hence $i_{ext} = 0$, Eqs (7.39) and (7.40) becomes $i_\lambda = i_d + i_{sh}$ and $V_{ex} = V_d$, respectively. Substituting for i_d and i_{sh} in the former, one obtains:

$$i_\lambda = i_o \left[\exp\left(\frac{eV_d}{kT}\right) - 1 \right] + \frac{V_d}{R_{sh}}$$

Rearranging, one obtains:

$$\exp\left(\frac{eV_d}{kT}\right) = 1 + \frac{i_\lambda}{i_o} - \frac{V_d}{i_o R_{sh}}$$

Under normal operating conditions, $i_\lambda > i_o$ whereas V_d is usually the same order of magnitude as iR_{sh}. For example, typical values might be $i_o \approx 10^{-8}$ A, $R_{sh} \approx 10^8$ Ω and $V_d \approx 0.6$ V. Thus, to first approximation, one finds:

$$\exp\left(\frac{eV_d}{kT}\right) = \frac{i_\lambda}{i_o} \quad (7.41)$$

It then follows:

$$V_d (= V_{ext}) = \frac{kT}{e} \ln\left(\frac{i_\lambda}{i_o}\right)$$

Substituting for i_λ from Eq. (7.38), one obtains:

$$V_{ext} = \frac{kT}{e} \ln\left(\frac{\eta i_o e \lambda A}{hci_o}\right) \quad (7.42)$$

Obviously, external voltage should be a *logarithmic function of incident light irradiance*.

A solar cell is basically a *pn* junction detector operated under conditions such that it can deliver power into an external load.

In photoconductive mode, a relatively small value of reverse bias is applied (i.e. few tenths of a volt). In Eq. (7.39):

$$i_\lambda = i_o + i_{sh} + i_{ext}$$

Now, $i_o = 10$ nA, while from Eq. (7.4), $i_{sh} = V_d/R_{sh} \approx 10$ V/100 MΩ ≈ 100 nA, so that when i_λ is of the order of microamps and above, one may write $i_{ext} = i_\lambda$.

Substituting from Eq. (7.38), one obtains:

$$i_{ext} = \frac{\eta I_o A e \lambda}{hc} \quad (7.43)$$

Thus, in *photoconductive mode, external current flowing is directly proportional to incident light irradiance*.

In addition to its inherently linear response, the photoconductive mode usually offers the advantages of **faster response, better stability** and **greater dynamic range**. The main drawback of this mode is the presence of dark current ($i_o + i_{sh}$) which as in a photomultiplier gives rise to shot noise and limits the ultimate sensitivity of the device. Both modes of operation are subject to noise generation but recombination noise is absent since charge carriers are separated in the depletion region before they can recombine.

The most common semiconductor material used in photodiodes is silicon. It has an energy gap of 1.14 eV and provides excellent photodiodes with quantum efficiency upto 80% at wavelengths between 0.8–0.9 μm.

7.5 PIN PHOTODIODES

A *p-i-n* (PIN) photodetector is a *p-n* junction with an intrinsic (*i*) layer sandwiched between the *p* and *n* layers. In practice, the *i*-layer is either a p^- or n^- layer (i.e., lightly doped) which is inserted between the p^+ and n^+ layers. This structure is illustrated in Fig. 7.20 along with the energy band diagram under an applied bias. The PIN photodiode is operated in the reverse-bias mode and, because the *i*-region has a low concentration of free carriers, it can be depleted with a minimum amount of voltage. Therefore, the depletion region extends through the entire *i*-region. When photons with energy greater than or equal to the bandgap energy are incident on the photodiode, electron-hole pairs are created. Carriers generated within a diffusion length of the depletion region diffuse into the *i*-region. These carriers along with all the carriers generated in the depletion region are transported by drift due to the applied reverse bias and are collected. The electric field in the depletion region is high and sufficiently uniform so that the carriers travel at saturation velocities. If all of the photogenerated carriers are collected, the *quantum efficiency* is given by Eq. (7.13) with *L* being the thickness of the region where the light is absorbed and $\zeta = 1$. *L* is usually assumed to be thickness of the *i*-region. As mentioned earlier, *r* can be made negligible by utilizing an antireflection coating to obtain good quantum efficiency. The internal quantum efficiencies $(1 - \exp(-\alpha L))$ for InGaAs and GaAs at various operating wavelength are shown in Fig. 7.21. Because there is no internal optical gain associated with the PIN diode, the maximum internal quantum efficiency that can be expected is 100%. The external quantum efficiency is given by Eq. (7.13) with $\zeta = 1$, and the responsivity is given by Eq. (7.16) with unity gain.

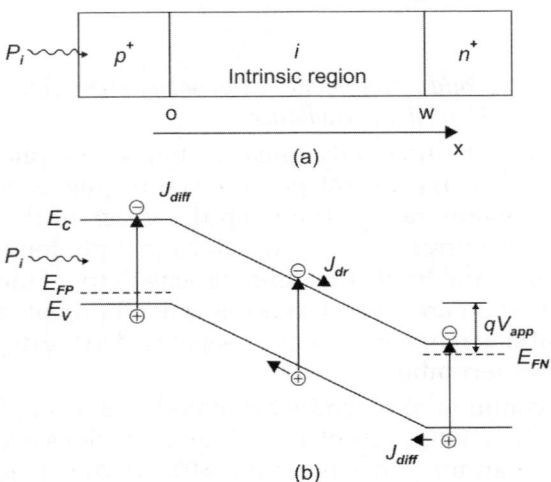

Fig. 7.20. The *p-i-n* photodiode: (a) the layer structure, and (b) the associated energy band diagram under an applied bias

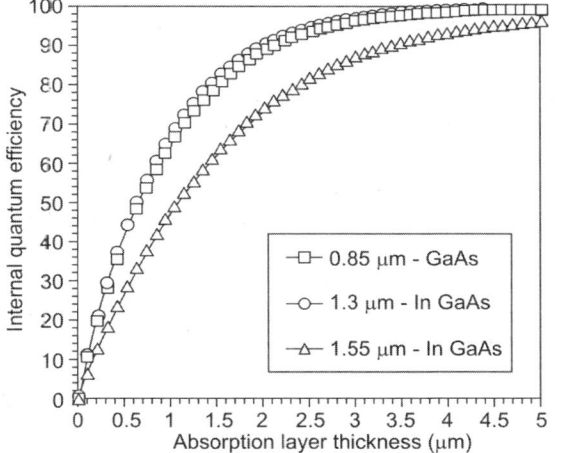

Fig. 7.21. Internal quantum efficiency for GaAs and InGaAs at certain wavelengths and various absorption layer thickness

A schematic of a typical *I-V* characteristic of a PIN photodiode is shown in Fig. 7.22. This is the usual *I-V* characteristic of a *p-n* junction but with an added current $-I_p$ which is proportional to the incoming photon flux. The *I-V* relationship is given by:

$$I = I_s [e^{eV/kT} - 1] - I_p \tag{7.44}$$

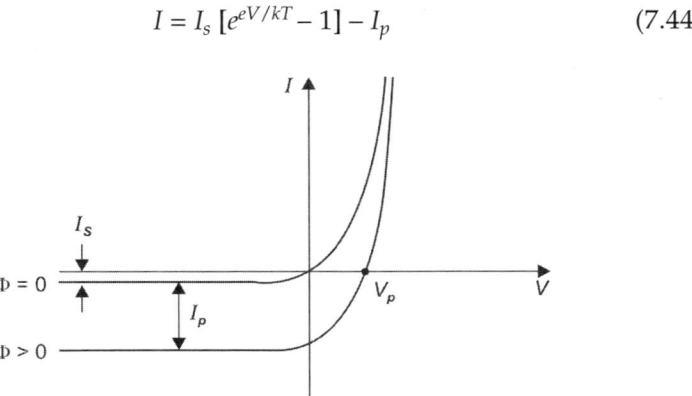

Fig. 7.22. An *I-V* relationship of a typical photodiode. The two different curves represent the *I-V* relationship when the incident photon flux is equal to zero and greater than zero

The applied reverse bias (V_A) needed to deplete the *i*-region of thickness L is given by:

$$L = \frac{\sqrt{2\varepsilon(V_0 + V_A)}}{qN_c} \tag{7.45}$$

where ε is the dielectric constant, V_0 is the built-in voltage, and N_C is the reduced carrier concentration, $N_C = N_A N_D / N_A + N_D$.

For high quantum efficiency or responsivity, it is necessary that $\alpha L \gg 1$. Of course, if L is large, the transit time of carriers across the i-region becomes large and device speed suffers. It is desirable to make α large, but it is a material property which cannot be changed. Therefore, there are many factors that must be considered to realize high-performance PIN photodiodes.

The response speed of PIN photodiodes can be limited by (a) the transit time of the photogenerated carriers across the depletion region, (b) the diffusion time of carriers generated outside the depletion region, and (c) the RC time constant with C consisting of junction capacitance and any other parasitic capacitances. The transit time is limited by the slower carriers, usually the holes. Another factor which can limit speed is change trapping at heterojunctions. For the RC time constant, the junction capacitance of the photodiode is given by:

$$C = \frac{\varepsilon A}{L} \qquad (7.46)$$

where A is the cross-sectional area of the device. As seen here, the device capacitance can be minimized by increasing the thickness of the depletion region and by reducing the diameter of the device. Other parasitic capacitances associated with packaging can also be reduced. As mentioned earlier, transit-time considerations suggest the use of a very thin i-layer to obtain high speed response. However, that increases parasitic junction capacitance which, in turn, reduces bandwidth. So, for a given detector area, an optimum i-layer thickness exists to obtain the highest speed possible. The transit-time limited frequency response of a PIN detector is given by:

$$\frac{i(\omega)}{i(0)} = \frac{1}{(1-e^{-\alpha L})} \left[\frac{e^{j\omega\tau_n - \alpha L} - 1}{j\omega\tau_n - \alpha L} - e^{-\alpha L} \frac{(e^{j\omega\tau_n} - 1)}{j\omega\tau_n} \right.$$

$$\left. + \frac{e^{j\omega\tau_p} - 1}{j\omega\tau_p} - e^{\alpha L} \frac{(e^{j\omega\tau_p + \alpha L} - 1)}{j\omega\tau_p + \alpha L} \right] \qquad (7.47)$$

where $i(\omega)$ is the detected current at an angular modulation frequency of ω, $i(0)$ is the dc current, L is the thickness of the i-layer, and $\tau_n = L/v_n$ and $\tau_p = L/v_p 6$ are the electron hole transit times, and v_n and v_p are the electron and hole saturation velocities, respectively. For InGaAs, the saturation electron and hole velocities are 6.5×10^6 and 4.8×10^6 cm/s, respectively. Using Eq. (7.47) and the frequency response of a parallel RC network, the theoretical 3 dB bandwidth of InGaAs/InP PIN detectors of several diameters are plotted against i-layer thickness in Fig. 7.23. The results were calculated for 1.3 μm operation

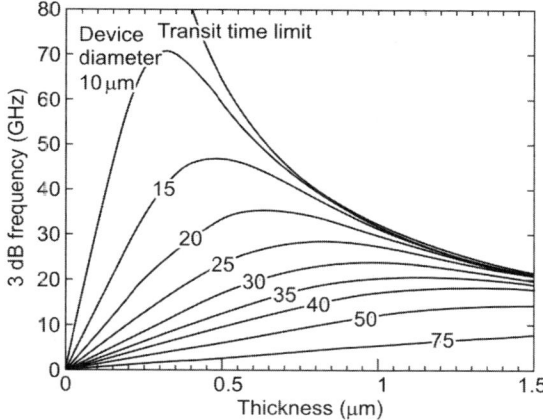

Fig. 7.23. The theoretical frequency response of a *p-i-n* photodiode for various active layer thicknesses and dimensions

(i.e. $\alpha = 1.16 \ \mu m^{-1}$) and for a load resistance of 50 Ω. In practice, there are parasitic resistances (the shunt or junction resistance and the series resistance) and a stray capacitance that should be taken into account for higher accuracy. The transit-time limited response is also shown in Fig. 7.23, which clearly demonstrates the compromises that are needed to design PIN photodiodes and also shows that devices with bandwidths over 60 GHz can be realized if the small diameter needed is not limiting for the particular application in hand.

An approximate expression for bandwidth for a detector with a very thin *i*-layer is:

$$B = \frac{0.45 \, v}{L} \quad (7.48)$$

for which $v_n = v_p = v$ has been assumed. Therefore, for a high speed detector where $\alpha L \ll 1$, the quantum efficiency from Eq. (7.13) is $(1-r)\alpha L$, and consequently, the bandwidth-efficiency product is:

$$B \cdot \eta = 0.45 \, \alpha v \, (1-r) \quad (7.49)$$

Once again, the critical importance of a large α is clear. The large α for InGaAs at long wavelengths (1.3 and 1.55 μm) gives it an advantage over Si and Ge as a material of choice for PIN photodetectors for optical communication applications.

The sources of noise in a PIN photodiode are from (a) the current due to the photocurrent I_p, (b) the dark current due to generation in the depletion region I_D, and (c) the current due to background radiation I_B. Due to the random generation of these currents, they contribute to shot noise which can be expressed as:

$$\langle i_s^2 \rangle = 2q(I_p + I_D + I_B)B \quad (7.50)$$

where B is the bandwidth. In addition, there is thermal or Johnson noise contributed with the shunt resistance of the diode and the input resistance of the following preamplifier stage. The Johnson noise for the PIN photodiode is given by Eq. (7.17). From these considerations, the NEP in units of watts for a PIN photodiode is given by:

$$\text{NEP} = \frac{h\nu}{q\eta}\left[2q(I_P + I_D + I_B) + \frac{4kT}{R_{\text{eff}}}\right]^{1/2} \quad (7.51)$$

It is seen that the key to increasing the sensitivity of the photodiode is to make R_{eff} and η as large as possible and make the I_B and I_D as small as possible. I_B, which is due to background radiation is usually very small, and I_D is also very small in a PIN photodiode because the device operates in the reverse-bias mode. The shot noise contributed by I_D is very small compared to that in photoconductors. Johnson noise is usually dominant in PIN detectors, but it can be minimized by optimizing the device and the circuit parameters. Therefore, PIN detectors are very useful for high speed, low noise applications such as encountered in optical communications.

In these applications which require operation as long wavelengths, photodiodes are made of InP-based materials, such as $In_{0.53}Ga_{0.47}As$ and InGaAsP. Heterojunctions of InP/InGaAs and InAlAs/InGaAs are needed for photodiodes to obtain higher breakdown and lower reverse leakage current than can be obtained using homojunctions. Practical heterostructures for PIN devices operating at infrared and long wavelengths are shown in Fig. 7.24(a) and (b).

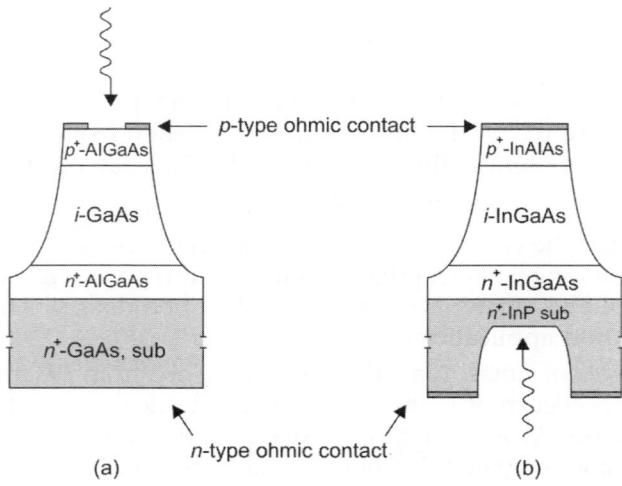

Fig. 7.24. Schematics of p-i-n photodiodes utilizing (a) front side illumination, and (b) back side illumination

Limitation of *p-i-n* photodiodes: The frequency response of a *p-i-n* photodiode is limited transit time effects or circuit parameters. The most important parameter which affects carrier transit time is the width of the intrinsic region. As the width of intrinsic region increases, the total distance travelled by photogenerated carriers also increases and due to this, the device bandwidth reduces. However, if intrinsic width is made too small, quantum efficiency is reduced. Obviously, one should select optimal intrinsic width.

Typically, a *p-i-n* photodiode is designed in such a way that transit time through the intrinsic region is equal to one-half of modulation period of the optical signal:

$$f = \frac{1}{t_{tr}} \qquad (7.52)$$

where t_{tr} is transit time within the intrinsic region.

7.6 RESPONSE TIME OF PHOTODIODES

Following three main factors limit the speed of response:

Diffusion time of carriers to the depletion region: Carrier diffusion is inherently a relatively slow process. Time taken for carriers to diffuse a distance d is found to be:

$$T_{diff} = \frac{d^2}{2D_e} \qquad (7.53)$$

where D_e is minority carrier diffusion coefficient.

To ensure that few carriers are generated outside the depletion region, we require that, at wavelengths being used:

$$W = X_n + X_p \geq \frac{1}{\alpha(\lambda)}$$

where $\alpha(\lambda)$ is **absorption coefficient** at wavelength λ. At wavelengths near the bandgap limit and at fairly low reverse-bias levels, the above inequality may not hold and the speed of detection of some optically generated carriers will then be limited by diffusion. In these circumstances, the response to a narrow optical pulse will be as illustrated in Fig. 7.25. The carriers generated within the depletion region respond rapidly, whereas those generated outside give rise to a slow tail. The relative importance of the tail may be decreased by increasing the reverse bias or by using the PIN structure where the depletion region is relatively wide.

Drift time of carriers through the depletion region: In very high electric fields, drift velocities of carriers in semiconductors tend to saturate. Since the field within the depletion region exceeds the saturation value for most of its length (which is usually the case in practice), then one may assume that carriers move with a constant

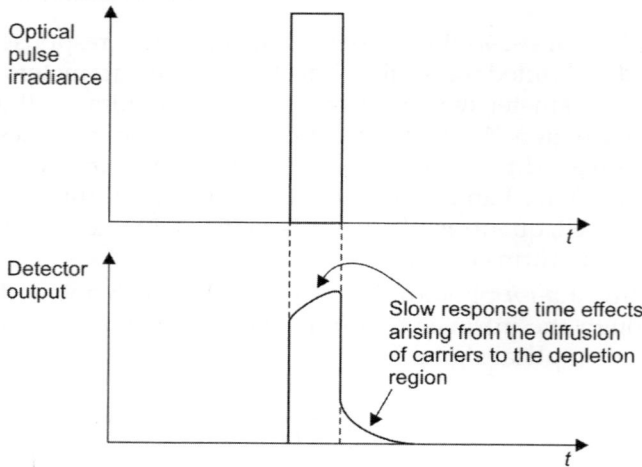

Fig. 7.25. Response of a *pn* junction detector to a narrow optical pulse. The effects arising from the relatively long time taken by carriers generated away from the depletion region to diffuse to the junction is shown

velocity v_{sat}. The longest transit time will result when carriers are generated near to one edge of the depletion region. In this case, one of the carriers will have to traverse the full depletion layer width W. The transit time is as follows:

$$T_{\text{drift}} = \frac{W}{v_{\text{sat}}} \quad (7.54)$$

Junction capacitance effects: A diode under reverse-bias exhibits a voltage-dependent capacitance caused by variation is stored charge at the junction. For example, an abrupt junction diode has capacitance given by:

$$C_1 = \frac{A}{2}\left[\frac{2e\varepsilon_r\varepsilon_o}{(V_o - V)}\left(\frac{N_d N_a}{N_d + N_a}\right)^{1/2}\right] \quad (7.55)$$

In we assume external bias potential V is large compared to zero bias junction potential V_o and that we have a p^+-n^- junction, then the above expression reduces to:

$$C_1 = \frac{A}{2}(2e\varepsilon_r\varepsilon_o N_d)V^{1/2} \quad (7.56)$$

In practice, junctions are rarely abrupt. However, it still remains true that capacitance decreases with increasing reverse-bias. For example, in a linearly graded junction, one finds:

$$C_1 \propto V^{-1/3}$$

One can refer to Fig. 7.17 that at high frequencies diode capacitance acts as a shunt across the output resistance network and reduces the output.

Usually $R_{sh} > R_L$ and $R_s \ll R_L$, hence as far as diode current i_λ is concerned, diode capacitance C_1 is parallel with load resistor R_L. One can easily show that output voltage of the device as a function of modulation frequency f is obtained as:

$$V_o = \frac{i_\lambda R_1}{(1 + 4\pi^2 f^2 C_1^2 R_L^2)^{1/2}} \quad (7.57)$$

Electrical bandwidth Δf_{at} is defined as the frequency range over which the output is above $\sqrt{\frac{1}{2}}$ of its maximum value. Thus, in the present case, we have $4\pi^2 f_{at}^2 C^2 R_L^2 = 1$, i.e.:

$$\Delta f_{at} = \frac{1}{2\pi R_L C_1} \quad (7.58)$$

The bandwidth may obviously be improved by reducing C_1. Eq. (7.56) reveals that this may be achieved by reducing the diode area A, reducing the doping level N_d or increasing the reverse-bias voltage V. To achieve the largest bandwidths, response time associated with diode capacitance is made equal to that associated with the drift of carriers across the depletion region. Silicon photodiodes with bandwidths greater than 1 GHz (i.e. response times less than 1 ns) are available.

The silicon photodiode is found to be one of the most popular all-purpose radiation detectors in the wavelength range 0.4 to 1 µm. It has the virtues of high quantum efficiency, small size, good linearity of response, large bandwidth, simple biasing requirements and relatively low cost.

Photodiodes made from materials other than silicon are also available, but they have not found suitable. Germanium photodiodes can be used up to a wavelength of 1.8 µm but they have rather low quantum efficiencies and somewhat high leakage currents (and hence high noise currents), although cooling to 77 K greatly improved their performance. This is unfortunate since the wavelength region from 1 to 1.6 µm is of great interest from the point of view of optical communications. Recently, diodes made from both ternary (e.g. GaInAs and HgCdTe) and quaternary (e.g. GaInAsP) compounds are gaining importance.

7.7 AVALANCHE PHOTODIODES

An avalanche photodiode (APD) is essentially a *p-n* junction operated at high reverse-bias voltages close to the breakdown voltage. At such high voltages, photogenerated carriers in the depletion region gain kinetic energy from the induced electric field and travel at their saturation velocities in the host material.

The carriers can acquire sufficient kinetic energies to undergo inelastic collisions with the lattice to create secondary electron-hole pairs. These secondary electron-hole pairs along with the primary carriers continue to drift and produce tertiary electron-hole pairs. To have an ionizing collision, a carrier must gain a threshold energy greater than the bandgap energy. Therefore, the critical field required to create ionizing collisions is material dependent and ranges from 10^4 to 10^5 V/cm. The process of creating carriers through ionizing collisions is called *impact ionization*, and the subsequent increase in the number of generated carriers is termed *avalanche multiplication*. This process is illustrated in the band diagram of Fig. 7.26 showing the direction of electron and hole injection via optical illumination. The avalanche multiplication process is an internal gain mechanism. The gain is typically much higher than that associated with a photoconductor. The quantum efficiency of an APD is given by Eq. (7.13) with L being the thickness of the absorption region.

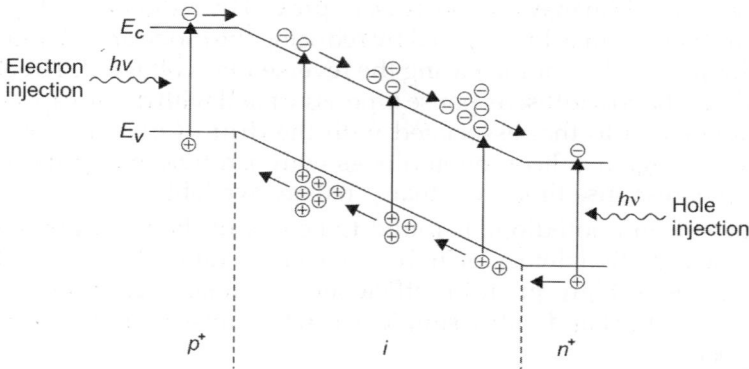

Fig. 7.26. The energy band diagram of an APD illustrating the avalanche gain process

As described above, the avalanche process is characterized by impact ionization. The electron-hole pairs create drift in opposite directions, and with sufficient energies, each carrier type can undergo impact ionization. The number of ionization collisions per unit length is designated α_n and α_p, the ionization coefficients for holes and electrons, respectively. Among other things, α_n and α_p depend on materials and their band structures. Ionization coefficients are also dependent on the applied electric field, and an approximate relationship is given by:

$$\alpha_n \alpha_p = a \exp\left[-\frac{b}{E}\right] \quad (7.59)$$

where a and b are constants dependent on the type of material and doping density and E is the applied electric field. The field-dependent impact ionization coefficients for various semiconductor materials have been determined and reported in the literature. In general, the two rates approach each other at fields higher than 3×10^5 V/cm for many semiconductors. Table 7.4 shows the values of a and b of Eq. (7.59) for various semiconductors.

The avalanche process produces a carrier multiplication factor M or an avalanche gain which is dependent on α_n and α_p. The responsivity of an APD is expressed by Eq. (7.16) where G is the gain and it is equivalent to the multiplication factor M. The low-frequency gain for electrons is given by:

$$M = \left\{1 - \int_0^W \alpha_n \exp\left[-\int_0^x (\alpha_n - \alpha_p)dx'\right]dx\right\}^{-1} \quad (7.60)$$

Table 7.4. The impact ionization coefficients for various semiconductors

Semiconductor material	$a\ (cm^{-1})$		$b\ (V\ cm^{-1})$	
	Electrons	Holes	Electrons	Holes
GaAs	1.1×10^7	5.5×10^6	2.2×10^6	2.2×10^6
GaAs	3.82×10^4	4.50×10^4	3.80×10^5	3.10×10^5
$In_{0.15}Ga_{0.63}Al_{10.22}As$	3.19×10^4	4.03×10^4	4.03×10^5	3.25×10^5
$In_{0.2}Ga_{0.8}As$	5.9×10^4	6.8×10^4	3.0×10^5	3.02×10^5
Si	9.2×10^5	2.4×10^5	1.45×10^6	1.64×10^6
InP	5.5×10^6	1.98×10^6	3.10×10^6	2.29×10^6
$In_{0.53}Ga_{0.47}As$	2.27×10^6	3.95×10^6	1.13×10^6	1.45×10^6
$In_{0.67}Ga_{0.53}As_{0.70}P_{0.30}$	3.37×10^6	2.94×10^6	2.29×10^6	2.40×10^6

where W is the width of the depletion region. When $\alpha_n \neq \alpha_p$ and the values of both ionization rates are independent of position as p-i-n diodes, then:

$$M = \frac{(1 - \alpha_p/\alpha_n)\exp[\alpha_n W(1 - \alpha_p/\alpha_n)]}{(1 - \alpha_p/\alpha_n)\exp[\alpha_n W(1 - \alpha_p/\alpha_n)]} \quad (7.61)$$

For $\alpha_n = \alpha_p = \alpha$, the multiplication factor M takes the form:

$$M = \frac{1}{(1 - \alpha W)} \quad (7.62)$$

It is observed that M goes to ∞ when $\alpha W = 1$, which signifies the condition for junction breakdown. Therefore, high values of M can be obtained for photodiodes biased near the breakdown voltage. A schematic of the avalanche gain in an APD is illustrated in Fig. 7.27 showing the sharp increase in gain as the reverse bias approaches the breakdown voltage. The critical field at which breakdown is initiated depends on impurity concentration in semiconductors and be calculated from the one-sided abrupt p-n junction approximation.

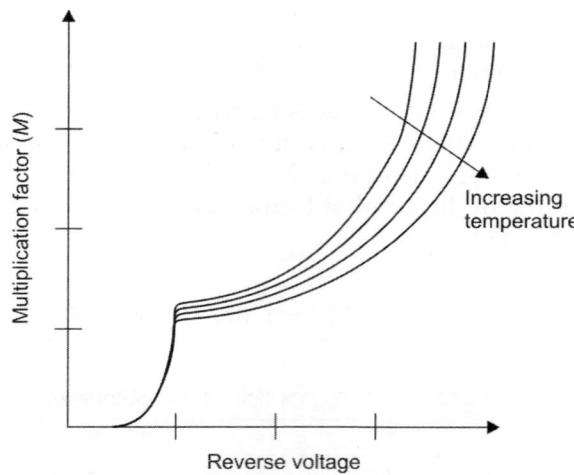

Fig. 7.27. A representation of the multiplication factor as a function of reverse bias voltage and temperature for an APD

The steady-state photocurrent in an APD in the presence of an avalanche gain is given by:

$$I_p = \frac{q\eta P_i}{h\nu} M \tag{7.63}$$

where P_i is the incident optical power. For the case of an intensity-modulated incident optical signal, the root mean square (rms) signal with a modulation index m is expressed as:

$$i_p = \frac{q\eta P_i m}{\sqrt{2}h\nu} M \tag{7.64}$$

The avalanche process is a regenerative process and it can be time-consuming. It takes time to build up which implies that the higher the avalanche gain, the longer the avalanche process persists. This results in the presence of a large number of secondary carriers in the depletion region after the primary carriers have been collected. Therefore, the gain and bandwidth of an APD are inextricably linked together. This results in a finite gain-bandwidth product fixed for a given material and device structure. The gain-bandwidth product is given by:

$$BM = \frac{1}{2\pi\tau_1} \tag{7.65}$$

where $\tau_1 = (\tau_n + \tau_p)/2$, τ_n is the electron transit time, and τ_p is the hole transit time. τ_n and τ_p are given by W/v_p where v_n and v_p are electron and hole saturation velocities, respectively. The equation above has only considered the transit-time effects on the response speed of a photodiode. Other usual effects are diffusion and RC limitations. Another effect particular to APDs is the time it takes carriers to complete the avalanche process, which is called the *avalanche build-up time*. Due to the randomness of the multiplication process, the build-up time is itself random. The dependence of the multiplication factor on frequency $M_0 > \alpha_n/\alpha_p$ is of the form:

$$M(\omega) \approx \frac{M_0}{\sqrt{1 + \omega^2 N_0^2 \tau_t^2}} \qquad (7.66)$$

where M_0 is the dc value of the multiplication factor and τ_t is an effective transit time through the avalanche region. A functional form that describes the effective transit time is:

$$\tau_t = N\tau_1(\alpha_n/\alpha_p) \qquad (9.67)$$

where τ_1 is the real transit time through the avalanche region and N is a number slowly varying from $N = 1/3$ to 2 as α_p/α_n varies from 1 to 0.001.

Apart from the usual shot noise limitations in photodiodes, the randomness associated with the avalanche or multiplication process makes it a principal source of noise in ADPs. These random fluctuations create a distribution of gain which causes excess noise in the device. For the condition where the avalanche is initiated by electrons (i.e. $\alpha_n > \alpha_p$), the excess noise factor is given by:

$$F = M\left[1 - (1-k)\left(\frac{M-1}{M}\right)^2\right] \qquad (7.68)$$

where k is the ratio of the ionization factor, i.e. α_p/α_n. An equivalent expression also exists for a hole-induced avalanche gain in which k is replaced by $k' = \alpha_n/\alpha_p$. There are two special cases, $k = 0$ and $k = 1$. The first case results in:

$$F = 2 - \frac{1}{M} \qquad (7.69)$$

whereas the second, denoting where both ionization coefficients are equal, results in:

$$F = M \qquad (7.70)$$

The noise factor for various multiplication gains and ratios of ionization coefficient is shown in Fig. 7.28.

In an APD, the dark current and the current due to background radiation are all enhanced by the multiplication gain similar to that of the signal current in Eq. (7.63). Therefore, the root mean shot noise is given by:

$$\langle i_s^2 \rangle = 2q(I_p + I_D + I_B)FBM^2 \qquad (7.71)$$

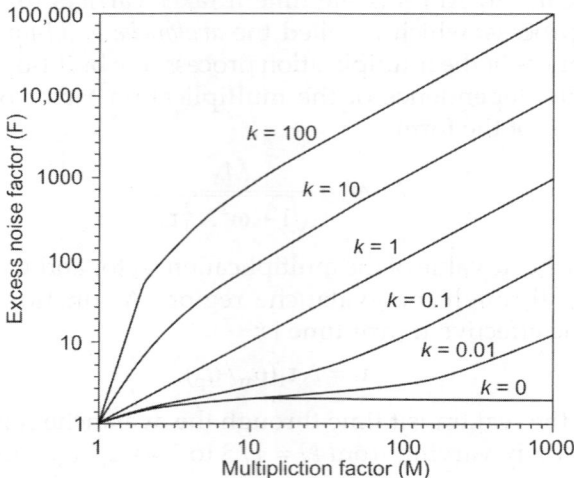

Fig. 7.28. Excess noise factor as a function of multiplication gain and the ratio of ionization coefficients

The *Johnson or thermal noise* for an APD is identical to that of a PIN photodiode. Therefore, the resulting NEP for an APD is:

$$\text{NEP} = \frac{h\nu}{q\eta M}\left[2q(I_P + I_D + I_B)FM^2 + \frac{4kT}{R_{\text{eff}}}\right] \quad (7.72)$$

It is observed that the NEP of an APD is almost identical to that of a PIN photodiode except for an extra gain factor M which appears in the denominator. It is evident, therefore, that the gain factor does reduce the NEP of an APD, making the device more sensitive than PIN photodiodes. Indeed, APDs have shown superior sensitivity compared to other photodiodes.

Important issues for consideration in designing and fabricating APDs concern *dark current* and *high-speed performance*. The contributions to the dark current of an APD include generation-recombination in the depletion region, tunneling of carriers across the bandgap, and leakage across junctions. Various methods have been developed to alleviate these problems. For example, guard rings and dielectric film deposition are used in Si APDs to minimize various forms of leakage currents. Modern APDs utilized for fiber-optic communications depend on InP related compounds. Lightwave operation at 1.55 μm relies on $In_{0.47}Ga_{0.53}As$ which has an energy bandgap of 0.75 eV. Large tunneling current is a problem if a material with such a low bandgap is used in homojunction devices. Therefore, a separate absorption and multiplication (SAM) structure, in which a low-field InGaAs region is utilized as the absorption region and a high-fiield InP region is the avalanching area, is now used for APDs. Fig. 7.29 shows a typical

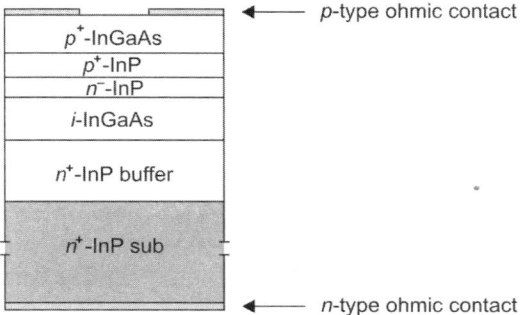

Fig. 7.29. Schematic diagram of a separate avalanche and multiplication avalanche photodiode (SAM-APD)

InP/InGaAs SAM-APD heterostructure where the p^+-n^- junction is located in the high-bandgap (1.35 eV) InP material. In practice, a graded bandgap material (InGaAsP) is placed at the InP/InGaAs heterojunction to prevent a sharp energy bandgap discontinuity which can trap carriers and result in slow device response. This modification results in a structure called *separate absorption-graded multiplication* APD (or SAGM-APD). These devices combine low dark current and high gain-bandwidth operations.

Other types of APDs have been designed to ensure preferential multiplication of one carrier type (i.e., k or k' approaches zero) over the other. These are the *superlattice* and the *multiquantum-well APDs*.

While using avalanche photodiode, one will have to take number of precautionary measures: (i) A very stable power supply for maintaining constant gain is required. A typical variation of current gain with reverse bias is shown in Fig. 7.22. (ii) Gain is highly dependent on temperature. One will have to overcome this by having two identical diodes. One diode is masked from the incident radiation and operated near breakdown at constant current. A change in temperature causes altering bias voltage at the control diode and one may use this change to regulate the operating bias applied to the signal diode.

Avalanche photodiodes are found quite suitable candidates for use in fiber optical communication systems. However, suitable materials giving low noise performance in the region beyond 1.1 μm (limit of silicon detectors) will have to be developed.

7.8 SCHOTTKY PHOTODIODE

In this photodiode a thin metal coating (usually gold) is applied to an n type silicon substrate (Fig. 7.30(a)). The energy band structure in the region of the junction is as shown in Fig. 7.30(b). One can see that when an electron-hole pair is generated within the depletion region, the electron and hole will be separated by the action of the internal field as

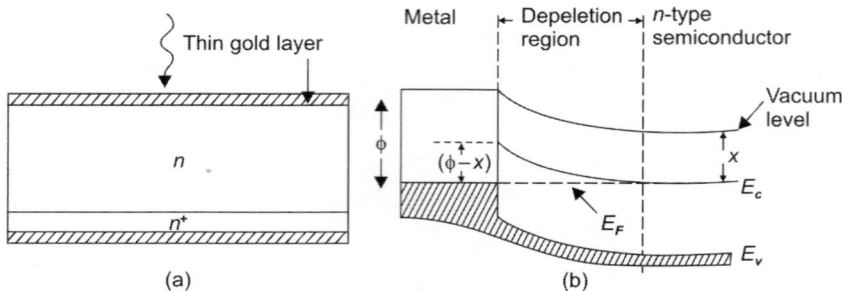

Fig. 7.30. Schottky photodiode (a) energy level behaviour in the region of the junction, (b) a potential barrier of height φ-x is formed between the metal and semiconductor

in *pn* junction photodiode. The main advantage of this photodiode is that the surface metal layer can be made sufficiently thin to transmit blue and near ultraviolet radiation, thus giving enhanced sensitivity in this region.

7.8.1 Metal-Semiconductor-Metal (MSM) Photodiode

A metal-semiconductor-metal (MSM) photodiode consists of two interdigitated electrodes which form back-to-back Schottky diodes on a semiconductor absorbing layer. The MSM is a planar device which can be fabricated easily. This ease in fabrication along with the ease in integrating the MSM with conventional field-effect-transistor (FET) technology have provided the underlying impetus to gradually improve the MSM's efficiency, dark current, and bandwidth to the point where it can now compete with the PIN photodiode. The schematic of the MSM is shown in Fig. 7.31. The interdigitated electrodes are similar to those of the photoconductors of Fig. 7.31(a) but differ in the type of contacts made to the semiconductor. The contacts of photoconductors are ohmic. Photogenerated electrons and holes in MSMs are transported along the electric field lines to the oppositely biased contacts as shown in Fig. 7.31(b). The electric field distribution is nonuniform due to the electrode geometry, with fields being strongest near the semiconductor's surface. Therefore, the applied bias must be sufficiently large to ensure the depletion of the entire absorbing region for the carriers to travel at their saturation velocities. The energy band diagram in Fig. 7.31(c) shows the flow of the generated electrons and holes and also shows how the Schottky barriers limit the injection of carriers from the metal contacts to the semiconductor.

The external quantum efficiency is basically given by Eq. (7.73), but taking the opaque interdigitated electrodes into account, we have:

$$\eta = (1-r)\frac{d}{d+w}\zeta[1-\exp(-\alpha L)] \qquad (7.73)$$

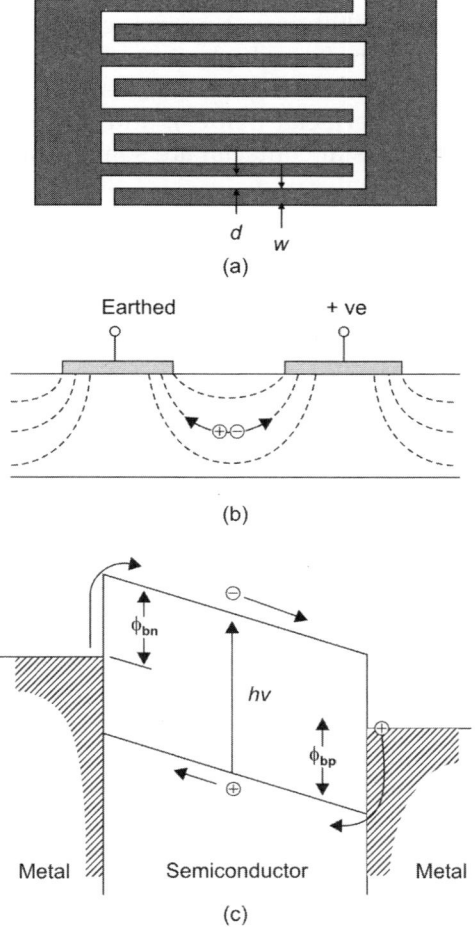

Fig. 7.31. (a) Schematic of a MSM photodiode, (b) cross-sectional view of the device displaying the direction of carriers travel in response to an electric field, and (c) the associated energy band diagram

where w is the electrode finger width and d is the finger spacing. It is seen that η is reduced by the opacity of the electrodes. The use of transparent Schottky contacts, such as indium tin oxide and cadmium tin oxide, to avoid the opacity of metal electrodes has been demonstrated. The shortcoming of this method is that carriers generated in the low-field region under the contacts degrade the bandwidth of the device. The responsivity of MSMs is given by Eq. (7.15). Ideally, the gain G should be unity. However, two possible gain mechanisms have been identified in MSMs. These are (i) the gain due to avalanche multiplication in high-field region of the device, for example, around the sharp edges of the contacts, and (ii) the gain due to trapping of carriers. Trapping of carriers can occur at trapping centers on semiconductor surface (between fingers), can have a detrimental influence on the dark current and high-speed response. The use of high quality epitaxial materials (low trap density), fabrication of high quality Schottky diodes (no interfacial oxides), and appropriate biasing of devices (avoid high fields at contacts) are some means of minimizing these effects.

The dark current of an epitaxial GaAs MSM is shown in Fig. 7.32. The photodiode has an area of 50 × 50 μm, a finger width of 1 μm, and an absorbing layer thickness of 1 μm. A symmetrical trace of the above result should be obtained for negative voltages due to the back-to-back Schottky diode configuration. The dark current of an MSM is dominated by thermionic emission over the barrier as illustrated in Fig. 7.32. A simple one-dimensional model gives the flat band voltage V_{fb}, at the forward-biased contract as:

$$V_{fb} = \frac{qN_D d^2}{2\varepsilon} \quad (7.74)$$

where q is the electron charge, ε is the dielectric constant of the semiconductor, N_D is the residual doping in the semiconductor, and d is the electrode spacing. The maximum V_{fb} is limited by the avalanche breakdown voltage near the reverse-biased contact. At low bias voltage compared to V_{fb}, electron injection at the reverse-biased contact is the dominant factor in carrier transport. At higher bias, hole injection at the forward-biased contact become dominant. A higher voltages, there is a reach-through condition, when the edges of the depletion regions of the two Schottky diodes merge. The total current density for a voltage higher than the reach-through voltage is given by:

$$J_{tot} = A_n^* T^2 \exp\left[\frac{-q(\phi_{bn} - \Delta\phi_{bn})}{kT}\right] + A_p^* T^2 \exp\left[\frac{-q(\phi_{bp} - \Delta\phi_{bp})}{kT}\right] \quad (7.75)$$

where A_n^* and A_p^* are Richardson constants for electrons and holes, ϕ_{bn} and ϕ_{bp} are the barrier heights for electrons and holes, and finally, $\Delta\phi_{bn}$ and $\Delta\phi_{bp}$ are the respective image force barrier lowerings. A_n^* and A_p^* are 8.15 and 74.4 A/cm²/K² for GaAs, respectively. The dark current in Fig. 7.32 compares to that of a PIN photodiode. Improvements in MSM dark current have been shown to be possible bychoosing an appropriate metal system for Schottky contacts and also by using surface epitaxial layers with higher bandgap than the absorbing layer. The later example is demonstrated for an AlGaAs/GaAs MSM in Fig. 7.32, where the Schottky contacts were made to an $Al_{0.3}Ga_{0.7}$, as layer grown on the GaAs absorbing layer.

The bandwidth of a MSM photodiode is determined by the carrier transit time and the RC time constant, and both depend on the physical

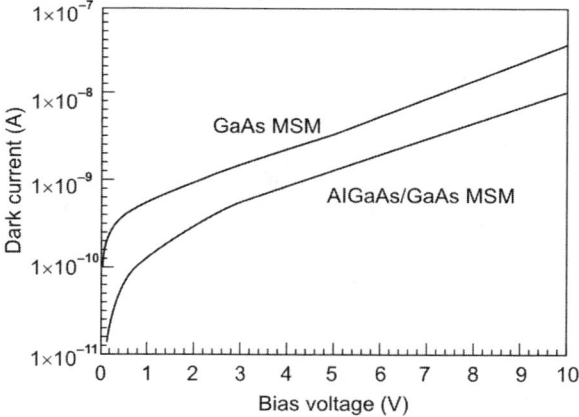

Fig. 7.32. Comparison of the dark current of 1 mm by 1 mm GaAs MSM photodiode with and without an AlGaAs cap layer

dimensions of the device. The carrier transit time is proportional to the distance between the electrodes if the effects of velocity overshoot and fringing electric field are neglected. The time is limited by the slowest carriers (holes). Using a dc model, the resistance of a MSM photodiode is given by:

$$R_{tot} = \frac{2R_0}{N} \quad (7.76)$$

where R_0 is the resistance of a single electrode and N is the number of electrodes on each side of the two sets of interdigitated electrodes. The resistance of the electrodes is usually less than the 50 Ω load resistance. A major advantage of an interdigitated MSM is its low capacitance. For an undoped and infinitely thick semiconductor, the capacitance is given by:

$$C = \frac{\kappa(k)}{\kappa(k')} \varepsilon_0 (1 + \varepsilon_r) \frac{A}{w + d} \quad (7.77)$$

where ε_r is the relative dielectric constant of the semiconductor, A is the active area of the MSM, w and d are the finger width and spacing, respectively, and $\kappa(k)$ is the complete elliptic integral of the first kind given by:

$$\kappa(k) = \int_0^{\pi/2} \frac{d\Phi}{\sqrt{1 - k^2 \sin^2 \Phi}} \quad (7.78)$$

$$k = \tan^2 \left[\frac{\pi w}{4(w + d)} \right] \quad (7.79)$$

and

$$k' = \sqrt{1 - k^2}. \quad (7.80)$$

Using these expressions, it has been shown that the capacitance of a MSM is significantly lower than that of a PIN photodiode of corresponding size.

The transit-time limited response of a surface-illuminated PIN photodiode can be adapted to the treatment of MSMs. The upper bound of the transit-time limited bandwidth of a MSM is given by Eq. (7.48), where the interdigitated spacing of the MSM is equal to the intrinsic layer thickness of the PIN. Because the effects of the carrier transit time and the RC time constant on the bandwidth are cumulative, then, the bandwidth of a MSM is given by:

$$B = \left(\frac{d}{0.45v} + R_L C\right)^{-1} \quad (7.81)$$

where R_L is the MSM load resistance and $v_n = v_p = v$. Although, the finger with no finger spacing ratio is a design parameter for MSM, if a ratio of unity is assumed for practical reasons, then, Eq. (7.81) becomes:

$$B = \left(\frac{d}{0.45v} + \frac{\pi K R_L}{d}\right)^{-1} \quad (7.82)$$

where K is a constant given by:

$$K = \frac{K(k_0)}{K(k_0')} \varepsilon_0 (1 + \varepsilon_r) A \quad (7.83)$$

with

$$k_0 = \tan^2\left(\frac{\pi}{8}\right) \quad (7.84)$$

The elliptic integrals in Eq. (7.83) are $K(k_0) = 1.58$ and $K(k_0') = 3.17$ for a device with $d = w$. Fig. 7.33 shows a plot of bandwidth as a function of finger spacing for an InGaAs MSM with the active area as a variable parameter. As shown in the curve, for small values of d the response in RC-limited for larger values of d the response is transit-time limited.

It is possible to obtain the finger spacing d_{max}, which gives a maximum available bandwidth, by equating the derivative of Eq. (7.82) with respect to d to zero, yielding:

$$d_{max} = \sqrt{0.45\pi v K R_L} \quad (7.85)$$

Therefore, for a device with a 50 × 50 μm² active area, an $In_{0.47}Ga_{0.53}As$ absorbing layer, and a 50 Ω load resistance, we obtain a d_{max} and B_{max} of 1 μm and 21.4 GHz, respectively. Fig. 7.34 shows the pulse response waveforms of InAlAs/InGaAs MSMs with Ti/Au Schottky contacts at a bias of 10 V for various device finger widths/finger spacings and a 50 × 50 μm² active area. The full-width-at-half-maximums (FWHMs) of the response pulses for the devices are 42, 47, and 52 ps, respectively.

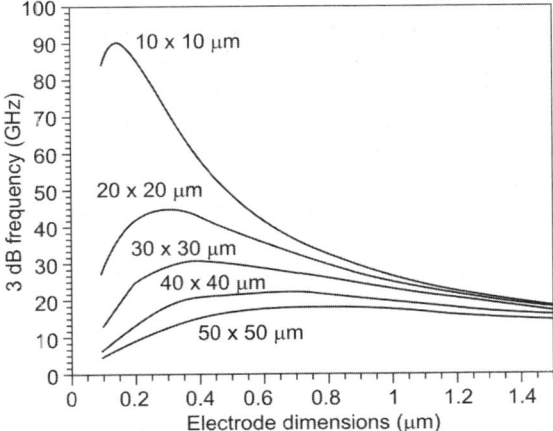

Fig. 7.33. The theoretical frequency response of an InGaAs MSM photodiode as a function of electrode and active area dimensions for a device with equal electrode width and spacing

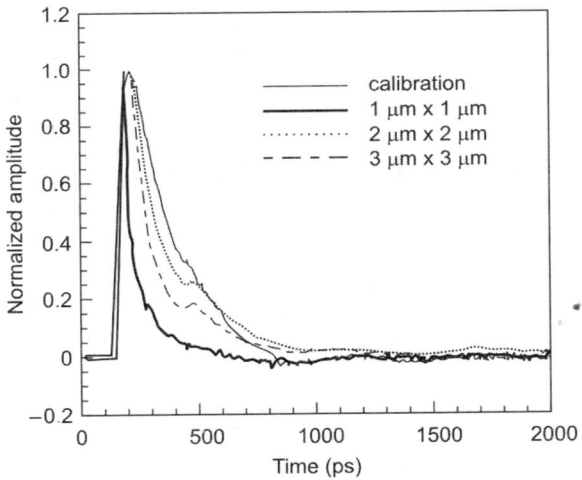

Fig. 7.34. A typical time-domain pulse response of an InGaAs MSM for devices with different electrode dimensions

Corresponding estimated bandwidths are 20, 14 and 11 GHz, respectively.

Various schemes have been adopted to improve the high speed response of MSMs. The use of thin absorbing layers and small finger spacing to finger width ratio will ensure very fast transit time but low quantum efficiency. The smaller ratio also increases the capacitance. Ion implantation in the absorbing region reduces transit time but also reduces sensitivity. This may contribute to excess noise and gain. In

GaAs MSM design, the use of an AlGaAs barrier layer in the buffer region has been shown to improve response with a slight reduction in the dc responsivity. The barrier impedes slow photogenerated carriers from deep in the substrate, where the electric field is week, from being collected by the electrodes. This leads to reduced transit time and a suppressed tail in the pulse response. To maximize charge collection and also reduce transit time, a graded layer is usually placed at the heterostructure interface to prevent a sharp, energy-band discontinuity.

Much like the PIN photodiode, the source of shot noise in MSM are the photocurrent, the dark current, and the current due to background radiation. There is also thermal noise component. The MSM photodiode has a lateral current flow with strong interactions with semiconductors with semiconductor interfaces like a FET device. Therefore, it is expected that MSM will exhibit a low-frequency, excess noise component with a $1/f^\alpha$-like dependence. This excess $1/f^\alpha$ noise has been observed at low frequency and high biases close to breakdown. Evidence shows that this excess noise is very dependent on the presence of traps. A significant excess noise was obtained at frequencies below 35 MHz for an InGaAs MSM which has a strained GaAs Schottky barrier enhancement layer. However, for a high quality InGaAs MSM with an InAlAs enhancement layer, the device was shot-noise dominated at frequencies as low as 1 MHz. Presumably, the excess noise for the GaAs/InGaAs MSMs is due to the large number of traps present at the strained interface between GaAs and InGaAs. Therefore, for MSMs fabricated on high quality epitaxial layers, the dominant noise properties will be similar to those of PIN photodiodes.

7.9 LIGHT-EMITTING DIODES (LEDs)

A LED is a *pn* junction diode of special construction. Like other semiconductor diodes, its behaviour is influenced by biasing conditions. Fig. 7.35 shows the basic operation of a LED. The *pn* junction is forward-biased when the cathode (*n* type region) is negative with respect to the anode. Under forward-bias, electrons from the *n* type region and holes in the *p* type material move towards the *pn* junction.

At the junction, the holes and electrons combine. Electrons are at a higher energy level than are holes. The electrons literally fall into holes as they combine, and

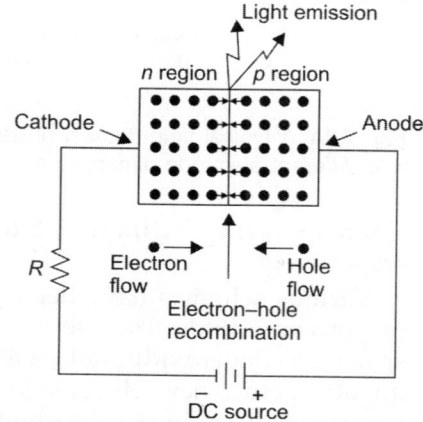

Fig. 7.35. Basic operation of a light emitting diode

in doing so, they release the excess amount of energy in the form of light. There is current through the LED in the forward direction, just as in the case of an ordinary diode.

When the *pn* junction is reverse-biased, the LED does not conduct and there is no release of light. Hence the LED emits light when forward-biased and does not emit light when reverse-biased. The schematic symbol of LED is shown in Fig. 7.36. The LED operates at forward voltages typically ranging from 1.2 V to 4.3 V. The reverse breakdown voltage is in the range of –3 to –10 V.

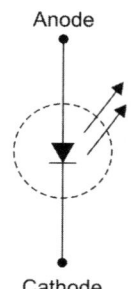

Fig. 7.36. Schematic symbol of LED

The main advantages of LEDs are:
  (i) LEDs have very long life.
 (ii) LEDs are highly durable and reliable.
(iii) The efficiency of LEDs is higher than that of incandescent light bulb and can possibly exceed fluorescence bulbs.
(iv) White-light LED sources can provide significant energy saving in comparison with conventional lighting.

7.9.1 LED Construction and Working

LED emits more light than an ordinary diode because ordinary diodes are made of silicon, which is opaque or impenetrable as far. Energy in ordinary diodes is, therefore, released in the form of heat.

On the other hand, LEDs are made of semiconductor materials that are semitransparent. LEDs are normally made of gallium arsenide phosphide, which emits visible red light. Gallium phosphide produces visible green light. LEDs that produce yellow and blue light are also available.

Figure 7.37 shows a LED chip and a typical LED package. The LED chip is attached to cathode and anode leads through thin wires. The

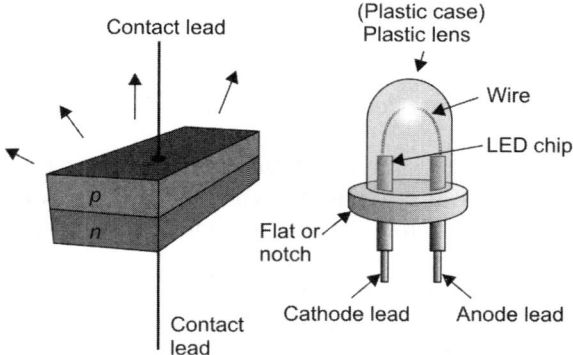

Fig. 7.37. The LED: (a) chip, (b) typical package

plastic case serves as a lens that conducts light away from the LED and also acts as a magnifier. The entire case may be dyed or tinted with a colour that enhances the on/off contrast of the LED. LED leads are identified by one of the following three ways (Fig. 7.38):
1. The leads may have different lengths, and the shorter of the two leads is the cathode (Fig. 7.38(a)).
2. The cathode lead is flattened (Fig. 7.38(b)).
3. One side of the case may be flattened. The lead closest to the flattened side is the cathode (Fig. 7.38).

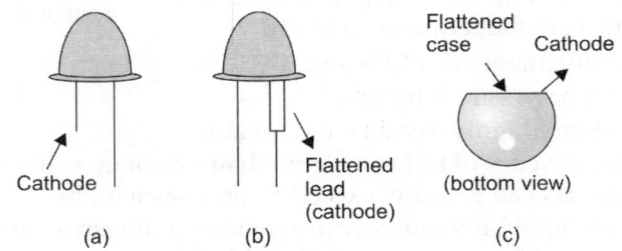

Fig. 7.38. Identification of LED leads

A cross-sectional view of a typical diffused LED is shown in Fig. 7.39.

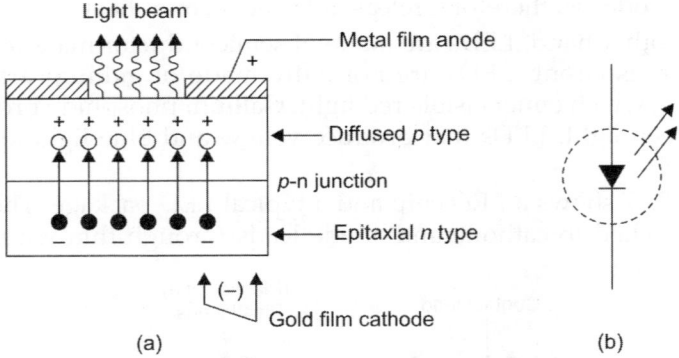

Fig. 7.39. (a) Structure of LED, (b) symbol (LED)

The semiconductor material employed may be GaAs, GaAsP or GaP. Here an *n* type epitaxial layer is grown upon the substrate, and the region is created by diffusion. When the *pn* junction is forward-biased, charge carrier recombination occurs in the *p* region, so this region is kept uppermost. The *p* region, therefore, becomes the surface of the device, and the metal film anode connection must be patterned to allow most of light to be emitted outside it. This is done by making a connection to the outside edges of the *p* layer or by depositing a comb-shaped pattern in the mid-region of p material. A gold film is applied at the bottom of the substrate to reflect light as much as possible towards the upper surface of the device and to provide a cathode connection. Materials like GaAs emit infrared (invisible) radiations. However, GaAsP provides either red or yellow light, while red or green light is emitted by GaP. Its symbol is shown in Fig. 7.39(b) where the direction of arrows indicates emitted light.

Normally a single LED is used to indicate current in the circuit but generally a seven-segment display is used where seven LEDs are combined to show a numerical display (Fig. 7.40(a)). Any desired number between 0 and 9 can be displayed. The seven segments are labelled a, b, c, d, e, f, g. We can easily follow that a display for number 2 will be shown by the glow of a, b, g, e, d segments. Similarly, for a display of number 8 all the segments shall grow. In this way we have the following series:

Numerals	0	1	2	3
Display segments	a, b, c, d, e, f	b, c	a, b, g, e, d	a, b, g, c, d

The LED device is usually very small, so to enhance the lighted segments surface, solid plastic light pipes are often used. The LEDs is seven-segment display may be connected in common anode configuration (Fig. 7.40(b)) or common cathode configuration (Fig. 7.40(c)), so when selecting a display, it is important to choose the configuration.

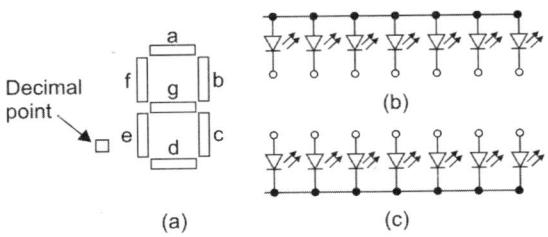

Fig. 7.40. (a) Seven-segment display, (b) common anode connection, (c) common cathode connection

The forward voltage for a LED is typically 1.2 V for forward current 20 mA. The relatively large amounts of current consumed by LEDs are their major disadvantage. Excluding the display, a typical electronic circuitry in LEDs may require as high as 10 mA, necessitating a bulky power supply. However, they have the big advantage of long life and ruggedness. Often LEDs are usually switched ON and OFF by means of a transistor circuit. In such cases, a resistance is connected in series with it.

With the introduction of liquid crystal diodes (LCD), its main disadvantage of requiring high current of the order of 20 mA has been offset. At present, LCDs are very popular and used mostly in measuring meters.

Luminous performance of a LED is related to its optical yield. Optical yield measures the efficiency of LED in the conversion of electric current into output photon. Luminous performance is expressed as the ratio of total output flux in lumens to input power in watts.

The quality of emitted photons reduces by the following three factors:
  (i) Absorption within the LED material.
  (ii) Reflection loss when light passes from semiconductor material to air due to differences in refractive index.
  (iii) Total internal reflection of light at an angle greater than critical angle (θ_c).

In LEDs, one might apply voltage across the *pn* junction which results in a diode current, which in turn produces photon and light output. This inverse mechanism is termed as **injection electroluminescence**.

When a voltage is applied across *pn* junction, electron and holes are injected across the space charge region, where they became excess minority carriers. These excess minority carriers diffuse into regions of neutral semiconductor and recombine with majority carriers. If this recombination is a direct band-to-band process, then photons are emitted. The intensity of output photon is directly proportional to ideal diffusion current.

The forward current voltage characteristics of LED is shown in Fig. 7.41, which is same as that of GaAs *pn* junction. One finds that at

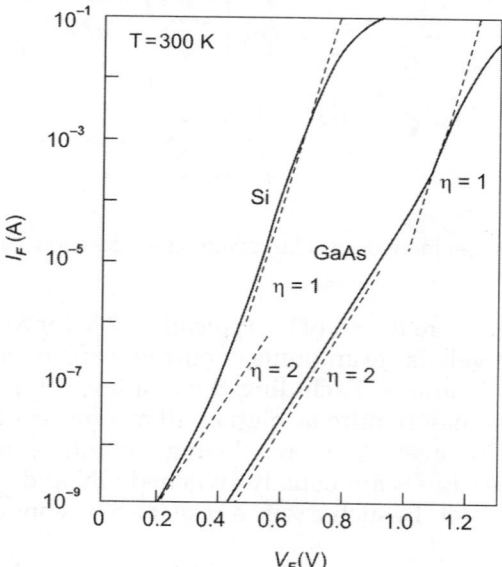

Fig. 7.41: Forward-bias current-voltage characteristic of LED

low forward voltages, diode current is dominated by non-radiative recombination current mainly due to surface recombination near the perimeter of LED chip. Diode current is dominated by radiative diffusion current at higher voltages. At even higher voltages, the series resistance will limit diode current. The total diode current can be expressed as the sum of saturation current due to diffusion (I_d) and saturation current due to recombination (I_r) as:

$$I = I_d \exp\left[\frac{q(V - IR_s)}{kT}\right] + I_r \exp\left[\frac{q(V - IR_s)}{2kT}\right] \quad (7.86)$$

where R_s is device series resistance.

7.9.2 Internal and External Quantum Efficiency of an LED

The internal quantum efficiency of a LED is a function of diode that will produce luminescence and is also a function of injection efficiency as well as of percentage of radiative recombination events compared with the total number of recombination events. The internal quantum efficiency of homo junction LED is influenced by the following factors:
  (i) Quality and purity of the material and substrate.
  (ii) Existence of a direct energy gap spectrally matched to the desired wavelength.

The presence of traps and impurities reduce the internal quantum efficiency of a LED. The internal and external quantum efficiency bears the following relationship:

$$\eta_{ext} = \eta_{int} \, C_{ox} \quad (7.87)$$

where C_{ox} represents extraction efficiency. Extraction efficiency is defined as the fraction of generated photons that escapes from the device package and emerges from the chip.

Extraction efficiency is determined by taking care of the following issues:
  (i) Absorption coefficient of the material at emission wavelength.
  (ii) Radiation geometry of the LED.
  (iii) Substrate.

We may note that LED is an **incoherent light source** because photons are emitted randomly from the junction in all directions and not in phase with each other. This is why, LED is less suitable for the transmission of digital information than lasers.

Recently certain organic semiconductors (e.g. 8-hydroxy-quinolinato aluminium) have been found useful for electroluminescent applications. The organic light emitting diode (OLED) is particularly useful for multicolour, large area, flat-panel display due to its attribute of low power consumption and excellent emissive quality with wide viewing angle.

For designing OLEDs, the criteria are:
(i) Ultrathin layer for low biasing voltage (typically 150 nm).
(ii) Low injection barriers.
(iii) Proper band gaps for the required colour.

There are two general types of OLED devices: those made with small organic molecules and those made with organic polymers. Low cost of OLEDs is the main advantage over standard LEDs. Since OLEDs are non-crystalline materials, they can be easily deposited. However, there are some problems with OLEDs, e.g., small lifetime of OLEDs; the organic materials used in LEDs are quite sensitive to water and oxygen, and therefore, they must be deposited on glass substrates and covered with a second sheet of glass. The usage of glass enhances the cost of manufacturing.

Applications: LEDs have replaced incandescent light bulbs as display devices. Long life expectancy, ruggedness and low power consumption have contributed to the choice of LEDs over incandescent light bulbs. Different types of liquid crystal displays have now replaced LEDs in some applications because they use much less energy than LEDs.

7.9.3 LED Indicator Circuits

Figure 7.42 shows how a LED can be used as a power level indicator in a computer, CD player, stereo amplifier or other such devices. The LED is connected in parallel to the system's internal power supply along with series resistor R_s. When the power switch is on, the diode is forward-biased and emits a light.

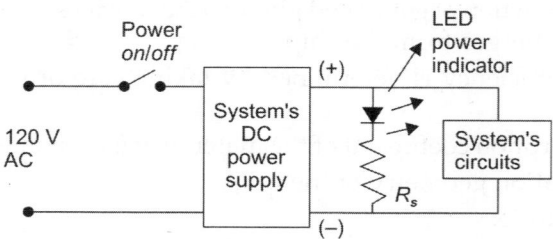

Fig. 7.42. LED power indicator

Series resistor R_s limits current to the LED and is often referred to as a "current limiter". Most LEDs have a forward-biased voltage drop between 1.2 V and 2.5 V although high intensity LEDs can have higher-biased voltage drops. Note that this is much higher than for a normal silicon diode (0.7 V) or germanium diode (0.3 V). The resistance value and wattage rating of R_s can be calculated using Ohm's law. R_s value is found by subtracting the LED's voltage drop from the source and then dividing by the LED's desired current. To find wattage rating for R_s, simply apply the power formula $P = I^2 R$.

Example

What is the resistance and wattage rating of a limiting resistor in a 12 V DC circuit for a LED that requires 2 V forward-bias with 20 mA current rating?

Solution

$$R_s = \frac{V_S - V_D}{I_D} = \frac{12\text{ V} - 2\text{ V}}{20\text{ mA}} = 500\ \Omega$$

and $\quad P = I^2 R = (20\text{ mA})^2 \times 500\ \Omega = 200\text{ mW}$

Therefore, a standard-size 0.25 watt (= 250 mW) resistor is required:

7.9.4 Multicolour LEDs

Multicolour LEDs are available that will:
 (i) emit one colour when supply voltage is one polarity.
 (ii) emit a second colour when polarity is reversed.
 (iii) emit a third colour when bias polarity is rapidly switched.

Figure 7.43 shows the schematic of multicolour LEDs. Multicolour LEDs are usually two LEDs connected in antiparallel. That is, the anode of each diode is connected to the cathode of the other. Each LED can emit light only when forward-biased. Thus, when voltage of either polarity is applied, one LED is forward-biased and emits its native colour.

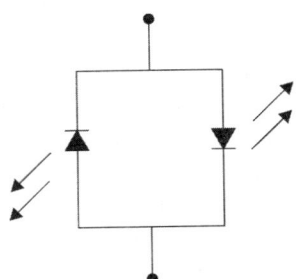

Fig. 7.43. Multicolour LEDs

The most commonly used are red and green in colour. The green LED is normally used to indicate whether something is functioning properly, and the red LED is used to indicate that there is a problem. If a multicolour LED is rapidly switched between two polarities, the red/green LED appears to produce a third colour (yellow).

Infrared-Emitting Diode

We know that the infrared band (30 THz to 400 THz) falls below the frequencies the human eye can detect. Diodes made of gallium arsenide release energy by way of heat and infrared light. Such a diode is called infrared-emitting diode (IRED). IREDs are used in remote controls, fiber-optic communication, discriminating organic solvents in the field of medicine and other such applications.

Multisegment LED Displays

LEDS are widely used in multisegment displays. Fig. 7.44 shows the most commonly used multisegment display. Its seven segments are labelled a, b, c, d, e, f and g. The LED labelled *dp* is used to display the

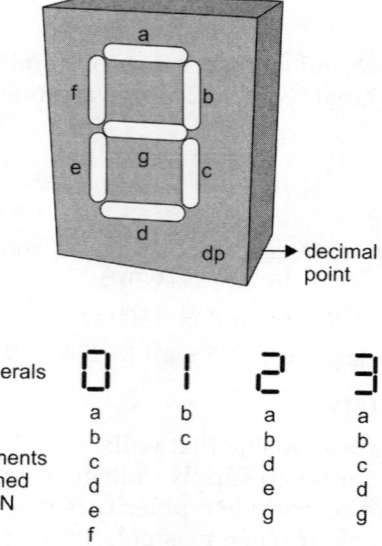

Fig. 7.44. Seven-segment display

decimal point. By lighting a combination of different LEDs, any number from 0 to 9 can be displayed.

The seven-segment display cannot be used effectively to display the alphabet, and thus other multisegment displays have been developed. The 16-segment display is illustrated in Fig. 7.45, and the 5 × 7 dotmatrix display is shown in Fig. 7.46.

7.9.5 LED Testing

LEDs are usually damaged by excessive current flowing through them. Such LEDs can be identified by discolouration on their casing, due to the burned junction. The LED can be tested with a DMM (digital multimeter). In the diode check position, the DMM supplies about 2.5 V at very low current.

If at this setting the cathode of LED is connected to the negative lead of DMM and the anode of LED is connected to the positive lead of DMM, the supplied 2.5 V should be sufficient to forward-bias the LED and cause it to grow dimly, and the forward breakover voltage should be indicated on the meter

Fig. 7.45. Sixteen-segment display

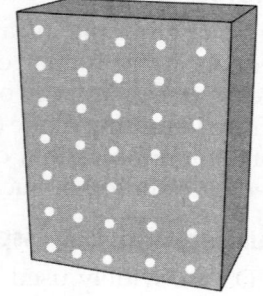

Fig. 7.46. 5 × 7 dotmatrix

display. Reversing the meter leads should, as in conventional dioded, read open (∞).

7.9.6 Quantum Efficiency and LED Power

An excess of electrons and holes in *p*- and *n*-type material, respectively (referred to as *minority carriers*) is created in a semiconductor light source by carrier injection at the device contacts. The excess densities of electrons n and holes p are equal, since the injected carriers are formed and recombine in pairs in accordance with the requirement for charge neutrality in the crystal. When carrier injection stops, the carrier density returns to the equilibrium value. In general, the excess carrier density decays exponentially with time according to the relation:

$$n = n_0 e^{-t/\tau} \tag{7.88}$$

where n_0 is the initial injected excess electron density and the time constant τ is the carrier lifetime. This lifetime is one of the most important operating parameters of an electro-optic device. Its value can range from milliseconds to fractions of a nanosecond depending on material composition and device defects.

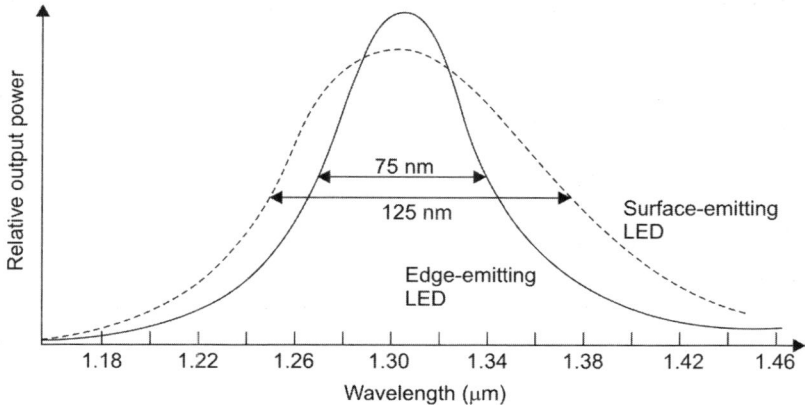

Fig. 7.47. Typical spectral patterns for edge-emitting and surface-emitting LEDs at 1310 nm. The patterns broaden with increasing wavelength and are wider for surface emitters

The excess carriers can recombine either radiatively or nonradiatively. In radiative recombination, a photon of energy $h\nu$, which is approximately equal to the band-gap energy, is emitted. Nonradiative recombination effects include optical absorption in the active region (self-absorption), carrier recombination at the heterostructure interfaces, and the Auger process in which the energy released during an electron-hole recombination is transferred to another carrier in the form of kinetic energy.

When there is a constant current flow into a LED, an equilibrium condition is established. That is, the excess density of electrons n and holes p is equal since the injected carriers are created and recombined in pairs such that charge neutrality is maintained within the device. The total rate at which carriers are generated is the sum of the externally supplied and the thermally generates rates. The externally supplied rate is given by J/qd, where J is the current density in A/cm^2, q is the electron charge, and d is the thickness of the recombination region. The thermal generation rate is given by n/τ. Hence, the rate equation for carrier recombination in a LED can be written as:

$$\frac{dn}{dt} = \frac{J}{qd} - \frac{n}{\tau} \qquad (7.89)$$

The equilibrium condition is found by setting Eq. (7.89) equal to zero, yielding:

$$n = \frac{J\tau}{qd} \qquad (7.90)$$

This relationship reveals the steady-state electron density in the active region when a constant current is flowing through it.

The *internal quantum efficiency* in the active region is the fraction of the electron-hole pairs that recombine radiatively. If the radiative recombination rate is R_r and the nonradiative recombination rate is R_{nr}, then the internal quantum efficiency η_{int} is the ratio of the radiative recombination rate to the total recombination rate:

$$\eta_{int} = \frac{R_r}{R_r + R_{nr}} \qquad (7.91)$$

For exponential decay of excess carriers, the radiative recombination lifetime is $\tau_r = n/R_r$ and the nonradiative recombination lifetime is $\tau_{nr} = n/R_{nr}$. Thus, the internal quantum efficiency can be expressed as:

$$\eta_{int} = \frac{1}{1 + \tau_r/\tau_{nr}} = \frac{\tau}{\tau_r} \qquad (7.92)$$

where the bulk recombination lifetime t is:

$$\frac{1}{\tau} = \frac{1}{\tau_r} + \frac{1}{\tau_{nr}} \qquad (7.93)$$

In general, τ_r and τ_{nr} are comparable for direct-band-gap semiconductors, such as GaAlAs and InGaAsP. This also means that R_r and R_{nr} are similar in magnitude, so that the internal quantum efficiency is about 50 percent for simple homojunction LEDs. However, LEDs having double-heterojunction structures can have quantum efficiencies of 60–80 percent. This high efficiency is achieved because the thin active regions of these devices mitigate the self-absorption effects, which reduces the nonradiative recombination rate.

If the current injected into the LED is I, then the total number of recombinations per second is:

$$R_r + R_{nr} = I/q \tag{7.94}$$

Substituting Eq. (7.94) into Eq. (7.91), then yields $R_r = \eta_{int} I/q$. Noting that R_r is the total number of photons generated per second and that each photon has an energy $h\nu$, then the optical power generated internally to the LED is:

$$P_{int} = \eta_{int} \frac{I}{q} h\nu = \eta_{int} \frac{hcI}{q\lambda} \tag{7.95}$$

Assuming the outside medium is air and letting $n_1 = n$, we have $T(0) = 4n/(n+1)^2$. The external quantum efficiency is then approximately given by:

$$\eta_{int} \approx \frac{1}{n(n+1)^2} \tag{7.96}$$

From this, it follows that the optical power emitted from the LED (Fig. 7.47) is:

$$P = \eta_{ext} P_{int} \approx \frac{P_{int}}{n(n+1)} \tag{7.97}$$

We may note that not all internally generated photons will exit the device. To find the emitted power, one needs to consider the *external quantum efficiency* η_{ext}. This is defined as the ratio of the photons emitted from the LED to the number of internally generated photons. To find the external quantum efficiency, we need to take into account reflection effects at the surface of the LED. As shown in Fig. 7.48, at the interface of a material boundary only that fraction of light falling within a cone defined by the critical angle $\phi_c = \pi/2 - \theta_c$ will cross the interface. We have that $\phi_c = \sin^{-1}(n_2/n_1)$. Here, n_1 is the refractive index of the semiconductor material and n_2 is the refractive index of the outside material,

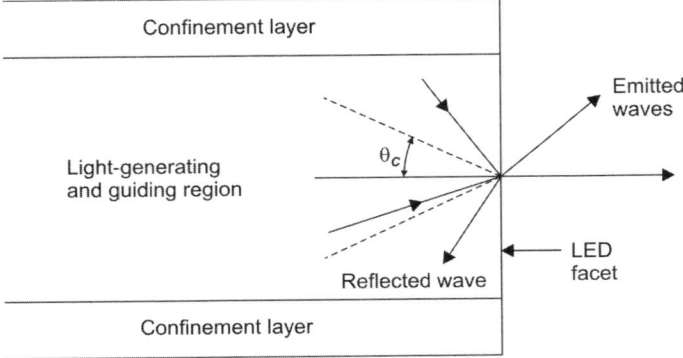

Fig. 7.48. Only light falling within a cone defined by the critical angle will be emitted from an optical source

which nominally is air with $n_2 = 1.0$. The external quantum efficiency can then be calculated from the expression:

$$\eta_{ext} = \frac{1}{4\pi} \int_0^\phi T(\phi)(2\pi \sin \phi)\, d\phi \qquad (7.98)$$

where $T(\phi)$ is the *Fresnel transmission coefficient* or *Fresnel transmissivity*. This factor depends on the incidence angle ϕ, but, for simplicity, we can use the expression for normal incidence, which is:

$$T(0) = \frac{4n_1 n_2}{(n_1 + n_2)^2} \qquad (7.99)$$

Assuming the outside medium is air letting $n_1 = n$, we have $T(0) = 4n/(n+1)^2$. The external quantum efficiency is then approximately given by:

$$\eta_{ext} \approx \frac{1}{n(n+1)^2} \qquad (7.99a)$$

From this, it follows that the optical power emitted from the LED is:

$$P = \eta_{ext} P_{int} \approx \frac{P_{int}}{n(n+1)^2} \qquad (7.99b)$$

7.9.7 Modulation of an LED

The *response time* or *frequency response* of an optical source dictates how fast an electrical input drive signal can vary the light output level. The following three factors largely determine the response time: the doping level in the active region, the injected carrier lifetime t_i in the recombination region, and the parasitic capacitance of the LED. If the drive current is modulated at a frequency ω, the optical output power of the device will vary as:

$$P(\omega) = P_0 [1 + (\omega t_i)^2]^{1\backslash 2} \qquad (7.100)$$

where P_0 is the power emitted at zero modulation frequency. The parasitic capacitance can cause a delay of the carrier injection into the active junction, and consequently, could delay the optical output. This delay is negligible if a small, constant forward bias is applied to the diode. Under this condition, Eq. (7.100) is valid and modulation response is limited only by the carrier recombination time.

The modulation bandwidth of an LED can be defined in either electrical or optical terms. Normally, electrical terms are used since the bandwidth is actually determined via the associated electrical circuitry. Thus, the modulation bandwidth is defined as the point where the electrical signal power, designated by $p(\omega)$, has dropped to half its constant value resulting from the modulated portion of the optical signal. This is the electrical 3-dB point; that is, the frequency at which the output electrical power is reduced by 3 dB with respect to the input electrical power, as shown in Fig. 7.49.

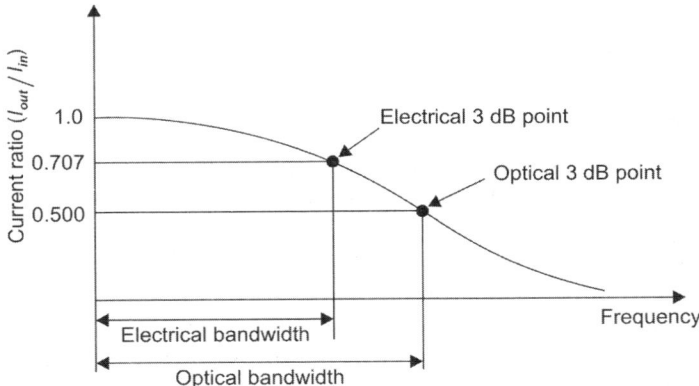

Fig. 7.49. Frequency response of an optical source showing the electrical and optical 3-dB-bandwidth points

Since an optical source exhibits a linear relationship between light power and current, currents rather than voltages (which are used in electrical systems) are compared in optical systems. Thus, since $p(\omega) = I^2(\omega)/R$, the ratio of the output electrical power at the frequency w to the power at zero modulation is:

$$\text{Ratio}_{\text{elec}} = 10 \log \left[\frac{p(\omega)}{p(0)} \right] = 10 \log \left[\frac{I^2(\omega)}{I^2(0)} \right] \quad (7.101)$$

where $I(\omega)$ is the electrical current in the detection circuitry. The electrical 3-dB point occurs at that frequency point where the detected electrical power $p(\omega) = p(0)/2$. This happens when:

$$\frac{I^2(\omega)}{I^2(0)} = \frac{1}{2} \quad (7.102)$$

or $I(\omega)/I(0) = 1/\sqrt{2} = 0.707$.

We may note that sometimes, the modulation bandwidth of an LED is given in terms of the 3-dB bandwidth of the modulated optical power $P(\omega)$; that is, it is specified at the frequency where $P(\omega) = P_0/2$. In this case, the 3-dB bandwidth is determined from the ratio of the optical power at frequency ω to the unmodulated value of the optical power. Since the detected current is directly proportional to the optical power, this ratio is:

$$\text{Ratio}_{\text{optical}} = 10 \log \left[\frac{P(\omega)}{P(0)} \right] = 10 \log \left[\frac{I(\omega)}{I(0)} \right] \quad (7.103)$$

The optical 3-dB point occurs at that frequency where the ratio of the currents is equal to $1/2$. As shown in Fig. 7.49, this gives an inflated value of the modulation bandwidth, which corresponds to an electrical power attenuation of 6 dB.

7.10 LASER DIODE

In a light emitting diode, the emitted light is incoherent. Obviously, when we require coherent light, LED is not useful. For coherent light, we use a laser beam.

Principle: A *pn* junction is manufactured with precisely defined length L from GaAs or GaAs combined with other materials. L (junction length) is related to λ (wavelength) to be emitted (Fig. 7.50(a)). The ends of the junction are polished to a mirror surface and possibly may have additional reflective coating, so that generated light may reflect back internally into the junction. One end is kept partially reflective so that light can pass through when lasing occurs.

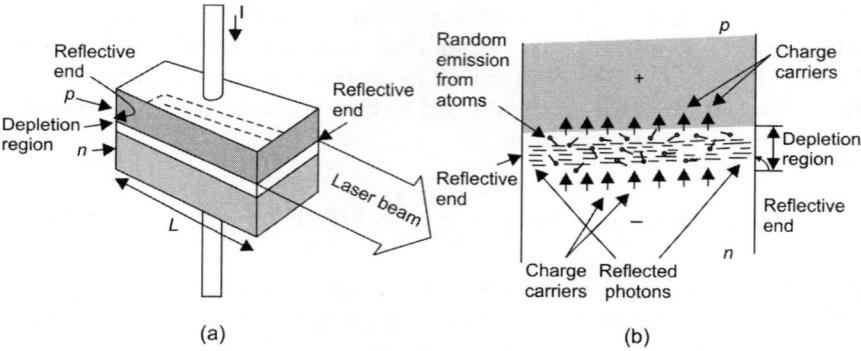

Fig. 7.50. (a) Laser diode, (b) illustration of random emission and laser action within depletion region

Function: Let us consider the effect of charge carrier injection into the depletion region when the junction is forward-biased (Fig. 7.50(b)). As forward current excites the atoms that strike, first emit photons (radiant energy) of energy randomly, as electrons are raised to a high energy level and then fall back to a lower level. Consequently, several photons strike the reflective ends of the junction perpendicularly so that they are reflected back from the other end of the junction. Obviously, photons have a to-and-fro motion several times and their number goes on increasing as they cause other atoms in their path to emit photons by striking them. This activity of reflection and generation of photons causes amplification of initial reflected photons. A laser beam emerges through the partially reflected surface of the junction. GaAs laser diodes normally require high forward current levels, ranging from 10 mA to tens of amperes. At low current levels, this device emits similar as in LED. Beyond a threshold current level, light intensity increases sharply and its bandwidth decreases as lasing commences. Due to high energy density, laser beam can be quite harmful to eyes.

Laser diodes operating in a pulsed manner are called **junction laser diodes**, whereas lasers producing continuous output are called **continuous wave** (CW) laser diodes. Each type of laser emits a particular light wavelength depending upon the material and junction dimensions. Each type of laser has a threshold input current level.

Basic Operation and Construction

Laser diodes are light emitting devices, capable of emitting laser beams. One of the commonly used laser diodes is *injection laser diode* (ILD). A typical ILD package is shown in Fig. 7.51. The laser diode works exactly like an ordinary diode. Fig. 7.52 shows the structures of an injection laser. The p type and n type materials are made of AlGaAs.

When forward-biased, light is emitted at the pn junction. However, unlike in the LED, the emitted light is coherent and monochromatic. In the LED, light emitted is scattered in all directions. In the laser diode, however, light emitted can escape only from the end faces because the edges are roughened. In most laser diodes, one of the end faces is coated with a reflective material so that radiation of laser is emitted only in one direction.

Fig. 7.51. A typical injection laser diode (ILD) package

The laser diode emits light when forward-biased and can withstand only relatively small reverse-biased voltages. A high reverse voltage can damage or destroy the laser diode. The symbol for laser diode is shown in Fig. 7.53. Note that light emission is shown zigzagged rather than straight, as in the symbol for an ordinary LED.

Applications of Laser Diode

Laser diodes are used primarily in fiber optic communications. It is the only device capable of producing optical energy high enough in concentration to pass through lengthy fiber optic cables. However, it also has some major disadvantages as compared to LEDs.

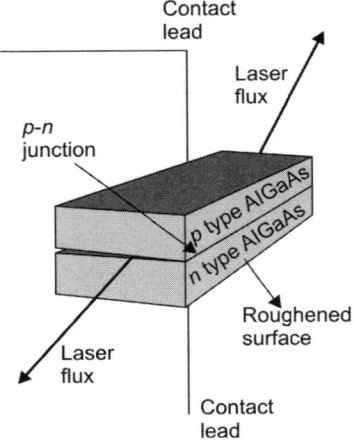

Fig. 7.52. Structure of injection laser diode (ILD)

(i) They cost 10 times more than LEDs.
(ii) Their life expectancy is 10 times shorter than that of LEDs.

(iii) They require elaborate power supplies and consume much more power than LEDs.

Laser diodes do not compete with LEDs on cost-reliability basis. Therefore, LEDs are used as much as possible in fiber optic systems and laser diodes are used only when absolutely necessary.

Some applications of laser diodes are:
(i) Fixed product scanners seen in grocery stores.
(ii) Handheld barcode scanners used for inventory control in warehouses.
(iii) Handheld laser pointers used in presentations.
(iv) Laser projection devices used by surgeons, mechanics, aviators and technicians. Laser scanners project low-power images directly onto the user's retina as an overlay in conjunction with the real-world image.
(v) Land survey range-finding has been revolutionized by the use of laser diodes. Laser transmission from Total Station (reflected from a distant point) is detected, time-delay measured, and converted to an equivalent distance. Fig. 7.54 shows a Leica model TC407 Total Station that incorporates laser range-finding.

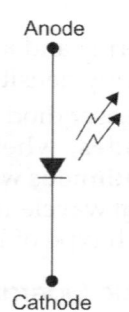

Fig. 7.53. Symbol for a laser diode

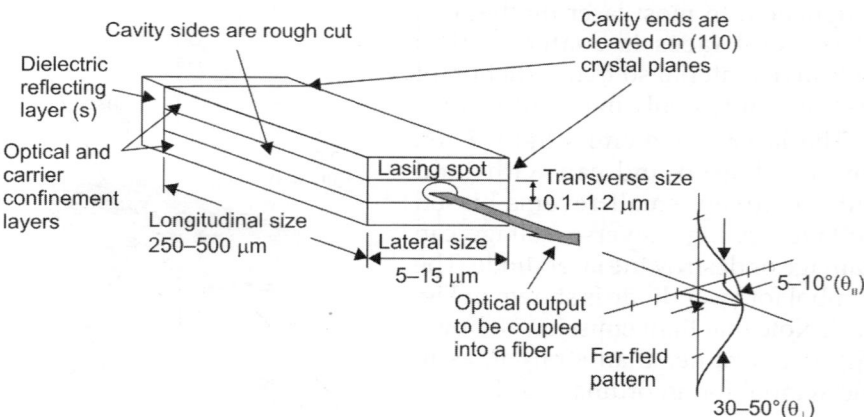

Fig. 7.54. Fabry-Perot resonator cavity for a laser diode. The cleaved crystal ends function as partially reflecting mirrors. The unused end (the rear facet) can be coated with a dielectric reflector to reduce optical loss in the cavity. The light beam emerging from the laser forms a vertical ellipse, even though the lasing spot at the active-area facet is a horizontal ellipse

(vi) Document printers and scanners often incorporate lasers.
(vii) In construction, lasers are used to paint line and angle projection, nearly eliminating the potential for manual measurement errors.

7.10.1 Laser Diode Modes and Threshold Condition

The radiation in one type of laser diode configuration is generated within a Fabry-Perot resonator cavity (Fig. 7.54), as in most other types of lasers. However, this cavity is much smaller (~ 250–500 μm long, 5–15 μm wide, and 0.1–0.2 μm thick). These dimensions commonly are referred to as longitudinal, lateral, and transverse dimensions of the cavity, respectively.

As illustrated in Fig. 7.55, two flat, reflecting mirrors are directed towards each other to enclose the Fabry-Perot resonator cavity. The mirror facets are constructed by making two parallel clefts along with natural cleavage planes of the semiconductor crystal. The purpose of the mirrors is to establish a strong optical feedback in the longitudinal direction. The feedback mechanism converts the device into an oscillator (and hence a light emitter) with a gain mechanism that compensates for optical losses in the cavity at certain resonant optical frequencies. The sides of the cavity are simply formed by roughing the edges of the device to reduce unwanted emissions in the lateral directions.

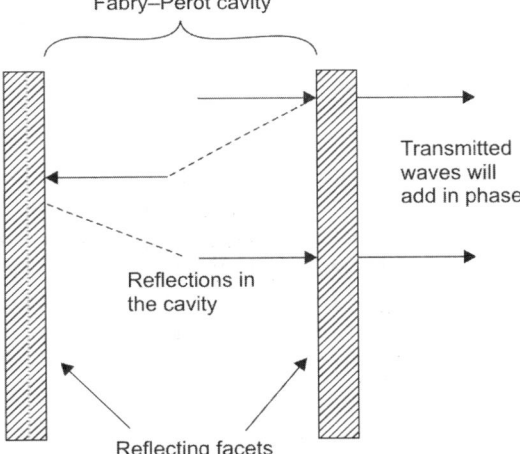

Fig. 7.55. Two parallel light-reflecting mirrored surfaces define a Fabry-Perot resonator cavity

As the light reflects back and forth within the Fabry-Perot cavity, the electric fields of the light interfere on successive round trips. Those wavelengths that are integer multiples of the cavity length interfere constructively so that their amplitudes add when they exit the device through the right-hand facet. All other wavelengths interfere destructively and thus cancel themselves out. The optical frequencies at which constructive interference occurs are the *resonant frequencies* of the cavity. Consequently, spontaneously emitted photons that have wavelengths at these resonant frequencies reinforce themselves after multiple trips through the cavity so that their optical field becomes very strong. The resonant wavelengths are called the *longitudinal modes* of the cavity, since they resonate along the length of the cavity.

Figure 7.56 shows the behaviour of the resonant wavelength for three values of the mirror reflectivity. The plots give the relative intensity as a function of the wavelength relative to the cavity length. As can be seen from Fig. 7.56, the width of the resonances depends on the value of the reflectivity. That is, the resonances become sharper as the reflectivity increases.

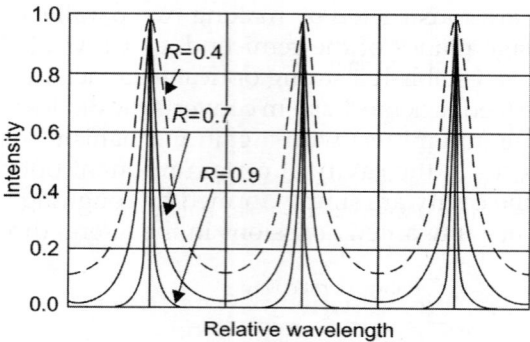

Fig. 7.56. Behaviour of the resonant wavelengths in a Fabry-Perot cavity for three values of the mirror reflectivity

In another laser diode type, commonly referred to as the *distributed-feedback* (DFB) *laser*, the cleaved facets are not required for optical feedback. A typical DFB laser configuration is shown in Fig. 7.57. The fabrication of this device is similar to the Fabry-Perot types, except that the lasing action is obtained from Bragg reflectors (gratings) or periodic variations of the refractive index (called *distributed-feedback corrugations*), which are incorporated into the multilayer structure along the length of the diode.

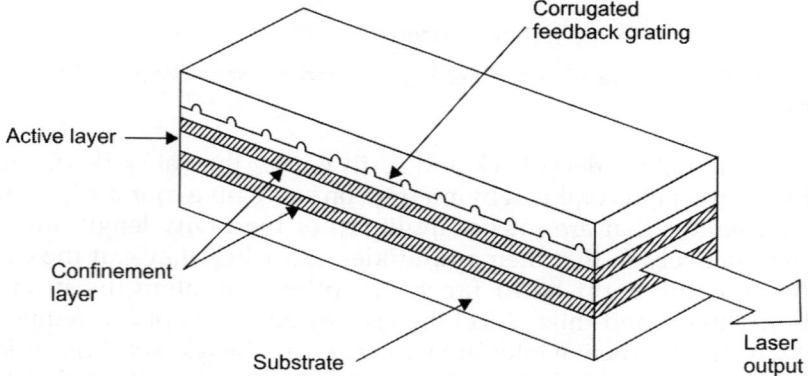

Fig. 7.57. Structure of a distributed-feedback (DFB) laser diode

In general, the full optical output is needed only from the front facet of the laser—that is, the one to be aligned with an optical fiber. In this case, a dielectric reflector can be deposited on the rear laser facet to reduce the optical loss in the cavity, to reduce the threshold current density (the point at which lasing starts), and to increase the external quantum efficiency. Reflectivities greater than 98 percent have been achieved with a six-layer reflector.

The optical radiation within the resonance cavity of a laser diode sets up a pattern of electric and magnetic field lines called the *modes of the cavity*. These can conveniently be separated into two independent sets of transverse electric (TE) and transverse magnetic (TM) modes. Each set of modes can be described in terms of the longitudinal, lateral, and transverse half-sinusoidal variations of the electromagnetic fields along the major axes of the cavity. The *longitudinal modes* are related to the length L of the cavity and determine the principal structure of the frequency spectrum of the emitted optical radiation. Since L is much larger than the lasing wavelength of approximately 1 μm, many longitudinal modes can exist. *Lateral modes* lie in the plane of the *pn* junction. These modes depend on the side wall preparation and the width of the cavity, and determine the shape of the lateral profile of the laser beam. *Transverse modes* are associated with the electromagnetic field and beam profile in the direction perpendicular to the plane of the *pn* junction. These modes are of great importance, since they largely determine such laser characteristics as the radiation pattern (the angular distribution of the optical output power) and the threshold current density.

In order to determine the lasing conditions and the resonant frequencies, we express the electromagnetic wave propagating in the longitudinal direction (along the axis normal to the mirrors) in terms of the electric field phasor.

$$E(z, t) = I(z)\, e^{j(\omega t - \beta z)} \qquad (7.104)$$

where $I(z)$ is the optical field intensity, ω is the optical radian frequency, and β is the propagation constant.

Lasing is the condition at which light amplification becomes possible in the laser diode. The requirement for lasing is that a population inversion be achieved. This condition can be understood by considering the fundamental relationship between the optical field intensity I, the absorption coefficient α_λ, the gain coefficient g is the Fabry-Perot cavity. The stimulated emission rate into a given mode is proportional to the intensity of the radiation in that mode. The radiation intensity at a photon energy $h\nu$ varies exponentially with the distance z that it traverses along the lasing cavity according to the relationship:

$$I(x) = I(0)\, \exp\{[\Gamma g\,(h\nu) - \bar{\alpha}(h\nu)]z\} \qquad (7.105)$$

where $\bar{\alpha}$ is the effective absorption coefficient of the material in the optical path and Γ is the *optical-field confinement factor*—that is, the fraction of optical power in the active layer.

Optical amplification of selected modes is provided by the feedback mechanism of the optical cavity. In the repeated passes between the two partially reflecting parallel mirrors, a portion of the radiation associated with those modes that have the highest optical gain coefficient is retained and further amplified during each trip through the cavity.

Lasing occurs when the gain of one or several guided modes is sufficient to exceed the optical loss during one roundtrip through the cavity; that is, $z = 2L$. During this roundtrip, only the fractions R_1 and R_2 of the optical radiation are reflected from the two laser ends 1 and 2, respectively, where R_1 and R_2 are the mirror reflectivities or Fresnel reflection coefficients, which are given by:

$$R = \left(\frac{n_1 - n_2}{n_1 + n_2}\right)^2 \tag{7.106}$$

for the reflection of light at an interface between two materials having refractive indices n_1 and n_2. From this lasing condition, Eq. (7.105) becomes:

$$I(2L) = I(0)\, R_1 R_2 \exp\{2L\,[\Gamma g\,(hv) - \alpha\,(hv)]\} \tag{7.107}$$

For an uncoated cleaved facet the reflectivity is only about 30 percent. To reduce the loss in the cavity and to make the optical feedback stronger, the facets typically are coated with a dielectric material. This can produce a reflectivity of about 99 percent for the rear facet and 90 percent for the front facet through which the lasing light emerges.

At the lasing threshold, a steady-state oscillation takes place, and the magnitude and phase of the returned wave must be equal to those of the original wave. This yields:

$$I(2L) = I(0) \tag{7.108}$$

for the amplitude and

$$e^{-j2\beta L} = 1 \tag{7.109}$$

for the phase. Equation (7.109) gives information concerning the resonant frequencies of the Fabry-Perot cavity. From Eq. (7.108), one can find which modes have sufficient gain for sustained oscillation and we can find the amplitudes of these modes. The condition to just reach the lasting threshold is the point at which the optical gain is equal to the total loss α_t, in the cavity. From Eq. (7.108), this condition is obtained as:

$$\Gamma g_{th} = \alpha_t = \bar{\alpha} + \frac{1}{2L} \ln\left(\frac{1}{R_1 R_2}\right) = \bar{\alpha} + \alpha_{end} \tag{7.110}$$

where α_{end} is the mirror loss in the lasing cavity. Thus, for lasing to occur, we must have the gain $g \geq g_{th}$. This means that the pumping source that maintains the population inversion must be sufficiently strong to support or exceed all the energy-consuming mechanisms within the lasing cavity.

The mode that satisfies Eq. (7.110) reaches threshold first. Theoretically, at the onset of this condition, all additional energy introduced into the laser should argument the growth of this particular mode. In practice, various phenomena lead to the excitation of more than one mode. Studies on the conditions of longitudinal single-mode operation have revealed that important factors are thin active regions and a high degree of temperature stability.

The relationship between optical output power and diode drive current is presented in Fig. 7.58. At low diode currents, only spontaneous radiation is emitted. Both the spectral range and the lateral beam width of this emission are broad like that of an LED. A dramatic and sharply defined increase in the power output occurs at the lasing threshold. As this transition point is approached, the spectral range and the beam width both narrow with increasing drive current. The final spectral width of approximately 1 nm and the fully narrowed lateral beam width of nominally 5–10° are reached just past the threshold point. The *threshold current* I_{th} is conventionally defined by extrapolation of the lasing region of the power-versus-current curve; as shown in Fig. 7.58. At high power outputs, the slope of the curve decreases because of junction heating.

For laser structures that have strong carrier confinement, the *threshold current density* for stimulated emission J_{th} can, to a good approximation, be related to the lasing-threshold optical gain by:

$$g_{th} = \beta J_{th} \tag{7.111}$$

where β is a constant that depends on the specific device construction.

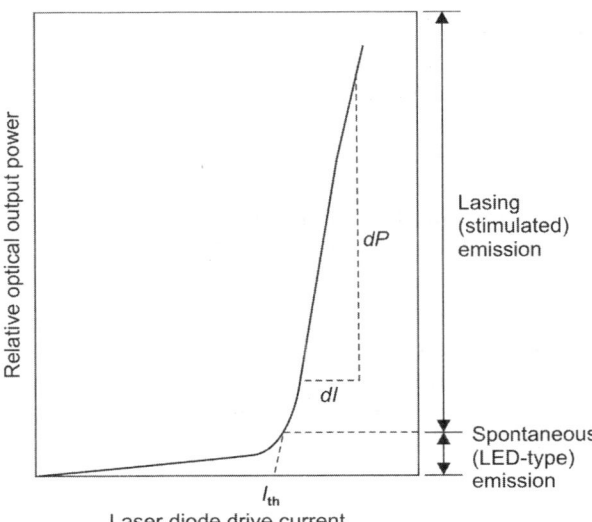

Fig. 7.58. Relationship between optical output power and laser diode drive current. Below the lasing threshold the optical output is a spontaneous LED-type emission

7.10.2 Laser Diode Rate Equations

The relationship between optical output power and the diode drive current can be determined by examining the rate equations that govern the interaction of photons and electrons in the active region. As noted earlier, the total carrier population is determined by carrier injection, spontaneous recombination, and stimulated emission. For a *pn* junction with a carrier-confinement region of depth d, the *rate equations* are given by:

$$\frac{d\Phi}{dt} = Cn\Phi + R_{sp} - \frac{\Phi}{\tau_{ph}} \qquad (7.112)$$

$$= \text{stimulated emission}$$
$$+ \text{spontaneous emission} + \text{photon gas}$$

which governs the number of photons Φ, and

$$\frac{dn}{dt} = \frac{J}{qd} - \frac{n}{\tau_{sp}} - Cn\Phi \qquad (7.113)$$

$$= \text{injection} + \text{spontaneous recombination}$$
$$+ \text{stimulated emission}$$

which governs the number of *electrons* n. Here, C is a coefficient describing the strength of the *optical absorption and emission* interactions, R_{sp} is the rate of spontaneous emission into the lasing mode (which is much smaller than the total spontaneous-emission rate), τ_{ph} is the photon lifetime, τ_s is the spontaneous-recombination lifetime, and J is the injection-current density.

Equations (7.112) and (7.113) may be balanced by considering all the factors that affect the number of carriers in the laser cavity. The first term in Eq. (7.112) is a source of photons resulting from stimulated emission. The second term, describing the number of photons produced by spontaneous emission, is relatively small compared with the first term. The third term in Eq. (7.112) indicates the decay in the number of photons caused by loss mechanisms in the lasing cavity. In Eq. (7.113), the first term represents the increase in the electron concentration in the conduction band as current flows into the device. The second and third terms give the number of electrons lost from the conduction band owing to spontaneous and stimulated transitions, respectively.

Solving these two equations for a steady-state condition, one obtains an expression for the output power. The steady state is characterized by the left-hand sides of Eqs (7.112) and (7.113) being equal to zero. First, from Eq. (7.112), assuming R_{sp} is negligible and noting that $d\Phi/dt$ must be positive when Φ is small, we have:

$$Cn - \frac{1}{\tau_{ph}} \geq 0 \qquad (7.114)$$

This reveals that n must exceed a threshold value n_{th} in order for Φ to increase. Using Eq. (7.113), this threshold value can be expressed in terms of the threshold current J_{th} needed to maintain an inversion level $n = n_{th}$ in the steady state when the number of photons $\Phi = 0$:

$$\frac{n_{th}}{\phi_{sp}} = \frac{J_{th}}{qd} \quad (7.115)$$

This expression *defines the current required to sustain an excess electron density in the laser* when spontaneous emission is the only decay mechanism.

Next, consider the photon and electron rate equations in the steady-state condition at the lasing threshold, respectively, Eqs (7.112) and (7.113) become:

$$0 = Cn_{th}\Phi_s + R_{sp} - \frac{\Phi}{\tau_{ph}} \quad (7.116)$$

and

$$0 = \frac{J}{qd} - \frac{n_{th}}{\tau_{sp}} - Cn_{th}\Phi_s \quad (7.117)$$

where Φ_s is the steady-state photon density. Adding Eqs (7.116) and (7.117), using Eq. (7.115) for the term n_{th}/τ_{sp}, and solving for Φ_s, one obtains the number of photons per unit volume:

$$\Phi_s = \frac{\tau_{ph}}{qd}(J - J_{th}) + \tau_{ph} R_{sp} \quad (7.118)$$

The first term in Eq. (7.118) is the number of photons resulting from stimulated emission. The power from these photons is generally concentrated in one or a few modes. The second term gives the spontaneously generated photons. The power resulting from these photons is not mode-selective, but is spread over all the possible modes of the volume, which are on the order of 10^8 modes.

7.10.3 External Differential Quantum Efficiency

The *external differential quantum efficiency* η_{ext} is defined as the number of photons emitted per radiative electron-hole pair recombination above threshold. Under the assumption that above threshold the gain coefficient remains fixed at g_{th}, η_{ext} is obtained as:

$$\eta_{ext} = \frac{\eta_i(g_{th} - \alpha)}{g_{th}} \quad (7.119)$$

Here, η_i is the internal quantum efficiency. This is not a well-defined quantity in laser diodes, but most measurements show that $\eta_i \approx 0.6$–0.7 at room temperature. Experimentally, η_{ext} is calculated from the straight-line portion of the curve for the emitted optical power P versus drive current I, one obtains:

$$\eta_{ext} = \frac{q}{E_g}\frac{dP}{dI} = 0.8056\lambda\ (\mu m)\frac{dP\ (mW)}{dI\ (mA)} \quad (7.120)$$

where E_g is the band-gap energy in electron volts, dP is the incremental change in the emitted optical power in milliwatts for an incremental change dI in the drive current (in milliamperes), and λ is the emission wavelength in micrometers. For standard semiconductor lasers, external differential quantum efficiencies of 15–30% per facet are typical. High-quality devices have differential quantum efficiencies of 30–40%.

7.10.4 Resonant Frequencies

Now, we again consider Eq. (7.109) to examine the resonant frequencies of the laser. The condition in Eq. (7.109) holds when:

$$2\beta L = 2\pi m \quad (7.121)$$

where m is an integer. Using $\beta = 2\pi n/\lambda$ for the propagation constant, we have:

$$m = \frac{L}{\lambda/2n} = \frac{2Ln}{c}\nu \quad (7.122)$$

where $c = \nu\lambda$. This states that the cavity resonates (i.e., a standing-wave pattern exists within it) when a integer number m of half-wavelengths spans the region between the mirrors.

Since in all lasers the gain is a function of frequency (or wavelength, since $c = \nu\lambda$), there will be a range of frequencies (or wavelengths) for which Eq. (7.122) holds. Each of these frequencies corresponds to a mode of oscillation of the laser. Depending on the laser structure, any number of frequencies can satisfy Eqs (7.108) and (7.109). Thus, some lasers are *single-mode* and some are *multimode*. The relationship between gain and frequency can be assumed to have the *Gaussian form*:

$$g(\lambda) = g(0)\exp\left[-\frac{(\lambda-\lambda_0)^2}{2\sigma^2}\right] \quad (7.123)$$

where λ_0 is the wavelength at the center of the spectrum, σ is the spectral width of the gain, and the maximum gain $g(0)$ is proportional to the population inversion.

We now look at the frequency, or wavelength, spacing between the modes of a multimode laser. Here, we consider only the longitudinal modes. Note, however, that for each longitudinal mode there may be several transverse modes that arise from one or more reflections of the propagating wave at the sides of the resonator cavity. To find the frequency spacing, consider two successive modes of frequencies ν_{m-1} and ν_m represented by the integers $m-1$ and m. From Eq. (7.122), one finds:

$$m - 1 = \frac{2Ln}{c}\nu_{m-1} \quad (7.124)$$

and
$$m = \frac{2Ln}{c} v_m \tag{7.125}$$

Subtracting these two equations yields:
$$1 = \frac{2Ln}{c}(v_m - v_{m-1}) = \frac{2Ln}{c} \Delta v \tag{7.126}$$

from which we have the frequency spacing:
$$\Delta v = \frac{\lambda^2}{2Ln} \tag{7.127}$$

This can be related to the wavelength spacing $\Delta\lambda$ through the relationship $\Delta v/v = \Delta\lambda/\lambda$, yielding:
$$\Delta\lambda = \frac{\lambda^2}{2Ln} \tag{7.128}$$

Thus, given Eqs (7.123) and (7.128), the output spectrum of a multimode laser follows the typical gain-versus-frequency shown in Fig. 7.59, where the exact number of modes, their heights, and their spacings depend on the laser construction.

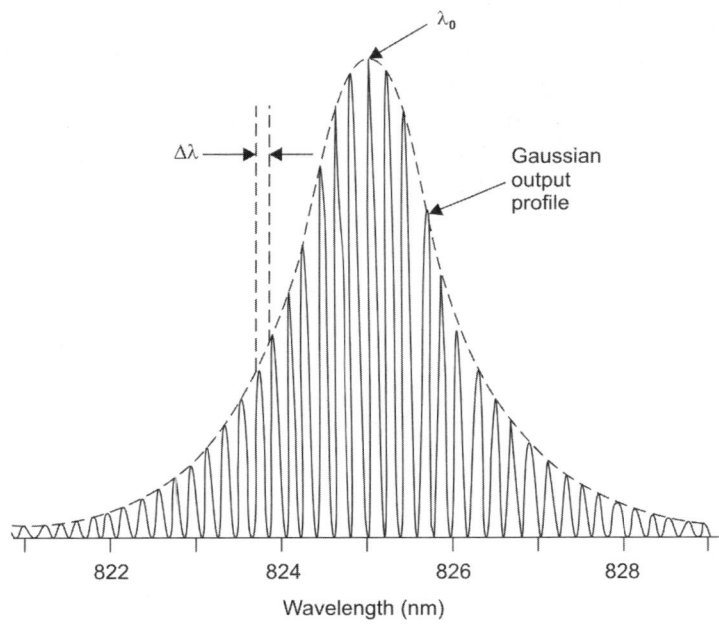

Fig. 7.59. Spectrum from a Fabry-Perot GaAlAs/GaAs laser diode

7.10.5 Laser Diode Structures and Radiation Patterns

A basic requirement for efficient operation of laser diodes is that, in addition to transverse optical and carrier confinement between heterojunction layers, the current flow must be restricted laterally to a narrow stripe along the length of the laser. Numerous novel methods of achieving this, with varying degrees of success, have been proposed, but all strive for the same goals of limiting the number of lateral modes so that lasing is confined to a single filament, stabilizing the lateral gain, and ensuring a relatively low threshold current.

Figure 7.60 shows the three basic *optical-confinement methods* used for bounding laser light in the lateral direction. In the first structure, a narrow electrode stripe (less than 8 μm wide) runs along the length of the diode. The injection of electrons and holes into the device alters the refractive index of the active layer directly below the stripe. The profile of these injected carriers creates a weak, complex waveguide that confines the light laterally. This type of device is commonly referred to as a *gain-guided laser*. Although, these lasers can emit optical power exceeding 100 mW, they have strong instabilities and can have highly astigmatic, two-peaked beams as shown in Fig. 7.60(a).

More stable structures use the configurations shown in Fig. 7.60(b) and (c). Here, dielectric waveguide structures are fabricated in the lateral direction. The variations in the real refractive index of the various materials in these structures control the lateral modes in the laser. Thus, these devices are called *index-guided lasers*. If a particular index-guided laser supports only the fundamental transverse mode and the fundamental longitudinal mode, it is known as a *single-mode laser*. Such a device emits a single, well-collimated beam of light that has an intensity profile which is a bell-shaped gaussian curve.

Index-guided lasers can have either positive-index or negative-index wave-confining structures. In a *positive-index waveguide*, the central

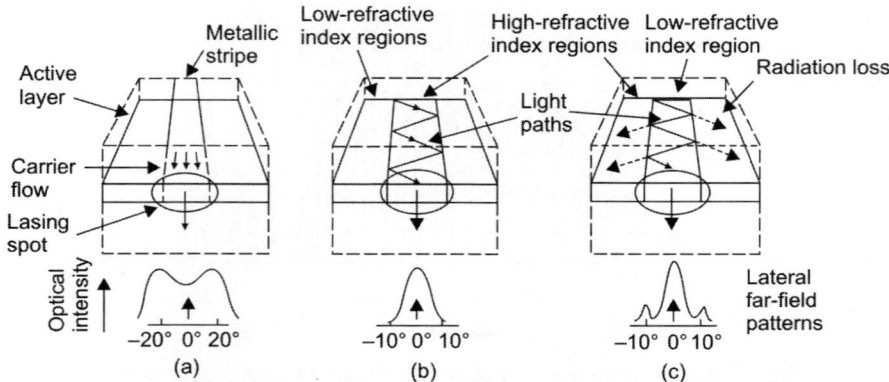

Fig. 7.60. Three fundamental structures for confining optical waves in the lateral direction: (a) in the gain-induced guide, electrons injected via a metallic stripe contact alter the index of refraction of the active layer; (b) the positive-index waveguide has a higher refractive index in the central portion of the active region; (c) the negative-index waveguide has a lower refractive index in the central portion of the active region

Optoelectronic Devices

region has a higher refractive index than the outer regions. Thus, all of the guided light is reflected at the dielectric boundary, just as it is at the core-cladding interface in an optical fiber. By proper choice of the change in refractive index and the width of the higher-index region, one can make a device that supports only the fundamental lateral mode.

In a *negative-index waveguide*, the central region of the active layer has a lower refractive index than the outer regions. At the dielectric boundaries, part of the light is reflected and the rest is refracted into the surrounding material and is thus lost. This radiation loss appears in the far-field radiation pattern as narrow side lobes to the main beam, as shown in Fig. 7.60(c). Since the fundamental mode in this device has less radiation loss than any other mode, it is the first to lase. The positive-index laser is the more popular of these two structures.

Index-guided lasers can be made using any one of four fundamental structures. These are the buried heterostructure, a selectively diffused construction, a varying-thickness structure, and a bent-layer configuration. To make the *buried heterostructure* (BH) laser shown in Fig. 7.61, one etches a narrow mesa stripe (1–2 µm wide) in double-heterostructure material. The mesa is then embedded in high-resistivitry lattice-matched n-type material with an appropriate band gap and low refractive index. This material is GaAlAs in 800 to 900 nm lasers with a GaAs active layer, and is InP for 1300 to 1600 nm lasers with an InGaAsP active layer. This configuration, thus strongly traps generated light in a lateral waveguide. A number of variations of this fundamental structure have been used to fabricate high-performing laser diodes.

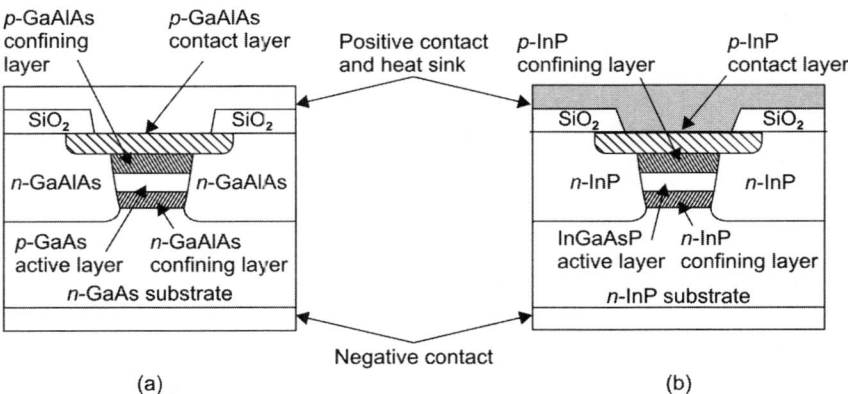

Fig. 7.61. (a) Short-wavelength (800–900 nm) GaAlAs, and (b) long-wavelength (1300–1600 nm) InGaAsP buried-heterostructure laser diodes

The *selectively diffused construction* is shown in Fig. 7.61(a). Here, a chemical dopant, such as zinc for GaAlAs lasers and cadmium for InGaAsP lasers, is diffused into the active layer immediately below the metallic contact stripe. The dopant changes the refractive index of the active layer to form a lateral waveguide channel. In the *varying-thickness structure* shown in Fig. 7.61(b), a channel (or other topological configuration, such as a mesa or terrace) is etched into the substrate. Layers of crystal are then regrown into the channel using liquid-phase epitaxy. This process fills in the depressions and partially dissolves the protrusions, thereby creating variations in the thickness, the thicker area acts as a positive-index waveguide of higher-index material. In the *bent-layer structure*, a mesa is etched into the substrate as shown in Fig. 7.61(c). Semiconductor material layers are grown onto this structure using vapour-phase epitaxy to exactly replicate the mesa configuration. The active layer has a constant thickness with lateral bends. As an optical wave travels along the flat top of the mesa in the active area, the lower-index material outside of the bends confines the light along this lateral channel (Fig. 7.62).

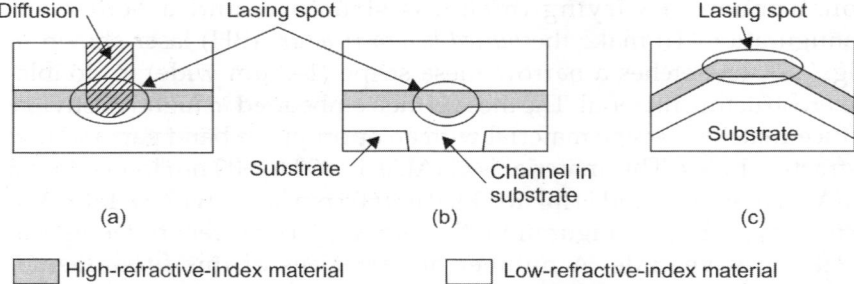

Fig. 7.62. Positive-index optical-wave-confining structure of the (a) selectively diffused, (b) varying thickness, and (c) bent-layer types

In addition to confining the optical wave to a narrow lateral stripe to achieve continuous high optical output power, one also needs to restrict the drive current tightly to the active layer so more than 60 percent of the current contributes to lasing. Fig. 7.63 shows the four basic *current-confinement methods*. In each method, the device architecture blocks current on both sides of the lasing region. This is achieved either by high-resistivity regions or by reverse-biased *pn* junctions, which prevent the current from flowing while the device is forward-biased under normal conditions. For structures with a continuous active layer, the current can be confined either above or below the lasing region. The diodes are forward-biased so that current travel from *p*-type to the *n*-type regions. In the *preferential-dopant diffusion* method, partially diffusing a *p*-type dopant (Zn or Cd) through an *n*-type capping layer establishes a narrow path for the current, since back-biased *pn* junctions block the current outside the diffused region. The *proton implantation method* creates regions of high resistivity, thus restricting the current to

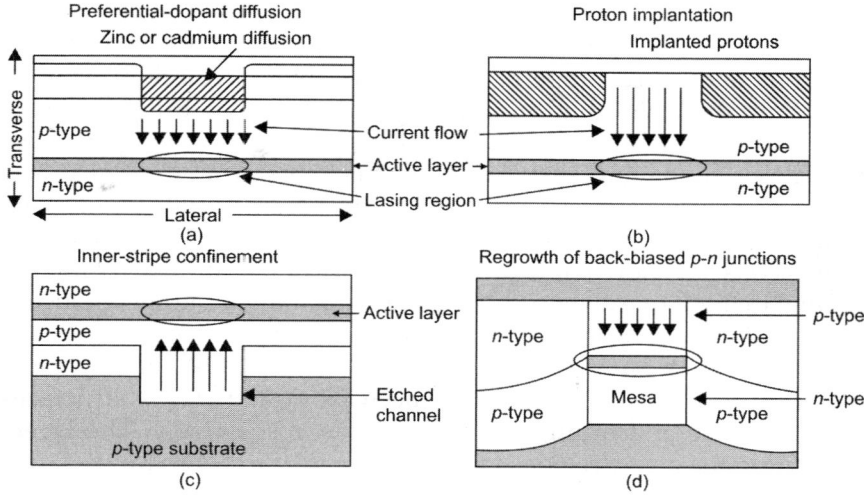

Fig. 7.63. Four basic methods for achieving current confinement in laser diodes: (a) preferential-dopant diffusion, (b) proton implantation, (c) inner-stripe confinement, and (d) regrowth of back-biased *pn* junctions

a narrow path between these regions. The *inner-stripe confinement* technique grows the lasing structure above a channel etched into planar material. Back-biased *pn* junctions restrict the current on both sides of the channel. When the active layer is discontinuous, as in a buried heterostructure, current can be blocked on both sides of the mesa by growing *pn* junctions that are reverse-biased when the device is operating. A laser diode can use more than one current-confining technique.

In a *double-heterojunction laser*, the highest-order transverse mode that can be excited depends on the waveguide thickness and on the refractive-index differentials at the waveguide boundaries. If the refractive-index differentials are kept at approximately 0.08, then only the fundamental transverse mode will propagate if the active area is thinner than 1 μm.

When designing the width and thickness of the optical cavity, a tradeoff must be made between current density and output beam width. As either the width or the thickness of the active region is increased, a narrowing occurs of the lateral or transverse beam widths, respectively, but at the expense of an increase in the threshold current density. Most positive-index waveguide devices have a lasing spot 3 μm wide by 0.6 μm high. This is significantly greater than the active-layer thickness, since about half the light travels in the confining layers. Such lasers can operate reliably only upto continuous-wave (CW) output powers of 3–5 mW. Here, the transverse and lateral half-power beam widths shown in Fig. 7.54 are about $\theta_\perp \simeq$ 30–50° and $\theta_\parallel \simeq$ 5–10°, respectively.

Although, the active layer in a standard double-heterostructure laser is thin enough (1–3 µm) to confine electrons and the optical field, the electronic and optical properties remain the same as in the bulk material. This limits the achievable threshold current density, modulation speed, and linewidth of the device. *Quantum-well lasers* overcome these limitations by having an active-layer thickness around 10 nm. This changes the electronic and optical properties dramatically, because the dimensionality of the free-electron motion is reduced from three to two dimensions. As shown in Fig. 7.64, the restriction of the carrier motion normal to the active layer results in a quantization of the energy levels. The possible energy-level transitions which lead to photon emission are designated by ΔE_{ij}. Both single quantum-well (SQW) and multiple quantum-well (MQW) lasers have been fabricated. These structures contain single and multiple active regions, respectively. The layers separating the active regions are called *barrier layers*. The MQW lasers have a better optical-mode confinement, which results in a lower threshold current density. The wavelength of the output light can be changed by adjusting the laser thickness d, e.g. in an InGaAs quantum well laser, the peak output wavelength moves from 1550 nm when d = 10 nm to 1500 nm for d = 8 nm.

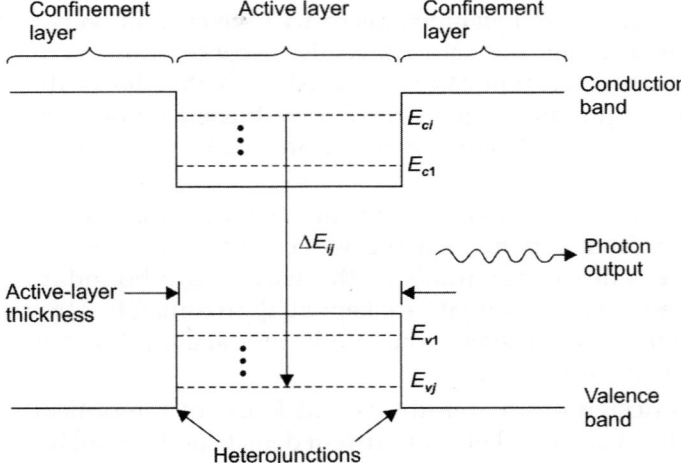

Fig. 7.64. Energy-band diagram for a quantum layer in a multiple quantum-well (MQW) laser. The parameter ΔE_{ij} represents the allowed energy-level transitions

Example 1

Calculate the ratio of rates of spontaneous and stimulated emission for a tungsten filament lamp operating at a temperature of 2000 K. Take average frequency = 5×10^{14} Hz.

Solution

$$R = \exp\left(\frac{h\nu}{kT}\right) - 1$$

$$R = \exp\left(\frac{6.6 \times 10^{-34} \times 5 \times 10^{14}}{1.38 \times 10^{-23} \times 2000}\right) - 1$$

or
$$= e^{12} - 1 = 1.5 \times 10^5.$$

Example 2

Estimate the relative populations of two energy levels such that a transition from higher to lower will give a visible radiation.

Solution

Average wavelength of visible radiation = 550 nm

$$\therefore \quad E_2 - E_1 = \frac{hc}{\lambda} = \frac{6.6 \times 10^{-34} \times 3 \times 10^8}{550 \times 10^{-9}}$$

$$= 3.6 \times 10^{-19} \, J = 2.25 \text{ eV}$$

Assuming room temperature = 300 K and $g_1 = g_2$, we have:

$$\frac{N_2}{N_1} = \exp\left[\frac{-3.6 \times 10^{-19}}{1.38 \times 10^{-23} \times 300}\right]$$

$$\approx e^{-87} = 10^{-37}.$$

Example 3

A single crystal of silicon sample having width of 0.3 µm is illuminated with a monochromatic light of energy 3 eV. The incident power is 10 MW. Calculate: (i) total energy absorbed by material per second, (ii) the rate of excess thermal energy dissipated to the lattice, and (iii) the number of photons per second given off from the recombination of intrinsic transition. Given $\alpha = 4 \times 10^4$/cm. [B.Tech.]

Solution

(i) Energy absorbed/sec is equal to:

$$\phi_o [1 - \exp(4 \times 10^4 \times 0.3 \times 10^{-4})]$$
$$= 10^{-12} (1 - \exp(4 \times 10^4 \times 0.3 \times 10^{-4}))$$
$$= 10^{-2} [1 - e^{1.2}] = 6.3 \text{ mW}$$

(ii) The portion of each photon energy that is converted into heat is obtained as:

$$\frac{h\nu - E_g}{h\nu} = \frac{3 - 1.12}{3} = 62\%$$

Obviously, the amount of energy dissipated/sec to lattice is $62\% \times 6.3 = 3.9$ mW.

(iii) Number of photons/sec from recombination is:

$$\frac{2.4}{1.6 \times 10^{-19} \times 1.12} = 1.3 \times 10^{16} \text{ photon/sec}$$

∴ Recombination radiation $= 6.3 - 3.9$
$= 2.4$ mW.

Example 4

Calculate the time taken for electrons to diffuse through a 5 μm thick layer of *p* type silicon. Given minority carrier diffusion coefficient $D_c = 3.4 \times 10^{-3}$ m²s⁻¹.

Solution

$$\tau_{\text{Diff}} = \frac{d^2}{2D_c}$$

$$= \frac{(5 \times 10^{-6})^2}{2 \times 3.4 \times 10^{-3}} = 3.7 \times 10^{-9} \text{ s}$$

Example 5

Estimate frequency bandwidth limitations arising from diode capacitance. Assume the typical values: $A = (1 \text{ mm}^2)$, $\varepsilon_r = 11.7$, $N_d = 10^{21}$ m⁻³ and $V = 10$ volt.

Solution

We have:

$$C_J = \frac{A}{2}(2e\varepsilon_r\varepsilon_o N_d)^{1/2} V^{-1/2}$$

$$= \frac{10^{-6}}{2}[2(1.6 \times 10^{-19})(11.7)(8.85 \times 10^{-12})10^{21}]^{1/2}(10)^{-1/2} \approx 30 \text{ pF}$$

Example 6

The reflection coefficients for a semiconductor laser are equal and have the value 0.5. The length of cavity is 1 μm. Assume that there are no additional loss terms. Calculate the gain of the laser. [B.Tech]

Solution

No internal loss means $d_i = 0$; we have:

$$g = \frac{1}{2L}\ln\left(\frac{1}{r_1 r_2}\right)$$

$$= \frac{1}{2 \times 1} \ln\left(\frac{1}{0.5 \times 0.5}\right)$$

$$= 6.93 \times 10^3 / \text{cm}.$$

Example 7

A cadmium sulphide photodetector is irradiated over a receiving area of 4×10^{-6} m² by a light of wavelength 0.4×10^{-6} m and intensity 200 Wm⁻². Assuming that each quantum generates an electron-hole pair, calculate the number of pairs generated per second. [B.Tech]

Solution

Here: $\lambda = 0.4 \times 10^{-6}$ m,

Area of the crystal = 4×10^{-6} m²,

Intensity of light = 200 Wm⁻²

Now, the number of photons $= \dfrac{\text{Intensity}}{h\nu} \times \text{Area}$

$$= \frac{200 \times 4 \times 10^{-6} \lambda}{hc}$$

$$= \frac{200 \times 4 \times 10^{-6} \times 0.4 \times 10^{-6}}{6.626 \times 10^{-34} \times 3 \times 10^8}$$

$$= 1.609 \times 10^{15}$$

As each quantum of energy will generate an electron-hole pair and hence the number of pairs generated per second = 1.6×10^{15}.

Example 8

A photodiode has a quantum efficiency of 70% when photons with energy 2.2×10^{-19} J are incident on it. Calculate: (a) operating wavelength of photodiode, (b) the incident power required to obtain a photocurrent of 0.2 mA when the diode is operating in the above condition. [B.Tech]

Solution

(a) $E = h\nu = \dfrac{hc}{\lambda}$ or $\lambda = \dfrac{hc}{E}$

$$\therefore \lambda = \frac{6.62 \times 10^{-34} \times 3 \times 10^8}{2.2 \times 10^{-19}}$$

$$= 0.9 \text{ μm}$$

$\eta = 70\%$ of photoenergy = 0.70

$$E = 2.2 \times 10^{-19} \text{ J}$$
$$I_p = 0.2 \text{ mA}$$
$$\text{and } v = \frac{c}{\lambda} = \frac{3 \times 10^8}{0.9 \times 10^{-6}}$$
$$= 3.33 \times 10^{14} \text{ Hz}$$

(b) $R = \dfrac{\eta e}{h v} = \dfrac{\eta e \lambda}{hc} = \dfrac{\eta e}{hv} \quad (\because c = v\lambda)$

$$= \frac{0.70 \times 1.6 \times 10^{-19}}{6.62 \times 10^{-34} \times 3.33 \times 10^{14}}$$
$$= 0.51 \text{ AW}^{-1}$$

$$R_o = \frac{I_p}{P_o} \text{ or } R_o = \frac{I_p}{P} = \frac{2.0 \times 10^{-6}}{0.51} = 3.92 \text{ }\mu\text{W}$$

Example 9

3×10^{11} photons each with a wavelength of 0.85 are incident on a photodiode. On an average 1.5×10^{11} electrons are collected at the terminals of the device. Calculate: (i) Quantum efficiency (η), and (ii) responsivity (R) of the photodiode at the wavelength of 0.85 mm.

[B.Tech]

Solution

(i) $\eta = \dfrac{r_e}{r_p} = \dfrac{\text{number of hole - pairs generated}}{\text{number of incident photon}}$

$$= \frac{1.5 \times 10^{11}}{3 \times 10^{11}} = 0.5$$

(ii) Responsivity, $R = \eta \dfrac{e\lambda}{hc}$

$$= \frac{0.5 \times 1.6 \times 10^{-19} \times 0.85 \times 10^{-6}}{6.62 \times 10^{-34} \times 3 \times 10^8}$$
$$= 0.34 \text{ AW}^{-1}$$

Example 10

A germanium p-i-n photodiode with an active area dimensions of 100×50 µm has a quantum efficiency (η) of 60% when operating at a wavelength 1.2 µm. The measured dark current is 10 nA. Assuming that the dark current is the main source of noise, calculate the noise equivalent power and specific directivity of the device.

[M.Sc. (Elec.)]

Solution

We have, noise equivalent power:

$$\text{NEP} = \frac{hc\,(2eI_D)^{1/2}}{\eta e^{\lambda}} \qquad (i)$$

Here

$$I_D = 10 \text{ nA}$$
$$\eta = 0.6$$
$$\lambda = 1.2 \times 10^{-6}$$

Substituting these values in Eq. (i), we obtain

$$\text{NEP} = \frac{6.62 \times 10^{-34} \times 3 \times 10^8 \,(2 \times 1.6 \times 10^8) \times (2 \times 1.6 \times 10^{-19} \times 10 \times 10^{-9})^{1/2}}{0.6 \times 1.0 \times 10^{-19} \times 1.2 \times 10^{-6}}$$

$$= 9.8 \times 10^{-14} \text{ W}$$

$$D^* = \frac{A^{1/2}}{\text{NEP}} = \frac{(100 \times 10^{-6} \times 50 \times 10^{-6})^{1/2}}{9.80 \times 10^{-14}}$$

$$= 7 \times 10^8 \text{ m} \cdot \text{Hz}^{1/2} \cdot \text{W}^{-1}$$

Example 11

GaAs$_{0.88}$Sb$_{0.12}$ is a direct band gap semiconductor and has a band gap energy of 1.115 eV. Determine the critical wavelength above which an intrinsic photodetector from this material can be operated. [B.Tech.]

Solution

The critical wavelength is given by:

$$\frac{hc}{\lambda_c} = E_g$$

or

$$\lambda_c = \frac{hc}{E_g}$$

$$h = 6.62 \times 10^{-34}$$
$$c = 3 \times 10^8 \text{ m/s}$$

$$E_g = 1.15 \times 1.6 \times 10^{-19} \text{ V}$$

or
$$\lambda_c = \frac{6.62 \times 10^{-34} \times 3 \times 10^8}{1.15 \times 1.6 \times 10^{-19}}$$

$$= 1.06 \text{ mm} = 1060 \text{ nm}$$

Example 12

Photons of wavelength 0.90 µm are incident on a *pn* photodiode at a rate of 5×10^{10} s^{-1} and, on an average, the electrons are collected at the terminals of the diode at the rate of 2×10^{10} s^{-1}. Calculate the quantum efficiency (η) and the responsibility (R) of the diode at this wavelength.

[B.Tech.]

Solution

$$\eta = \frac{r_e}{r_p} = \frac{2 \times 10^{10}}{5 \times 10^{10}} = 0.40$$

Here:

$\lambda = 0.90$ µm $= 0.90 \times 10^{-6}$ m
Rate of electron (r_e) $= 2 \times 10^{10}$ s^{-1}
Rate of photons (r_p) $= 5 \times 10^{10}$ s^{-1}

$$R = \frac{\eta e \lambda}{hc} = \frac{0.40 \times 1.6 \times 10^{-19} \times 0.90 \times 10^{-6}}{6.626 \times 10^{-34} \times 3 \times 10^8} = 0.29 \text{ AW}^{-1}$$

Example 13

A *pn* photodiode has a quantum efficiency of 70% for photon of energy 1.52×10^{-19} J. Calculate (i) the wavelength at which the diode is operating and (ii) the optical power required to achieve a photocurrent of 3 µA when the wavelength of the incident photons is that calculated in part (i).

[B.Tech.]

Solution

(i) Photon energy, $E = h\nu = \dfrac{hc}{\lambda}$

or
$$\lambda = \frac{hc}{E} = \frac{6.626 \times 10^{-34} \times 3 \times 10^8}{1.52 \times 10^{-19}}$$

$$= 1.30 \times 10^{-6} \text{ m} = 1.30 \text{ µm}$$

(ii) $\quad R = \dfrac{\eta e}{h\nu} = \dfrac{0.70 \times 1.6 \times 10^{-19}}{1.52 \times 10^{-19}} = 0.736 \text{ AW}^{-1}$

$\therefore \quad R = \dfrac{I_p}{P_{in}} \quad \text{or} \quad P_{in} = \dfrac{I_p}{R} = \dfrac{3 \times 10^{-6}}{0.736} = 0.736 \text{ AW}^{-1}$

or $\quad P_{in} = 4.07 \times 10^{-6}$ W = 4.07 µW.

Example 14

A silicon RAPD (reach through avalanche photodiode), operating at a wavelength of 0.80 µm, exhibits a quantum efficiency of 90%, a multiplication factor of 800, and a dark current of 2 nA. Calculate the rate at which photons should be incident on the device so that the output current (after avalanche gain) is greater than the dark current.

[B.Tech.}

Solution

We have, $I = I_p M = P_{in}\, RM = P_{in} \left(\dfrac{\eta e \lambda}{hc}\right) M = \left(\dfrac{P_{in}}{hc/\lambda}\right) \eta e M$

$= [(\text{photon rate})\, e]\, \eta M = \dfrac{2 \times 10^{-9}}{1.6 \times 10^{-19} \times 0.90 \times 800}$

$= 1.736 \times 10^7 \text{ s}^{-1}$

Thus, for $I > 2$ nA, photon rate $\approx 1.74 \times 10^7 \text{ s}^{-1}$.

SUGGESTED READINGS

1. Kakani SL and Bhandari KC, 'Electronics: Theory and Applications', New Age Int. Publishers, New Delhi (5th ed., 2017).
2. Kakani SL and Bhandari KC, 'Electronic Devices and Circuits', Viva Books, New Delhi-2 (2nd ed., 2017).
3. Kakani SL and Bhandari KC, 'Objective Electronics', Himalaya Publishing House, Mumbai-4 (2015).
4. Adesida I and Coleman J in 'The Handbook of Photonics', Gupta MC and Ballato J (Eds.), CRC Press (2007).
5. Bhattacharya P, 'Semiconductor Optoelectronic Devices', Prentice Hall (1994).
6. Saleh BEA and Teich MC, 'Fundamentals of Photonics', Wiley (1991).
7. Wilson J and Hawkes, JFB, 'Optoelectronics: An Introduction, Prentice Hall (2nd ed., 1989).

GLIMPSES

1. Optoelectronics is a branch of electronics which deals with light sensitive devices. When light energy is incident upon these devices, electric current is generated. These devices are called photonic or optoelectronic devices.
2. There are two types of optoelectronic devices: (i) light (emitted) source type, e.g., LEDs, and laser diodes, (ii) light detector type, e.g., photodiodes, phototransistors, photoresistors, photovoltaic cell and LCDs.
3. The phenomenon of emission of electrons from a metal surface under the effect of light radiations is termed a photoelectric effect. The minimum frequency of light that can cause photoelectric effect is called threshold frequency.
4. A photoconductive cell is a two-terminal semiconductor device, whose resistance decreases with the intensity of light falling on it, and as a result free electrons are available. It is also called light (photo) *dependent resistance* (LDR).
5. An LED is a specially fabricated semiconductor *pn* junction diode that emits monochromatic light when forward-biased. LED has voltage drop-rating of about 2 V and current rating of about 20 mA.
6. The internal quantum efficiency of an LED is the fraction of diode current that will produce luminescence.
7. The external quantum efficiency of an LED is defined as the ratio of the number of photons emitted from the device per incident injected electron.
8. The brightness of light emitted by an LED depends upon current flowing through it.
9. LEDs have replaced small incandescent light bulbs as display devices. The long life expectancy, ruggedness, and much lower power consumption have contributed to the choice of LEDs over incandescent light bulbs. Some of the applications of LEDs are seven-segment display, polarity tester, and continuity tester.
10. Laser is an acronym of light amplification by stimulated emission of radiation.
11. Laser diodes are light-emitting devices capable of emitting laser beams. One of the commonly used laser diodes is the injection laser diode (ILD).
12. Laser light sources are monochromatic. All the light waves emitted by a laser diode are in phase with each other. Due to resonance, it emits a narrow beam of coherent, intense, focussed and monochromatic light.

13. A photodiode is a *pn* junction which generates significant current when light is incident upon it.
14. A phototransistor is like a photodiode except that photocurrent produced in a transistor is more than that produced in the photodiode, because it is multiplied in transistor by a factor β.
15. The sensitivity of photodiodes and phototransistors is expressed as the ratio of photocurrent output (in amperes) to radiant energy (in watts).
16. A photoresistor is a semiconductor light-detector device without any *pn* junction diode.
17. A solar cell (also called photovoltaic cell) converts light energy into electric energy.
18. A photomultiplier tube uses secondary emission to give multiplied output.
19. An optocoupler or optoisolator is an arrangement of coupling (or isolating) a light emitting source from a light dependent device (light detector) through a transparent insulating medium.
20. An LCD consists of a thin film (about 15 µm) of a conductive crystal in liquid form. Normally the crystal is transparent. However, when energized by an AC source in the presence of light, it changes to provide black-tone or gray readout.

REVIEW QUESTIONS

1. What is optical light spectrum?
   (a) Frequency and wavelength
   (b) Visible versus invisible
   (c) Infrared and ultraviolet
2. The infrared band is given as 30 to 400 THz. What is its value in terms of wavelength?
3. Mention the difference between coherent and incoherent light.
4. Discuss the operation of LED.
   (a) How does it emit light? (b) When does it emit light.
5. A certain piece of equipment shows a green light when it is working properly and red light when it is not working properly. The light appears to come from the same device. How might this work?
6. A particular fiber-optic application will use a very long length of fiber-optic cable. Would this application use an LED or a laser diode? Why?
7. Discuss the two modes of operation of photodiode.
8. A certain control circuit is being used to control the operation of a high-power motor starter. The motor starter generates enormous amounts of noise that might interfere with the control system.

Discuss how the circuit might be designed to isolate noise from the control.

9. What is an LED? Explain the modulation of an LED?
10. You wish to operate a door opener with a light source. Draw a block diagram to show how this might be done.
11. What are optoelectronic devices? Illustrate your answer with suitable examples.
12. What is photoelectric effect? Write names of its types and explain each.
13. Describe the working of a photoconductive cell and mention its important applications.
14. What is an LCD? Write its applications, make a comparison of LCD and LED.
15. Explain the principle of solar cell and describe its working.
16. Describe the working of an optocoupler. Mention its uses.
17. Explain the working of a photomultiplier tube and mention its uses.
18. What is the minimum frequency to cause photoconduction?
19. Mention where the liquid crystal diode is practically used?
20. Show that when the simple bias circuit (Fig. 7.65) is used with a photoconductive detector to detect small signal levels, then maximum voltage output signals across R_L are obtained when R_L is equal to detector resistance R_D.

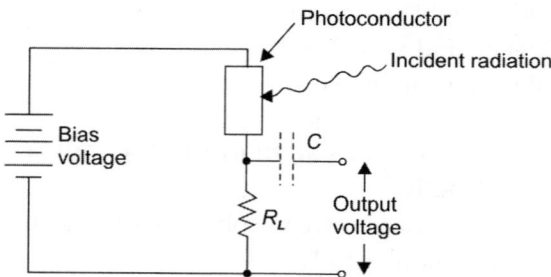

Fig. 7.65. Photoconductor bias circuit. The photoconductor is placed in a series circuit comprising a voltage source, a load resistor R_L, and the photoconductor. Changes in the resistance of photoconductor cause changes in voltage appearing across R_L. If only the AC component of this voltage is required, then a blocking capacitor C may be placed as shown

21. Show that gain G of a photoconductive detector can be written as $Ge = \tau_{c_c}/\tau_d$ where t_c is minority carrier lifetime, and τ_d is given by:

$$\frac{1}{\tau_d} = \frac{1}{\tau_n} + \frac{1}{\tau_p}$$

τ_n and τ_p being the times taken for electrons and holes to drift across the photodetector.

22. Show the responsivity of a photodiode detector can be written as:

$$R(f) = \frac{R(o)}{(1 + 4\pi^2 f^2 C^2 R_L^2)}$$

where C is diode capacitance and R_L load resistance.

PROBLEMS

1. A photo cell cathode is coated with materials having a work function of 3.5 eV. A photo cell is irradiated of frequency 4×10^{15} Hz. Calculate the velocity of emitted electrons.
   [**Ans.** 2.08×10^6 ms^{-1}]

2. The electrical conductivity of a semiconductor increases when electromagnetic radiation of wavelength shorter than 2480 nm is incident on it. Calculate the bandgap of the semiconductor.
   [**Ans.** $\Delta E_g = 0.5$ eV]

3. A *pn* photodiode has a quantum efficiency of 50% at $\lambda = 0.90$ μm. Calculate (a) its responsivity at this wavelength, (b) if the mean photocurrent is 10^{-6} A, then the optical power received, and (c) the corresponding photons received at this wavelength.
   [**Ans.** (a) 0.362 AW^{-1}, (b) 2.76 mW, (c) 1.25×10^{13} s^{-1}]

4. In a photocell, a copper surface was irradiated by light of wavelength 1849Å, the stopping potential was found to be 2.72 eV. Show that the maximum energy (I_{max}) of photoelectrons is 2.72 eV.

5. Find the current through LED in the circuit shown (Fig. 7.66). Assume that voltage drop across the LED is 2 V.
   [**Ans.** $I_F = 5.91$ mA]

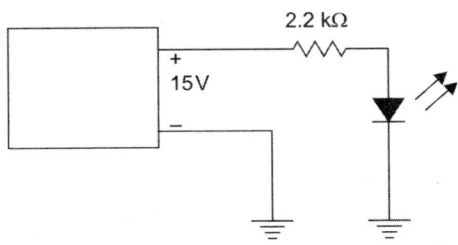

Fig. 7.66. LED circuit

6. Calculate the responsivity of an ideal photodiode at the following wavelengths: (a) 1.55 μm, (b) 1.30 μm, (c) 0.85 μm. [B.Tech.]
   [**Ans.** (a) 1.248 AW^{-1}, (b) 1.046 AW^{-1}, (c) 0.684 AW^{-1}]

7. The quantum efficiency of an APD is 50% at 1.3 μm. It produces an output current of 8 μA, after avalanche gain when illuminated with

optical power of 0.4 µW at this wavelength. Show that the multiplication factor of the diode is 38.
8. The responsivity of a typical photodiode is 0.40 AW^{-1} for a He–Ne laser source (λ = 632.8 nm). If the active area of the photodiode is 2 mm^2, calculate the output photocurrent when the incident flux is 100 λW/mm^2? **[Ans. 80 µA]**

SHORT ANSWER QUESTIONS

1. Do all optoelectronic devices produce light, either visible or invisible?

Ans. No.

2. Is an LED a specially fabricated *pn* junction diode that works in forward-bias?

Ans. Yes.

3. Do laser diodes find application in compact disk players and laser printers?

Ans. Yes.

4. Can a photovoltaic cell be used to measure the intensity of illumination?

Ans. Yes.

5. What is a photovoltaic cell?

Ans. This is a device that converts light directly into electric energy.

6. What is a autocoupler?

Ans. This is an assembly in a single package that contains a light emitting diode and a photosensitive device.

7. What do you understand by photodiode sensitivity?

Ans. It is the ratio of photocurrent output (in amperes) to radiant energy (in watts) and expressed in amps/watts.

8. What is a laser diode?

Ans. A laser diode is a heavily doped *pn* junction diode.

9. What is photoemissive or photoelectric effect?

Ans. When radiation with a wavelength less than a critical value is incident upon a particular metal (sodium, potassium, etc.) surface, electrons are found to be emitted, this is called photoemissive or photoelectric effect.

10. What is a solar cell?

Ans. The solar cell is basically a *pn* junction detector operated under conditions such that it can deliver power into an external load. It is used to convert light energy to electrical energy and thereby serves as a DC voltage source.

11. **What is a phototransistor?**
Ans. A phototransistor is a light detecting device, which is also called a photo-sensor.
12. **Why there has been interest in the development of white LEDs for general illumination?**
Ans. LEDs are three times as efficient as incandescent lamps and can last ten times longer. A white LED requires LEDs of red, green and blue colours.
13. **What is the advantage of using heterojunction LEDs?**
Ans. Since a photon emitted to a radiative recombination process of direct bandgap material in one part of the device and reabsorbed prior to it reaches the surface, one obvious choice is to put the *pn* junction close to the surface to avoid reabsorption. However, the presence of surface states, producing non-radiative recombination channels, can result in a reduction in radiative efficiency of the device. To solve this problem, heterojunction is used instead of avoiding reabsorption. A commonly used heterojunction is that of GaAs and AlGaAs. A heterojunction LED is designed in such a way that recombination occurs on the narrow gap semiconductor. Light is subsequently emitted through the AlGaAs window into the external environment. The bandgap of AlGaAs is greater than that of GaAs and therefore, most of the light will transmit through the AlGaAs without being absorbed.
14. **Explain the importance of infrared LEDs.**
Ans. Under infrared LEDs fall GaAs LEDs which can emit light of wavelength near 0.9 μm and many III–IV compounds such as quaternary $Ga_x In_{1-x} As_y P_{1-y}$ LEDs, which emit light from 1.1 to 1.6 μm. Infrared LEDs have an important application in opto-isolators where an input or control signal is decoupled from the output. Infrared LEDs also find application for transmission of optical signal through an optical fiber.

MULTIPLE CHOICE QUESTIONS

1. The densities of electrons and holes are same in:
   (a) a *p-n* junction in equilibrium
   (b) a forward biased *p-n* junction
   (c) an extrinsic semiconductor
   (d) an intrinsic semiconductor
2. Which of the following material is not suitable for making an LED?
   (a) GaAlAs (b) Silicon
   (c) GaAs   (d) InGaAsP

3. To make an efficient LED, the material should be:
   (a) a metal
   (b) an insulator
   (c) an indirect band gap type semiconductor
   (d) a direct band gap type semiconductor
4. An LED with an external quantum efficiency of 0.012 is coupled to an optical fiber having NA = 0.15 (with air between them). The overall source-fiber coupling efficiency is:
   (a) 3.2×10^{-4}
   (b) 1.8×10^{-4}
   (c) 1.1×10^{-4}
   (d) 2.7×10^{-4}
5. In a p-n homojunction, the majority concentration are almost equal to the dopant concentrations at:
   (a) high temperature
   (b) absolute zero temperature
   (c) normal temperature
   (d) all temperatures
6. In an LED, which of the following factors affects severely the efficiency of the diode and also cannot be eliminated in principle?
   (a) Back emission
   (b) Total internal reflection
   (c) Absorption
   (d) Fresnel reflection
7. The internal quantum efficiency (η_{int}) of an LED is given by:
   (a) $\eta_{int} = \dfrac{1}{1 + \tau_{rr}\tau_{nr}}$
   (b) $\eta_{int} = \dfrac{1}{\tau_{rr} + \tau_{nr}}$
   (c) $\eta_{int} = \dfrac{1}{1 + \dfrac{\tau_{rr}}{\tau_{nr}}}$
   (d) $\eta_{int} = \dfrac{\tau_{nr}}{\tau_{rr}}$
8. For LEDs, the overall source-fiber coupling efficiency (η_T) is given by:
   (a) $\eta_T = \eta_{ext} \dfrac{(NA)^2}{n_a^2}$
   (b) $\eta_T = \dfrac{n_a}{n_{ext}}(NA)^2$
   (c) $\eta_T = \dfrac{n_{ext}}{n_a^2}(NA)$
   (d) $\eta_T = \dfrac{n_{ext}}{n_a^2}(NA)^{1/2}$
9. Which of the following pairs are suitable for making a heterojunction?
   (a) GaAs and GaAlAs
   (b) Si and GaAs
   (c) Si and Ge
   (d) GaAs and AlAs
10. A photodetector is a device that converts:
    (a) heat energy to optical energy
    (b) sound energy to electrical energy
    (c) optical energy to electrical energy
    (d) mechanical energy to electrical energy

11. The principal mechanism of converting optical energy to electrical energy in a photodetector is:
    (a) photoconductivity
    (b) photovoltaics
    (c) polarization
    (d) photoelasticity
12. A *p-n* junction photodiode is:
    (a) forward biased
    (b) a very fast photodetector
    (c) embedded in an opaque capsule
    (d) dependent on thermally generated minority carriers
13. A *pn* junction that radiates energy as light instead of heat is called as:
    (a) Zener diode
    (b) LED
    (c) photocell
    (d) photodiode

ANSWERS

1. (d)   2. (b)   3. (d)   4. (d)   5. (c)   6. (b)   7. (c)
8. (a)   9. (a)   10. (c)   11. (a)   12. (b)   13. (b)

8

Photovoltaic Devices

8.1 INTRODUCTION

Photovoltaic (PV) *devices*, i.e. PV *cells* directly convert the incident solar radiation energy into electrical energy. Incident photons are absorbed to photographic charge carriers that pass through an external load to do electrical work. Photovoltaic device applications cover a wider range from smaller consumer electronics, such as a *solar cell*, calculator using less than a few milliwatts to photovoltaic power generation by a central power plant (generating a few megawatts).

World consumption of electric energy in 2009 was around 12–13 TW. This energy was produced by various methods, e.g., nuclear power reactors, thermal power plants, hydroelectric power plants, solar energy, etc. We look briefly at two of them: nuclear power and solar energy.

Assuming that a single nuclear plant produces 1 GW of power, creation of 10 TW of power would require 10,000 nuclear power plants. This huge number will certainly create environmental and social problems. Further, the uranium as a fuel needed for these plants could be diminished in less than two decades.

As regarded the solar energy, the sun supplies more solar energy on earth in one hour than we use globally in one year. However, by increasing efficiency of solar cells, for example by using *multijunction devices* and increasing concentration of solar energy, the area of semiconductor solar cells can be significantly reduced.

Those rough estimates indicate that solar energy can be a practical alternative as a source of electric energy, and also that increasing the efficiency of solar cells is important. One should, however, remember that the incidence of solar energy is not uniform.

Solar irradiation spectrum (spectrum of solar energy) is shown in Fig. 8.1. In the spectrum, it is shown that spectral intensity is of the intensity of solar energy per unit wagvelength. Above the Earth's atmosphere, it is denoted as AM0 (air-mass zero). The integrated (over

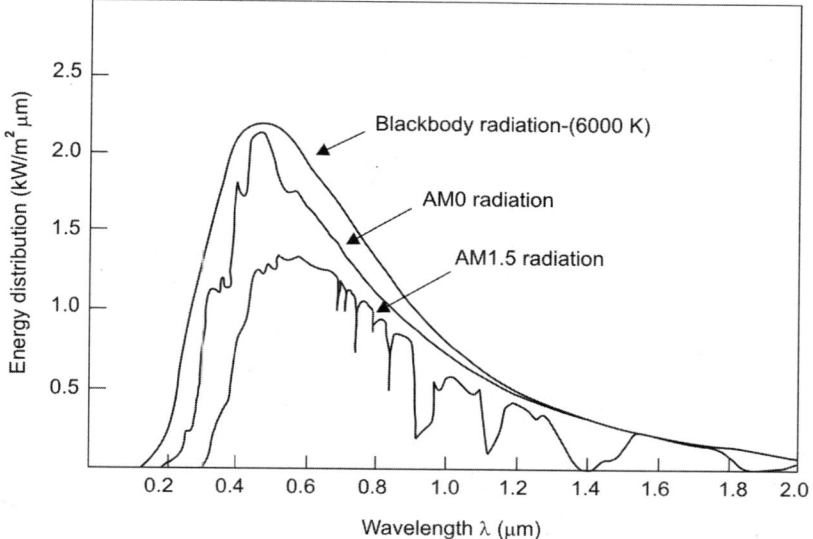

Fig. 8.1. Solar irradiation spectrum

the whole spectrum) intensity above the Earth's atmosphere is constant and its approximate value is 1.353 kW·m². It gives the total power flow through a unit area perpendicular to the direction of the Sun.

The Earth's atmosphere significantly absorbs radiation from the Sun. In Fig. 8.1, the radiation at the Earth's surface is labelled as AM 1.5. In the figure, several absorption bands and lines due to various effects are shown. For comparison, the spectrum of black body radiation at the temperature of 6000 K is also shown. One can notice that the solar spectrum above the Earth's atmosphere (AM0) resembles blackbody radiation.

The air mass coefficient defines the direct optical path length through the Earth's atmosphere, expressed as a ratio relative to the path length vertically upwards, i.e. at the zenith.

Generally, air mass (AM), m is defined as the ratio of the actual radiation path h to the shortest path h_0, $m = h/h_0 \cos\theta$ (Fig. 8.2).

Changing the light intensity incident on a polar cell changes all solar cell parameters, including the short-circuit current, the open-circuit voltage, the efficiency and the impact of series and shunt resistances.

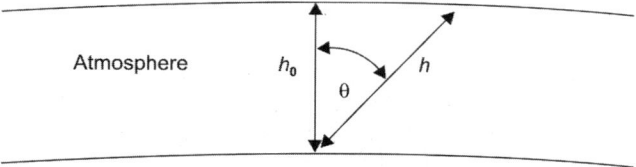

Fig. 8.2. Schematic of air-mass definition

The light intensity on a solar cell is measured in the number of 'suns', where 1 sun corresponds to standard illumination at AM 1.5, or 1 kW·m^2. For example, a system with 10 kW·m^2 incident on the solar cell would be operating at 10 suns, or at 10X.

Generations of Photovoltaic Devices

In general, photovoltaic cells can be classified into three "generations". The *first generation* technology of cells are based on *silicon wafers* that are thick enough to be self-supporting, typically about 300 μm thick. Production of these single-crystal and multicrystalline wafer technologies comprise nearly 91% of the current PV market. These photovoltaic cells typically have an efficiency of 12–17%. The primary limitation of first-generation technologies is their cost, which is dominant by the cost of the high-purity silicon wafer.

The *second generation* represents *thin-film* technologies that aim for lower cost at the expense of efficiency by using thin layers of semiconducting material. Because the active layer is only a few microns thick, a rigid substrate or transparent superstrate is required for mechanical support. The thin films are deposited rapidly over large areas of the sub/superstate and are subsequently patterned to form multiple cells. Therefore, no soldering is required to connect the individual cells and the basic production unit is the module rather than the cell. The four most important thin-film technologies are *amorphous silicon* (a-SI), *polycrystalline silicon, cadmium telluride* (CdTe), and *copper indium diselenide* (CIS), with current production efficiencies ranging from 4% to 10%. While thin-film cells have been expected to overtake the first-generation cells' dominance for decades, this has not yet occurred for a variety of reasons. More recently, organic thin-film devices have emerged as a possible path for achieving truly low-cost PV devices, albeit with a lower conversion efficiency. These devices, which only exist in laboratories have low efficiencies of 0.1–0.5%.

Third-generation approaches are eventually intended to achieve both high-efficiency and low cost. One example of a third-generation photovoltaic cell that is already commercially produced is the *tandem cell*. Two types of tandem cells exist. The first, a *thin-film technology* based on layers of a-Si and microcrystalline silicon (μc-Si) and amorphous germanium (a-Ge) is a low-cost and low-efficiency device. The second, made from elements of groups III and V of the periodic table (known simply as III–V cells), achieves ultra-high efficiency, but at a high cost premium.

In general, several thin-film technologies are meeting the target for second-generation devices, whereas the first-generation PV modules still lie in the upper price range defined by region I.

The final device presented here is the *dye-sensitized solar cell* (DSSC). This solar cell does not rely on the photovoltaic effect to convert sunlight into electricity, but rather the same photoelectrochemical effect as observed by Becquerel in 1839. Consequently, this device is not included in the generation of PV technology outlined above.

8.2 PRINCIPLE OF PHOTOVOLTAIC CELL

The simplest solar or photovoltaic (PV) cell is a single *p-i-n* junction operating under forward bias. It is formed by using, e.g. properly doped silicon (Fig. 8.3). It can produce potential difference, thus acting as a battery. It contains a very narrow and heavily doped *n*-region. It is covered with thin antireflecting coating to reduce sun reflection which penetrates *p-n* junction from *n*-side (Fig. 8.4) for more details.

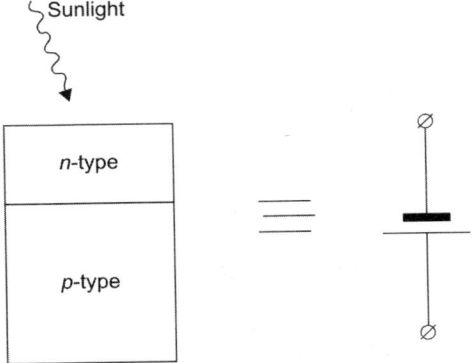

Fig. 8.3. A *p-n* junction operating as a solar cell

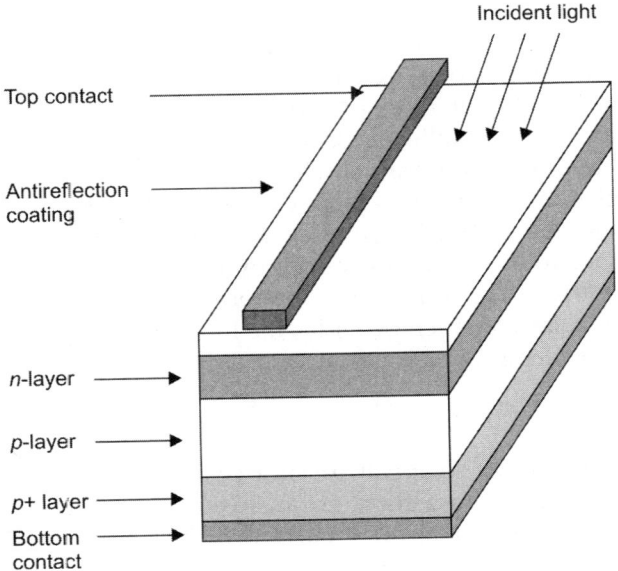

Fig. 8.4. Solar cell perspective

Short wavelengths penetrate only the *n*-region, longer wavelengths penetrate depletion region of width W, and finally the longest wavelengths reach the *p*-region, (Fig. 8.5).

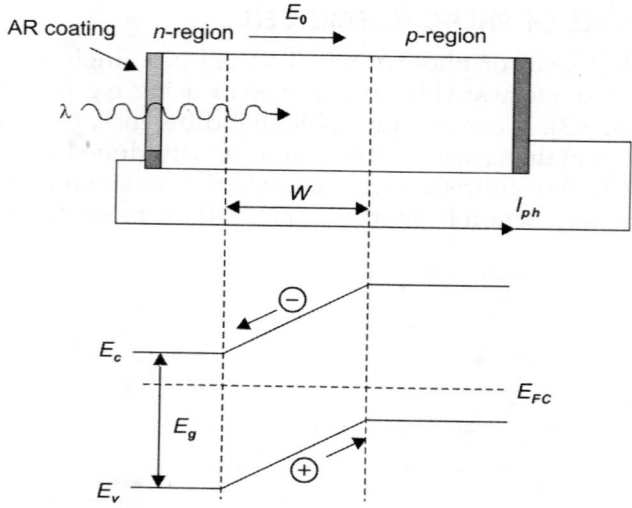

Fig. 8.5. The principle of photocurrent generation in a solar cell in short circuit

Photons of frequency ν having energies $h\nu \geq E_g$, where E_g is the bandgap energy, generate electron-hole (*e-h*) pairs. Due to a built-in electric field E_0 in the depletion region, electrons move towards metal contact in the *n*-region whereas holes travel in opposite direction towards metallic contact in the *p*-region. Since the *n*-region is very narrow, most of the photons are absorbed in the depletion region. Each generated electron increases charge in the *n*-region by $-e$; similarly each hole makes the *p*-region more positive by $+e$. Thus, potential difference is created between metallic electrodes.

A semiconductor can only efficiently convert photons into current with energies equal to the bandgap. Photons with energies smaller than the bandgap are not absorbed, and photons with higher energies reduce their energies to the bandgap energy by thermalization of the photo-generated carriers; a process which involves losses.

The basic relation which provides the link between photons' energy and their wavelength is:

$$\lambda[\mu m] = \frac{1.24}{E[eV]} \tag{8.1}$$

If both terminals of solar cell are shorted, the excess electrons on the *n*-side will start flowing through an external wire contributing to electrical current known as photocurrent, I_{ph} (Fig. 8.6). The short circuit conventional current I_{sc} flows opposite to photocurrent. If I_{light} (I_{ph}) is the light intensity, one can write:

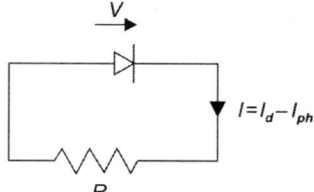

Fig. 8.6. Solar cell driving an external load resistance R

$$I_{sc} = -I_{ph} = -K \cdot I_{light}$$

where K is some (device dependent) constant.

When the illuminated solar cell is loaded with resistance R, the conventional current $I = I_d - I_{ph}$ (Fig. 8.6), where I_d is a forward diode current:

$$I_d = I_0 \left[\exp \frac{eV}{\eta k_B T} - 1 \right] \quad (8.2)$$

and where I_0 is a constant, V is the voltage across diode and η is known as diode fidelity factor. It is equal to 1 for diffusion controlled and 2 for space charge layer recombination controlled characteristics.

The I–V characteristic of solar cell is shown in Fig. 8.7. It is obtained from a dark diode by shifting its I–V curve downward by I_{ph}. The analytical expression for conventional current is:

$$I = -I_{ph} + I_0 \left[\exp \left(\frac{eV}{\eta k_B T} \right) - 1 \right] \quad (8.3)$$

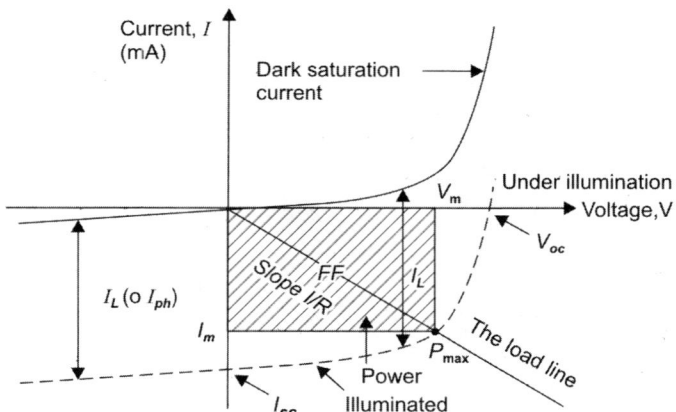

Fig. 8.7. Typical I–V characteristic for a photovoltaic cell

With the load resistance R connected to the solar cell (Fig. 8.6), the current-voltage relation is:

$$I = -\frac{V}{R} \qquad (8.4)$$

Operating point of solar cell with load resistance R is obtained by solving Eqs (8.3) and (8.4). The graphical solution is illustrated in Fig. 8.8, where we have plotted Eqs (8.3) and (8.4a). Crossing of both lines determines operating point (V_1, I_1) of a solar cell.

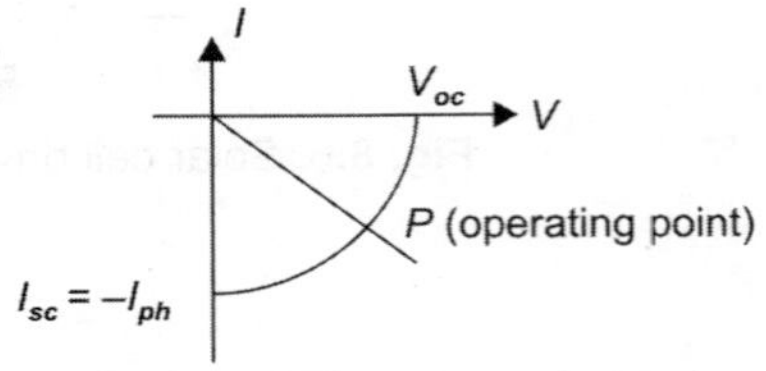

Fig. 8.8. Finding an operating point of a solar cell. V_{oc} is an open-circuit voltage

The point at which a characteristic intersects the vertical current axis is known as a short circuit condition and the corresponding current as short-circuit current I_{sc}. Similarly, the point at which I–V characteristic intersects the horizontal voltage axis is known as the open circuit condition. It defines the open-circuit voltage V_{oc}, which is the maximum voltage which can be drawn from the solar cell. It corresponds to zero current.

Power delivered to the load R is $P_{\text{load}} = I_1 V_1$, which is the area of the rectangle (Fig. 8.7). The goal of the solar cell design is to maximize that power.

The main processes of solar-cell operation are shown in Fig. 8.9. The sun's photons strike the front surface of the solar cell. High-energy (ultraviolet or blue) photons are absorbed strongly near the front surface of the device, but as the photon energy decreases towards the infrared, these photons are more weakly absorbed and penetrate deeper into the device. The absorption of a photon results in an electron in the valence band of the semiconductor being excited up to the conduction

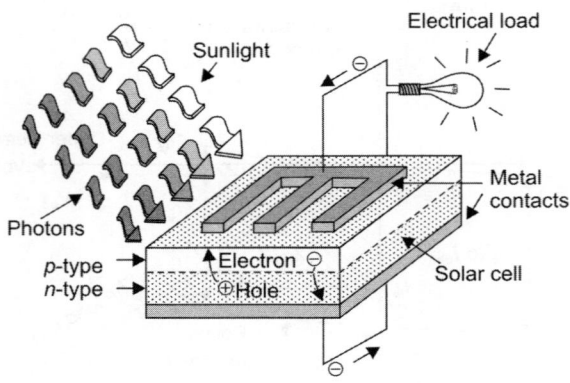

Fig. 8.9. Operation of a *p-n* junction solar cell

band, leaving behind a positively charged electron vacancy or "hole" in some materials, known as *direct bandgap semiconductors*, the absorption involves just the photon and the created carrier pair; however, in indirect bandgap semiconductors, the process also requires interaction with one or more photons. Indirect bandgap semiconductors, the process also requires interaction with one or more photons. Indirect bandgap semiconductors, such as silicon (Si), exhibit weaker absorption due to the required presence of this third "particle" for carrier generation, while gallium arsenide (GaAs) is a good example of a direct bandgap semiconductor.

Once created, the free electrons and holes are mobile and they will move due to two mechanisms; electric field (drift), or diffusion, to regions of lower carrier concentration. The flow of minority carriers (electrons in *p*-type material and holes in an *n*-type material) principally determined the performance of the majority of photovoltaic cells. The minority carriers from electron-hole pairs generated near the junction are swept across it by the strong electric field. As each crosses the junction to become a majority carrier, a real contribution to the cell's output current is made. Minority carriers that are generated far away from the junction can be transported to the junction by diffusion, as a lower minority carrier concentration exists near the junction due to the field action there. However, until a minority carrier is collected across the junction, there is always the possibility of it recombining with a surrounding majority carrier. Recombination can be either radiative, which is the inverse of the optical generation process that produced the electron-hole pair and energy is lost in the production and emission of a new photon, or nonradiative, where energy is dissipated as heat in the cell. Nonradiative recombination mechanisms dominate in silicon solar cells, with crystal defects, unwanted impurities, and the abrupt termination of the crystal structure at the surfaces all serving as strong nonradiative recombination sites. Metal contacts at the front and rear of the device allow physical connection of the photovoltaic cell to an electrical load and the photogenerated current to flow. The front contact is normally in the form of a fine metallic grid to minimize the fraction of the front surface area that is shaded, preventing light from passing through into the semiconductor. The rear side of the device is typically covered with the second metal contact.

The light-generated current flows in what is normally regarded as the reverse direction in diode theory, with electrons flowing out of the cell into the circuit from the *n*-type contact and back into the cell through the *p*-type contact, even though the voltage across the cell is in the forward bias direction (i.e., positive at the *p*-type contact). An ideal solar cell can be considered as being a current source connected in parallel with a rectifying diode.

The current-voltage (I–V) relationship of an ideal device can be described by Eq. (8.3).

The light-generated current can be regarded as being roughly independent of the voltage across the device. The dark current can be thought of as simultaneously flowing in the opposite direction to the light-generated current. As indicated by the solid line in Fig. 8.7, the dark current increases strongly with the voltage; at the open circuit voltage (V_{oc}), it completely cancels the light-generated current. Thus, a typical I–V characteristic for a photovoltaic cell appears as a dashed line in Fig. 8.7. The (forward) dark current is added to the (reverse) light-generated current (I_L) to make the resultant current negative for a range of positive voltages. In this quadrant, the cell can generate power. The power curve (not shown) is zero at both zero volts and V_{oc}, with maximum power (P_{max}) being produced near the "knee" of the I–V curve. The voltage (V_m) and current (I_m) at this point are used to define the fill factor (FF) of the solar cell:

$$FF = \frac{I_m V_m}{I_{sc} V_{oc}} = \frac{P_{max}}{I_{sc} V_{oc}} \qquad (8.4a)$$

where the short-circuit current (I_{sc}) is equal to the light-generated current I_L in the ideal case. Ideally, electrical loads connected to the solar cell should be chosen to keep the operating point close to the optimum knee-point during normal operation. The conversion efficiency (η) of a solar cell is defined as the ratio of generated electrical power to the incident solar radiation, P_{in}:

$$\eta = \frac{P_{max}}{P_{in}} \qquad (8.4b)$$

8.3 BASIC MODEL OF A SOLAR CELL

The equivalent circuit of solar cell which includes parasitic effects is shown in Fig. 8.10. It consists of current source which generates photoelectric current I_{ph}, ideal diode, shunt resistor and series resistor. Based on this model, the goal is to obtain current-voltage relation. From Kirchoff rule one obtains (from now on we reverse direction of flow of conventional current, which is a common practice):

$$I = I_{ph} - I_d \qquad (8.5)$$

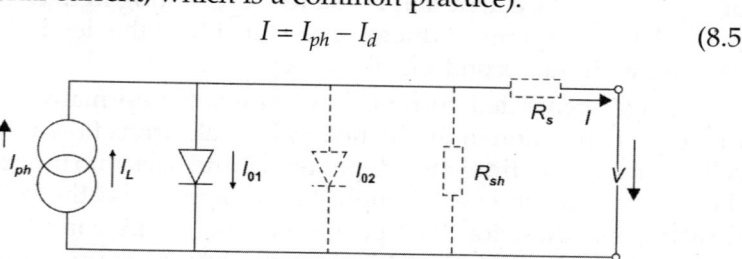

Fig. 8.10. The equivalent circuit of a solar cell, showing the configuration for both an ideal (solid lines) and nonideal device (dotted lines).

where I is the cell total current and I_d is the diode current. Combining the above relation [Eq. (8.3)], one finds:

$$I = I_{ph} - I_0 \left[\exp\left(\frac{eV_d}{\eta k_B T}\right) - 1 \right] \tag{8.6}$$

This equation describes an ideal situation without parasitic effects. In a real solar cell there exists leakage current flowing through shunt resistance R_{sh}. Potential drop across the device is represented by series resistance R_s. The inclusion of shunt resistance modifies solar current I as:

$$I = I_{ph} - I_d - \frac{V_d}{R_{sh}} \tag{8.7}$$

The impact of series resistance is included as:

$$V_d = V + I \cdot R_s \tag{8.8}$$

With the parasitic effects included, the solar current I takes the form:

$$I = I_{ph} - I_0 \left[\exp\left(\frac{eV_d}{\eta k_B T}\right) - 1 \right] - \frac{V + I \cdot R_s}{R_{sh}} \tag{8.9}$$

The photocurrent I_{ph} depends on the solar insolation and cell's temperature and it is given by:

$$I_{ph} = [I_{sc} + K_I (T_c - T_{ref})] S_{in} \tag{8.10}$$

where I_{sc} is the cell's short-circuit current at a 25°C and 1 kW m², K_I is the cell's short-circuit current temperature coefficient, T_{ref} is the cell's reference temperature and S_{in} is the solar insolation in kW·m².

The expression for I_0 is:

$$I_0 = I_{RS} \left(\frac{T_c}{T_{ref}}\right)^{3/n} \exp\left[\frac{E_{gap}}{Ak_B \left(\frac{1}{T_{ref}} - \frac{1}{T_c}\right)}\right] \tag{8.11}$$

where I_{RS} is cell's reverse saturation at a reference temperature and a solar radiation, E_{gap} is the bandgap energy of the semiconductor used in the cell and A is an ideal factor which depends on PV technology. Some typical values are: $A = 1.2$ for Si-mono, $A = 1.3$ for Si-poly and $A = 1.5$ for CdTe.

Other Models

There exist other models of solar cells which are based on a basic model. Here, we will only mention two such models: the *double exponential model* and *ideal PV cell model*.

Double exponential model is shown in Fig. 8.11. This model contains a light generated current source, two dioded D_1 and D_2 and a series and parallel resistances. The model is derived from the physics of the *p-n* junction and can describe cells constructed from polycrystalline silicon.

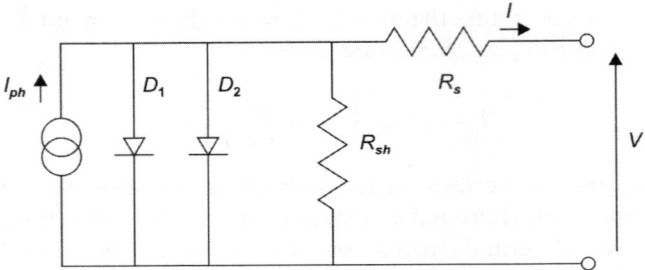

Fig. 8.11. Double exponential model of a solar cell

Simple models: The shunt resistance R_{sh} is inversely related to shunt leakage current to the ground. PV efficiency shows little sensitivity to the variation of R_{sh} and one can assume that $R_{sh} = \infty$. The resulting model is similar to the basic one, shown in Fig. 8.10, without R_{sh}. The current-voltage expression for this model is:

$$I = I_{ph} - I_0 \left[\exp\left(\frac{e(V + IR_s)}{\eta k_B T} \right) - 1 \right] \quad (8.12)$$

An ideal PV cell model assumes no series loss and no leakage to ground; i.e. $R_s = 0$ and $R_{sh} = \infty$. The equivalent circuit consists of only current source and one diode. Current is expressed as:

$$I = I_{ph} - I_0 \left[\exp\left(\frac{eV}{\eta k_B T} \right) - 1 \right] \quad (8.13)$$

We conducted analysis using typical parameters, which are summarized in Table 8.1. Current-voltage characteristics for several values of series resistance are shown in Fig. 8.12 and in Fig. 8.13, current-voltage characteristics at three different values of temperatures are shown.

Table 8.1. Parameters for equivalent circuit model of Si solar cell

Description of parameter	Symbol	Value
Illumination	I_{ph}	10 mA
eta	η	1.5
Current	I_0	3×10^{-6} mA
Series resistance	R_s	0 Ω, 20 Ω, 50 Ω
Shunt resistance	R_{sh}	415 Ω

8.4 MULTIJUNCTIONS

As determined by Shockley and Queisser in 1971, the maximum theoretical limit for a single junction solar cell is about 31%. In order to increase theoretical efficiency, one needs to increase number of *p-n* junctions in the cell. Henry in 1980, found that maximum theoretical

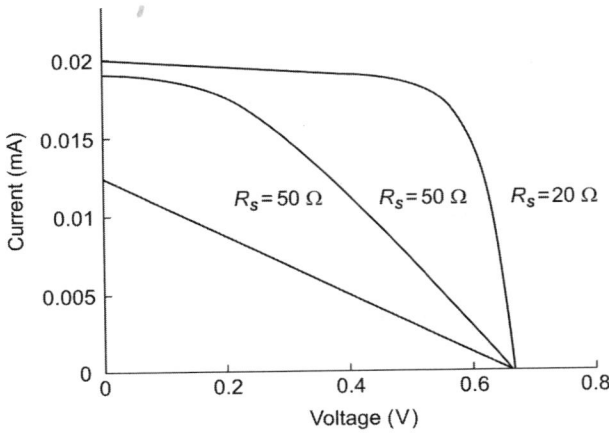

Fig. 8.12. Current-voltage characteristics of an ideal solar cell based on Si for several values of series resistance

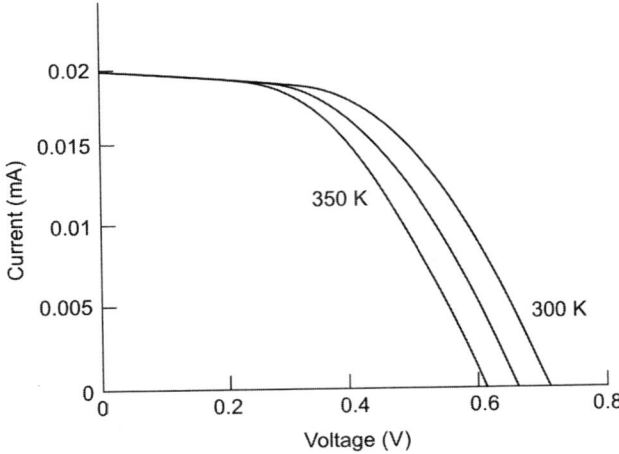

Fig. 8.13. Current-voltage characteristics of an ideal solar cell based on Si for several values of temperature

efficiency at a concentration of 1 sun is 31%. At a concentration of 1000 suns with the cell at 300 K, the maximum efficiencies are 37%, 50%, 56% and 72% for cells with 1, 2, 3, and 36 p-n junctions (with different, properly chosen energy gaps), respectively.

Increase in efficiencies can be obtained by splitting solar spectrum into several parts and using different materials for conversion in various parts. The principle is illustrated in Fig. 8.14 for a triple solar cell built from $Ga_{0.49}In_{0.51}P$ (1.9 eV), $Ga_{0.99}In_{0.01}As$ (1.4 eV) and Ge (0.7 eV). Germanium is used mainly due to its robustness and its low cost of production. Those different semiconductor materials are grown on one

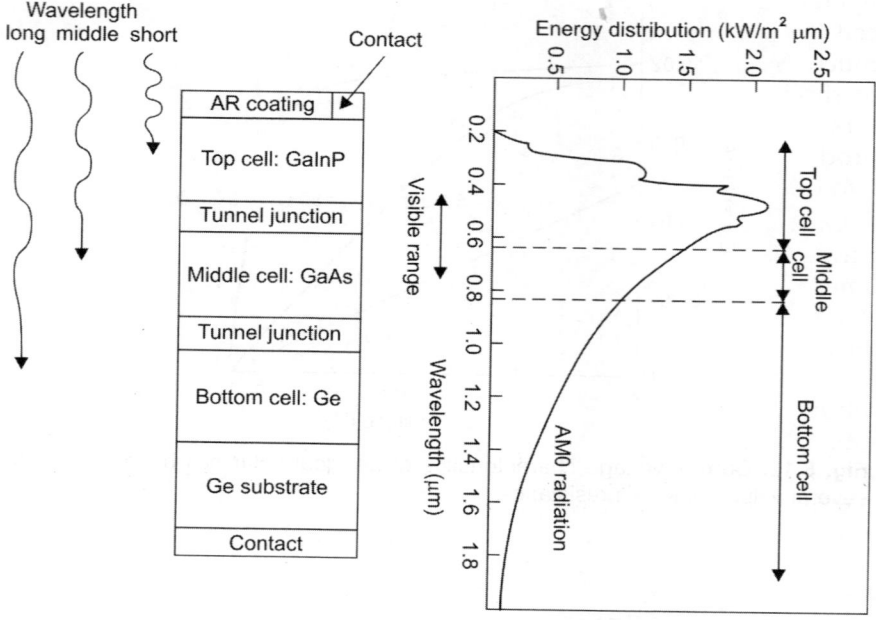

Fig. 8.14. Triple-junction solar cell

substrate. Tunnel junctions are used to form an electrical series connection of the subcells.

Suitable materials must be chosen for each cell so that photons of appropriate wavelengths are absorbed. The bandgap of each material is determined using Eq. (8.1). Energy bandgap E_g must decrease from top cell to bottom cell in order not to absorb all photons by the top cell. Also, all layers of semiconductors must be lattice matched, so that there is no built-in strain.

The anti-reflective (AR) coating typically consists of several layers. Design of AR layers can be performed.

The presence of tunnel junctions is to provide low electrical resistance between subcells. Modelling of those layers specifically for the purpose of MJ solar cells has been described by Baudrit and Algora in 2010.

8.4.1 Quantum Dots in Multijunctions

The triple-junction subcells shown are connected in series and their voltages add up, but the current flowing through the structure is determined by the smallest current produced by subcells. Total power generated by the structure is greater than that of the single junction cell.

The thickness of each subcell is determined by its absorption and the light power available in the spectral range where the subcell operates. As shown in Fig. 8.15, the bottom cell (formed by Ge, including substrate) is much thicker than two upper subcells. Due to this and also because it covers much larger spectral range, the Ge subcell produces much larger current compared to two upper subcells.

One of the possible solutions of reducing current of the Ge subcell and at the same time increasing current of the GaAs subcell, is to introduce InAs-based quantum dots into the GaAs subcell. These quantum dots are designed to have an effective bandgap slightly smaller than that of GaAs. Therefore, they can 'steal' current from bottom Ga subcell roughly by about 15%. Increasing also the thickness of top GaInP subcell to match the increased current of GaAs results in an overall current increase by about 15%. With the minimal change in voltage over GaAs subcell, this concept resulted in an increase of triple-junction conversion efficiency by about 13%.

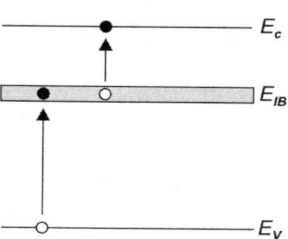

Fig. 8.15. Triple-junction solar cell

8.4.2 Intermediate Band Solar Cells (IBSC)

The concept of *intermediate band solar cells* (IBSC) was first described by Luque and Marti in 1997 and by Wolf in 1960. The main idea was to create an additional band between conduction and valence band (Fig. 8.15). The additional intermediate band can be created using several approaches, like for example, minibands.

The standard cell is only able to absorb photons with energy equal to or greater than $E_{gap} = E_c - E_v$. Within this new scheme, it is possible to have additional absorption via the IB.

The electronic states in the IB should be accessible via direct transitions. Therefore, photon of energy E_{IV} can transfer an electron from the valence band into IB whereas another photon with energy E_{CI} will be able to pump electrons from IB to conduction band. Thus, the performance of solar cell is increased.

Solar cells based on this design are attractive because of their predicted photon conversion efficiency of up to about 60%. Shao and Coworkers in 2007 show that the ordered three-dimensional arrays of quantum dots, i.e. quantum dot supracrystals, can be used to implement the intermediate-band solar cell with the efficiency significantly exceeding the Shockley-Queisser limit for a single junction cell. The increase is due to the utilization of photogenerated hot carriers which can produce higher voltages and higher photocurrents.

Modelling of such structures has been conducted using the CFD Research Corporation 3D device simulator, NanoTCAD. They performed accurate simulations of quantum dot solar cell performance and degradation due to effects of space radiation.

8.5 TEMPERATURE EFFECTS

The output voltage and the efficiency of a solar cell increases with decreasing temperature; solar cells operate best at lower temperatures. Consider the open circuit voltage V_{oc} of the device in Fig. 8.7. As the total cell current is zero, the photocurrent I_{ph} generated by light must be balanced by I_d which is generated by the photovoltaic voltage V_{oc}, that is, $I_d = I_o \exp(eV_{oc}/nk_BT)$. If n_i is the intrinsic concentration then I_o is proportional to n_i^2 which means that I_o decreases rapidly with decreasing temperature. Consequently, a greater voltage is developed to generate the necessary I_d that balances I_{ph}.

The output voltage, V_{oc}, when $V_{oc} \gg nk_BT/e$ is given by:

$$V_{oc} = \frac{nk_BT}{e} \ln\left(\frac{I_{ph}}{I_o}\right) \tag{8.14}$$

In Eq. (8.14), I_o is the reverse saturation current which is *strongly temperature dependent* because it depends on n_i^2 where n_i is the intrinsic concentration. Further, since $I_{ph} = KI$ where K is a constant and I is the light intensity, we can write Eq. (8.14) as:

$$V_{oc} = \frac{nk_BT}{e} \ln\left(\frac{KI}{I_o}\right) \quad \text{or} \quad \frac{eV_{oc}}{nk_BT} = \ln\left(\frac{KI}{I_o}\right)$$

Assuming $n = 1$, then at two different temperatures T_1 and T_2 but at the same illumination level, by subtraction:

$$\frac{eV_{oc2}}{k_BT_2} - \frac{eV_{oc1}}{k_BT_1} = \ln\left(\frac{KI}{I_{o2}}\right) = \ln\left(\frac{I_{o1}}{I_{o2}}\right) \approx \ln\left(\frac{n_{i1}^2}{n_{i2}^2}\right)$$

where the subscripts 1 and 2 refer to the temperatures T_1 or T_2 respectively.

We can substitute $n_i^2 = N_c N_v \exp(-E_g/k_BT)$ and neglect the temperature dependences of N_c and N_v compared with the exponential part to obtain:

$$\frac{eV_{oc2}}{k_BT_2} - \frac{eV_{oc1}}{k_BT_1} = \frac{E_g}{k_B}\left(\frac{1}{T_2} - \frac{1}{T_1}\right)$$

Rearranging for V_{oc2} in terms of other parameters, we find:

$$V_{oc2} = V_{oc1}\left(\frac{T_2}{T_1}\right) + \frac{E_g}{e}\left(1 - \frac{T_2}{T_1}\right) \quad \text{Open circuit-voltage vs temperature} \tag{8.15}$$

For example, a Si solar cell that has $V_{oc1} = 0.55$ V at $20°C$ ($T_1 = 293$ K) will have V_{oc2} at $60°C$ ($T_2 = 333$ K) given by:

$$V_{oc2} = (0.55 \text{ V})\left(\frac{333}{293}\right) + (1.1 \text{ V})\left(1 - \frac{333}{293}\right) = 0.475 \text{ V} \tag{8.16}$$

If we assume to first order that the absorption characteristics are unaltered (E_g, diffusion lengths etc. remaining roughly the same), so that I_{ph} remains the same, the efficiency decreases at least by this factor.

8.6 OTHER TECHNOLOGIES: DYE-SENSITIZED CELLS

In 1991, O'Regan and Gräztel developed a new dye-sensitized non-crystalline photovoltaic cell. These cells are now commonly referred to as *Gräztel cells* and are fundamentally different from the others discussed here in that they do not rely on semiconductor *p-n* junctions. Instead, they are *electrochemical devices* in which the optical absorption and carrier-collection processes are separate. As shown in Fig. 8.16, a porous film of a wide-bandgap semiconductor, usually titanium dioxide (TiO_2), is coated with a redox charge-transfer dye. Figure 8.17 shows the basic operating and chemical principles of a Gräztel cell using TiO_2, an organic dye, and an iodide solution as an electrolyte. The dye is excited by the absorption of a photon (a), an electron is injected into an excited energy level of the oxide (b), and the dye is regenerated by a liquid electrolyte (c). The excited electron in the oxide reaches an electrode and is then able to migrate through an external circuit, perform useful work in the load, then pass to the other electrode where it regenerates the electrolyte (d). The thick TiO_2 film is deposited onto TCO-coated glass that forms the front contact. They work despite the high defect densities at the interface because of the very fast (picosecond) spatial separation of electrons from ions at the dye–TiO_2 interface; this makes them effectively majority-carrier devices.

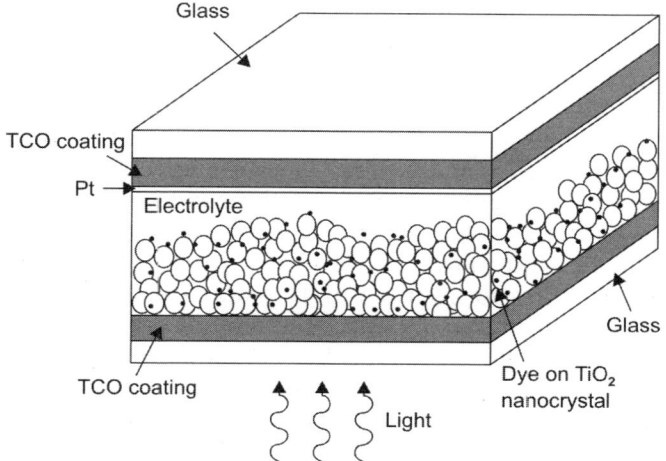

Fig. 8.16. Structure of the nanocrystalline dye-sensitized solar cell

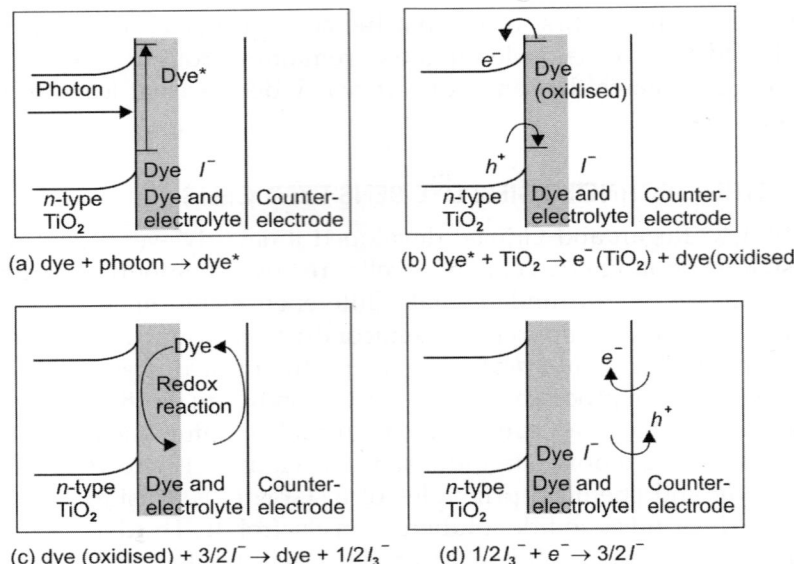

Fig. 8.17. Operating and chemical principles of a Gräztel cell. (a) Photons excite the dye. (b) Electrons are transferred to the *n*-type material from the dye where they can be collected by an electrode. (c) The dye is regenerated (redox reaction) by the electrolyte. (d) An electron from the counter-electrode regenerates the electrolyte

The underlying operating principles of Gräztel cells are not considered difficult. Simple devices using natural dyes from berries can quite easily be made and used to demonstrate chemical and physical topics that are taught at a senior-high-school level.

The best-demonstrated efficiency for a laboratory device is 11% and reasonable stability under illumination has been demonstrated. Module efficiencies using dye-sensitized cells are similar to those of commercially available thin-film amorphous silicon modules. Dye-sensitized cells are coming into small-scale commercial production in Australia and the US, and the installation of the first wall panel has been performed in Australia. Dye-sensitized cells exhibit promise as potentially low-cost cells, in terms of both finance and energy, with the added advantage that their temperature coefficient can be positive, increasing the output power at elevated operating temperatures. The main challenges are to improve high-temperature stability, scale-up production, and, preferably, replace the liquid electrolyte with a solid. Tandem dye-sensitized cells are particularly aimed at the direct splitting of water for hydrogen production rather than at electricity production. In one experimental system, yielding 4.5% conversion efficiency from sunlight to chemical energy, two semiconductors, TiO_2 and WO_3 or Fe_2O_3, are used, with water being the electrolyte.

8.7 APPLICATIONS OF PHOTOVOLTAIC SYSTEM

Photovoltaic modules are used in many cost-effective applications today, supplying both direct current (DC) and alternating current (AC) electrical loads. For supplying DC electricity, a photovoltaic system usually consists of the solar panel itself, an appropriately sized battery to supply a continuous load, and the necessary control electronics to keep the photovoltaic cell operating at its maximum power point and maintaining the battery state of charge. In AC systems, an inverter is required to transform the DC output of the solar panel to 120 to 240 V AC. The final possible system component is a solar tracker that mechanically moves the solar panels through one-axis or two-axes to keep them normal to the incident snlight, thus maximizing the output power. The increase in system complexity and the decrease in reliability means that these trackers are not often used. The primary applications of photovoltaic cells is described below.

Many of the highest-efficiency photovoltaic cells produced over the years have been destined for space applications, especially for powering satellites. In this market, high specific power (kW/kg) and hardness to cosmic radiation are more important than price. High efficiency is important because if fewer devices are required to power a satellite, then more mass can be used for scientific instruments. A major advantage of space operation is the constant presence of the sun, however, the cells are not protected by the Earth's atmosphere from high-energy particles and radiation; they consequently have a life expectancy of only about 7 years. Two recent applications of photovoltaic cells in such applications are the Mars rover and an unmanned high-altitude solar plane that is intended to replace satellites.

The first terrestrial markets to use solar cells were typically located in remote areas and capitalized on the weatherproof nature of photovoltaic modules. Applications included powering navigation buoys, lighthouses, repeater stations for telecommunications, and weather-monitoring stations. Powering such applications from photovoltaics was economical due to the high cost of replacing batteries in these systems, the low power consumption (10–100 W) of the load, and the high system availability, or low loss-of-load probability (LOLP) required. Where a low LOLP was critical, upto 15 days of battery storage was included in the system. Another traditional market for solar cells include powering cathodic protection systems to provide an electrical current that counteracts the natural electrochemical currents that cause corrosion, such as oil and gas pipes.

Screen-printed monocrystalline silicon cells were used in the majority of the applications mentioned above. For many years, amorphous silicon solar cells have been used in consumer products such as calculators and watches. Using the photovoltaic technology obviates the need for batteries, as these devices can operate under extremely low (indoor) light levels.

Photovoltaic cells have played a role in bringing power to some of the estimated 2 billion people in the world that do not have access to electricity. Solar home systems (SHS) are specifically designed for application in developing countries, and typically include one 50 W solar panel, a battery, a charge controller, and some light emitting diode (LED)-based lamps. In addition, owners might typically add a radiation or black and white television to the system. By providing light in the evenings, children are able to be better educated, adult community members can work to earn additional income for their family, and air pollution from kerosene lamps is diminished. Street lights that include an integrated solar panel and becoming more common in both developing and developed countries. In the latter case, a photovoltaic power light becomes more economical than a grid-connected light as soon as the necessary trenches for cables are more than a few meters in length.

Photovoltaic systems are used to power water pumping systems in a remote community, to bring water from the local river or well. This water is either used for human consumption or irrigation. Where the water quality is too poor to drink, photovoltaics have been used to power the pumps in desalination systems that force the water through membranes to filter out the unwanted particulates, bacteria, viruses, and salts. The source water for such systems is typically brackish, but the desalination of seawater is commonplace in the Middle East.

The biggest growth market for photovoltaics over the past few years are for on-grid residential rooftop systems. On-grid systems do not reuire batteries as the electricity grid is used as "storage", with excess power being exported to the grid during the day, and brought back at night as required. The installation of such systems is being driven by astute policy-making decisions in the places such as Japan, Germany, and California, with rebates for offsetting installation costs or attractive buyback rates for generated electricity being offered. Although, they have yet to become a common sight in our everyday lives, photovoltaics have been used to power many vehicles, including solar boats and racing care.

The final measure of the performance of a PV module or system is its energy yield ratio (EYR)—the ratio of the electrical energy generated over the 25-year lifetime of a PV module to the energy used in the fabrication of a PV module from raw materials. The EYR of PV modules is high; today's PV modules generate between 5.6 to 12.4 times more energy than went into their fabrication. The EYR is not significantly reduced for a rooftop grid-connected PV system because there is minimal framing and the embodied energy of the inverter is low. However, for stand-alone PV systems that rely on battery storage, the

Photovoltaic Devices

EYR is significantly lower, although the value is still nearly three times greater—even for a region with a poor solar-radiation resource.

Example 1

Suppose that a particular family house in a sunny geographic location over a year consumes a daily average electrical power of 500 W. If the annual average solar intensity incident per day is about 6 kW·h·m^{-2}, and a photovoltaic device that converts solar energy to electrical energy has an efficiency of 15%, calculate the required device area.

Solution

We know the average light intensity incident:

Total energy available for 1 day = Incident solar energy in 1 day per unit area × Area × Efficiency

which must equal to the average consumed per house in 1 day.

$$\therefore \quad \text{Area} = \frac{\text{Energy per house}}{\text{Incident solar energy per unit area} \times \text{Efficiency}}$$

$$= \frac{50 \text{ W} \times 60 \frac{\text{s}}{\text{min}} \times 60 \frac{\text{min}}{\text{hr}} \times 24 \text{ hrs}}{(6 \times 10^3 \text{ W} \times \text{hr} \times \text{m}^{-2} \text{ day}^{-1})\left(60 \frac{\text{s}}{\text{min}} 60 \times \frac{\text{min}}{\text{hr}}\right) \times 0.15}$$

$= 13.3 \text{ m}^2$ or a panel $(3.6 \times 3.6) \text{ m}^2$.

The problem is that this area is based on averages over the years. Such a panel cannot supply the peak power when a number of electrical appliances are running at the same time (consuming several kilowatts). An energy storage devices can be used to store the surplus energy generated during low-power consumption periods but this adds to the cost and complexity of the system.

Example 2

Consider two identical solar cells with the properties $I_o = 25 \times 10^{-6}$ mA, $n = 1.5$, $R_f = 20$ Ω, subjected to the same illumination so that $I_{ph} = 10$ mA. Explain the characteristics of two solar cells connected in parallel. Find the maximum power that can be delivered by one cell and two cells in series and also find the corresponding voltage and current at the maximum power point (assume $R_p = \infty$).

Solution

Consider one individual solar cell as shown in Fig. 8.18. The voltage V_d across the diode is $V - R_s I$ so that the external current I is:

$$I = -I_{ph} + I_o \exp\left(\frac{eV_d}{nk_BT}\right) - I_o + - I_{ph} + I_o \exp\left(\frac{e(V - IR_s)}{nk_BT}\right) - I_o \quad (1)$$

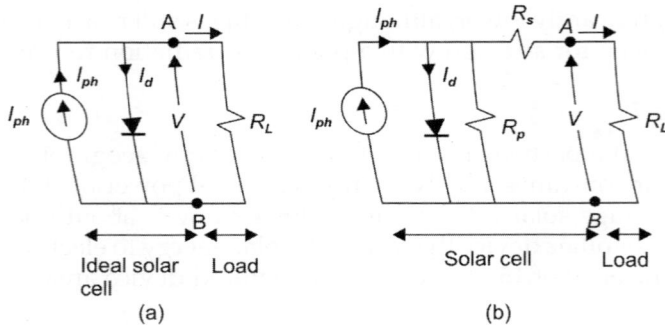

Fig. 8.18. The equivalent circuit of a solar cell (a) Ideal *pn* junction solar cell, (b) Parallel and series resistances

Equation (1) gives the *I–V* characteristic of 1 cell and is plotted in Fig. 8.19. The output power *P* is simply *IV* which is also plotted in Fig. 8.19. The power is maximum at 2.2 mW when the current is about 8 mA and the voltage is 0.27 V. The load must be 34 Ω.

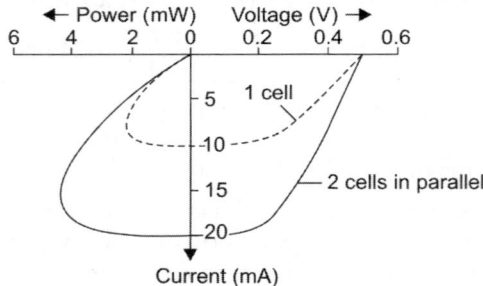

Fig. 8.19. Current vs voltage and power vs current characteristics of one cell and two cells in parallel. The two parallel devices have $R_s/2$ and $2I_{ph}$

Figure 8.20 shows the equivalent circuit of the two solar cells in parallel running, a load R_L. *I* and *V* now refer to the whole system of two devices in parallel. Each device is now delivering a current $I/2$. The diode voltage for one cell is $V_d - R(I/2)$. Thus:

$$\frac{1}{2}I = -I_{ph} + I_o \exp\left(\frac{V - \frac{1}{2}IR_s}{nk_BT}\right) - I_o$$

or

$$I = -2I_{ph} + 2I_o \exp\left(\frac{V - \frac{1}{2}IR_s}{nk_BT}\right) - 2I_o \qquad (2)$$

Comparing Eq. (2) with Eq. (1), we see that the parallel combination has halved the series resistance, doubled the photocurrent, and also doubled the diode reverse saturation current I_o. All these are in line

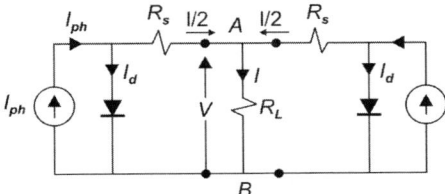

Fig. 8.20. Two identical solar cells in parallel under the same illumination and driving a load R_L

with intuitive expectation as the device area has now been effectively doubled. Figure 8.19 shows the I–V and P vs. I characteristics of the combined device. The power is maximum at about 4.4 mW when $I \approx 16$ mA, $V \approx 0.27$ V. The corresponding load is 17 Ω. It is clear that the parallel combination increases the available current and allows a lower resistance load to be driven.

It we were to use the two solar cells in series, then V_{oc} would be at 4.4 mW available at about 8 mA and 0.55 V. This output power would need a load of about 34 Ω. These simple ideas, however, do not work when the cells are not identical. The connections of such mismatched cells can lead to much poor performance than idealized predictions based on parallel and series connections of matched devices.

Example 3

Consider a solar cell driving a 30 Ω resistive load as shown in Fig. 8.21(a). Suppose that the cell has an area of 1 cm × 1 cm and is illuminated with light of intensity 600 W·m^{-2} and has the I–V characteristics in Fig. 8.21(b). What are the current and voltage in the circuit? What is the power delivered to the load? What is the efficiency of the solar cell in this circuit?

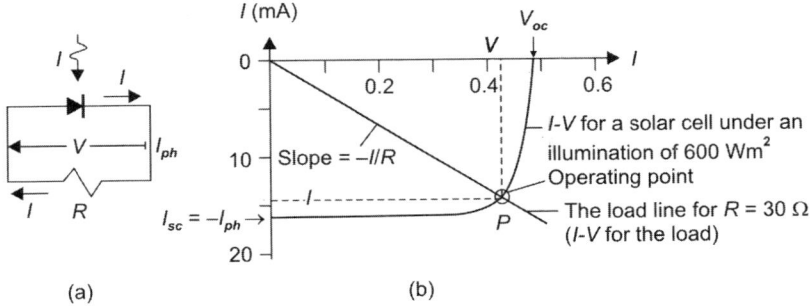

Fig. 8.21. (a) When a solar cell drives a load R, R has the same voltage as the solar cell but the current through it is in the opposite direction to the convention that current flows from high to low potential. (b) The current I' and voltage V' in the circuit of (a) can be found from a load line construction. Point P is the operating point (I', V'). The load line is for R = 30 Ω

Solution

The I–V characteristic of the load is the load line:

$$I = -\frac{V}{30\,\Omega}$$

The line is drawn in Fig. 8.21(b) with a slope $1/30\,\Omega$. It cuts the I–V characteristics of the solar cells at $I' = 14.2$ mA and $V' = 0.425$ V which are the current and voltage in the photovoltaic circuit of Fig. 8.21(a) respectively. The power delivered to the load is:

$$P_{out} = I'V' = (14.2 \times 10^{-3})(0.425\text{ V}) = 0.006035\text{ W, or } 6.035\text{ mW}.$$

This is not necessarily the maximum power available from the solar cell. The input sun-light power is:

P_{in} = (Light Intensity) (Surface Area) = $(600\text{ W·m}^{-2})(0.01)^2 = 0.060$ W

The efficiency is:

$$\eta = 100\,\frac{P_{out}}{P_{in}} = 100\,\frac{0.006035}{0.060} = 10.06\%$$

This will increase, if the load is adjusted to extract the maximum power from the solar cell but the increase will be small as the rectangular area $I'V'$ in Fig. 8.21(b) is already close to the maximum.

Example 4

A solar cell under an illumination of 600 W·m^{-2} has a short circuit I_{sc} of 16.1 mA and an open circuit output voltage V_{oc} of 0.485 V. What are the short circuit current and the open circuit voltage when the light intensity is doubled?

Solution

The general I–V characteristics under illumination is given by Eq. (2) (Example 2). Setting $I = 0$ for open circuit, we have:

$$I = -I_{ph} + I_o\,[\exp(eV/nk_BT) - 1] = 0$$

Assuming that $V_{oc} \gg nk_BT/e$, rearranging the above equation, we can find V_{oc}:

$$V_{oc} = \frac{nk_BT}{e}\ln\left(\frac{I_{ph}}{I_o}\right) \qquad \text{Open circuit output voltage (1)}$$

In Eq. (1), the photocurrent, I_{ph}, depends on the light intensity I via, $I_{ph} = KI$. At a given temperature, the change in V_{oc} is:

$$V_{oc2} - V_{oc1} = \frac{nk_BT}{e}\ln\left(\frac{I_{ph2}}{I_{ph1}}\right) = \frac{nk_BT}{e}\ln\left(\frac{I_2}{I_1}\right)$$

The short circuit current

$$I_{sc2} = I_{sc1}\left(\frac{I_2}{I_1}\right) = (16.1 \text{ mA})(2) = 32.2 \text{ mA}$$

GLIMPSES

1. *Photovoltaic cells* directly convert light to electricity.
2. The most common source of light on earth is the *sun*. The photosphere, or the external region of the sun, emits radiation closely approaching that of a thermodynamic "black body" or perfect radiator at a temperature of 6000 K, with a spectral distribution governed by Planck's radiation law.
3. The radiant power per unit area perpendicular to the direction of the sun just outside the earth's atmosphere is essentially constant and is known as *solar constant*. This is also known as the *air-mass zero* (AM0) spectrum, which has the value of 1.367 kW·m^{-2}.
4. In general *photovoltaic cells* can be classified into "*three generations*".

   The *first generation* technology of cells are based on silicon wafers that are thick enough to be self-supporting, typically about 300 mm thick.

   The *second generation* represents *thin-film* technologies that aim for lower cost at the expense of efficiency by using thin layers of semiconducting material. Because the active layer is only a few microns thick, a rigid substrate or transparent substrate is required for the mechanical support.

   Third generation approaches are intended to achieve both high-efficiency and low cost, e.g. *tandem cell*. Two types of tandem cells exists: (i) a thin film technology based on layers of a-Si and microcrystalline silicon (μc-Si) or a Si and amorphous germanium (a-Ge), (ii) made from elements of group III and V of the periodic table (known simply as III–V cells), achieves ultra-high efficiency, but at a high cost premium.
5. *Solar cells* can be regarded as a large-area illuminated semiconductor diode. The semiconductor's properties are exploited in two ways: (i) light is absorbed and generated free charge carriers within the material, and (ii) a junction in the semiconductor separates the negative and positive charge carriers, producing a unidirectional electrical current through the two contacts that have a voltage difference between them. The I–V relationship of an ideal device can be described by:

$$I = I_L - I_o [\exp(eV/kT) - 1]$$

where I_L is photogenerated current, and I_o is dark saturation current.

6. *Organic semiconducting materials* offer significant advantages compared with traditional inorganic semiconducting materials in terms of cost and versatility.

SUGGESTED READINGS

1. Anderson EE, 'Fundamentals of Solar Energy Conversion, Addison-Wesley (1983).
2. Rabl A, 'Active Solar Collectors and their Applications', Oxford University Press (1985).
3. Moeller HJ, 'Semiconductor for Solar Cells', Artech House (1993).
4. Green MA, 'Power to the People: Sunlight to Electricity Using Solar Cells", University of New South Wales Press (2000).
5. Richards BS and Shalav A, 'Photovoltaic Devices', in 'The Handbook of Photonics', CRC Press (2nd ed., 2007).
6. Julian Chen C, Physics of Solar Energy, John Wiley (2012).

REVIEW QUESTIONS

1. Explain the principles of photovoltaics and the *p-n* junction operating as a solar cell.
2. What is a solar cell? Draw a figure for the operation of a *p-n* junction solar cell.
3. Explain the three generations of photovoltaic devices.
4. Write current-voltage (*I–V*) relationship for an ideal photovoltaic device and draw *I–V* characteristic for a photovoltaic cell.
5. Draw equivalent circuits of solar cells which includes parasitic effects. Obtain expression for I_o (constant) in terms of I_{RS} (cell's reverse saturation current at a reference temperature.
6. What is the function of multijunctions? How it helps to enhance the efficiency of a photovoltaic cell?
7. What is the role of quantum dots in multijunctions.
8. Describe the principle and working of intermediate band solar cells.
9. How efficiency limits of single-material cells are exceeded in tandem cells?

SHORT ANSWER QUESTIONS

1. What is the importance of solar energy for us?

Ans. At current consumption rates, all of the world's readily exploitable *fossil* fuel reserves are expected to be depleted within the next 50 (gas) to 200 (coal) years. In 2002, electricity generated from renewable energy sources such as hydro, wind, biomass, geothermal, solar thermal, and photovoltaics amounts to 14% of total electricity generated.

As far as the solar energy is concerned, the sun supplies more energy in one hour than we use globally in one year. This shows that solar energy can be a practical alternative as a source of electric energy.

2. What is a dye-sensitized solar cell?

Ans. This solar cell does not rely on photovoltaic effect to convert sunlight into electricity but it works on the principle of photo-electrochemical effect. These cells are now commonly referred to as Gratzel cells. In these cells optical absorption and carrier collection processes are separate.

3. What are organic solar cells?

Ans. Inorganic thin-film technology requires high-temperature processing in high-vacuum environments. Manufacturing costs are still quite high due to energy-intensive processes and the use of relatively large, expensive substrates that can withstand the high temperature during processing. Organic semiconducting materials are being investigated for photovoltaic effect due to their lower cost and versatility.

Organic materials including conjugated polymers, dyes, and organic glasses, can show p-type or n-type semiconducting properties. These materials have very high absorption coefficients that make them ideal for thin-film PV technology. Film thicknesses on the order of only a few hundred nms would be required. The development of organic solar cells could benefit from already available commercial organic photoconductive materials currently used in *laser printing* and *organic light emitting* diode (OLED) technology. Unfortunately, organic PV devices still have very low power conversion efficiencies and stabilities (compared with inorganic PV devices).

4. What is a quantum dot?

Ans. A quantum mechanical system, usually made from a semiconductor, in which electrons can be confined into a small region a few nms in size containing a few thousand atoms. Such systems act as *'artificial atoms'* with their own sets of quantum states for the electrons.

5. What is intermediate band solar cells (IBSC)?

Ans. In IBSC and additional band between conduction and valence band is created using several approaches, e.g. minibands. Solar cells based on this design are attractive because of their predicted photon conversion efficiency upto about 60%.

6. **What is the main advantage of tandem cell?**

Ans. Tandem cell is one approach to exceed the efficiency limits of single-material cell. They reduce the two main losses: (i) the thermalization of the excess energy of high-energy photons, and (ii) transparency to low energy photons. Tandem cells are stacks of *p-n* junctions each of which is formed from a semiconductor of different bandgap energy. Each responds to a different section of the solar spectrum, yielding higher overall efficiency. The cells are stacked in order of decreasing bandgaps, such that the light is automatically filtered as each cell extracts photons that exceeds its bandgap. This technique is commercially used to fabricate triple-junction amorphous cells.

Tandem cells are also fabricated from compounds of elements from group III and V of the periodic table, such as GaAs. GaAs is used extensively in the optoelectronics, etc.

7. **Mention some applications of photovoltaic system.**

Ans. Photovoltaic modules are used in many cost effective applications, e.g., supplying both direct current (DC) and alternating current (AC) electrical loads. For supplying DC electricity, a photovoltaic system usually consists of the solar panel itself, an approximately sized battery to supply a continuous load, and the necessary control electronics to keep the photovoltaic cell operating at its maximum power point and maintaining the battery state of change. In AC systems, an inverter is required to transform the DC output of the solar panel to AC.

Highest efficiency photovoltaic cells produced over the years have been destined for *space applications*, especially for powering satellites.

Photovoltaic cells are used in powering navigation buoys, light houses, repeater stations for telecommunications, and weather-monitoring stations.

Photovoltaic cellshave played a role in bringing power to some 2 billion people in the world that do not have access to electricity. *Solar home systems* (SHS) are specifically designed for application in developing countries.

Photovolatic systems are used to power water pumping systems in remote community, to bring water from the local river or well.

The biggest growth market for the photovoltaics over the past few years are for *on-grid residential rooftop systems*.

8. **What is a quantum well?**

Ans. A linear conductor that is narrow enough for quantum effects to affect the resistance. If the wire diameter is small, then the electrons have quantized energy levels for motion perpendicular to the axis (as for a particle in a box). Consequently, the resistance of the wire is also quantized. *Carbon nanotubes* (CNTs) have been investigated as practical implementations of quantum wires. *Quantum well solar cells* using quantum transport are also under development. Such approaches combine elements from semiconductor optics and quantum transport in nanostructures.

9. Draw a diagram of solar cell and explain its principle.

Ans. A simplified schematic diagram of a solar cell is shown in Fig. 8.22. Consider a *pn* junction with a very narrow and more heavily doped *n*-region. The illumination is through the thin *n*-side. The depletion region (*W*) or the space charge layer (SCL) extends primarily into the *p*-side. There is a **built-in field** E_o in this depletion layer. The electrodes attached to the *n*-side must allow illumination to enter the device and at the same time result in small *series resistance*. They are deposited on to *n*-side to form an array of *finger electrodes* on the surface. A thin antireflection coating on the surface (not shown in the figure) reduces reflections and allows more light to enter the surface.

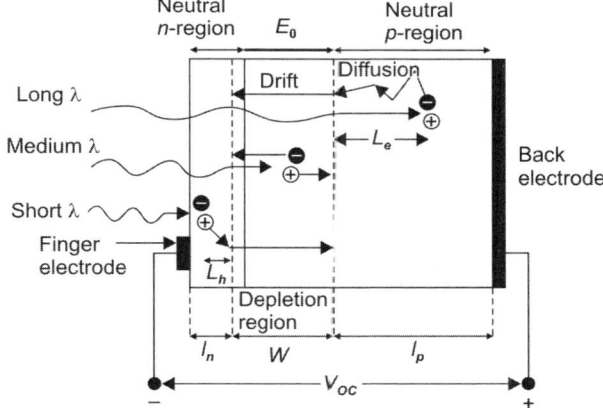

Fig. 8.22. The principle of operation of the solar cell

If the terminals of the device are shortened, then the excess electron in the very narrow *n*-side can flow through the external circuit to neutralize the excess hole in the *p*-side. This current due to the flow of the photodegenerated carriers is called the *photocurrent*.

MULTIPLE CHOICE QUESTIONS

1. Photovoltaic cells directly convert:
   (a) light to heat (b) light to electricity
   (c) light to sound (d) None of the above

2. The value of solar constant is:
   (a) 1.367 kW·m^{-2}
   (b) 0.367 kW·m^{-2}
   (c) 3.367 kW·m^{-2}
   (d) 0.5
3. Dye-sensitized solar cell rely on:
   (a) photovoltaic effect
   (b) photoelectrochemical effect
   (c) Raman effect
   (d) None of the above
4. The current-voltage (I–V) relationship of an ideal p-n junction photovoltaic cell is (symbols have usual meanings).
   (a) $I = I_L - I_o$
   (b) $I = I_L - I_o (\exp (qV/kT)]$
   (c) $I = I_L - I_o (\exp (qV/kT) - 1]$
   (d) $I = I_o (\exp (qV/kT) - 1]$
5. First-generation photovoltaic technology is based upon:
   (a) thin silicon wafter
   (b) thick silicon wafers
   (c) stacks of p-n junctions
   (d) due to sensitized nanocrystalline materials

ANSWERS

1. (b)   2. (a)   3. (b)   4. (c)   5. (b)

9

Optical Receivers

9.1 INTRODUCTION

Optical receivers are the essential and important parts of optical communication system. An optical receiver converts an optical signal, transmitted through an optical fiber cable into an electrical signal suitable for a receiving device installed at the other end of the communication system. The conversion process in the receiver is performed by two essential parts: a *detector* and an *electronic signal processor*. The detector converts the optical signal into an electrical signal. The electronic signal processor converts the raw detector signal into a form *decipherable* by the receiving device, e.g. a telephone, camera, or scanner.

A block diagram of an optical detector is shown in Fig. 9.1.

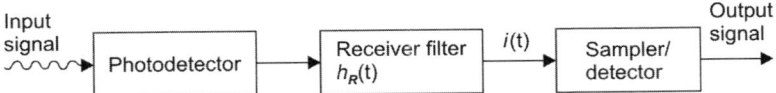

Fig. 9.1. Block diagram of an ideal optical receiver

The receiver consists of a photodetector which converts the optical signal into electrical current. A good light detector should generate a large photocurrent at a given incident light power. They should also respond fast to the input changes and add minimal noise to the output signal. This last requirement is of crucial importance since the received signal is typically very weak. In digital optical communication system the detection process is often conducted with a PIN photodiode.

There are generally two types of detection: *direct detection* (also called *incoherent detection*) and *coherent detection*.

Direct detection detects only the intensity of the incident light. It is used mainly for intensity or amplitude modulation schemes. It can only detect an amplitude modulated (AM) signal.

Coherent detection can detect both the power and phase of the incident light. It is, therefore, used when phase modulation (PM) or frequency modulation (FM) is preferred. Coherent detection is also important in applications such as WDM.

The coherent detection requires a local oscillator to coherently down-convert the modulated signal from optical frequency to intermediate frequency (IF). The incoherent detection which dominates in currently deployed systems is based on square-law envelope detection of the optical signals.

After detection, the electrical current is often amplified (not shown in Fig. 9.1) and then passes through an electrical filter which is normally of the Bessel type. At that point, electrical eye diagrams are typically observed for the assessment of signal quality. Next, sampling of electrically filtered received signal is performed. The received electrical signal is corrupted with noise of various origins. The noise sources will be discussed in the following sections.

Performance evaluation of an optical transmission system is done by evaluating optical signal-to-noise ratio, eye opening and bit error rate (BER) which is the ultimate indicator of the system's performance.

As stated above, the function of the optical receiver is to pick up an optical signal and convert it into an electrical signal suitable for a receiving device at the end of the communication line. Optical signals can be data, video, or audio. Optical detectors perform this conversion of an optical signal into an electrical signal. This is why *optical receivers are sometimes called optical detectors*. Optical detectors perform the opposite function of optical transmitters, such as light-emitting diodes (LEDs), and semiconductor lasers.

The detector requirements in optical fiber communication system are very similar to those of optoelectronic sources, i.e. they should have *very high sensitivity* at operating wavelengths, *high fidelity, fast response, high reliability, low noise* and *low cost*. Moreover, the size of the detector should be comparable with that of fiber employed in the link. These requirements are easily met by detectors made of semiconducting materials.

The most common type of optical detector is the *semiconductor photodiode*, which produces current in response to incident light. Detectors operate based on the principle of the semiconductor diode (*p-n* junction). An incident photon striking the diode gives an electron in the valence band of the atom. If the photon has sufficient energy to move the electron to the conduction band, this creates a free-moving electron and a hole. If the creation of these carriers occurs in a depleted region, the charge carriers (electrons and holes) will quickly separate and create a current. As they reach the edge of the depleted area, the electrical forces diminish and current cases.

Figure 9.2 shows one application of an optical receiver as conversion of an optical signal into a digital form. The incoming optical signal is

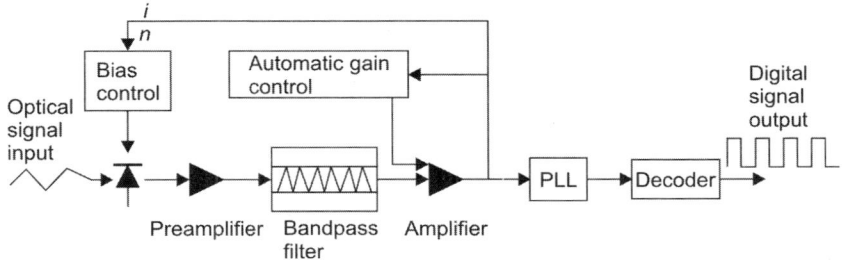

Fig. 9.2. Optical signal conversion process

converted to an electronic signal using a photodetector, such as a *p-i-n* (PIN) photodiode or an avalanche photodiode (APD). The signal is then amplified by a preamplifier, and passed through a bandpass filter that removes unwanted wavelengths. Further amplification, with a gain feedback control circuit, provides stable signal levels for the rest of the process. This control circuit controls the bias current, and thus the sensitivity of the photodiode.

A phase-locked loop (PLL) recovers the data bit stream and the timing information. The stream of bits needs to be decoded, from the coding used on the line, into its data format coding. This decoding process varies, depending on the encoding, and is occasionally integrated with the PLL, depending on the code in use.

9.2 PRINCIPLE OF OPTOELECTRONIC DETECTION

Figure 9.3 shows a reverse-biased *p-n* junction. A bias voltage V is applied to the *p-n* junction. Electrons diffuse away from the *n*-region into the *p*-region, leaving behind positively-charged ionized atoms called "donors".

In the *p*-region, these electrons recombine with the abundant holes. Similarly, holes diffuse away from the *p*-region, leaving behind negatively-charged ionized atoms (called "acceptors"). In the *n*-region, the holes recombine with the abundant mobile electrons.

This diffuse process cannot continue indefinitely, however, because it causes a disruption of the charge balance in the two regions. As a result, a narrow region on both sides of the junction becomes almost totally depleted of mobile charge carriers (electrons and holes). This region is called the *depletion region*. This region has a built-in electric field E_o, due to the acceptors and donors beyond its edges. This occurs even without an applied voltage.

When a photon of energy greater than the band gap of the semi-conductor material (i.e. $hv \geq E_g$) is incident on or near the depletion region of the device, it excites an electron from the valence band into the conduction band. The vacancy of an electron creates a hole in the

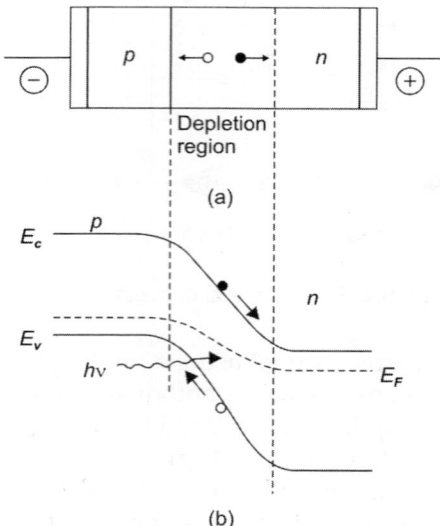

Fig. 9.3. (a) A reverse-biased *p-n* junction and depletion region. (b) Energy band diagram showing carrier generation and their drift

valence band. Electrons and holes so generated experience a strong electric field and drift rapidly towards the *n* and *p* sides, respectively. The resulting flow of current is proportional to the number of incident photons. Such a reverse-biased *p-n* junction, therefore, acts as a photodetector and is normally referred to as a *p-n* photodiode.

The fundamental principle behind the photodetection process is *optical absorption*. If the energy ($h\nu$) of an incident photon exceeds the energy of the band gap (between the conduction and valence bands) and is absorbed in the depletion region, then an electron moves upto the conduction band leaving a hole in the valence band. Thus an electron-hole pair is generated each time a photon is absorbed by the semiconductor. Under the influence of the electric field, electrons and holes are swept across the semiconductor in opposite directions. This flow of carriers results in the flow of electrical current called "generated photocurrent" when connected to an electric circuit. An applied voltage serves to speed up the carrier movement, increasing the current.

The fraction of light absorbed by the photodiode depends on:
(i) Wavelength (λ) of the light, determined by the photon energy ($\varepsilon = h\nu = hc/\lambda$).
(ii) The thickness of the absorption material (depletion region width or depletion layer thickness).

9.3 OPTICAL RECEIVER

The main characteristics of optical receiver are:

9.3.1 Sensitivity

This property is a measure of the minimum level of optical power P_{sens} at the receiver required for a reliable operation. Specifically, one expects that the BER is smaller than a specified level. Typically, that level is established to be equal to 10^{-9}.

The receiver sensitivity is a function of both the *signal* and also the *noise* parameters of photodetector and a preamplifier. It is a measure of the operating limit of the optical receiver, which, however, rarely operates close to that limit. There always exists a possibility of degradation of the system (temperature, ageing, etc.), so a typical margin (normally 3–6 dB).

Receiver sensitivity is a fundamental parameter of an optical receiver. It is directly responsible for the spacing between two points in any optical link, e.g. between transmitter and receiver or between repeaters.

9.3.2 Dynamic Range

It (expressed in dB) is the difference betwen the maximum allowable power and minimum power determined by receiver sensiti-vity. The maximum allowable power on the receiver is determined by nonlinearity and saturation.

Large dynamic range is important because it allows for more flexibility in the design of an optical network. The design of every network should take into account the wide range of possible changes of received optical powers due to change in temperature, ageing or various types of losses (in the fiber, connectors, etc.)

9.3.3 Bit-Rate Transparency

It refers to the ability of the optical receiver to operate over a range of bit rates. It describes the ability of the same receiver to be used for several networks operating at different bit rates.

9.3.4 Bit-Pattern Independency

This is the property of an optical receiver determining its operation for various data formats. The main constraint is imposed by non-return-to-zero (NRZ) code.

9.4 PHOTODETECTORS

At the end of travel through an optical fiber, the optical signal reaches a photodetector (PD), where it is converted into an electrical signal. In PD photon of energy $h\nu$ is absorbed and produces photocurrent i_p:

$$i_p = \eta \frac{P}{h\nu} e \qquad (9.1)$$

where P is the power of the incoming light, η is the quantum efficiency, h is Planck's constant and $\nu = c/\lambda$ is the frequency of the absorbed light. PD is similar to a semiconductor diode polarized in the reverse direction. Therefore, PD can be *modelled as a current source*.

Another very popular detector is a *human eye*. It is the most popular natural detector of light but it has several disadvantages: *it is slow, has bad sensitivity for low-level signals, has no natural connection to electronic amplifiers* and its *spectral response is limited* to the 0.4–0.7 µm range.

Artificial (human made) optical detectors are based on two physical mechanisms: external photoelectric effect and internal photoelectric effect. In the external photoelectric effect, electrons are removed from the metal surface of electrode known as a cathode by absorbing energy from incident light. Then, under an electric field due to the potential difference between both electrodes, they travel to another electrode known as an anode, thus producing electrical current. Vacuum photodiodes and photomultiplier tubes operate on that principle.

Main choices for photodetectors are the *p-i-n* (*p*-type intrinsic *n*-type) photodiode and avalanche photodiode (APD). APD provides gain which increases system sensitivity but introduces more noise.

In internal photoelectric effect, physical processes take place inside semiconductor junction devices. There, free carriers (electrons and holes) are generated by absorption of incoming photons, and as a result an electrical current is produced. These devices can be viewed as the inverse of a light emitting diode (LED).

9.5 PRINCIPLES OF PHOTODETECTION

Photodetection using semiconductors is possible because of optical absorption. When light is incident on the semiconductor surface, it may or may not be absorbed depending on the wavelength.

Absorbed optical power is described as:

$$\frac{dp}{dx} = \alpha(\lambda)\, P \qquad (9.2)$$

where $\alpha(\lambda)$ is the absorption coefficient which is wavelength dependent. The spectra of optical absorption for several semiconductors and semiconductor compounds which are commonly used is shown in Fig. 9.4. As can be seen the absorption coefficient $\alpha(\lambda)$ strongly depends on wavelength.

We now consider a photodiode. Photodiode absorbs photons of a specific wavelength to produce electron-hole pairs and thus a photocurrent depends on the absorption coefficient $\alpha(\lambda)$ (hereafter α).

We assume that the total optical power incident on the photodiode is P_{in} and the Fresnel reflection coefficient at the air-semiconductor interface is R. Then the optical power entering the semiconductor will be, $P = P_{in}(1 - R)$. If d is the absorption region of the semiconductor,

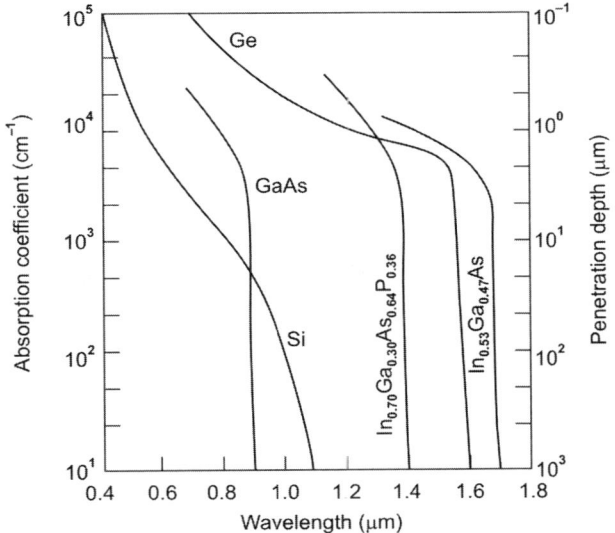

Fig. 9.4. Wavelength dependence of the absorption coefficient α for some semiconductors

then the power absorbed by the semiconductor can be obtained by integrating Eq. (9.2) (in accordance with Beer's law), we have:

$$P_{abs}(d) = P_{in}(1-R)[1-\exp(-\alpha d)] \qquad (9.3)$$

Let us assume that the incident light is monochromatic and energy of each photon in $h\nu$. Then the rate of photon absorption will be given by:

$$\frac{P_{abs}(d)}{h\nu} = \frac{P_{in}(1-R)}{h\nu}[1-\exp(-\alpha d)]$$

$$= \frac{P}{h\nu}[1-\exp(-\alpha d)] \qquad (9.4)$$

Incident light penetrates the semiconductor surface and generates electrical current I_p (rate of flow of charge carriers) is given by:

$$I_p = \frac{P_e}{h\nu}[1-\exp(-\alpha d)] \qquad (9.5)$$

where e is electronic charge. Here, we have assumed that (i) the semiconductor is an intrinsic absorper, i.e., the absorption of photons excites the electrons from the valence band directly to the conduction band, (ii) each photon produces an electron-hole pair, and (iii) all charge carriers are collected at the electrodes.

A schematic of a photodetector is shown in Fig. 9.5.

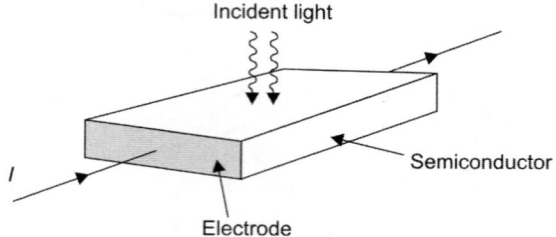

Fig. 9.5. Schematic of a photodetector

We may note that the intensity of the optical signal at the receiver is very low, the detector has to meet high performance specifications:
  (i) The conversion efficiency should be high at the operating wavelength.
  (ii) The speed of response should be high enough to ensure that signal distortion does not occur.
  (iii) The detection process introduce the minimum amount of noise.
  (iv) The detector size should be compatible with the fiber dimensions.
  (v) It must be possible to operate continuously over a wide range of temperature for several years.

9.6 PROPERTIES OF SEMICONDUCTOR PHOTODETECTORS

9.6.1 Quantum Efficiency (η)

The quantum efficiency (QE) of a photodetector is a measure of how effectively the detector converts light into electrical current. QE denoted by η, is defined as the ratio of the flux of generated electron-hole pairs (EHPs) that contribute to the detector current, to the flux of the incident photons. The quantum efficiency of a detector is the ratio of the number of photons actually detected, to the number of incident photons. The QE range is $0 \leq \eta \geq 1$. The quantum efficiency of the photodiode is defined as:

$$\eta = \frac{\text{Number of free EHP generated and collected}}{\text{Number of incident photons}} = \frac{r_e}{r_p} \quad (9.6)$$

Since QE is a function of photon energy, the QE is calculated at a particular photon energy. The measured photocurrent (I_{ph}) in the external circuit is due to the flow of electrons to the terminals of the photodiode. The number of electrons collected at the terminal per second is I_{ph}/e, where e is the charge of an electron. For incident optical power P_o, the number of incident photons arriving per second is $P_o/h\nu$. Thus, the QE (η) can also be defined as:

$$\eta = \frac{I_{ph}/e}{P_o/h\nu} \quad (9.7)$$

One of the major factors which determines η is the absorption coefficient (α) of semiconductor material used within the photodetector. η is generally less than unity as not all of the incident photons are absorbed to create electron-hole (e-h) pairs. η is often quoted as a percentage, e.g. 75% is equivalent to 75 electrons collected per 100 incident photons. Further, in common with α, η is also a function of photon wavelength and must, therefore, only be quoted for a specific wavelength.

9.6.2 Responsivity (R)

The responsivity (R) of a photodetector is defined as the ratio of the photocurrent (I_{ph}) flowing in the device, to the incident optical power (P_o). Responsivity is measured in amps per watt. Thus:

$$R = \frac{\text{Photocurrent}}{\text{Incident optical power}} = \frac{I_{ph}}{P_o} = \frac{I_p}{P_o} \ (AW^{-1}) \quad (9.8)$$

Since R is a function of photon wavelength, R is calculated at a particular wavelength λ.

The output photocurrent (I_{ph} or I_p) may be expressed in terms of the rate, r_e of the electrons collected as follows:

$$I_p = e\, r_e \quad (9.9)$$

where e is the electronic charge. Combining Eqs (9.6) and (9.9), one obtains:

$$I_p = e\, \eta\, r_p \quad (9.10)$$

The rate of incident photon is given by:

$$r_p = \frac{\text{Incident optical power}}{\text{Energy of the photon}} = \frac{P_{in}}{h\nu} \quad (9.11)$$

Thus,
$$I_p = \frac{\eta e P_{in}}{h\nu} \quad (9.12)$$

Substituting for I_p from Eq. (9.12) in Eq. (9.8), we get expression for R in terms of η as follows:

$$R = \eta \frac{e}{h\nu} = \eta \frac{e\lambda}{hc} \quad (9.13)$$

Also given in Eq. (9.7) and Eq. (9.13), the efficiency and responsivity depend on wavelength, R is also called the *spectral responsivity* or *radiant sensitivity*. The R vs. λ characteristics represent the spectral response of the photodiode. Spectral response curves are generally provided by the manufacturer. The spectral response characteristics for various quantum efficiencies are shown in Fig. 9.6, and can be calculated using Eq. (9.13). The outer area of the detector has a higher responsivity than the centre area, which can cause problems when aligning the fiber cable to the detector.

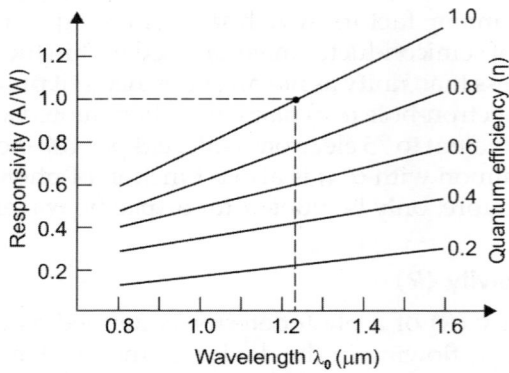

Fig. 9.6. Responsivity vs wavelength for various quantum efficiencies

Equation (9.13) also shows that the responsivity (R) is directly proportional to η at a particular wavelength and in the ideal case, when $\eta = 1$, R is directly proportional to λ (Fig. 9.6). This means that ions responsivity is a linear function of wavelength.

In case of a practical diode, as the wavelength of the incident photon becomes longer, its energy becomes smaller than that required for exciting the electron from the valence band to the conduction band. The responsivity (R), thus falls of the cut-off wavelength (λ_c) as shown in Fig. 9.7(b).

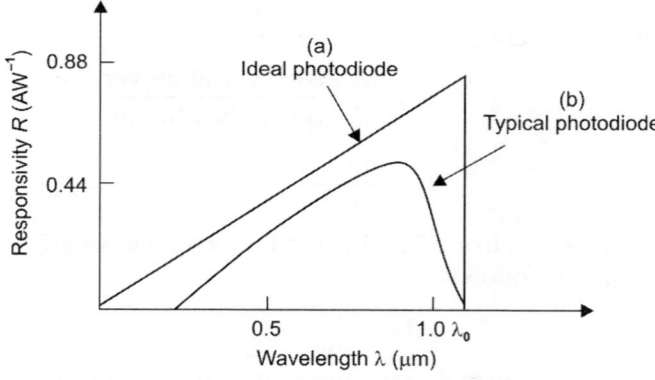

Fig. 9.7. Responsivity (R) as a function of wavelength (λ) for (a) an ideal Si photodiode and (b) a practical Si diode

We may note that responsivity gives transfer characteristics of detector, i.e., photocurrent per unit incident optical power.

Typical responsivities of *p-i-n* photodiodes are as follows:
- Silicon pin photodiode at 900 nm $\rightarrow$ 0.65 AW^{-1}
- Germanium pin photodiode at 1.3 µm $\rightarrow$ 0.45 AW^{-1}
- In GaAs pin photodiode at 1.3 µm $\rightarrow$ 0.9 AW^{-1}

In most photodiodes the quantum efficiency (η) is independent of the power level falling on the detector at a given photon energy. This means that ideal responsivity (R) is a linear function of the optical power, i.e., the photocurrent I_p is directly proportional to the optical power P_i incident upon the photodetector, so that R is constant at a given wavelength. However, η is not a constant at all wavelengths, since, it varies according to photon energy ($h\nu$). Consequently, R is a function of λ and of the photodiode material.

9.6.3 Long-Wavelength Cut-off

It is essential when considering the intrinsic absorption process that the energy of incident photons be greater than or equal to the band gap energy E_g of the material used to fabricate the photoconductor. Therefore, the photon energy:

$$\frac{hc}{\lambda} \geq E_g \qquad (9.14)$$

giving:

$$\lambda \leq \frac{hc}{E_g} \qquad (9.15)$$

Thus, the threshold for detection, commonly known as the long-wavelength cutoff point λ_c, above which photons are simply not absorbed by the semiconductor, given by:

$$\lambda_c = \frac{hc}{E_g} \qquad (9.16)$$

The expression given in Eq. (9.16) allows the calculation of the longest wavelength of light to give photodetection for the various semiconductor material used in the fabrication of detectors.

We may note that the above criterion is only applicable to intrinsic photodetectors. Extrinsic semiconductors violate the condition given by Eq. (9.16), but are not currently used in optical fiber communication.

9.6.4 Response Time

Response time is defined as the time needed for the photodiode to respond to an optical input by producing photocurrent. When light incident on the photodiode generates an electron-hole pair in a photodetector material, an electrical charge is generated in an external circuit, as shown in Fig. 9.8. This electrical charge, due to the electron and hole, equals $2e$ (e is the charge of an electron).

The charge delivered to the external circuit, by the movement of carriers in the photodetector material, is not provided instantaneously. The charge is delivered over an extended period. It is as if the motion of the charged carriers in the material draws charge slowly away from

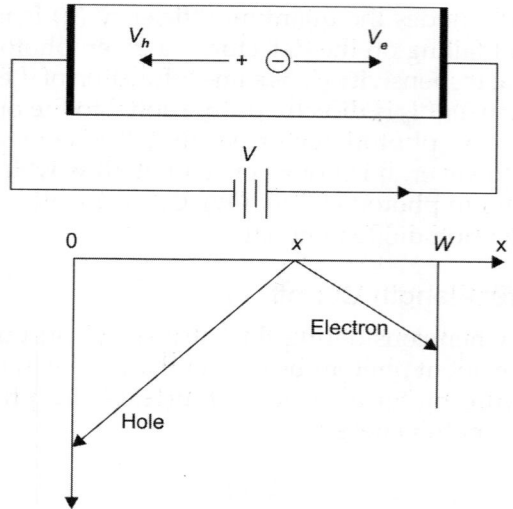

Fig. 9.8. Generated electron-hole pair at position *x*

the wire on one side of the device and then pushes it slowly into the wire at the other side. In this way, each charge passing through the external circuit is spread out in time. This phenomenon is called *transit-time spread*. It is an important limiting factor for the speed of operation of all semiconductor photodetectors.

Consider an electron-hole pair generated (by photon absorption, for example) at position x in a semiconductor material. The semiconductor material has width W, to which a voltage V is applied, as shown in Fig. 9.8. When an electron-hole pair is generated at position x, the hole moves to the left with velocity v_h, and the electron moves to the right with velocity v_e. This movement terminates when the carriers reach the edge of the material. The current (i) in the external circuit, generated by this movement, is given by:

$$i(t) = \frac{Q}{W} v(t) \qquad (9.17)$$

where
    t is the time
    Q is the total charge of the photo-generated electrons

If the voltage is increased, the electron velocity increases, and thus current increases. This means that for an input light pulse, the output current pulse will have a faster response time, for a higher applied voltage.

Response time can be affected by dark current, noise, responsivity linearity, back-reflection, and detector edge effect. Edge effect occurs because detectors only provide a fast response in their centre area.

9.6.5 Sensitivity

A photodetector is a device that converts photon energy into an electrical signal. A photodetector usually detects the energy of some photons better than others. Detection sensitivity is a function of the photon's energy being detected. The sensitivity is usually given as a function of the wavelength and expressed as the quantum efficiency. A high sensitivity allows a low level of light to be detected.

9.7 PERFORMANCE PARAMETERS OF PHOTODETECTORS

Main parameters which determine characteristics of photodetector are: dark current, spectral response, quantum efficiency, noise, detectivity, linearity and dynamic range, speed and frequency range. We have:

9.7.1 Dark Current or Leakage Current

The current resulting from absorbed incident light is called "generated photocurrent". *The current passing through the detector in the absence of light is called "dark current" or "leakage current."* Low leakage current is an important measure of device quality. If the dark current is high, the generated photocurrent needs to be larger, in order to provide a good signal. Otherwise, the leakage current will dominate the detector current. Therefore, it is important to control the dark current.

Sources of leakage current: The three fundamental sources of leakage current are:

(i) *Generation-recombination (g-r) current*: Arises from the generation and recombination of electron-hole pairs in the diode depletion region. The *g-r* current dominates the leakage current at low temperature.

(ii) *Diffusion current*: Arises from the diffusion of the minority carriers, toward or away from the junction, in the diode neutral region. In the case of p^+-n junction with the intrinsic region width larger than the hole diffusion length, the intrinsic region alone may be considered. Then diffuse current dominates the leakage current at high temperature.

(iii) *Tunneling current*: Refers to the band-to-band tunneling in the presence of high electric fields. A high field reduces the effective band gap barrier, allowing carriers to cross the band gap.

Typical values of dark current for popular semiconductors are shown in Table 9.1.

Table 9.1. Dark currents for different semiconductors and their compounds

Semiconductor	Dark current (nA)
Si	0.1–1
Ge	100
InGaAsP	1–10

9.7.2 Quantum Efficiency

See Section 9.6.

9.7.3 Responsivity (R)

See Section 9.6.

9.7.4 Speed of Response

It determines how photodetector responds to an optical signal. Fig. 9.9 shows a typical response to a pulse.

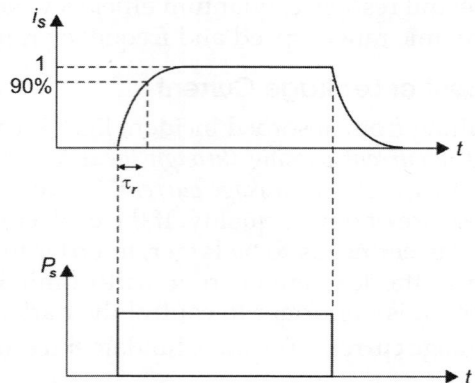

Fig. 9.9. Typical response of a photodetector to a square-pulse signal

Speed of response is determined by the *RC* time constant. In terms of parameters of previously introduced equivalent circuit, the *rise time* τ_r is given by:

$$\tau_r = 2.19 \cdot R_L \cdot C_j \tag{9.18}$$

The evaluation of τ_r is left as an exercise. Time response is directly related to the frequency response. The 3 dB bandwidth is given by:

$$f_{s-dB} = \frac{1}{2\pi R_L C_j} \tag{9.19}$$

9.8 TYPES OF OPTICAL DETECTORS

There are many types of optical detectors including the phototransistor, photovoltaic, metal-semiconductor-metal (MSM), *pin* photodiode, and avalanche photodiode (APD). These detectors are explained in the following sections.

9.8.1 Phototransistors

Phototransistors are the simplest type of photodetector. The basic operating principle of a phototransistor is shown in Fig. 9.10. The device

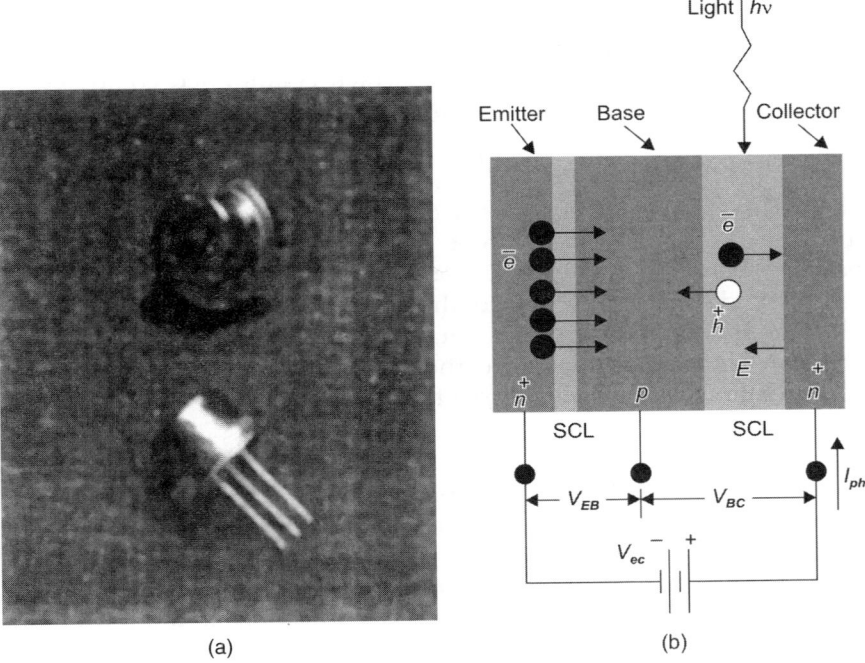

Fig. 9.10. Phototransistor

consists of an *n-p-n* junction, in which *n* is the emitter, *p* is the base, and the other *n* is the collector. The base terminal is normally open, and there is a voltage applied between the collector and emitter terminals (just as in the normal operation of common bipolar junction transistor (BJT).

A large space charge layer (SCL) forms between the base and collector. The SCL region is called the absorption region. The operation of this device begins when an incident photon is absorbed in the SCL, and generates an electron-hole pair. The electrical field E_o drifts the electron and hole in opposite directions. Phototransistors operate as a photodetector that amplifies the photocurrent. An applied voltage V will increase E to become E_o plus V. When the drifting electron reaches the collector, it gets collected (and thereby neutralized) by the power supply (applied voltage). On the other hand, when the hole enters the neutral base region, it can only be neutralized by injecting a large number of electrons into the base. It forces a large number of electrons to be injected from the emitter.

Normally, the electron recombination time in the base is very long, compared with the time it takes for electrons to diffuse across the base. The means that only a small fraction of electrons injected from the emitter can recombine with holes in the base. Thus, the emitter has to inject a large number of electrons to neutralize this extra hole in the base. These electrons diffuse across the base and reach the collector, and thereby, create a photocurrent, which is amplified compared to the original electron. Thus, phototransistors have photocurrent gain.

9.8.2 Photovoltaics

Photovoltaic panels, or solar cells, convert the incident solar radiation, through the photovoltaic effect, into electrical current. The basic principle behind this effect relies on the small energy gap between the valence and conduction bands of the photovoltaic material. When light photons incident on a photovoltaic have enough energy to excite electrons from the valence to the conduction band, the resulting accumulation of charge leads to a flow of current.

Figure 9.11 shows a typical solar panel and its cross-section. Consider a p-n junction with a very narrow and more heavily doped n-region. Solar radiation is incident on the thin n-side. The electrodes attached to the n-side must allow illumination to enter the cell and at the same time have a small series resistance. The electrodes are deposited on the n-side to form an array of finger electrodes on the surface. A thin anti-reflection coating on the surface reduces reflections and allows more light to enter the cells.

The width (W) of the depletion region or the space charge layer (SCL) extends primary into the p-side. Most photons are absorbed within the n-region and depletion region. Thus, short and medium wavelengths are absorbed. The generated electron-hole pairs are swept away by the built-in field E_o in the depletion layer. This creates an open circuit voltage

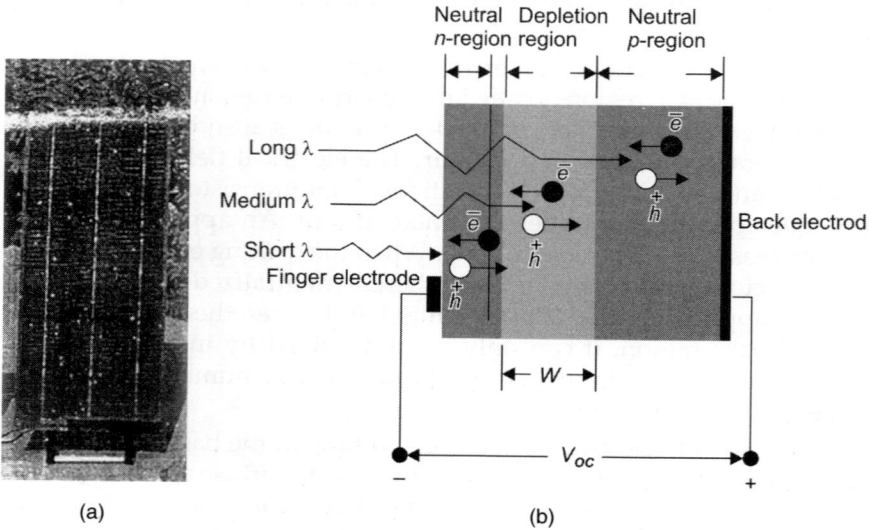

Fig. 9.11. Photovoltaic panel

V_{oc} between the electrodes. If an external load is connected, a photocurrent results.

The efficiency of a solar cell is one of its most important characteristics; it allows the device to be accessed economically, in comparison to other energy conversion devices. The solar cell efficiency refers to the fraction of incident light energy converted to electrical energy. This conversion efficiency depends on the semiconductor material properties, the device structure, and the incident light wavelength spectrum, which is mostly solar radiation. The efficiency of a solar cell decreases with increasing temperature. Therefore, the temperature of solar cells must be controlled for maximum efficiency.

Most solar cells are silicon based; silicon based semiconductor fabrication is a very developed technology, enabling cost effective devices for energy production in remote applications. A solar cell fabricated by making a *p-n* junction in the same crystal is called "homojunction". A silicon homojunction solar cell is called a *"single crystal passivated emitter rear locally diffused"* (PERLD) *cell*. It has higher efficiency than other types of semiconductor solar cells.

9.8.3 Metal-Semiconductor-Metal (MSM) Detectors

Metal-semiconductor-metal (MSM) detectors are probably the fastest and simplest optical detector to fabricate. The basic idea is to create a Shottky barrier, which forces the material at the surface to be depleted. This barrier is created by contacting a metal to the semiconductor surface. Fig. 9.12(a) shows the cross-section of a metal area. Fig. 9.12(b) shows top view of MSM structure and its cross-sectional view is shown in Fig. 9.12(c). The barriers are often in the form of inter-digitated metal fingers separated by a small distance, typically on the order of microns. The metal is usually opaque to the incoming light; the remainder of the

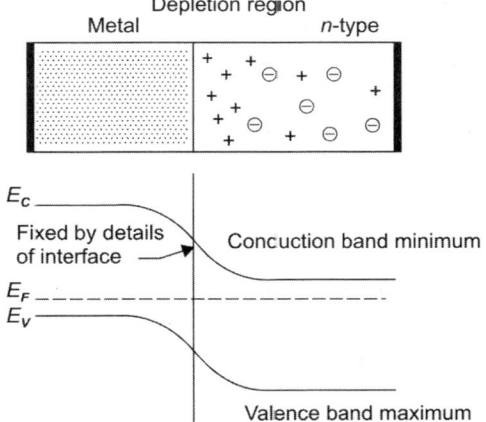

Fig. 9.12. (a) Shottky barrier and energy diagram

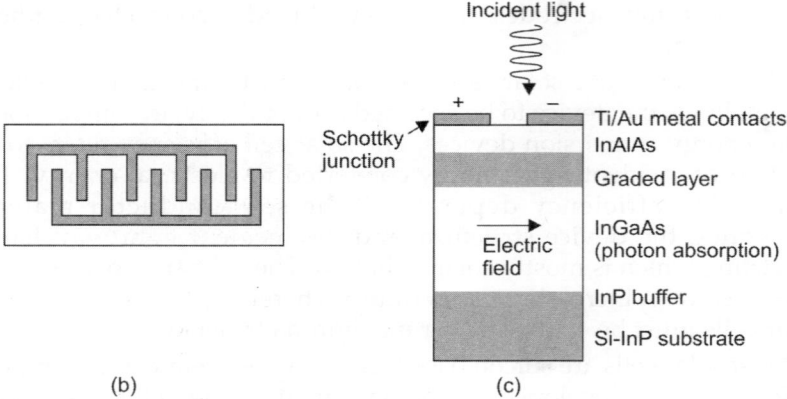

Fig. 9.12. (b) Top view of the structure of MSM (c) and its cross-section

surface area absorbs the light. All the depletion layers are connected together. Any absorbed light generates electron-hole pairs, which are quickly swept out to the contacts. Full-width at half maximum (FWHM) pulses are measured in picoseconds for such structures, since the response time is so quick.

When using the MSM for the detection of 1300–1500 nm ranges, the MSM suffers from following two serious drawbacks:

 (i) Shottky barriers on indium phosphate (InP) tend to have high dark current, and therefore, low receiver sensitivity.

 (ii) Low quantum efficiency results, because the metal fingers prevent some of the incoming light from reaching the absorption layer.

Salient Features of MSM Photodetectors

- MSM photodetector uses a sandwitched semiconductor between two metals. The middle semionductor layer acts as optical absorbing layer. A Schottky barrier is formed at each metal semi-conductor interface (junction), which prevents flow of electrons.
- When optical power is incident on it, *e-h* pairs generated through photoabsorption flow towards metal contacts and causes photo-current.
- These photodetectors are manufactured using different combinations of semiconductors such as GaAs, InGaAs, InP, InAlAs. Each MSM photodetector has distinct features, e.g. R, η, W, etc.
- With InAlAs based MSM photodetector, 92% η can be achieved at 1.3 μm with low dark current. An *inverted* MSM photodetector shows high R when illuminated from top.
- A GaAs based MSM photodetector with travelling wave structure gives a bandwidth (W) beyond 500 GHz.

- MSM structure (an interdigital pattern of metal fingers deposited on a semiconductor substrate and a typical *p-i-n* detector) shows several improvements compared to traditional designs, e.g. sensitivity. Most of the improvements result from the lateral design.

9.8.4 p-n Photodiode

Figure 9.13(a) shows the simplest structure of a *p-n* photodiode. Incident photons of energy (say *h*v) are absorbed not only inside the depletion region but also outside it (Fig. 9.13(b)). Photons absorbed within the depletion region generate electron-hole (*e-h*) pairs. Due to built-in strong electric field (Fig. 9.13(c)), electrons and holes generated within this region get accelerated in opposite directions, and thereby, drift to the *n*-side and *p*-side respectively. The resulting flow of photocurrent constitutes the response of the photodiode to the incident optical power. The response time is governed by the transit time τ_{drift} in accordance with the relation:

$$\tau_{drift} = \frac{W}{V_{drift}} \qquad (9.20)$$

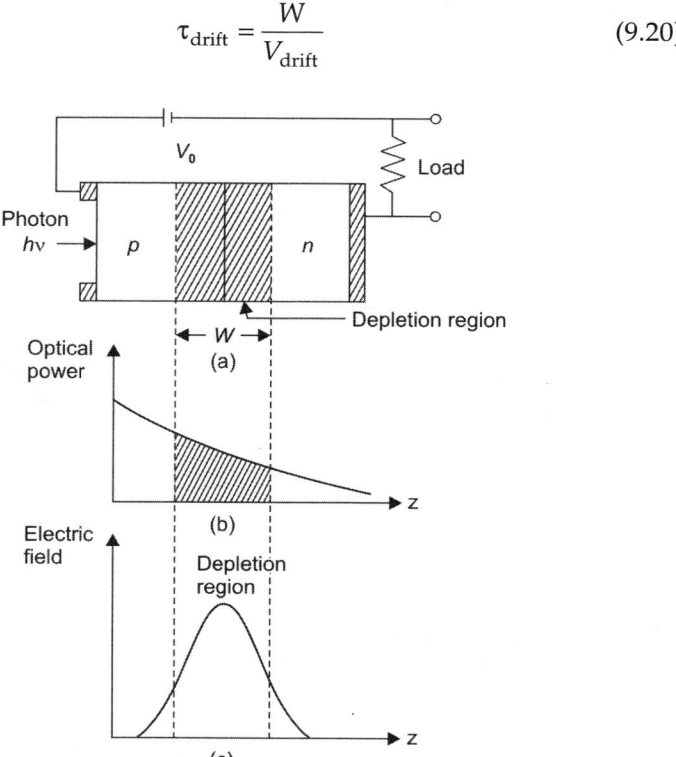

Fig. 9.13. (a) Schematic structure of a *p-n* photodiode and the associated depletion region under reverse bias. (b) Variation of optical power within the diode. (c) Variation of electric field inside the diode

where W is the width of the depletion region and V_{drift} is the average drift velocity. τ_{drift} is of the order of 100 ps, which is small enough for the photodiode to operate upto a bit rate of about 1 G bits/s. The depletion layers width W of a p-n photodiode is given by:

$$W = \left[\frac{2\varepsilon}{e}(V_{bi} + V_o)\left(\frac{1}{N_a} + \frac{1}{N_d}\right)\right]^{1/2} \tag{9.21}$$

Here ε is the dielectric constant, V_{bi} is built in voltage and depends on semiconductor, V_o is the applied bias voltage, and N_a and N_d are the acceptor and donor concentrations respectively used to fabricate the p-n junction. V_{drift} depends on the bias voltage but attains a saturation value depending on the semiconductor of the p-n diode.

Incident photons are absorbed outside the depletion region also (Fig. 9.13(b)). The electrons generated in the p-side have to diffuse to the depletion region boundary, prior to they can drift (under the built in electric field) to the n-side. In a similar way, holes generated in the n-side have to diffuse to the depletion region boundary for their drift towards the p-side. However, the diffusion process is inherently slow, and therefore, the presence of a diffusive component may distort the temporal response of a photodiode (Fig. 9.14).

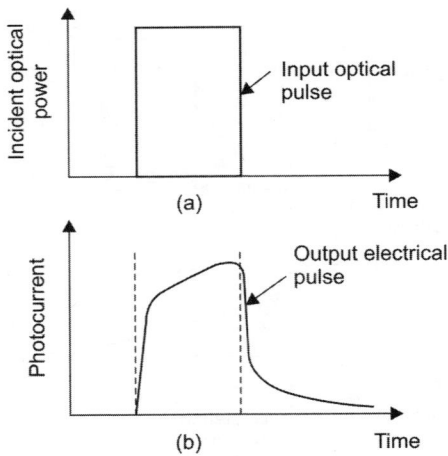

Fig. 9.14. Schematic response of a typical p-n photodiode to a rectangular optical pulse when both drift and diffusion contribute to the photocurrent

In reverse biased photodiode, the semiconductor material absorbs a photon of energy ($h\nu$), which excites an electron from the valence band to the conduction band (opposite to photon emission). The photo generated electron leaves behind it a hole (h), and thus, each photon generates two charge carriers. This enhances the material conductivity, so called *photoconductivity* resulting in an increase in the diode current (I_{diode}). Obviously, the diode equation is modified as:

$$I_{\text{diode}} = (I_d + I_s)(e^{V_q/nkT} - 1) \qquad (9.22)$$

where I_d is dark or leakage current that flows when no signal is present and I_s the photo generated current due to incident optical signal. A plot of Eq. (9.22) for varying amounts of incident optical power is shown in Fig. 9.15.

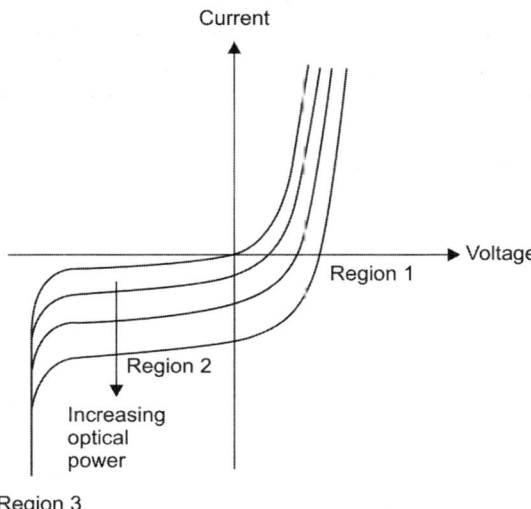

Fig. 9.15. *V-I* characteristics of photodiode

We see that there are three regions: Fowards bias, reverse bias and avalanche breakdown.

(i) **Region 1: Forward bias:** A change in incident power causes a change in terminal voltage, it is called as *photovoltaic mode*. Now, if the diode is operated in this mode, the frequency response of the diode is poor and this is why photovoltaic operation is rarely used in optical links.

(ii) **Region 2: Reverse bias:** We see that a change in optical power produces a proportional change in diode current and it is called as *photoconductive mode* of operation which most optical detector use. Under these conditions, the exponential term in Eq. (9.22) becomes insignificant and one obtains the reverse bias current as:

$$I_{\text{diode}} = (I_d - I_s) \qquad (9.23)$$

(iii) **Region 3: Avalanche breakdown:** When photodiode biased in this region, a photogenerated electron-hole pair causes avalanche breakdown, resulting in large diode current for a single incident photon. *Avalanche photodiode* (APD) operate in this region. APDs exhibits carrier multiplication. APDs are usually very sensitive detectors. However, V-I characteristic is very steep in this region, and hence, the bias voltage have to be tightly controlled to prevent spontaneous breakdown.

9.8.5 The *p-i-n* (PIN) Photodiodes

In *p-i-n* photodiodes, the conversion of light into electrical current is achieved by the creation of free electron hole pairs by the absorption of photons. This absorption process creates electrons in the conduction band and holes in the valence band. Figure 9.16 shows the simplified structure of a typical *p-i-n* junction photodiode. The structure of the photodiode is the p^+-intrinsic-n^+ junction. The intrinsic silicon (*i*-Si) layer has much less doping than both the p^+ and n^+ regions, and it is much wider than these regions. At long wavelengths, where penetration depth is large, the photons can be absorbed in the wide depletion region. Thus, the width depends on the particular wavelength used in the application. In contrast, a *p-n* junction has a narrow depletion region and fewer photons are absorbed.

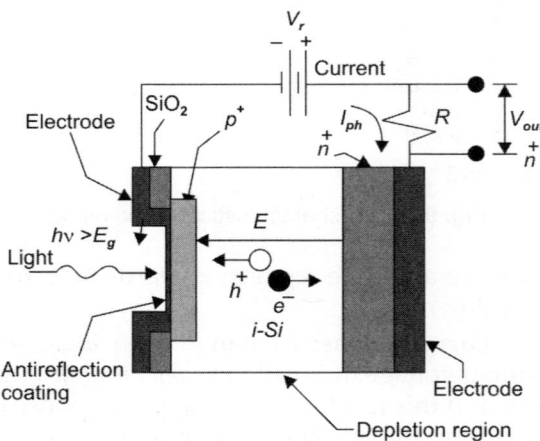

Fig. 9.16. A *p-i-n* junction photodiode

When the structure is formed, holes diffuse from the p^+ side, and electrons from the n^+ side, into the *i*-Si layer, also called the depletion region. In this region, they recombine (with other holes and electrons) and disappear. This leaves behind a thin layer of exposed, negatively-charged acceptor ions in the p^+ side, and a thin layer of exposed, positively-charged donor ions in the n^+ side. The two charges are separated and create the built-in electric field E in the *i*-Si layer. An exterior voltage increases E, which increase the response speed. While the photogenerated carriers are drifting through the *i*-Si layer, they create an external photocurrent, when a voltage is applied. The photocurrent can be detected as the voltage across a small external resistor R, as shown in Fig. 9.16. A larger thickness of the *i*-Si layer increases QE, but slows the response time since carriers have further to travel.

In some photodiodes, such as pyroelectric detectors, the energy conversion generates heat, which increases the temperature of the detector. The temperature increase changes the polarization and relative permitivity of the photodiode.

The *p-i-n* photodetectors offer high bandwidth, high quantum efficiency, and low dark current. High bandwidth and low dark current are important characteristics for good receiver sensitivity. However, the device has no gap, which places a lower limit on the sensitivity achievable, before dark current becomes significant. The *p-i-n* photodetectors are small devices with small capacitance, thus can detect high-speed signals with high sensitivity. The most important applications of the photodiodes are in optical communications.

A double heterostructure (Fig. 9.17) improves the performance of *p-i-n* photodiode. In this structure, the middle *i*-region of a material with lower bandgap is sandwiched between *p*- and *n*-type materials of higher bandgap, so that incident light is absorbed only in *i*-region (Fig. 9.17). The bandgap of InP is 1.35 eV and hence, transparent to light for λ > 0.2 μm, whereas the bandgap of lattice matched InGaAs is about 0.75 eV, which corres-ponds to λ_c of 1.65 μm. This means that the intrinsic layer of InGaAs absorbs strongly in the wavelength range 1.3–1.6 μm. In heterostructure, the diffusive component of the photocurrent is completely eliminated, because the incident photons are absorbed only within the depletion region. Heterostructure *p-i-n* photodiodes are very useful for fiber-optic communication systems operating in the range 1.3–1.6 μm.

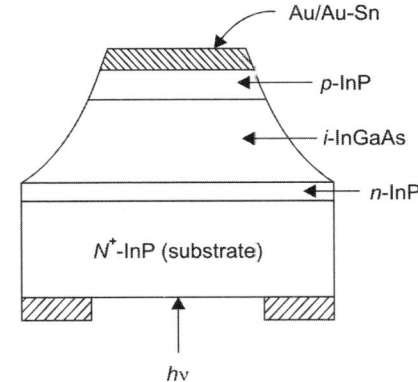

Fig. 9.17. Schematic of double heterostructure of *p-i-n* photodiode using InGaA/InP

Salient features of p-i-n photodiodes

(i) In the absence of light, *p-i-n* photodiodes behave electrically just like an ordinary rectifier diode. When forward biased, they conduct large amount of current.

(ii) *p-i-n* detectors can be operated in two modes: *photovoltaic* and *photoconductive*. In photovoltaic mode, no bias is applied to the detector. In this mode, the detector works very slow, and output of the detector is approximately *logarithmic* to the input level. In practice, fiber optic receivers never use this mode.
In *photoconductive mode*, the detector is reverse biased. In this case, the output is a current that is very linear with the input light power.

(iii) The intrinsic region somewhat improves the sensitivity of the *p-i-n* photodetector. It does not provide internal gain. The combination of different semiconductors operating at different wavelength permits the selection of material capable of responding to the desired operating wavelength. Table 9.2 summarizes the characteristics of common *p-i-n* photodiodes.

Table 9.2. Characteristics of common *p-i-n* photodiodes

S. No.	Parameters	Symbols	Unit	Si	Ge	InGaAs
1.	Wavelength	λ	μm	0.4–1.1	0.8–1.8	1.0–1.7
2.	Responsivity	R	A/W	0.4–0.6	0.5–0.7	0.6–0.9
3.	Quantum efficiency	η	%	75–90	50–55	60–70
4.	Dark current	I_d	nA	1–10	50–500	1–20
5.	Rise time	T_r	nS	0.5–1	0.1–0.5	0.02–0.5
6.	Bandwidth	B	GHz	0.3–0.6	0.5–3	1–10
7.	Bias voltage	V_b	V	50–100	5–10	5–6

Depletion Layer Photocurrent

Figure 9.18 shows a reverse biased *p-i-n* photodiode and electric field distribution under reverse bias. The total current density through depletion layer is:

$$J_{\text{total}} = J_{\text{drift}} + J_{\text{diffusion}} \quad (9.24)$$
$$(J_{\text{tot}}) \quad (J_{\text{dr}}) \quad (J_{\text{diff}})$$

where, J_{dr} is drift current density due to carriers generated in depletion region and J_{diff} is diffusion current density due to carriers generated outside depletion region. One can express the drift current density as:

$$J_{\text{dr}} = \frac{I_p}{A}$$

or
$$J_{\text{dr}} = e\phi_o (1 - e^{\alpha W}) \quad (9.25)$$

where A is photodiode area and ϕ_o is incident photon flux per unit area.

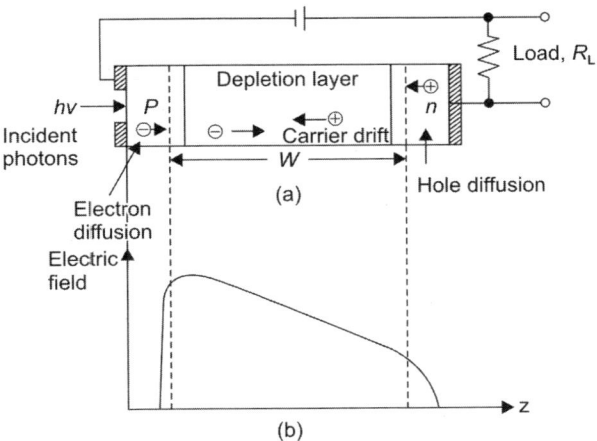

Fig. 9.18. (a) Schematic structure of a reverse biased *p-i-n* photodiode. (b) Electric-field distribution inside the device under reverse bias

The diffusion current density is given as:

$$J_{\text{diff}} = e\phi_o \frac{\alpha L_p}{1+\alpha L_p} e^{-\alpha W} + eP_{no} \frac{D_p}{L_p} \quad (9.26)$$

where D_p is hole diffusion coefficient, P_{no} is equilibrium hole density. Substituting (9.25) and (9.26) in (9.24), one obtains:

$$J_{\text{tot}} = e\phi_o \left[1 + \frac{\alpha W}{1+\alpha L_p}\right] + eP_{no} \frac{D_p}{L_p} \quad (9.27)$$

Response Time

Response time of a photodiode depends upon the following factors:
 (a) Transit time of photocarriers within the depletion region.
 (b) Diffusion of photocarriers outside the depletion region.
 (c) RC time constant of diode and external circuit.

The transit time (t_d) is given by:

$$t_d = \frac{W}{V_d} \quad (9.28)$$

The diffusion process is slow and diffusion times are less than carrier drift time. One can calculate the effect of diffusion by considering the photodiode response time. Fig. 9.19 shows the response time of photodiode which is not fully depleted.

If R_T is the combination of the load and amplifier input resistance and C_T is the sum of the photodiode and amplifier capacitances, the detector behaves like a simple RC low-pass filter with a passband given by:

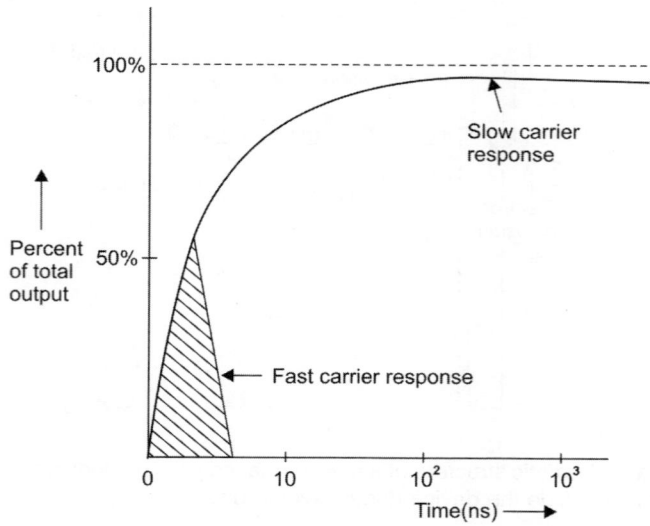

Fig. 9.19. Response time of photodiode not fully depleted

$$B = \frac{1}{2\pi R_T C_T} \tag{9.29}$$

9.8.6 Avalanche Photodiodes

Avalanche photodiodes (APDs) are high performance devices and are widely used in many applications, such as optical communication, due to their high speed and internal gain. Although not as fast as *p-i-n* photodiodes, the devices offer superior receiver sensitivity in their own bandwidth range. The device bandwidth at high gain is limited by the gain-bandwidth product (GBW), for InGaAs-InP APDs. However, there are difficulties in fabricating the device; this process requires stringent process control. For this reason, commercial high performance APDs cost more than similar photodetectors.

Figure 9.20 shows the cross-section of the structure of an InGaAs-InP avalanche photodiode with separate absorption and multiplication (SAM) regions. The InP multiplication (avalanche) layer has a wider bandgap than InGaAs absorption layer. The *p*-type and *n*-type multiplication (avalanche) layer has a wider bandgap than InGaAs absorption layer. The *p*-type and *n*-type doping of InP is indicated by capital letters, *P* and *N*. Fig. 9.20(b) shows the configuration for achieving carrier multiplication with little excess noise and is called a *reach through avalanche photodiode* (RAPD).

The main depletion layer forms between the P^+-InP and the N-InP layers and is within the N-InP (Fig. 9.20(a)). The electric field is greatest in this N-InP layer, this is, therefore, where avalanche multiplication

Optical Receivers 475

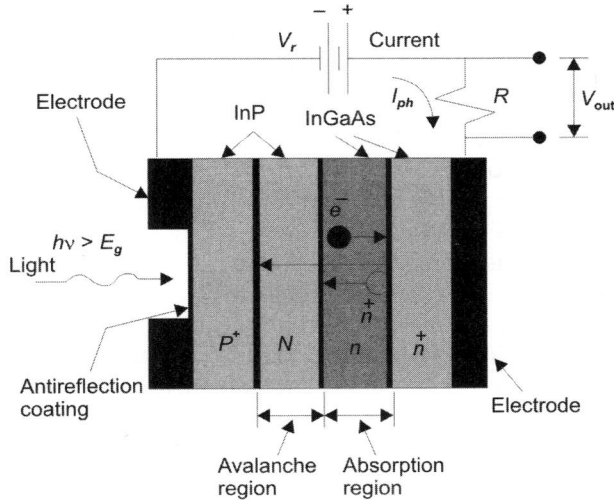

Fig. 9.20. (a) Avalanche photodiode

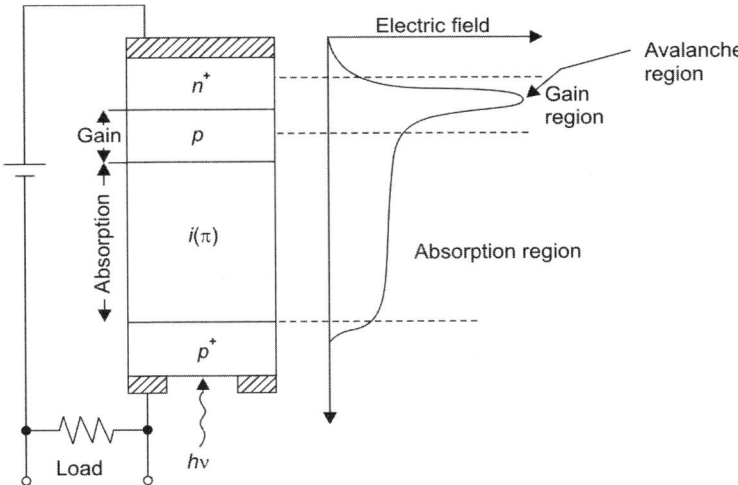

Fig. 9.20. (b) Schematic configuration of RAPD and the variation of electric field in the depletion and multiplication regions

takes place (Fig. 9.20(b)). With sufficient reverse voltage bias, the depletion layer in the n-InGaAs extends into the N-InP layer.

The electric field in the n-InGaAs depletion layer is not as great as that in the N-InP. Although long wavelength photons are incident on the InP side, they are not absorbed by InP, since the photon energy is less than the bandgap energy of InP ($E_g = 1.35$ eV). Long wavelength photons pass through the InP layer and are absorbed in the n-InGaAs layers.

The electric field E in the n-InGaAs layer drifts the holes to the multiplication region, where impact ionization multiplies the carriers. The impact ionization, from the physics point of view, is the mechanism that creates the internal current gain. Primary electrons and/or holes (carriers) are generated through the absorption of photons. Carriers can acquire large amounts of energy from a high E field, when the device has a strong reverse-voltage bias. This can be translated into high-speed motion. When a collision between a carrier and the lattice occurs, the energy from the carrier can be transferred to the lattice. Sufficient energy can be absorbed by the lattice for an electron to be promoted from the valence to the conduction band, creating an electron-hole pair. This process is called "impact ionization". These new carriers are swept out by E and can acquire high energy, causing further electron-hole pairs to be created. The entire process in which many carriers are created from one initial carrier is called "avalanche multiplication".

Salient features of APD

- APD uses the avalanche breakdown phenomena for its operation. The APD has its internal gain which increases its responsivity (R).
- When APD is biased close to breakdown, it will result in reverse leakage current. This is why, APDs are usually biased just below breakdown, with the bias voltage being tightly controlled.
- The multiplication (M) for all carriers generated in the photodiode is given as:

$$M = \frac{I_M}{I_P}$$

where I_M is average value of total multiplied output current, and I_P is primary unmultiplied photocurrent.

- Responsivity of APD is given by:

$$R_{APD} = \frac{\eta e}{h\nu} M = \frac{n e \lambda}{hc} M \quad (\because \nu = c/\lambda) \qquad (9.30)$$

$$\therefore \qquad R_{APD} = R_o M \qquad (9.31)$$

where $R_o = \dfrac{n e \lambda}{hc}$ is unity gain responsivity.

Characteristics of common APD are given in Table 9.3.

A comparison of *p-i-n* and APD photodetectors is given in Table 9.4.

9.9 PHOTODETECTOR NOISE

Typical optical signals arriving at the receiver front end are very weak. They are, therefore, significantly affected by various noise sources.

There are several physical processes which contribute to noise. For example, the APD generates noise in the avalanche process and optical

Table 9.3. Characteristics of common APDs

S. No.	Parameters	Symbols	Unit	Si	Ge	InGaAs
1.	Wavelength	λ	μm	0.4–1.1	0.8–1.8	1.0–1.7
2.	Responsivity	R_{APD}	A/W	80–130	3–30	5–20
3.	APD gain	M	–	100–500	50–200	10–40
4.	k-factor	kA	–	0.02–0.05	0.7–1.0	0.5–0.7
5.	Dark current	I_d	nA	0.1–1	50–500	1–5
6.	Rise time	T_r	ns	0.1–2	0.5–0.8	0.1–0.5
7.	Bandwidth	B	GHz	0.02–1	0.4–0.7	1–10
8.	Bias voltage	V_b	V	200–250	20–40	20–30

Table 9.4. Comparison of *p-i-n* and APD photodetectors

p-i-n detectors	*APD detectors*
• Fast	Not as fast
• High bandwidth, upto 40 GHZ at quantum efficiency > 80%	Significantly less bandwidth
• Low dark current	High dark current
• No gain, which leds to lower sensitivity	Built in gain, which extends the sensitivity to lower levels of received light
• Conversion efficiency (responsivity) from 0.5 to 1.0 A/W	Conversion efficiency (responsivity) from 0.5 to 100.0 A/W
• Less expensive	More expensive
• Sensitivity less (0–12 dB)	More sensitivity (5 to 15 dB)
• Low reverse biasing voltage (5 to 10 V)	High reverse bias voltage (20 to 40 V)
• Wavelength region 300 to 1100 nm	Wavelength region 400–100 nm
• S/N ratio poor	S/N ratio better
• Simple detector circuit	More complex detector circuit

amplifiers produce noise due to amplified spontaneous emission (ASE). In addition, there is always a *quantum noise* which exists in all devices and sets a fundamental lower limit on noise power.

The working parameter characterizing the photodetector is the signal-to-noise ratio (SNR), which is defined as:

$$\frac{S}{N} = \frac{\text{signal power from photocurrent}}{\text{photodetector noise power + amplifier noise power}} \quad (9.32)$$

Equation (9.32) clearly reveals that to achieve high S/N, (i) the photodetector should have high quantum efficiency and low noise so that it generate large signal power, and (ii) the amplifier noise must be kept low. The sensitivity of a detector is defined as the power necessary to generate a photocurrent equal in magnitude to the *rms* value at the total noise current. Noise appears as random fluctuations in a signal. A typical measure of noise is associated with the variance or the mean square deviation of signal s, defined as:

$$a_s^2 = \overline{(s-\bar{s})^2} = \overline{s^2} - \bar{s}^2$$

The mean value (average) of s is defined as:

$$\bar{s} = \sum_s p(s)s$$

where $p(s)$ is the probability of the measured signal having a value s and the sum is evaluated over all possible values obtained from measuring the signal.

Noise in a signal s can be represented by random variable s_n:

$$s_n = s - \bar{s}$$

If in the device (or system), there are two or more simultaneously present noise sources $s_{n1}, s_{n2}, \ldots$ which are independent, their combined effect is found by adding their mean square values (or their powers):

$$\overline{s_n^2} = \overline{s_{n1}^2} + \overline{s_{n2}^2} + \ldots$$

In terms of the above quantities, the SNR is expressed as:

$$\text{SNR} = \frac{\overline{s^2}}{\overline{s_n^2}} = \frac{\overline{s^2}}{\sigma_s^2} \quad \text{or} \quad \text{SNR} = 10 \log \frac{\overline{s^2}}{\sigma_s^2} \text{[dB]} \qquad (9.33)$$

Main types of noise arising in a photodetector are as follows:

9.9.1 Shot Noise

Shot noise is due to random distribution of electrons generated in the photodetector. It is associated with the quantum nature of photons arriving at the photodetector which generates carriers. Photons arrive at the photodetector randomly in time due to their quantum-mechanical nature. Their randomness is described by Poisson statistics. The resulting expression for shot noise of current in the photodetector is:

$$\overline{i_s^2} = 2eB\bar{i_s} \qquad (9.34)$$

where e is the electron charge, B is the bandwidth and $\bar{i_s}$ is the average value of signal current.

9.9.2 Thermal Noise

Thermal noise or *Johnson noise* is due to the random motion of electrons in the resistor R. It is modelled as a Gaussian random process with zero mean and autocorrelation function given by:

$$\frac{4k_B T}{R} \delta(\tau) \qquad (9.35)$$

where $\delta(\tau)$ is the Dirac delta function. Thermal noise powder in a bandwidth B is:

$$p_{s \cdot th} = 3k_B T B \tag{9.36}$$

where $k_B = 1.38 \times 10^{-23}$ J/K is the Boltzmann's constant and T is the absolute temperature. Thermal noise can be expressed as a current source:

$$\overline{i_{s.th}^2} = \frac{4k_B T}{R} B \equiv I_T^2 B \tag{9.37}$$

Here I_T is the parameter used to specify standard deviation in units of pA/$\sqrt{\text{Hz}}$. Its typical value is 1pA/$\sqrt{\text{Hz}}$.

Now, we can write a current generated by a *p-n* or *p-i-n* photodiode in response to an instantaneous optical signal as:

$$I(t) = <i_p(t)> + i_s(t) + i_d(t) + i_T(t) \tag{9.38}$$

where $\langle i_p(r) \rangle = I_p = RP_{in}$ is the average photocurrent, and $i_s(t)$, $i_d(t)$, and $i_T(t)$ are the current fluctuations related to *shot noise*, *dark current noise*, and *thermal noise*, respectively.

Quantum or shot noise arises

$$\langle i_s^2(t) \rangle = 2eI_p \, \Delta f \, M^2 F(M) \tag{9.39}$$

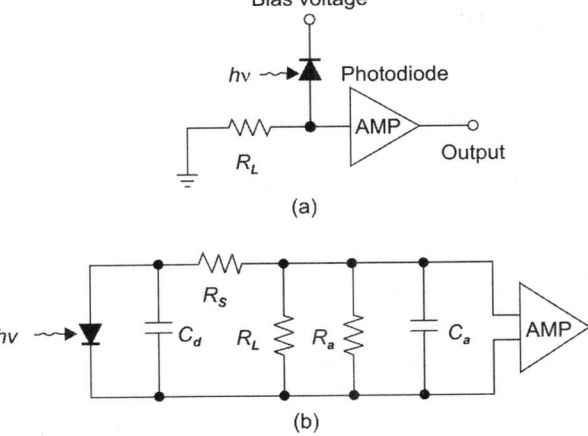

Fig. 9.21. (a) A schematic of simplest model of an optical receiver; (b) equivalent circuit

where $F(M)$ is a noise factor related to the random nature of an avalanche process and Δf is the effective noise bandwidth. Experimentally, it has been reported that $F(M) = M^x$, where x depends on the material and varies from 0 to 1. For *p-n* and *p-i-n* photodiodes, $F(M)$ and M are unity.

Dark current or leakage current is a reverse leakage current that continues to flow through the device when no light is incident on the photodetector. Normally, it arises from the electrons or holes which are thermally generated near the *p-n* junction of a photodiode. In an APD, these carriers get multiplied by the avalanche gain mechanism. Thus, the mean square value of the dark current is given by:

$$\overline{i_d^2(t)} = \langle i_d^2(t) \rangle = 2eI_d M^2 F(M)\, \Delta f \qquad (9.40)$$

where I_d is the average primary dark current (before multiplication) of the detector.

Thermal (or *Johnson*) noise is a random fluctiation in current due to the thermally induced random motion of electrons in a conductor. The load resistance R_L adds such fluctuations to the current generated by the photodiode. The mean square value of this current is given by:

$$\langle i_T^2 \rangle = \frac{4kT}{R_L} \Delta f \qquad (9.41)$$

where k is Boltzmann's constant and T is the absolute temperature of the load resistor.

The noise power associated with the amplifier following the detector will depend on the active elements of the amplifier circuit. For the present discussion, let us assume that its mean square noise current is $\langle i_{amp}^2 \rangle$.

In general, therefore, the signal to noise ratio (S/N) of an optical receiver may be written as:

$$\frac{S}{N} = \frac{\langle i_p^2 \rangle M^2}{2e(I_p + I_d)\,\Delta f\, M^2 F(M) + \dfrac{2kT\Delta f}{R_L} + \langle i_{amp}^2 \rangle} \qquad (9.42)$$

When *p-n* and *p-i-n* photodiodes are used in the receiver, M and $F(M)$ become unity.

9.10 RECEIVER ANALYSIS

This involves detection of signals with noise. In optical systems the information is transmitted using light which consists of photons (Fig. 9.21). Due to their statistical nature, the transmitted information will always show random fluctuations. Those fluctuations determine the lowest limit on transmitted power. In addition, there exist other noise contributions which originate from various processes. Some of them have already been discussed.

Digital signal under consideration operates at a bit rate B. The time slot (or bit interval) T is:

$$T = \frac{1}{B}$$

and it is the inverse of the bit rate. The input data sequence in the communication system is denoted by $\{b_k\}$. The optical power $p(t)$ falling on the photodetector is the sequence of pulses and it is written as:

$$p(t) = \sum_{k=-\infty}^{k=+\infty} b_k h_p(t - k \cdot T) \tag{9.43}$$

where k is a parameter denoting the k-th time slot and $h_p(t)$ represents the pulse shape of an isolated optical pulse at the photodetector input.

It is assumed that:

$$\frac{1}{T} \int_{-\infty}^{+\infty} h_p(t) dt = 1 \tag{9.44}$$

so that b_k represents the received optical power in the k-th time slot.

Equation (9.43) is based on the assumption that the system consisting of transmitter and optical fiber is a linear one and time-invariant. b_k in Eq. (9.43) can take two values b_0 and b_1, which correspond to logical values in the k-th time slot being zero or one. In an ideal case, one would expect the b_0 to be zero so no optical power is transmitted for the logical zero. However, semiconductor lasers always operate at the nonzero bias current, so there is always some small optical power transmitted.

9.11 BER OF AN IDEAL OPTICAL RECEIVER

There are several ways to measure *bit error rate* (BER), that is the rate of error occurrence in a digital data stream. It is equal to number of errors occurring over some time interval divided by total number of pulses (both ones and zeros).

The simplest method to define BER is:

$$\text{BER} = \frac{N_e}{N_p} = \frac{N_e}{B \cdot t} \tag{9.45}$$

where N_e is the number of errors appearing over time interval t, N_p is the number of pulses transmitted during that interval, $B = 1/T$ is the bit rate and T is the bit interval.

The required BER for high-speed optical communication systems today is typically 10^{-12}, which means that on average one bit error is allowed for every terabit of data transmitted. BER depends on the various signal-to-noise ratio (SNR) of the fiber system, like the receiver noise level.

Direct optical detection is a process of determining the presence or absence of light during a bit interval. No light is interpreted as logical zero; some of the light present signals is logical one.

In a real life, the detection process is not so simple because of the random nature of photons arriving at the receiver. Their arrival is modelled as a Poisson random process. The random process, in time arrivals of photons at the photodetector is shown in Fig. 9.22.

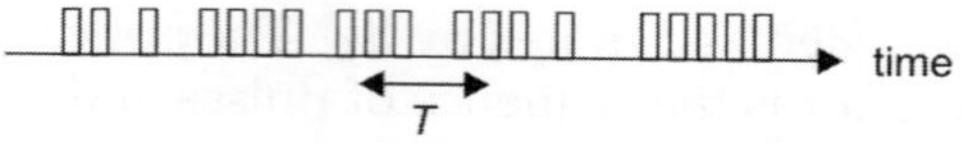

482 Photonics | Optoelectronics

```
__ΠΠ_Π_ΠΠΠΠ_ΠΠΠ_ΠΠΠ_Π_ΠΠΠΠΠ___→ time
         ←——→
           T
```

Fig. 9.22. Random arrivals of photons at photodetector are described by the Poisson process. Each photon is represented by a box and they all have the same amplitude

For an ideal optical receiver, we will assume that there are no noise sources in the system. The average number of photons arriving at the photodetector, with $h\nu_c$ being the energy of a single photon, is thus:

$$N = \frac{p(t)}{h\nu_c} \qquad (9.46)$$

where $p(t)$ is power of light signal, h is Planck's constant and ν_c is the carrier frequency. The output power impinging on the photodetector is expressed by Eq. (9.43)

A simple expression for BER for ideal receiver (not noise) can be obtained as follows:

The probability that n photons are received during a bit interval T is:

$$e^{-\frac{N}{S}} \frac{\left(\frac{N}{B}\right)^n}{n!}$$

where N is the average number of photons given by Eq. (9.46). Probability of not receiving any photons ($n = 0$) is $\exp(-N/B)$. Assume equal probabilities of receiving zero and one. The BER of an ideal receiver is thus:

$$\text{BER} = \frac{1}{2}e^{-N/B} \equiv \frac{1}{2}e^{-M} \qquad (9.47)$$

where $M = \frac{N}{B} = \frac{p}{h\nu_c B}$ represents the average number of photons received during one bit.

Equation (9.47) represents BER for an ideal receiver and it is called the *quantum limit*. To get a typical bit rate of 10^{-12}, the average number of photons is $M = 27$ per one bit.

Example 1

When 3×10^{11} photons each with a wavelength of 0.85 µm are incident on a photodiode, on an average 1.2×10^{11} electrons are collected at the terminals of the device. Calculate the quantum efficiency (η) and the responsivity (R) of the photodiode at 0.85 µm.

Solution

$$\eta = \frac{\text{Number of electrons collected}}{\text{Number of incident photons}} = \frac{1.2 \times 10^{11}}{3 \times 10^{11}} = 0.4$$

Thus, η of the photodiode at 0.85 μm is 40%.

$$R = \frac{\eta e \lambda}{hc} = \frac{0.4 \times 1.602 \times 10^{-19} \times 0.85 \times 10^{-6}}{6.626 \times 10^{-34} \times 3 \times 10^8} = 0.274 \text{ AW}^{-1}$$

Thus, R of the photodiode at 0.85 μm is 0.27 AW^{-1}.

Example 2

GaAs has a bandgap energy (E_g) of 1.43 eV at 300 K. Determine the wavelength (λ_c) above which an intrinsic photodetector fabricated from this material will cease to operate.

Solution

The long wavelength cutoff:

$$\lambda_c = \frac{hc}{E_g} = \frac{6.626 \times 10^{-34} \times 3 \times 10^8}{1.43 \times 1.602 \times 10^{-19}} = 0.867 \text{ μm}$$

Obviously, GaAs photoconductor will cease to operate above 0.867 μm.

Example 3

A p-n photodiode has a quantum efficiency (η) of 70% for photons of energy 1.52×10^{-19} J. Calculate (a) the wavelength at which the diode is operating, and (b) the optical power required to achieve a photocurrent of 3 μA when the wavelength of incident photons is that calculated in part (a). [B.Tech.]

Solution

(a) The photon energy

$$E = h\nu = \frac{hc}{\lambda}$$

Therefore, $\lambda = \dfrac{hc}{E} = \dfrac{6.626 \times 10^{-34} \times 3 \times 10^8}{1.52 \times 10^{-19}} = 1.30 \times 10^{-6}$ m

$= 1.30$ μm

(b) $R = \dfrac{\eta e}{h\nu} = \dfrac{0.70 \times 1.6 \times 10^{-19}}{1.52 \times 10^{-19}} = 0.736 \text{ AW}^{-1}$

Since $R = \dfrac{I_p}{P_{in}}$, $P_{in} = \dfrac{I_p}{R} = \dfrac{3 \times 10^{-6}}{0.736}$

or $P_{in} = 4.07 \times 10^{-6}$ W = 4.07 μW.

Example 4

A photodiode has a quantum efficiency (η) of 65% when photons of energy 1.5×10^{-19} J are incident upon it. Calculate: (a) at what wavelength is the photodiode operating? (b) the incident optical power required to obtain a photocurrent of 2.5 µA when the photodiode is operating as stated above. [B.Tech]

Solution

We have:

$$\text{Photon energy } h\nu = hc/\lambda$$

$$\therefore \quad \lambda = \frac{hc}{E} = \frac{6.626 \times 10^{-34} \times 3 \times 10^8}{1.5 \times 10^{-19}} = 1.32 \; \mu m$$

Thus, the photodiode is operating at a wavelength of 1.32 µm.

$$\text{Responsivity } R = \frac{\eta e}{h\nu} = \frac{0.65 \times 1.602 \times 10^{-19}}{1.5 \times 10^{-19}} = 0.964 \; AW^{-1}$$

We have $\quad R = \dfrac{I_o}{P_o}$

$$\therefore \quad P_o = \frac{I_o}{R} = \frac{2.5 \times 10^{-6}}{0.694} = 3.60 \; \mu W$$

Clearly, the incident optical power required is 3.60 µW.

Example 5

Photons of wavelength 0.90 µm are incident on a *p-n* photodiode at a rate of 5×10^{10} s^{-1} and, on an average, the electrons are collected at the terminals of the diode at the rate of 2×10^{10} s^{-1}. Calculate (a) the quantum efficiency (η), and (b) the responsivity (R) of the diode at this wavelength. [B.Tech.]

Solution

(a) $\eta = \dfrac{2 \times 10^{10}}{5 \times 10^{10}} = 0.40$

(b) $R = \dfrac{\eta e \lambda}{hc} = \dfrac{0.40 \times 1.6 \times 10^{-19} \times 0.90 \times 10^{-6}}{6.626 \times 10^{-34} \times 3 \times 10^8} = 0.29 \; AW^{-1}$

Example 6

A *p-i-n* photodiode has an intrinsic region with a width of 20 µm and a radius of 500 µm in which the drift velocity of electrons is 10^5 ms^{-1}. When the permittivity of the device material is 10.5×10^{-13} F cm^{-1}, calculate: (a) the drift time of the carriers across the depletion region; (b) the junction capacitance of the photodiode. [B.Tech.]

Solution

We have the drift time for the carriers across the depletion region for the photodiode:

$$t_{drift} = \frac{W}{V_d} = \frac{20 \times 10^{-6}}{1 \times 10^5} = 2 \times 10^{-10} \text{ s}$$

Thus, the drift time for the carriers across the depletion region is 200 ps.

The junction capacitance:

$$C_j = \frac{\varepsilon_s A}{W} = \frac{10.5 \times 10^{-13} \times 0.79 \times 10^{-6}}{20 \times 10^{-6}}$$
$$= 0.41 \times 10^{-13} = 4 \text{ pF}.$$

Here, $A = \pi r^2 = 3.14 \times (500 \times 10^{-6})^2 \simeq 0.79 \times 10^{-6} \text{ m}^2$

Example 7

A p-i-n photodiode, on an average, generates one electron-hole pair per two incident photons at a wavelength of 0.85 μm. Assuming all the photo-generated electrons are collected, calculate (a) the quantum efficiency of the diode: (b) the maximum possible band gap energy (in eV) of the semiconductor, assuming the incident wavelength to be a long-wavelength cut-off; and (c) the mean output photocurrent when the incident optical power is 10 μW.

Solution

(a) $\eta = \dfrac{1}{2} = 0.5 = 50\%$

(b) $E_g = \dfrac{hc}{\lambda_c} = \dfrac{6.626 \times 10^{-34} \times 3 \times 10^8}{0.85 \times 10^{-6}} = 2.33 \times 10^{-19} \text{ J} = 1.46 \text{ eV}$

(c) $I_P = RP_{in} = \dfrac{\eta e}{h\nu} P_{in} = \dfrac{0.5 \times 1.6 \times 10^{-19}}{2.33 \times 10^{-19}} \times 10 \times 10^{-6} = 3.43 \times 10^{-6} \text{ A}$

$= 3.43 \text{ μA}$

Example 8

The carrier velocity in a silicon photodiode with a 25 μm depletion layer width is 3×10^4 ms^{-1}. Calculate the maximum response time for the device. [B.Tech.]

Solution

The maximum 3 dB bandwidth for the photodiode may be obtained from:

$$B_m = \frac{V_d}{2\pi w} = \frac{3 \times 10^4}{2\pi \times 25 \times 10^{-6}} = 1.91 \times 10^8 \text{ Hz}$$

Maximum response time for the device

$$= \frac{1}{B_m} = \frac{1}{1.91 \times 10^8} = 5.2 \text{ ns}$$

Example 9

A germanium p-i-n photodiode with an active area dimensions of 100 × 50 μm has a quantum efficiency (η) of 60% when operating at a wavelength 1.2 μm. The measured dark current is 10 nA. Calculate the noise equivalent power and specific directivity of the device. Assume that dark current is the main source of noise. [B.Tech.]

Solution

We have, the noise equivalent power.

Here:
$I_B = 10 \text{ nA} = 10 \times 10^{-9}$ A
$\eta = 0.6$
$\lambda = 1.2 \times 10^{-6}$ m

$$\text{NEP} = \frac{hc\,(2eI_D)^{1/2}}{\eta e \lambda}$$

$$= \frac{6.62 \times 10^{-34} \times 3 \times 10^8 \,(2 \times 1.6 \times 10^{-19} \times 10 \times 10^{-9})^{1/2}}{0.6 \times 1.6 \times 10^{-19} \times 1.2 \times 10^{-6}}$$

$$= 9.8 \times 10^{-14} \text{ W}$$

Specific directivity:

$$D^* = DA^{1/2} = \frac{\eta e \lambda}{hc \left(\frac{2eI_D}{A}\right)^{1/2}} = \frac{A^{1/2}}{\text{NEP}}$$

$$= \frac{(100 \times 10^{-6} \times 50 \times 10^{-6})^{1/2}}{9.8 \times 10^{-14}} = 7 \times 10^8 \text{ mHz}^{1/2} \cdot \text{W}^{-1}$$

Example 10

A silicon RAPD, operating at a wavelength of 0.80 μm, exhibits a quantum efficiency of 90%, a multiplication factor of 800, and a dark current of 2 nA. Calculate the rate at which photons should be incident on the device so that the output current (after avalanche gain) is greater than the dark current. [B.Tech.]

Solution

$$I = I_P M = P_{in} R M = P_{in} \left(\frac{\eta e \lambda}{hc}\right) M = \left(\frac{P_{in}}{hc/\lambda}\right) \eta e M = [(photon\ rate)e] \eta M$$

For $I = 2$ nA:

$$\text{Photon rate} = \frac{I}{e\eta M} = \frac{2 \times 10^{-9}}{1.6 \times 10^{-19} \times 0.90 \times 800} = 1736 \times 10^7 \text{ s}^{-1}$$

For $I > 2$ nA:
$$\text{Photon rate} \approx 1.74 \times 10^7 \text{ s}^{-1}$$

Example 11

In a 100 ns pulse, 6×10^6 photons at a wavelength of 1300 nm fall on an InGaAs photodetector. On an average, 5.4×10^6 electron-hole (e-h) pairs are generated. Calculate the quantum efficiency (η). [M.Sc. (Ele.)]

Solution

$$\eta = \frac{\text{number of } e\text{-}h \text{ pairs generated}}{\text{number of incident photons}} = \frac{5.4 \times 10^6}{6 \times 10^6} = 0.90$$

Thus, at 1300 nm, $\eta = 90\%$.

Example 12

The quantum efficiency (η) for the wavelength range 1300 nm < λ < 1600 nm for InGaAs is around 90%. Calculate the responsivity (R) and cutoff wavelength. [B.Tech.]

Solution

$$R = \frac{\eta e}{h\nu} = \frac{\eta e \lambda}{hc} = \frac{(0.90 \times 1.6 \times 10^{-19}) \lambda}{(6.62 \times 10^{-34})(3 \times 10^8)} = 7.25 \times 10^5 \lambda$$

At 1300 nm, we have:

$$R = 7.25 \times 10^5 \left(\frac{AW}{m}\right)(1.30 \times 10^{-6} \text{ m}) = 0.92 \text{ AW}^{-1}$$

At wavelength higher than 1600 nm, the photon energy is not sufficient to excite an electron from the valence band to the conduction band, e.g. $In_{0.53}Ga_{0.47}As$ has an energy gap $E_g = 0.73$ eV. Thus, the cutoff wavelength is:

$$\lambda_c = \frac{1.24}{E_g} = \frac{1.24}{0.73} = 1.73 \text{ μm}$$

We may note that at wavelength less than 1100 nm, the photons are absorbed very close to the photodetector surface, where the combination rate of the generated electron-hole pairs is very short. The responsivity (R) thus decreases rapidly for smaller wavelengths, since many of the generated carriers do not contribute to the photocurrent.

Example 13

An APD has a quantum efficiency (η) of 40% and 1.3 μm. When illuminated with optical power of 0.3 μW at this wavelength, it produces an output photocurrent of 6 μA, after avalanche gain. Calculate the multiplication factor (M) of the diode. [B.Tech.]

Solution

The

$$M = \frac{I}{I_P} = \frac{I}{P_{in}R} = \frac{I}{P_{in}\left(\frac{\eta e \lambda}{hc}\right)} = \frac{I(hc)}{P_{in}(\eta e \lambda)}$$

$$= \frac{6 \times 10^{-6} \times (6.626 \times 10^{-34} \times 3 \times 10^{8})}{0.3 \times 10^{-6} \times (0.4 \times 1.6 \times 10^{-19} \times 1.3 \times 10^{-6})} = 47.6.$$

Example 14

A silicon avalanche photodiode has a quantum efficiency of 65% at a wavelength of 900 nm. Suppose 0.5 μW of optical power produces a multiplied photocurrent of 10 μA. Find the multiplication M.

Solution

We have the primary photocurrent

$$I_p = R P_{in} = \frac{\eta e}{h\nu} P_{in} = \frac{\eta e \lambda}{hc} P_{in}$$

$$= \frac{(0.65)(1.6 \times 10^{-19})(9 \times 10^{-7})}{(6.62 \times 10^{-34})(3 \times 10^{8})} \times 5 \times 10^{-7} = 0.235 \, \mu A$$

∴ Multiplication factor:

$$M = \frac{I_M}{I_p} = \frac{10 \, \mu A}{0.235 \, \mu A} = 43$$

This means that the primary photocurrent is multiplied by a factor of 43.

Example 15

An InGaAs p-i-n photodiode has the following parameters at a wavelength of 1300 nm: $I_D = 4$ nA, $\eta = 0.90$, $R_L = 1000 \, \Omega$ and the surface leakage current is negligible. The incident optical power is 300 nW (−35 dBm), and the receiver bandwidth is 20 MHz. Calculate the various noise terms of the receiver.

Solution

We have:

$$I_p = R P_{in} = \frac{\eta e}{h\nu} P_{in} = \frac{\eta e \lambda}{hc} P_{in}$$

Substituting

$$I_p = \frac{(0.90)(1.6 \times 10^{-19})(1.3 \times 10^{-6})}{(6.62 \times 10^{-34})(3 \times 10^{8})} \times 3 \times 10^{-7} = 0.282 \, \mu A$$

Mean square shot noise current for p-i-n photodiode is given by:

$$\langle I_{shot}^2 \rangle = 2eI_p Be = 2 \times 1.6 \times 10^{-19} \times 0.282 \times 10^{-6} \times 20 \times 10^6$$
$$= 1.80 \times 10^{-18} \, A^2$$

or $\langle I_{shot}^2 \rangle^{1/2} = 1.34$ mA

Mean square dark current:

$$\langle I_{DB}^2 \rangle = 2eI_D B$$
$$= 2 \times 1.6 \times 10^{-19} \times 4 \times 10^{-9} \times 20 \times 10^6 = 2.56 \times 10^{-20} \, A^2$$

∴ $\langle I_{DB}^2 \rangle^{1/2} = 0.16$ nA

Mean square thermal noise current: We have:

$$\langle I_T^2 \rangle = \frac{4k_B T}{R_L} B = \frac{4 \times 1.38 \times 10^{-23} \times 293 \, K \times 20 \times 10^6}{1 \, k\Omega}$$
$$= 323 \times 10^{-18} \, A^2$$

∴ $\langle I_T^2 \rangle^{1/2} = 18$ nA

We see that for this receiver, the *rms* thermal noise current is about 14 times greater than the *rms* shot noise current and about 100 times greater than *rms* dark current.

SUGGESTED READINGS

1. Donati S, 'Photodetectors', Prentice Hall (2000).
2. Yariv V, 'Optical Electronics', Wiley (1997)
3. Kasap, SO, 'Optoelectronics and Photonics: Principles and Practices, Prentice Hall (2001).
4. Razavi, B, 'Design of Integrated Circuits for Optical Communications', McGraw Hill Higher Education (2003).
5. Agrawal GP and Dutta NK, 'Semiconductor Lasers', Van Nostrand (1993).
6. Setian L, 'Applications in Electro-Optics', Prentice Hall (2002).
7. Degiorgio V and Cristian I, 'Photonics', Springer (2014).
8. Kumar S and Jamal Deen M, 'Fiber Optic Communications', Wiley (2014).

GLIMPSES

- In an optical fiber system, it is required to convert the optical signals at the receiver end back into electrical signal. This task is performed by an optical receiver system. Optical receiver converts optical energy into electrical signal, amplify the signal and process it.

- The important blocks of an optical receiver are:
    (i) Photodetector/Front end
    (ii) Amplifier/Linear channel
    (iii) Signal processing circuitary/Data recovery.
- The aim of a receiver is the recovery of the transmitted data. This process involves two steps:
    (i) the recovery of bit clock
    (ii) the recovery of transmitted bit within each bit interval.
- A good light detector should generate a large photocurrent at a given incident light power. They should also respond fast to the input changes and add minimal noise to the output signal. This last requirement is of crucial importance since the received signal is typically very weak. In digital optical communication systems, the detection process is often conducted with a PIN photodiode.
- There are generally two types of detection: direct detection (also called incoherent detection) and coherent detection.
- *Direct detection* detects only the *intensity* of the incident light. It is used mainly for intensity or amplitude modulation schemes. It can only detect an amplitude modulated (AM) signal.
- *Coherent detection* can detect both the power and phase of the incident light. It is, therefore, used when phase modulation (PM) or frequency modulation (FM) is preferred. Coherent detection is also important in applications such as WDM.
- The *coherent detection* requires a local oscillator to coherently down-convert the modulated signal from optical frequency to intermediate frequency.
- The *incoherent detection* which dominates in currently deployed systems is based on square-law envelope detection of the optical signals.
- A reverse biased *p-n junction* is used for conversion of the optical signals at the receiver end back into electrical signals. An incidence photon of energy ($h\nu$) greater than the band gap of the semi-conductor creates an electron-hole pair. The two charge carriers are swept in opposite directions by the applied bias, and photo-current flows in the external circuit.
- The photocurrent (I_p) depends on the absorption coefficient (α) of the semiconductor for incident wavelength. The relation between I_p and α is:

$$I_p = \frac{P_{in}(1-R)e}{h\nu}[1-\exp(-\alpha d)]$$

where P_{in} is power incident on the photodiode, R is Fresnel reflection coefficient at the air-semiconductor interface, d is the width of the absorption region.

- The *quantum efficiency* (η) of a device is the ratio of the rate of electrons collected at the diode terminals to the rate of photons incident on it. η is related to the responsivity (R) of the detector by the relation:

$$R = \frac{\eta e}{h\upsilon} = \frac{\eta e \lambda}{hc}$$

- The absorption of photon of energy ($h\upsilon$) is possible only when its energy is greater than or equal to the energy gap (E_g) of the semiconductor, i.e. $hc/\lambda \geq E_g$. Therefore, there is a long wavelength cutoff λ_c, above which photons are not absorbed by the semiconductor. λ_c is given by:

$$\lambda_c \,(\mu m) = \frac{hc}{E_g} = \frac{1.24}{E_g \,(eV)}$$

- Photodiodes are of three types: (i) *p-n*, (ii) *p-i-n*, and (iii) Avalanche photodiodes. The first two diodes produce current without gain, whereas the third one produces current with gain. One can use photoconductivity detectors for long wavelength operations.
- Main characteristics of optical receivers are:
   (i) *Receiver sensitivity*: This property is a measure of the minimum level of optical power P_{sens} at the receiver required for a reliable operation.
   (ii) *Dynamic range*: Dynamic range (in dB) is the difference between the maximum allowable power and minimum power determined by receiver sensitivity.
   (iii) *Bit-rate transparency*: This refers to the ability of the optical receiver to operate over a range of bit rates.
   (iv) *Bit-pattern independency*: This determines the operation of an optical receiver for various data formats. The main constraint is imposed by non-return-to-zero (NRZ) code.
- Performance evaluation of an optical transmission system is done by evaluating optical signal-to-noise ratio, eye opening and bit error rate (BER) which is the ultimate indicator.

REVIEW QUESTIONS

1. Explain the principle and working of optical receivers.
2. Explain the detection process in a *p-n* photodiode. Compare the device with *p-i-n* photodiode.
3. Explain the quantum efficiency (η) and responsivity (R) of a photodiode. Derive the relation between η and R.
4. Draw the layer diagram and explain the operation of *p-i-n* diode. Draw diagrams for three practical photodiodes and show that the detector current is given by:

$$I_p = \frac{eP_o}{\eta f}[1 - \exp(-\alpha(\lambda)W)(1 - R_f)$$

where R_f is reflectivity, W is the width of the depletion layer, η is quantum efficiency and P_o is optical power.

5. What is the difference between p-n diode, p-i-n diode and an APO? Can we make these types of photodiodes using the same semiconductor?

6. Define the quantum efficiency (η) of photodiode and show that:

$$\eta = \frac{I_p}{P_o}\frac{h\nu}{e}[1 - \exp\{(-\alpha(\lambda)W\}](1 - R_f)$$

Also define responsivity (R) and show that:

$$R = \eta\frac{e}{h\nu} = \frac{e}{h\nu}(1 - R_f)[1 - \exp\{(-\alpha(\lambda)W)\}]$$

where I_p is average photocurrent, P_o is optical power, R_f is effect of reflectivity, $\alpha(\lambda)$ is absorption coefficient at the opening wavelength and W is width of depletion layer.

7. Explain, why the responsivity (R) versus wavelength (λ) curve for a practical Si diode deviate from an ideal curve? How, one can improve the quantum efficiency of such a diode?

8. Describe with the help of a relevant diagram the operation of a silicon RAPD and explain how it differs from a p-i-n photodiode. Outline the advantages and drawbacks of RAPD as detector in optical fiber communication.

9. The avalanche photodiode and photoconducting detector both provide gain. Compare their merits for their use in optical fiber communication and other applications.

10. Describe the different types of noise encountered in a photodetector.

PROBLEMS

1. Calculate the cutoff wavelength for Si and Ge p-i-n photodiodes. Their bandgap energies are 1.1 eV and 0.67 eV respectively.

   [Hint. $\lambda_c = \frac{hc}{E_g} = \frac{1.24}{E_g\,(eV)}$

   $h = 6.62 \times 10^{-34}$
   $c = 3 \times 10^8$

   (i) For Si, $\lambda_c = \dfrac{6.62 \times 10^{-34} \times 3 \times 10^8}{1.1 \times 10^{-19}} = 1.8 \times 10^{-6}\,m = 1.8\,\mu m$

   (ii) For Ge, $\lambda_c = \dfrac{6.62 \times 10^{-34} \times 3 \times 10^8}{0.67 \times 10^{-19}} = 2.96 \times 10^{-6}\,m = 2.96\,\mu m$]

2. A *p-i-n* photodiode is fabricated by GaAs which has bandgap energy 1.43 eV at 300 K. Calculate its upper cutoff wavelength.

   [Hind. $\lambda_c = \dfrac{hc}{E_g} = \dfrac{6.62 \times 10^{-34} \times 3 \times 10^8}{1.43 \times 10^{-19}} = 1.33 \times 10^{-6}$ m $= 1.33$ m]

3. A *p-n* photodiode has a quantum efficiency (η) = 50% at $\lambda = 0.90$ μm. Calculate (a) its responsivity (R) at this wavelength, (b) the optical power received for mean photocurrent 10^{-6} A, and (c) the corresponding number of photon received at this wavelength.
   [B.Tech.]
   [Ans. (a) 0.36 AW^{-1}, (b) 2.76 μW, (c) $\eta_p = 1.25 \times 10^{13}$ s^{-1}]

4. A pulse of 85 ns emits 6×10^6 photons at 1300 nm wavelength from an InGaAs photoconductor. Average number of *e-h* pairs generated are 5.4×10^6. Calculate quantum efficiency (η) of the detector.
   (M.Sc. (Ele.))

   [Hint. $\eta = \dfrac{\text{Number of } e-h \text{ pairs generated}}{\text{Number of incident photons}}$
   $= \dfrac{5.4 \times 10^6}{6 \times 10^6} = 0.9 = 90\%$]

5. Calculate the responsivity of an ideal *p-n* photodiode at the following wavelengths: (a) 0.85 μm, (b) 1.35 μm, and (c) 1.55 μm.
   [B.Sc. (Ele.)]
   [Ans. (a) 0.684 AW^{-1}, (b) 1.046 AW^{-1}, (c) 1.248 AW^{-1}]

6. Photons having energy 1.53×10^{-19} J are incident on a photodiode having responsivity (R) = 0.65 AW^{-1}. If output power (P_o) = 10 μW, calculate the generated photocurrent. [B.Tech.]

   [Hint. $R = \dfrac{I_P}{P_o}$ or $I_P = RP_o = 0.65 \times 10 = 6.5$ μA]

7. The quantum efficiency (η) of an APD is 50% at 1.3 μm. When this APD is illuminated with optical power of 0.4 μW, this wavelength, it produces an output photocurrent of 0.8 μA, after avalanch gain. Show that the multiplication factor of diode is 30.

8. Compute the bandwidth (W) of a photodetectors having parameters as: (a) Photodiode capacitance = 3 pF, (ii) Amplifier capacitance = 4 pF, (iii) Load resistance (R_L) = 50 Ω, (iv) Amplifier input resistance = 1 MΩ.

   [Hint. Total capacitance of photodiode and amplifier:
   $$C_T = 3 + 4 = 7 \text{ pF}$$
   Combination of load resistance and amplifier input resistance:
   $$R_L = 50 \text{ Ω} \,||\, 1 \text{ MΩ} \approx 50 \text{ Ω}$$

Bandwidth of photodetector:
$$W = \frac{1}{2\pi R_L C_T} = \frac{1}{2 \times 3.14 \times 50 \times 7 \times 10^{-12}} = 454.95 \text{ MHz}]$$

9. A typical photodiode has a responsivity $(R) = 0.40$ AW^{-1} for a He-Ne laser source ($\lambda = 632.8$ nm). The active area of the photodiode is 2 mm^2. If the incident flux is 100 µW/mm^2, calculate the output photocurrent. **[Ans. 80 µA]**

10. The maximum 3-dB bandwidth permitted by an InGaAs photoconducting detector is 450 MHz when the electron transit time in the device is 6 ps. Calculate (a) the gain G, and (b) the output photocurrent when an optical power of 5 µW at a wavelength of 1.30 µm is incident on it, assuming quantum efficiency of 75%.

    [**Hint.** We know that the current response in the photoconductor decays exponentially with time once the incident optical pulse is removed. The time constant of this decay is equal to the slow carrier transit time t_s. Therefore, the maximum 3 dB bandwidth (Δf_m) of the device will be given by:
    $$(\Delta f)_m = \frac{1}{2\pi t_s} = \frac{1}{2\pi t_f G} \quad \left(\because G = \frac{t_s}{t_f} \right)$$

    (a) $G = \dfrac{1}{2\pi t_f (\Delta f)_m} = \dfrac{1}{2\pi \times 6 \times 10^{-12} \times 450 \times 10^6} = 58.94$

    (b) $I = G I_p = \dfrac{G \eta P_{in} e \lambda}{hc}$
    $$= \frac{58.94 \times 0.75 \times 5 \times 10^{-6} \times 1.6 \times 10^{-19} \times 1.3 \times 10^{-6}}{6.626 \times 10^{-34} \times 3 \times 10^8}$$
    $$= 232.1 \text{ µA} = 2.321 \times 10^{-4} \text{ A}]$$

11. A given APD has a quantum efficiency (η) of 65% at wavelength of 900 nm. If a 0.5 µW of optical power produced a multiplied photocurrent of 10 µA, calculate the multiplication factor (m).

    [**Hint.** $R = \dfrac{\eta e \lambda}{hc} = \dfrac{0.65 \times 1.6 \times 10^{-19} \times 900 \times 10^{-9}}{6.62 \times 10^{-34} \times 3 \times 10^8} = 0.47$ AW^{-1}

    Photocurrent, $I_p = P_{in} \times R = 0.5 \times 10^{-6} \times 0.4705 = 0.235$ µA

    Multiplication factor $(M) = \dfrac{I_M}{I_p} = \dfrac{10 \times 10^{-6}}{0.235 \times 10^{-6}} = 42.55$]

12. An InGaAs p-i-n photodiode is operating at room temperature (300 K) at a wavelength of 1.3 µm. Its quantum efficiency (η) is 70% and the incident optical power is 500 nW. Take the primary dark current I_d of the device is 5 nA, R_L is a kΩ, and the effective bandwidth is 25

MHz. Calculate (a) the rms values of shot noise current, dark current, and thermal noise current; and (b) S/N at the input end of an amplifier of the receiver. [B.Tech.]

[Hint.

(a) $\quad I_P = RP_{in} = \dfrac{\eta e \lambda}{hc} P_{in}$

$$I_P = \dfrac{0.70 \times 1.6 \times 10^{-19} \times 1.3 \times 10^{-6}}{6.626 \times 10^{34} \times 3 \times 10^8} \times 500 \times 10^{-9}$$

$\qquad = 3.663 \times 10^{-7}$ A $= 0.3662$ µA

$\langle i_s^2 \rangle = 2eI_p(\Delta f)M^2 \, F(M)$

$\qquad = 2 \times 1.6 \times 10^{-19} \times 3.662 \times 10^{-7} \times 25 \times 10^6 \times 1 \times 1$

$\qquad = 293.03 \times 10^{-20}$ A^2

$\langle i_s^2 \rangle^{1/2} = 17.15 \times 10^{-10}$ A $= 1.715$ nA

$\langle i_d^2 \rangle = 2eI_d(\Delta f)$ (M and $F(M)$ are unity)

$\qquad = 2 \times 1.6 \times 10^{-19} \times 5 \times 10^{-19} \times 25 \times 10^6$

$\qquad = 400 \times 10^{-22}$ A^2 $= 4 \times 10^{-20}$ A^2

$\langle i_d^2 \rangle^{1/2} = 20 \times 10^{-11}$ A $= 0.2$ nA

$\langle i_T^2 \rangle = \dfrac{4kT(\Delta f)}{R_L} = \dfrac{4 \times 1.38 \times 10^{-23} (J/K) \times 300(K)}{1,000} \times 25 \times 10^6$

$\qquad = 414 \times 10^{-18}$ A^2

(b) Sum of mean square noise current $= 41{,}698.16 \times 10^{-20}$ A^2

$\qquad\qquad\qquad\qquad\qquad\qquad\qquad = 4.17 \times 10^{-16}$ A^2

and $I_p^2 = 1.352 \times 10^{-13}$ A^2

$\dfrac{S}{N} = \dfrac{1.352 \times 10^{-13}}{4.17 \times 10^{-16}} = 0.324 \times 10^3 = 324$]

SHORT ANSWER QUESTIONS

1. What is an optical receiver?

Ans. An optical receiver is an essential part of an optical fiber communication system. An optical receiver converts an optical signal, transmitted through an optical fiber cable into an electrical signal suitable for a receiving device installed at the other end of the communication system. The conversion process in the receiver is performed by two essential parts: (i) *photodetector*, and (ii) an *electronic signal processor*.

2. **What is the function of a photodetector?**

Ans. The photodetector senses the luminescent power falling upon it and converts the variation of this optical power into a correspondingly varying electric current.

3. **How many types of photodetectors are in existence?**

Ans. Several different types of photodetectors are in existence. Among these are photomultipliers, pyroelectric detectors, and semiconductor based photoconductors, phototransistors, and photodiodes.

4. **What is the function of electronic signal processor?**

Ans. The electronic signal processor converts the raw detector signal into a form decipherable by the receiving device, such as a telephone, camera, or a scanner.

5. **Of the semiconductor-based photodetectors, which is used almost exclusively for fiber optic systems and why?**

Ans. Photodiode is used almost exclusively for fiber optic communication system because of its small size, suitable material, high sensitivity, and fast response time.

6. **For efficient operation should a detector have a high or low responsivity?**

Ans. High.

7. **Which two types of photodiodes are used in fiber optic communication?**

Ans. (i) *p-i-n* photodiode, (ii) Avalanche photodiode (APD).

8. **What is photocurrent?**

Ans. The current produced when photons are incident on the active area of the detector.

9. **What is responsivity?**

Ans. The ratio of optical detector's output photocurrent in amperes to the incident power in watts.

$$R = \frac{I_p \text{ (A)}}{P_{in} \text{ (W)}} \text{ (in Aw}^{-1}\text{)}$$

10. **How are *p-i-n* photodiodes usually biased?**

Ans. Usually reverse-biased.

11. **What is dark current or leakage current?**

Ans. The leakage current that continuous to flow through a photodetector when there is no incident light.

12. **Whether dark current increase or decrease as the temperature of photodiode increases?**

Ans. Increases.

13. How upper wavelength cutoff (λ_c) is related to band-gap energy?

Ans. $\lambda_c\ (\mu m) = \dfrac{hc}{E_g} = \dfrac{1.24}{E_g\ (eV)}$

14. On what factors the fraction of light absorbed by the photodiode depends.

Ans. (i) Wavelength (λ) of the light, determined by the photon energy ($\varepsilon = h\nu = hc/\lambda$). (ii) The thickness of the absorption material (depletion region width or depletion layer thickness).

15. Why it is important to control the dark current below a certain value?

Ans. Low leakage current is an important measure of device quality. If the dark current is high, the generated photocurrent needs to be larger, in order to provide a good signal. Otherwise, the leakage current will dominate the detector current. This is why, it is important to control the dark current below the dark current.

16. Should the capacitance of photodetector be small or large? Explain why?

Ans. Small. This prevents RC time constant from limiting the response time.

17. What is quantum efficiency (η) of a device?

Ans. The ratio of the rate of electrons collected at the diode terminals to the rate of photons incident on it is called the quantum efficiency of a device.

18. How the quantum efficiency (η) of a device is related to its responsivity (R).

Ans. $R = \dfrac{\eta e \lambda}{hc} = \dfrac{\eta e}{h\nu}$

19. What are the sources of leakage current?

Ans. 1. *Generation-recombination current*: Arises from the generation and recombination of electron-hole pair in the diode depletion region. 2. *Diffusion current*: Arises from the diffusion on the minority carriers towards or away from the junction, in the diode neutral region. 3. *Tunneling current*: Refers to the band-to-band tunneling in the presence of high electric field.

20. How the gain of APD can be increased?

Ans. By increasing reverse-bias voltage.

21. What are the main types of noises in receiver?

Ans. (i) Thermal noise, (ii) Dark current noise, and (iii) Quantum noise.

22. The sensitivity of receiver is decided by which parameter?

Ans. Noise.

498 Photonics | Optoelectronics

23. What is response time?

Ans. Response time is defined as the time needed for the photodiode to respond to an optical input by producing photocurrent.

24. What are phototransistors?

Ans. These are simplex type of photodetectors. The device consists of an n-p-n transistor. A large *space charge layer* (SCL) is formed between the base and collector. The SCL region is called the absorption region. Phototransistors operate as a photodetector that amplifies the photocurrent.

25. What are Metal-Semiconductor-Metal (MSM) detectors?

Ans. MSM detectors are probably the fastest and simplest optical detector to fabricate. The basic idea is to create a Shottky barrier, which forces the material at the surface to be depleted.

26. What factors limit the speed of response of a photodiode?

Ans. (i) Drift time of carriers through the depletion region:

$$t_{drift} = \frac{\text{Depletion layers width } (W)}{\text{Drift velocity } (v_d)}$$

(ii) Diffusion time of carriers generated outside the depletion region:

$$t_{drift} = \frac{d^2}{2D_c}$$

where d is distance for carriers to diffuse and D_c is the minority carriers diffusion time.

(iii) Time constant incurred by the capacitance of the photodiode with its load. The junction capacitance C_j is given by:

$$C_j = \frac{\varepsilon_s A}{W}$$

$\varepsilon_s \rightarrow$ permittivity of semiconductor, A $\rightarrow$ diode junction area and $\omega \rightarrow$ depletion layer width.

27. What are important requirements of an optical detector?

Ans. (i) High responsivity (R), (ii) High quantum efficiency (η), (iii) Least response time, (iv) Zero dark current.

28. What is 'bulk dark' current?

Ans. The flow of current through a detector in the absence of light.

29. Explain, how long wavelength cutoff related to photodiode?

Ans. Photodiode of long cutoff wavelength can emit optical power in wide range that is used for a fiber optic communication.

MULTIPLE CHOICE QUESTIONS

1. If r_e is the rate of electrons collected at the detector terminals and r_p is the rate of photons on the device, then quantum efficiency (η) of a optoelectronic detector is:

   (a) $\eta = \sqrt{\dfrac{r_e}{r_p}}$
   (b) $\eta = \dfrac{r_e}{r_p}$
   (c) $\eta = \dfrac{r_e^2}{r_p^2}$
   (d) $\eta = \left(\dfrac{r_e}{r_p}\right)^{1/3}$

2. If I_p is the output photocurrent in amperes and P_{in} is the incident optical power in watts, then:

   (a) $R = \dfrac{I_p(A)}{P_{in}(W)}$
   (b) $R = \sqrt{\dfrac{I_p(A)}{P_{in}(W)}}$
   (c) $R = \dfrac{P_{in}(W)}{I_p(A)}$
   (d) $R = \dfrac{P_{in}^2(W)}{I_p(A)}$

3. The cutoff wavelength (λ_c) for intrinsic absorption process bears a following relation between bandgap energy (E_g) of the material and energy of incident photons ($h\nu = hc/\lambda$):

   (a) $\lambda_c = \dfrac{h\nu}{E_g}$
   (b) $\lambda_c = \dfrac{h\lambda}{E_g}$
   (c) $\lambda_c = \dfrac{hc}{E_g}$
   (d) $\lambda_c = \dfrac{h}{E_g}$

4. The performance parameters of a photodetector are:
   (a) responsivity (R) only
   (b) quantum efficiency (η) only
   (c) response time and dark current only
   (d) All of the above

5. The photodiode which produce current with gain is:
   (a) p-n
   (b) p-i-n
   (c) Avalanche photodiode (APD)
   (d) p-n and p-i-n

6. The photodiodes produce current with gain are:
   (a) p-n, p-i-n and APD
   (b) p-n diode only
   (c) p-i-n diode only
   (d) p-n and p-i-n both

7. In a fiber-optic communication system, it is required to convert the optical signals at the receiver end into electrical signals. This task is performed by:
   (a) detector
   (b) transmitter
   (c) optic connector
   (d) none of the above

8. Practically, in order to create an electron-hole pair in a *p-n* diode, the energy of the incident photon should be:
   (a) less than E_g
   (b) equal to E_g
   (c) greater than E_g
   (d) much greater than E_g

9. The responsivity of a given *p-i-n* diode is 0.5 AW^{-1} for a wavelength of 1 µm. What is the output photocurrent when optical power of 0.2 µW at this wavelength is incident on it?
   (a) 0.1 µA
   (b) 1 µA
   (c) 10 µA
   (d) 1 A

10. Which of the following is an inherent property of an optical signal and cannot be eliminated even in principle?
    (a) Thermal noise
    (b) Shot noise
    (c) Environmental noise
    (d) Background noise

11. A photoconducting detector can be constructed from:
    (a) an intrinsic semiconductor
    (b) an extrinsic semiconductor
    (c) polycrystalline material
    (d) All of the above

12. Given the germanium (Ge) has a bandgap of 09.67 eV, what is the maximum wavelength that will be absorbed by it?
    (a) 7,080 nm
    (b) 4,560 nm
    (c) 1,850 nm
    (d) 1,100 nm

13. The highest wavelength that silicon (Si) can absorb is 1.12 µm. What is the approximate band gap of Si?
    (a) 1.1 eV
    (b) 1.4 eV
    (c) 1.74 eV
    (d) 2.3 eV

14. Which one of the following material is more suitable for making a *p-n* diode?
    (a) A direct band gap semiconductor
    (b) An indirect band gap semiconductor
    (c) A metal
    (d) An insulator

15. A *p-n* photodiode, on an average, generates one electron-hole pair per five incident photons at a wavelength of 0.90 µm. Assuming all the photogenerated electrons are collected, what is the quantum efficiency of the diode?
    (a) 20%
    (b) 60%
    (c) 40%
    (d) 50%

16. Photons of wavelength 0.85 µm are incident on a *p-i-n* photodiode at the rate of 4×10^{10} s^{-1} and, on an average, electrons are

collected at the terminals of the diode at the rate of 2×10^{10} s^{-1}. What is the responsivity of the diode at this wavelength?
(a) 0.15 AW^{-1} (b) 0.23 AW^{-1}
(c) 0.34 AW^{-1} (d) 0.50 AW^{-1}

17. Which of the following detectors give amplified output?
    (a) *p-n* photodiode (b) *p-i-n* photodiode
    (c) Avalanche photodiode (d) Photovoltaic detector

18. A photodiode has a quantum efficiency (η) of 65% when photons of energy 1.5×10^{-19} J are incident upon it, the photodiode is operating at wavelength (λ).
    (a) 1.32 µm (b) 0.66 µm
    (c) 2.64 µm (d) 0.33 µm

    [**Hint.** $\eta = \dfrac{\text{No. of electrons collected}}{\text{No. of incident photons}}$

    $= \dfrac{1.2 \times 10^{11}}{3 \times 10^{11}} = 0.4$]

19. When 3×10^{11} photons each with a wavelength of 0.85 µm are incident on a photodiode, on an average 1.2×10^{11} electrons are collected at the terminals of the device. The quantum efficiency of photodiode at 0.85 µm is:
    (a) 0.8 (b) 0.4
    (c) 0.6 (d) 0.3

    [**Hint.** $R = \dfrac{ne\lambda}{hc} = \dfrac{0.4 \times 1.6 \times 10^{-19} \times 0.85 \times 10^{-6}}{6.62 \times 10^{-34} \times 3 \times 10^{8}}$

    $= 0.27$ AW^{-1}]

20. In Q.No. 19, the responsivity of photodiode is:
    (a) 0.624 AW^{-1} (b) 0.424 AW^{-1}
    (c) 0.274 AW^{-1} (d) 0.324 AW^{-1}

    [**Hint.** $R = \dfrac{\eta e \lambda}{hc} = \dfrac{0.4 \times 1.6 \times 10^{-19} \times 0.85 \times 10^{-6}}{6.62 \times 10^{-34} \times 3 \times 10^{8}} = 0.27$ AW^{-1}]

ANSWERS

1. (b)	2. (a)	3. (c)	4. (d)	5. (c)	6. (d)	7. (a)
8. (c)	9. (a)	10. (b)	11. (d)	12. (c)	13. (a)	14. (b)
15. (a)	16. (c)	17. (c)	18. (a)	19. (b)	20. (c)	

10

Optoelectronic Modulation

10.1 INTRODUCTION

Information can be carried out by an optical beam only if the beam is modulated. In optical fiber communication system, data are encoded in the form of variation of some property of the optical output of light emitting diode (LED) or an injection laser diode (ILD). This property of the optical signal may be its *amplitude* or the *phase* or the width of the pulses being transmitted. There are many ways to modulate the LED or ILD emission: output wavelength, frequency, intensity, etc. Intensity modulation is used most often because it is well adapted to *digital* communications and relatively simple to implement. Normally, there are two ways of modulating an optical signal from a LED or ILD. These are: *External modulation* and *internal modulation*.

Internal modulation is accomplished by modulating the drive current of LED or ILD. For this purpose, a circuit is designed to modulate the current injected into the device. As the output of the source, i.e. LED/ILD is controlled by the injected current, one can achieve the desired modulation. Internal (i.e. current) modulation has the advantage that it is both simple and economical to implement. The disadvantage of internal modulation are related primarily to transient effects associated with turning on or turning off the LED/ILD. In this modulation process, the upper modulation frequencies are limited to about 40 GHz, and the emission frequency may change as the drive current is changed. Thus, only amplitude modulation is possible with ease and for phase or frequency modulation, one will have to take additional care to design the drivers. The disadvantages of internal modulation are related primarily to transient effect associated with turning on or turning off the LED/ILD. Some of the main difficulties are *chirping* and *self-pulsations*. The laser chirp refers to the change of the laser output wavelength with time as the laser is being pulsed on or off. The *chirp* may be large enough to increase the communications error rate beyond acceptable limits. Self-pulsations are the result of a resonant coupling between the population of photons and the population of excited carriers in the laser structure. The presence of self-pulsation or relaxation oscillations put a limit on the modulation bandwidth of the laser.

For external modulation, an optical signal from the source is passed through a material (or device) whose optical properties can be altered by external means. This can be achieved by a mechanical wheel such as a compact disc, or by an electro-optic modulator that changes the optical density or index of refraction of the propagation path. In this scheme of modulation, the device speed is controlled by the modulation property and may be quite fast; the emission frequency remains unaffected and both amplitude and phase modulation are possible with ease. The main disadvantage of this modulator is normally large on the scale of microelectronic devices, and therefore, cannot become a part of the same integrated circuit (IC). In the present chapter, we will discuss external modulations only. The design of these modulators is based either on the *electro-optic effect* or on the *acoustic-optic* effect. In high speed communications, devices based on electro-optic effect are most common. Interestingly, the technology used in these devices is also becoming compatible with that of the semiconductor devices.

10.2 ELECTRO-OPTIC EFFECT

When an electric field is applied across a crystal, it may change its refractive indices. This means that the field may induce *birefringence* in an otherwise isotropic crystal or change the birefringent property of a doubly refracting crystal. This is called the electro-optic effect. If the refractive index varies linearly with the applied electric field, it is called as the *Pockel's effect*, and if the variation of refractive index is proportional to the square of the applied electric field, the effect is termed as the *Kerr effect*.

In general, one can express the change in the refractive index (n) as a function of electric field (E) by the following relation:

$$\Delta\left(\frac{1}{n^2}\right) = rE + PE^2 \qquad (10.1)$$

where r is linear electro-optic coefficient and P is the quadratic electro-optic coefficient.

10.3 LINEAR ELECTRO-OPTIC MODULATOR

The precise effects of an applied electric field depend on the crystal structure and symmetry. We now consider the case of KDP crystal, which is one of the most widely used electro-optic crystals. As stated earlier, KDP is a uniaxial birefringent crystal. It possesses a fourfold axis of symmetry (i.e. the rotation of the crystal structure about this axis by an angle $2\pi/4$ leaves it invariant), which is chosen as the optic axis of the crystal, normally marked as z-axis. Further, it also possesses two mutually orthogonal two-fold axes of symmetry, marked as the x-axis and y-axis.

We now consider a longitudinal configuration as shown in Fig. 10.1. We can see that a plane-polarized light is propagating along the optic axis (marked as z-axis) of a KDP crystal which is being acted upon by an applied electric field (E) directed along the z-axis. In the absence of external field, the incident wave polarized normal to the z-axis (i.e., in the x-y plane) will propagate as a principal wave with an ordinary refracting index n_o, because KDP is a uniaxial crystal and the optic axis is along the z-axis. Upon the application of electric field E_z along the z-axis, the crystal no longer remains uniaxial but becomes biaxial. The principal x-axis and y-axis of the crystal are rotated through 45° into new principal axes x' and y'. One can show that the refractive index $n_{x'}$ for a wave propagating along the z-direction and polarized along the x'-direction is given by:

$$n_{x'} = n_o + \frac{1}{2} n_o^3 r_{63} E_z \tag{10.2}$$

where r_{63} is an *appropriate electro-optic coefficient*. Similarly, the refractive index $n_{y'}$ for a wave polarized in the y'-direction is given by the relation:

$$n_{y'} = n_o + \frac{1}{2} n_o^3 r_{63} E_z \tag{10.3}$$

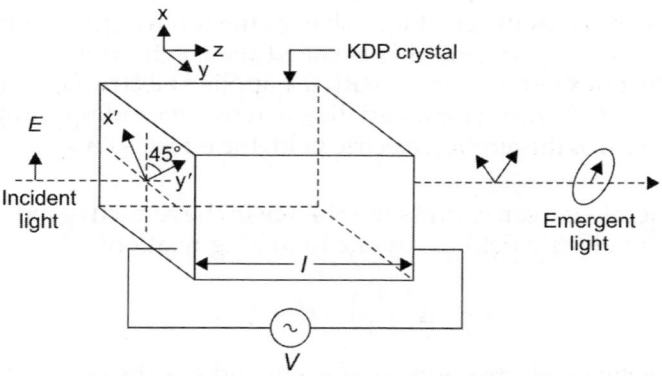

Fig. 10.1. A beam of plane-polarized light propagating along the z-axis (optic axis) of a KDP crystal subject to an external electric field applied in the direction of the applied field

If the incident wave is represented by the following equation:
$$E = E_o \cos(\omega t - kz)$$

the components along the x'- and y'-direction will be, respectively, given by:

$$E_{x'} = \frac{E_o}{\sqrt{2}} \cos(\omega t - kz) \tag{10.4}$$

$$E_{y'} = \frac{E_o}{\sqrt{2}} \cos(\omega t - kz) \qquad (10.5)$$

However, components having refractive indices $n_{x'}$ and $n_{y'}$ given, respectively by Eqs (10.4) and (10.5), will become increasingly out of phase as they propagate through KDP crystal. Let us consider that the thickness of the crystal along the direction of propagation is l, the phase change experienced by these said two components (at $z = l$) may be expressed by the following relations:

$$\phi_{x'} = kn_{x'} l = \frac{2\pi}{\lambda} n_{x'} l \qquad (10.6)$$

and
$$\phi_{y'} = kn_{y'} l = \frac{2\pi}{l} n_{y'} l \qquad (10.7)$$

Substituting the value of $n_{x'}$ from Eq. (10.2) in Eq. (10.6), one obtains:

$$\phi_{x'} = \frac{2\pi}{\lambda} ln_o \left(1 + \frac{1}{2} r_{63} n_o^2 E_z \right) \qquad (10.8a)$$

Assuming $2\pi l n_o = \phi_o$ and $\pi / n_o^3 r_{63} E_z / \lambda = \Delta\phi$, one finds:

$$\phi_{x'} = \phi_o + \Delta\phi \qquad (10.8b)$$

Similarly, substituting $n_{y'}$ from Eq. (10.3) in Eq. (10.7), one obtains:

$$\phi_{y'} = \frac{2\pi}{\lambda} ln_o \left(1 - \frac{1}{2} r_{63} n_o^2 E_z \right) \qquad (10.9a)$$

or
$$\phi_{y'} = \phi_o + \Delta\phi \qquad (10.9b)$$

where
$$\Delta\phi = \frac{\pi}{\lambda} l r_{63} n_o^3 E_z = \frac{\pi}{\lambda} r_{63} n_o^3 V \qquad (10.10)$$

and $V = E_z l$ is the applied voltage. One finds that an extra phase shift $\Delta\phi$ (due to the application of the electric field) for each component is directly proportional to the applied voltage V. This, if V is made to oscillate with frequency ω_m, that is, if $V = V_0 \sin \omega_m t$, the phase shift $\Delta\phi$ will also vary sinusoidally and the peak value will be $\pi r_{63} n_o^3 V_0 / \lambda$. Thus, the *electro-optic effect* may be used for *phase modulation*. The net phase shift or total retardation at $(z = l)$ between the two waves polarized in the x' and y'-direction as a result of the application of voltage V to the crystal will, therefore, be obtained as:

$$\Phi = \phi_{x'} - \phi_{y'} = 2\Delta\phi = \frac{2\pi}{\lambda} r_{63} n_o^3 V \qquad (10.11)$$

In general, the superposition of two plane-polarized waves that are perpendicular to each other produces an elliptically polarized wave. Thus, inspecting Eq. (10.11), it appears in the present case that, in general, the wave emerging at $z = l$ will be elliptically polarized. However, if the superposition gives a phase difference which is an integral multiple of π, the emergent beam will be plane-polarized, and if the phase difference is an odd integer multiple of $\pi/2$, the emergent beam will be circularly polarized.

The voltage $V = V_\pi$ required to introduce a phase shift of π between the two polarization components is called *half-wave voltage* and usually obtained as follows:

$$\Phi = \pi = \frac{2\pi}{\lambda} r_{63} n_o^3 V_\pi$$

or
$$V_\pi = \frac{\lambda}{2 n_o^3 r_{63}} \qquad (10.12)$$

The *half-wave voltage* is one of the important parameters of an electro-optic modulator.

Using Eqs (10.4), (10.5), (10.8b), and (10.9b), the equations for the components of the wave emerging from the crystal polarized in the x'- and y'-direction may be expressed (omitting common phase factors, ϕ_o) as:

$$E_{x'(z=l)} = \frac{E_o}{\sqrt{2}} \cos(\omega t + \Delta\phi) \qquad (10.13)$$

and
$$E_{y'(z=l)} = \frac{E_o}{\sqrt{2}} \cos(\omega t + \Delta\phi) \qquad (10.14)$$

Let us put a plane polarizer at the output end of the KDP crystal and orient it at right angles to the polarizer producing the original plane-polarized beam as shown in Fig. 10.2.

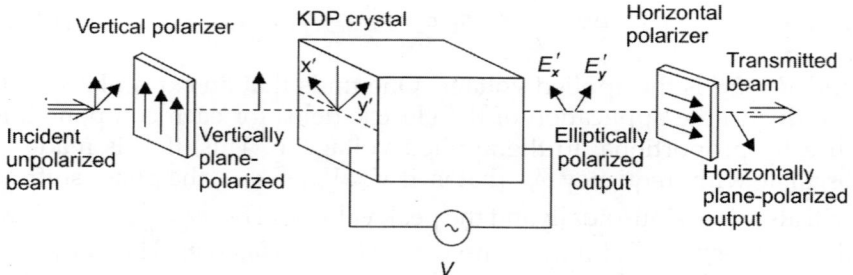

Fig. 10.2. Schematic diagram of a longitudinal electro-optic amplitude modulator using KDP crystal

Then the transmitted electric-field components will be given by $-E_{x'}/\sqrt{2}$ and $E_{y'}/\sqrt{2}$. Now, using Eqs (10.13) and (10.14), one can write the following expression for the transmitted electric field:

$$E = \frac{E_o}{2}[-\cos(\omega t + \Delta\phi) + \cos(\omega t - \Delta\phi)]$$

or
$$E = E_o \sin\delta\phi \sin\omega t \qquad (10.15)$$

The intensity of the transmitted beam may be obtained by averaging E^2 [E being given by Eq. (10.15), over a complete period $T = 2\pi/\omega$]. Thus, the intensity:

$$I = \frac{1}{T}\int_0^T E^2\, dt = \frac{\omega}{2\pi}\int_{t=0}^{2\pi/\omega} E_o^2\,(\sin^2 \Delta\phi)(\sin^2 \omega t)\, dt$$

$$= \frac{E_c^2}{2}\sin^2 \Delta\phi$$

or $\quad I = I_o \sin^2 \Delta\phi = I_o \sin^2\left(\dfrac{\Phi}{2}\right) \qquad (10.16)$

where $I_o = E_o^2/2$ is the amplitude of the intensity of the incident beam. Substituting for DF from Eq. (10.10) in Eq. (10.16), the *transmittance* I/I_o of the modulator, shown in Fig. 10.2, will be given by:

$$\frac{I}{I_o} = \sin^2\left(\frac{\pi}{\lambda} r_{63} n_o^3 V\right) \qquad (10.17)$$

Using Eq. (10.12), this expression may be written as:

$$\frac{I}{I_o} = \sin^2\left(\frac{\pi}{2}\frac{V}{V_o}\right) \qquad (10.18)$$

One can now define V_π as the voltage required for maximum transmission, i.e., $I = I_o$. In general, the transmittance of the modulator can be altered by changing the voltage applied across the crystal. The variation of I/I_o as a function of applied voltage V is shown in Fig. 10.3. Such a system is named as the *Pockels electro-optic amplitude modulator*.

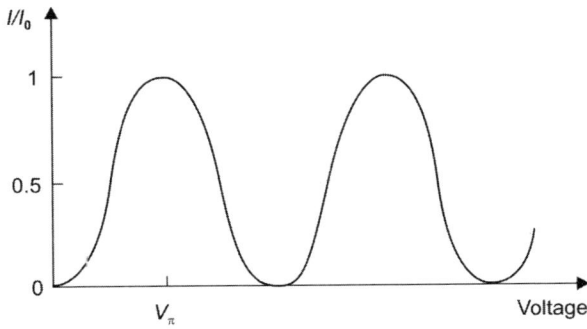

Fig. 10.3. The variation of the transmittance of (I/I_o) of an amplitude modulator as a function of applied voltage (*V*)

We see that, if such a modulator is operated around $V = 0$, the output intensity of the modulated beam does not vary linearly with the input signal. In fact, from Eq. (10.18), one finds that for $V \ll V_\pi$, the transmitted intensity is proportional to V^2.

To overcome this problem, a common practice is to introduce an external bias, so that with no signal, the transmittance of this modulator is 1/2. In general, it is more convenient to bias the modulator optically to the 50% transmittance point Q, by introducing a quarter-wave plate with its fast and slow axes parallel to the x' and y' axes of the modulator crystal, respectively, as shown in Fig. 10.4. This retarder plate introduces a phase difference of $\pi/2$ between $E_{x'}$ and $E_{y'}$.

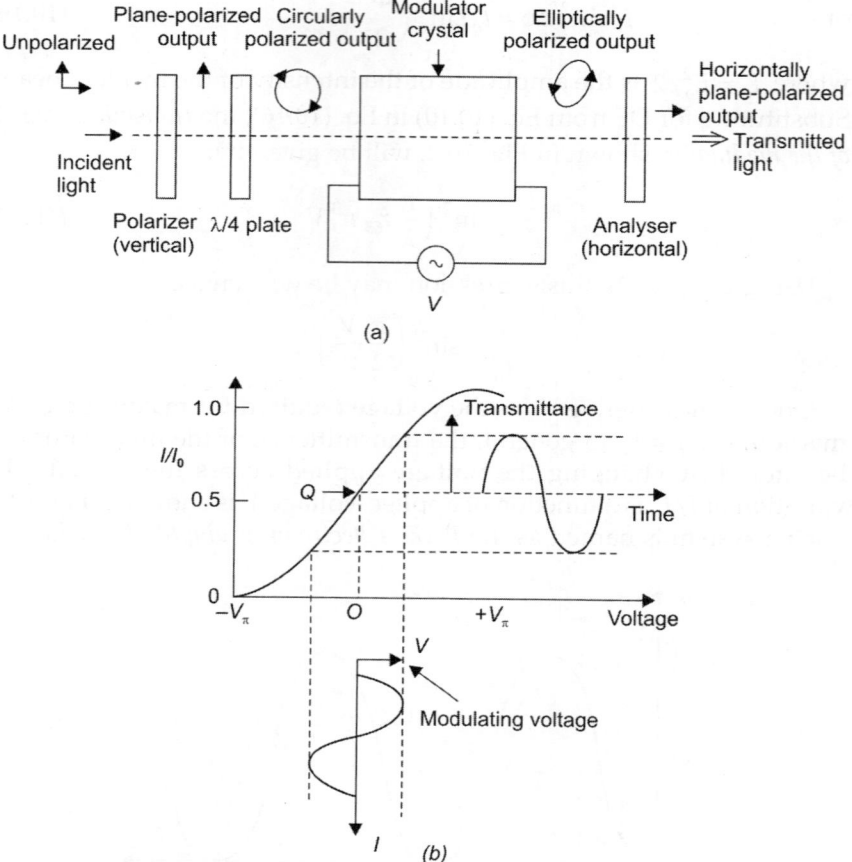

Fig. 10.4. (a) Schematic of Pockels electro-optic amplitude modulator biased with a quarter-wave plate. (b) An almost linear variation of transmittance with applied voltage, when the modulator is operated about Q

With this bias, the net retardation Φ between the two polarized components is obtained as:

$$\Phi = \frac{\pi}{2} + 2\Delta\phi = \frac{\pi}{2} + \pi \frac{V}{V_\pi} \qquad (10.19)$$

Substituting the value of Φ in Eq. (10.16), one finds:

$$\frac{I}{I_0} = \sin^2\left(\frac{\Phi}{2}\right) = \sin^2\left(\frac{\pi}{4} + \frac{\pi}{2}\frac{V}{V_\pi}\right)$$

For $V \ll V_p$:

$$\frac{I}{I_0} \approx \frac{1}{2}\left(1 + \frac{\pi V}{V_\pi}\right) \quad (10.20)$$

which shows that the transmitted intensity varies almost linearly with the applied voltage V. If a small sinusoidally varying voltage of amplitude V_o and frequency w_m is applied to the modulator, then the intensity of the transmitted beam will also vary sinusoidally with frequency ω_m as shown in Fig. 10.4(b). This variation, for small V_o, may be expressed as:

$$\frac{I}{I_0} \approx \frac{1}{2}\left[1 + \frac{\pi V}{V_\pi} V_0 \sin\omega_m t\right] \quad (10.21)$$

In this amplitude modulator, the voltage, and hence, the electric field, is applied along the direction of propagation of the optical beam, and hence, it is called a *longitudinal electro-optic modulator*. To achieve uniform transmission across the effective aperture of the device, a *cylindrical crystal* is taken and *ring electrodes* are used for applying the electric field.

10.4 TRANSVERSE ELECTRO-OPTIC MODULATOR

An electro-optic modulator in the transverse mode of operation, where the direction of propagation of light is perpendicular to the direction of the applied electric field is shown in Fig. 10.5.

This configuration is advantageous as the electrodes do not obstruct the beam as the case of the longitudinal modulator and the retardation

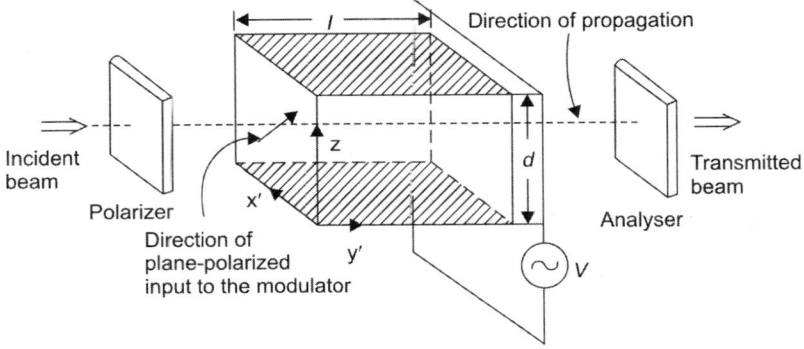

Fig. 10.5. Schematic of transverse electro-optic modulator. Herein, the input wave is plane-polarized at an angle of 45° to the x'-direction (in the x'-z plane) and propagated in the y'-direction. The electric field is applied along the z-direction. The analyser is placed with its pass axis normal to that of the polarizer

(or phase difference), which is proportional to the electric field and the crystal length, can be increased by using longer crystals. We may note that retardation in the case of longitudinal modulators is independent of crystal length.

Let us assume that electric field along the z-direction, and the direction of propagation is along the y' induced principal axis (Fig 10.5). Further, we assume that the incident light, after passing through the polarizer, is plane-polarized in the x'-z plane at 45° to the x'-principal axis. In the presence of electric field E_z in the z-direction, the refractive indices for a wave propagating along the y'-direction and polarized along the x'- and z-direction are, respectively, given by:

$$n_{x'} = n_o + \frac{1}{2} n_o^3 r_{63} E_z \tag{10.22}$$

and

$$n_z = n_e \tag{10.23}$$

Thus the phase difference between the emergent field components along the x'- and y'-direction after traversing a length l of the crystal will be given by:

$$\Delta\phi = \phi_{x'} - \phi_z = \frac{2\pi}{\lambda} l(n_{x'} - n_z)$$

$$= \frac{2\pi}{\lambda} l \left[n_o + \frac{1}{2} n_o^3 r_{63} E_z - n_e \right]$$

$$= \frac{2\pi}{\lambda} l (n_o - n_e) + \frac{\pi}{\lambda} r_{63} n_o^3 \left(\frac{V}{d}\right) l \tag{10.24}$$

where V is the voltage applied across the width d of the crystal. It is important to note that even when the applied voltage $V = 0$, there is finite retardation given by:

$$(\Delta\phi)_{V=0} = \frac{2\pi}{\lambda} l (n_o - n_e)$$

This is due to the intrinsic birefringence of the crystal. Thus, the retardation induced by the external voltage is given by:

$$\Delta\phi = \frac{\pi}{\lambda} r_{63} n_o^3 \left(\frac{V}{d}\right) l \tag{10.25}$$

We define a *half-wave voltage* V_π for this configuration as the voltage required to produce a phase difference of π between the two polarization components (in addition to that produced by intrinsic birefringence). From Eq. (10.25), one obtains:

$$\Delta\phi = \pi = \frac{\pi}{\lambda} r_{63} n_o^3 \left(\frac{V_\pi}{d}\right) l$$

or

$$V_\pi = \frac{\lambda}{n_o^3 r_{63}} \left(\frac{d}{l}\right) \tag{10.26}$$

Contrary to the longitudinal modulator, V_π in this case is *not independent of the length l of the modulator crystal, but depends on the ratio d/l*. This means half-wave voltage may be reduced by employing long, thin crystals.

Table 10.1 contains some materials that are useful for making electro-optic modulators.

Table 10.1. Physical parameters of some electro-optic crystals used in Pockels modulators (the values quoted below are near λ = 550 nm)

Material	Refractive index		Relevant electro-optic coefficient r (10^{-12} m/V)
	n_o	n_e	
KDP (KH_2PO_4)	1.51	1.47	r_{63} = 10.5
KD*P (KD_2PO_4)	1.51	1.47	r_{63} = 26.4
ADP ($NH_4H_2PO_4$)	1.52	1.48	r_{63} = 8.5
Lithium niobate ($LiNbO_3$)	2.29	2.20	r_{33} = 30.8
Lithium tantalate ($LiTaO_3$)	2.175	2.18	r_{33} = 30.3
Gallium arsenide (GaAs)	3.6	–	r_{41} = 1.6

10.5 ACOUSTO-OPTIC EFFECT AND ACOUSTO-OPTIC MODULATORS

The change in the refractive index of a medium caused by the mechanical strain produced due to the passage of an acoustic wave through the medium is called as acoustic optic effect.

It is reported that the refractive index of a medium, in general, varies with mechanical strain in a complicated manner. Let us consider a simple case (Fig. 10.6), where a monochromatic light of wavelength λ is incident normally on an acousto-optic medium, in which the periodic variations in the refractive index of the medium. When the light enters the medium, the portion of the incident wavefront near the acoustic wave crests (or pressure maxima) encounter high refractive index, and hence, advance with a lower velocity than those portions of the wavefront that encounter acoustic wave troughs (or pressure minima). As a consequence, the wavefront in the medium soon acquires a wave-like appearance, as shown in Fig. 10.6. As the velocity of the acoustic wave is much lesser than that of the optical wave, the refractive index in the medium may be considered to have almost no variation. In effect, the acoustic wave sets up a refractive index grating within the medium, so that when a light beam falls on it, either multiple-order or single-order diffraction takes place. The first one is called *Raman-Nath diffraction*, normally observed at low acoustic frequencies, and the second one is referred to as *Bragg diffraction*, usually observed at high acoustic frequencies. *Acousto-optic modulators operating in the Raman-Nath and Bragg regimes* are discussed in the following sections:

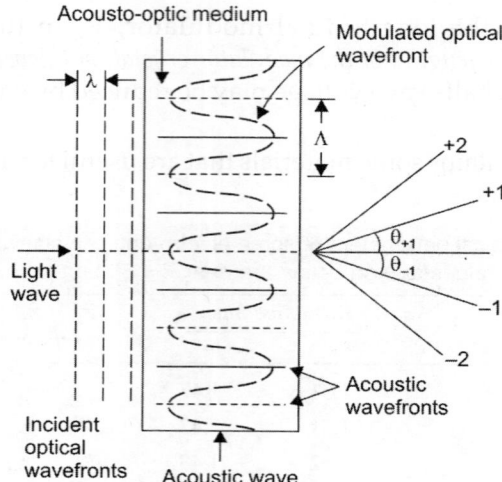

Fig. 10.6. Schematic simplified illustration of acousto-optic modulation. Crests (pressure maxima) and troughs (pressure minima) in the acoustic wave are represented by solid and dashed horizontal lines, respectively

10.5.1 Raman-Nath Modulator

In the Raman-Nath regime, the acousto-optic diffraction grating is so thin that it behaves almost like a *plane transmission grating*. In general, the mth-order diffracted wave propagates along a direction making an angle θ_m with the direction of the incident beam, given by:

$$\sin \theta_m = m \left(\frac{\lambda}{n_0 \Lambda} \right) \tag{10.27}$$

where n_0 is the refractive index of the medium in the absence of the acoustic wave and:

$$m = 0, \pm 1, \pm 2, \pm 3, \pm 4, \ldots$$

is the order number.

In this configuration, (Fig. 10.7), the signal carrying the information modulates the amplitude of the acoustic wave propagating through the medium. The light beam incident on the acousto-optic medium gets diffracted and the zeroth-order beam of the diffracted output is blocked using a stop. For small acoustic powers, the relative intensity in the first order is obtained as:

$$\eta \approx \frac{\pi^2 (\Delta n)^2 L^2}{\lambda^2} \tag{10.28}$$

where Δn is the peak change in refractive index of the medium due to the acoustic wave and L is width of the acoustic beam, normally equal to the length of the medium. It can be shown that $(\Delta n)^2$ is proportional

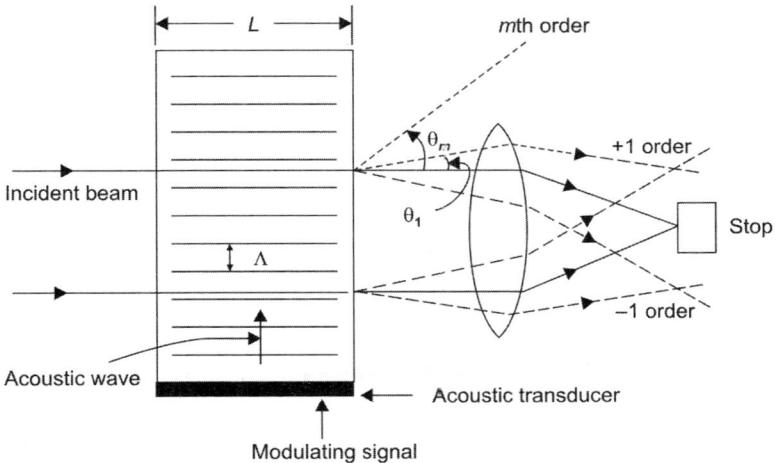

Fig. 10.7. Acousto-optic modulator based on Raman-Nath diffraction

to the acoustic power. Thus, if the acoustic wave is amplitude-modulated, the first-order diffracted beam (corresponding to $m = \pm 1$) will be intensity-modulated.

10.5.2 Bragg Modulator

The configuration of a Bragg modulator is as shown in Fig. 10.8. In the Bragg regime, the interaction length L is larger, so the acoustic field creates a thick grating inside the medium. The situation here is analogous to Bragg's crystal grating in which the x-rays reflected by the different planes of the crystal produce diffraction. Thus, when the light beam is incident at an angle θ, it is reflected by successive layers of the acoustic grating. Diffraction occurs for an angle of incidence $θ = θ_B$ (known as the Bragg angle) under the condition:

$$\sin θ_B = \frac{\lambda}{2 n_0 \Lambda} \qquad (10.29)$$

For small acoustic power P_a, the diffraction efficiency for an angle of incidence $θ_B$ may be given by:

$$\eta \approx \frac{\pi^2 M}{2 \lambda^2 \cos^2 θ_B} \left(\frac{L}{H}\right) P_a \qquad (10.30)$$

where M is the figure of merit of the acousto-optic device, and L and H are the length and height, respectively, of the acoustic transducer. Thus, the intensity of the diffracted beam is directly proportional to the acoustic power, and hence, modulation of the acoustic power will lead to corresponding modulation of the diffracted beam.

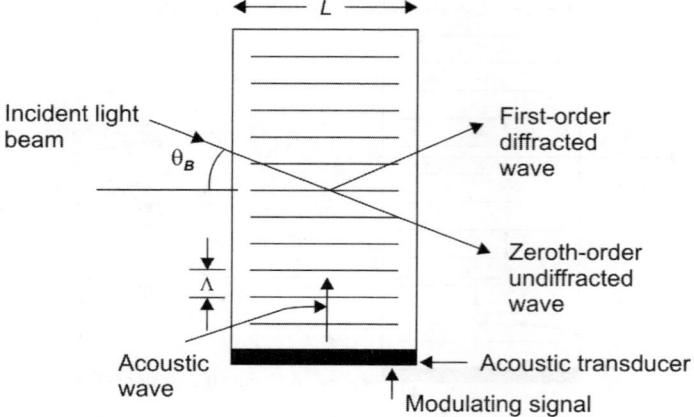

Fig. 10.8. An acousto-optic modulator based on Bragg diffraction

10.5.3 Applications of Optoelectronic Modulators

Electro-optic effect in fiber optic systems is used to convert phase modulation into intensity variation. Figure 10.9 shows the typical structures for achieving this. We can see that these consists of two planar single-mode waveguides made from a similar electro-optic material.

The input beam is split equally between two waveguides using a 3 dB coupler. The split beams after travelling through the waveguides are recombined through a second 3 dB coupler. In one scheme (Fig. 10.9(a)), the modulation voltage is applied to one waveguide but not to the other. If the two waveguides are equal in length and voltage

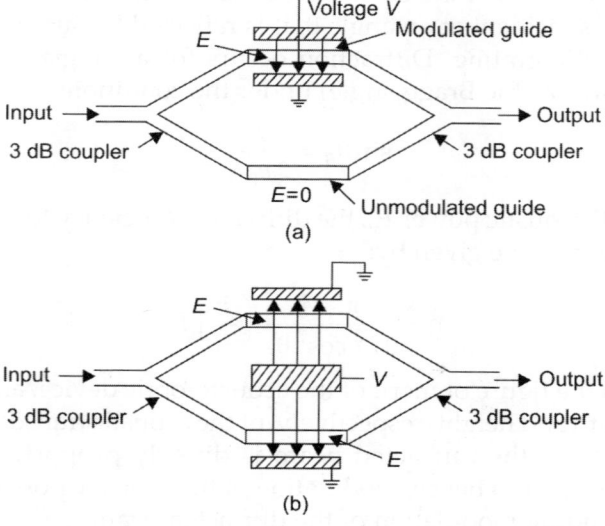

Fig. 10.9. Application of electro-optic planar waveguides in digital modulation

is zero, the optical path lengths of the two arms are equal and the two waves arrive at the second coupler in phase and constructively interfere (producing a 'high' signal). However, if we apply the voltage needed to delay the phase by π, the two waves get out of phase when they recombine. They interfere destructively (producing a 'low' signal). In the second scheme (Fig. 10.9(b)), voltages are applied across both the waveguides, but with opposite polarities, so that the voltage increases the refractive index in one arm and decreased it in the other. This differential change in the refractive index may be used to achieve the same modulation as in the first scheme but with lower voltage. Lithium niobate modulators are now widely used for this purpose; modulation at 10 Gbits/s is possible with these.

The acousto-optic effect may be used in the spectral analysis of radio-frequency (*rf*) signals. A typical integrated optic spectrum analyzer consists of an antenna that picks up the *rf* signal and sends it to an amplifier that drives an acousto-optic planar waveguide to periodically constrict or expand. The spatial period of the acoustic wave helps in getting the signature of the *rf* signal.

Optoelectronic modulators can also be used as *switches* or *routers* in optical networks.

Example 1

Calculate the change in refractive index due to longitudinal electro-optic effect for a 1-cm-wide KDP crystal for an applied voltage of 5 kV. If the wavelength of light being propagated through the crystal is 550 nm, calculate the net phase shift between the two polarization components after they emerge from the crystal. Also calculate V_π for the crystal. [B.Tech.]

Solution

For a voltage of 5 kV across a 1-cm crystal, the electric field produced will be:

$$E_z = \frac{V}{l} = \frac{5000}{1 \times 10^{-2}} = 5 \times 10^5 \text{ V/m}$$

From Table 10.1, for KPD, $n_o = 1.51$ and $r_{63} = 10.5 \times 10^{-12}$ m/V. Thus the change Δn in the refractive index (that is, the difference between n_o and $n_{x'}$ or n_o and $n_{y'}$) will be:

$$\Delta n = \frac{1}{2} N_o^3 r_{63} E_z = \frac{1}{2} \times (1.51)^3 \times 10.5 \times 10^{-12} \times 5 \times 10^5$$

or $\quad \Delta n = 9.04 \times 10^{-6}$

$\therefore \quad \Delta\phi = \frac{2\pi}{\lambda} \Delta nl = \frac{2\pi}{550 \times 10^{-9}} \times 9.04 \times 10^{-6} \times 1 \times 10^{-2} = 0.33 \pi$

Clearly, the net phase shift suffered by the two polarization components will be:
$$\Phi = 2\Delta\phi = 0.66\pi$$

Now, the half-wave voltage V_π may be calculated from Eq. (10.12)

$$V_\pi = \frac{\lambda}{2n_o^3 r_{63}} = \frac{550 \times 10^{-9}}{2 \times (1.51)^3 \times 10.5 \times 10^{-12}} = 7606 \text{ V} = 7.6 \text{ kV}$$

Example 2
A transverse electro-optic modulator with a KDP crystal is operating at a wavelength $\lambda = 550$ nm. The crystal has length $l = 3$ cm and width $d = 0.25$ cm. The optical constants of the crystal may be taken from Table 10.1. Calculate (a) the phase difference between the emergent field components with applied voltage $V = 0$, (b) the additional phase difference between the emergent field components with $V = 2$ V, and (c) the half-wave voltage V_π for the crystal. [B.Tech.]

Solution
(a) $\Delta\phi$ (due to intrinsic birefringence):

$$= \frac{2\pi l}{\lambda}(n_o - n_e) = \frac{2 \times 3 \times 10^{-2}}{550 \times 10^{-9}}(1.51 - 1.47)\pi = 4.363 \times 10^3 \pi$$

(b) $\Delta\phi$ (due to external field):

$$\frac{\pi}{\lambda} r_{63} n_o^3 \left(\frac{V}{d}\right) l = \frac{26.4 \times 10^{-12} \times (1.51)^3}{550 \times 10^{-9}} \times \frac{200 \times 3 \times 10^{-2}}{0.25 \times 10^{-2}} \pi = 0.396 \pi$$

(c) $V_\pi = \frac{\lambda}{n_o^3 r_{63}}\left(\frac{d}{l}\right) = \frac{550 \times 10^{-9}}{(1.51)^2 \times 26.4 \times 10^{-12}} \times \left(\frac{25 \; 10^{-2}}{3 \times 10^{-2}}\right) = 504 \text{ V}$

Example 3
A typical acousto-optic cell of a Raman-Nath modulator contains water. A piezoelectric crystal (an acoustic transducer shown in Fig. 10.7) bonded to the cell generates an acoustic wave of frequency 5 MHz in water. The velocity of the acoustic wave in water is 1500 ms^{-1} and the thickness of the cell is 1 cm. If a He-Ne laser beam ($\lambda = 633$ nm) is incident on the cell, calculate the (a) angle between the first-order diffracted beam and the direct beam, and (b) relative intensity of the diffracted beam in the first order if $\Delta n = 10^{-5}$.

Solution
(a) We have:
$$\sin\theta_m = m\left(\frac{\lambda}{n_0 \Lambda}\right)$$

Here, $m = 1$, $\lambda = 633$ nm, $n_0 = 1.33$ (for water).

$$\Lambda = \frac{\text{velocity of acoustic waves}}{\text{frequency (Hz)}} = \frac{1{,}500 \text{ ms}^{-1}}{5 \times 10^6 \text{ s}^{-1}} = 300 \times 10^{-6} \text{ m}$$
$$= 300 \text{ μm}$$

$$\theta_1 = \sin^{-1}\left[\frac{633 \times 10^{-9}}{1.33 \times 300 \times 10^{-6}}\right] = 0.09°$$

(b) $\eta = \dfrac{\pi^2 (\Delta n)^2 L^2}{\lambda^2} = \dfrac{\pi^2 (10^{-5})^2 \times (1 \times 10^{-2})^2}{(633 \times 10^{-9})^2} = 0.246.$

SUGGESTED READINGS

1. Liu JM, 'Photonic Devices', Cambridge University Press (2005).
2. Chuang SL, 'Physics of Photonic Devices', Wiley (2009).
3. Liu MMK, 'Principles and Applications of Optical Communications, Irwin, Chicago (1966).
4. Yariv A and Yeh P, 'Photonics: Opto-Electronics in Modern Communications', Oxford University Press (6th ed.) (2007).
5. Kakani SL and Bhandari KC, 'Optics, Sultan Chand & Sons, New Delhi-2 (2nd ed.).
6. Degiorgio V and Cristiani J, 'Photonics', Springer (2014).

GLIMPSES

- One can achieve the modulation of an optical signal at the source either internally or externally. Internal or direct modulation has its limitation, whereas external modulation is faster and permits both amplitude as well as phase modulations with ease.
- In optically anisotropic medium, e.g., calcite, KDP etc., the velocity of propagation depends on the direction as well as the state of polarization. Clearly, there are two principal refractive indices: (i) corresponding to the ordinary (*o*) ray (which follows Snell's law), and (ii) the extraordinary (*E*) ray (which does not follow Snell's law). Calcite and quartz crystals have two principal refractive indices and one optic axis (the direction along which the velocity of O and E rays is same) are called *uniaxial crystals.*
- When the electric field is applied across a birefringent crystal, say KPD crystal, it may change its refractive indices. If this change is linearly proportional to the applied field, the effect is called the *Pockels electro-optic effect*, which may be used for phase modulation. When a voltage *V* is applied along the optic axis of such a birefrigent crystal, the incident plane polarized light splits into two components and the net phase shift between them is as follows:

$$\phi = \frac{2\pi}{\lambda} r_{63} n_o^3 V$$

Obviously, ϕ can be changed by changing *V*.

- For a *longitudinal modulator*, the voltage (V) required to introduce a phase shift of π is called as *half-wave voltage* (V_π). We have

$$V_\pi = \frac{\lambda}{2n_o^3 r_{63}}$$

  In case of a transverse modulator:

$$V_\pi = \frac{\lambda}{n_o^3 r_{63}}\left(\frac{d}{l}\right)$$

- One can also change the refractive index of a crystal by passing an acoustic wave through it. This effect is known as *acoustic-optic effect*. It produces a grating within a crystal, so that when a light beam is incident on it, either multiple order (*Raman-Nath regime*) or single-order (*Bragg regime*) diffraction takes place. Acousto-optic modulator based on these effects and operating in the two regime can be constructed.

REVIEW QUESTIONS

1. Explain electro-optic effect. How this effect can be used for modulating the phase of an optical signal? How one can modulate the amplitude of the optical signal.
2. Describe the principle and working of a longitudinal electro-optic modulator. Show that the transmittance I/I_o of the modulator is given by:

$$\frac{I}{I_o} = \sin^2\left(\frac{V}{V_\pi}\right)$$

   where V is applied voltage and V_π is the voltage required for maximum transmittance, i.e. $I = I_o$.
3. Explain the principle and working of transverse electro-optic modulator. Show that half-wave voltage for this configuration is given by:

$$V_\pi = \frac{\lambda}{n_o^3 r_{63}}\left(\frac{d}{l}\right)$$

   where symbols have usual meanings. How this differs from longitudinal electro-optic modulator.
4. What is acoustic optic effect? Distinguish between modulators based on this effect and operating in the Raman-Nath and Bragg regime.

PROBLEMS

1. Calculate the change in refractive index due to longitudinal electro-optic effect for a 5 mm long crystal of lithium niobate when the applied voltage is 100 V. Take the wavelength of light propagating through the crystal as 550 nm. Calculate the net phase shift between

the two polarization components after they emerge from the crystal. Take the values of necessary constants from Table 10.1.

[**Ans.** 3.698×10^{-6}, 0.134 p and 743.5 V]

2. A transverse electro-optic modulator uses a lithium niobate crystal and operates at 550 nm (for constants of lithium biobate crystal (Table 10.1)).

Calculate the length of the crystal required to produce a phase difference of $\pi/2$ between the emergent field components with zero applied field.

Also calculate the width of the crystal required to produce an additional phase difference of $\pi/2$ between these components with an applied voltage of 20 V.

What will be the value of half wave voltage V_π for the crystal.

[**Ans.** 1.528 µm, 0.041 µm and 40 V]

3. A Raman-Nath modulator employs a cell containing water. An acoustic transducer bonded to cell generated a wave of frequency in water of frequency 4 MHz. A He-Ne laser beam ($\lambda = 633$) incident on the cell. Taking the following data: $n_o = 1.33$ for water, thickness of the cell = 1.2 cm, and the velocity of the acoustic wave in water = 1500 ms^{-1}, show that the observed first-order diffracted beam is 0.727°.

SHORT ANSWER QUESTIONS

1. What are the two ways of modulating an optical signal from a LED or ILD?

Ans. (i) Internal or direct modulation in which a circuit is designed to modulate the current injected into the device. (ii) External modulation: In this scheme an optical signal from the source is passed through a material (or device) whose optical properties can be altered by external means.

2. What are the disadvantages of direct modulation?

Ans. (i) The upper frequencies are limited to about 40 GHz. (ii) The emission frequency may change as the drive current is changed. (iii) Only amplitude modulation is possible with ease, i.e., for phase or frequency modulation, additional care is required to design the drivers.

3. What is the disadvantage of external modulation?

Ans. The modulator is normally large on the scale of microelectronic devices and hence cannot be a part of same integrated circuit.

4. What do you understand by optical anisotrophy?

Ans. Calcite, quartz, KPD (potassium dihydrogen phosphate), etc. are few examples which exhibit optical anisotrophy. In these materials, the velocity of propagation of light, in general, depends on the direction of propagation and also the state of polarization of polarization of light, i.e., the refractive index of these crystals varies with direction. Such crystals are called birefringent or doubly refracting.

5. What is electro-optic effect?

Ans. When we apply an electric field across a crystal, it may change its refractive indices. Obviously, the field may induce birefringence in an otherwise isotropic crystal or change the birefringent property of the doubly refracting crystal. This is known as electro-optic effect.

6. What is acusto-optic effect?

Ans. The change in the refractive index of a medium caused by the mechanical strain produced due to the passage of an acoustic wave through the medium is termed as the acousto-optic effect.

MULTIPLE CHOICE QUESTIONS

1. Which one is birefringent crystal?
   - (a) NaCl
   - (b) Diamond
   - (c) Calcite
   - (d) Nickel

2. If r is the linear electro-optic coefficient and P is the quadratic electro-optic coefficient, then the change index n as a function of applied field E can be given by the equation of the form:
   - (a) $\Delta\left(\dfrac{1}{n^2}\right) = rE + PE$
   - (b) $\Delta\left(\dfrac{1}{n^2}\right) = rE^2 + PE$
   - (c) $\Delta\left(\dfrac{1}{n^2}\right) = rE + PE^2$
   - (d) $\Delta\left(\dfrac{1}{n^2}\right) = rE^2 + PE^2$

3. Half-wave voltage (V_π) of an electro-optic modulator is given by:
   - (a) $V_\pi = \dfrac{\lambda}{2n_o^3 r_{63}}$
   - (b) $V_\pi = \dfrac{\lambda}{2n_o r_{63}}$
   - (c) $V_\pi = \dfrac{\lambda}{2n_o^2 r_{63}}$
   - (d) $V_\pi = \dfrac{\lambda}{2\sqrt{n_o} r_{63}}$

4. Half-wave voltage V_π for transverse electro-optic modulator is given by:
   - (a) $V_\pi = \dfrac{\lambda}{n_o r_{63}}\left(\dfrac{d}{l}\right)$
   - (b) $Vn_o = \dfrac{\lambda}{r_{63}}$
   - (c) $V_\pi = \dfrac{\lambda}{n_o^3 r_{63}}\left(\dfrac{d}{l}\right)$
   - (d) $V_\pi = \dfrac{\lambda}{n_o r_{63}^2}\left(\dfrac{d}{l}\right)^2$

5. In a Raman-Nath regime, the acoustic-optic diffraction gratting behaves almost like a:
   (a) prism
   (b) plane transmission grating
   (c) convex lens
   (d) concave lens
6. Consider an electromagnetic wave travelling in the z-direction. The orthogonal components of its E-vector are along the x- and y-direction. Assume that the two components have same amplitudes but a phase difference of $\pi/2$. The resultant E-vector at any point in space:
   (a) is constant in amplitude but rotates with an angular frequency ω
   (b) changes in amplitude but rotates with an angular frequency ω
   (c) remains stationary
   (d) varies randomly
7. The electromagnetic wave of Question 6 is said to be:
   (a) unpolarized
   (b) plane-polarized
   (c) circularly polarized
   (d) elliptically polarized
8. In a birefringent crystal:
   (a) the o-ray follows Snell's law but the e-ray does not
   (b) the e-ray follows Snell's law but the o-ray does not
   (c) both the o-ray and e-ray follow Snell's law
   (d) both the o-ray and e-ray do not follow Snell's law
9. In a longitudinal electro-optic modulator, half-wave voltage is that voltage which introduces the following phase shift between two polarization components:
   (a) $\pi/4$
   (b) $\pi/2$
   (c) π
   (d) 2π
10. In a transverse electro-optic modulator:
    (a) V_π is dependent of the length l and width d of the modulator crystal
    (b) V_π is dependent on the length l but not on the width d of the crystal
    (c) V_π is dependent on the width d but not on the length l of the crystal
    (d) V_π is dependent on the ratio d/l
11. In a Raman-Nath modulator, the acousto-optic grating is:
    (a) so thin that it behaves almost like a plane transmission grating
    (b) so thick that it behaves almost like Bragg's crystal grating
    (c) analogous to a concave Rowland's grating
    (d) quite complicated

12. The application of an electric field across a birefringent crystal, e.g., KDP, may change its refractive indices. If the change is linearly proportional to the electric field, it is called:
    (a) Pockel's electro-optic effect
    (b) Raman-Nath effect
    (c) Snell's law
    (d) Acousto-optic effect
13. When a voltage V is applied along the optic axis of a birefringent crystal, e.g., KDP, the incident plane polarized light splits into two components and the net phase shift between them is given by:
    (a) $\phi = \dfrac{2\pi}{\lambda} V$
    (b) $\phi = \dfrac{2\pi}{\lambda} r_{63} n_o^3 V$
    (c) $\phi = \dfrac{2\pi}{\lambda} n_o^3 V$
    (d) $\phi = \dfrac{2\pi}{\lambda} r_{63} V$

ANSWERS

1. (c) 2. (c) 3. (a) 4. (c) 5. (b) 6. (a) 7. (c)
8. (a) 9. (c) 10. (d) 11. (a) 12. (a) 13. (b)

11

Optical Amplifiers

11.1 INTRODUCTION

An *optical amplifier* (OA) is a device that amplifies the optical signal directly, without converting it to an electrical signal and then to an optical signal again. OAs are used for amplifying a weak signal in order to increase the distance, the signal can be transmitted down the transmission lines. In comparison, repeaters and generators convert the signal to electrical form, regenerate or amplify the signal, and then convert it to optical form again. The conversion of the signal from one form to another is a complex process, subject to high losses, slow speed, and more costlier than simple optical amplifiers.

An optical amplifier operates solely in the optical domain, i.e. it takes in a weak optical signal from one segment of the link, amplifies it optically to produce a strong optical signal (without recourse to photon-to-electron conversion and *vice versa*), and couples it to the next segment of the link. Obviously, such devices offer several advantages over regenerators, e.g. (i) they are insensitive to data rate or signal format, and (ii) they have large gain widths. Clearly, a single OA can simultaneously amplify many WDM signals, propagating through the same fiber. On the other hand, if the system employs regenerators, it will need a regenerator for each wavelength. A major disadvantage of the present OAs is that they cannot regenerate signals, i.e. they cannot clean up noise or compensate for dispersion. However, if one can take appropriate steps to reduce noise or compensate for dispersion, such OAs are much simpler, less expensive and widely applicable. This clearly reveals that OAs have become essential components in high performance, long-hand and multi-channel fiber optic communication systems.

Some of the basic applications of OAs are listed below:
(a) as in-line amplifiers for power boosting
(b) as per-amplifiers to increase the received power at the receiver
(c) as power amplifiers to increase transmitted power
(d) as a power booster in a local area network.

The main types of optical amplifiers are:
- erbium doped fiber amplifiers (EDFAs),
- Raman and Brillouin amplifiers, and
- semiconductor optical amplifiers (SOAs).

The operation and characteristics of all types of amplifiers are:
1. population inversion is created, which means that more systems (atoms, molecules) are in a high energy state than in a lower one,
2. the incoming pulse of signal induce stimulated emission,
3. amplifiers saturate above a certain signal power, and
4. amplifiers add noise to the signal.

The general characteristics of EDFA and SOA are compared in Table 11.1.

Table 11.1. Some properties of EDFA and SOA

Properties of amplifiers	EDTA	SOA
Active medium	Er^{3+} ion in silica	Electron-hole in semiconductors
Typical length	Few meters	500 µm
Pumping	Optical	Electrical
Gain spectrum	1.5–1.6 µm	1.3–1.5 µm
Gain bandwidth	24–35 nm	100 nm
Relaxation time	0.1–1 ms	< 10–100 ps
Maximum gain	3–50 dB	25–30 dB
Saturation power	> 10 dBm	0–10 dBm
Crosstalk	–	For bit rate 10 GHz
Polarization	Insensitive	Sensitive
Noise figure	3–4 dB	6–8 dB
Insertion loss	< 1 dB	4–6 dB
Optics	Pump laser diode couplers, fiber splice	Antireflection coatings, fiber-waveguide coupling
Integration	No	Yes

These technologies and applications are discussed in the following sections.

11.2 SEMICONDUCTOR OPTICAL AMPLIFIERS

The functional applications of *semiconductor optical amplifiers* (SOAs) were first studied in the early 1990s. Since then, the diversity and scope of such applications have been steadily growing. SOAs are another common type of in-line amplifiers that are developed to support *dense wavelength division multiplexing* (DWDM) and to expand to the other wavelength bands supported by fiber optics. They have many applications in optical fiber communication, switching, and signal processing systems.

SOAs are based on the same technology as basic-semiconductor Fabry-Perot diodes, but they have anti-reflection (AR) coating at the endfaces. Fabry-Perot laser diodes are generally presented in the laser theory. The structure of the SOAs is much the same as the diode, with two stacked slabs of specially designed semiconductor material, with another material between them that forms the active layer, as shown in Fig. 11.1(a). The schematic diagram of an SOA is shown in Fig. 11.1(b) and Fig. 11.1(c) shows SOA in an amplifier configuration. An electrical current is passed through the device in order to excite electrons to high-state level. The electrons then fall back to the non-excited ground state, emitting photons by stimulated emission. An incoming optical signal stimulates emission of photons at its own wavelength. This is accomplished by blocking the cavity reflectors using an antireflection (AR) coating on both end faces. Fiber optic cables are attached to both ends. As explained in EDFAs, optical isolators are commonly used at

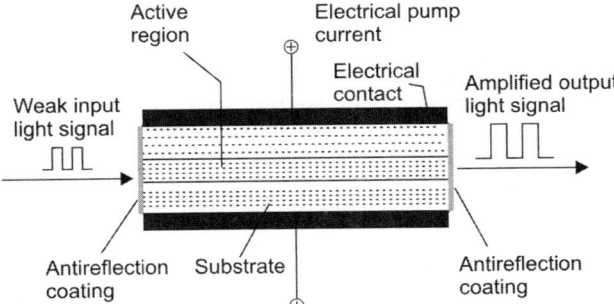

Fig. 11.1. (a) Semiconductor optical amplifier (SOA)

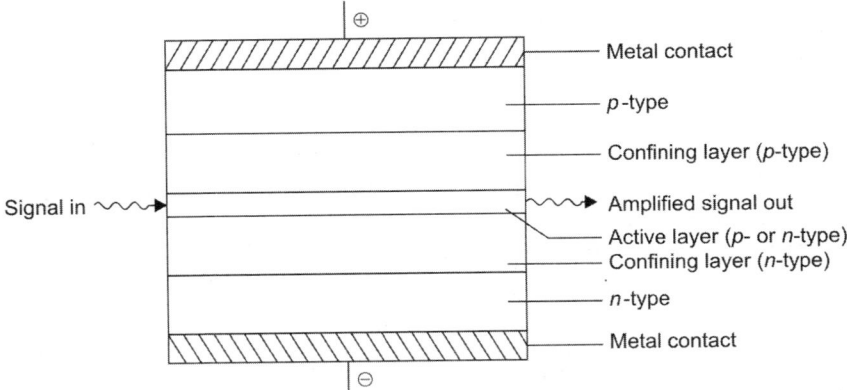

Fig. 11.1. (b) Schematic diagram of an SOA

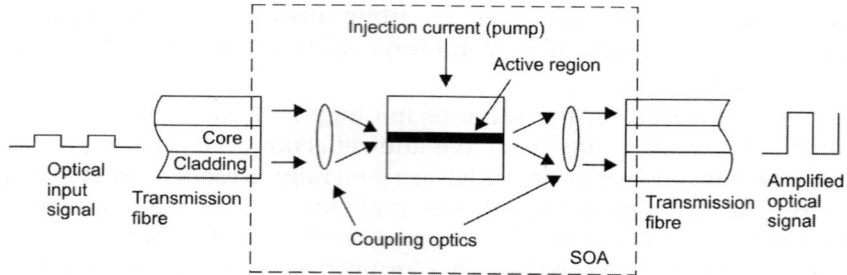

Fig. 11.1. (c) SOA in amplifier configuration

both ends of the SOA to prevent light signals from returning back. Depending on the material of the active layer, they operate from 1310 to 1550 nm in telecommunication systems.

SOAs are typically constructed in a small package. In addition, they transmit bi-directionally, making the reduced size of the device an advantage over EDFAs. They can be integrated with optical devices, such as semiconductor lasers, modulators, and DWDM. But the actual performance is still not compatible with EDFAs. They have high noise, less gain, medium polarization dependence, and high optical gain non-linearity with fast transient time. High nonlinearity makes the SOAs attractive for optical signal processing, such as all-optic switching, wavelength conversion and regeneration, time demultiplexing, clock recovery, and pattern recognition. A number of SOA chips can be integrated on the same substrate to create high-density switching matrices.

SOAs are classified into two groups: *Fabry-Perot Cavity Amplifiers* (EPA) and *Travelling Wave Amplifiers* (TWA). The difference depends on the efficiency of the reflection value of the antireflection coating material used.

We may note that in SOA, the light signal passes through the amplifying medium only once. Fig. 11.1(b) clearly shows that SOA is a familiar double-heterostructure (DH).

The material of the active layer is chosen such that it has a band gap lower than that of the confining layers. When a forward bias is applied to this DH, electrons from the n-type semiconductor and holes from the p-type semiconductor travel towards the active layer where they get trapped in a low-band-gap potential well. If the biasing current is large enough, large concentrations of electrons and holes built up in the active layer, leading to *population inversion*. Signal photons passing through the active layer can stimulate radiative recombination of electrons and holes, resulting in the amplification of signal power. This is the basic principle underlying the functioning of this structure as an optical amplifier. It is also possible that the carriers (electrons and holes) recombine spontaneously, leading to amplified spontaneous emission (ASE), or even decay non-radiatively.

In order that a DH functions efficiently as an OA, at least following requirements have to be met:

(i) The active layer must have a band gap lower than that of the surrounding layers so that the carriers are confined in this layer and population inversion is achieved. This implies that the lifetime of the carriers should also be sufficiently long.

(ii) The active layer should also confine the light passing through the structure. Its lower band gap with respect to the confining layers implies a large refractive index of this layer, leading to waveguiding within this region.

(iii) The energy of signal photons should match with that of the inverted active layer in order to achieve optical gain. Further amplification should be independent of the polarization of the signal beam.

(iv) The signal beam must be coupled efficiently into and out of the SOA chip, usually to a single-mode optical fiber. This implies that the SOA should function as a single-mode waveguide with a circular beam waist matching the mode field diameter of the single-mode fiber.

(v) Finally, the optical feedback have to be suppressed. This means that all measures must be taken to reduce optical reflections at the facets of the active layer to less than 0.01%.

Basic parameters of SOA are summarized in Table 11.2.

Table 11.2. Basic parameters of SOA

Description	Symbol	Value
Differential gain	a	2.7×10^{-16} cm^2
Gain coefficient	a_2	0.15 cm^{-1} nm^{-2}
Gain coefficient	a_3	2.7×10^{-17} nm·cm^{-3}
Transparency density	n_o	1.1×10^{18} cm^{-3}
Amplifier length	L	350 µm
Density at threshold	n_{th}	1.8×10^{18} cm^{-3}

11.2.1 General Properties of Optical Amplifiers

Amplification in optical amplifiers is through stimulated emission (the same as in lasers). One can, therefore, consider similar characteristic parameters, like gain and its spectrum, bandwidth, etc. We will discuss them in some detail.

11.2.2 Gain Spectrum and Bandwidth

The model gain, one typically starts with a homogeneously broadened two-level system. Local gain coefficient for such system is:

$$g(\nu, P) = \frac{g_0}{1 + \frac{(\nu - \nu_0)^2}{\Delta \nu_0^2} + \frac{P}{P_{sut}}} \qquad (11.1)$$

where g_0 is the *peak value of the unsaturated gain*, ν_0 atomic transition frequency, $\Delta \nu_0$ 3-dB local gain bandwidth, P_{sat} saturation power and P and ν are optical power and frequency of the amplified signal.

Local gain can also be expressed as:

$$g(\omega, P) = \frac{g_0}{1 + (\omega - \omega_0)^2 T_2^2 + P/P_{sat}} \qquad (11.2)$$

where $\omega = 2\pi \nu$ is the angular frequency and T_2 is known as dipole relaxation time.

For small signal power one can consider a unsaturated regime defined as $P \ll P_{sat}$. In this limit local gain is:

$$g(\omega) = \frac{g_0}{1 + (\omega - \omega_0)^2 T_2^2} \qquad (11.3)$$

The following conclusions can be derived from the above equation:
  (i) maximum gain corresponds to transitions with angular frequency $\omega = \omega_0$,
  (ii) for $\omega \neq \omega_0$ gain spectrum is described by Lorentzian profile,
  (iii) local gain bandwidth, which is defined as the *full width at half maximum* (FWHM), is:

$$\Delta \omega_0 = \frac{2}{T_2} \qquad (11.4)$$

Local gain bandwidth $\Delta \omega_0$ is defined by points in frequency where local gain takes half the value at the maximum. In terms of frequency it can be written as:

$$\Delta \nu_0 = \frac{\Delta \omega_0}{2\pi} = \frac{1}{\pi T_2} \qquad (11.5)$$

Let $P(z)$ be be the optical power at a distance z from the input end. Its change is described as:

$$\frac{dP(z)}{dz} = g(\nu, P) \cdot P(z) \qquad (11.6)$$

Assume a linear device (here for power levels $P \ll P_{sat}$), where local gain is independent of the signal power. Integration of the above equation gives:

$$P(z) = P(0) \, e^{g \cdot z} \qquad (11.7)$$

where $P(0) = P_{in}$ is the signal input power. Linear amplifier gain is defined as:

$$G = \frac{P(L)}{P(0)} \tag{11.8}$$

where $P(L) = P_{out}$ is the output power. From the above solution, one obtains:

$$G = \frac{P(L)}{P(0)} = e^{gL} = \exp\left[\frac{g_0 \cdot L}{1 - \frac{(\nu - \nu_0)^2}{\Delta\nu_0^2}}\right] \tag{11.9}$$

$$= \exp\left[g_0 L - \left(\frac{P(L) - P(0)}{P_{sat}}\right)\right]$$

where P_{sat} is saturation power.

If $P(L) \gg P(0)$ and the small signal gain of the amplifier is expressed by $G_0 = \exp(g_0 L)$, one may write to a good approximation:

$$\log G = \log G_0 - \frac{P(L)}{P(0)} \tag{11.9a}$$

Amplifier bandwidth B_0 is evaluated using the above solution. It is defined by two frequency points where power drops by 50%, i.e. $P_{3dB} = \frac{1}{2} P_{max}(L)$, which translates into $G_{3dB} = \frac{1}{2} G_{max}$. Here G_{max} is the maximum value of gain evaluated at $\nu = \nu_0$. One can determine following from Eq. 11.9(a).

$$\ln(G_0/2) = \ln G_0 - \frac{(P_{sat})_{3dB}}{P_{sat}} \tag{11.9b}$$

or

$$(P_{sat})_{3dB} = \ln(2) P_{sat} = \ln(2) \frac{h\nu A}{\Gamma \sigma_g \tau_c} \tag{11.9c}$$

where Γ is confinement factor, τ_0 being life time, $h\nu$ is photon energy, A is cross-section area of active region and σ_g is gain cross-section.

In detail:

$$\exp\left[\frac{g_0 \cdot L}{1 + \frac{B_0^2}{\Delta\nu_0^2}}\right] = \frac{1}{2} \exp\left[\frac{g_0 \cdot L}{1 + \frac{0}{\Delta\nu_0^2}}\right] = \frac{1}{2} \exp(g_0 \cdot L)$$

where we have introduced 3 dB bandwidth B_0 as $B_0 = \nu - \nu_0$. By the straightforward algebra, from the above relation, one finds:

$$B_0 = \Delta\nu_0 \sqrt{\frac{\ln 2}{g_0 L - \ln 2}} \tag{11.10}$$

Macroscopic bandwidth of the amplifier B_0 is smaller than the local gain bandwidth $\Delta\nu_0$.

11.2.3 Gain Saturation

Let us now analyse the gain when signal power has large value and saturation effects are becoming important. Assume that $\omega = \omega_0$. Substituting Eq. (11.2) into (11.6), one obtains:

$$\frac{dP}{dz} = \frac{g_0 P}{1 + P/P_{sat}} \qquad (11.11)$$

We introduce new variable $u = P/P_{sat}$ and use separation of variables method to integrate the above equation. One obtains:

$$\int_{u_{in}}^{u_{out}} \frac{1+u}{u} du = \int_0^L g_0 dz$$

where $u_{in} = P_{in}/P_{sat}$, $u_{out} = P_{out}/P_{sat}$ and P_{in}, P_{out} are input and output powers, respectively, L is the length of an amplifier. Gain G is defined as:

$$G = \frac{P_{out}}{P_{in}} \qquad (11.12)$$

One obtains:

$$G = G_0 \exp\left(-\frac{G-1}{G}\frac{P_{out}}{P_{sat}}\right) \qquad (11.13)$$

where $G_0 = \exp(g_0 \cdot L)$.

Fig. 11.2 shows saturation gain dependence from Eq. (11.13).

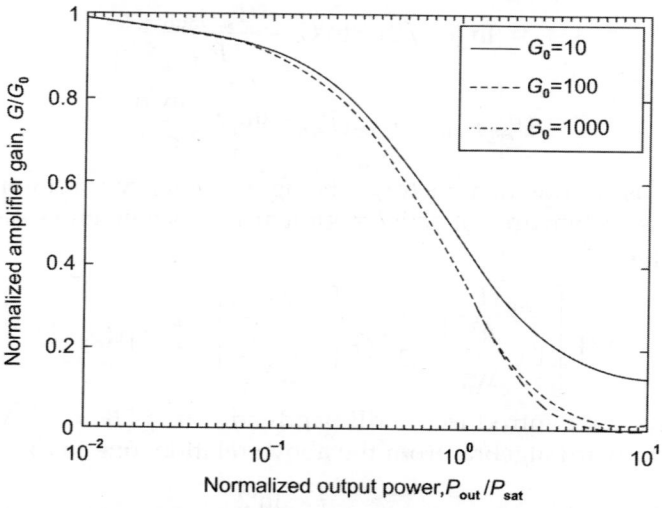

Fig. 11.2. Saturated normalized amplifier gain G/G_0 as a function of the normalized output power for three values of the unsaturated amplifier gain G_0

11.2.4 Amplifier Noise

Signal-to-noise ratio (SNR) of optical amplifiers is degraded because spontaneous emission adds to the signal during its amplification. Amplifier noise figure F_n is defined as:

$$F_n = \frac{(SNR)_{in}}{(SNR)_{out}} \qquad (11.14)$$

SNR refers to the electrical power generated when the signal is converted to electric current by using a photodetector. We model F_n by considering an ideal detector limited only by a shot noise:

$$(SNR)_{in} = \frac{\langle I \rangle^2}{\sigma_s^2} \qquad (11.15)$$

where $\langle I \rangle = RP_{in}$ is the average photocurrent, $R = \frac{q}{h\nu}$ is the responsivity of an ideal detector with unit quantum efficiency $\sigma_s^2 = 2q(RP_{in})\Delta f$ is the variance from shot noise and Δf is the detector bandwidth. At the output, we should add spontaneous emission to the receiver noise.

$$S_{sp}(\nu) = (G - 1) n_{sp} h\nu \qquad (11.16)$$

Here, S_{sp} is the spectral density of the noise induced by spontaneous emission, ν is optical frequency and n_{sp} is the spontaneous-emission factor or population inversion factor. The value of n_{sp} is $n_{sp} = 1$ for amplifiers with complete population inversion (all atoms in the upper state) and $n_{sp} > 1$ for incomplete population inversion.

For a two-level system:

$$n_{sp} = \frac{N_2}{N_2 - N_1} \qquad (11.17)$$

where N_1 and N_2 are the atomic populations in the lower and upper states, respectively.

Total variance of the shot noise plus spontaneous emission noise is thus:

$$\sigma^2 = 2q(RGP_{in})\Delta f + 4(GRP_{in})(RS_{sp})\Delta f \qquad (11.18)$$

All other contributions to the receiver noise are neglected. At the output, the SNR of the amplified signal is:

$$(SNR)_{out} = \frac{\langle I \rangle^2}{\sigma^2} = \frac{(RGP_{in})^2}{\sigma^2} \approx \frac{GP_{in}}{4S_{sp}\Delta f} \qquad (11.19)$$

assuming $G \gg 1$ and we neglected first term in (11.18).

Using definition of F_n, one finds:

$$F_n = 2n_{sp}\frac{G-1}{G} \approx 2n_{sp} \qquad (11.20)$$

It shows that even for an ideal amplifier ($n_{sp} = 1$), amplified signal is degraded by a factor of 2 (3 dB). In practice, F_n is in the range 6–8 dB.

SOA is very similar to a semiconductor laser. There are two categories of SOA (Fig. 11.3): (a) *Fabry-Perot* (FP) *amplifier*, and (b) *travelling-wave amplifier* (TWA). The FP amplifier displays high gain but has a non-uniform gain spectrum, whereas TWA has broadband gain but requires very low facet reflectivities. The FP amplifier has large reflectivities at both ends which results in resonant amplification, and also has large gain at the wavelength corresponding to longitudinal modes of the FP cavity.

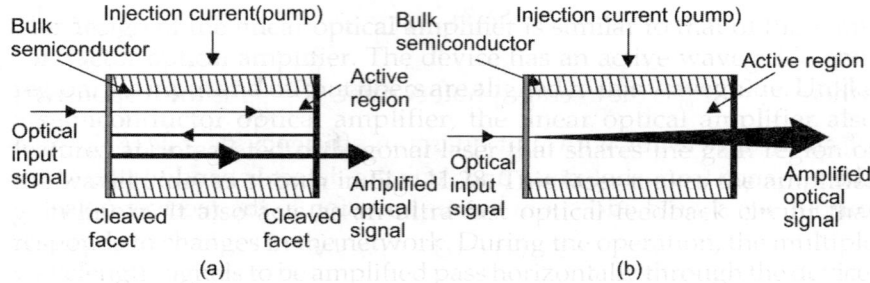

Fig. 11.3. Types of SOA: Fabry-Perot (left) and travelling-wave (right)

TWA has very small reflectivities, achieved by AR (anti-reflection) coating; its gain spectrum is broad but small ripples exist in gain spectrum, resulting from residual facet reflectivity. It is more suitable for system applications but the gain must be polarization independent. The phenomenological expression for gain of SOA is written as:

$$g_m = a(n - n_0) - a_2 (\lambda - \lambda_p)^2 \qquad (11.21)$$

and the wavelength peak values as:

$$\lambda_p = \lambda_0 + a_3 (n - n_0) \qquad (11.22)$$

Typical values of the parameters which appear in the above relations are given in Table 11.2.

From the previous equations, the 3 dB gain bandwidth is determined as:

$$2\Delta\lambda = 2\sqrt{\frac{a(n - n_0)}{2n_2}}$$

Substituting typical values from Table 11.2, one obtains SOA bandwidth equal to 54 nm.

11.2.5 Gain Formula for SOA with Facet Reflectivities

Consider a typical *Fabry-Perot ethalon* with a gain medium in-between (Fig. 11.4). Total output electric field E_{out} consists of all transmitted contributions:

$$E_{out} = E_1 + E_2 + E_3 + \ldots$$

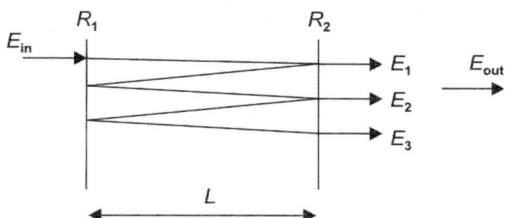

Fig. 11.4. Basic model of a Fabry-Perot amplifier

Here R_1 and R_2 are coefficients of internal reflections for electric field, and $1 - R_1$ and $1 - R_2$ corresponding coefficients of transmission. The single-pass gain G_s for field is:

$$G_s = e^{\Gamma(g-a) \cdot L} \qquad (11.23)$$

where Γ is optical confinement factor, g is gain coefficient and α are internal losses. The longitudinal propagation constant β_2 is:

$$\beta_z = k_0 \cdot n_g \qquad (11.24)$$

where $k_0 = \dfrac{2\pi}{\lambda}$ and n_g effective group index:

$$n_g = \frac{c}{v_g} \qquad (11.25)$$

Here v_g is the group velocity. Applying the standing wave condition to an electromagnetic wave of wavelength λ is a resonator of length L gives:

$$L = m\frac{\lambda}{2} \qquad (11.26)$$

Expressions for transmitted components are:

$$E_1 = E_0 e^{-j\beta_2 L} \sqrt{G_s} \sqrt{1-R_1} \sqrt{1-R_2}$$

$$E_2 = E_0 e^{-j\beta_2 \cdot 3L} \left(\sqrt{G_s}\right)^3 \sqrt{1-R_1} \sqrt{1-R_2} \sqrt{R_1} \sqrt{1-R_2}$$

$$E_3 = E_0 e^{-j\beta_2 \cdot 5L} \left(\sqrt{G_s}\right)^5 \sqrt{1-R_1} R_2 R_1 \sqrt{1-R_2}$$

Total field, which is determined as a sum of all the above components, is therefore:

$$E_{out} = E_0 \sqrt{(1-R_1)(1-R_2) G_s} \, e^{-j\beta_2 L} \{1 + G_s \sqrt{R_1 R_2} \, e^{-j\beta_2 \cdot 2L}$$
$$+ G_s^2 R_1 R_2 e^{-j\beta_2 \cdot 4L} + \ldots\}$$

Summing geometrical series, one finally obtains:

$$E_{out} = \frac{\sqrt{1-R_1}\sqrt{1-R_2}\; G_s E_0 e^{-j\beta_2 L}}{1 - G_s \sqrt{R_1 R_2}\; e^{-2j\beta \cdot L}}$$

Finally, gain of SOA with facet reflectivities R_1 and R_2 is:

$$G = \frac{(1-R_1)(1-R_2)G_s}{(1-\sqrt{R_1 R_2}\; G_s)^2 + 4\sqrt{R_1 R_2}\; G_s \sin^2 \phi} \quad (11.27)$$

Phase shift ϕ is obtained by assuming that the phase of the incident wave is taken as zero. The relation for phase change ϕ at the output is thus:

$$\phi = \frac{2\pi}{\lambda} L \quad (11.28)$$

Using Eq. (11.25), one can write the above relation for frequency as:

$$\nu = \phi \frac{v_g}{2\pi L} \quad (11.29)$$

The above relation can be interpreted on gain spectrum graph as shown in Fig. 11.5, where frequency ν_0 corresponds to gain peak G_{max}, which is obtained when $\sin \phi = 0$, or $\phi = k\pi$. Using an expression for phase (11.28), one finds its value corresponding to ν_0 (using relation $1/\lambda_0 = \nu_0/v_g$):

$$\phi = \frac{2\pi}{\lambda_0} L = \frac{2\pi}{v_g} \nu_0 L$$

or

$$\nu_0 = \frac{v_g}{2L} \cdot k = \frac{v_g}{2L}$$

assuming $k = 1$. The difference in frequencies $\nu - \nu_0$ is determined as:

$$\nu - \nu_0 = \frac{v_g}{2\pi L} \phi - \frac{v_g}{2L} = \frac{v_g}{2L}\left(\frac{\phi}{\pi} - 1\right)$$

or

$$\frac{2\pi(\nu - \nu_0)L}{v_g} = \pi\left(\frac{\phi}{\pi} - 1\right) = \phi - \pi$$

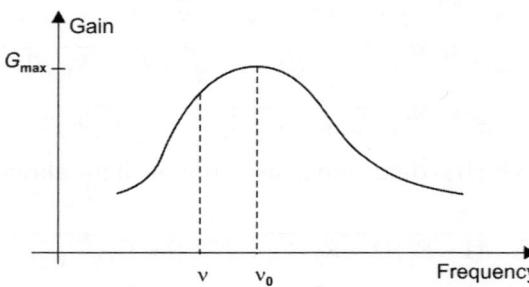

Fig. 11.5. Gain spectra showing frequency corresponding to gain peak

Finally:

$$\phi = \frac{2\pi (\nu - \nu_0) L}{v_g} \quad (11.30)$$

where we have neglected π since $\sin \pi = 0$.

11.2.6 The Effect of Facet Reflectivities

An uncoated SOA has facet reflectivities (due to FP reflections) determined by taking the values of refractive indices of a typical semiconductor and the air and is approximately equal to 0.32. Even with the AR coatings, there are some residual reflectivities which result in the appearance of the so-called gain ripples (Fig. 11.6). Ripples are superimposed on gain spectrum. The peak-to-valley ratio between the resonant and non-resonant gains is known as the *amplifier gain ripple* G_r. From Eq. (11.27), one obtains:

$$G_r = \frac{1 + G_s \sqrt{R_1 R_2}}{1 - G_s \sqrt{R_1 R_2}} \quad (11.31)$$

where G_s is the *single-pass amplifier gain*.

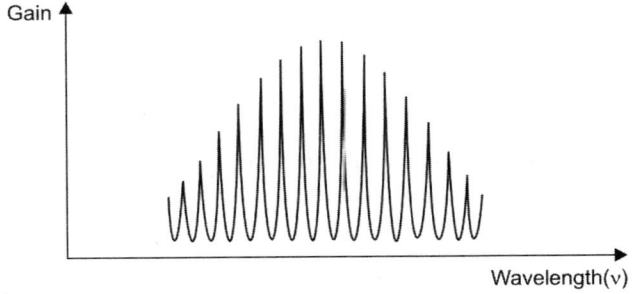

Fig. 11.6. Typical gain spectra of Fabry-Perot amplifier showing gain ripples

For an ideal TWA both R_1 and R_2 are zero. In this case, $G_r = 1$, i.e., no ripple occurs at the cavity mode frequencies. The quantity G_r is plotted in Fig. 11.7 as a function of reflectivity (R) (assuming that $R_1 = R_2$) for two values of gain. One observes that gain ripple increases with increasing gain and also increasing facet reflectivity.

11.2.7 SOA Rate Equations for Pulse Propagation

Let us discuss basic rate equations describing pulse propagation in SOA. Discussion involves electromagnetic field and carriers. We start with the development of electromagnetic equation.

Pulse propagation inside SOA is described by the wave equation:

$$\nabla^2 E(r, t) - \frac{\varepsilon}{c^2} \frac{\partial^2 E(r, t)}{\partial t^2} = 0 \quad (11.32)$$

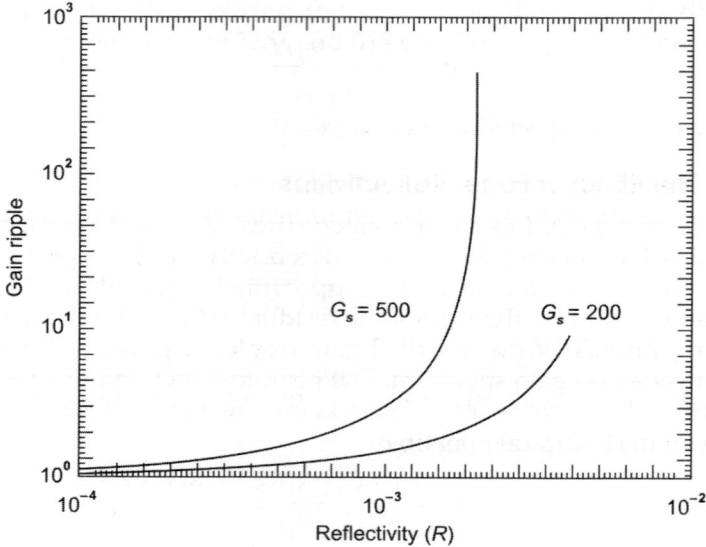

Fig. 11.7. Gain ripple as a function of reflectivity for two values of gain

where $E(r, t)$ is the electric field vector, c is the light velocity and ε is the dielectric constant of the amplifier medium which is expressed as:

$$\varepsilon = n_b^2 + \chi \qquad (11.33)$$

Here n_b is the background refractive index of the semiconductor and susceptibility χ which represents the effect of charges inside an active region in the phenomenological model is:

$$\chi = -\frac{cn}{w_0}(\alpha_H + i)g(n) \qquad (11.34)$$

where n is the effective mode index, α_H is the linewidth enhancement factor (Henry factor) and $g(n)$ is the optical gain approximated here (for bulk devices) as:

$$g(n) = a(n - n_0) \qquad (11.35)$$

where a is differential gain, n is the injected carrier density and n_0 is the carrier density at transparency.

In the following, we assume a travelling wave semiconductor wave amplifier which supports single mode propagation with a perpendicular electric profile described by $F(x, y)$. The electric field $E(r, t)$, which obeys wave equation (11.32) is expressed as:

$$E(r, t) = \hat{n}\frac{1}{2}\{F(x, y) A(z, t) e^{i(k_0 z - \omega_0 t)}\} \qquad (11.36)$$

when $\hat{n}$ is the polarization vector, $k_0 = n\omega_0/c$ and $A(z, t)$ is the slowly varying amplitude of the propagating wave. We introduce the following notation:

$$\nabla^2 = \nabla_\perp^2 + \frac{\partial^2}{\partial z^2}, \quad \text{where} \quad \nabla_\perp^2 = \frac{\partial^2}{\partial x^2} + \frac{\partial^2}{\partial y^2}$$

and evaluate derivatives:

$$\nabla^2 E \sim (\nabla_\perp^2 F) A e^{ik_0 z} + F \frac{\partial^2 A}{\partial z^2} e^{ik_0 z} + 2F \frac{\partial A}{\partial z} ik_0 e^{ik_0 z} + FA (ik_0)^2 e^{ik_0 z} \quad (11.37)$$

In the slowly varying envelope approximation (SVEA), the term with second derivative with respect to z and r is neglected. Substituting (11.37) into (11.32), applying SVEA with respect to z and t and integrating over the transverse direction gives:

$$\nabla_\perp^2 F + \frac{\omega_0^2}{c^2}(n_b^2 - n^2)F = 0 \qquad (11.38)$$

and

$$\frac{\partial A}{\partial z} + \frac{1}{v_g}\frac{\partial A}{\partial t} = \frac{i\omega_0 \Gamma}{2cn}\chi A - \frac{1}{2}\alpha_{loss}A \qquad (11.39)$$

where we have accounted for losses described by α_{loss}. The group velocity is defined as c/n_g and group index n_g is:

$$n_g = n + \omega_0 \frac{\partial n}{\partial \omega} \qquad (11.40)$$

The confinement factor Γ is:

$$\Gamma = \frac{\int_0^\omega dx \int_0^d dy\, |F(x,y)|^2}{\int_{-\infty}^{+\infty} dx \int_{-\infty}^{+\infty} dy\, |F(x,y)|^2} \qquad (11.41)$$

where ω and d are the width and thickness of the amplifier active region. At this stage, one simplifies Eq. (11.39) by introducing transformation to a reference frame moving with pulse as:

$$\tau = t - \frac{z}{v_g}$$
$$z' = z$$

In the new reference frame (Eq. 11.39) takes the form:

$$\frac{\partial A}{\partial z} = \frac{i\omega_0 \Gamma}{2cn}\chi A - \frac{1}{2}\alpha_{loss}A \qquad (11.42)$$

Rate equation for carrier density n is in the following form:

$$\frac{dn}{dt} = \frac{1}{qV} - \frac{n}{\tau_c} - \frac{\Gamma g(n)}{\hbar\omega_0 \sigma_m}|A|^2 \qquad (11.43)$$

where l is the injection current, V is the active volume, q is the electron charge, τ_c is the carrier lifetime and σ_m is the cross-section of the active region. In the new reference frame, the above equation is:

$$\frac{dn}{d\tau} = \frac{1}{qV} - \frac{n}{\tau_c} - \frac{\Gamma g(n)}{\hbar\omega_0\sigma_m}|A|^2 \qquad (11.44)$$

Slowly varying amplitude $A(z, t)$ of the propagating wave is expressed as:

$$A = \sqrt{P}\, e^{i\phi} \qquad (11.45)$$

where $P(z, \tau)$ and $\phi(z, \tau)$ are the instantaneous power and the phase of the propagating pulse. Using the above equations, one obtains:

$$\frac{\partial A}{\partial z} = \frac{1}{2}(1 + i\alpha_H)\, g \cdot A \qquad (11.46)$$

$$\frac{dg}{d\tau} = \frac{g - g_0}{\tau_c} - \frac{gP}{E_{sat}} \qquad (11.47)$$

$$\frac{\partial P}{\partial z} = (g - \alpha_{loss})\, P \qquad (11.48)$$

$$\frac{\partial \phi}{\partial z} = \frac{1}{2}\alpha_H g \qquad (11.49)$$

The quantity E_{sat} is defined as $E_{sat} = \tau_c P_s$, where P_s is the saturation power of the amplifier:

$$P_s = \frac{\hbar\omega_0\sigma_m}{a\Gamma\tau_c} \qquad (11.50)$$

In the above g_0 is the small signal gain:

$$g_0 = \Gamma a\left(\frac{I\tau_c}{eV} - n_0\right) \qquad (11.51)$$

Finally, the cross section σ_m of the active region is $\sigma_m = wd$.

Pulse Amplification

Using previously defined equations, let us now analyse pulse propagation assuming zero losses, i.e. $\alpha_{loss} = 0$, and also assuming that $\tau_p \ll \tau_c$, where τ_p is the width of the input pulse. Under this approximation pulse is so short that gain has no time to recover. Observing that $\tau_c = 0.2$–0.3 ns for typical SOA, this approximation works for τ_p equal to about 50 ps. Under the above approximations, one can obtain the analytical solution of the amplifier equations. One first integrates Eq. (11.49) to obtain output power as:

$$P_{out}(\tau) = P_{in}(\tau)\, e^{h(\tau)} \equiv P_{in}(\tau)\, G(\tau) \qquad (11.52)$$

where $P_{in}(\tau)$ is the input power, and

$$h(\tau) = \int_0^L g(z, \tau)\, dz \qquad (11.53)$$

Quantity $h(\tau)$ is known as the *total integrated net gain*. Replacing the last term in Eq. (11.47) using (11.48) gives:

$$\frac{dg}{d\tau} = \frac{g_0 - g}{\tau_c} - \frac{1}{E_{sat}} \frac{dP}{dz}$$

Integrating the above over amplifier length and using (11.52) gives:

$$\frac{dh(\tau)}{d\tau} = \frac{g_0 L - h(\tau)}{\tau_c} - \frac{1}{E_{sat}} P_{in}(\tau)[G(\tau) - 1] \qquad (11.54)$$

The solution of the previous equation is:

$$G(\tau) = e^{h(\tau)} = \frac{G_0}{G_0 - (G_0 - 1)\exp[-E_0(\tau)/E_{sat}]} \qquad (11.55)$$

where G_0 is the unsaturated gain of the amplifier and the quantity $E_0(t)$ is given by:

$$E_0(\tau) = \int_{-\infty}^{t} P_{in}(\tau')\tau' \qquad (11.56)$$

$E_0(\tau)$ represents the fraction of the pulse energy contained in the leading part of the pulse to $\tau' \leq \tau$.

The above solution shows that due to time dependence of the gain, different parts of input pulse experience different amplification which leads to a modification of pulse shape after being amplified by SOA.

In the following, we will restrict our analysis to the Gaussian input pulse:

$$P_{in}(\tau) = P_0 \exp(-\tau^2) \qquad (11.57)$$

For the Gaussian input pulse, the expression for the quantity $E_0(\tau)$ can be found in a closed form as:

$$E_0(\tau) = P_0 \tau_0 \frac{1}{2} \sqrt{\pi}\,[1 + erf(\tau)] \qquad (11.58)$$

where $erf(\tau)$ is the error function defined as:

$$erf(\tau) = \frac{2}{\sqrt{\pi}} \int_{\tau}^{\infty} e^{-x^2} dx \qquad (11.59)$$

We have also used the following property of error function:

$$1 - erf(\tau) = \frac{2}{\sqrt{\pi}} \int_{\tau}^{\infty} e^{-x^2} dx \qquad (11.60)$$

Fig. 11.8 shows pulse shape for several values of unsaturated gain G_0. One can observe that the amplified pulse becomes asymmetric, i.e., its leading edge is sharper compared to its trailing edge.

11.3.4 Design of SOA

Let us now briefly describe the main effects which must be considered when designing SOA with proper characteristics. Out of many issues

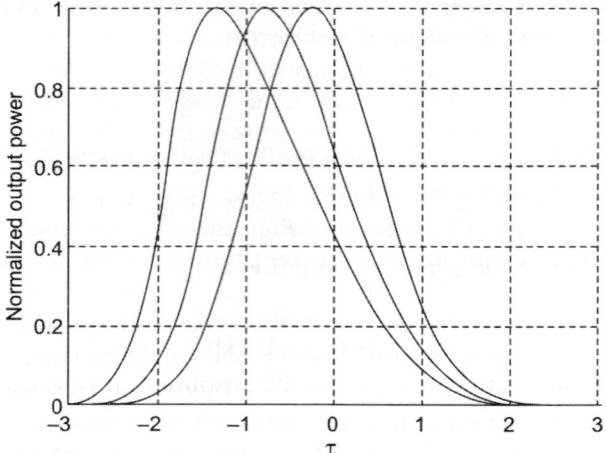

Fig. 11.8. Output pulse shapes for several values of the unsaturated gain $G_0 = 10$, 100, 1000 (increasing values to the left)

related to the design of SOA, one should briefly concentrate here on two: *suppression of cavity resonances* and *polarization insensitivity*.

Suppression of Cavity Resonance

To fabricate a travelling wave SOA, the Fabry-Perot cavity resonances must be suppressed. To accomplish this, the reflectivities at both facets must be reduced. Three approaches were used to achieve this goal (Fig. 11.9): (a) to put anti-reflection (AR) coating at both facets, (b) to tilt the active region, and (c) to use transparent window regions.

AR coating only works for a particular wavelength and it is not suitable for a wide bandwidth. The analysis shows that this appropriate combination of the previously discussed methods, it is possible to obtain an effective facet reflectivity of less than 10^{-4}. Multilayer coatings can broaden wavelength range where there is low reflectivity.

Polarization Insensitive Structures

The state of polarization of an electric field during propagation in optical fiber changes randomly. Therefore, after propagation, when signal is ready for amplification (say, in SOA), its state of polarization is unknown. For amplifications of such signals, it is therefore, desirable that SOA has polarization independent amplification. The main factor responsible for polarization sensitivity is the difference between confinement factors for TE and TM modes. Proper design of polarization insensitivities SOA involves several techniques.

Additionally, in modern SOAs which are based on quantum well (QW) designs instead of bulk structures, due to their significantly reduced threshold current and increased efficiency, there exist

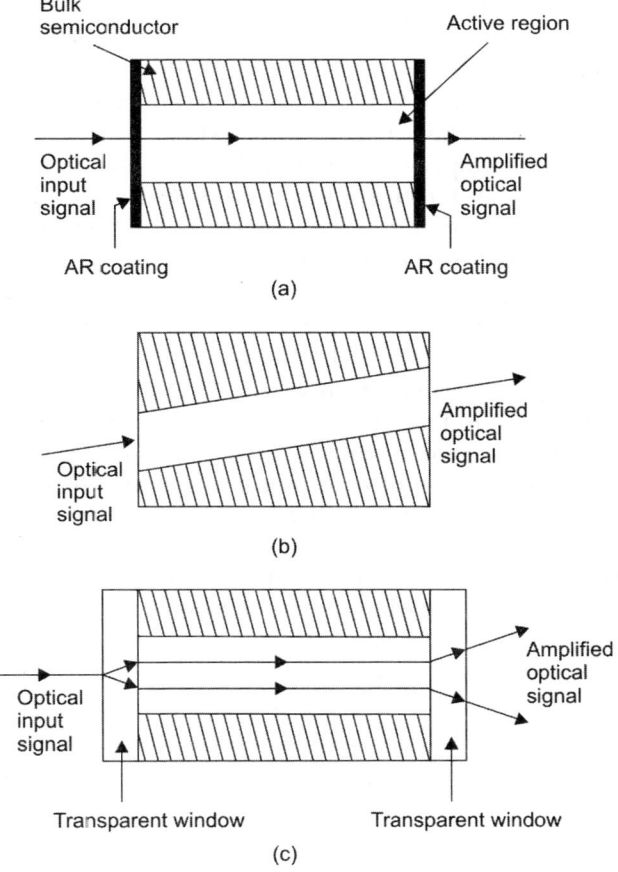

Fig. 11.9. Three main ways to make travelling wave SOA

additional polarization related effects associated with QW. QW significantly suffers from polarization sensitivity, that is, a significant difference in gain between the transverse-electric (TE) and transverse-magnetic (TM) polarization modes. This is a major concern as the polarization of a signal cannot be always controlled. This issue is of significant importance for SOA built using quantum wells.

11.2.8 Some Amplifications of SOA

(i) Wavelength Conversion

One of the important applications of SOA is for wavelength conversion (WC). Three main methods of WC have been analysed: *cross-gain modulation* (XGM), *cross-phase modulation* (XPM) and *four-wave mixing* (FWM). XGM will be discussed in detail in the following section.

Other methods include those based on the nonlinear optical loop mirror (NOLM) with the nonlinearity achieved by using fiber or SOA.

Cross-gain modulation

The reduction of gain, known as gain saturation, typically occurs for input powers of the order of 100 µW or higher. To understand this effect, one must remember that the amplification in SOA is the result of stimulated emission. The rate of stimulated emission in turn depends on the optical input power. At high optical injection, the carrier concentration in the active region is depleted through stimulated emission to such an extent that the gain of SOA is reduced.

WC based on XGM can operate in either copropagation or counter-propagation configurations. Counter-propagation configuration does not require optical filtering of the target wavelength. However, counter-propagation suffers from speed limitations.

The schematic representation of the XGM used as wavelength conversion is shown in Fig. 11.10 (copropagation configuration) and in Fig. 11.11 (counter-propagation configuration). The principle of wavelength conversion employing nonlinear characteristics of SOA is explained in Fig. 11.12. Two optical signals enter a single SOA. One of the signals known as the probe beam) at wavelength λ_{CW} is injected continuously (CW); the other (known as the signal beam) injected at wavelength λ_S is carrying digital information.

Fig. 11.10. Wavelength conversion using XGM in SOA in copropagation configuration

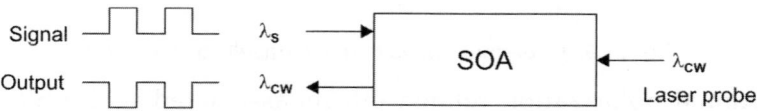

Fig. 11.11. Wavelength conversion using XGM in SOA in a counter-propagation configuration

If the peak optical power is the modulated signal is near the saturation power of the SOA, the gain will be modulated synchronously with the power. When the data signal is at high level (a logical **one**), the gain is depleted and vice versa. This gain modulation is imposed on the unmodulated (CW) input probe beam. Thus, an inverted replica of the input data is created at the probe beam wavelength. This form of wavelength conversion is one of the simplest all-optical wavelength conversion mechanisms available today.

Very fast wavelength conversion can be achieved with speeds in the order of 40 Gbs^{-1} with a small bit error ratio penalty. Previously, it was thought that the speed of WC is limited by the intrinsic carrier lifetime,

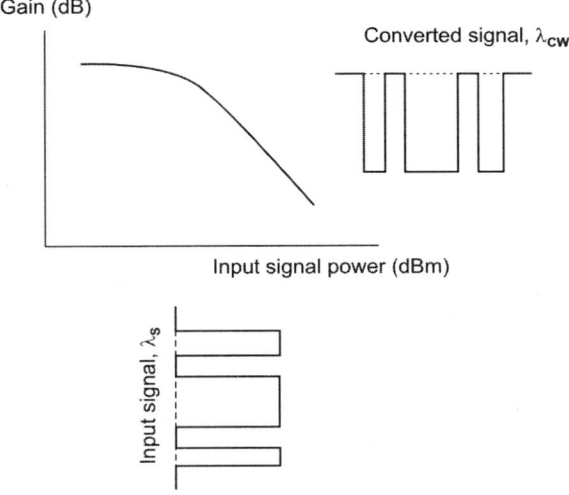

Fig. 11.12. Principle of wavelength conversion in SOA using XGM

which is around 0.5 ns. This is because the effective carrier lifetime can be decreased by the use of high optical injection to values as low as 10 ps.

Another effect associated with pulse propagation in long SOA (>1 mm) is that *pulse distortion* of the input data due to gain saturation effects can tend to sharpen the leading edge of the data pulse at the probe wavelength.

(ii) All-optical Logic Based on Interferometric Principles

For the future high-speed optical networks, several all-optical signal processing functionalities will be required to avoid cumbersome and power consuming electro-optic conversion. The central role in this process is played by all-optical high-speed logic gates.

In recent years several schemes of implementation have been investigated. Some of these schemes exploit gain saturation of SOA, and the other methods employ the interferometric configurations.

Optical logic devices are constructed using the *Mach-Zehnder (MZI) interferometer* (Fig. 11.13) for an introduction to the notation and symbols $\Delta\phi$ is the phase difference between signals propagating in the upper and lower arms of MZI. Each arm of the interferometer contains SOA where the effect of cross-phase modulation (XPM) is utilized to change phase of the transmitted light. The XPM effect is based on the physical effect of the dependence of refractive index on the carrier density in the active region. Depending on operating conditions which are controlled by driving currents within each SOA and direction and intensity of external light, this configuration can perform like an all-

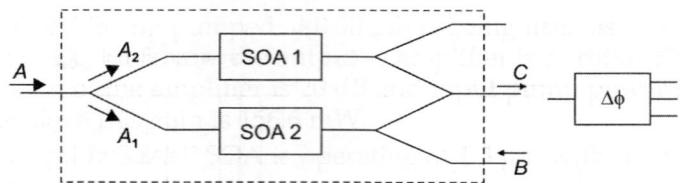

Fig. 11.13. The MZ interferometric configuration and its equivalent symbol

optical gain. In what follows, for several logic states, we schematically show operating conditions of MZI, the resulting phase difference and the equivalent logical table.

A and not B

Here, bias conditions are set in such way that, when a second signal B is input in a counter-propagating scheme, the phase difference is zero. The configuration and logical table are shown in Fig. 11.14.

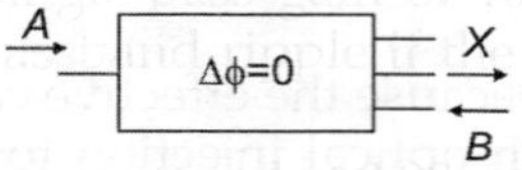

A	B	X
0	0	0
0	1	0
1	0	1
1	1	0

Fig. 11.14. Block diagram of A not B and truth table

A and B

Here, both SOA are biased such that in the absence of counterpropagating signal B, a phase difference of π exists at the output X. The relevant configuration and the truth table are shown in Fig. 11.15.

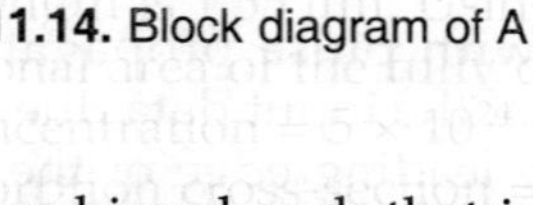

A	B	X
0	0	0
0	1	0
1	0	0
1	1	1

Fig. 11.15. Block diagram of A and B and truth table

Alternative Method

Optical gain

Let us consider the schematic structure of an SOA, shown in Fig. 11.1(b). The optical signal power P propagating through such an amplifier may be described by:

$$\frac{dP}{dz} = gP - \alpha_{\text{eff}} P \tag{11.61}$$

where g is the gain coefficient (per unit length) and α_{eff} is the effective loss coefficient (per unit length). If N is the carrier concentration (per unit volume), N_u is the carrier concentration (per unit volume) at transparency (i.e., when the gain is unity), σ_g is the gain cross section (also known as differential gain coefficient and normally expressed as dg/dN), and Γ is the confinement factor, the gain coefficient can be expressed as:

$$g = \Gamma \sigma_g (N - N_{tr}) \tag{11.62}$$

The rate equation can be written by considering various physical phenomena through which the carrier population N changes with the injection current I and the signal power P. Thus,

$$\frac{dN}{dt} = \frac{I}{eV} - \frac{N}{\tau_c} - \frac{gP}{h\nu A} \tag{11.63}$$

The first term on the right-hand side of Eq. (11.63) gives the total number of carriers (per unit volume) pumped into the active region by the injection current I. Here, e is the electronic charge and V is the volume of the active region. The second term describes the carrier loss (per unit volume) through non-radiative processes, τ_c being the carrier lifetime. The third term gives the carrier loss (per unit volume) through the stimulated emission process. Here, $h\nu$ is the photon energy and A is the cross-sectional area of the active region.

Under steady-state conditions, $dN/dt = 0$ and the solution for N may be obtained. Setting the left-hand side (LHS) of Eq. (11.63) to be zero and solving for N, we get:

$$N = \frac{I\tau_c}{eV} - \frac{gP\tau_c}{h\nu A} \tag{11.64}$$

Substituting this value of N in Eq. (11.62) gives:

$$g = \frac{\Gamma \sigma_g \left[\dfrac{I\tau_c}{eV} - N_{tr}\right]}{\left[1 + \dfrac{P}{\dfrac{h\nu A}{\Gamma \sigma_g \tau_c}}\right]} \tag{11.65}$$

It the signal power P is small, the second term in the denominator of Eq. (11.65) may be neglected, and hence, a small-signal gain coefficient g_0 is obtained, which is expressed by the following relation:

$$g_0 = \Gamma \sigma_g \left[\frac{I\tau_c}{eV} - N_{tr}\right]$$

The term $(h\nu A/\Gamma\sigma_g\tau_c)$ gives the saturation power P_{sat} of the amplifier. Thus, in terms of g_0 and P_{sat}, Eq. (11.65) may be written as:

$$g = \frac{g_0}{(1 + P/P_{sat})} \tag{11.66}$$

Substituting the value of g from Eq. (11.66) in Eq. ((11.61), one obtains:

$$\frac{dP}{dz} = \frac{g_0}{\left(1 + \dfrac{P}{P_{sat}}\right)} P - \alpha_{eff} P \qquad (11.67)$$

Neglecting α_{eff} and integrating Eq. (11.67), one obtains:

$$\int_{P=P_{in}}^{P_{out}} \frac{dP}{\left[\dfrac{P}{1+P/P_{sat}}\right]} = \int_{z=0}^{L} g_0\, dz$$

where P_{in} and P_{out} are the input and output signal powers, respectively, and L is the length of the active region. Solving this, one obtains:

$$\int_{P=P_{in}}^{P_{out}} \frac{dP}{P} + \frac{1}{P_{sat}} \int_{P_{in}}^{P_{out}} dP = g_0 \int_0^L dz$$

or
$$[\ln P_{out} - \ln P_{in}] + \frac{1}{P_{sat}}[P_{out} - P_{in}] = g_0 L$$

or
$$\left[\ln \frac{P_{out}}{P_{in}}\right] = g_0 L - \left(\frac{P_{out} - P_{in}}{P_{sat}}\right) \qquad (11.68)$$

Therefore, the amplifier gain G may be expressed as:

$$G = \frac{P_{out}}{P_{in}} = \exp\left[g_0 L - \left(\frac{P_{out} - P_{in}}{P_{sat}}\right)\right] \qquad (11.69)$$

If $P_{out} \gg P_{in}$ and the small-signal gain of the amplifier is expressed by $G_0 = \exp(g_0 L)$, we may write to a good approximation:

$$\ln G = \ln G_0 - \frac{P_{out}}{P_{sat}} \qquad (11.70)$$

The 3-dB saturation power $(P_{sat})_{3dB}$ is defined as the output power P_{out} at which the amplifier gain G has dropped to $G_0/2$. By putting $G = G_0/2$ and $P_{out} = (P_{sat})_{3dB}$, we can determine the following from Eq. (11.70):

$$\ln(G_0/2) = \ln G_0 - \frac{(P_{sat})_{3dB}}{P_{sat}}$$

or
$$(P_{sat})_{3dB} = \ln(2) P_{sat} = \ln(2) \frac{h\nu A}{\Gamma \sigma_g \tau_c} \qquad (11.71)$$

Thus amplifier saturation is governed by the material properties σ_g and τ_c.

Effect of Reflections

The performance of amplifier can severely affected by the optical reflections at the facets of active region, especially when the single pass gain is high. At high signal powers, the reflections at the facet create a

Fabry-Perot type of resonator, which may lead to oscillations in the amplifier gain versus wavelength curve. This is termed as *gain ripple*.

If we assume that the reflections at the facets are independent of wavelength, then one can obtain the equation governing the transmitted optical power $P(L)$ relative to the input power $P(O)$. This leads to a well-known expression for a Fabry-Perot resonator (FPR) with optical gain:

$$\frac{P(L)}{P(O)} = G = \frac{(1-R_1)(1-R_2)G_s}{(1-G_s\sqrt{R_1R_2})^2 + 4G_s\sqrt{R_1R_2}\sin^2\phi} \qquad (11.72)$$

where R_1 and R_2 are input and output reflectivities, G is the real (i.e., measured) gain, G_s is the single pass gain, and ϕ is the phase shift that the light wave L of the amplifier once (i.e., the single-pass phase shift). We have:

$$\phi = \frac{2\pi n(v-v_o)L}{c} \qquad (11.73)$$

where n is the refractive index of the active region material, v is the incident signal frequency, v_o is the frequency of the resonant mode of the amplifier, and c is the speed of light in vacuum ($= 3 \times 10^8$ m·s^{-1}).

Using Eqs (11.72) and (11.73), one may calculate the 3 dB spectral bandwidth $\Delta v = 2(v-v_o)$ of a single longitudinal mode of a *Fabry-Perot amplifier* (FPA). One obtains:

$$\Delta v = \frac{c}{\pi n L}\sin^{-1}\frac{1-G_s\sqrt{R_1R_2}}{(4G_s\sqrt{R_1R_2})^{1/2}} \qquad (11.74)$$

We may note that the 3dB spectral bandwidth of a single-pass amplifier (with $R_1 = R_2 = 0$), also known as pure *travelling wave amplifier* (TWA), is determined by the full gain width of the amplifier medium itself (Fig. 11.16). For near TWAs, the pass band comprises peaks and troughs whose relative amplitudes are governed by R_1 and R_2, the single pass gain and input signal power. One can define the *peak-trough* ratio of the pass band ripple, ΔG as the difference between the resonant FPA and non-resonant TWA signal gain (Fig. 11.16). ΔG is given by:

$$\Delta G = \left(\frac{1+G_s\sqrt{R_1R_2}}{1-G_s\sqrt{R_1R_2}}\right)^2 \qquad (11.75)$$

Limitations

The theory of SOA discussed so far is valid under the assumption of continuous wave (CW) operation. However, optical signal carrying information through optical fibers are modulated in time. If one assumes that the signals are intensity-modulated, then according to Eq. (11.9b), the gain will adjust itself to the changing signal intensity, but only as

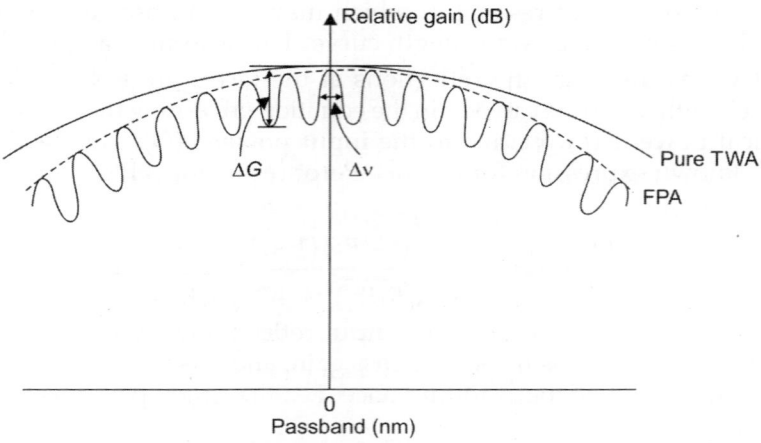

Fig. 11.16. Schematic passband characteristics of pure TWA and FPA showing large ripples in the latter case

long as the carriers can respond to the time-varying signal. However, in the case of an SOA, the carrier lifetime is of the order of 0.1 ns, hence, the gain recovery time is short with respect to the data rate of the order of gigahertz. Therefore, different levels of signal intensity (e.g., 0's and 1's in a digital signal) will experience different optical gains, leading to signal distortion. This effect gets significant when the SOA is operating close to saturation conditions. This poses an upper limit on the maximum amplifier output power. Moreover, in many applications, it is required that the gain be independent of the polarization of the input signal wave, but the semiconductor active layers are, in general, sensitive to polarization. In a multichannel operation, it is ideally expected that each channel should be amplified by the same amount and there is no crosstalk. In practice, however, the presence of several non-linear phenomena in SOAs leads to interchannel crosstalk, an undesirable feature. In spite of these drawbacks, SOAs have found application in wavelength conversion and fast-switching in WDM networks.

The said drawbacks of SOAs led to the developments of *rare-earth doped fiber amplifiers,* and *Raman fiber optical amplifiers.* These are discussed in subsequent sections.

11.3 RARE EARTH DOPED FIBER AMPLIFIERS

Doped fiber optical amplifiers are made from optical fiber whose core is doped with atoms of an element that can be excited by an external pump light to a state where stimulated emission can occur. Pump light from an external laser source is steadily pumped in one end or both ends of the fiber. The pump light is guided along the fiber length where it excites the doped atoms of the core. The core also guides the input light signal and the amplified light resulting from stimulated emission. The doped material types and doping concentrations in the fiber core depend on the wavelengths of light to be amplified. This is a general description of the operation of the doped fiber optical amplifiers. The following sections present details of each type.

11.3.1 Erbium-Doped Fiber Optical Amplifiers

In the late 1980s, a group of researchers at the University of Southampton in the United Kingdom successfully developed the *Erbium doped fiber amplifier* (EDFA). The EDFA then became the dominant type of optical amplifier. EDFAs combined with wavelength division multiplexing technology are widely used in long-distance optical communications, networks, and signal modulation. EDFAs operate in the 1540–1570 nm range, called the C-band by convention. Erbium ions have quantum levels that allow them to be stimulated to emit light in the C-band, which is the wavelength band having the least power loss in most silica-based fiber, where high-quality amplifiers are most needed.

Figure 11.17 shows the basic operation of the EDFA. An Erbium doped fiber amplifier consists of a 10–30 meter length of optical fiber, the core of which is doped with a rare-earth element of Erbium (Er^{3+}). The fiber is pumped with a laser light at 980 or 1480 nm to raise the Erbium ions to a high energy state. When an Erbium ion is in a high energy state, an incident photon of input light signal can stimulate the Erbium ions to give up some of its energy in the form of light and return to a lower energy state, which is more stable. This operation is called stimulated emission, and is generally presented in light production and laser theory.

The pump laser power supplies the optical energy for the amplifier. The pumped laser light is mixed with the input light signal via a coupler at the input fiber cable. The mixed light is guided into the fiber section with Erbium ions included in the core. A photon of the laser excites the Erbium ion to its higher-energy state. When the photon of the input light signal meets the excited Erbium atom, the Erbium atom gives up some energy in the form of a photon and returns to its lower-energy

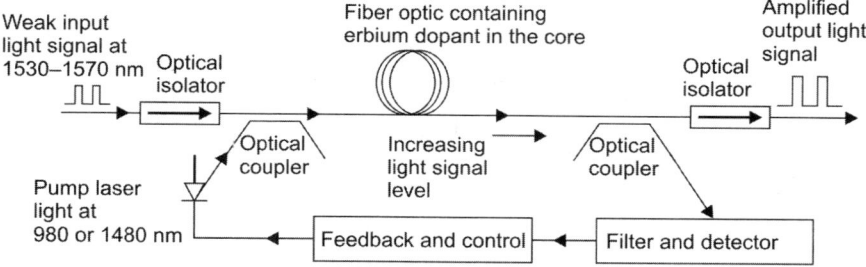

Fig. 11.17. Schematic of erbium-doped fiber optical amplifier (basic operation)

state. The new photon is in exactly the same phase and direction at the light signal that is being amplified. Thus, the light signal is amplified along the fiber core in a forward direction of travel only. Fig. 11.17 also shows the need to have a pump laser beam along the length of a fiber to provide the energy for EDFAs. This design requires power and optics, such as couplers and filters.

EDFAs also have gain that varies with a signal's wavelength, which creates problems in many WDM applications. This can be solved by using special optical passive filters that are designed to compensate for the gain variation of the EDFA.

Pumping power can be applied in a forward direction, as shown in Fig. 11.17, backward from the output end, or in both directions. Optical isolators are commonly used at the output end or both ends of the EDFA, to prevent the pump power signal and light signal from returning back down the fiber, or unwanted reflections that may affect laser stability.

As explained above, the pump laser light is supplied using a coupler at the inlet of the fiber cable, as shown in Fig. 11.17. Pump laser light can be pumped in the direction and/or opposite direction of the light signal. The pump signal can be coupled in various locations in the amplification system, as shown in Fig. 11.18. Fig. 11.18 also shows that the pump power can be (a) at the end of the fiber cable using a coupler, or (b) coupled on both ends. Fig. 11.18 shows a design for the remote pumping of the power, used where the pump laser is a long distance from the amplifier, such as in undersea systems.

EDFAs have a number of main technical characteristics, such as efficient pumping, wavelength selection, minimal polarization sensitivity, low insertion loss, low distortion and interchannel crosstalk, high power output, low noise, very high sensitivity, low power consumption, and low cost.

An optical fiber amplifier consists of an optical fiber where amplification takes place which is doped with a rare-earth element, and a pumping light supply system for supplying pumping light to the optical fiber for amplification (Fig. 11.19). The pumping light supply system usually includes a semiconductor laser and an optical coupler for guiding the pumping light into the optical fiber for amplification.

Erbium-doped amplifiers are made by doping a segment of the fiber with erbium and then exciting the erbium atoms to a high energy level through the introduction of pumping light. The energy is transferred gradually to signal light passing through the fiber segment during excitation, resulting in an amplification of the signal light upon exit from the amplifier. Fiber optic amplifiers can amplify signal light including one or more wavelengths within a predetermined wavelength band without converting them into an electrical signal.

Illustration of the amplification process in EDFA based on the three-level model is shown in Fig. 11.20 for the case of a 980 nm pump. A

Optical Amplifiers 551

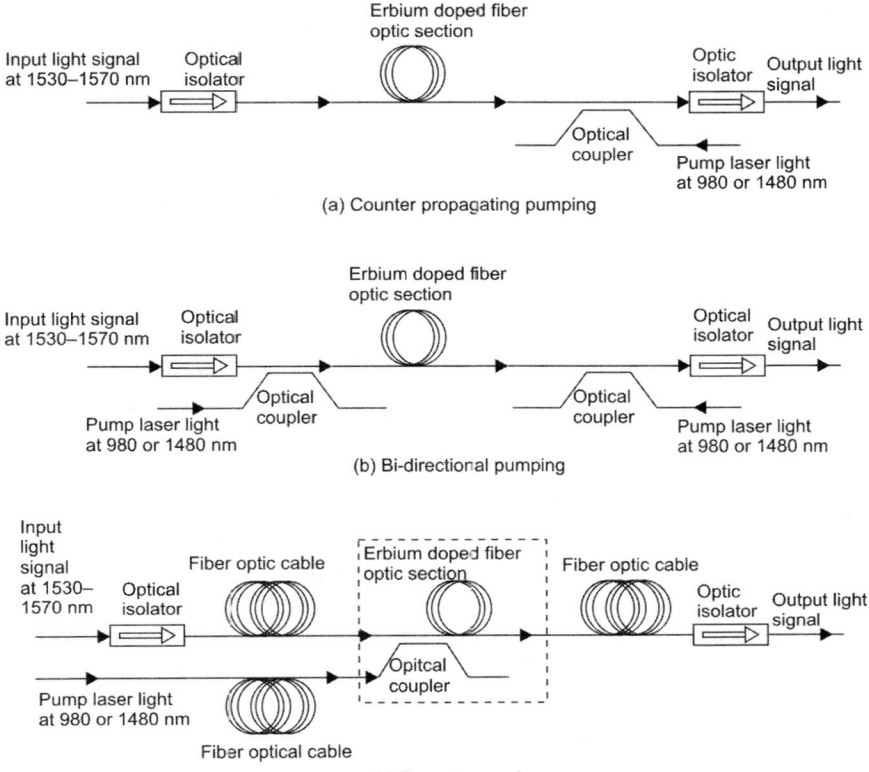

Fig. 11.18. Erbium-doped fiber optical amplifier basic operation

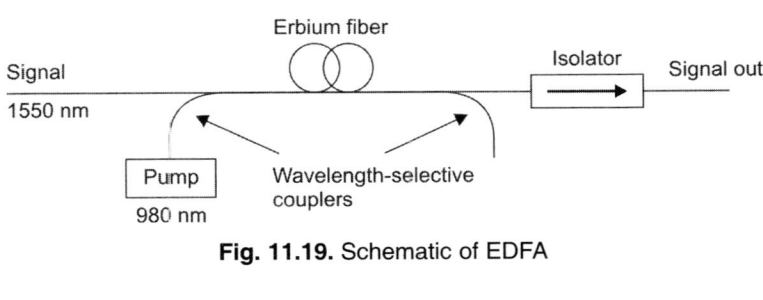

Fig. 11.19. Schematic of EDFA

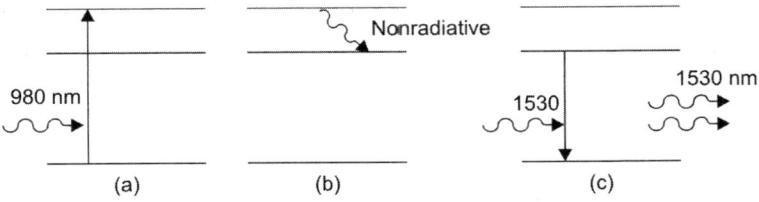

Fig. 11.20. Illustration of amplification in EDFA using a three-level model

pump photon at a 980 nm wavelength is absorbed by an erbium ion in the ground state and jumps into the highest energy level. Then, through a non-radiative decay the ion loses its energy and arrives into a metastable state. Once there, a photon having wavelength of 1530 nm can force a stimulated transition of the ion into its ground state, creating one additional photon and an amplification.

In this section, we will describe EDFA using a three-level model. The model is shown in Fig. 11.21. Level 1 is a ground state, level 2 is known as a metastable one (it has a long lifetime) and level 3 is an intermediate state. The population of levels are introduced as N_1, N_2, N_3. The spontaneous transition rates of the ion (transition probabilities) which include radiative and also non-radiative contributions are denoted as Γ_{32} and Γ_{21} and correspond to transitions between levels $3 \to 2$ and $2 \to 1$, respectively. σ_p is the pump absorption cross section and σ_s is the signal emission cross section. The incident light intensity fluxes of pump and signal are denoted by ϕ_p and ϕ_s, respectively. They are defined as number of photons per unit time per unit area.

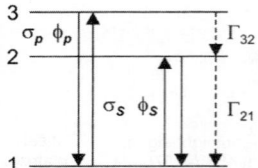

Fig. 11.21. Illustration of notation for a three-level system

This three-level model represents energy level structure of Er^+ which is involved in the amplification process. To obtain amplification, we need a population inversion between levels 1 and 2. Here, we only cosnider a one-dimensional model where we assume that the pump and signal intensities and also distribution of Er ions are constant in the transverse direction.

Based on the above observations, the rate equations for the changes of populations for all levels are postulated as:

$$\frac{dN_1}{dt} = \Gamma_{21} N_2 - (N_1 - N_3)\sigma_p \phi_p + (N_2 - N_1)\sigma_s \phi_s \quad (11.76)$$

$$\frac{dN_2}{dt} = -\Gamma_{21} N_2 + \Gamma_{32} N_3 - (N_2 - N_1)\sigma_s \phi_s \quad (11.77)$$

$$\frac{dN_3}{dt} = -\Gamma_{32} N_3 + (N_1 - N_3)\sigma_p \phi_p \quad (11.78)$$

11.3.2 Steady-State Analysis

In the steady-state conditions:

$$\frac{dN_1}{dt} = \frac{dN_2}{dt} = \frac{dN_3}{dt} = 0 \quad (11.79)$$

Also, total population N is assumed to be constant:

$$N = N_1 + N_2 + N_3 \quad (11.80)$$

From Eq. (11.78), one obtains:

$$N_3 = N_1 \frac{1}{1 - \dfrac{\Gamma_{32}}{\sigma_p \phi_p}} \tag{11.81}$$

In what follows, we will assume fast decay from level 3 to level 2; i.e. decay from level 3 is dominant compared to pump rate. Mathematically, the assumption corresponds to the following condition $\Gamma_{32} \gg \sigma_p \phi_p$. Lifetime of level 3, τ_{32} is related to transition probability as $\tau_{32} = 1/\Gamma_{32}$. In this limit, there is almost no population of level 3, and therefore, $N_3 \approx 0$. With those assumptions the system can be effectively considered as consisting of two levels only which are described by the following equations:

$$\frac{\sigma_p \phi_p + \sigma_s \phi_s}{\Gamma_{21} + \sigma_s \phi_s} N_1 - N_2 = 0 \tag{11.82}$$

$$N_1 + N_2 = N \tag{11.83}$$

One obtains the solution for population inversion as:

$$N_2 - N_1 = N \frac{\sigma_p o_p - \Gamma_{21}}{\Gamma_{21} + 2\sigma_s \phi_s + \sigma_p \phi_p} \tag{11.84}$$

11.3.3 Effective Two-Level Approach

Keeping the above assumptions, i.e. $\Gamma_{32} \gg \sigma_p \phi_p$ which allows us to neglect level 3, from Eq. (11.83), one has:

$$\frac{dN_1}{dt} = -\frac{dN_2}{dt}$$

If is, therefore, enough to consider only one equation, say for N_2; the other population N_1 can be found from $N_1 = N - N_2$. We postulate the following equations:

$$\frac{dN_2}{dt} = -\Gamma_{21} N_2 + [\sigma_s^{(a)} N_1 - \sigma_s^{(e)} N_2] \phi_s - [\sigma_p^{(e)} N_2 - \sigma_p^{(a)} N_1] \phi_p \tag{11.85}$$

$$\frac{dN_1}{dt} = -\Gamma_{21} N_2 + [\sigma_s^{(e)} N_2 - \sigma_s^{(a)} N_1] \phi_s - [\sigma_p^{(a)} N_1 - \sigma_p^{(e)} N_2] \phi_p \tag{11.86}$$

where $\sigma_s^{(a)}, \sigma_s^{(e)}, \sigma_p^{(e)}$ represent signal and pumping cross sections for absorption and emission, respectively. Assume steady-state and from Eq. (11.85) determine N_2.

$$N_2 = N \frac{\sigma_s^{(a)} \phi_s + \sigma_p^{(a)} \phi_p}{\dfrac{1}{\tau} + \left[\sigma_s^{(a)} + \sigma_s^{(e)}\right] \phi_s + \left[\sigma_p^{(a)} + \sigma_p^{(e)}\right] \phi_p}$$

where we have introduced $\tau = 1/G_{21}$. Further, introduce signal I_s and pump I_p intensities as:

$$\phi_s = \frac{I_s}{h\nu_s}, \quad \phi_p = \frac{I_p}{h\nu_p}$$

where h is the Planck constant and ν_s, ν_p are frequencies of the signal and pump, respectively. This allows us to write:

$$N_2 = N \frac{\tau \dfrac{\sigma_s^{(a)}}{h\nu_s} I_s(z) + \tau \dfrac{\sigma_p^{(a)}}{h\nu_p} I_p(z)}{\tau \dfrac{\sigma_s^{(a)} + \sigma_s^{(e)}}{h\nu_s} I_s(z) + \tau \dfrac{\sigma_p^{(a)} + \sigma_p^{(e)}}{h\nu_p} I_p(z) + 1}$$

We further assume that N is independent of distance along fiber z. The variation of signal and pump intensities is described as:

$$\frac{dI_s(z)}{dz} = \left[\sigma_s^{(e)} N_2 - \sigma_s^{(a)} N_1\right] I_s(z)$$

$$\frac{dI_p(z)}{dz} = \left[\sigma_p^{(e)} N_2 - \sigma_p^{(a)} N_1\right] I_p(z)$$

Table 11.3. Typical parameters of a low-noise in-line EDFA

Parameter	Symbol	Value	Unit
Signal mode area	πw^2	1.3×10^{-11}	m²
Erbium concentration	N_{tot}	5.4×10^{24}	m⁻³
Signal overlapping integral	Γ_s	0.4	–
Pump overlapping integral	Γ_s	0.4	–
Signal emission cross section	s_{se}	5.3×10^{-25}	m²
Signal absorption cross section	s_{sa}	3.5×10^{-25}	m²
Pump absorption cross section	s_p	3.2×10^{-25}	m²
Signal local saturation power	P_{ss}	1.3	mW
Pump local saturation power	P_{sp}	1.6	mW

11.3.4 Gain Characteristics of Erbium-Doped Fiber Amplifiers

We outline a basic approach to evaluate the performance of EDFA. Neglecting amplified spontaneous emission (ASE) and assuming copropagation configuration, the equations describing steady-state are:

$$\frac{dI_s(z)}{dz} = 2\pi \Gamma_s \left\{\sigma_{se} N_{me}(z) I_s(z) - \sigma_{sa} N_{gr}(z) I_s(z)\right\} \quad (11.87)$$

$$\frac{dI_p(z)}{dz} = 2\pi \Gamma_p \sigma_{sa} N_{gr}(z) I_p(z) \quad (11.88)$$

$$N_{me}(z) = N_{tot} \frac{I_s(z)/I_{ss} + I_p(z)/I_{sp}}{1 = I_p(z)/I_{sp} + 2I_s(z)/I_{ss}} \quad (11.89)$$

$$N_{me}(z) + N_{gr}(z) = N_{tot} \quad (11.90)$$

In the previous equation $N_{gr}(z)$ is the population of the ground state, $N_{me}(z)$ is the population of the metastable state, $I_p(z)$ is the intensity of the pump wave propagating at the wavelength λ_p and $I_s(z)$ is the intensity of the signal wave propagating at the wavelength λ_s in the positive z-direction.

Finally, the overall *amplifier gain G* is obtained using the following relation:

$$G = \frac{I_s(L)}{I_s(0)} \qquad (11.91)$$

where L is the doped fiber length.

11.3.5 Typical EDFA Characteristics

The variation of gain with fiber length summarized in Fig. 11.22 for different values of pump power. Constant signal input power of 10 µW and constant erbium doping density have been assumed. Gain was evaluated for four different pump power levels equal to 3, 5, 7 and 9 mW. As it is shown, gain increases up to a certain fiber's length and then begins to decrease after reaching a maximum point.

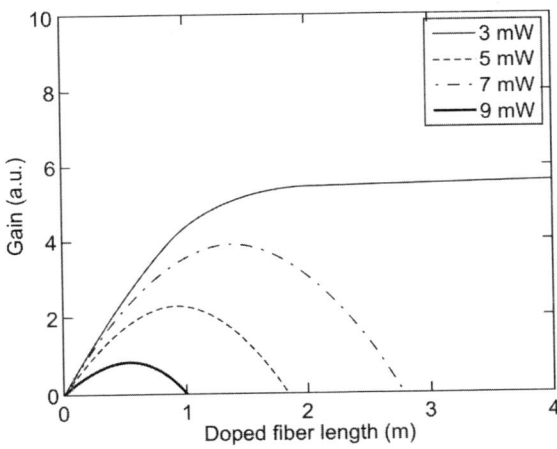

Fig. 11.22. Variation of gain vs fiber length for four different values of pump power

The optimum fiber length (the one which corresponds to a maximum gain) is a few metres and it increases with the pump power. The reason for the decrease in gain is insufficient population inversion due to excessive pump depletion.

Figure 11.23 shows the variation of gain with pump power for three different values of fiber lengths of 5, 10 and 15 m and a constant signal input power equal to 1 mW. Constant *Er* doping density was also assumed. As it is seen, gain of the EDFA increases with the increasing pump

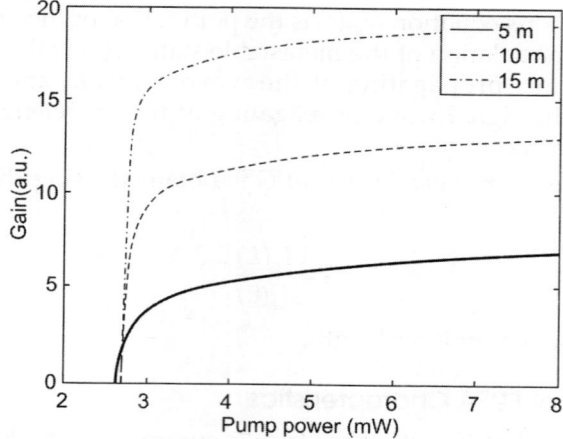

Fig. 11.23. Variation of gain vs pump power for three different values of fiber length

power and then saturates after a certain level of pump power. The pump saturation effect occurs for input powers in the range 3–6 mW.

11.3.6 Alternative Model of EDFA

The energy bands of an Er^{3+} ion in a silica matrix, shown in Fig. 11.24, can be approximately described as a non-degenerate three-level system,

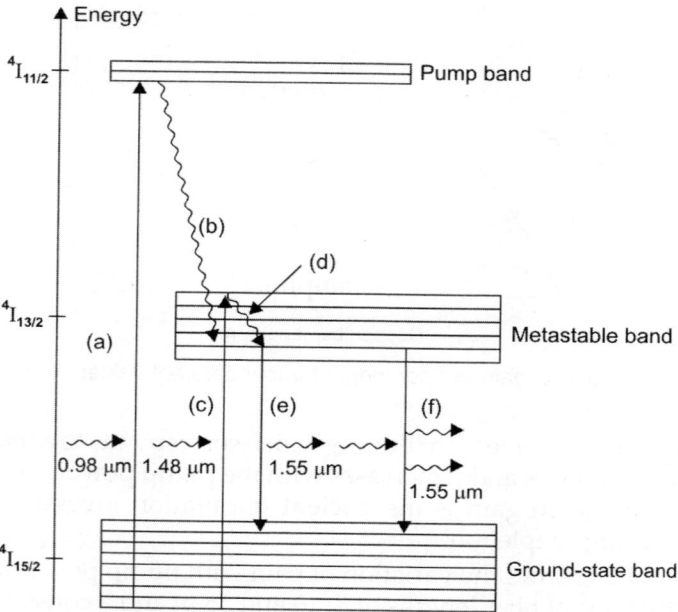

Fig. 11.24. Simplified energy-level diagram of Er^{3+} ion in silica fiber exhibiting various possible transitions

provided the transition is characterized by different absorption and emission cross section. Let us assume that the core of the erbium-doped silica fiber has erbium ion density of N_t and that the fiber is single-moded at both pump wavelength ($\lambda_p = 0.98$ µm) and signal wavelength ($\lambda_s = 1.55$ µm). Further, assume that the population density (number of ions per unit volume) of Er^{3+} ion in the ground state $^4I_{15/2}$ (with energy E_1) is N_1 and that in the upper amplifier level (metastable level) $^4I_{13/2}$ (with energy E_2) is N_2. Pumping this system by $\lambda_p = 0.98$ µm takes the ground-state Er^{3+} ions from E_1 to the pump level $^4I_{11/2}$ (with energy E_3), from which the ions rapidly relax to the metastable level (energy E_2). Since the relaxation rate of the pump level is very fast, we may assume that this top level (energy E_3) remains almost empty. Thus, one may write:

$$N_1 + N_2 = N_t \tag{11.92}$$

Let P_p and P_s represent the optical powers for the pump and signal waves, and σ_{pa}, σ_{sa}, and σ_{se} denote the absorption cross section at the pump frequency ($v_p = c/\lambda_p$), absorption cross section at the signal frequency ($v_s = c/\lambda_s$), and the emission cross section at the signal, respectively. Then the rate of change of population of the ground level (energy E_1) may be expressed as:

$$\frac{dN_1}{dt} = -\frac{\sigma_{pa} P_p}{a_p h v_p} N_1 - \frac{\sigma_{sa} P_s}{a_s h v_s} N_1 + \frac{\sigma_{se} P_s}{a_s h v_s} N_2 + \frac{N_2}{\tau_{sp}} \tag{11.93}$$

where a_p and a_s are the cross-sectional areas of the fiber modes for λ_p and λ_s, and τ_{sp} is the spontaneous emission lifetime for the transition from E_2 to E_1. Further, ($\sigma_{pa} P_p / a_p h v_p$) N_1 is the rate of absorption per unit volume from the ground level E_1 to the pump level E_3 due to the pump at v_p ($\sigma_{sa} P_s / a_s h v_s$), N_1 is the rate of absorption per unit volume from level E_1 to the metastable level E_2 due to the signal at v_s, ($\sigma_{pa} P_s / a_s h v_s$), N_2 is the rate of stimulated emission (per unit volume) from level E_2 to level E_1 due to the signal at v_s, and (N_2 / τ_{sp}) is the rate of spon-taneous emission (per unit volume) from level E_2 to level E_1.

Similarly, the rate for the upper amplifier level N_2 may be written as:

$$\frac{dN_2}{dt} = \frac{\sigma_{pa} P_p}{a_p h v_p} N_1 + \frac{\sigma_{sa} P_s}{a_s h v_s} N_1 - \frac{\sigma_{se} P_s}{a_s h v_s} N_2 - \frac{N_2}{\tau_{sp}} \tag{11.94}$$

Equations (11.93) and (11.94) can be reduced to the following compact forms:

$$\frac{dN_1}{dt} = \frac{-\sigma_{pa} P_p}{a_p h v_p} N_1 + \frac{P_s}{a_s h v_s}[\sigma_{se} N_2 - \sigma_{se} N_1] + \frac{N_2}{\tau_{sp}} \tag{11.95}$$

and

$$\frac{dN_2}{dt} = \frac{\sigma_{pa} P_p}{a_p h v_p} N_1 + \frac{P_s}{a_s h v_s}[\sigma_{sa} N_1 - \sigma_{se} N_1] - \frac{N_2}{\tau_{sp}} \tag{11.96}$$

The pump power P_p and the signal power P_s vary along the length of the amplifier due to absorption, stimulated emission, and spontaneous emission. If we neglect the contribution of spontaneous emission, the variation of P_s and P_p along the amplifier length z is given by:

$$\frac{dP_s}{dz} = \Gamma_s (\sigma_{se} N_2 - \sigma_{sa} N_1) P_s - \alpha P_s \quad (11.97)$$

$$\pm \frac{dP_p}{dz} = \Gamma_p (-\sigma_{pa} N_1) P_p - \alpha' P_p \quad (11.98)$$

where α and α' take into account fiber losses at the signal and pump wavelengths, respectively. Such losses can be neglected for small amplifier lengths.

The confinement factors Γ_s and Γ_p take into account the fact that the doped region within the core provides the gain for the entire fiber mode. The $\pm$ sign in the LHS of Eq. (11.98) indicates the direction of propagation of the pump wave (positive for the forward direction and negative for the backward direction).

For lumped amplifiers, the fiber length is small (10–30 m) and hence, both the absorption coefficients α and α' can be assumed to be zero. Because N_1 and N_2 are related through Eq. (11.92), one only need to solve either Eq. (11.95) or Eq. (11.96). Let us consider Eq. (11.96). Under steady-state conditions, we have:

$$\frac{dN_2}{dt} = 0$$

$$\therefore \quad N_2(z) = -\frac{\tau_{sp}}{a_d h \nu_s} \frac{dP_s}{dz} - \frac{\tau_{sp}}{a_d h \nu_p} \frac{dP_p}{dz} \quad (11.99)$$

assuming pump propagation in the forward direction. Hence $a_d = \Gamma_s a_s = \Gamma_p a_p$ is the cross-sectional area of the doped portion of the fiber core. Substituting $N_2(z)$ from Eq. (11.99) into Eqs (11.97) and (11.98) and integrating over the fiber length, one can get the pump power P_p and signal power P_s in the analytical form at the output end of the doped fiber.

The total gain G for an EDTA of length L can be obtained using the expression:

$$G = \Gamma_s \exp\left[\int_0^L (\sigma_{se} N_2 - \sigma_{sa} N_1) dz\right] \quad (11.100)$$

where $N_1 = N_1 - N_2$ and N_2 is given Eq. (11.47). The variation of gain with fiber length L and the pump power are same as shown in Fig. 11.22 and Fig. 11.23 respectively.

In the above analysis, we have assumed that the pump and signal beams are continuous waves. However, in practice, the EDFA is pumped by cw ILD and the signal is in the form of pulses of 1's and 0's, whose duration is inversely proportional to the bit rate. Fortunately, owing to a relatively longer lifetime of the excited state (≈ 10 ms) for Er^{3+} ions, the gain does not vary from pulse to pulse.

11.3.7 Other Rare Earth Doped Fiber Optical Amplifiers

(a) Praseodymium-doped Fluoride Optical Amplifiers

EDFAs have shifted the optical telecommunication emphasis towards the third transmission window, called long wavelength band in the 1510–1600 nm range. There is still great interest in 1300 nm in O-band amplifiers. This is mainly because a substantial part of the fiber optic network worldwide is designed for operation in the second transmission window of about 1310 nm. Praseodymium-doped fluoride fiber amplifiers (PDFFAs) can provide substantial gain in this region. However, to compete with EDFAs, the quantum efficiency of the 1310 nm transition of Pr^{3+} should be increased. Low-photon-energy glass hosts are needed for this purpose. Other alternatives are directed towards Gallium-Lanthanum-Sulfide (GLS) and Gallium-Sulfide-Iodide (GSI) glasses.

(b) Neodymium-doped Optical Amplifiers

Neodymium (Nd)-doped optical amplifiers amplify in the 1310 nm band. Nd will amplify over the 1310–1360 nm range when doped into Fluoro-zirconate (ZBLAN) glass and over the 1360–1400 nm range when doped into silica. The most efficient pump wavelengths are at 795 and 810 nm.

(c) Telluride-based, Erbium-doped Fiber Optical Amplifiers

Telluride-based, Erbium-doped optical amplifiers offer the potential optical bandwidth of over 76 nm in the 1532–1608 nm band, thus increasing the potential bandwidth of an Erbium doped optical amplifier from 30 to over 110 nm.

(d) Thulium-doped Optical Amplifiers

Thulium-doped optical amplifiers amplify between about 1450–1500 nm in the S-band.

(e) Other Doped Fiber Optical Amplifiers

Other fiber optical amplifiers use doping materials, such as Ytterbium (Yb). The host fiber material can be silica, a fluoride-based glass, or a multi-component glass. Some plastic fiber amplifiers are under research and development. Modern plastics have characteristics similar to doped glass.

11.4 RAMAN FIBER OPTICAL AMPLIFIERS OR FIBER RAMAN AMPLIFIERS (FRA)

Raman optical amplifiers differ in principle from EDFAs. They utilize *stimulated Raman scattering* (SRS) to create optical gain. Stimulated Raman scattering occurs when light waves interact with molecular vibrations in a material having a solid lattice structure. In Raman scattering, the molecule absorbs the light, then quickly re-emits a photon with energy equal to the original photon, plus or minus the energy of a molecular vibration mode. This has the effect of both scattering light and shifting its wavelength.

When a fiber transmits two suitably spaced wavelengths, stimulated Raman scattering can transfer energy from one wavelength to the other. In this case, one wavelength excites the molecular vibration; then light of the second wavelength stimulates the molecule to emit energy at the second wavelength.

Fig. 11.25 shows the topology of a typical Raman optical amplifier. The pump laser and circulator comprise the two key elements of the Raman optical amplifier. The pump laser, in this case, has a wavelength of 1535 nm. Raman amplifiers work in the 1550 nm window. The circulator provides a convenient means of injecting light backwards into the transmission fiber with minimal optical loss. The pump laser is coupled into the transmission fiber either in the same direction as the trans-mission signal, which is called "co-directional pumping", or is coupled into the transmission fiber in the opposite direction, which is called "contra-directional pumping". Contra-directional pumping is more common because co-directional pumping has the problem of optical nonlinearity (nonlinear amplification)). In contra-directional pumping, the attenuation of the pump light is so small that it travels a great distance, several kilometers along the transmission fiber. It also

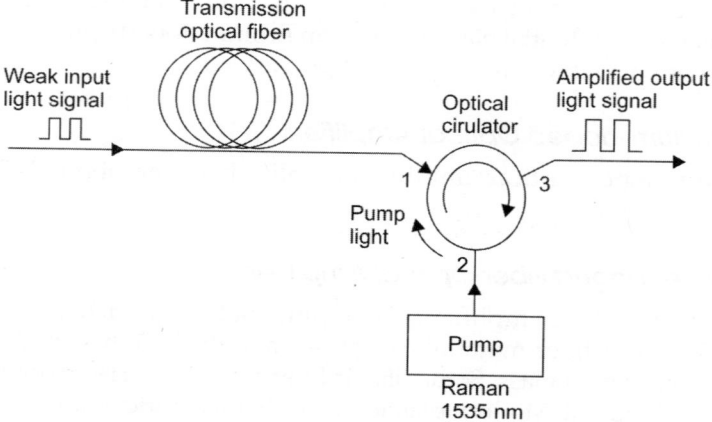

Fig. 11.25. Raman optical amplifier

keeps pump photons from reaching the receiver, where they could interfere with reception of the desired signal.

Optical isolators are commonly used at both ends of the Raman amplifier to prevent pump power and light signal from returning back down in the fiber or unwanted reflections that may affect laser stability. Raman amplifiers are used as pre-amplifiers, power amplifiers, and distributed amplifiers in digital and analogue transmissions in communication systems.

Figure 11.26 shows another technique for pumping light into Raman amplifiers. The amplification bandwidth can be extended, by using multiple pump light sources along with a wavelength division multiplexor (WDM). This can be done by using more than two pump light sources, producing a broadband amplifier over bands of more than 100 nm, for example in the range of 1500–1600 nm.

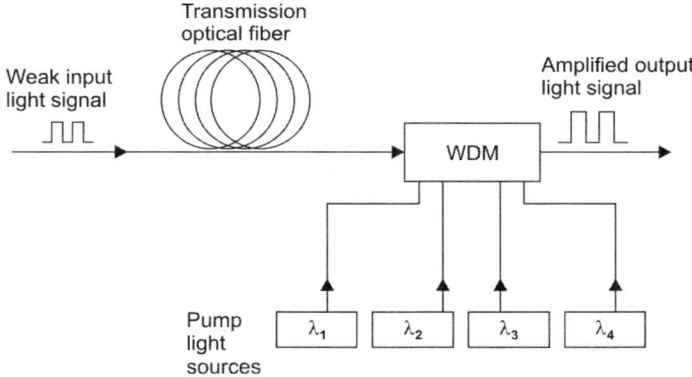

Fig. 11.26. Raman optical amplifier with multiple pump light sources

Raman amplifiers have a number of main technical characteristics, such as efficient pumping, simplicity, wider wavelength coverage, minimal polarization sensitivity, low insertion loss, high gain, low noise, fast reaction to changes of the pump power, low power consumption, and low cost. These amplifiers are used in long-haul and ultra-haul DWDM transmission systems.

The basic configuration of an FRA is shown in Fig. 11.27. Both the pump beam at a frequency v_p and the input signal beam at frequency v_s are injected into a specific optical fiber serving as an optical amplifier, through an optical coupler. The pump wavelength λ_p $(= c/v_p)$ is converted into a signal wavelength λ_s $(= c/v_s)$ by SRS, thereby increasing the power at λ_s. In other words, if a suitable optical fiber is optically pumped by an appropriate source, the signal beam will get amplified as the two beams co-propagate along the fiber. In practice, both forward pumping (i.e., the pump beam in the direction of propagation of the signal beam) and backward pumping (i.e., the pump beam in the

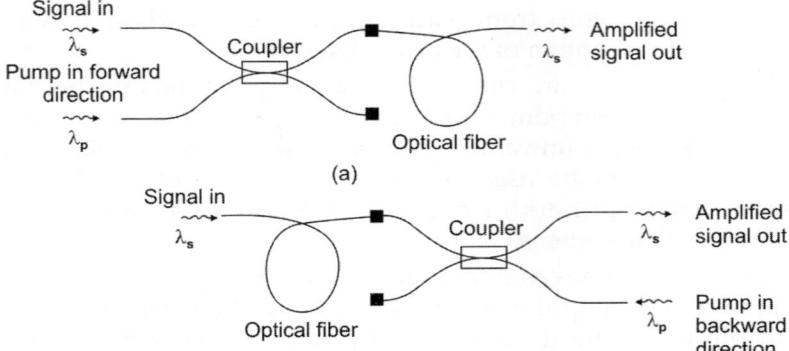

Fig. 11.27. Basic configuration of an FRA with (a) forward pumping, and (b) backward pumping

direction opposite to that of the signal beam) are possible. Since SRS is not a resonant phenomenon, it does not require population inversion.

In the case of a forward pumping, the variation of pump and signal powers along the FRA for small-signal amplification may be studied by solving the following two equations:

$$\frac{dP_s}{dz} = -\alpha_s P_s + \left(\frac{g_R}{a_p}\right) P_p P_s \quad (11.101)$$

and

$$\frac{dP_s}{dz} = -\alpha_s P_s \quad (11.102)$$

where α_s and α_p represent fiber losses (per unit length) at the signal and pump frequencies ν_s and ν_p, respectively, P_s and P_p are the signal and pump powers, respectively, that vary along the length z of the fiber, g_R is the Raman gain coefficient, and α_p is the cross-sectional area of the pump beam inside the fiber.

Solving Eq. (11.102), one obtains the expression for pump power $P_p(z)$ at any point z along the length of the fiber. Thus

$$\int_{P_{p,\,in}}^{P_p(z)} \frac{dP_p}{P_p} = -\alpha_p \int_0^z dz$$

or

$$P_p(z) = P_{p,\,in} \exp(-\alpha_p z) \quad (11.103)$$

where $P_{p,\,in}$ is the input pump power (at $z = 0$). Substituting for P_p in Eq. (11.101) from Eq. (11.103), one obtains:

$$\frac{dP_s}{dz} = -\alpha_s P_s + \left(\frac{g_R}{a_p}\right) P_s P_{p,\,in} \exp(-\alpha_p z)$$

$$= \left[-\alpha_s P_s + \left(\frac{g_R}{a_p}\right) P_s P_{p,\,in} \exp(-\alpha_p z)\right] P_s \quad (11.104)$$

If we assume that the signal power at the input end of the FRA is $P_{s,\,in}$ and that at the output end of the total fiber length L is $P_s(L)$, solving Eq. (11.104), we have:

$$\int_{P_{s,\,in}}^{P_s(L)} \frac{dP_s}{P_s} = \int_0^L \left[-\alpha_s + \left(\frac{g_R}{a_p}\right) P_{p,\,in} \exp(-\alpha_p z) \right] dz$$

or

$$\ln\left[\frac{P_s(L)}{P_{s,\,in}}\right] = \left[\frac{g_R}{a_p} P_{p,\,in} \frac{(1-e^{-\alpha_p L})}{\alpha_p} - \alpha_s L\right]$$

or

$$P_s(L) = P_{s,\,in} \exp\left[\frac{g_R}{a_p} P_{p,\,in} \frac{(1-e^{-\alpha_p L})}{\alpha_p} - \alpha_s L\right]$$

$$= P_{s,\,in} \exp\left[\frac{g_R}{a_p} P_{p,\,in} L_{\text{eff}} - \alpha_s L\right] \tag{11.105}$$

where L_{eff} is called the effective length of the fiber and is given by:

$$L_{\text{eff}} = \left[\frac{1 - \exp(-\alpha_p L)}{\alpha_p}\right] \tag{11.106}$$

If $\alpha_p L \gg 1$, $L_{\text{eff}} \approx 1/\alpha_p$. Thus the overall net gain, for small-signal amplification, will be given by:

$$G_{FRA} = \frac{P_s(L)}{P_{s,\,in}} = \exp\left[\frac{g_R}{a_p} P_{p,\,in} L_{\text{eff}} - \alpha_s L\right] \tag{11.107}$$

This can also be written as:

$$G_{FRA} = \exp[(g_0 - \alpha_s)L] \tag{11.108}$$

where

$$g_0 = \frac{g_R P_{p,\,in} L_{\text{eff}}}{a_p L} \tag{11.109}$$

Gain (in dB) may be obtained as follows:

$$\text{Gain (dB)} = 10 \log_{10}(G_{FRA}) = 4.343\,[g_0 - \alpha_s]L \tag{11.110}$$

In this case of backward pumping, for small-signal amplification, Eq. (11.101) for signal variation will be modified. Therein, dP_s/dz will be replaced by $-dP_s/dz$. Other things will remain the same. In this case the gain of the amplifier can be shown to be given by:

$$G_{FRA} = \frac{P_{s,\,in}}{P_s(L)} = \exp[(g_0 - \alpha_s)L] \tag{11.111}$$

where the symbols have their usual meaning.

For *fiber Raman amplifiers* used either in the forward or backward configuration, gains exceeding 20 dB have been achieved experimentally in a silica fiber, which in principle exhibits a broad spectral bandwidth of upto 50 nm with suitable doping. Such a broad bandwidth is attractive for WDM-based system applications. The main drawback with the FRA is that it requires high power lasers for pumping.

11.5 PLANER WAVEGUIDE OPTICAL AMPLIFIERS

Rare-earth-doped planar waveguide optical amplifiers are becoming increasingly important and provide compact and inexpensive alternatives to fiber amplifiers. In addition, planar technology is quite suitable for *optical integration* and will be essential to the development of fully integrated advanced optical devices.

11.6 LINEAR OPTICAL AMPLIFIERS

The design of the linear optical amplifier is similar to that of the semiconductor optical amplifier. The device has an active waveguide gain region; the input and output fibers are aligned to this waveguide. Unlike a semiconductor optical amplifier, the linear optical amplifier also features an integrated orthogonal laser that shares the gain region of the waveguide, as shown in Fig. 11.28. This laser makes the amplifier gain linear. It also acts as an ultra-fast optical feedback circuit that responds to changes in the network. During the operation, the multiple wavelength signals to be amplified pass horizontally through the device, directly through the path of the laser, which is pumping photons of light vertically in the same device. The linear optical amplifiers are designed to be small and low cost. They are used in high data telecommunication systems.

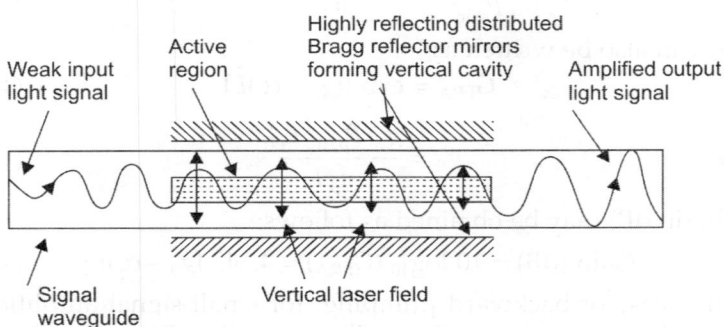

Fig. 11.28. Linear optical amplifier

11.7 BASIC APPLICATIONS OF OPTICAL AMPLIFIERS

Optical amplifiers (OAs) are used to boost signals over long distances in network systems. A schematic diagram illustrating the components of a basic communication system is shown in Fig. 11.29.

The characteristics and advantages of the OAs have led to many applications. OAs can be used to boost signal power after multiplexing, or prior to multiplexing, or at any point in modern optical networks.

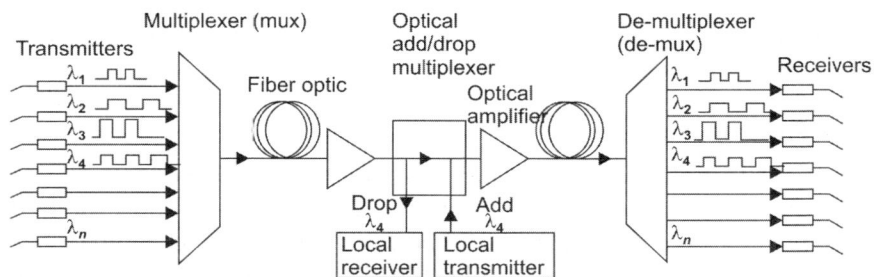

Fig. 11.29. A schematic diagram of basic communication system

OAs are ideal for metro and long-hour dense multiplexing (DWDM) as well as single wavelength applications. The optical design, coupled with sophisticated control circuitry, permit these OAs to provide constant gain even with signals being added to the network, such as λ_4.

(i) In-line Optical Amplifiers

An optical amplifier is used as a in-line amplifier allowing signals to be amplified within the optical signal path, as shown in Fig. 11.30. Optical amplifiers can compensate for transmission loss and thus increase the distance between the transmitters and receivers. This application enables the signal to travel through lines hundreds of kilometres long.

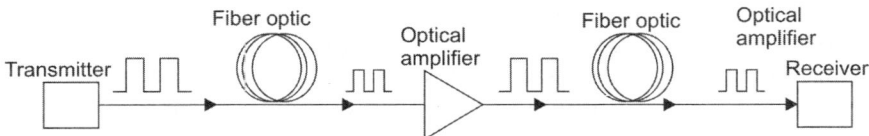

Fig. 11.30. In-line optical amplifier

(ii) Postamplifier

An optical amplifier is used as a postamplifier when placed immediately after the transmitter, as shown in Fig. 11.31. An optical amplifier will boost the light signal to the required power levgel at the beginning of a fiber line. This arrangement enables the signal to travel hundreds of kilometres down the fiber cable. A common application of the post-amplification technique, together with an optical preamplifier at the

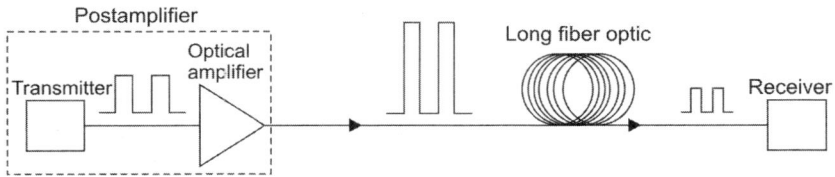

Fig. 11.31: Postamplifier application

receiving end, enables continuous underwater transmission distances up to a few kilometers.

(iii) Preamplifier

An optical amplifier is used as a preamplifier when placed immediately before the receiver, as shown in Fig. 11.32. An OA will boost the light signal to the required power level before being received by an optical receiver. The preamplifier enables the signal to be processed directly by the receiver. The preamplification technique is commonly used before optical herodyne detectors or avalanche photodiodes.

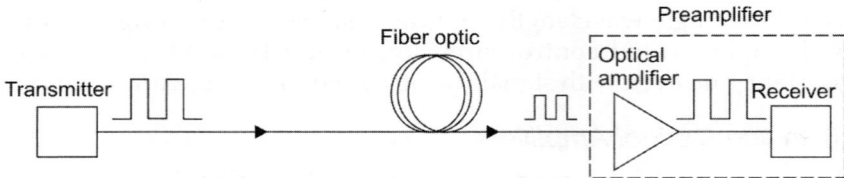

Fig. 11.32. Preamplifier application

(iv) Local Area Network

Optical amplifiers are used to boost signals in local area networks, when placed in sub-centres within the transmission lines and branches, as shown in Fig. 11.33. This type of arrangement also enables optical amplifiers to have the characteristics suitable for analogue transmission. The video content multiplexed by head-end equipment is converted

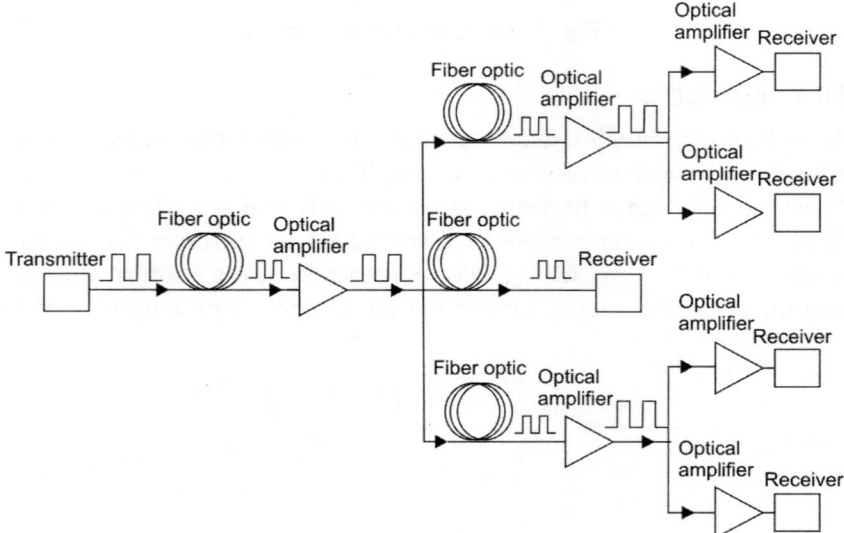

Fig. 11.33. In local area network application

from optical/electrical transmission into the 1550 nm band-wavelength optical video signal that is suitable for optical communication. The optical video signals are optically amplified to compensate the losses of splitting and transmission. We may note that the signals repeatedly undergo amplification, splitting and transmission.

Example 1

Consider an InGaAsP SOP with amplifier width (W) = 5 µm and thickness (d) = 0.5 µm. Given that group velocity (v_g) = 2 × 10^8 m·s^{-1}, if a 1.0 µW optical signal at 1500 nm enters the device, then calculate the photon density.

Solution

We have:

$$N_{ph} = \frac{P_s}{v_g (h\nu)(Wd)}$$

$$= \frac{1 \times 10^{-6} \text{ W}}{(2 \times 10^8 \text{ ms}^{-1}) \frac{(6.626 \times 10^{-34} \text{ Js}) \times 3 \times 10^8 \text{ m·s}^{-1}}{1.55 \times 10^{-6} \text{ m}} (5 \text{ µm})(0.5 \text{ µm})}$$

$$= 1.56 \times 10^6 \text{ photons m}^{-3}$$

Here

$P_s = 1 \times 10^{-6}$ W
$v_g = 2 \times 10^8$ m·s^{-1}
$\nu = \frac{c}{\lambda} = \frac{3 \times 10^8 \text{ m·s}^{-1}}{1.55 \times 10^{-6} \text{ m}}$
$W = 5$ µm
$d = 0.5$ µm

Example 2

A SOA has uncoated facet reflectivities of 30% and a single-pass gain of 5 dB. The device has an active region of length 320 µm, a mode spacing of 1 nm, and a peak gain wavelength of 1.55 µm. Calculate the refractive index of the active region and the spectral bandwidth of the amplifier.

[B.Tech., M.Sc. (Ele.)]

Solution

The refractive index n of the active region at the peak gain wavelength λ is given by:

$$n = \frac{\lambda^2}{2(\delta\lambda)L}$$

where $\delta\lambda$ is the mode spacing and L is the length of the active region. Substituting the values of λ, $\delta\lambda$ and L in the above equation, one obtains:

$$n = \frac{(1.55 \times 10^{-6})^2}{2 \times 1 \times 10^{-9} \times 320 \times 10^{-6}} = 3.75$$

Now, using the expression for the 3 dB spectral bandwidth of the amplifier as follows and substituting the values, one obtains:

$$\Delta v = \frac{c}{\pi n L} \sin^{-1} \frac{1 - G_s \sqrt{R_1 R_2}}{(4 G_s \sqrt{R_1 R_2})^{1/2}}$$

$$= \frac{3 \times 10^8 \, (\text{m} \cdot \text{s}^{-1})}{\pi \times 3.75 \times 320 \times 10^{-6} \, \text{m}} \sin^{-1} \left[\frac{1 - 3.16 \sqrt{0.30 \times 0.30}}{(4 \times 3.16 \sqrt{0.30 \times 0.30})^{1/2}} \right]$$

(5 dB gain is equivalent to 3.16)

$= 2.125 \times 10^9$ Hz

$= 2.125$ GHz

Example 3

Consider an EDFA being pumped at 980 nm with 30 mW pump power. If the gain at 1550 nm is 20 dB, then calculate the maximum input and output powers. [B.Tech.]

Solution

Here, $\dfrac{\lambda_p}{\lambda_s} = \dfrac{980}{1550}$, $P_{p,\text{in}} = 30$ mW, $G = 100$

We have input signal power:

$$P_{s,\text{in}} \leq \frac{(\lambda_p / \lambda_s) P_{p,\text{in}}}{G - 1} \leq \frac{(980/1550)(300 \text{ mW})}{100 - 1} = 190 \, \mu W$$

$$P_{s,\text{out}} \leq P_{s,\text{in}} + \frac{\lambda_p}{\lambda_s} P_{p,\text{in}} = 190 \, \mu W + 0.03 \, (30 \text{ mW})$$

$= 190 \, \mu W + 0.03 \, (30 \text{ mW}) = 19.1$ mW $= 12.8$ dB $\cdot$ m

Example 4

Consider an EDFA which is used as a power amplifier with a 10 dB gain. Assume the amplifier input is 0-dB m level from a laser diode transmitter. If the pump wavelength is 980 nm, then calculate the least pump power for a 10 dB m output at 1540 nm. [M.Sc. (Ele.)]

Solution

We have:

$$P_{p,\text{in}} \geq \frac{\lambda_s}{\lambda_p} (P_{s,\text{out}} - P_{s,\text{in}}) = \frac{1540}{980} (10 \text{ mW} - 1 \text{ mW}) = 14 \text{ mW}$$

Example 5

Consider a typical EDFA with the following parameters: doping concentration 6×10^{24} m^{-3}, signal wavelength $\lambda_s = 1.536$ µm, absorption cross section at λ_s, $\sigma_{sa} = 4.644 \times 10^{-25}$ m^2, emission cross section at λ_s, $\sigma_{sc} = 4.644 \times 10^{-25}$ m^2, lifetime for spontaneous emission, $\tau_{sp} = 1.2 \times 10^{-2}$ s, length of the doped fiber $L = 7$ m, and $\Gamma_s = 0.80$. Assume that (N_2/N_1) is nearly constant over the length of the EDFA and is equal to 0.70. (In actual practice, N_1 and N_2 vary with z). Calculate the small-signal gain of EDFA and the maximum possible achievable gain. [B.Tech.]

Solution

The total gain G for a lumped EDFA of length L can be obtained using following expression for total gain:

$$G = \Gamma_s \exp\left[\int_0^L (\sigma_{se} N_2 - \sigma_{sa} N_1)\, dz\right]$$

where Γ_s is confinement factor taking into account the fact that the doped region within the core provides the gain for the entire mode fiber. σ_{se} and σ_{sa} represent the absorption cross-section at the pump frequency $(v_p = c/\lambda_p)$ and absorption cross section at the signal frequency $(v_s = c/\lambda_s)$ respectively. N_1 and N_2 represent the population density of Er^{3+} ion in the ground state and that in the upper amplifier level (metastable level) respectively. With the above assumptions, simple mathematical manipulation of this equation gives:

$$G = \Gamma_s \exp\left[\sigma_{sa} N_t \left\{\left(\frac{\sigma_{se}}{\sigma_{sa}} + 1\right)\frac{N_2}{N_t} - 1\right\} L\right]$$

Here N_1 has been replaced by $N_t - N_2$. Thus, one finds:

$$\left[G = 0.80 \exp (4.644 \times 10^{-25}\ m^2)(6 \times 10^{24}\ m^{-3})\right.$$

$$\left.\times \left\{\left(\frac{4.644 \times 10^{-25}}{4.644 \times 10^{-25}} + 1\right) \times 0.70 - 1\right\} \times 7\ m\right]$$

or $\qquad G = 0.80 \exp 97.80) = 1956 = 32.9$ dB

In the above expression for G, if we substitute $N_2 = N_t$ and $\Gamma_s = 1$, we get the expression for maximum possible achievable gain:

$$G_{max} = \exp (s_{se} N_t L)$$

Substituting the values of relevant parameters, one obtains for the present case:

$$G_{max} = \exp [(4.644 \times 10^{-25}\ m^2) \times (6 \times 10^{24}\ m^{-3}) \times 7\ (m)]$$
$$= 2.9568 \times 10^8 = 84.70\ dB$$

Indeed this value of G_{max} is quite high. A more realistic estimate of G_{max} can be obtained using the principle of conservation of energy. Thus, if $P_{s,\,in}$ and $P_{s,\,out}$ are the signal powers at the input and output ends of the erbium-doped fiber at the signal wavelength λ_s, and $P_{p,\,in}$ is the input pump power at wavelength λ_p, the following inequality should hold true:

$$P_{s,\,out} \leq P_{s,\,in} + \frac{\lambda_p}{\lambda_s} P_{p,\,in}$$

Assuming that there is no spontaneous emission, the gain may be written as:

$$G = \frac{P_{s,\,out}}{P_{s,\,in}} \leq 1 + \frac{\lambda_p}{\lambda_s} \frac{P_{p,\,in}}{P_{s,\,in}}$$

This yields:

$$G_{max} = 1 + \frac{\lambda_p}{\lambda_s} \frac{P_{p,\,in}}{P_{s,\,in}}$$

Example 6

Consider an optical transmission path containing N cascaded optical amplifiers each having a 30 dB gain. If a fiber has a loss of 0.2 dB/km, then the span between optical amplifiers is 150 km if there are no other system impairments. A 900 km link require five amplifiers. Calculate the noise penalty factor over the total path. [M.Sc. (Ele.)]

Solution

We have:

$$F_{path}(G) = \frac{1}{G}\left(\frac{G-1}{\log G}\right)^2$$

Substituting the values, one obtains:

$$10 \log F_{path}(G) = 10 \log\left[\frac{1}{1000}\left(\frac{1000-1}{\log 1000}\right)^2\right] = 10 \log 20.9 = 13.2 \text{ dB}$$

Now, if we reduce the gain to 20 dB, then the impairment free transmission distance is 100 km for which one need eight amplifiers. Now, the noise penalty factor is obtained as:

$$10 \log F_{path}(G) = 10 \log\left[\frac{1}{100}\left(\frac{100-1}{\log 100}\right)^2\right] = 10 \log 4.62 = 6.6 \text{ dB}$$

Example 7

A fiber Raman amplifier has a length of 2 km. The attenuation coefficients α_s and α_p for signal and pump wavelengths for this fiber are 0.15 and 0.20 dB/km, respectively. Assume that $a_p = 60$ µm² and

$g_R = 5 \times 10^{-14}$ m/W. The amplifier is pumped by a laser of 1 W power. If the input signal power is 1 μW, calculate (a) the output signal power for forward pumping, and (b) the overall gain in dB.

Solution

In the case of forward pumping, we can use the following relation:

$$P_s(L) = P_{s,\text{in}} \exp\left[\frac{g_R}{a_p} P_{p,\text{in}} \left\{\frac{1-\exp(-\alpha_p L)}{\alpha_p}\right\} - \alpha_s L\right]$$

$P_{s,\text{in}} = 1$ μW $= 1 \times 10^{-6}$ W, $P_{p,\text{in}} = 1$ W

$g_R = 5 \times 10^{-14}$ mW^{-1}, $a_p = 60$ μm$^2 = 60 \times 10^{-12}$ m^2

$L = 2$ km $= 2000$ m, $a_s = 0.15$ dB/km ($= 3.39 \times 10^{-5}$ m^{-1})

$\alpha_p = 0.20$ dB/km ($= 4.50 \times 10^{-5}$ m^{-1})

$$P_s(L) = 1 \times 10^{-6}\,(W) \exp\left[\frac{5 \times 10^{-14}\,\text{mW}^{-1}}{6 \times 10^{-1}\,\text{m}^2} \times 1\,(W)\right.$$

$$\left.\times \left\{\frac{1-\exp(-3.40 \times 10^{-5}\,(\text{m}^{-1}) \times 2000(\text{m}))}{4.5 \times 10^{-5}\,\text{m}^{-1}}\right\} - 3.39 \times 10^{-5}\,(\text{m}^{-1}) \times 2000(\text{m})\right]$$

Therefore, the net gain of the amplifier will be:

$$G_{\text{FRA}} = \frac{4.582}{1\,\mu W} = 4.582$$

Gain (in dB) $= 10 \log_{10} G_{\text{FRA}} = 10 \log_{10}(4.582) = 6.61$ dB.

SUGGESTED READINGS

1. Agrawal GP, 'Fiber Optic Communication Systems', 2nd ed., Wiley (1997).
2. Connelly MJ, 'Semiconductor Optical Amplifiers', 1st ed., Kulwer Academic Publishers (2002).
3. Dutta NK and Wang Q,. 'Semiconductor Optical Amplifiers, 1st ed., World Scientific (2005).
4. Kolinbiris H, 'Fiber Optic Communications', Prentice Hall (2004).
5. Salah BEA and Teich MC, 'Fundamentals of Photonics', Wiley (1991).
6. Yeh C, 'Applied Photonics', Academic Press (1991).
7. Degiorgio V and Cristiani I, 'Photonics', Springer (2014).
8. Kumar S and Jamal Deen M, 'Fiber Optic Communications', Wiley (2014).

GLIMPSES

- The optical signal gets attenuated as it propagates along the fiber in a fiber-optic communication system. Therefore, signal strength

have to be restored at appropriate points along the link. This can be achieved using either an amplifier or a regenerator. A regenerator requires conversion of the optical signal into the electrical domain and reconversion again into the optical domain. The use of regenerators is limited to some specific systems.
- An optical amplifier is a laser without feedback. An optical amplifier operates solely in the optical domain. Most amplifiers amplify incident light through stimulated emission. Optical gain is achieved when the amplifier is pumped optically or electrically to achieve *population inversion*.
- Optical amplification depends on:
    (i) Frequency (or wavelength) of incident signal.
    (ii) Local beam intensity.
- In an optical amplifier, the external pump source energy is absorbed by the electrons in the active medium. The electrons shifts to the higher energy level producing population inversion. Photons of incoming signal triggers these excited electrons to lower energy level through a stimulated emission process, producing amplified optical signal.
- Types of optical amplifiers:
    (a) *Doped fiber amplifiers* (DFA); (i) In these amplifiers, active medium is created by erbium (Er), ytterbium (Yb), neodymium (Nd), and praseodymium (Pr), (ii) these amplifiers can pump device at several different wavelength, (iii) coupling loss is low, and (iv) there is constant gain which is provided either by rare-earth dopants or stimulated Raman scattering.
    (b) *Semiconductor optical amplifiers* (SOA): These amplifiers utilize stimulated emission from injected carriers. In these amplifiers active medium consists of alloy semiconductor (P, Ga, In, As). SOA works in both low attenuation windows, i.e., 1300 nm and 1550 nm. The 3 dB bandwidth is about 70 nm because of very broad gain spectrum. SOA consumes less power and has fewer components
        – SOAs are mainly of two types: (i) Fabry-Perot amplifier (FPA), and (ii) Travelling wave amplifier (TWA).
        – SOA has rapid gain response 1 ps to 0.1 ns.
    (c) *Raman fiber optical amplifiers* (RFOA): These amplifiers differ in principle from Erbium-doped fiber optical amplifiers (EDFOA). They utilize stimulated Raman scattering (SRS) to create optical gain. SRS occurs when light waves interact with molecular vibrations in a material having a solid lattice structure.
- Light signal in an SOA passes through the active layer of the forward-biased semiconductor DH and in the process gets amplified. The amplifier gain (G) is given by:

$$G = \frac{P_{out}}{P_{in}} = \exp\left[g_o L - \frac{P_{out} - P_{in}}{P_{sat}}\right]$$

where P_{sat} is saturation power of amplifier, g is the gain coefficient (per unit length), P_{in} and P_{out} are the input and output powers respectively, and L is the length of the active region.
- At the facets of the active region, the optical reflections can severely affect the performance of the amplifier, especially when the single pass gain is high. Moreover, several nonlinear phenomena in SOAs, lead to interchannel crosstalk in multichannel operation.
- An EDFA operates on the principle of optical pumping of an Er^{3+} ion in silica fiber by either a 0.98 μm or a 1.48 μm source and stimulated emission at 1.55 μm, which is the low-attenuation window of silica based fibers. EDFAs combined with WDM technology are widely used in long-distance optical communications, networks, and signal modulation. The gain of EDFA depends on several factors, e.g. doping concentration, fiber length, pump power, etc. EDFAs are widely used in multichannel systems.
- FRA utilize stimulated Raman scattering (SRS) to create optical gain. Herein, the pump energy at λ_p is transferred to the signal energy at λ_s in a non-resonant process to provide gain at λ_s. The overall gain for small signal amplification is given by:

$$G_{FRA} = \exp\left[(g_o - \alpha_s) L\right]$$

where α_s represents fiber losses per unit lengths at the signal, frequency v_s, L is total fiber length, and:

$$g_o = \frac{g_R P_{p,in} L_{eff}}{a_p L}$$

where $L_{eff} \approx 1/\alpha_p$, where α_p resent fiber losses at the pump frequencies v_p and L_{eff} is called the effective length of the fiber, and g_R is the Raman gain coefficient.
- FRA have a number of main technical characteristics, e.g. efficient pumping, simplicity, wider wavelength coverage, minimal polarization sensitivity, low insertion loss, high gain, low noise, fast reaction to changes of the pump power, low power consumption, and low cost. FRA are used in long-haul and ultra-haul DWDM transmission systems.
- The design of the *linear optical amplifier* is similar to that of the semiconductor optical amplifier. The device has an active waveguide gain region; the input and output fibers are aligned to this waveguide. They are used in high data rate telecommunication systems.
- Optical amplifiers (OAs) are used to boost signals transmitted over long distances in network systems. OAs are ideal for Metro and

Long-Haul dense wavelength multiplexing (DWDM) as well as single wavelength applications. The optical design, coupled with sophisticated control circuitry, allows these OAs to provide constant gain even with signals being added to the network.

REVIEW QUESTIONS

1. Explain the principle of the optical amplifiers. Write their characteristics and important applications.
2. Describe the basic operation of EDTA. Mention its main technical characteristics.
3. Derive an expression for amplifier gain of SOA. Mention limitations of SOA.
4. Derive an expression for total gain G fo an EDTA of length L.
5. Explain the principle of FRA. Distinguish between the amplification processes in (a) EDFA, and (b) FRA.
6. What requirements should be met so that semiconductor DH functions efficiently as an optical amplifier?
7. Derive an expression for gain ripple G_r of an SOA.
8. Using steady-state amplifier equations and appropriate approximation, derive an analytical expression for an optimum length of fiber which gives maximum amplication for EDFA.
9. Distinguish between pure non-resonant TWA and resonant FPA signal gain.
10. The power conversion efficiency (PCE) of an EDFA is defined as:

$$PCE = \frac{P_{s,\,out} - P_{s,\,in}}{P_{p,\,in}}$$

Show that PCE is less than unity and the maximum value of its quantum conversion efficiency (QCE), which is defined by:

$$QCE = \frac{\lambda_s}{\lambda_p} (PCE)$$

is unity.

11. Explain the origin of gain saturation in FRAs. Derive an expression (approximate) for the saturated amplifier gain.
12. Write the flexibilities available in FRAs that are not available in SOAs and EDFAs.
13. Briefly mention the application of optical amplifiers.
14. Write the limitations of SOAs.

PROBLEMS

1. Taking the following parameters for a 1300 nm InGaAsP SOA:

W	active area width	3 μm
d	active area thickness	0.3 μm
L	amplifier length	500 μm
T	confinement factor	0.3
τ_r	time constant	1 ns
a	gain coefficient	2×10^{-20} m²
n_{th}	threshold density	1.0×10^{24} m⁻³

   (a) Calculate the pumping rate if a 100 mA bias current is applied to the device.
   (b) Calculate the zero signal gain.

   [**Ans.** (a) 1.39×10^{33} (electron/m³), (b) 2340 m⁻¹ = 23.4 cm⁻¹]

   [**Hint:**

   (a) $R_p(t) = \dfrac{J(t)}{qd} = \dfrac{1}{qdWL}$

   $= \dfrac{0.1 \, A}{(1.6 \times 10^{-19} C)(0.3 \, \mu m)(3 \, \mu m)(500 \, \mu m)}$

   $= 1.39 \times 10^{33}$ (electrons/m³)/s

   (b) $g_o = a\tau_r \left(\dfrac{J}{qd} - \dfrac{n_{th}}{\tau_r} \right)$

   $= 0.3 \, (2.0 \times 10^{-20} \, m^2)(1 \, ns)$

   $\times \left(1.39 \times 10^{33} \, m^{-3} s^{-1} - \dfrac{1.0 \times 10^{24} \, m^3}{1.0 \, ns} \right)$

   $= 2340$ m⁻¹ = 23.4 cm⁻¹]

2. To achieve a bit error rate (BER) = 10^{-9}, the factor Q specifying the receiver performance is 6. Calculate the optical signal-to-noise ratio (OSNR). Comment on your result.

   [**Ans.** OSNR (BER = 10^{-9}) ≈ 13.5 dB]

   [**Hint:** OSNR $= \dfrac{1}{2} Q(Q + \sqrt{2})$

   or   OSNR = (BER = 10^{-9})
   $= 0.5(6)(6 + \sqrt{2})$
   $= 22.24 \approx 13.5$ dB

   This reveals that if an OSA measures an OSNR ≤ 13.5 dB, then the corresponding error rate are equal to or higher than BER = 10^{-9}.]

3. An EDFA being pumped at 0.98 μm is being used as a power of 0 dB·m at λ_s = 1.55 μm, the output of amplifier is 20 dB·m. Show that the gain of the amplifier is 20 dB and input pump power required to achieve this gain is 156.6 mW.

4. A typical InGaAsP SOA is operating at 1.3 μm with the following parameters:
active region thickness = 0.5 mm
active region length = 200 mm
confinement factor (Γ) = 0.4
Time constant (τ_c) = 1 ns
σ_g (gain cross section or differential gain coeffiient = dg/dN) = 3×10^{-20} m^2
N_{tr} (carrier concentration per unit volume) = 1.0×10^{24} m^{-3}
Bias current (I) = 100 mA
Calculate (i) P_{sat}, (ii) the zero-signal gain coefficient, and (iii) the zero signal net gain. [**Ans.** (i) 31.85 mW, (ii) 3000 m^{-1}, (iii) 1.82]

5. A SOA has single-pass gain of 10 dB. Calculate the peak-trough ratio of the passband ripple if the facet reflectivities are (i) 0.01%, (ii) 1%. [**Ans.** (i) 1.004, (ii) 1.4938]

6. An EDFA being pumped at 0.98 μm with 30 mW pump power. The signal wavelength is 1.55 μm. Using the following data:
  - cross-sectional area of the fully doper fiber cone = 8.50 μm^2
  - doping concentration = 5×10^{24} m^{-3}
  - pump absorption cross-section = 2.17×10^{-25} m^2
  - signal absorption cross-section = 2.57×10^{-25} m^2
  - signal emission cross section = 3.41×10^{-25} m^2
  - input signal power = 200 μW

  and assuming that the fiber modes for λ_p and λ_s are fully confined, calculate (a) the rate of absorption per unit volume from the Er^{3+} level E_1 to level E_3 due to pump at λ_p (assuming $N_2 \approx 0$); and (b) the rate of absorption per unit volume from level E_1 to the metastable level E_2 and the rate of stimulated emission per unit volume from level E_2 to level E_1, both due to the signal at λ_s (assuming $N_2 \approx N_1$).
  [**Ans.** (a) 1.888×10^{28} m^{-3} s^{-1}, (b) 1.178×10^{26} m^{-3} s^{-1}, 1.5640×10^{26} m^{-3} s^{-1}]

SHORT ANSWER QUESTIONS

1. How the development of optical amplifiers revolutionized communication system?

Ans. Optical amplifiers had an important impact similar to the invention of laser in early 1960s. Both devices contributed to the development of communication systems and other applications, such as lower pump optical power, single pixel multicolour displays, and light emitting devices.

2. How a regenerator is an optoelectronic device?

Ans. Yes, a regenerator is an optoelectronic device. It amplifies and cleans up the optical signal in three steps: (i) The first step is to convert an optical signal into an electrical signal and then amplify it electronically, (ii) In the second step, to clean up the signal pulses using re-timing and pulse-shaping circuits, and (iii) In the third step, to reconvert the amplified electrical signal into an optical signal. This signal is then coupled into the next segment of the optical fiber.

3. What is an optical amplifier (OA)?

Ans. An OA is a device that amplifies the optical signal directly, without converting it to an electrical signal and then to an optical signal again. An OA operates solely in the optical domain, i.e., it takes in a weak optical signal from one segment of the link, amplifies it optically to produce a strong optical signal (without recourse to photon-to-electron conversion and vice versa), and couples it to the next segment of the link, or the OA is ideally a transparent box which provides gain and is also insensitive to the bit rate, modulation format, power and wavelengths of the signal(s) passing through it. The signals remain in optical form during amplification. Much of the most relevant recent advances in OAs (i.e. long distance NRZ and soliton systems, and wide area and broadcast multi-channel systems) can be traced to the incorporation of OAs.

4. How many types of optical amplifiers are there?

Ans. There are mainly two classes of amplifiers, namely (i) semiconductor optical amplifiers, which utilize stimulated emission from injected carriers, and (ii) fiber optical amplifiers, in which the gain is provided by either rare-earth dopants or stimulated Raman scattering.

5. What is planar waveguide optical amplifiers?

Ans. Rare-earth-doped planar waveguide optical amplifiers provide compact and inexpensive alternatives to fiber amplifiers. Moreover, planar technology is quite suitable for optical integration and will be essential to the development of fully integrated advanced optical devices.

6. What is EDFA?

Ans. EDFA is a length of glass fiber which has been doped with the rare-earth metal Erbium ions. These ions act as an active medium with the potential to experience inversion of carriers and emit spontaneous and stimulated emission of light near a desirable signal wavelength. To produce the amplifier gain medium, the silica fiber core of a standard single-mode fiber is doped with Erbium ions.

578 Photonics | Optoelectronics

7. **What are semiconductor optical amplifiers (SOA)?**
Ans. SOA is nothing more than a semiconductor laser, with or without facet reflections (the anti-reflection coating reduces the reflections). An electrical current inverts the medium, e.g. electrons are transferred from the valence to the conduction band, which produces spontaneous emission (fluorescence) and the potential for stimulated emission if external optical field is present. The stimulated emission yields the signal gain.

8. **Why EDFAs have been used in multi-channel WDM systems?**
Ans. EDFAs have been used in multi-channel WDM systems to compensate for (i) fiber attenuation losses in transmission, (ii) component excess losses, and (iii) optical network splitting losses.

9. **What material properties govern the amplifier saturation in semiconductor optical amplifiers?**
Ans. (i) gain cross section (σ_g), (ii) carrier life time (τ_c).

10. **Give some examples of rare-earth elements which can be dissolved in glass to make optical amplifiers.**
Ans. A number of SOA chips can be integrated on the same substrate to create high-density switching matrices.

11. **How high-density switching matrices can be created using SOA chips?**
Ans. Neodymium, Praseodymium, Holonium, and Erbium. However, the laser transition in erbium-doped glass occurs at a wavelength that is very close to the wavelength of minimum attenuation of glass fibers, and this gives erbium special importance.

MULTIPLE CHOICE QUESTIONS

1. An optical amplifier operates:
   (a) solely in the optical domain
   (b) solely in the ultraviolet domain
   (c) solely in the infrared region
   (d) None of the above

2. Semiconductor optical amplifiers utilize:
   (a) stimulated Raman scattering
   (b) stimulated emission from injected carriers
   (c) spontaneous emission from injected carriers
   (d) None of the above

3. The amplifier gain (G) for SOA may be expressed as:
   (a) $G = \dfrac{P_{out}}{P_{in}} = \exp\left[g_o L - \left(\dfrac{1 - P_{in}}{P_{sat}}\right)\right]$
   (b) $G = \dfrac{P_{out}}{P_{in}} = \exp\left[g_o L - \dfrac{P_{in}}{P_{sat}}\right]$

(c) $G = \dfrac{P_{out}}{P_{in}} = \exp\left[g_0 L - \dfrac{1}{P_{sat}}\right]$

(d) $G = \dfrac{P_{out}}{P_{in}} = \exp\left[g_0 L - \dfrac{P_{out} - P_{in}}{P_{sat}}\right]$

4. Which of the following optical amplifiers is most suited for multi-channel bidirectional operation?
   (a) SOA
   (b) EDFA
   (c) FRA
   (d) None of these

5. Optical amplifiers can be used as:
   (a) in-line amplifiers to compensate for loss
   (b) power amplifiers to follow transmitter
   (c) pre-amplifiers to precede the receiver
   (d) All of the above

6. Erbium-doped fiber amplifiers operate at the following windows:
   (a) Low-dispersion window (around 1.30 μm)
   (b) Low-attenuation window (around 1.55 μm)
   (c) Both the windows
   (d) None of these

7. The structure of a semiconductor optical amplifier differs from a semiconductor laser in the following aspect:
   (a) The reflectivity of the end facets of the active region in the SOA is zero
   (b) The reflectivity of the end facets of the active region is 100%
   (c) The SOA is pumped electrically
   (d) There is no difference

8. An SOA differs from an EDFA in the following manner:
   (a) An SOA operates in the electrical domain while the EDFA operates in the optical domain
   (b) An SOA is pumped electrically while the EDFA amplifies 1.55 μm
   (c) An SOA amplifies 1.30 μm while the EDFA amplifies 1.55 μm
   (d) There is no difference

9. Gain in EDFA depends on the following factors:
   (a) Doping concentration
   (b) Length of the doped fiber
   (c) Pump power
   (d) All of these

10. The power conversion efficiency (PCE) of an EDFA defined as
    $PCE = \dfrac{P_{s,\,out} - P_{s,\,in}}{P_{p,\,in}}$ is:

(a) equal to 1   (b) less than 1
(c) greater than 1   (d) infinite

11. The difference between a regenerator and an optical amplifier is:
    (a) A regenerator amplifies as well as restores the signal
    (b) An optical amplifier compensate for transmission loss
    (c) A regenerator converts the optical signal into the electrical domain for amplification and then reconverts it into the optical domain, whereas an optical amplifier operates only in the optical domain
    (d) There is no difference between the two

12. The most suitable wavelength for pumping an EDFA is:
    (a) 0.85 μm   (b) 0.98 μm
    (c) 1.30 μm   (d) 1.55 μm

13. In what way does an EDFA differ from a fiber Raman amplifier?
    (a) An EDFA requires population inversion while the FRA does not
    (b) An FRA operates on the principle of stimulated Raman scattering
    (c) An EDFA operates on the principle of stimulated emission
    (d) There is no difference

14. In what way are EDFA and FRA similar?
    (a) Both of them operate in the all-optical domain
    (b) Both of them can be used around the 1.55 μm window
    (c) Both of them can be employed for multichannel operation
    (d) All of the above

15. A Raman optical amplifier is based on a nonlinear effect called:
    (a) spontaneous Raman scattering
    (b) stimulated Raman scattering
    (c) Rayleigh scattering
    (d) None of the above

16. In a long transmission system, optical amplifiers are needed to periodically restore the power level after it has decreased due to:
    (a) attenuation in the fiber   (b) reflection in the fiber
    (c) absorption in the fiber   (d) None of the above

ANSWERS

1. (a)   2. (b)   3. (d)   4. (b)   5. (d)   6. (b)   7. (a)
8. (b)   9. (d)   10. (b)   11. (c)   12. (b)   13. (a)   14. (d)
15. (b)   16. (a)

APPENDIX
OPTICAL SWITCHES

A11.1 INTRODUCITON

There are many optical networks which integrate optical switches into their design. *Opto-mechanical switches* redirect optical signals from one port to another by moving an optical fiber tube assembly or an optical component, such as a mirror or prism. There are several types of optical switches incorporated into networks. However, in practice, most optical switches are still operated mechanically and controlled by an electronic control circuit. Speed is a crucial parameter in network applications, since a high-speed data transmission of tenths of milli-seconds is required. In few years, we expect that dynamic optical routing will require much faster speeds. However, more technology exists for optical switches than any other functional component within the optical network. Efforts through research are there for developing optical switches to increase the number of outputs, and to reduce size, cost and switching time. Presently, optical switches include many types, e.g., *opto-mechanical switches, thermo-optic switches, electro-optic switches, micro-electro-mechanical switches* (MEMSs), and *micro-opto-mechanical switches* (MOMSs). Development of new types of optical switches are in the research and development ranges.

Opto-mechanical and electro-mechanical switches are the oldest type of optical switches and most widely deployed at this time. These devices achieve switching of moving fiber or other optical components by means of stepper motors or relay arms. This causes them to have relatively slow switching time; however, their reliability is excellent and they offer low insertion and crossmark losses.

This chapter presents a few optical switch designs, i.e. four cases in building opto-mechanical switches using a movable mirror or prism to switch between the input and output ports for use in communication systems.

A11.2 OPTO-MECHANICAL SWITCHES

Figure A11.1 illustrates common switch configuration. The input signal comes through the input fiber cable on the left side of the switch. A mechanical slider moves that fiber up and down, latching into one of the two output fiber cables on the right side of the switch. In OFF/ON positions, the switch directs light from the input fiber into one of the two outputs. This arrangement is called "1 × 2 switch configuration". As input at port 1, the signal can be switched to either port 2 or port 3.

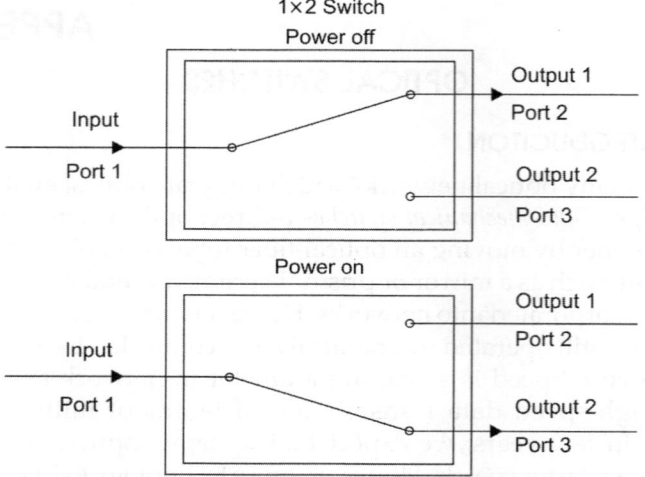

Fig. A11.1: A typical 1 × 2 switch configuration

For the following definitions, assume the switch is configured to couple to port 2. The insertion loss L_{IL} (in decibels) is defined by Eq. A11.1. Insertion loss depends on fiber cable alignment at the input and output ports. Low insertion loss value can be obtained on switches with good mechanical alignment. A good switch provides similar values of insertion loss for all switch positions.

$$L_{IL} = -10 \times \log_{10} \frac{P_2}{P_1} \qquad (A11.1)$$

where, P_1 is the power going into port 1 and P_2 is the power exiting from port 2.

Crosstalk loss L_{CT} is one of the important losses, which should be considered in opto-mechanical switches. Crosstalk loss is a measure of how well the uncoupled port is isolated. The crosstalk loss L_{CT} (in decibels) is defined by Eq. (A11.2). Crosstalk loss values depend on the particular design of the switch.

$$L_{CT} = -10 \times \log_{10} \frac{P_3}{P_1} \qquad (A11.2)$$

where, P_1 is the power going into port 1 and P_3 is the power exiting from port 3.

There are other important optical parameters that need to be specified for each switch type. These parameters include: *polarization dependent loss* (PDL), *return loss* (RL), and the *Etalon effect*. The PDL is defined as the maximum difference in insertion loss between any two-polarization states. It is caused by mechanical stress and temperature variation on optical components or fiber cables. This causes changes in the birefringence and a gradient of index of refraction (*n*) of the optical material. The RL is defined as the light reflected back into the input path. It is caused by scattering and reflection from optical surfaces like mirrors, lenses, and connectors or from defects, such as cracks and scratches. The back reflection is equal to the RL with a negative quantity. Elaton effect is defined as light resonance (ripple) at a certain wavelength. It is caused by reflection of light from parallel optical surfaces and interference between the signals. All the above losses are measured in decibels (dB). Special optical parameters can be specified by the customers.

Another important parameter of the optical switches is the repeatability—achieving the same insertion loss each time the switch is returned to the same position. Switching speed is also another important specification of a switch. The switching speed is defined as how fast the switch can charge the signal from one port to the other. It is an important factor in some switch applications in communication systems.

Figure A11.2 shows a schematic diagram of a mechanical switch configuration with two inputs and two outputs. The inputs are located on the one side and the outputs on the other side of the switch. This configuration is called a 2 × 2 switch. The signal enters port 1 and port 4, and exists from port 2 and port 3, respectively. This case is called the *bypass state*, in the OFF position. When the latching mechanism changes position between port 2 and port 3, signal enters port 1 and exists port 3 and from port 4 to port 2. This case is called the *operate state*, in the ON position.

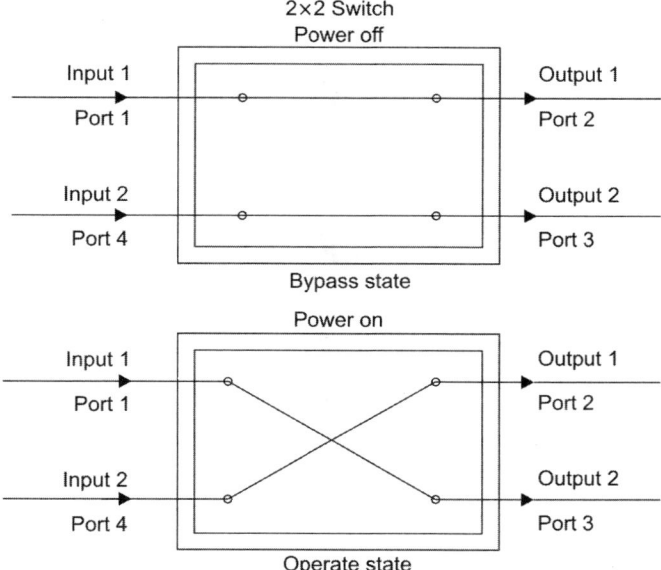

Fig. A11.2: A typical 2 × 2 switch configuration

584 Photonics | Optoelectronics

Opto-mechanical switches collimate the optical beam from each input and output fiber and move these collimated beams around inside the switch. This creates low optical loss, and allows distance between the input and output fiber. These switches have more bulky components compared to newer alternatives, such as the micro-opto-mechanical switches.

Figure A11.3 shows a schematic diagram of a two-position switch. The switch consists of a *sliding prism* and *quarter pitch graded index* (GRIN) lenses at the input and output ports. The components are assembled in a packaging base and sealed with a lid. Each GRIN lens is connected to the fiber tube assembly using an epoxy. Figure A11.3 illustrates the OFF/ON positions of a 1 × 2 switch. As explained above, the GRIN lens collimates the divergence beam exiting from the input fiber. The right angle prism deflects the light by total internal reflection (TRIN) at its two slanting surfaces. The GRIN lens refocuses the collimated beam onto a fiber cable at one of the output ports. To direct the signal from port 1 to port 3, the prism slides to a new position, as shown in Fig. A11.3 in the OFF position. Figure A11.3 also shows the

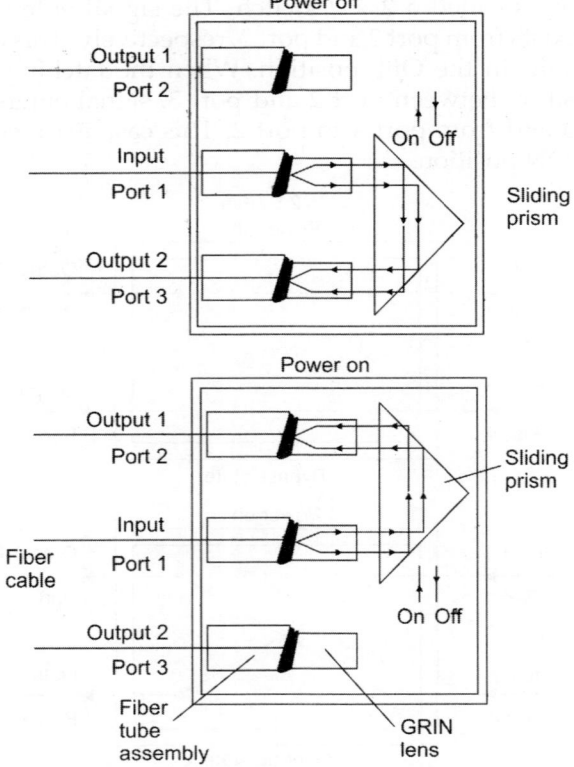

Fig. A11.3: A 1 × 2 opto-mechanical switch

signal directed between port 1 and port 2, in the ON position, when the prism changes position.

Opto-mechanical switches drive optical fiber networks mechanically. They can switch between light paths at high speed and with low insertion loss. They are widely used in rapidly developing areas of the fiber-optic field, such as optical cross connection and wavelength multiplexing. Figure A11.4 shows the design of a 2 × 4 opto-mechanical switch. The switch uses an electromagnetic actuator with a latching function to drive a movable block to change the light path between the ports. Figure A11.4 shows the switch in the OFF position. The light passes through from port 1 and port 2 to port 3 and port 4, respectively. When the power is turned ON, an electromagnetic actuator (with latching function) drives a movable block to change light path from port 1 and port 2 to port 5 and port 6, respectively, as shown in Fig. A11.4. The optical and mechanical components of a switch are

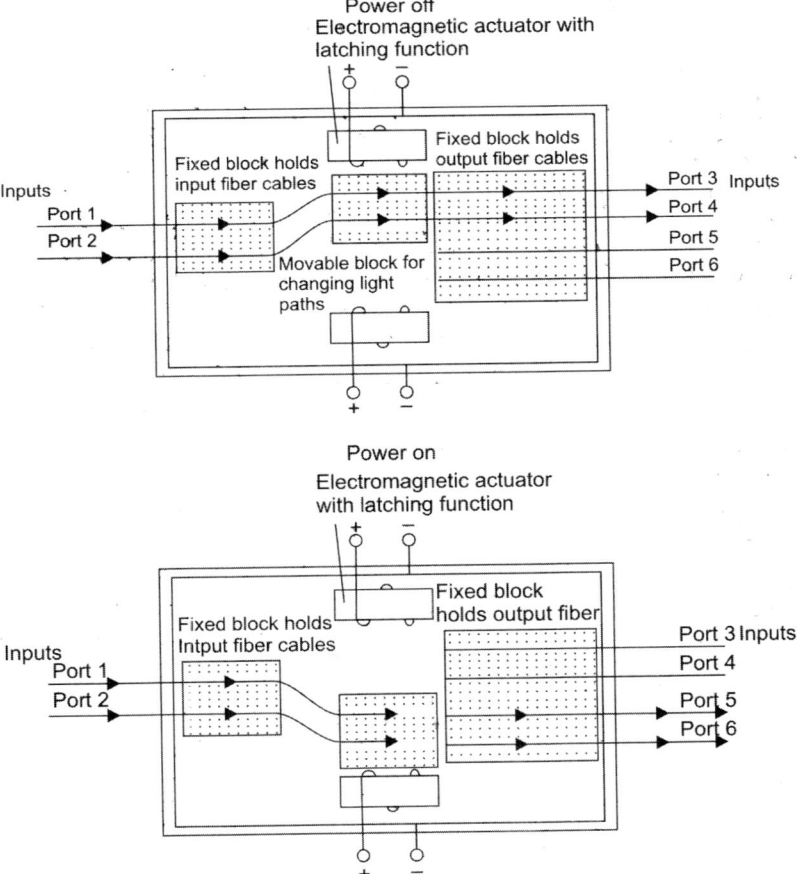

Fig. A11.4. A 2 × 4 opto-mechanical switch

assembled in a packaging box with minimal alignment work. There are three configurations of this switch: 1 × 2, 2 × 2, and 2 × 4.

A practical *electromagnetic bypass switch* is illustrated in Fig. A11.5. The switch contains a quarter pitch GRIN Lens connected to fiber tube

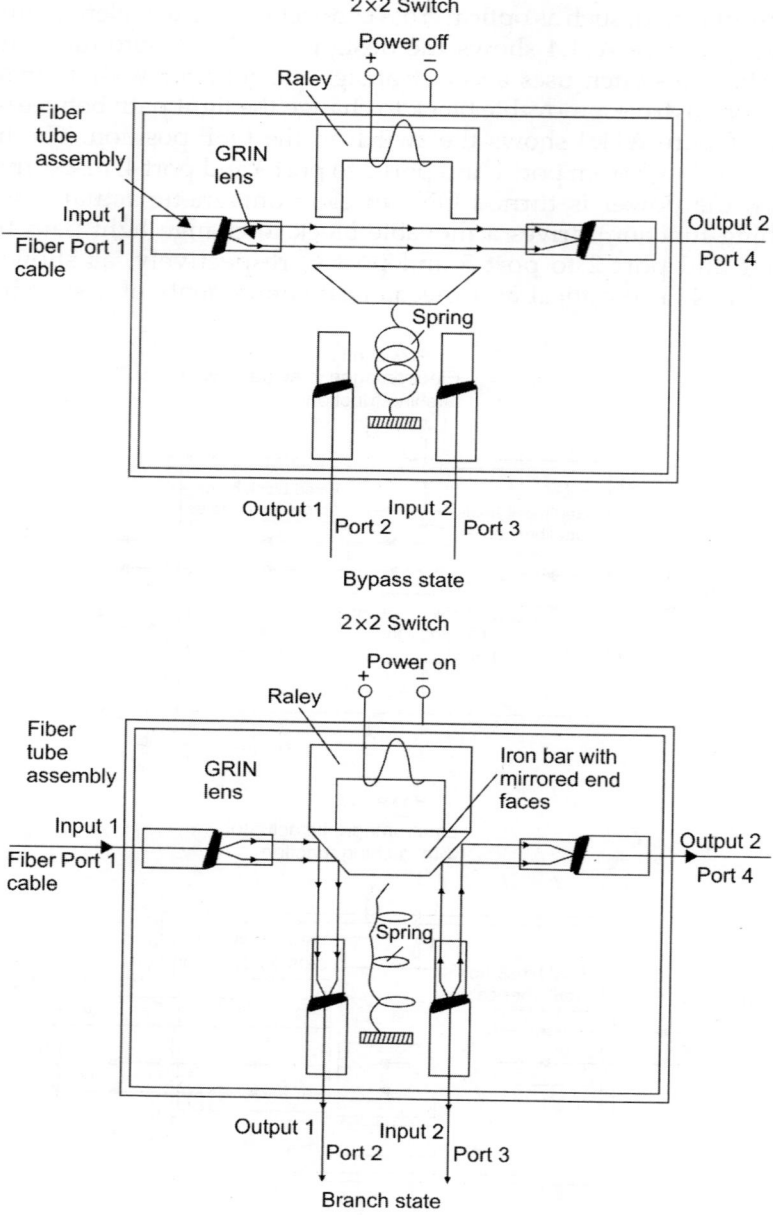

Fig. A11.5: An electromagnetic bypass switch

assembly at the input and output ports, a relay, and an iron bar with mirror end faces. The components are assembled in a packaging base, which is sealed with a lid. When the power is turned OFF, a spring pulls the iron bar out of the signal path, returning the switch to the bypass condition. This is called the bypass state. In the bypass state, the signal passes directly from port 1 and port 4. When the power is ON, the electromagnet is activated and the iron bar is raised. This is called the branch state. In the branch state, mirrors direct the signal between port 1 and port 2, and between port 3 and port 4.

Another type of *bypass switch* is also used in communication network. Figure A11.6 illustrates the function of this type of bypass switch. When the power is in the OFF position, the input signal comes through the input fiber cable on port 1 on the left side and leaves through the output port 4 on the right side of the switch. This is called the bypass state. When the power is ON, a mechanical slider moves two fiber connections to the up position, latching into two output fiber cables at port 2 and port 3 on the side of the switch. In this position, the input signals from port 1 and port 4 are launched into port 2 and port 3, respectively. This position is called *branch state*. As input at port 1, the signal can be directed to either port 4 or port 2. Also, an input signal at port 4 can be directed to output port 3.

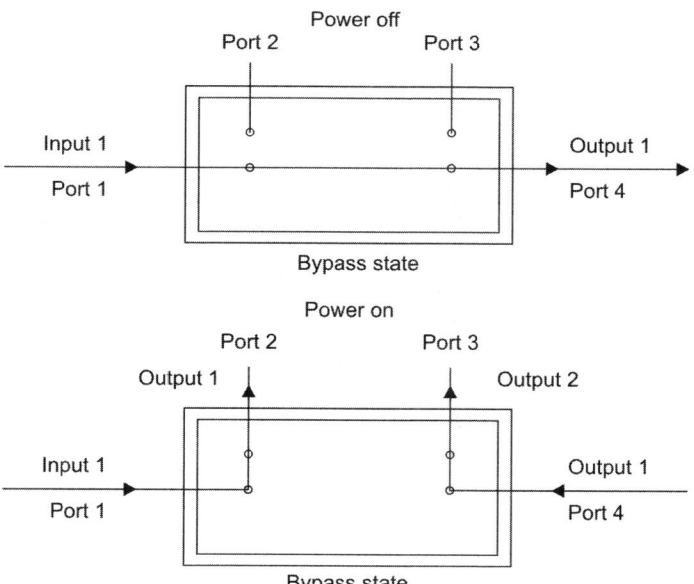

Fig. A11.6: A bypass switch

Now, we present new switch designs of a 1 × 8 latching switch configuration using prisms. These switches are commercially available in the market. There are two types of models: *the linear and triangular models*. The linear model directs the signal from the input to the outputs by arranging the prism linearly, as shown in Figs A11.7 and A11.8. The triangular model directs the input to the outputs by arranging the prisms triangularly, as shown in Figs A11.9 and A11.10. These models have come into wide use because they are simple, offer 8 outputs, and are cost effective. They are also used in back-up systems to re-route signals around broken fiber optic cable and in fiber optical instruments.

Figure A11.7 illustrates a schematic diagram for one configuration of a *linear model of a 1 × 8 latching switch*. The common element of this type of opto-mechanical switch is that their operation involves mechanical sliding motion of prisms in OFF/ON positions to direct the signal from one port to another. Figure A11.7 shows the input located on the one side and the outputs on the other side of the switch.

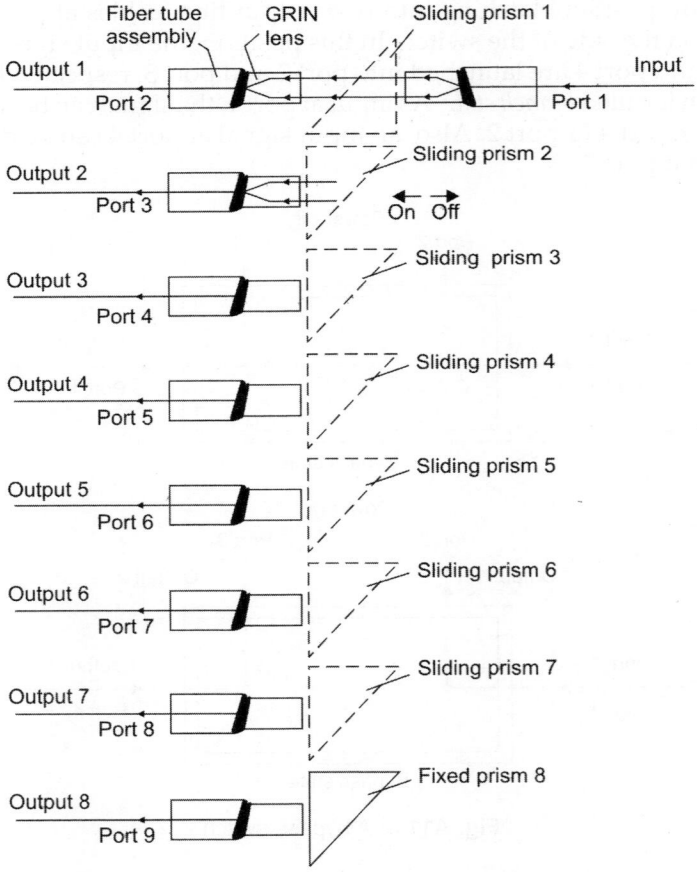

Fig. A11.7: Schematic configuration of the one type of linear model of a 1 × 8 latching switch

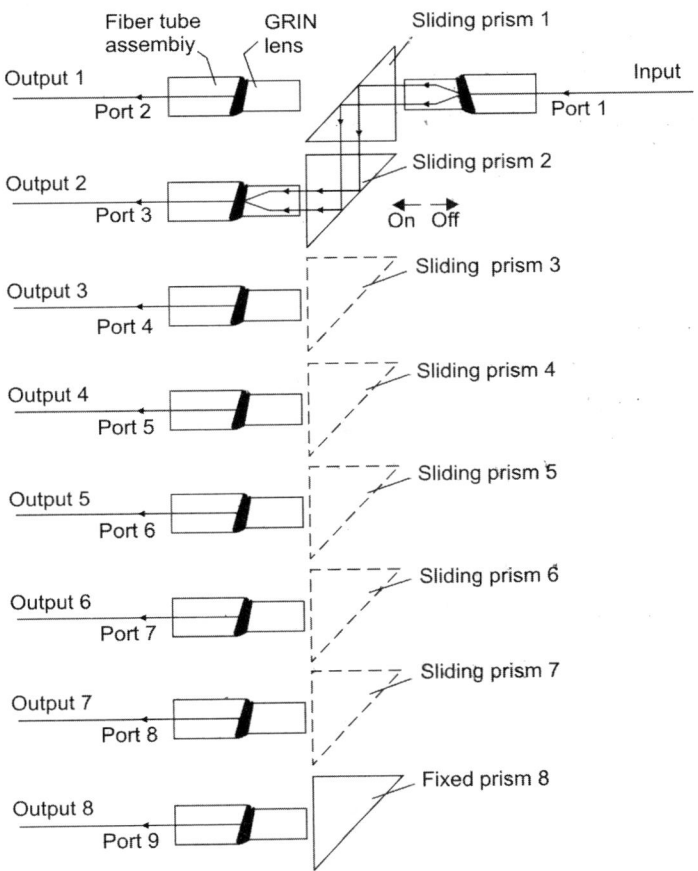

Fig. A11.8: Signal from port 1 to port 3 in the linear model of a 1 × 8 latching switch

This configuration is called a 1 × 8 switch in the linear model. Light enters port 1 and exits from port 2 when sliding prism 1 is in the OFF position. When the latching mechanism places the sliding prisms 1 and 2 in position, the light enters port 1 and exits from port 3, as shown in Fig. A11.8. Similarly, light enters port 1 and exits from port 4, when the sliding prism 2 is in the OFF position and sliding prism 3 is in the ON position. This switch configuration is more complicated than the second switch configuration because the input is located on one side and the outputs on the other side of the switch. This configuration includes a complex mechanism, controls, seven sliding prisms, and one fixed prism.

Figure A11.9 illustrates a schematic diagram of the second type of configuration of the linear model of a 1 × 8 latching switch. This configuration is different because the input and the outputs are located on the same side of the switch, as shown in Fig. A11.9. Prisms are also used in

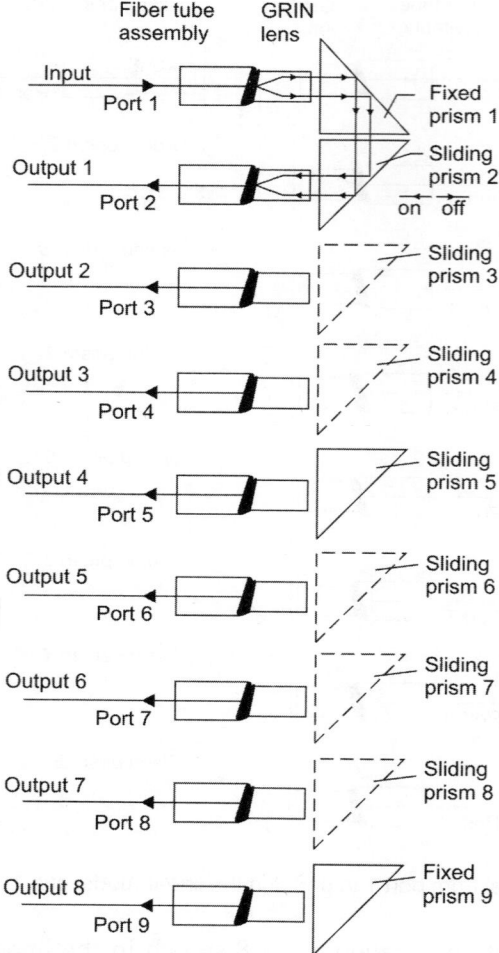

Fig. A11.9: Schematic of second configuration of the linear model of a 1 × 8 latching switch

the operation of this type of switch configuration. Light enters port 1 and exits from port 2 when fixed prism 1 and sliding prism 2 are in position. When the latching mechanism places the sliding prism 2 in the OFF position and sliding prism 3 in the ON position, the light enters port 1 and exits from port 3. Similarly, light enters port 1 and exits from port 4 when the sliding prism 3 is in the OFF position and sliding prism 4 is in the ON position. The same procedure is used for the signal exiting from other ports. This switch configuration is simpler than the first configuration because the input and outputs are located on the same side of the switch. This configuration includes less complex mechanism, controls, seven sliding prisms, and two fixed prisms.

Figure A11.10 illustrates a schematic diagram of a configuration of the triangular model of a 1 × 8 latching switch. The common element of this type of opto-mechanical switch is that the operation involves mechanical sliding motion of parallelogram prisms in OFF/ON positions to direct the signal from one port to another. Figure A11.10 shows the input located on one side and the outputs on the other side of the switch. This configuration is called a 1 × 8 switch in the triangular model. Light enters port 1 and exits from one of the outputs. When the latching mechanism places the sliding parallelogram prism 1 in the OFF position and sliding prism 2 into position, the light enters port 1 and exits from port 4, as shown in Fig. A11.10. Similarly, light enters port 1 and exits from port 3 when the sliding prisms 2 and 3 are in the ON position. Figure A11.11 shows light entering port 1 and exiting from port 5, when sliding prism 1 is in the OFF position. This switch configuration is more complicated than the linear model because the input is located on one side, and the outputs on the other side of the switch. Seven sliding parallelogram prisms with additional mechanisms and controls form this configuration. Both models have difficulty achieving precise alignment and low losses during the manufacturing processes.

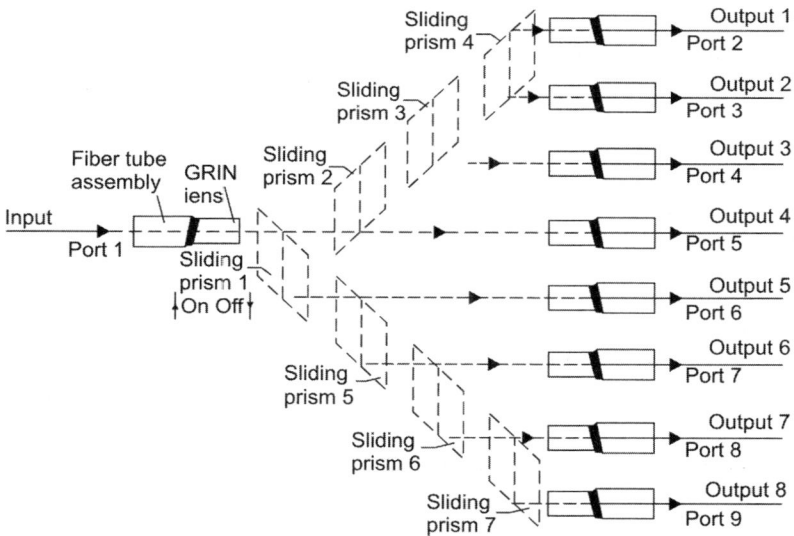

Fig. A11.10: Schematic of the triangular model of a 1 × 8 latching switch

Many other modern opto-mechanical switches are used in telecommunication networks management, monitoring, restoration, and protection. They have excellent optical performance and the high reliability necessary for network applications. They feature low insertion loss, high RL and channel isolation, excellent repeatability, and fast switching speeds. The switches are available in single-mode and multi-

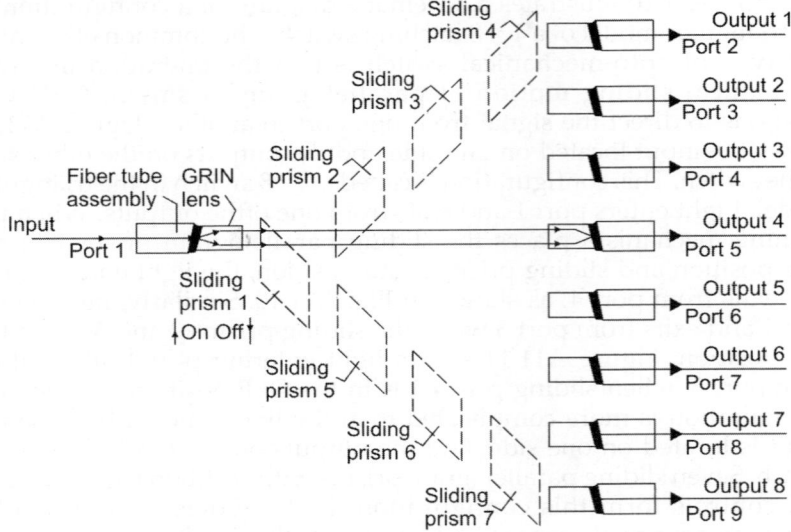

Fig. A11.11: Signal from port 1 to port 5 in a 1 × 8 latching switch

mode, and cover wide wavelength ranges. They are available in 1× 1, 1 × 2, and 2 × 2 configurations. The switching mechanism is latching and remains in its selected states following a loss of power. The switch consists of a quarter pitch GRIN lens glued to a double bore fiber tube assembly, relay, and mirror mounted on a shaft. The components are assembled inside a packaging base box and covered by a lid. Figure A11.12 illustrates a schematic diagram of a 2 × 2 switch. This figure illustrates the switch in the OFF/ON positions. When the power is off, light transmits from port 1 to port 2 and port 4 to port 3. This configuration is called the *transmission state*. When the power is ON, the mirror is in position, the light is reflected by the mirror, light exits port 1 and reflects to port 4 and similarly, port 2 reflects to port 3. This state is called the *reflection state*.

A11.3 ELECTRO-OPTIC SWITCHES

Switches with no moving parts can be built by using some of the passive devices, such as Mach–Zehnder interferometers (MZIs) and couplers. Some optical materials, such as lithium niobate crystal ($LiNiO_3$), Avalanche photo diode (APD) ($NH_4H_2PO_4$), and KDP (KH_2PO_4) exhibit an electro-optic effect. The index of refraction (RI) of the optical material changes in the presence of an electric field. These optical materials are used in building devices, such as the MZI, APD, and KDP. An electric field applied across the lithium niobate crystal causes a variation in the RI. This changes the transit time, creating a phase shift of the optical signal passing through the lithium niobate crystal.

Optical Amplifiers 593

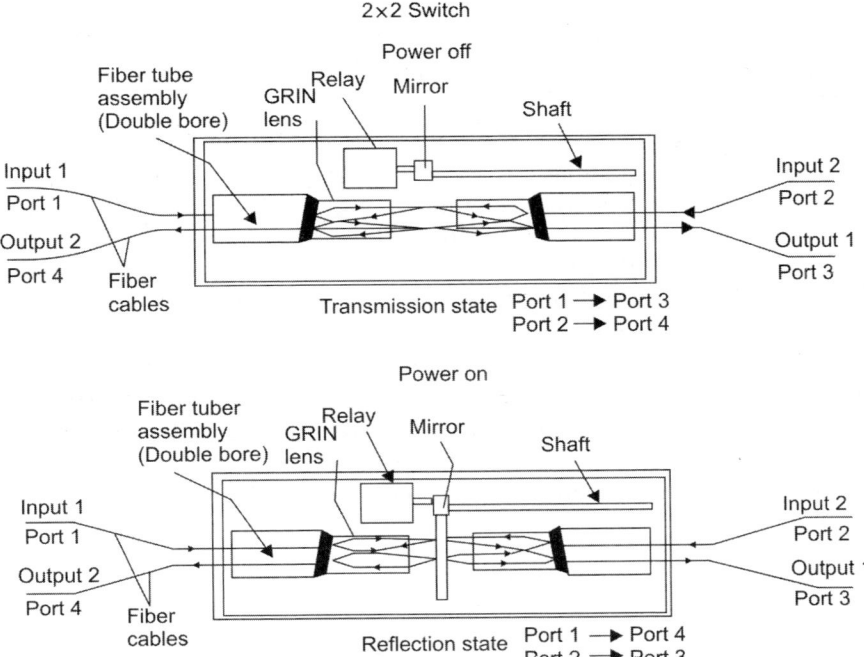

Fig. A11.12: A 2 × 2 opto-mechanical switch

Mach–Zehnder interferometers are used in building optical devices, which are used in a wide variety of applications in optic communication systems. The basic requirement of the Mach-Zehnder interferometer is to have a balanced configuration of a splitter and a combiner connected by a pair of optically matched waveguides, as shown in Fig. A11.13. The optical signal entering the *Mach-Zehnder interferometer* input port is split through a "Y" splitter section into two equal components. Each component goes to one of the two arms of the Y splitter. When there is no phase change in signal components after passing through both arms of the interferometer, the signal components is recombined at the "Y" coupler immediately before the optical signal exits the Mach–Zehnder

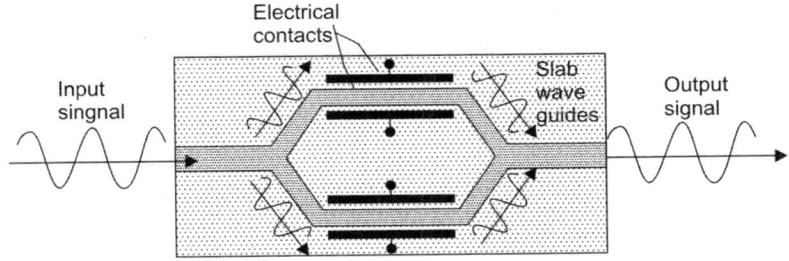

Fig. A11.13: Mach-Zehnder interferometer acts as a passive device

interferometer. The recombination of the two signal components takes place as constructive interference between two components and regenerates the original optical signal. In this case, the Mach-Zehnder Interferometer acts as a passive device.

When an electric field is applied to one arm of the Mach–Zehnder Interferometer, the RI changes and causes 180° shift in the phase of the signal component, due to the change in optical path length of this arm. As shown in Fig. A11.14, when there is a difference in phase at the destination "Y" coupler, the signal components will be out of phase with one another. The signal components recombination will be lost because the components will cancel each other in destructive interference. If the phase difference is a full 180°, then the output will be zero. In other words, applying the electric field to one of the arms of the Mach–Zehnder interferometer will make the phase shift one of the signal components. The Mach-Zehnder interferometer acts as an active device, when an external electric voltage is applied causing the switching.

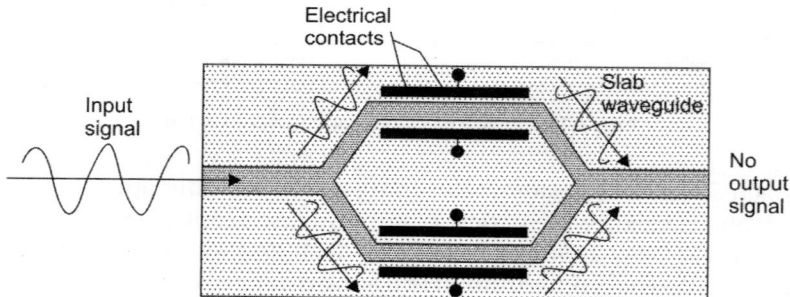

Fig. A11.14: An electro-optic switch using a Mach-Zehnder interferometer

Using the same principle as discussed above, one can built an *electro-optic switch* using two branching waveguides arranged like a 3 dB coupler to switch one input between two outputs. You can replace the one input with two parallel outputs coupled to the pair of switching waveguides by a combining coupler, as shown in Fig. A11.15. An electric field applied to one arm of the waveguide causes a 180° shift in the phase of the signal component. The electrical voltage is raised or lowered to shift the delay between waveguides by 180°. This directs the output from one waveguide on the right side to the other output. Because signal interference depends on phase shift, it is possible to further increase the voltage to switch the signal back to the other output. Table A1 presents the possible outcomes achieved by applying different voltages across the waveguide arms.

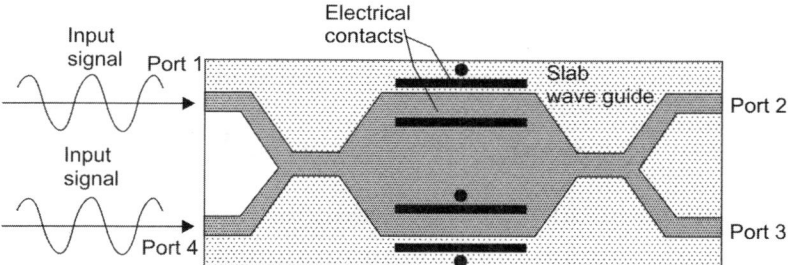

Fig. A11.15: A 2 × 2 electro-optic switch

Table A11.1. Input and output signals connections	
Voltage	Connections
V_1	Port 1 to Port 2 and Port 4 to Port 3
V_2	Port 1 to Port 3 and Port 4 to Port 2

A11.4 THERMO-OPTIC SWITCHES

A novel rib waveguide-integrated thermo-optic switch has appeared recently. The device is based on the TIR phenomenon and the thermo-optic effect (TOE) in hydrogenated amorphous silicon (a-Si:H) and crystalline silicon (c-Si). It takes advantage of a bandgap-engineered a-Si:H layer to explore the properties of an optical interface between materials showing similar refractive indexes but different thermo-optic coefficients. In particular, the modern plasma-enhanced chemical vapour deposition techniques, the refractive index of the amorphous film can be properly tailored to match that of c-Si at a given temperature. TIR may be achieved at the interface by acting on the temperature, because the two materials have different dermo-optic coefficients. The switch is integrated in a 4-pm-wide and 3-µm-thick single-mode rib waveguide, as shown in Fig. A11.16. The substrate is a silicon-on-insulator wafer with an oxide thickness of 500 nm. The active middle region has an optimal length of 282 µm. The device performance is analysed at a wavelength of 1.55 µm. As shown in Fig. A11.12, the optical waveguide-integrated switch consists of a 2 × 2 waveguide structure with an input Y branch and an output Y branch and an output Y branch. They are joined by a middle active region, although in this work, which guarantee both an effective optical confinement and low propagation losses. When properly designed, single mode operation can be achieved in the input and output of the Y branches.

As shown in the top-left insert of Fig. A11.16, the device structure is symmetric with respect to the (yz) plane. It consists of a core layer of c-Si in the upper half, and a core layer of a-Si:H in the lower half, both laying on a SiO_2 layer grown on a highly doped crystalline silicon

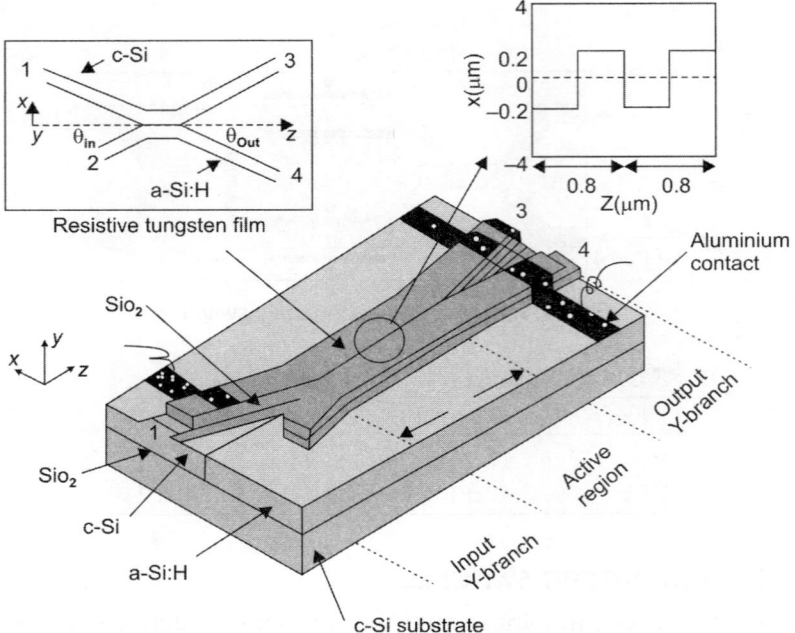

Fig. A11.16: Schematic of a waveguide-integrated thermo-optic switch

substrate. The thickness of the two guiding layers together is 3 µm. Due to the refractive index of SiO$_2$ (n_{SiO_2} = 1.48), a 500 mm thick under cladding layer ensures the optical confinement for both waveguides, as suggested by electromagnetic field propagation simulations. In the top-right inset of Fig. A11.16, a detail of the interface between the a-Si:H waveguide and the crystalline Si waveguide in the active region is also shown. The irregular profile at the TIR interface takes into account surface roughness that may result from the fabrication process.

We can exploit the TOE in a SI:H and c-Si by changing the refractive index of the core layers and thereby, switching the light beam at the output of the structure. A 300 nm thick tungsten heating film is introduced on top of the stacked structures. It is separated from the active region by a 100-nm-thick SiO$_2$ film. This reduces the optical absorption by the electrodes due to the evanescent field of the optical mode. Finally, the heating structure is completed by aluminium bonding pads.

The operating principle of the device is the TIR, which can be activated or dropped by exploiting the different thermo-optic coefficients in a-Si:H and c-Si. In particular, by choosing a proper gas phase composition during the deposition process of a-Si:H, the two materials develop the same refractive index at a given temperature.

By changing the device operating temperature, a refractive index discontinuity is created at the (yz) interface, producing the desired optical switching. At room temperature, an incident channel-guided light beam coming from port 1 will encounter a refractive index discontinuity between C-Si and a-Si:H, and the reflection (straight state) exists. Under these conditions TIR will occur and the incident light beam at port 1 will be reflected to port 3.

Another type of thermo-optical switch uses the Mach-Zehnder interferometer for the switching process. This type of thermo-optical switch is used in communication systems. Figure A11.17 illustrates the logic of an 8 × 8 thermo-optical switch. The 8 × 8 optical matrix switch employe Mach-Zehnder interferometer with a thermo-optic phase shifter as the switching mechanism. The small switch offers low loss, low crosswalk, low return loss, excellent stability, high reliability, and low power consumption. Applications of such switches include: space division switching systems (with analog and/or digital signals), wavelength routing (such as cross-connect and add-drop), protective switching, video switching, and inter-module connection.

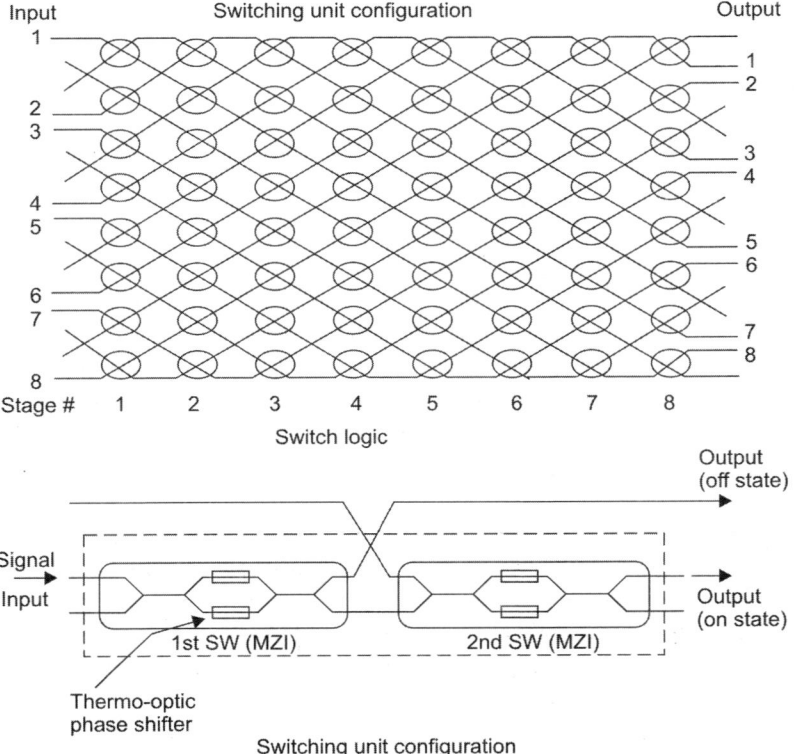

Fig. A11.17: An 8 × 8 thermo-optical switch using Mach-Zehnder interferometer

A11.4.1 Switch Logic

(i) Switching Unit Configuration

A high-speed all-optical switch using a fiber optic coupler and a light-sensitive variable-index material is illustrated in Fig. A11.18. *Refractive index variation with light is the principle of the switch operation.* Evanescent-wave coupling between two mono-mode fibres is extremely sensitive to the RI of the material surrounding the coupling region. Two ground and polished fibers, producing an evanescent field, can be brought in close proximity so that light in one fiber will couple into the other fiber in any desired ratio. Such polished couplers have been constructed to produce very low losses.

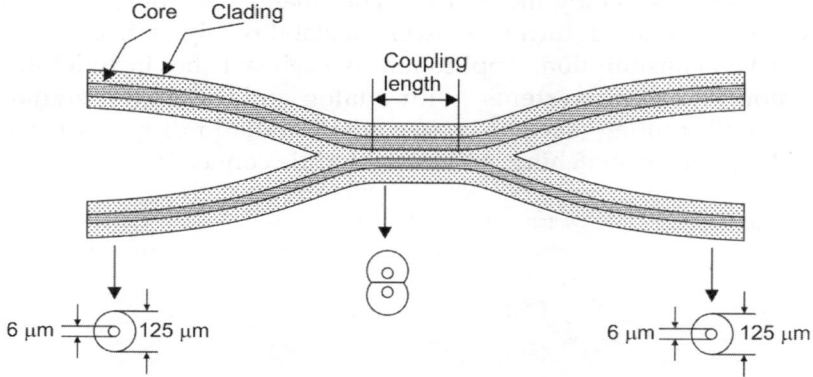

Fig. A11.18: A high-speed optical switch using a fiber optic coupler

A similar coupler may be made by timed etching of the fibers. Hydrofluoric acid may be used to remove as much cladding as desired; this exposes the core and produces an evanescent field. Within the etched region, fiber-to-fiber optical coupling will occur for fibers placed in close proximity. The coupling efficiency will depend on the RI of the surrounding medium, the core-to-core separation, the length of the interaction, and the amount of etching.

Another evanescent-field coupler is based on a non-etched, fused, and drawn coupler. This type of coupler is drawn to such an extent that the core is essentially lost and the cladding reaches a diameter near that of the undrawn core. The claddings become the core, and the evanescent field is forced outside the new core (waist region) into the air. Twisting two fibers together and fusing by using heat makes fused bi-conical tapered couplers.

The contribution of charge carriers (electrons) to the RI provides a simple way to modulate the index by the introduction or withdrawal of such carriers in the material. This can easily be done by the creation of electron-hole pairs if the material is also photoconductive (optical modulation).

One of the important parameters of the couplers is the coupling length. The coupling length is wavelength dependent. Thus the shifting of power between the two parallel waveguides will take place at different places along the coupler for different wavelengths. Fig. A11.19 shows two wavelengths entering at port 1 and port 2. When coupler length is made exactly to match the wavelength of the signal, the coupler works to combine wavelengths. Combined signals exit from port 2.

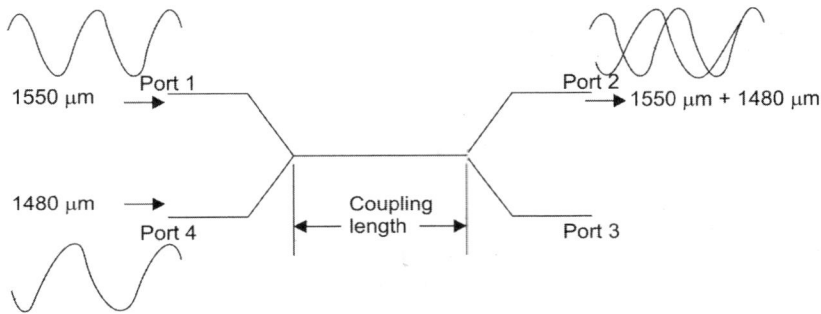

Fig. A11.19: Two wavelengths entering at port 1 and port 2

Figure A11.20 shows the reverse process where two different wavelengths arrive on the same input fiber at port 1. At a particular location along the coupler, the wavelengths will be in different waveguides. Then the wavelengths separate exactly and each wavelength exits from a different port. In this case, one wavelength exits from port 2 and the other from port 3. The processes described in Figs A11.19 and A11.20 are performed in the same coupler. This process is *bi-directional*. The coupler in Fig. A11.19 works as a *splitter*; the same coupler in Fig. A11.20 also works as a *combiner*.

There are other types of switches that use the same elements of either the micro-opto-mechanical switch or micro-electro-optical switch. They

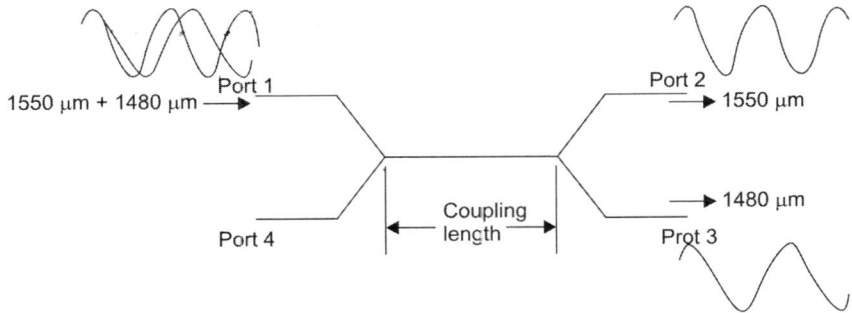

Fig. A11.20: Two wavelengths entering at port 1

employ couplers for switching between the inputs and outputs. These types of optical switches are used in communication systems. Figure A11.21 shows the configuration of a 4 × 4 optical space-division switch. The switch is designed to connect any input port to any output port as desired by the user. Any input may be switched to any output; however, two inputs may not go to the same output at the same time. The device is bi-directional such that once a connection has been established between an input port and an output port, that particular connection may be used in either one or both directions. The switches have no moving parts, are very stable, and reliable while exhibiting very low loss; thereby reducing the need for expensive amplifiers.

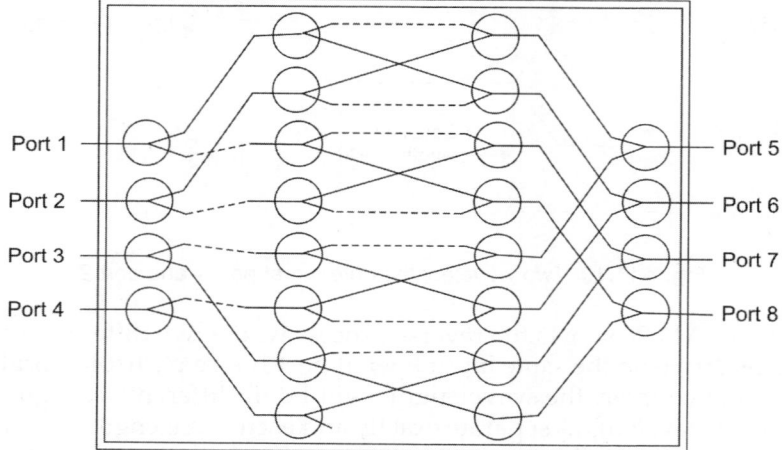

Fig. A11.21: A 4 × 4 optical space-division switch

Figure A11.22 shows the cross connect switch, which selects outputs by optical cross connecting. This results in a significant reduction of overall complexity and number of required elements. These switches are used in protection/backup switching, optical cross-connecting, network testing and monitoring, optical routing, and optical burst switching. A 4 × 4 switch configuration can be cascaded to built 16 × 16–256 × 256 switch configurations.

A11.5 ACOUSTO-OPTIC SWITCHES

Sound waves are generated when a material is in mechanical vibration mode. They can also be generated by acoustic transducers. Like any light wave, a sound wave is a moving wave, which has a frequency. Light waves travel at the speed of light; sound waves travel at the speed of sound, which is slower than light waves. Sound waves are used to control light transmission in acousto-optic switches and modulators. The refractive index of some optical material is altered by the presence

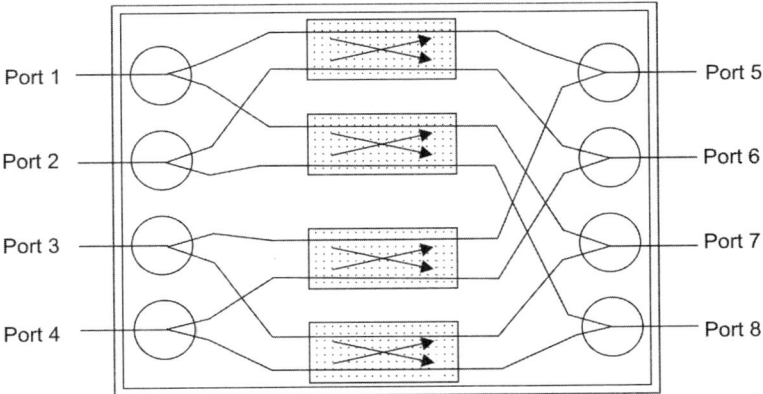

Fig. A11.22: A cross connect switch

of sound waves. The sound wave causes regular zones of compression and tension within the optical material. This creates a regular pattern of changes in index of refraction n of the optical material; this is called a *Bragg diffraction grating*. Within the optical material, there is interference between sound and light waves. The power of the deflected light is controlled by the intensity of the sound wave. The angle of deflection is controlled by the frequency of the sound wave. Figure A11.23 illustrates a design of an acoustic-optic switch. The figure shows that an incident light can be controlled by the frequency of the sound wave. The incident light exits the switch from one or more selected output ports depending on the sound wave intensity.

A11.6 MICRO ELECTROMECHANICAL SYSTEM (MEMS)

MEMS is a rapidly growing technology for the fabrication of miniature devices using processes similar to those used in the integrated circuit (IC) industry. MEMSs are widely used in optical switching in telecom networks. The appeal of MEMS goes beyond just switching applications in defense, aerospace, and medical industries. MEMS technology provides a way to integrate electrical, electronic, mechanical, optical, material, chemical, and fluids engineering on very small devices ranging in size from a few microns to one millimeter. MEMS devices have many important advantages over conventional opto-mechanical switches. First, like integrated circuits, they can be fabricated in large numbers, so that cost of production can be reduced substantially. Second, they can be directly incorporated into integrated circuits, so that far more complicated system can be made than with other technologies. Third, MEMS have small size, low cost, and high reliability and stability. Fourth, MEMS have the important capability of high-density digital transmission communication with different bandwidths.

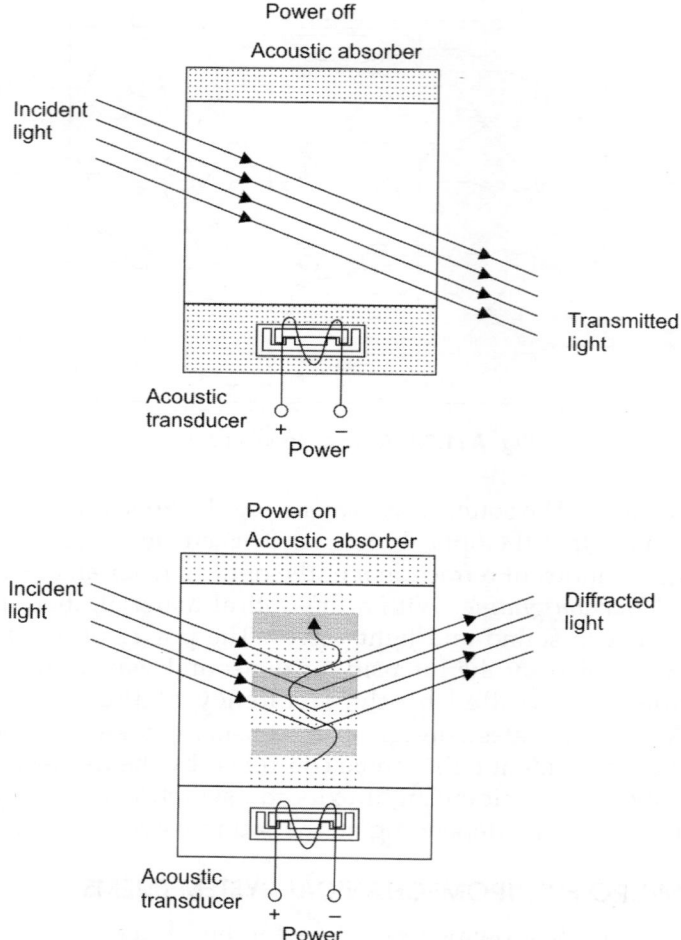

Fig. A11.23: Schematic design of an acoustic-optic switch

There are two categories of MEMS switches: MEMS 2D and 3D. They are typically fabricated onto a substrate that may also contain electronics needed to drive the MEMS switching element. MEMS-based optical switches route light from one fiber to another to enable equipment to switch traffic completely in the optical domain without requiring any optical-to-electrical conversion. At the core of MEMS 2D matrix switches is an array of micro mirrors capable of redirecting light either in free space or within a waveguide framework. The 2D switch architecture shown in Fig. A11.24 employs one mirror for every possible switched node in a matrix switch, and thus requires N^2 mirrors for an $N \times N$ array. 2D mirror arrays are characterized by two-state mirror positioning. One state is inactive and requires only that the mirror can be parked out of the optical path. During the switching state, the mirror

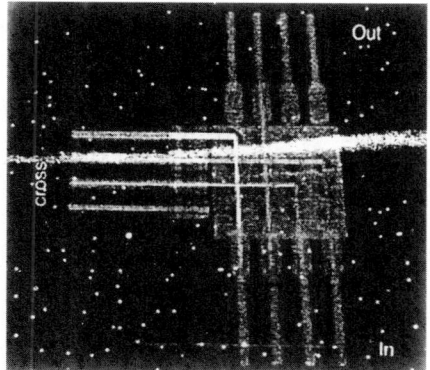

Fig. A11.24: MEMS 2D switch architecture

redirects the light path. Mirror positioning accuracy, repeatibility, and stability are critical in determining switch performance. Unlike 3D switch architectures that require servo positioning of individual mirrors, a 2D switch can rely on passive positioning control of the switched mirror, simplifying the control scheme. But a successful MEMS 2D approach must provide means for actuating the mirror into a highly predictable, stable state and hold it there indefinitely.

MEMS use an array of pop-up MEMS mirrors fabricated on the surface of a silicon wafer. The mirror is hinged to allow its rotation off the plane of the substrate to an angle of 90° where it redirects a light channel from the through to cross state, as shown in Fig. A11.25. An addressing scheme is required to select individual mirrors for actuation into the popped-up state and also for positioning them with sufficient accuracy for efficient coupling into the switched channel. The 2D MEMS array described here, called MagOXC, which stands for *"magnetically optical actuated cross-connect"*, uses a combination of magnetic and

Fig. A11.25: Pop-up mirror array passive mirror alignment

electrostatic actuation to rotate the mirrors and to select and deselect individual mirrors for clamping into the up or down state.

To rotate unclamped mirrors into the up state, magnetic actuation is implemented globally by applying an external field generated with a small electromagnet. The magnetic signal only needs to be applied momentarily. Using a global field avoids the need to fabricate indiviual magnetic actuators for each chip. Mirrors are fabricated with a layer of attached nickel to produce torque on the mirror hinge in response to the applied field. The nickel plate aligns with the magnetic field lines and generates a magnetostatic torque on the mirror. This lifts the mirror off the substrate and orients it near the desired vertical position, where electrostatic force can take over in setting and holding the final desired mirror position. An electrostatic field is applied mirror by mirror, either to hold the mirror down against the torque produced by the magnet or to hold the mirror in the up position against the restoring force of the elastic hinge. Since the magnetic field is applied globally, all mirrors will attempt to rotate when the field is turned on. Only the mirrors to be rotated into the up position are unclamped; all others are held down electrostatically. Similarly, mirrors clamped in the up state remain so until the electrostatic signal is removed; the magnet is no longer needed to hold the mirror up. The combination of magnetic and electrostatic actuation provides an effective means for configuring a mirror array without resorting to complex individual actuators for each mirror. Since electrostatic clamping of the mirrors requires virtually no current flow, the switch array consumes very little steady state power. Power is consumed only during transitions when the magnet is activated. The components of the switch are packed in a packaging base and lid.

A11.7 3D MEMS BASED OPTICAL SWITCHES

3D MEMS based optical switch route light from any of 80 input fibers to any of 80 output fibers. Designed for fiber-based test and measurement, 10 Gbit/s *Ethernet*, *high-definition video*, and *telecom* applications, this all-optical micro-photonic subsystem fits in the palm of a hand. The switch design is based on 3D MEMS mirror arrays, which is called *Reflexion*. It can switch signals within 10 ms, which is well within the telecommunications requirement for communications applications. 3D designs have switching elements accommodating hundreds, even thousands of ports, Fig. A11.26 shows a 3D MEMS optical switch. The design of the 3D MEMS is simple, solves mechanical and optical issues, is easy to fabricate, and achieves manufacturing tolerances that are accepted by the telecommunications industry.

A 3D MEMS design is shown in Fig. A11.27. A micro-morror array rests atop a single piece of silicon on a ceramic substrate. No bonding pads or other integrated electronics exist on the chip. All routing to the rest of the actuation electronics is done on the back end of the ceramic

Optical Amplifiers 605

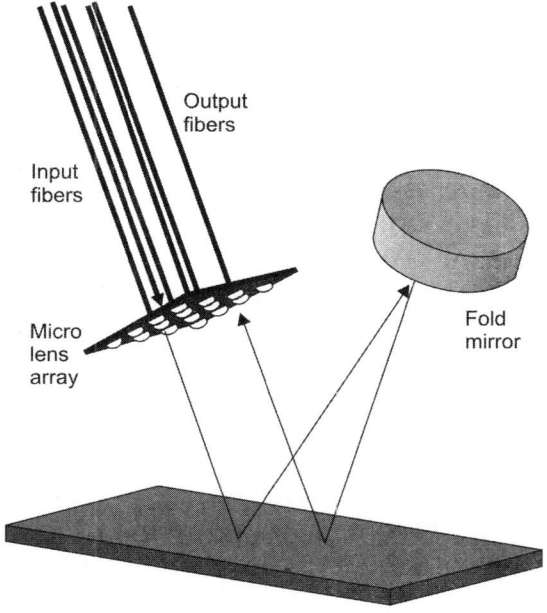

Fig. A11.26: A 3D MEMS optical switch architecture

substrate. Additional electronics is located on a photodetector card, along with constant-delay two-pole *Bessel filters*, and an *analog-to-digital converter* with a conversion-phasing time. Parallel-plate electrostatic actuation of the mirrors, with potentials of about 200 V, is provided by high voltage linear amplifiers.

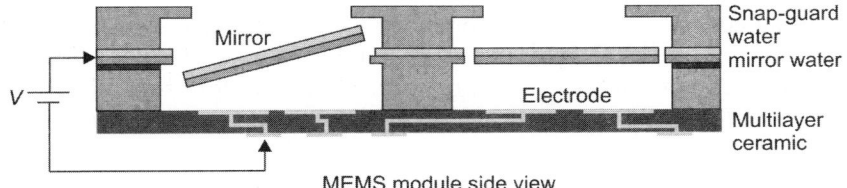

Fig. A11.27: A 3D MEMS optical switch design

Figure A11.28 shows a *torsional micro mirror*, which is driven by a vertical bi-directional comb driver (micro electrostatic actuator). Underneath the mirror plate is the substrate electrode. During operation, a voltage is applied to the electrode in order to generate an electrostatic attractive force on the mirror plate. The mirror plate rotates around the supporting axis in two dimensions. Such tilting mirror position can direct light from many distinct input ports to any of many distinct output ports.

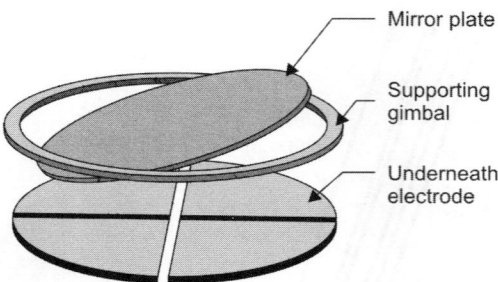

Fig. A11.28: A torsional micro mirror

A11.8 MICRO-OPTIC MECHANICAL SYSTEM (MOMS)

On-chip integrated MOMS were developed for a variety of applications for optical telecommunications. It is a new technology that allows for the integration of multiple passive and active components at the chip level. The technology is an extension of integrated optics and is based on suspended waveguides fabricated on chips, which are integrated with other optical components. It is used for a variety of optical solutions, including optical switching and cross-connect, signal dispersion correction, configurable optical add/drop multiplexers, and signal intensity equalizers. The technology bases its main technological concept on integrating optics at the chip level. The technology explores low-cost, high-performance, planar optical waveguide switches. Waveguides are used to channel light, rather than allowing it to propagate in free-space. The use of waveguides allow a degree of freedom in reducing size, while at the same time operating in a controlled physical environment. Switching time and losses are very low. By using waveguides, photons can be channeled in a controlled fashion, making it possible to lay out a photonic network within the chip with low losses. Since silicon is used as the propagating material, wavelength transparency for the 1300–1600 nm is utilized. The use of silicon for the waveguide material creates a tight confinement of light that allows the use of very small curvature radii, enabling a significant reduction of footprint chip size.

12
Wavelength Division Multiplexing (WDM) Technology

12.1 INTRODUCTION

Until the late 1980s, optical fiber communications was mainly confined to transmitting a single optical channel. Due to fiber attenuation, this channel required periodic regeneration which included *detection, electronic processing* and *optical retransmission*. Such regeneration causes a high-speed optoelectronic bottleneck, is bit-rate specific, and can only handle a single wavelength. The need for these single-channel regenerators (i.e. *repeaters*) was replaced when the *erbium-doper fiber amplifier* (EDFA) were developed, enabling high-speed repeaterless single channel transmission. One can think of this single ~Gbits/s channel as a single high speed "lane" in a highway in which the cars represent packets of optical data and highway represents the optical fiber. Single fiber has its low-loss window near 1.55 µm is approximately 25,000 GHz wide. The high-bandwidth characteristic of the optical fiber implies that a single optical carrier at 1.55 µm can be baseband modulated at ~25,000 Gbit/s, occupying this 25,000 GHz bandwidth. Obviously, this bit rate is impossible for present-day optical devices to achieve, given that heroic lasers, external-modulators, switches, or detectors have bandwidths < 100 GHz; we may note that practical data links today would be significantly slower, perhaps no more than 40 Gbit/s per channel. As such, a single high-speed channel takes advantage of an extremely small portion of the available fiber bandwidth. It seems natural to dramatically increase the system capacity by transmitting several different independent wavelengths simultaneously down a fiber in order to fully utilize this enormous fiber bandwidth. Therefore, the intent was to develop a *multiple-lane highway*, with each lane representing data travelling on a different wavelength. This *highway cartoon* scenario is illustrated in Fig. 12.1.

In addition to high-capacity transmissions, *WDM technology* also enables wavelength routing and switching of data paths in an optical network. By utilizing wavelength-selective component technologies,

608 Photonics | Optoelectronics

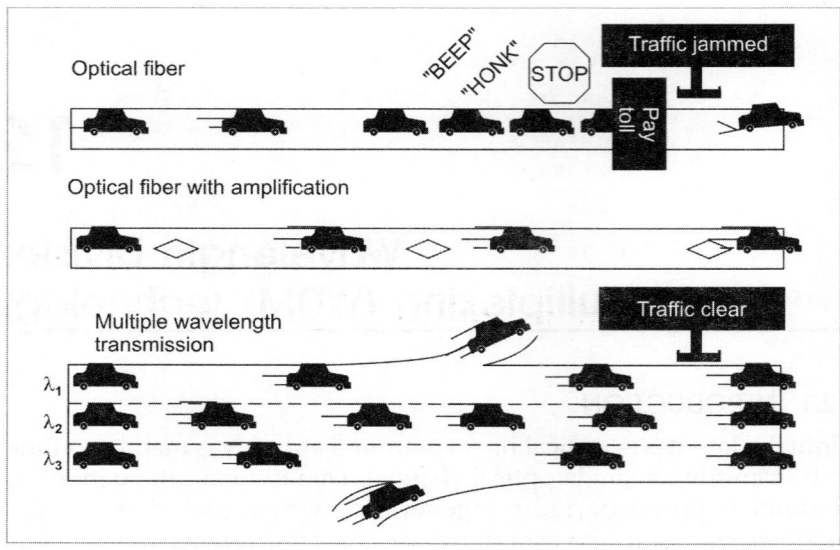

Fig. 12.1. Multiwavelength optical transmission as represented by a multiple-lane highway

each data channel's wavelength can be used to determine the routing through the network. Therefore, data can be thought of as travelling not on optical fiber but on wavelength-specific "light paths" from source to destination that can be arranged by a network controller to optimize throughput.

Basic Operation

The practicality of EDFA which can provide gain to many channels simultaneously over a ~THz wavelength range opened the door to multiplexing signals at many wavelengths onto the same optical fiber. This technique, known as WDM, wonderfully enhances an optical system's capacity. Along with the EDFA, WDM is the other technology that is the key to terabit-per-second optical systems. Conceptually, WDM is the same as the frequency-division multiplexing (FDM) used to place many radio channels on carrier waves of different frequencies. The carrier wave of each optical WDM channel, however, is a million times higher in frequency (terahertz verses megahertz).

In the most basic WDM arrangement as shown in Fig. 12.2, the desired number of lasers, each emitting a different wavelength, are multiplexed together by a wavelength multiplexer (or a combiner) into the same high-bandwidth fiber. After being transmitted through a high-bandwidth optical fiber, the combined optical signals must be multiplexed by a WDM (or a splitter) at the receiving end by distributing the total optical power to each output port and then requiring that each receiver

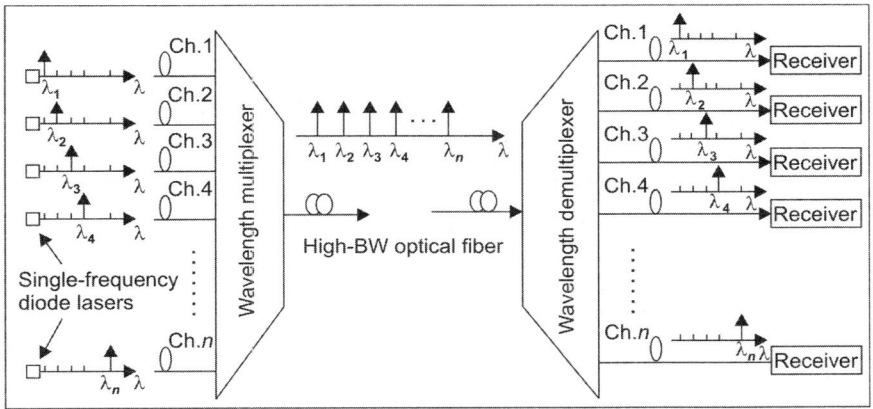

Fig. 12.2. Schematic diagram of a simple WDM system

selectively recover only one wavelength. At the receiver, a narrow-band optical filter is used to select just one of the incoming wavelength, so that only one signal is allowed to pass and establish a connection between source and destination.

It is important to space channel wavelengths an adequate distance apart. The goal is to minimize transmission of unwanted channels through the filter and to accommodate the drift in the wavelength characteristics of optoelectronic components over time—a change in temperature is just one possible cause. Typical channel spacings range from 0.4 nm to 4 nm (50–500 GHz).

Figure 12.3 illustrates the concept of *wavelength demultiplexing* using an optical fiber. In this example, four channels are input to an optical filter which has a nonideal transmission filtering function. The filter transmission peak is centered over the desired channel, in this case λ_3, thereby transmitting that channel and blocking all other channels. Due to the nonideal filter transmission function, some optical energy of the neighboring channels leaks through the filter causing inter-channel inter-wavelength crosstalk. This crosstalk has the effect of reducing the

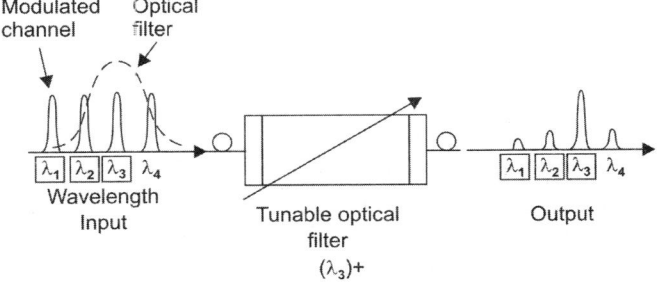

Fig. 12.3. Optical WDM being demultiplexed by an optical filter

selected signal's contrast ratio and can be minimized by increasing the spectral separation between channels. Although there is no set definition, a nonstandardized convention exists for defining optical WDM, dense-WDM, and FDM as encompassing a system for which the channel spacing is approximately 10, 1 and 0.1 nm, respectively. However, we will not make any distinction among these system labels.

Obviously, WDM is an optical technology that permits several wavelengths to be coupled into the same fiber cable, effectively increasing the aggregaiton bandwidth per fiber cable. Figure 12.4(a) illustrates the components of a basic communication system. The *de-multiplexer* (de-mux) decouples what the multiplexer has coupled. The de-mux separates several wavelength in a single fiber cable, and directs them individually onto many fiber cables, which are connected to the receiver channels.

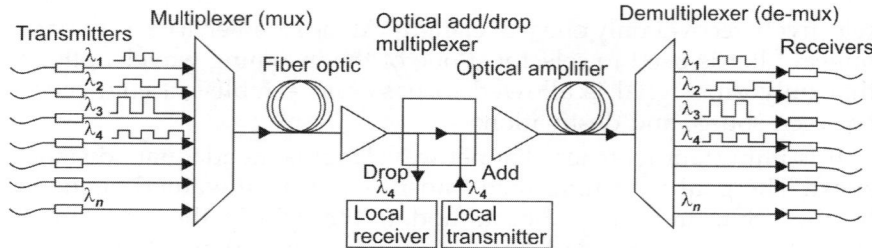

Fig. 12.4. (a) A schematic diagram of basic communication system

WDM systems are based on the ability of a fiber cable to carry many different wavelengths without mutual interference. Each wavelength represents an optical channel within the fiber cable. Several optical methods are available to combine individual channels within a fiber cable, and to extract them at appropriate points along a network. WDM technology has evolved to the point that channel wavelength separations can be as small as a few nanometers, giving rise to dense wavelength division multiplexing (DWDM) systems.

Key features:
- capacity upgrade
- transparency (each optical channel can carry any transmission format)
- wavelength routing
- wavelength switching

The output of each laser transmitter in a WDM system is set to one of the channel frequencies. These signals of various frequencies must be then multiplexed (superimposed or combined) and then inserted into a single fiber optic cable. These signals then travel through the cable from the multiplexer to an add-drop multiplexer. The add-drop multiplexer routes one wavelength, λ_4, to a point and picks up another signal at the same wavelength, also λ_4. Note that this is a different signal, as shown in Fig. 12.4a. A *demultiplexer* is used to extract the multiplexed channels at the receiver end. Multiplexing and demultiplexing devices employ narrowband filters. Multiplexer and demultiplexer devices can be cascaded and combined to achieve the desired results. Several devices exist to perform such filtering, including thin-film fiber *Bragg gratings*, *optic gratings, tapered fibers, liquid crystal filters,* and *integrated optical devices*. These multiplexing technologies are also used for other optical fiber applications. These applications include telephone and data communications, SONET/SDH networks, inter-exchange networks, and in links for trunk exchange and local exchange hubs.

Implementation of a typical WDM system employing N channels is shown in Fig. 12.4(b). In the shown system *three* wavelengths are multiplexed in one fiber to increase transmission capacity. The light of laser diodes with wavelengths recommended by the ITU is launched into the inputs of a wavelength multiplexer (MUX), where all wavelengths are combined and coupled into a single-mode fiber. When needed, propagating light can be amplified by an optical fiber amplifier and eventually imputed at the wavelength demultiplexer (DMUX), which separates all optical channels and sends them to different outputs.

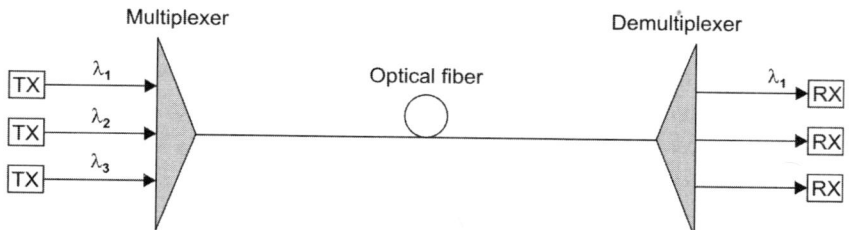

Fig. 12.4. (b) Implementation of a typical WDM link

In order to find the optical bandwidth corresponding to a spectral width in optical region, we use the relation $c = \lambda \cdot v$, where λ is wavelength and v carrier frequency and c velocity of light. Differentiating:

$$dv = c \frac{d}{d\lambda} \frac{1}{\lambda} d\lambda = -\frac{c}{\lambda^2} d\lambda \qquad (12.1)$$

or

$$|\Delta v| = \frac{c}{\lambda^2} |\Delta \lambda| \qquad (12.2)$$

Equation (12.2) describes the frequency change Δv which corresponds to the wavelength change $\Delta \lambda$ around λ. Using the above formula, we can estimate the usable wavelength range for a standard single-mode fiber. Assuming that telecommunication wavelength range extends from $\lambda_1 = 1280$ nm to $\lambda_2 = 1625$ nm, the ultimate bandwidth of optical fiber is 40 THz. Assuming 50 or 25 GHz channel spacing, there is the possibility to transmit 800–1600 wavelength channels.

12.2 TIME DIVISION MULTIPLEXING

Time division multiplexing (TDM) is a technique for transmitting digitized data, voice, and video signals simultaneously over one fiber cable. This is accomplished by inter-leaving pulses representing bits from different channels or time slots. The public-switched telephony network (PSTN) is based on the TDM technologies and is often called a *TDM access network*.

The time division multiplexer is a device that uses TDM techniques to combine several slower speed data streams into a single high speed data stream, as shown in Fig. 12.5. Data from multiple sources is broken into portions (bits or bit groups); these portions are transmitted in a defined sequence. The transmission order must be maintained so that the input streams can be reassembled at the destination. Typically, using the same TDM techniques, the same device can also perform the reverse process; de-compose the high-speed data streams into multiple low speed data streams, a process called demultiplexing. Therefore, a time division multiplexer and demultiplexer are very often packaged in the same box.

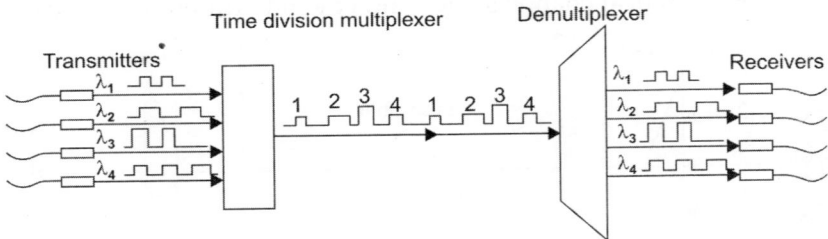

Fig. 12.5. A schematic diagram of time division multiplexing

12.3 FREQUENCY DIVISION MULTIPLEXING (FDM)

Frequency division multiplexing (FDM) is a scheme in which numerous analogue signals are combined for transmission on a single communications line or channel. Each signal is assigned a different frequency (sub-channel) within the main channel. This technology is used in broadcast radio, television, and cable division. Home local area network (LAN) use this technology to ensure compatibility between the different services sharing the same telephone wire, specifically voice, and the home network. To eliminate interference, each service has a frequency spectrum that is different from the others. Traditionally, frequency-division multiplexing is used for analogue signals, but it also can be used for digital signals.

When FDM is used in a communication network, each input signal is sent and received at maximum speed at all times. However, if many signals must be sent along a single long-distance line, the necessary bandwidth is large, and careful design is required to ensure that the system will perform properly. In some systems, time-division multiplexing is used instead.

12.4 DENSE WAVELENGTH DIVISION MULTIPLEXING (DWDM)

DWDM is an acronym for dense wavelength division multiplexing, an optical technology used to increase bandwidth over existing fiber optic backbones. DWDM is a fiber-optic transmission technique that employs light wavelengths to transmit data as parallel bits or a serial string of characters. Using DWDM, up to 80 (and theoretically more) separate wavelength or channels of data can be multiplexed into a single light stream, and then transmitted on a single fiber optic cable. Each channel carries a time division multiplexed (TDM) signal. In a system with each channel carrying data, billions of bits per second, can be delivered by the fiber optic cable. DWDM is also sometimes called wave division multiplexing (WDM). Since each channel is de-multiplexed at the end of the transmission back into the original source, different data formats can be transmitted together, at different rates. Specifically, internet data, synchronous optical network (SONET) data, and asynchronous transfer mode (ATM) data can all be transmitted at the same time within the same optical fiber. Utilizing DWDM technology is a suitable solution for high-speed data transmission, without the addition of more fiber cables.

12.5 COARSE WAVELENGTH DIVISION MULTIPLEXING (CWDM)

CWDM is an acronym for coarse wavelength division multiplexing, which is a technology that combines upto 16 wavelengths onto a single fiber. When there are just a few channels (upto 16 channels) and they are spaced more widely (10 nm or more) apart, the system is called CWDM. The coarse wavelength division multiplexer and de-multiplexer (CWDM) are designed to multiplex and demultiplex wavelength signals in metropolitan, access and enterprise networks, and for cable television applications. They are a low-cost approach for systems that use un-cooled laser sources, and are an alternative to more expensive DWDM components based on 100 or 200 GHz channel spacing. CWDMs are used to isolate a specific wavelength channel, whereas CWDM channel splitters are used to isolate a band channels.

12.6 PASSIVE COMPONENTS

For implementing WDM (or DWDM) various passive and active components are required to combine, distribute, isolate and amplify optical power at different wavelength. The components do not need any external control for their operation.

Passive components are mainly used to split or combine optical signals. These components operates in optical domains. Passive components don't need external control for their operation. Passive components are fabricated by using optical fibers or by planar optical waveguides. Commonly required passive components are: 1. N × N couplers, 2. *Power splitters*, 3. *Power taps*, 4. *Star couplers*.

Most passive components are derived from basic *star couplers*.

Star coupler can perform combining and splitting of optical power. Therefore, star coupler is a multiple input and multiple output port device.

12.6.1 Couplers

Couplers are the simplest passive fiber optic devices. Couplers direct multiple input light waves to multiple inputs. Normally, couplers split signals into two or more outputs, or combine two or more inputs into one output. Couplers can have more than two inputs or outputs. Couplers work as power splitters, power taps, and wavelength selectors.

(i) 2 × 2 Couplers

A simple 2 × 2 fiber coupler consists of two inputs and two output port as shown in Fig. 12.6.

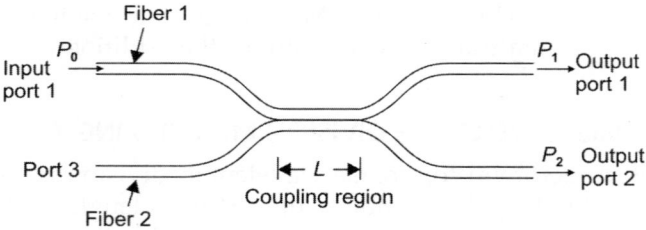

Fig. 12.6. A simple 2 × 2 fiber-optic coupler

The arrows indicate the power flow through the coupler. 2 × 2 coupler can be made by fusing two optical fibers together in the middle and then stretching them so that a coupling region is created. Such devices can be made wavelength independent over a wide spectral range.

When an optical signal launched at input port 1, it may split into two signals that can be collected at the output ports 1 and 2, based on a set splitting ratio. Ideally, no power will reach port 3, called the isolated port. By convention, the power P_1 emerging from output port 1 is equal to or greater than the power P_2 from output port 2, depending on the designed *splitting ratio* or *coupling ratio* of the coupler. The splitting ratio or coupling ratio is denoted by $P_1:P_2$ or else percentage of power e.g. 1:1 same as 50/50% splits power in half. By careful design, it is possible to achieve coupling ratio from 1:99 to 50:50. A device with a 50:50 coupling ratio is termed as a *3 dB coupler*, as 50% of the input power is coupled to each output port. One can use it as a *splitter*. We may note that a coupler with a coupling ratio of 1:99 can be used as an optical tap.

One can analyse the coupling mechanism using electromagnetic theory for dielectric waveguides. One can easily understand this mechanism in a simple manner also as follows. The V parameter of an optical fiber is given by:

$$V = \frac{2\pi a}{\lambda} n_1 \sqrt{2\Delta} \qquad (12.3)$$

where $2a$ is the core diameter, n_1 is the core index, λ is the wavelength of light propagating through the fiber, and Δ is the relative refractive index difference. In the process of manufacturing a coupler, the fibers are heated, fused together, and stretched. Stretching reduces the core diameter and so also the V-parameter. Thus, the optical power propagating through the core of a single-mode fiber (say) will be less confined. If two identical single-mode fibers are used to make a 2×2 coupler, the power in the single mode propagating through the core of the first fiber will couple to that in the core of the second adjacent fiber in the coupling region (fused portion). By controlling the distance between the fibers, it is possible to obtain a desired coupling ratio. Such couplers are called *directional couplers*, because the fibers allow the launched light to pass through them in one direction. It is also possible to make the coupling ratio wavelength selective. Such couplers are used to combine or separate two signals of different wavelengths.

Let us assume that the above-mentioned 2×2 coupler is *loss-less* and the *two-single-mode fibers* are identical, the power P_2 coupled from the first fiber into the second fiber over an axial length z is given by:

$$P_2 = P_0 \sin^2(\kappa z) \qquad (12.4)$$

where P_0 is the power launched at input port 1 (Fig. 12.6) and κ is the coupling coefficient describing the interaction between the propagating fields in the two fibers.

Assuming that the power is conserved, one can write the following expression for power P_1 delivered to output port 1:

$$P_1 = P_0 - P_2 = P_0 [1 - \sin^2(kz)] = P_0 \cos^2(\kappa z) \qquad (12.5)$$

From Eqs (12.4) and (12.5), one can easily infer that there is a periodic exchange of power between the two fibers. Thus, at $z = m\pi/\kappa$, where $m = 0, 1, 2, ...$, $P_1 = P_0$ and $P_2 = 0$, which means that the entire power is in the first fiber; and at $z = (m + 1/2)(\pi/\kappa)$, $m = 0, 1, 2, ...$, $P_1 = 0$ and $P_2 = P_0$; that is, the entire power is in the second fiber. The minimum interaction length over which the power is completely trans-ferred from the first fiber to the second fiber is given by:

$$z = L_c = \frac{\pi}{\kappa} \qquad (12.6)$$

This length L_c is called the *coupling length*.

The main parameters of couplers are optical power losses, i.e. the performance of a directional coupler may be specified in terms of coupling or splitting ratio, defined as follows:

(i) Coupling ratio (%) $= \left(\dfrac{P_2}{P_1 + P_2} \right) \times 100$ (12.7a)

(ii) Coupling ratio (dB) $= -10 \log \left(\dfrac{P_2}{P_1 + P_2} \right)$ (12.7b)

So far, we have assumed that the coupler is loss-less. However, in a practical device of this type, some power is always lost when the signal passes through it. There are two basic *parameters related to the loss*: (i) *excess loss*, defined as the ratio of the total output power to the input power, and (ii) *insertion loss* (for a specific port-to-port path), defined as the ratio of power at output port j to power at input port i. Thus, in decibels (dB):

$$\text{Excess loss (dB)} = -10 \log \left(\dfrac{P_1 + P_2}{P_0} \right) \quad (12.8)$$

(Power loss within the coupler)

and $\quad\quad$ Insertion loss (dB) $= -10 \log \left(\dfrac{P_j}{P_i} \right)$ (12.9)

For a 2 × 2 coupler, for a path from input port 1 to output port 2, using Eqs (12.7(b)) and (12.8), one can write:

Insertion loss $= -10 \log_{10} \left(\dfrac{P_2}{P_0} \right)$

$= -10 \log \left(\dfrac{P_2}{P_1 + P_2} \right) \times \left(\dfrac{P_1 + P_2}{P_0} \right)$

$= -10 \log \left(\dfrac{P_2}{P_1 + P_2} \right) - 10 \log_{10} \left(\dfrac{P_1 + P_2}{P_0} \right)$

$= \text{coupling ratio} + \text{excess loss}$ (12.10)

Table 12.1 lists splitting ratio values, throughput loss, and tap loss for several common couplers.

Table 12.1. Characteristics of several common couplers

Coupler description (dB)	Splitting ratio	L_{THP} (dB)	L_{TAP} (dB)
3	1:1	3	3
6	3:1	1.25	6
10	9:1	0.46	10
12	15:1	0.28	12

Here, L_{THP} (dB) is *throughput loss* specifies the transmission loss between the input power P_0 at input port 1 and transmission power P_1 at output port 1, and L_{TAP} (dB) is *tap loss* specifies the transmission loss between the input power P_0 at input port 1 and the tap power P_2 at output port 2.

(ii) 3 dB Couplers

A simple four-port coupler is often called a 3 dB coupler, if the input light splits into two equal portions at the output ports. The signal is split in half (3 dB = half). The 3 dB comes from the power loss formula: $[-10 \times \log_{10}(P_2/P_1) = -10 \log_{10}(0.5/1.0) = -10 \times (-0.3) = 3$ dB]. Half of the light entering at Port 1 will exit at Port 2 and half at Port 3, as shown in Fig. 12.7.

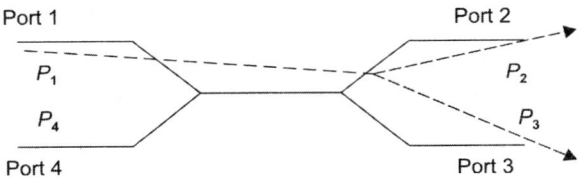

Fig. 12.7. 3 dB coupler

It is often useful to cascade many 3 dB couplers, as shown in Fig. 12.8. The configuration shown is called a *splitter*. This splitter divides a single input into four equal outputs and is denoted as a 1 × 4 coupler. As might be expected, if the device is perfect, each output port will contain one fourth of the input power.

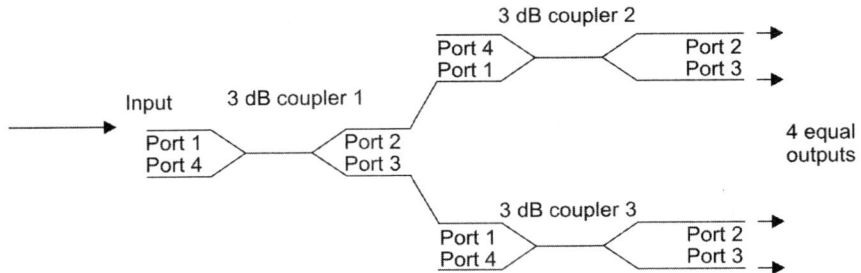

Fig. 12.8. Cascaded 3 dB couplers to produce a 1 × 4 coupler

Figure 12.9 shows how a 1 × 8 coupler can be constructed by cascading several 3 dB couplers in a tree configuration. The signal input power will divide into eight equal outputs. Each output port will contain one eighth of the input power.

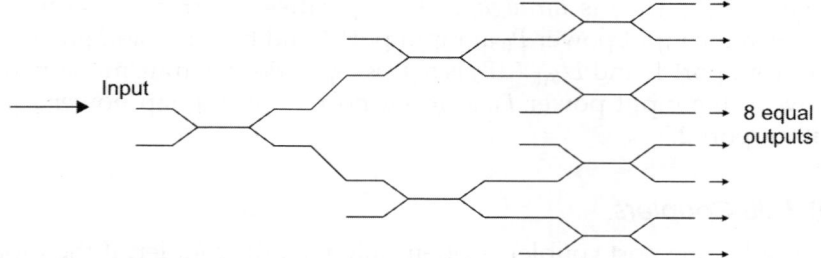

Fig. 12.9. Cascaded 3 dB couplers to produce a 1 × 8 coupler

(iii) Y-Couplers

Y-couplers or splitters, sometimes called 3 dB couplers, split the light equally. Y-couplers are 3 dB couplers in which Port 4 is not used. In the Y-coupler (splitter), as shown in Fig. 12.10, the light entering Port 1 will be split equally between Ports 2 and 3 with almost no loss. They are extremely efficient at splitting light with little loss. Y-couplers are difficult to construct in fiber optics, but they are easy to construct in planar waveguide systems. The power loss in the Y-coupler system can be calculated by:

$$\text{Loss (dB)} = -10 \log_{10}\left(\frac{P_{out}}{P_{in}}\right) \tag{12.11}$$

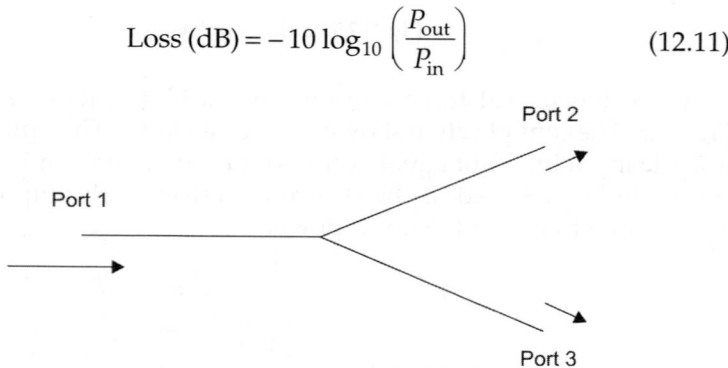

Fig. 12.10. Y-coupler (splitter)

Y-couplers are very seldom built as separate planar devices; instead they are manufactured on the same substrate as other devices. Connecting Y-couplers to a fiber optic cable is expensive; and significant loss is experienced in the connections. However, Y-couplers of this kind are used extensively in *complex planar devices*.

If is often useful to cascade Y-couplers, as shown in Fig. 12.11. The splitter configuration shown divides a single input into four equal outputs. If the device is perfect, each output port will contain one fourth of the input power. Fig. 12.11 shows how a 1 × 4 Y-coupler can be constructed by cascading three 1 × 2 Y-couplers in a tree arrangement.

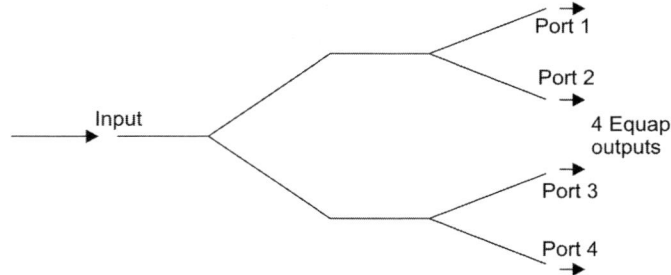

Fig. 12.11. Cascaded 1 × 2 Y-couplers to produce a 1 × 4 coupler

Figure 12.12 shows how a 1 × 8 Y-coupler can be constructed by cascading seven Y-couplers in a tree configuration. The power of the input signal will divide into eight equal outputs. Each output port will contain one eighth of the input power.

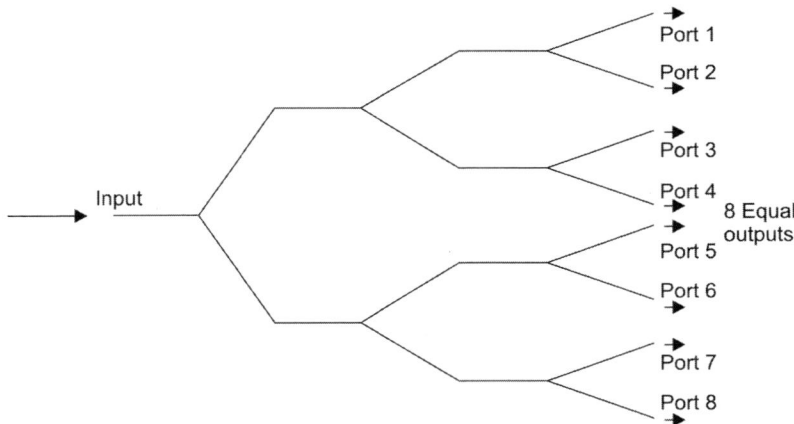

Fig. 12.12. Cascaded Y-couplers to produce a 1 × 8 Y-coupler

(iv) Star Couplers

A star coupler is simply a multiple output coupler in which each input signal is made available on every output fiber. There are two star coupler designs, as shown in Fig. 12.13. Fig. 12.13(a) shows an 8 × 8 coupler. *This coupler distributes the power from any input port to all the output ports, splitting equally among the output ports.* This type of coupler is called a *transmission star coupler*. Fig. 12.13(b) shows a reflection star coupler; any input is split equally and is reflected back among all fibers. Star couplers are typically used in *local area networks* (LAN) and *metropolitan area networks* (MAN).

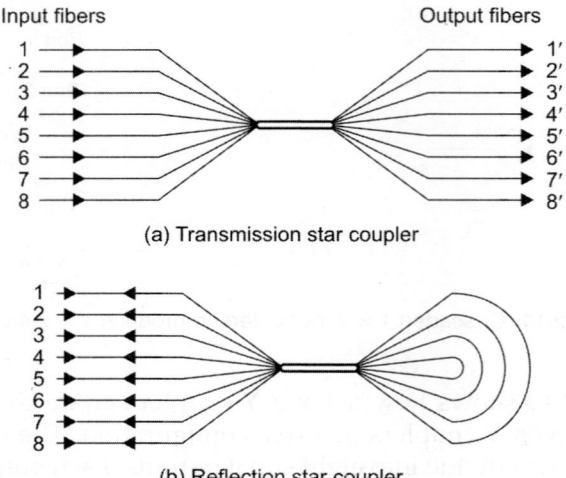

(a) Transmission star coupler

(b) Reflection star coupler

Fig. 12.13. Star couplers

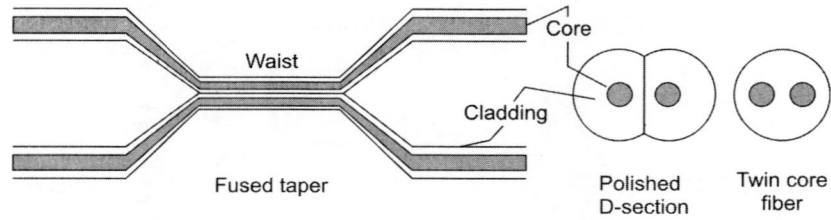

Fig. 12.14. Some coupler configurations

Salient features of a star coupler

(i) A star coupler is mainly used for combining optical powers from N-inputs and divide them equally at M-output ports.

(ii) For producing $N \times N$ star couplers, the fiber fusion technique is popularly used. Fig. 12.15 shows a 4×4 fused star coupler.

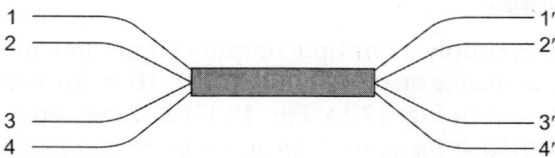

Fig. 12.15. 4×4 fused star coupler

(iii) The optical power put into any port on one side of coupler is equally divided among the output ports. Ports on same side of coupler are isolated from each other.

(iv) Total loss in star coupler is constituted by splitting loss and excess loss.

$$\text{Splitting loss} = 10 \log\left(\frac{1}{N}\right) = 10 \log N$$

$$\text{Excess loss} = 10 \log\left(\frac{P_{in}}{\sum_{i=1}^{N} P_{out,i}}\right)$$

(v) An 8 × 8 star coupler can be formed by interconnecting 2 × 2 couplers. It requires twelve 2 × 2 couplers (Fig. 12.16).

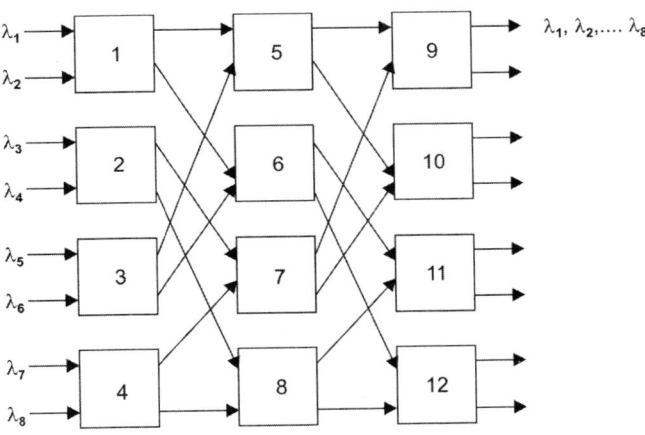

Fig. 12.16. 8 × 8 coupler

(vi) Excess loss in dB is given as:

$$\text{Excess loss} = 10 \log\left(F_T^{\log_2 N}\right)$$

where, F_T is fraction of power traversing each coupler element.
Splitting loss = $10 \log N$
Total loss = Splitting loss + Excess loss
= $10(1 - 3.32 \log F_T) \log N$

(v) Theory of Passive Couplers

Consider two waveguides 'a' and 'b' (Fig. 12.17). For a single waveguide, say 'a', one can express the field as:

$$E(x, y, z) = E^{(a)}(x^{(a)}, y) a(z)$$
$$H(x, y, z) = H^{(a)}(x^{(a)}, y) a(z)$$

and where

$$a(z) = a_0 e^{i\beta_a z}$$

where $E^{(a)}(z, y)$ and $H^{(a)}(x, y)$ are modal distributions in (x, y) plane. They are normalized as:

$$\frac{1}{2}\text{Re}\iint E^{(a)*}(x, y) \times H^{(a)*}\, dxdy \cdot \hat{z} = 1$$

Also

$$\frac{da(z)}{dz} = i\beta_a a(z)$$

Total guided power

$$P = \frac{1}{2}\text{Re}\iint E^{(a)*}(x, y, z) \times H^{(a)*}(x, y, z)\, \hat{z}dxdy$$
$$= |a(z)|^2$$

For two parallel waveguides, fields in each one are written as:

$$E(x, y, z) = a(z)E^{(a)}(x, y) + b(z)E^{(b)}(x, y) \quad (12.12a)$$
$$H(x, y, z) = a(z)H^{(a)}(x, y) + b(z)H^{(b)}(x, y) \quad (12.12b)$$

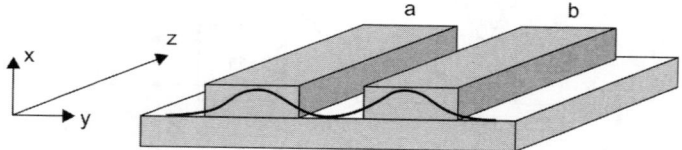

Fig. 12.17. Two coupled optical waveguides (distribution of electric fields is also shown in the figure)

Amplitudes $a(z)$ and $b(z)$ satisfy the following (coupled-mode) equations:

$$\frac{da(z)}{dz} = i\beta_a a(z) + i\kappa_{ab} b(z) \quad (12.13)$$

$$\frac{da(z)}{dz} = i\kappa_{ba} a(z) + i\beta_b b(z) \quad (12.14)$$

where κ_{ab} and κ_{ba} are coupling coefficients. Guided power is:

$$P = s_a|a(z)|^2 + s_b|b(z)|^2 + \text{Re}\{a(z)b^*(z)C_{ba} + b(z)a^*(z)C_{ab}\} \quad (12.15)$$

where
$$C_{pq} = \frac{1}{2}\int\int_{-\infty}^{+\infty} E^{(q)}(x, y) \times H^{(p)*}(x, y) \cdot \hat{z}dxdy \quad (12.16)$$

In the above, $s_a, s_b = +1$ is for propagation in the +z direction and $s_a, s_b = -1$ is for propagation in the $-z$ direction.

Coupled mode equations can be written in a matrix form as:

$$\frac{d}{dz}\begin{bmatrix} a(z) \\ b(z) \end{bmatrix} = i\overline{M}\begin{bmatrix} a(z) \\ b(z) \end{bmatrix} \quad (12.17)$$

where
$$\vec{M} = \begin{bmatrix} \beta_a & \kappa_{ab} \\ \kappa_{ba} & \beta_b \end{bmatrix} \quad (12.18)$$

The solution is assumed to be:
$$\begin{bmatrix} a(z) \\ b(z) \end{bmatrix} = \begin{bmatrix} A \\ B \end{bmatrix} e^{i\beta z} \quad (12.19)$$

After substituting into Eq. (12.17), one finds:
$$[\vec{M} - \beta \vec{1}] \begin{bmatrix} A \\ B \end{bmatrix} = 0 \quad (12.20)$$

where $\vec{1}$ is an identity matrix. In the full form:
$$\begin{bmatrix} \beta_a - \beta & \kappa_{ab} \\ \kappa_{ba} & \beta_b - \beta \end{bmatrix} \begin{bmatrix} A \\ B \end{bmatrix} = 0 \quad (12.21)$$

For non-trivial solutions, a determinant must vanish:
$$\det = (\beta_a - \beta)(\beta_b - \beta) - \kappa_{ab} \cdot \kappa_{ba} = 0 \quad (12.22)$$

From the above, two eigenvalues are found as:
$$\beta = \frac{1}{2}(\beta_a + \beta_b) \pm \gamma \equiv \begin{cases} \beta_+ \\ \beta_- \end{cases} \quad (12.23)$$

where
$$\gamma = \sqrt{\Delta^2 + \kappa_{ab} \cdot \kappa_{ba}},\ \Delta = \frac{1}{2}(\beta_b - \beta_a) \quad (12.24)$$

Eigenvectors are:
$$V_1 = \begin{bmatrix} \kappa_{ab} \\ \Delta - \gamma \end{bmatrix} \text{ or } \begin{bmatrix} \Delta - \gamma \\ \kappa_{ba} \end{bmatrix} \text{ for } \beta_+ \quad (12.25)$$

and
$$V_2 = \begin{bmatrix} \kappa_{ab} \\ \Delta - \gamma \end{bmatrix} \text{ or } \begin{bmatrix} \Delta - \gamma \\ \kappa_{ba} \end{bmatrix} \text{ for } \beta_- \quad (12.26)$$

The general solution is therefore:
$$\begin{bmatrix} a(z) \\ b(z) \end{bmatrix} = \vec{V} \begin{bmatrix} e^{i\beta_+ z} & 0 \\ 0 & e^{i\beta_- z} \end{bmatrix} \vec{V}^{-1} \begin{bmatrix} a(0) \\ b(0) \end{bmatrix} \quad (12.27)$$

where matrix $\vec{V}$ is formed from eigenvectors as:
$$\vec{V} = [V_1; V_2] \quad (12.28)$$

After some algebra, one finds the final solution as:
$$\begin{bmatrix} a(z) \\ b(z) \end{bmatrix} = \vec{S}(z) \begin{bmatrix} a(0) \\ b(0) \end{bmatrix} \quad (12.29)$$

with $\vec{S}(z) = \begin{bmatrix} \cos\gamma z - i\dfrac{\Delta}{\gamma}\sin\gamma z & i\dfrac{\kappa_{ba}}{\gamma}\sin\gamma z \\ i\dfrac{\kappa_{ba}}{\gamma}\sin\gamma z & \cos\gamma z = i\dfrac{\Delta}{\gamma}\sin\gamma z \end{bmatrix} \cdot e^{\dfrac{i}{2}(\beta_a + \beta_b)z}$ (12.30)

As a special case, consider a situation when at $z = 0$, the optical power is incident only in waveguide 1, $a(0) = 1$, $b(0) = 0$. One finds in this case:

$$|b(z)|^2 = \left|\dfrac{\kappa_{ba}}{\gamma}\right|^2 \sin^2 \gamma z$$

At $\gamma z = \dfrac{\pi}{2}, 3\dfrac{\pi}{2}, \ldots, (2n+1)\dfrac{\pi}{2}$, the power transfer from guide 'a' to guide 'b' is maximum. Since

$$\left|\dfrac{\kappa_{ba}}{\gamma}\right|^2 = \dfrac{|\kappa_{ba}|^2}{\left[\dfrac{1}{2}(\beta_b - \beta_a)\right]^2 + |\kappa_{ba}|^2} < 1$$

for $\beta_a \neq \beta_b$ the power transfer between waveguides is never complete.

12.7 MULTIPLEXERS AND DEMULTIPLEXERS

To implement a WDM-based system, a multiplexer is required at the transmitting end to combine optical signals from the several sources into a single fiber, and a demultiplexer is needed at the receiving end to separate the signals into appropriate channels. We may note that optical sources, e.g. an LED or ILD, do not emit significant optical power outside their designated channel width, interchannel cross talk is relatively unimportant at the transmitting end. As regards the design problem, the multiplexing should have low insertion loss. On the other hand, there exists a different requirement for demultiplexers, because photodetectors are generally sensitive over a broad range of wavelengths, which may include all the WDM channels. Obviously, a demultiplexer design must be such that it provides a good channel, isolation of the different wavelengths being used.

On the basis of the above study, we see that these devices are based on the reversible structure. Clearly, any wavelength-division demultiplexer (at least in principle) can also be used as multiplexer by simple exchanging the input and output directions. Therefore, we will restrict to the discussion of wavelength division demultiplexers.

One can classify the commonly used wavelength-division multiplexers (and multiplexers) into two categories: (i) *interference filter based devices*, and (ii) *angular dispersion based devices*.

Figure 12.18 shows the basic configuration of a two-wavelength (or two channel) interference filter demultiplexer.

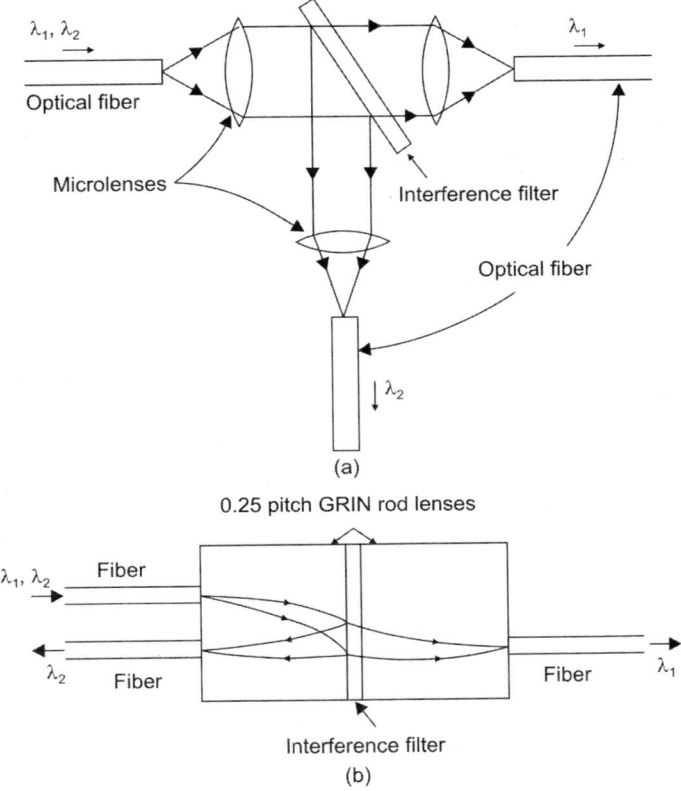

Fig. 12.18. The basic configuration of a two-wavelength (or two-channel) interference filter demultiplexer employing (a) conventional microlenses, and (b) GRIN rod lenses

An interference filter consists of a thin film obtained by depositing several dielectric layers of alternately low and high refractive index. When light propagates through such a structure, it undergoes multiple reflections, giving rise to either constructive or destructive interference depending on the wavelength. Therefore, a filter can be designed to produce *high transmittance in a given wavelength range and high reflectance outside this range*. In Fig. 12.18(a), appropriate conventional microlenses are used for collimating and focusing light. The incident beam consists of two wavelength, λ_1 and λ_2. The filter transmits the wavelength λ_1 and reflects the wavelength λ_2, thus demultiplexing the two channels. A compact low-loss two-channel demultiplexer (or multiplexer) may be implemented by employing two 0.25 pitch graded refractive index (GRIN) rod lenses as shown in Fig. 12.18(b). These lenses are used for collimating and focusing. The filter is deposited at the interface between these lenses (say, on either of the lens faces). Off-axis entry at the first lens makes it possible to easily separate the reflected beam.

Interference filters can (in principle) be used in series to separate N wavelength channels. However, the complexity involved in cascading the filters and the increase in signal loss that occurs with the addition of each filter generally limit the operation to four or five filters (i.e., four or five channels).

The second type of demultiplexers (or multiplexers) are mainly based on angular dispersion. Herein, the input beam (containing several wavelengths) is collimated onto a dispersive element which may be a prism or a grating. The latter angularly separates different wavelengths, the separation depending on the angular dispersion ($d\theta/d\lambda$) of the dispersion element. The separated output beams at different wavelengths are then focused using appropriate optics and collected by separate optical fibers. This type of demultiplexer is more suited to narrow line width sources, such as ILDs.

Three configurations of angular dispersion type demultiplexers are shown in Fig. 12.38 (Example 7). The first one (Fig. 12.38(a)) is a prism-type device. A Littrow prism (a half-prism, with its near surface serving as a reflector) has been used here for compact configuration. A multiwave-length signal from the input fiber is collimated onto the prism and the dispersed wavelengths are focused onto the output fibers by the same lens. Angular separation depends on the refractive index n of the material of the prism, which in turn depends on the wavelength. Suitable materials, giving a high value of $d\theta/dn$, for practical applications in the range 1.3–1.5 μm are not available. Therefore, blazed reflection gratings are normally employed for WDM applications. With a plane grating, the *Littrow mount* is often preferred because it allows the use of only one lens for collimating as well as focusing purposes, thus, reducing the device size. The basic structures of a Littrow grating demultiplexer employing a conventional lens and a 0.25 pitch GRIN rod lens are shown in Figs 12.38(b) and (c), respectively. Wavelength division multiplexing can also be achieved using the *Mach-Zehnder* interferometer.

Now, we briefly present some of the techniques that are used in multiplexing and de-multiplexing of many wavelengths:

12.7.1 Multiplexing and Demultiplexing Using a Prism

A simple way of multiplexing or demultiplexing wavelengths can be done using a prism. Figure 12.19 shows the de-multiplexing of multiple wavelengths exiting from the fiber cable. The first lens makes the diverging light beam become parallel and incident on the prism surface. Each component of the light is refracted differently when it exits from the prism. This spreading of the wavelengths produces the rainbow effect. Each wavelength is refracted by a different angle from the next wavelength. This angle is called the angle of refraction. The angle of refraction depends on the wavelength, the apex angle, and the refraction

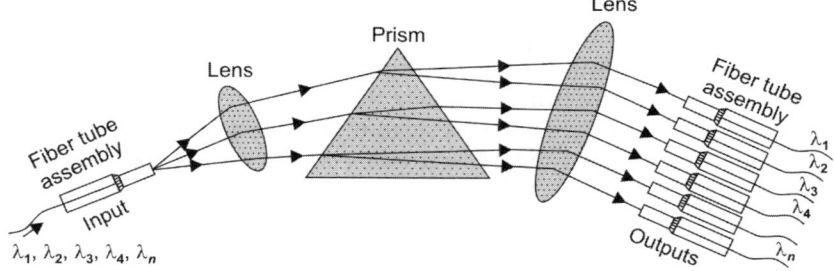

Fig. 12.19. Multiplexing and demultiplexing of wavelengths using a prism

index of the prism. The second lens focuses each wavelength to the designated output receiver via a fiber optic assembly. Therefore, this device is *bi-directional*.

12.7.2 Multiplexing and Demultiplexing Using a Diffraction Grating

Another technology based on the principles of diffraction uses a diffraction grating. When a light source is incident on a diffraction grating, each wavelength is diffracted at a different angle, and therefore, to a different point in space. It is necessary to use a lens to focus the wavelengths onto individual fibers, as shown in Fig. 12.20. Separate wavelengths can be combined onto the same output port, or a single mixed input may be split into multiple outputs, one per wavelength. This device is bi-directional.

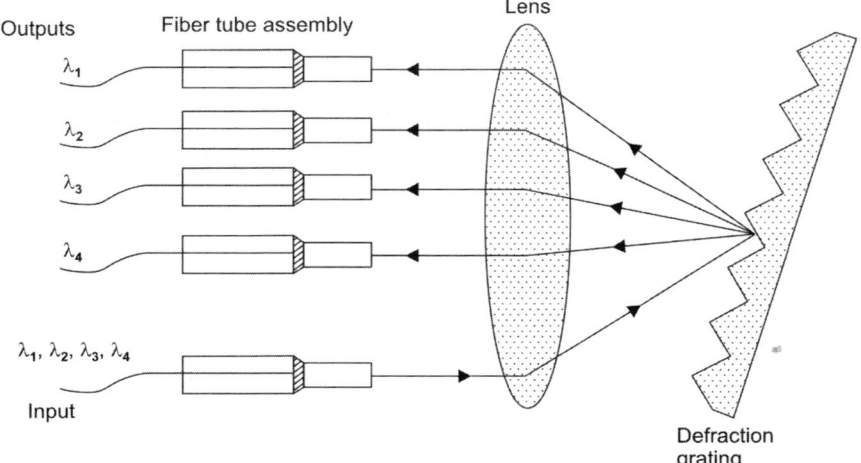

Fig. 12.20. Multiplexing and demultiplexing of wavelengths using a diffraction grating

12.7.3 Optical Add/Drop Multiplexers/Demultiplexers

Figure 12.21 illustrates a schematic representation of a design for an optical add/drop multiplexer/demultiplexer (OADM), which is widely used in communication systems. Between multiplexing and demultiplexing points in a DWDM system, there is a span where multiple wavelengths exist. An OADM removes or inserts one or more wavelengths, the OADM can remove some, while passing others on. OADMs are a key part of moving toward the goal of all-optical networks. The design shown in Fig. 12.21 includes both *pre-* and *post-amplification components* that may or may not be present in an OADM design.

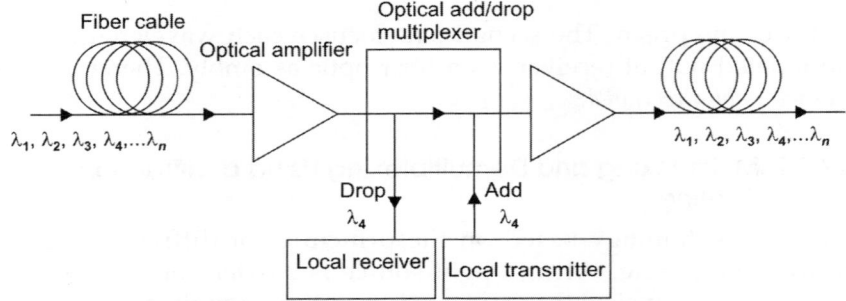

Fig. 12.21. An optical add/drop multiplexer

There are two general types of OADMs. The first generation is a fixed device, physically configured to drop specific present wavelengths while adding others. The second generation is reconfigurable and capable of dynamically selecting which wavelengths are added and dropped. *Thin-film filters* are used for OADMs in metropolitan DWDM systems, because of their low loss, low cost, and high stability. The new third generation of OADMs involves other technologies, such as tunable fiber gratings, fiber Bragg gratings, and circulators.

12.7.4 Arrayed Waveguide Gratings (AWGs)

Arrayed waveguide gratings (AWGs) are also based on the principles of diffraction. An AWG device is sometimes called an *optical wave-guide*, a *waveguide grating router*, a *phase array*, or a *phasar*. An AWG device consists of an array of curved-channel waveguides (W_1, W_2, W_3, ... W_n) with a fixed difference in the length of optical path between the adjacent channels, as shown in Fig. 12.22. The waveguides are connected to cavities at the input (S_1) and output (S_2). When light enters the input cavity, it is diffracted and enters the waveguide array. There the optical path length difference of each waveguide creates pulse delays in the output cavity, where an array of fibers is coupled. The process results in different wavelengths having constructive interference at different locations, where the output ports are aligned.

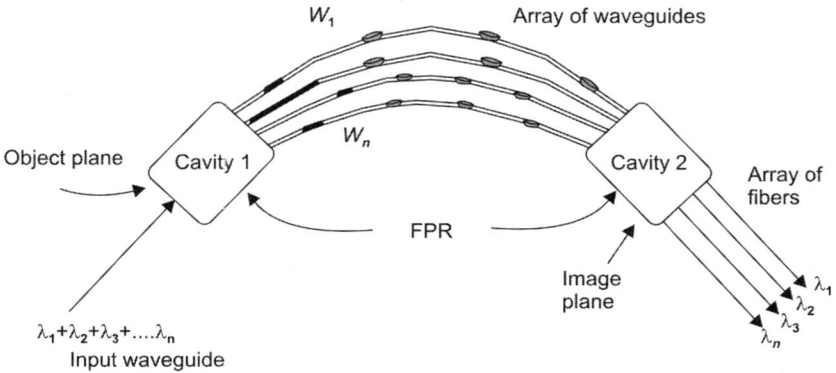

Fig. 12.22. Schematic of an arrayed waveguide grating device

One may understand its operation as follows:

When the optical beam (consisting of wavelengths $\lambda_1, \lambda_2, ..., \lambda_N$) propagating through the input waveguide enters the FPR, it is no longer laterally confined but becomes divergent. On arriving at the input aperture, this divergent beam is coupled into the array of waveguides and propagates through them to the output aperture. The length of the individual arrayed waveguides differs from its adjacent waveguides by ΔL, which is chosen such that ΔL is an integral multiple (m) of the central wavelength λ_c of the multiplexer, i.e.:

$$\Delta L = m \frac{\lambda_c}{n_g} \qquad (12.31)$$

where the integer m is called the order of the array and n_g is the group index of the guided mode. For this wavelength λ_c, the signals propagating through individual waveguides will arrive at the output aperture with equal phase (apart from an integral multiple of 2π), so that the image of the input field in the object plane will be formed at the centre of the image plane. The dispersion is caused by the length increment ΔL of the adjacent array waveguide to vary linearly with signal frequency. Here β is the phase propagation constant of the waveguide mode. As a result, the focal point for different frequencies shifts along the image plane. The spatial shift per unit frequency change (ds/dx) is called the *spatial dispersion D* of the device.

$$D = \frac{1}{v_c} \frac{\Delta L}{\Delta \alpha} \qquad (12.32)$$

where v_c is the central frequency of the PHASAR and $\Delta\alpha$ is the divergence angle between adjacent array waveguides nearthe input and output apertures. Substituting ΔL from Eq. (12.31) in Eq. (12.32), one obtains:

$$D = \frac{1}{v_c} \frac{m}{\Delta\alpha} \frac{\lambda_c}{n_g} = \frac{c}{n_g v_c^2} \frac{m}{\Delta a} \qquad (12.33)$$

where c is the speed of light in free space. It is clear from Eq. (12.33) that the dispersion is fully determined by the order m and the divergence angle $\Delta\alpha$ between adjacent array waveguides. Thus, by placing the output waveguides at appropriate positions along the image plane, the spatial separation of different wavelengths (λ_a, λ_2, ..., λ_N) can be obtained.

12.7.5 Fiber Bragg Grating (FBG)

A Bragg grating is made of a small section of fiber cable, which is modified by exposure to ultraviolet radiation to create periodic changes in the refractive index of the core of the fiber cable. Fig. 12.23 shows a fiber Bragg grating fiber cable. The Bragg grating reflects some of the light waves when travelling through it. The reflected waves usually occur at one particular wavelength. The reflected wavelength, known as the Bragg resonance wavelength, depends on the change in refractive index that is applied to the Bragg grating fiber. This also depends on the basic parameters of the grating (the grating period, the grating length, and the modulation depth). In such a grating, large coupling may occur between the forward and backward propagating modes if the following Bragg condition is satisfied:

$$\lambda_B = 2\Lambda n_{\text{eff}} \qquad (12.34)$$

where λ_B is called the Bragg wavelength, Λ is the grating period, and n_{eff} is the effective index of the mode. Thus, proper design can ensure that most of the power is effectively reflected, whereas signals with other wavelengths are transmitted. The advnatage of such gratings is that they are fiber-compatible so that the losses generated by connecting them to other fibers are very low.

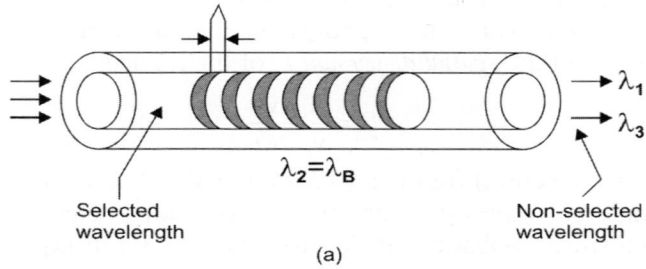

Fig. 12.23. (a) A fiber Bragg grating

Wavelength Division Multiplexing (WDM) Technology 631

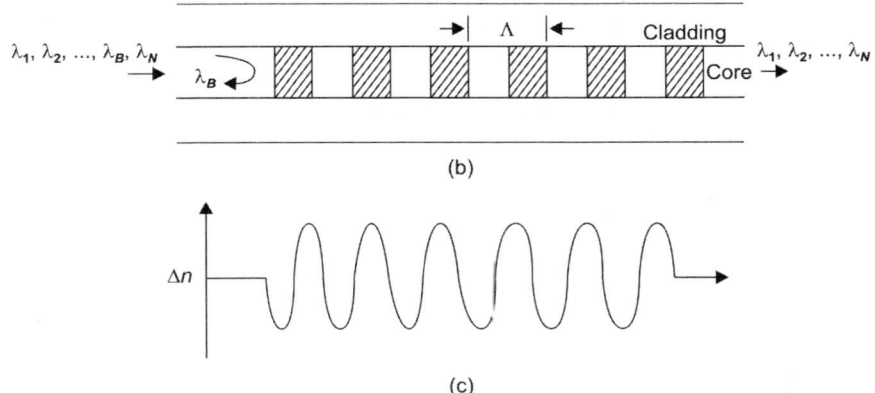

Fig. 12.23. (b) Fiber Bragg grating. (c) Periodic variation of the refractive index of the core

12.7.6 Thin Film Filters or Multilayer Interference Filters

Figure 12.24 shows one multiplexing technique that uses interference filters in devices called thin film filters, or multilayer interference filters. By positioning the thin filters in the optical path, wavelengths can be distributed. The property of each filter is such that it transmits one wavelength while reflecting others. By arranging the thin filters in a device, a de-multiplexer is created, and many wavelengths can be de-multiplexed.

There are several designs that use spectral filters positioned in the

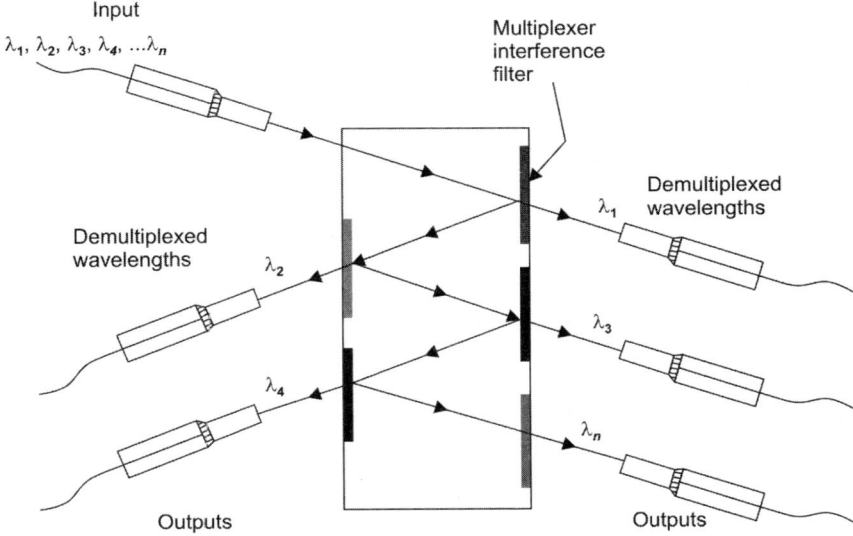

Fig. 12.24. Multilayer interference filters

optical path to sort out wavelengths. These designs can be used as de-multiplexers.

12.7.7 Periodic Filters, Frequency Slicers, Interleavers Multiplexing

Figure 12.25 is a schematic diagram of periodic filters, frequency slicers, and interleaver components that share the same functions, and are usually used together to make a multiplexer device. Stage 1 is a type of periodic filter, called an AWG. Stage 2 represents a frequency slicer. In this instance, stage 2 is another AWG; an interleaver function on the output is provided by six Bragg gratings. Six wavelengths are received at the input to the AWG at stage 1, which then breaks the wavelengths down into odd and even wavelengths. The odd and even wavelengths go to their respective stage 2 frequency slicer, and then are delivered by the interleaver in the form of six discrete, interference-free optical channels.

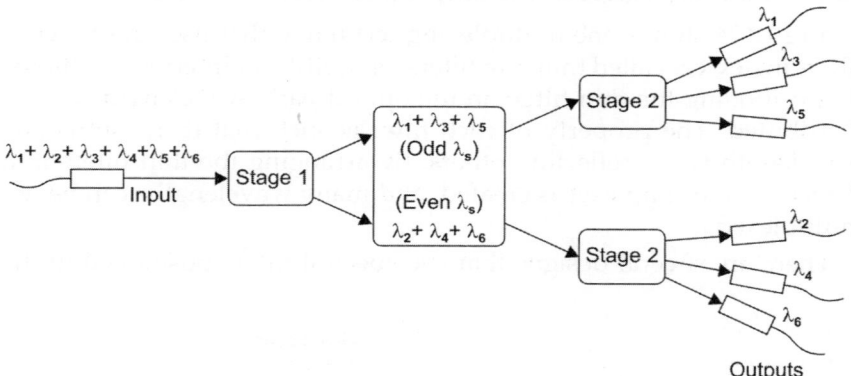

Fig. 12.25. Periodic filters, frequency slicers, and interleavers multiplexing

12.7.8 Mach-Zehnder Interferometer

Interferometers are based on the interferometric properties of light and the principles of Mach-Zehnder interferometer. Mach-Zehnder interferometers can be used to direct a specific wavelength to a specific output port, as shown in Fig. 12.26(a). The interferometer consists of parallel titanium-diffused waveguides on a lithium niobate substrate. The incoming signal is split evenly along the arms of the Mach-Zehnder interferometer, and then recombined at the output. Electrodes are fixed on the arms of the Mach-Zehnder interferometer, and two couplers are connected to the interferometer. While voltage is applied to the electrodes, a specific wavelength can be directed to either port 2 or 3.

This device consists of three parts: (a) 3dB directional coupler which splits the input signal equally and directs it along two paths having different lengths; (b) a central region consisting of two arms, one arm

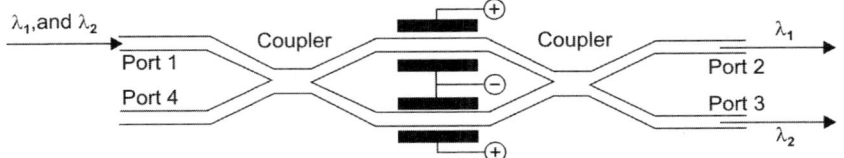

Fig. 12.26. (a) Mach-Zehnder interferometer

being longer by ΔL (say) than the other arm, which introduces a phase shift between two wavelengths; and (c) another 3 dB direction coupler which recombines the signals at the output. With this configuration, it is possible to introduce a phase shift in one of the paths so that the recombined signals will interfere constructively at one output port and destructively at the other. The combined (multiplexed) signals will emerge from the port at which constructive interference has occurred. It is possible to make any size $N \times N$ multiplexer (demultiplexer) using basic 2×2 Mach-Zehnder interferometers (Fig. 12.26b).

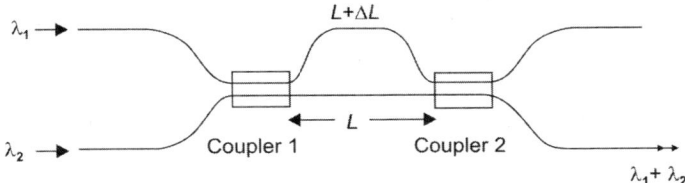

Fig. 12.26. (b) WDM using a 2×2 Mach-Zehnder interferometer

12.8 ACTIVE COMPONENTS

One can control the performance of active WDM components by electronic means. This provides a greater degree of flexibility in the design of optical networks. The prime components in this category are: (i) tunable sources, (ii) tunable filters, and (iii) optical amplifiers. We will discuss here only tunable sources and tunable filters.

(i) Tunable Sources

In implementing a WDM, it is required to generate several wavelengths (λ_1 to λ_N). One simple way is to use a series of discrete distributed feedback (DFB) or distributed Bragg reflector (DBR) lasers operating at different wavelengths and multiplex their outputs into one fiber using a power combiner or a wavelength multiplexer. But this solution requires a large number of lasers, each of which has to be controlled individually. Further, using a power combiner introduces a loss of at least $10 \log_{10} N$ (dB), where N is the number of wavelength channels. If a multiplexer is used, loss can be reduced, but at the cost of more stringent requirements on the control of emitted wavelengths. Therefore, *wavelength-tunable lasers* are used in modern WDM systems. These devices are based on DFB or DBR structures.

A controlled variation of emission wavelength is possible by changing the effective refractive index of the cavity or a part of it in accordance with the relation:

$$\Lambda = m\lambda_B/2n_e$$

where Λ is the grating period, n_e is the effective refractive index of the waveguide for that particular mode wavelength (i.e., mode index), and m is an integer representing the order of Bragg diffraction. The coupling is strongest for $m = 1$, i.e. first order diffraction.

At least two independent control currents are needed: (i) in the *active region* and (ii) in the *tuning region,* for the variation of the effective refractive index. Two configurations of lasers using this scheme are shown in Fig. 12.27. The first structure (Fig. 12.27a) has a cavity that is subdivided into two to three sections for independent current injection. By controlling the current properly, it is possible to vary the lasing wavelength without altering the output power. A tuning range of 2–3 nm is possible with three-section DFB lasers. The second structure (Fig. 12.27b) consists of three sections. Each section can be biased independently by different injection currents. The current injected into the Bragg section (section A) changes the Bragg wavelength (λ_B) through changes in the refractive index. The current injected into the phase control section (section B) changes the phase of the feedback from the

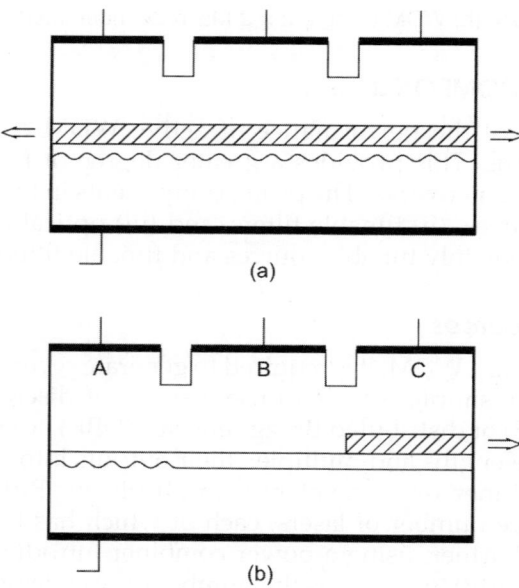

Fig. 12.27. Tunable lasers: (a) multisection DFB laser and (b) multisection DBR laser

Wavelength Division Multiplexing (WDM) Technology

DBR, and current injection in the gain section (section C) controls the output power. Continuous tuning in the range of 10–15 nm with an output power of the order of 100 mW is possible with such lasers. With superstructure grating DBR lasers, a tuning range of up to 100 nm is possible.

(ii) Tunable Optical Filters

Tunable optical filters are versatile devices that are used in many photonic applications. They are essential in wavelength-flexible WDM systems, and they can also play a key role in wavelength-tunable lasers for WDMs.

There is an extensive range of optical tunable filter types. The following filters are available in the market:
- Micron optic fiber Fabry-Perot (FFP) tunable filters.
- Digitally tunable optical filters based on dense wavelength division multiplexer (DWDM) thin film filters and semiconductor optical amplifiers.
- Narrowband tunable optical filters using fiber Bragg gratings.
- High-speed tunable optical filters using a semiconductor double-ring resonator.
- Micromachined in-plane tunable filters using the thermo-optic effect of crystalline silicon.
- Acousto-optic tunable filters (AOTFs).
- Liquid crystal tunable filters (LCTFs).
- Others not listed here.

We will explain some common types of tunable filters.

Optical tunable filter selection depends on various factors, such as fast tuning speed, simple control mechanism, being scalable without additional insertion loss, and long-term operation temperature stability.

(a) Fiber Fabry-Perot tunable filters

The FFP tunable filter principle is based on Fabry-Perot etalon technology. An FFP tunable filter passes wavelengths that are equal to integer fractions of the cavity (etalon) length; all other wavelengths are attenuated or reflected back. The key to the design of the FFP tunable filter is its lensless fiber construction. There are no collimating optics or lenses; thus, the FFP tunable filter achieves high precision, maintains low loss, and good transmission profile. Fig. 12.28 shows a cross-section of an FFP tunable filter design. The design has two pieces of fiber, the ends of which are polished and silvered, so that each end acts like a mirror. The ends are placed precisely opposite one another with a specific gap between them. The fiber assemblies are mounted on two piezo-electric crystals and packaged in a box. By applying a voltage across the crystals, the distance between the fiber ends changes, thus

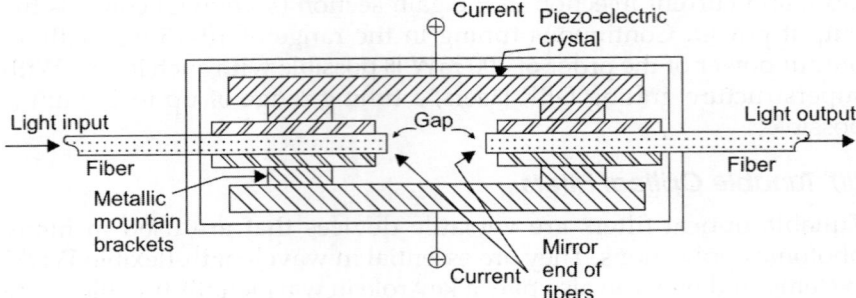

Fig. 12.28. Fiber Fabry-Perot (FFP) tunable filter

changing the resonant cavity length, and therefore, a change in the wavelength selection.

The design of tunable filter eliminates the *pitfalls* of other Fabry-Perot component technologies, including misalignment and environmental sensitivity. Fiber Fabry-Perot tunable filters have low loss, high isolation, long-term alignment stability, high reliability, and accurate power or wavelength measurements. Fiber Fabry-Perot tunable filters are used in optical performance monitoring, tunable optical noise filtering, dropping of a tunable channel for ultra dense WDM, etc.

The design of FFP tunable filter can be modified by putting a liquid crystal material into the gap between the ends of the two fibers. The index of refraction of the liquid crystal can be changed very quickly, by passing current through the liquid crystal. By changing the index of refraction of the crystal, a change in the wavelengths passing through the crystal can be achieved, thus eliminating unwanted wavelengths.

(b) Mach–Zehnder interferometer tunable filters

The Mach–Zehnder interferometer tunable filter has a ladderlike structure, in which each section resembles a Mach–Zehnder interferometer, as shown in Fig. 12.29. The output waveguide (across the top) and the input waveguide (across the bottom) are joined by regularly spaced linking waveguides, each longer than the previous one by ΔS,

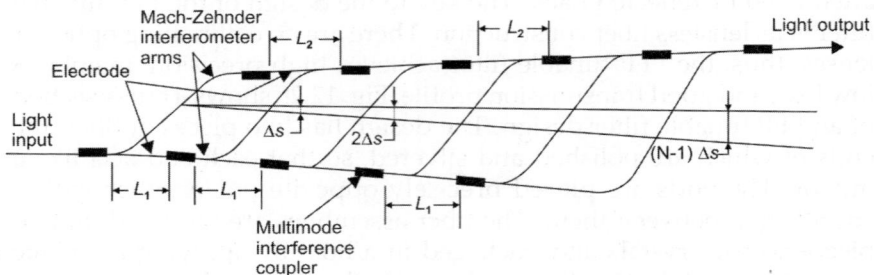

Fig. 12.29. Mach-Zehnder interferometer tunable filter

as illustrated in Fig. 12.29. For constructive interference to occur at each coupler in the output waveguide, ΔS must be equal to an integral number of wavelengths of the input light. However, similar to a Mach-Zehnder that can be tuned by adjusting the refractive index of one or both arms, this filter can be tuned by adjusting the refractive index of the arms with an injected current.

(c) Fiber grating tunable filters

Fiber grating tunable filters are used in a wide variety of applications within optics and fiber communication systems. They are an important element in wavelength division multiplexer systems for combining and separating individual wavelengths. Fiber Bragg grating (FBG) transmits one wavelength and reflects all others. Basically, a grating is a periodic structure within an optical material. This variation in the structure of the optical material reflects or transmits light in a certain direction depending on the light wavelength. Therefore, gratings can be categorized as either transmitting or reflecting.

Figure 12.30 shows a design of an in-fiber Bragg grating tunable filter. The filter contains two Bragg gratings and a four-port circulator. The

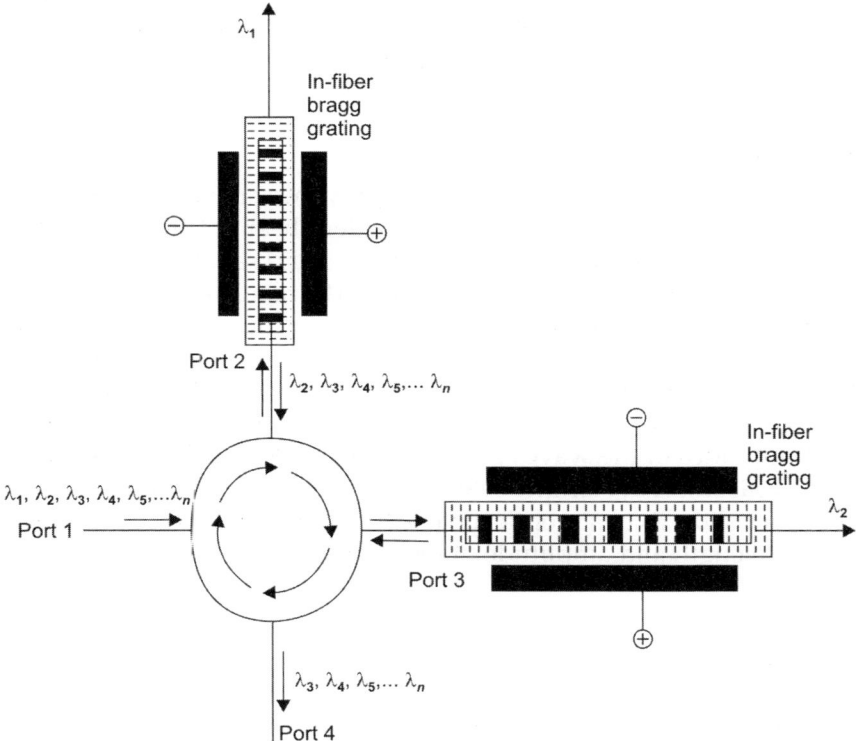

Fig. 12.30. Fiber grating tunable filter

gratings have high reflectance in wavelength bands at specified wavelengths. The Bragg grating fiber is glued to the piezoelectric crystal contacts. Current can be applied on one or two Bragg gratings. The current deforms the crystal, stretching the gratings to match the wavelength of the signals. The wavelength filtered depends on the current level. This type of filter is used in aircraft or spaceborne differential absorption systems that measure water vapour in Earth's atmosphere. It is also used for a unique optical receiver that couples a laser radar signal from a telescope to the in-fiber Bragg grating filter.

(d) Liquid grating tunable filters

Liquid crystal tunable filters (LCTFs) use electrically controlled liquid crystal elements, which select a specific visible wavelength of light for transmission through the filter. A typical wavelength-selective LCTFs is constructed from a stack of fixed filters that consist of interwoven birefringent crystal/liquid-crystal combinations and linear polarizers. The spectral region transmitted through an LCTF depends upon the choice of polarizers, optical coatings, and the liquid crystal characteristics (nematic, cholesteric, smectic etc.). In general, visible-wavelength devices of this type usually perform in the range of 400–700 nm. This type of filter is ideal for use with electronic imaging devices, such as CCDs, because it offers excellent imaging quality with a simple optical pathway.

(e) Acousto-optic tunable filters

Acousto-optic tunable filters (AOTFs) apply the same technology used in acousto-mechanical switches. An incident beam of light impacts a dioxide crystal of an AOTF. The dioxide crystal is sandwiched between an acoustic transducer and absorber that can be regulated by the acoustic power and acoustic frequency sliders. Upon encouraging the standing wave in the dioxide crystal, a portion of the incident light beam is diffracted into the output port, while the remainder of the beam passes through the crystal and is absorbed by a beam stop. As the slider moves, the amplitudes of the waves passing through the AOTF are increased or decreased. Wavelength selection is controlled by the acoustic frequency slide. Acousto-optic tunable filters are employed to modulate the wavelength and amplitude of incident laser light.

(f) Thermo-optic tunable filters

Thermo-optic tunable filters apply the thermo-optic effect on an optical material, such as crystalline silicon. Changing the temperature results in a change in the index of refraction of the material. For example, current applied to a resistive element creates heat which increases the temperature of the filter. Thus, by applying current, wavelength selection can be achieved. This type of filter is used for spectroscopy and in optical communication systems.

(g) Other types of tunable filters

There are many other types of tunable filters used in building various optical devices and systems. Many tunable filters perform specific functions. Tunable filters are also available in the market, such as wide-band-pass filters, gas tunable filters, active optical filters, volume holographic grating-based filters, digitally tunable filters based on dense wavelength division multiplexer (DWDM) thin film filters, and high-speed tunable filters using a semiconductor double-ring resonator.

12.9 TOPOLOGIES AND ARCHITECTURES

WDM technology enables the utilization of a significant portion of the available fiber bandwidth by allowing many independent signals to be transmitted simultaneously in one fiber. Additionally, high-bandwidth routing can be also facilitated through a multi-user network. The WDM channels can be routed and detected independently, with the wavelength determining the communication path by acting as the signature address of the origin, destination, or routing. Therefore, the basic system architecture that can take the full advantage of WDM technology is an important issue.

12.9.1 Point-to-point WDM Links

As shown in Fig. 12.31, in a simple point-to-point WDM system, several channels are multiplexed at one node, the combined signals are then transmitted across some distance of fiber, and the channels are demultiplexed at a destination node. This point-to-point WDM link facilitates the high bandwidth fiber transmission without routing or switching in the optical data path.

12.9.2 Wavelength-routed Networks

Figure 12.32 shows a more complex multi-user WDM network structure, where the wavelength is used as the signature address for either the transmitters or the receivers, and determines the routing path through an optical network. In order for each node to be able to communicate with any other node and facilitate proper link setup, either the transmitters or the receivers must be wavelength tunable; we have arbitrarily chosen the transmitters to be tunable in this network example. We may note that the wavelength are routed passively in wavelength-routed-networks.

Point-to-point systems

Optoelectronic routing node

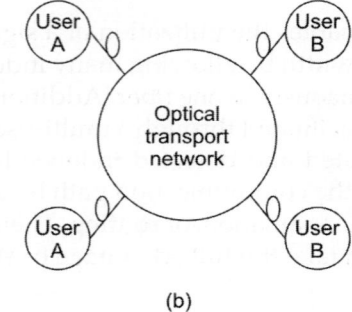

(a)

Point-to-multipoint systems

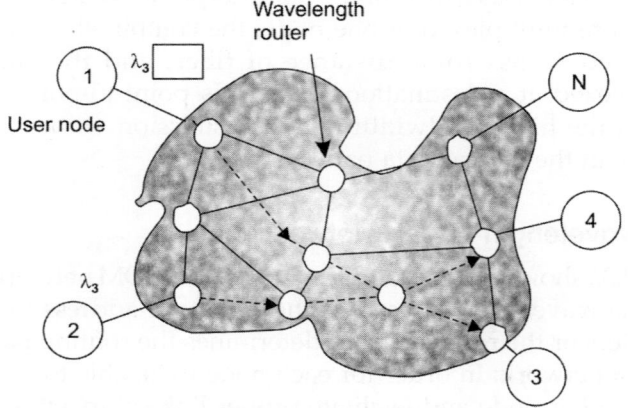

(b)

Fig. 12.31. Point-to-point and point-to-multipoint optical systems

Wavelength-determined routing

Fig. 12.32. A generic multiuser network in which the communication links and routing paths are determined by the wavelengths used within the optical switching fiber

12.9.3 WDM Star, Ring, and Meshes

Three common WDM network topologies are *star*, *ring*, and *mesh* networks. In the star topology, each node has a transmitter and receiver, with the transmitter connected to one of the passive central star's inputs and the receiver connected to one of the star's outputs, as is shown in Fig. 12.33(a). Rings, as shown in Fig. 12.33(b), are also popular because (1) many electrical networks use this topology, and (2) rings are easy to implement for any geographical network configuration. In this example, each node in the unidirectional ring can transmit on a specific signature wavelength, and each node can recover any other node's wavelength signal by means of a wavelength-tunable receiver. Although not depicted in the figure, each node must recover a specific channel. This can be performed (1) where a small portion of the combined traffic is tapped off by a passive optical coupler, thereby allowing a tunable filter to recover a specific channel; or (2) in which a channel-dropping filter completely removes only the desired signal and allows all other channels to continue propagating around the ring. Furthermore, a *synchronous optical network* (SONET) dual-ring architecture, with one ring provoding service and the other protection, can provide automatic fault detection and protection switching.

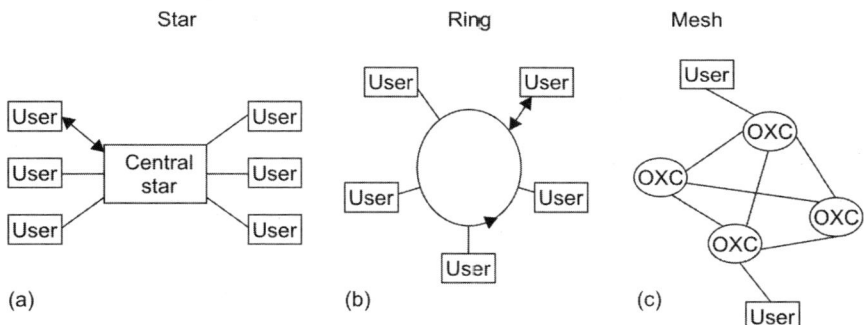

Fig. 12.33. WDM stars, rings, and meshes

In both the star and ring topologies, each node has a signature wavelength, and any two nodes can communicate with each other by transmitting and recovering that wavelength. This implies that N wavelengths are required to connect N nodes. The obvious advantage of this configuration, known as a single-hop network, is that data transfer occurs with an uninterrupted optical path between the origin and destination; the optical data starts at the originating node and reaches the destination node without stopping at any other intermediate node. A disadvantage of this single-hop WDM network is that the network and all its components must accommodate N wavelengths, which may be difficult (or impossible) to achieve in a large network, i.e. present fabrication technology cannot provide and transmission capability cannot accommodate 1000 distinct wavelengths for a 1000-user network!

It is important to mention that reliability is a problem in fiber ring. If a station is disabled or if a fiber breaks, the whole network goes down. To address this problem, a double-ring optical network, also called a *"self healing"* ring, is used to bypass the defective stations an loops back around a fiber break, as shown in Fig. 12.34. Each station has two inputs and two outputs connected to two rings that operate in opposite directions.

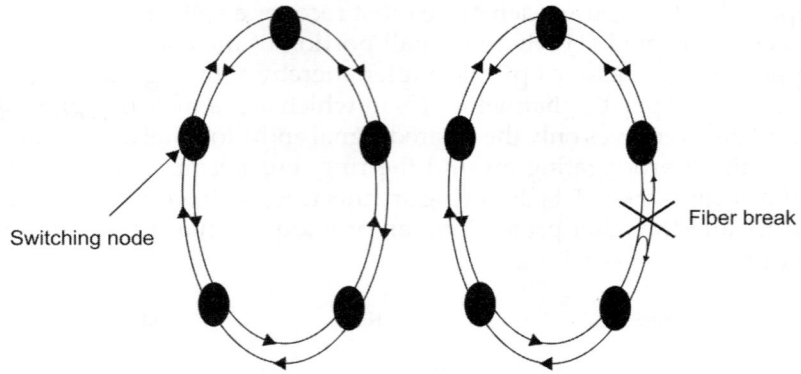

Fig. 12.34. A self-healing ring network

An alternative to requiring N wavelengths to accommodate N modes is to have a multi-hop network (mesh network) in which two nodes can communicate with each other by sending data through a third node,, with many such intermediate hops possible, shown in Fig. 12.33(c). In the mesh network, the nodes are connected by reconfigurable optical crossconnects (OXCs). The wavelength can be dynamically switched and routed by controlling the OXCs. Therefore, the required number of wavelengths and the wavelength tunable range of the components can be reduced in this topology. Moreover, the mesh topology can also provide multiple-paths between two nodes to make network protection and restoration easier to realize. If a failure occurs in one of the paths, the system can automatically find another path and restore communications between any two nodes. However, OXCs with large numbers of ports are extremely difficult to obtain, which limits the scalability of the mesh network.

In addition, there exist several other network topologies, such as a tree network, which is a favourite of broadcast, or distribution, systems. At the "base" of the tree is the source transmitter from which emanates the signal to be broadcast throughout the network. From this base, the tree splits many times into different "branches", with each branch either having nodes connected to it or further dividing into sub-branches. This continues until all the nodes in the network can access the base transmitter. Whereas the other topologies are intended to support bi-directional communication among the nodes, this topology is useful for distributing information uni-directionally from a central point to a multitude of users. This is a very straightforward topology and is in use in many electrical systems, most notably cable television (CATV).

By introducing Fig. 12.35 in which a large network is composed of smaller ones, we have also introduced the subject of the architecture of the network which depends on the network's geographical extent. The three main architectural types are the local-, metropolitan-, and wide-area networks, denoted by LAN, MAN, and WAN, respectively. Although no rule exists, the generally accepted understanding is that a LAN interconnects a small number of users covering a few km (i.e., intra- and inter-building). a MAN interconnects users inside a city and its outlying regions, and a WAN interconnects significant portions of a country (100s of km). Based on Fig. 12.35, the smaller network represents LANs, the larger ones MANs, and the entire figure would represent a WAN. In other words, a WAN is composed of smaller MANs, and a MAN is composed of smaller LANs. Hybrid systems exist, and typically a wide-area network will consist of smaller LANs, with mixing and matching between the most practical topologies for a given system. For example, stars and rings may be desirable for LANs, whereas buses may be the only practical solution for WANs. It is, at present unclear

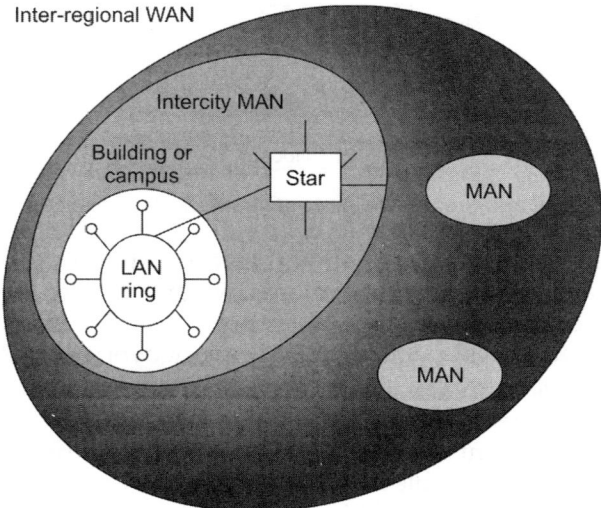

Fig. 12.35. Hybrid network topologies and architectures woven together to form a large network.

12.9.4 Network Reconfigurability

Figure 12.36 shows an example of fixed and reconfigurable optical add/drop nodes that can add and drop wavelengths at intermediate nodes in a communications network. A fixed add/drop multiplexing node can only process the signal(s) at a given wavelength or a group of wavelengths. This added flexibility would save operating and maintenance costs and would improve network efficiency. In general, a network is reconfigurable if it can provide the following functionality for multi-channel operations: (1) channel add/drop, and (2) path reconfiguration for bandwidth allocation or restoration. It appears that a reconfigurable network is highly desirable to meet the requirements of high bandwidth and bursty traffic in future networks.

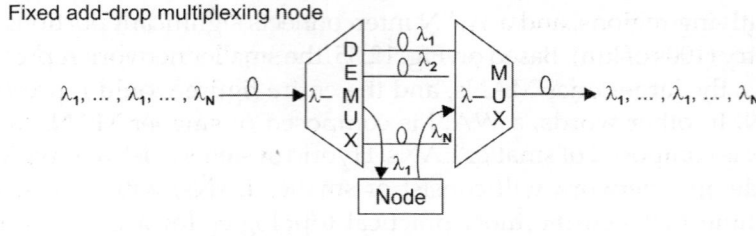

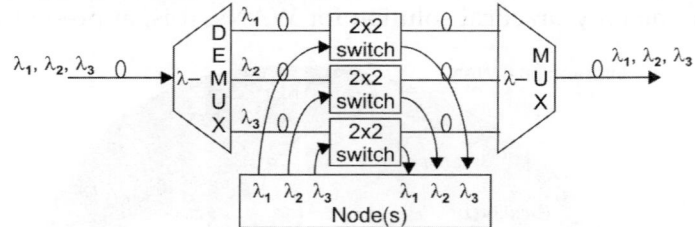

Fig. 12.36. Optical add-drop multiplexing (OADM) systms; fixed vs reconfigurable

A reconfigurable network allows dynamic network optimization to accommodate changing traffic patterns, which provides more efficient use of network resources. Fig. 12.37 shows blocking probability as a function of call arrival rate in a WDM ring network with 20 nodes. A configurable topology can support 6 times the traffic of a fixed WDM topology for the same blocking probability.

The key component technologies enabling network reconfigurability include wavelength tunable lasers and laser arrays, wavelength routers, optical switches, OXCs, OADMs, and tunable optical filters, etc.

Although huge benefits are possible with a reconfigurable topology, the path to reconfigurability is paved with various degrading effects.

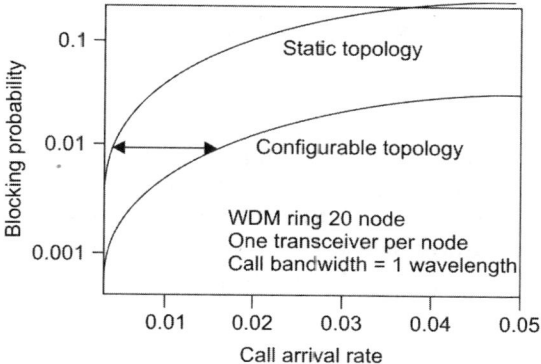

Fig. 12.37. Blocking probability as a function of call arrival rate in a WDM ring

Fig. 12.32 shows that the signal may pass through different lengths of fiber links due to the dynamic routing, causing some degrading effects in reconfigurable networks to be more critical than in static networks, such as nonstatic dispersion and nonlinearity accumulation due to reconfigurable paths, EDFA gain transients, channel power nonuniformity.

Example 1

For a 2 × 2 fiber coupler, input power is 200 μW, throughput power is 90 μW, coupled power is 85 μW and cross talk power is 6.3 μW. Compute the performance parameter of the fiber coupler. [B.Tech.]

Solution

Here $P_0 = 200$ μW, $P_1 = 90$ μW, $P_2 = 85$ μW, and $P_3 = 6.3$ μW:

(i) Coupling ratio = $\dfrac{P_2}{(P_1 + P_2)} \times 100\% = \dfrac{85}{(90 + 85)} \times 100\% = 48.57\%$

(ii) Excess ratio = $10 \log \left(\dfrac{P_0}{P_1 + P_2} \right)$ dB

$= 10 \log \left(\dfrac{200}{90 + 85} \right)$ dB $= 0.5799$ dB

(iii) Insertion loss (for port 0 to port 1) = $10 \log \left(\dfrac{P_0}{P_1} \right)$ dB

$= 10 \log \left(\dfrac{200}{90} \right) = 3.46$ dB

Insertion loss (for port 0 to port 2) = $10 \log \left(\dfrac{200}{85} \right) = 3.71$ dB

(iv) Cross talk $= 10 \log \left(\dfrac{P_3}{P_0} \right) = 10 \log \left(\dfrac{6.3 \times 10^{-3}}{200} \right) = -45 \text{ dB}$

Example 2
Design a broadband WDM 3dB coupler which splits at $\lambda = 1310$ nm and 1550 nm. The two step-index fibers used to make the coupler are identical and single-moded with a core diameter of 8.2 μm, core index $n_1 = 1.45$, and cladding index $n_2 = 1.446$. Calculate the position of the output ports with respect to the input port for the two wavelengths.
[B.Tech.]

Solution
We know that the interaction length L required to make a 3dB coupler is $\pi/(4\kappa)$, where κ is the coupling coefficient. Using simple empirical relationship given below may be used to calculate the value of λ:

$$\kappa = \dfrac{\pi}{2} \dfrac{\sqrt{\delta}}{a} \exp[-(A + Bd + Cd^2)]$$

where $A = 5.2789 - 3.663 V + 0.3841 V^2$
$B = -0.7769 + 1.2252 V - 0.0152 V^2$
$C = -0.0175 - 0.0064 V - 0.0009 V^2$

$$\delta = \dfrac{n_1^2 - n_2^2}{n_1^2}, \quad d = \dfrac{d}{a}$$

n_1 is the core refractive index of the fiber, n_2 is the cladding refractive index of the fiber, a is the fiber core radius, and d is the separation between the fiber axis.

Let us take $d = 10$ μm. With the given parameters, we obtain for $\lambda_1 = 1.31$ μm:

$$V_1 = \dfrac{2\pi a}{\lambda}(n_1^2 - n_2^2)^{1/2} = \dfrac{\pi \times (8.2 \text{ μm})}{(0.31 \text{ μm})}[(1.45)^2 - (1.446)^2]^{1/2}$$

or $\quad V_1 = 2.115$

and for $\lambda_2 = 1.55$ mm:

$$V_2 = 1.787$$
$$d = 5.5096 \times 10^{-3} \text{ and } d = 2.439$$

For coupling coefficient for λ_1 will be:
$$\kappa_1 = 1.0483 \text{ mm}^{-1}$$

And that for λ_2 will be:
$$\kappa_2 = 1.2938 \text{ mm}^{-1}$$

Therefore, the interaction length L_1 and L_2 for $\lambda_1 = 1.31$ μm and $\lambda_2 = 1.55$ μm, respectively, are obtained as:

$$L_1 = \frac{\pi}{4\kappa_1} = \frac{\pi}{4 \times 1.0483} = 0.7488 \text{ mm}$$

and
$$L_2 = \frac{\pi}{4\kappa_2} = \frac{\pi}{4 \times 1.2839} = 0.6114 \text{ mm}$$

Obviously, the output port positioned at 0.6114 mm with respect to the input port will gather signals at $\lambda_1 = 1310$ nm and that positioned at 0.7488 mm will gather signals at $\lambda_2 = 1550$ nm. Therefore, the coupler can be used as a WDM device.

Example 3

A 32 × 32 star coupler is formed by interconnecting 2 × 2 couplers. If 5% of power is lost in each coupler element, calculate total loss in the coupler. [M.Sc. (Ele.)]

Solution

Here $N = 32$

$$\therefore \quad F_T = \frac{100 - 5}{100} = 0.95 \; (\because 5\% \text{ power is lost})$$

Total loss = $10 (1 - 3.322 \log F_T) \log N$
= $10(1 - 3.32 \log 0.95) \log 32 = 16.16$ dB

Example 4

A directional coupler uses two identical single-mode fibers. Determine the interaction length so that the input power P_0 is divided equally at the two output ports. [B.Tech.]

Solution

We have
$$P_1 = P_2 = \frac{P_0}{2}$$

or
$$\sin^2(\kappa L) = \cos^2(\kappa L) = \frac{1}{2}$$

or
$$\kappa L = \frac{\pi}{4}$$

This gives the interaction length L to be equal to $\pi/2\kappa$. We may note that such a coupler can act as a power divider.

Example 5

Make a PHASAR-based demultiplexer for 16 channels with a channel spacing of 100 GHz. The channels are centred around 1.55 µm. Calculate the required order of the arrayed waveguides. [B.Tech.]

Solution

We know that the dispersion of the PHASAR is due to the difference ΔL in the optical path length of adjacent arrayed waveguides, which causes a phase difference $\Delta \phi = \beta \Delta L$ [where $b = 2\pi n_g/\lambda_c = (2\pi n_g/c)(v_c)$]. Thus, $\Delta \phi$ increases with frequency. If the change in frequency is such that $\Delta \phi$ increases by 2π, the transfer will be the same as before. Hence, the response of the PHASAR is periodical. This period Δv in the frequency domain is called the *free spectral range* (FSR), and it can be calculated as follows:

$$\Delta \beta \Delta L = 2\pi$$

or

$$\left[\frac{2\pi n_g}{c}(\Delta v_c)_{FSR}\right]\Delta L = 2\pi$$

or

$$(\Delta v_c)_{FSR} = \frac{c}{n_g \Delta L} = \frac{v_c}{m} \qquad (i)$$

Now, a demultiplexer for 16 channels with a channel spacing of 100 GHz should have an FSR of at least 1600 GHz. Since the centre wavelength is 1.55 μm, the corresponding frequency is obtained as:

$$v_c = \frac{c}{\lambda_c} = \frac{3 \times 10^8}{1.55 \times 10^{-6}} = 1.935 \times 10^{14} \text{ Hz}$$

Using Eq. (i), we obtain:

$$m = \frac{v_c}{\Delta v_{FSR}} = \frac{1.935 \times 10^{14}}{1600 \times 10^9} \approx 121$$

This reveals that the PHASAR-based demultiplexer would require an array with an order of at least 121.

Example 6

The LED used in 100 Mb·s^{-1} silica fiber link has a rise time of 8 ns. The wavelength emitted by the LED is 830 nm with a spectral width of 40 nm. The *p-i-n* photodiode has a rise time of 10 ns. The value of $\lambda^2 \frac{d^2 n}{d\lambda^2} = 0.024$ for silica fiber.

Calculate the system rise time provided the length of link is 2.5 km. The inter modal dispersion of the fiber is 3.5 nm/km. Estimate the maximum bit rate that may be achieved in the link using RZ and NRZ for solution. [B.Tech.]

Solution

We have:
$$\Delta t_{mat} = \left(\frac{-L}{c}\right)\left(\frac{\Delta \lambda}{\lambda}\right)\left(\lambda^2 \frac{d^2 n}{d\lambda^2}\right)$$

$$= -\frac{2.5 \times 10^3}{3 \times 10^8} \times \frac{40}{830} \times (0.024)$$

$$= -9.64 \times 10^{-9} \text{ s} = -9.64 \text{ ns}$$

For a 2.5 km link, we have $\Delta t_{\text{modal}} = 2.5 \times 3.5 = 8.8$ ns

We have $\Delta t_{\text{sys}} = 1.1 \sqrt{(\Delta t_s)^2 + (\Delta t_r)^2 + (\Delta t_{\text{max}})^2 + (\Delta t_{\text{modal}})^2}$

$$= 1.1 \sqrt{(8)^2 + (10)^2 + (9.6)^2 + (8.8)^2} \text{ ns}$$

$$= 19.8 \times 10^{-9} \text{ s} = 19.8 \text{ ns}$$

(i) B_T (max) $\dfrac{0.7}{\Delta t_{\text{sys}}} = \dfrac{0.7}{19.8 \times 10^{-9}} = 34 \times 10^6$

With RZ format, B_T (max) $= 34 \times 10^6$

(ii) With NRZ format is used: B_T (max) $= \dfrac{0.35}{19.8 \times 10^{-9}} = 17 \times 10^6$

Example 7

A mulfiplexer that uses a plane blazed reflection grating [Fig. 12.38(b)] has to be made. It is required to achieve a channel spacing of 10 nm in the wavelength range of 1500–1600 nm with a center wavelength of 1550 nm. (a) What should be the grating element if the angle of blaze of the grating is 10°? (b) What should be the focal length of the lens? Assume that the output fibers have a spacing of 150 µm. [B.Tech.]

Solution

A reflection grating has on its surface an array of parallel grooves which are identical in depth and shape, equally spaced, and are provided with a highly reflecting coating (Fig. 12.39). These grooves are inclined at a specific angle (called the angle of blaze, β) with respect to the grating surface. The grating is highly efficient in diffracting wavelengths close to those for which specular reflection occurs. The wavelength for which maximum efficiency is observed is called the blazed wavelength λ_B. When such a grating is used in the Littrow mode, the angle of incidence (α) and angle of diffraction (θ) are nearly equal to the angle of blaze as shown in Fig. 12.39; that is, $\alpha \approx \theta \approx \beta$. The fundamental grating equation, therefore, modifies to, taking the blazed wavelength $\lambda = \lambda_B$.

$$m\lambda_B = d[\sin \alpha + \sin \theta] \approx 2d \sin \beta \quad \text{(i)}$$

where m is the order of diffraction and d is the grating element (i.e., the distance between two grooves).

With such a demultiplexer using a reflection grating in the Littrow mode, the wavelength channel spacing is given by:

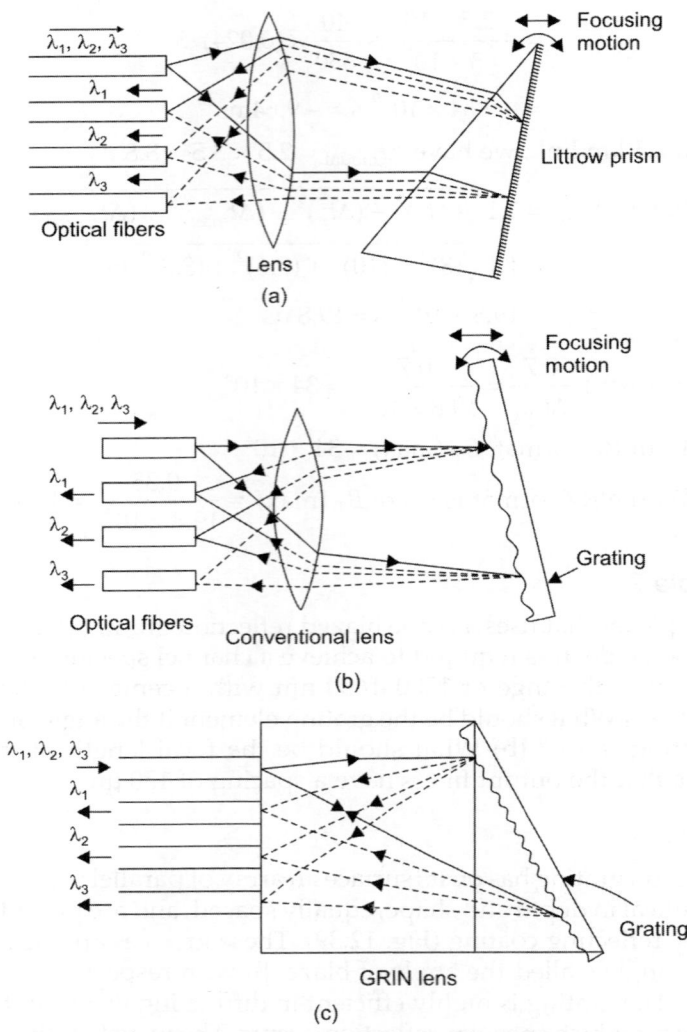

Fig. 12.38. Angular dispersion demultiplexers: (a) Littrow prism type (b) reflection grating type (with a conventional lens), and (c) reflection grating type (with a GRIN rod lens)

$$\Delta\lambda = \frac{x}{f}\left(\frac{d\lambda}{d\theta}\right) = \frac{x}{f}\sqrt{d^2 - \left(\frac{\lambda_2}{2}\right)^2} \qquad \text{(ii)}$$

where λ_c is central wavelength corresponding to λ_B, x is the spacing of core centres of the output fibers, and f is the local length of the lens.

Here, the required λ_c = 1.55 μm, so that λ_B = 1.55 μm, β = 10°, and m = 1 (for first-order diffraction). Using Eq. (i), we get d = 4.463 μm. The

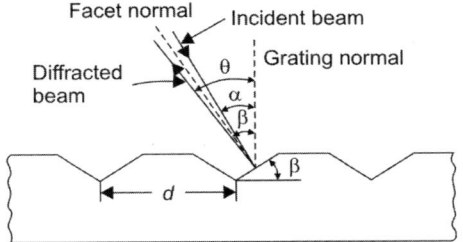

Fig. 12.39. Blazed reflection grating in the Littrow mode

focal length f of the lens may be obtained by putting $x = 150$ λm, $\Delta\lambda = 10$ nm, and $\lambda_c = \lambda_B = 1.55$ μm in Eq. (ii). This gives $f = 65.92$ mm.

Example 8
Consider an $N \times N$ waveguide grating multiplexer having $L_f = 10$ mm, $x = d = 5$ μm, $n_c = 1.45$, and a central design wavelength $\lambda_c = 1550$ nm. Calculate the waveguide length difference for $m = 1$. [B.Tech.]

Solution
We have:
$$\Delta L = m \frac{\lambda_c}{n_c}$$

$$\therefore \quad \Delta L = 1 \times \frac{1550}{1.45} = 1.069 \text{ μm}$$

Here, $m = 1$, $l_c = 1550$ nm, $n_c = 1.45$.
If $n_s = 1.45$ and $n_g = 1.47$, then:

$$\Delta\lambda = \frac{x}{L_f} \frac{n_s d}{m} \frac{n_c}{n_g} = \frac{5}{10^4} \times \frac{1.45 \times 5}{1} \times \frac{1.45}{1.47} \text{ μm} = 3.58 \; nm$$

Example 9
(a) Assume that the input wavelengths of a 2×2 silicon MZI are separated by 10 GHz (i.e. $\Delta\lambda = 0.08$ nm at 1550 nm). With $n_{\text{eff}} = 1.5$ in a silicon waveguide, calculate the waveguide length difference. (b) If the frequency difference is 130 GHz (i.e. $\Delta\lambda = 1$ nm), then calculate ΔL.

Solution
(a) $\Delta L = \left[2n_{\text{eff}}\left(\frac{1}{\lambda_1} - \frac{1}{\lambda_2}\right)\right] = \frac{c}{2n_{\text{eff}} \Delta v} = \frac{3 \times 10^8 \text{ m/s}}{2 \times 1.5 \times 10^{10}/\text{s}} = 10^{-2}$ m

(b) $\Delta L = \dfrac{3 \times 10^8 \text{ m/s}}{2 \times 1.5 \times (13) \times 10^{10}/\text{s}} = 0.77$ mm

Example 10

The maximum index change of a particular DBR laser operating at 1550 nm is 0.65%. Calculate the tuning rate.

If the source spectral width $\Delta\lambda_{signal} = 0.02$ for a 2.5 Gb/s signal, then find the number of channels that can operate in this tuning range.

Solution

We have:

$$\Delta L_{tune} = \lambda \frac{\Delta n_{eff}}{n_{eff}} = (1550 \text{ nm})(0.0065) = 10 \text{ nm}$$

$$N = \frac{\Delta\lambda_{tune}}{\Delta\lambda_{channel}} = \frac{10 \text{ nm}}{10 \times 0.02 \text{ nm}} = 50$$

SUGGESTED READINGS

1. GP Agrawal, 'Fibre Optic Communication Systems', 2nd ed., John Wiley (1997).
2. HJR Dutton, 'Understanding Optical Communications', IBM, Prentice Hall (1998).
3. G Keiser, 'Optical Communication Essentials', McGraw Hill (2003).
4. JC Palais, 'Fibre Optic Communications', 4th ed., Prentice Hall (1998).
5. JP Laude, 'DWDM Fundamentals, Components and Applications', Artech House, Boston (2002).
6. V Degiorgio and I Cristiani, 'Photonics', Springer (2014).
7. S Kumar and M Jamal Deen, 'Fiber Optic Communications', Wiley (2014).

GLIMPSES

- Optical signals of different wavelength (1300–1600 nm) can propagate without interfering with each other. The scheme of combining a number of wavelengths over a single fiber is called *wavelength division multiplexing* (WDM), i.e., the oncept of WDM involves simultaneous transmission of several wavelengths (say, $\lambda_1, \lambda_2, \lambda_3, ..., \lambda_N$) over the same single optical fiber, effectively increasing the aggregation bandwidth per fiber cable.
- WDM technology has evolved to the point that channel wavelength separations can be as small as a few nms, giving rise to *dense wavelength division multiplexing* (DWDM) *systems.*
- The implementations of WDM or DWDM requires a number of passive and active components. The performance of passive components is fixed whereas that of active components can be controlled electronically.

- To prevent spurious signals to enter into receiving channel, the *demultiplexer* must have narrow spectral operation with sharp wavelength cutoff. The acceptable limit of crosstalk is ~ 30 dB.
- *Passive components* are mainly used to split or combine optical signals. These components operates in optical domains. Passive components don't need external control for their operation. Passive components are fabricated by using optical fibers or by planar optical waveguides. Commonly used passive components are: (i) NXN couplers, (ii) Power splitters, (iii) Power taps, and (iv) Star couplers. Most passive components are derived from basic star couplers.
- Star coupler can perform combining and splitting of optical power. Therefore, star coupler is a multiple input and multiple output port device.
- A device with two inputs and two outputs is called as 2 × 2 coupler.
- Star coupler is mainly used for combining optical powers from N-inputs and divide them equally at M-output ports.
- *Directional couplers* are used to split or combine two signals at different wavelengths. In a directional coupler made up of two identical signal-mode fibers, there is a periodic exchange of power between the two optical fibers. The minimum interaction length over which the power is completely transferred from the first fiber to the second fiber is termed as the *coupling length*.
- The optical power put into any port on one side of coupler is equally divided among the output ports. However, ports on same side of coupler are isolated from each other.
- The performance of a coupler is expressed in terms of the following parameters:

  (i) Coupling ratio (%) = $\dfrac{P_2}{P_1 + P_2} \times 100$

  (ii) Splitting loss = $-10 \log \left(\dfrac{1}{N}\right) = 10 \log N$

  (iii) Excess loss (dB) = $+10 \log_{10} \left(\dfrac{P_0}{P_1 + P_2}\right) - 10_{10} \left(\dfrac{P_1 + P_2}{P_0}\right)$

  $= -10_{10} \left(\dfrac{P_{in}}{\sum_{i=1}^{N} P_{out, i}}\right)$

  (iv) Insertion loss (dB) = $-10 \log_{10} \left(\dfrac{P_j}{P_i}\right)$

- To implement WDM, a multiplexer is required to combine several wavelengths at the transmitting end, and a demultiplexer is required to isolate the different wavelengths at the receiver end. There are several mechanisms of multiplexing (demultiplexing) studies. Some of these are:
    (i) Interference filter devices
    (ii) Angular dispersion using a *Littrow* prism or Littrow grating.
    (iii) Match-Zehnder interferrometer
    (iv) Arranged wavelength grating
    (v) Fiber Bragg grating.
- Many different lasers designs have been proposed to generate the spectrum of wavelength needed of DWDM networks. The use of discrete single-wavelength DFB or DBR lasers is the simplest method. A tuning wavelength range of 10–15 nm is possible with DFB and DBR lasers. For wider ranges, an array of tunable lasers has to be used.
- *Tunable optical filters* are key components for dense WDM optical networks. Two key technologies to make a tunable filters are *MEMS*-based and *Bragg grating* based devices. Tunable filters can be used as add/drop multiplexers in optical networks or demultiplexes in the receiver module.
- *Time division multiplexing* (TDM) is a technique for transmitting multiple digital data, voice, and video signals simultaneously over one fiber cable. This is accomplished by interleaving pulses representing bits from different channels or time slots. The public-switched telephony network (PSTN) is based on the TDM technologies and is often called a TDM access network.
- *Frequency-division multiplexing* (FDM) is a scheme in which numerous analogue signals are combined for transmission on a single communications line or channel. Each signal is assigned a different frequency (sub-channel) within the main channel. This technology is used in broadcast, radio, television, and cable television.
- *Dense wavelength division multiplexing* (DWDM) is an optical technology used to increase bandwidth over existing fiber optic backbones. DWDM is a fiber-optic transmission technique that employs light wavelength to transmit data as parallel bits or a serial string of characters.
- *Coarse wavelength division multiplexing* (CWDM) is a technology that combines upto 16 wavelengths onto a single fiber. When there are just a few channels (upto 16 channels) and they are spaced more widely (10 nm or more) apart, the system is called CWDM.

REVIEW QUESTIONS

1. Draw a neat sketch of WDM scheme and explain it.
2. What is the significance of passive components in WDM?
3. Distinguish between WDM and DWDM. Explain the base frequency and channel spacing specified by ITU for DWDM.
4. Explain the construction and working of 2 × 2 directional fiber coupler and an $N \times N$ star coupler.
5. Explain various performance parameters of optical coupler.
6. How one can change the coupling ratio of a 2 × 2 coupler?
7. How many major types of devices are there for multiplexing/demultiplexing? Explain them and compare with merits and demerits.
8. Why tunable sources needed? Explain the principle of at least two types of tunable sources.
9. What are tunable filters? Where they are used?
10. Explain the principle of operation of an AOTF.
11. Mention and explain the possible applications of fiber optic gratings.
12. Write the definitions of following parameters: (a) Insertion loss, (B) Channel width, (c) Cross talk of a WDM system.
13. Explain the operation with diagram of a unidirectional WDM. Also explain bidirectional WDM system.
14. Write short notes on:
    (a) Time-division multiplexing (TDM)
    (b) Frequency-division multiplexing (FDM)
    (c) Dense wavelength division multiplexing (DWDM)
    (d) Coarse wavelength division multiplexing (CWDM).
15. Briefly explain the techniques for multiplexing and demultiplexing.

PROBLEMS

1. A 2 × 2 loss-fiber coupler is using identical single mode fibers. Show that the interaction length required to achieve a splitting ratio of 10:90 is $L \approx 1.25/k$. [B.Tech.]
2. How many fibers are required for full duplex communications between two nodes? [B.Tech.]
   [**Hint.** Generally, two fibers are rquired for traffic transmission in each direction. However, it is possible to reduce this to one fiber by using DWDM, WDM or bi-directional coupler technology.]
3. A PHASAR-based demultiplexer with 32 channels spaced at 50 GHz and a central wavelength of 1.55 µm is to be designed. Calculate the FSR and the order of array. [**Ans.** 1600 GHz, 121] [B.Tech.]
4. How one can increase bandwidth (BW, i.e. traffic capacity) economically? [B.Tech.]

656 Photonics | Optoelectronics

[**Hint.** There are number of different ways through which one can economically double fiber capacity. One may use bi-directional couplers that allow traffic transmission in both directions on one fiber. However, installing of bi-directional couplers does not require end mode equipment to be modified.

To enhance the capacity further, one may be able to install passive *Cross Band WDMS*, DWDMs or CWDMs.

5. Can you interchange multimode and single mode equipment?

   [**Solution.** Generally no, since multimode and single mode equipment are not interchangeable and also not compatible.]

6. A fiber system uses multimode optical fiber. Can one enhance the capacity of multimode fiber by using WDM or DWDM technology?

   [**Solution.** One can enhance the multimode fiber technology as WDM and DWDM technology is not available for multimode fiber.

7. Using DWDMs, find the maximum number of wavelengths that can be placed onto a fiber. [NET-SET]

   [**Solution.** This basically depends on the use of the technology of equipment by manufacturer. With the presently available technology, the maximum number of wavelengths that can be placed onto a fiber using DWDMs lies anywhere from 4 to 80.]

8. A 2 × 2 biconical fiber coupler has an input optical power level of $P_0 = 200$ μW. The output powers at the other three ports are $P_1 = 90$ μW, $P_2 = 85$ μW, and $P_3 = 6.3$ μW. Calculate coupling ratio, excess loss, insertion loss: port 0 to 1 port, port 0 to port 2, return loss.

   [B.Tech.]

   [**Hint.** We have splitting ratio $= \dfrac{P_2}{P_1 + P_2} \times 100\%$

   (i) Coupling ratio $= \left(\dfrac{85}{90 + 85}\right) \times 100\% = 48.6\%$

   (ii) Excess loss $= 10 \log \dfrac{P_0}{P_1 + P_2} = 10 \log \left(\dfrac{200}{90 + 85}\right) = 0.58$ dB

   (iii) Insertion loss $= 10 \log \left(\dfrac{P_i}{P_j}\right)$

   (a) Port 0 to Port 1 $= 10 \log \left(\dfrac{200}{90}\right) = 3.47$ dB

   (b) Port 0 to Port 2 $= 10 \log \left(\dfrac{200}{85}\right) = 3.72$ dB

   (iv) Return loss $= 10 \log \left(\dfrac{P_3}{P_0}\right)$

$$= 10 \log \left(\frac{6.3 \times 10^{-3}}{200} \right) = -45 \text{ dB}$$

9. A symmetric waveguide coupler has a coupling coefficient $\kappa = 0.6$ mm^{-1}. Find the coupling length for $m = 1$. [B.Tech.]

 [**Hint.** We have $L = \dfrac{\pi}{2\kappa}(m+1)$ with $m = 0, 1, 2, ...$

 Here $\kappa = 0.6$ mm^{-1}, m = 1

 $\therefore$ $L = 5.24$ mm.

10. A 32 × 32 single-mode coupler is made from a cascade of 3 dB fused fiber 2 × 2 couplers, where 5% of the power is lost in each element. Calculate the total loss. [B.Tech.]

 [**Hint.**
 (i) Excess loss $= -10 \log (F_T \log_2 N)$
 $= -10 \log (0.95 \log 32 / \log 2) = 1.1$ dB
 (ii) Splitting loss $= -10 \log \dfrac{1}{N} = -10 \log 32 = 15$ dB
 $\therefore$ Total loss $= 16.1$ dB.

SHORT ANSWER QUESTIONS

1. What is WDM?

Ans. Wavelength division multiplexing (WDM) is an optical technology that permits several wavelengths to be coupled into the same fiber cable, effectively increasing the aggregation bandwidth per fiber fable.

2. What is time-division multiplexing (TDM)?

Ans. TDM is a technique for transmitting multiple digitized data, voice, and video signals simultaneously over one fiber cable. This is accomplished by interleaving pulses representing bits from channels or time slots.

3. What is frequency-division multiplexing (FDM)?

Ans. FDM is a scheme in which numerous signals are combined for transmission on a single communication line or channel. Each signal is assigned a different frequency (sub-channel) within main channel. This technology is used in broadcast, radio, television and cable television.

4. What is dense wavelength division multiplexing (DWDM)?

Ans. DWDM is an optical technology used to increase bandwidth (BW) over existing fiber-optic backbones.

 DWDM works by combining and transmitting multiple signals simultaneously at different wavelengths on the same fibers.

5. What is coarse wavelength division multiplexing (CWDM)?

Ans. CWDM is a technology that combine upto 16 wavelengths onto a single fiber. When there are just a few channels (upto 16 channels) and they are spaced more widely (10 nm or more) apart, the system is called CWDM.

6. What method allows large number of independent, selectable channels to exist on a single fiber?

Ans. Frequency division multiplexing (FDM).

7. Mention the advantages of WDM?

Ans. (i) Capacity upgradation, (ii) Transparency, (iii) Wavelength routing, and (iv) Wavelength switching.

8. What are passive optical couplers?

Ans. Passive devices operate completely in the optical domain to split and combine light streams. They include $N \times N$ couplers (with $N \geq 2$), power splitters, power taps, and star couplers. These components can be fabricated either from optical fibers or by means of planar optical waveguides using material such as lithium niobate ($LiNbO_3$), InP, silica, silicon oxynitride, or various polymers.

9. What is splitting ratio or coupling ratio?

Ans. In order to specify the performance of an optical coupler, one usually indicates the percentage division of optical power between the output ports by means of splitting ratio or coupling ratio, splitting ratio or coupling ratio $= \left(\dfrac{P_2}{P_1 + P_2}\right) \times 100\%$ where P_1 and P_2 are output powers and P_0 being the input power.

10. What is excess loss?

Ans. The excess loss is defined as the ratio of the input power to the total output power. In decibels, the excess loss for a 2×2 coupler is:
$$\text{Excess loss} = 10 \log \left(\dfrac{P_0}{P_1 + P_2}\right)$$

11. What is insertion loss?

Ans. The insertion loss refers to the loss for a particular port-to-port path, e.g. for the path from port i to port j, we have in decibels:
$$\text{Insertion loss} = 10 \log \left(\dfrac{P_i}{P_j}\right)$$

12. What is cross talk or return loss?

Ans. Return loss measures the degree of isolation between the input at one port and the optical power scattered or reflected back into the other input port, i.e., it is a measure of the optical power level P_3. Thus:
$$\text{Return loss} = 10 \log \left(\dfrac{P_3}{P_0}\right)$$

Wavelength Division Multiplexing (WDM) Technology

13. What are star couplers?

Ans. A star coupler is simply a multiple output coupler in which each input signal is made available on every output fiber. The principal role of all star couplers is to combine the powers from *N*-inputs and divide them equally (usually) among *M* output ports. Usual techniques for creating star couplers include fused fibers, gratings, micro-optic technologies, and integrated-optics schemes.

14. What are optical isolators?

Ans. These are devices that allow light to pass through them in only one direction. This is important in a number of instances to prevent scattered or reflected light from travelling in the reverse direction. One common application of an optical isolator is to keep such backward-travelling light from entering a laser diode and possibly causing instabilities in the optical output.

15. What are optical circulators?

Ans. An optical circulator is a non-reciprocal multi-port passive device that directs light sequentially from port to port in one direction only. This device is used in optical amplifiers, add/drop multi-plexers, and dispersion compensation modules.

16. What are the three common network topologies?

Ans. Star, ring and mesh networks. In the star topology, each node has a transmitter and receiver, with the transmitter connected to one of the passive central star's inputs and receiver connected to one of the star's outputs.

Many electrical networks use ring topology because rings are easy to implement for any geographical network configuration.

In both star and ring topologies, each node has a signature wavelength, and any two nodes can communicate with each other by transmitting and recovering that wavelength.

In the mesh network, the nodes are connected by reconfigurable optical crossconnects (OXCs). The wavelength can be dynamically switched and routed by controlling the OXCs. Therefore, the required number of wavelengths and wavelength tunable range of the components can be reduced in this topology.

17. Mention some generic laudable goals which a WDM-device technologist aims to achieve:

Ans.   (i) Large wavelength tuning range
 (ii) Multi-user capability
 (iii) Wavelength stability and repeatability
 (iv) Low crosstalk
 (v) High extinction ratio
 (vi) Minimum excess losses
 (vii) Fast wavelength tunability (especially for packet switching)

660 Photonics | Optoelectronics

  (viii) High speed modulation bandwidth
  (ix) Low residual chirp
  (x) High finesse
  (xi) Low noise
  (xii) Robustness
  (xiii) High yield
  (xiv) Potential low cost

18. Write advantages of WDM.

Ans. Important advantages of WDM are as follows:
  (i) *Capacity upgrade*: Since each wavelength supports independent data rate in Gbps.
  (ii) *Transparency*: WDM can carry fast asynchronous, slow synchronous, synchronous analog and digital data.
  (iii) *Wavelength routing*: Link capacity and flexibility can be inserted by using multiple wavelength.
  (iv) *Wavelength switching*: WDM can add or drop multiplexers, cross connects and wavelength converters.

19. What are optical couplers?

Ans. Optical couplers are passive devices which either split optical signal into multiple paths, or combine several signals into one pair. A prime characteristics of couplers is the number of input and output ports, which is typically expressed as an $N \times M$ configuration, where N represents the number of inputs and M represents the number of outputs.

MULTIPLE CHOICE QUESTIONS

1. The technology that combines a number of wavelengths into the same fiber cable is known as:
   (a) WDM   (b) DWDM
   (c) Optical amplifier (d) Optical detector

2. Passive optical couplers include:
   (a) only NXN couplers (with $N \geq 2$)
   (b) only power splitters
   (c) only power taps and star couplers
   (d) All of the above

3. In general, an NXM couplers has:
   (a) N inputs only
   (b) M outputs only
   (c) N inputs and M outputs
   (d) None of the above

4. Couplers are devices that are used to:
   (a) combine and split optical signals
   (b) block communication message

(c) transfer information
(d) None of the above
5. The principal role of all star couplers is to combine the powers from N inputs and divide them usually:
   (a) unequally among M output ports
   (b) equally among M output ports
   (c) sometimes equally sometimes unequally among M output ports
   (d) None of the above
6. Which of the following schemes is most suitable for DWDM?
   (a) Arrayed waveguide grating multiplexer
   (b) Mach–Zehnder interferometer
   (c) Fiber Bragg gratings
   (d) Blazed reflection gratings
7. Which of the following tunable filters is most suitable for DWDM?
   (a) Mach–Zehnder interferometer
   (b) Fabry–Perot filters
   (c) Acousto-optic tunable filters
   (d) Fiber Bragg gratings
8. A 2 × 2 durectuibak ciyoker gas an input power level of 100 µW. The power available at output ports 1 and 2 are, respectively, 45 µW and 45 µW. What is the coupling ratio?
   (a) 45% (b) 50%
   (c) 90% (d) 100%
9. For the coupler of Question 8, what is the excess loss?
   (a) 3 dB (b) 1 dB
   (c) 0.5 dB (d) 0.46 dB
10. For the coupler of Question 8, what is the insertion loss for the path from input port 1 to output port 2?
    (a) 3.46 dB (b) 5.23 dB
    (c) 6.92 dB (d) 10 dB
11. A 1 × 10 coupler has an input signal 0 dB·m. What is the power level at each output port?
    (a) 0 dB·m (b) −1 dB·m
    (c) −3 dB·m (d) −10 dB·m
12. The function of wavelength-division multiplexer is to:
    (a) separate signals at different wavelengths and couple of them to different detectors
    (b) combine signals at different wavelengths to pass through a single fiber

(c) tap off part of the energy of the incoming signal
(d) change the transmission speed of the input signal

13. The scheme of WDM is similar to:
    (a) FDM for *rf* transmission (b) TDM
    (c) SDM   (d) OTDM

14. What is the channel spacing (in nm) specified by the ITU-T recommendation G.692 for DWDM?
    (a) 1.6 nm   (b) 0.8 nm
    (c) 0.4 nm   (d) 0.2 nm

15. What is the channel spacing (in GHz) corresponding to the wavelength of Question 14?
    (a) 200 GHz   (b) 100 GHz
    (c) 50 GHz   (d) 25 GHz

16. An optical circulator is a non-reciprocal ....
    (a) single-port passive device
    (b) two-port passive device
    (c) 2 × 2 port passive device
    (d) multiport passive device

ANSWERS

1. (a)   2. (d)   3. (c)   4. (a)   5. (b)   6. (a)   7. (a)
8. (b)   9. (d)   10. (a) 11. (d) 12. (b) 13. (a) 14. (b)
15. (b) 16. (d)

13

Fiber Optic Communication

13.1 INTRODUCTION

Optical communications is very old form of data transfer. Line-of-sight primitive digital systems have included lighting bone fires on mountain tops to send a simple one-bit message, smoke signals to send a multiple-bit message, and ship-to-ship road incoherent beam transmission of Morse-code messages. The inventions of *low-loss optical fiber* and the *high-speed semiconductor* laser have caused an explosion in the transmission capacity of optical systems. The availability of information depends on the transmission speed of data, voice, and multimedia across telecommunication networks. Despite new technologies that enable legacy copper telephone lines to carry information more efficiently, optical networks remain the most ideal medium for *high-bandwidth* communications. There are two distinct modes of communications:

  (i) *fiber optics* (*fiber optic cable*), and
  (ii) *optical wireless* based on free-space optic technology. For long-distance network deployments, nothing is better than fiber. When coupled with the *wavelength division multiplexing* technologies, fiber optic cables are capable of carrying information more densely across the globe.

One of the most basic and undisputed applications of *optical* rather than *electrical communication* is data transmission on long-distance links. Optics is an obvious choice because of the ultra-wide bandwidth (> 25 THz) of low-loss (< 0.2 dB/km) transmission properties of an optical fiber, while the impedance of the electrical cable increases substantially at GHz speeds to > 100 dB/mile. In addition to *ultra-wide* bandwidth transmissions, a *wavelength division multiplexing* (WDM) optical network also enables straightforward routing and switching of optical data paths, in which the data travels on wavelength-specific "light-paths" from source to destination that can be arranged by *network controller* to optimize throughput.

The progress in optical communications over the past three decades has been asounding. The advances of optical fiber systems have progressed in five *generation* of technology in the past three decades:

i. The *first generation* of optical fiber system used 0.8 μm GaAs semiconductor lasers and multimode fibers. Other specification of this generation are as under:
Bit rate: 50–100 Mbit/s
Repeater spacing: 10 km in length

ii. *Second generation* employed single mode fibers and 1.3 μm InGaAs and 1.55 μm InGaAsP lasers. Other specifications are:
Bit rate: 100 Mbit/s to 1.7 Gb/s
Repeater spacing: 50 km
Operating wavelength: 1.3 μm

iii. *Third generation* also employed single mode fibers and 1.3 μm InGaAs and 1.55 μm InGaAsP lasers. Other specifications are:
Bit rate: 10 Gb/s
Repeater spacing: 100 km
Operating wavelength: 1.55 μm

iv. *Fourth generation*: The fourth generation systems employed *optical amplifiers* and WDM technologies:
Bit rate: 10 Tbs^{-1}
Repeater spacing: > 10,000 km
Operating wavelength: 1.45 to 1.62 μm

v. *Fifth generation*: Fifth generation uses Roman amplification technique and optical solitons.
Bit rate: 40–160 Gbs^{-1}
Repeater spacing: 24000 km–35000 km
Operating wavelength: 1.53 to 1.57 μm

Today, the high-capacity amplified WDM systems have hundreds of channels at 10 Gbs^{-1} with channel spacing as low as 50 GHz and distances to a few thousand kilometers. Systems operate at 40 Gbs^{-1} channel^{-1} rate already in operation.

The *next generation* systems will be beyond fiber transmission links to optical networks. There are major efforts to promote the concept of Fiber-to-the X (FTTX:X = Curb, Home, Desktop, etc.) to support the demand for voice, data, and internet application across metropolitan areas.

Thus, optical communications is a fairly large and still rapidly-advancing scientific field. The use of wavelength-division multiplexing (WDM) have boosted in transmission capacity. *The basic principle of WDM is to use multiple sources at slightly different wavelengths* to send several independent packets of information data over the same fiber cable. For example, one of many of the world's WDM optical networks the SEA-ME-WE3-cable system. This undersea network runs from

Germany to Singapore, connecting many countries in between; hence, the name SEA-ME-WE, which refers to Southeast Asia (SEA), the Middle East (ME), and the West Europe (WE). The network has two pairs of undersea fibers with a capacity of eight STM-16 wavelengths per fiber (equivalent to eight OC-48 which is 8×2.5 Gb/s).

Another submarine cable system, the undersea network connects many countries between Portugal (Sesimbra) to Malaysia (Penang). The route distance between these two countries is 23,455 km.

13.2 ESSENTIAL COMPONENTS OF FIBER COMMUNICATION SYSTEM

The basic components of optical fiber communication system are shown in Fig. 13.1(a). The three major components are:
  (i) Transmitter,
  (ii) Medium (optical fiber cable) or information channel, and
  (iii) Receiver.

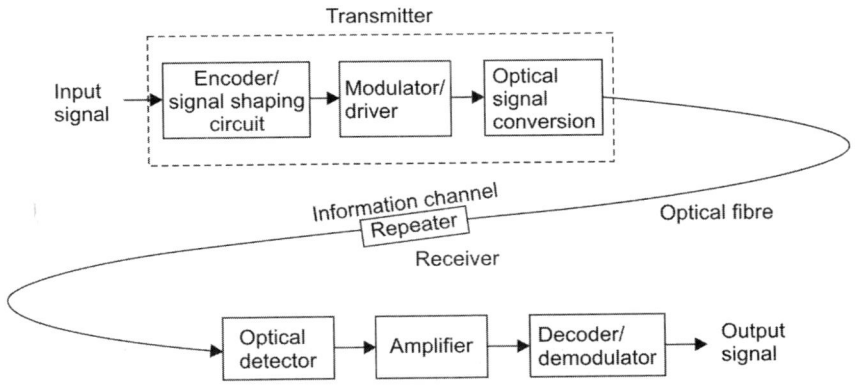

Fig. 13.1 (a) Schematic diagram of optical fiber communication

The transmission system consists of encoder, modulator/driver and optical signal converter, the medium or informal channel is optical fiber and receiving system consists of optical detector, amplifier decoder/demodulator circuit.

System design considerations for point to point links: The problems (most) associated with the design of fiber optic communication systems are due to the unique properties of the optical fiber serving as a transmission medium.

Information can be classified in two general forms, *analog* and *digital*. *Analog* signals (Fig. 13.1(b)) represent quantities which can take on any value within an infinite continuum of amplitude and time. This kind of signal is typical of many measured quantities, such as temperature,

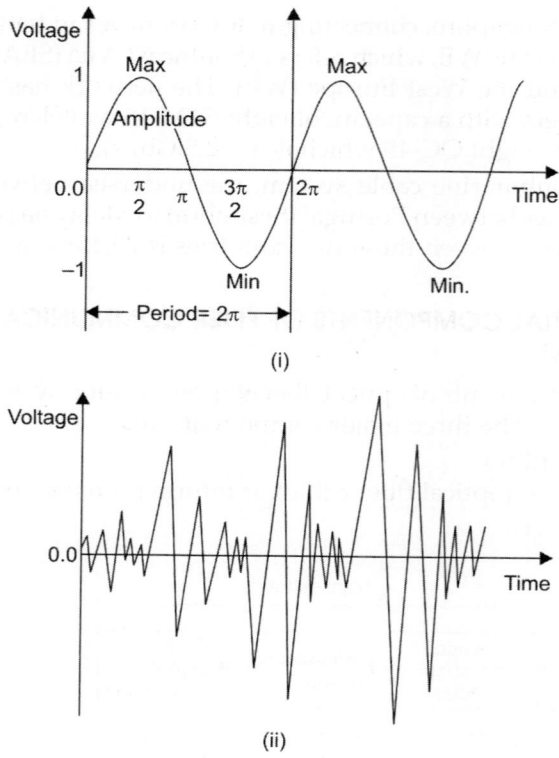

Fig. 13.1. (b) Analog signals. (i) Sine wave ($y = \sin\theta$), (ii) Analog voice wave

volume, frequency, dimensions, etc. Analog signals, in theory ultimately accurate. However, to achieve a high-degree of accuracy in signal recovery, the noise must be kept extremely low compared to the signal power ($\sim -50d$). Alternatively, *digital signals* can represent many quantities which may only require measuring of a limiting number of discrete levels at specific bit-time interval. It is only necessary to transmit and detect the difference between discrete levels (i.e., between "O" and "I" levels). Consequently the noise in relation to the signal that can be tolerated for error-free transmission is much higher than in analog transmission and is typically ~ -20 dB.

We may note that in common with other communication systems, the major design criteria for a specific application using either digital or analog techniques of transmission are the required transmission distance and also the speed of information transfer. These criteria are directly related to two important parameters of optical fibers: (i) *Fiber attenuation*, and (ii) *Fiber dispersion*. Keeping these facts in mind, we will discuss system design consideration for *digital systems* and for analog systems.

(a) Digital Systems

The simplest kind of *fiber-optic link* is the simplex (one-directional) point-to-point link having an optical transmitter at one end and an optical receiver at the other (Fig. 13.1(c)). The key system parameters needed to analyse this link are (i) the desired (or possible) transmission distance (without any repeaters), (ii) the data rate, and (iii) a specified bit-error rate (BER). In order to fulfil these requirements, the system designer has many choices. The major ones are as follows:

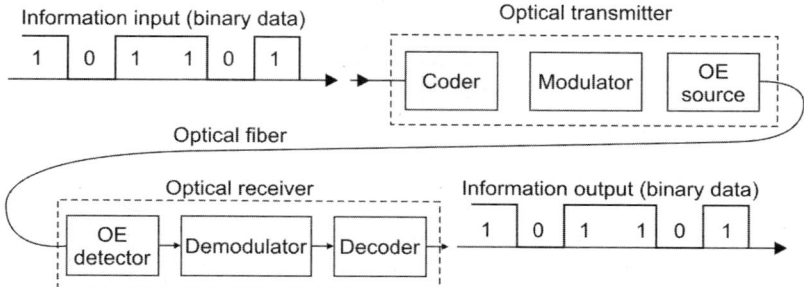

Fig. 13.1. (c) A typical simple *point-to-point digital fiber-optic link*. The information input is binary data (shown as a series of 1's and 0's). A coder in the optical transmitter organizes the 1's and 0's and the modulator acts on this data by producing a current that turns the OE source (LED or ILD) on and off. The resulting pulses of light containing the information are transmitted through the optical fiber. At the receiver end, the OE detector converts the pulses of light into pulses of current. A demodulator-decoder combination extracts the information from the electrical pulses

Converting Analog Signal to Digital Signal

The term digital is used frequently because digital circuits are becoming so widely used in computation, robotics, medical science and technology, communications, transportation, etc. Digital electronics developed from the principle that the circuitry of a transistor could be designed

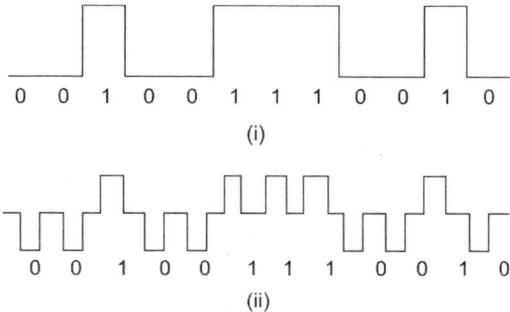

Fig. 13.1. (d) Common digital signal configuration (i) Digital pulse stream. (ii) Dipolar return-to-zero pulse

and easily fabricated to have an output of one or two voltage levels, based on its input voltage (Fig. 13.1d).

Figure (13.1(e)) shows an *analog-to-digital converter* (ADC) used to convert an analog signal into a digital signal. The ADC can be a single chip, or can be one circuit within a chip. The two voltage levels are usually 5 V (high) and 0 V (low); the levels can be represented by 1 and 0.

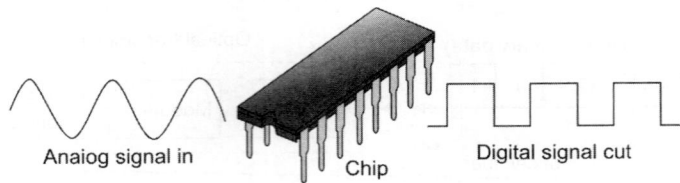

Fig. 13.1. (e) A chip used to convert an analog to digital signal

The binary numbering system (base-2 numbering system) is the main numbering system used in digital electronics. A digital value is represented by a combination of on and off voltage levels, and expressed as a string of 1s and 0s (Fig. 13.1d). Signals that have theoretically infinite number of levels are converted into signals that have two defined levels.

To convert an analog signal to a digital form, one starts by taking instantaneous measures of the height of the analog signal wave at regular interval, this is called sampling the signal. One way to convert these analog samples to a digital format is to simply divide the amplitude of the analog signal into N equally spaced levels, which are designated by integers, and to assign values to one of these integers. This process is called *quantization*. Since the signal varies continuously in time, this process generates a sequence of real numbers.

Figure 13.1(f)) shows the equally spaced levels that are the simplest method of quantization produced by a uniform quantizer. If the digitization samples are taken frequently enough relative to the rate at which the signal varies, then a good approximation of the signal can be recovered from the samples by drawing a straight line between the sample points. The resemblance of the reproduced signal to the original signal depends on the fineness of the quantizing process, and on the effect of noise and distortion added into the transmission system. If the sampling rate is at least two times the highest frequency, then the receiving device can easily reconstruct the analog signal. Thus, if a signal is limited to bandwidth of B Hz, then the signal can be reproduced without distortion if it is sampled at a rate of $2B$ times per second. These data samples are represented by a binary code.

As shown in Fig. 13.1(f), eight quantized levels having upper bounds, $V_1, V_2, V_3, \ldots V$, can be described by 4 binary digits (for example 8 in

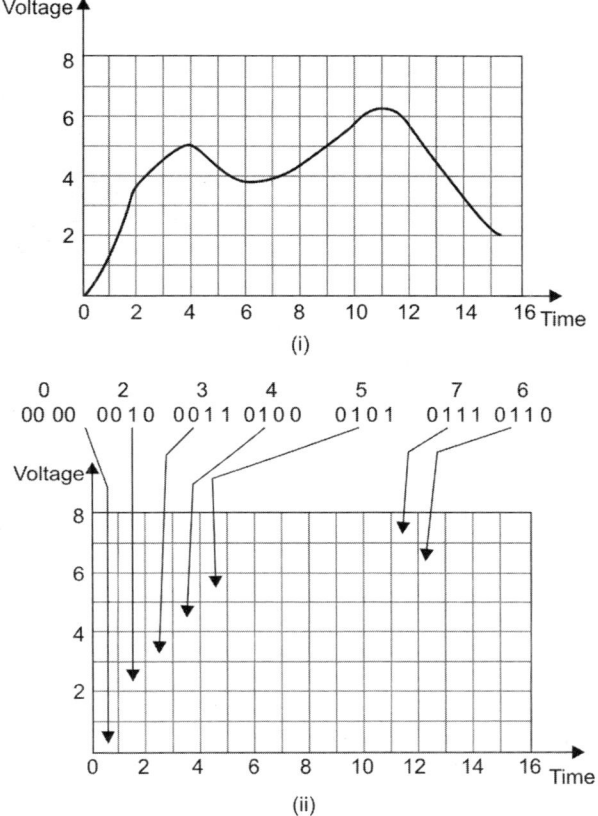

Fig. 13.1. (f) Digitization of analog waveforms. (i) Analog signal varying between 0 to V volts. (ii) Quantized and sampled digital signal

binary becomes $2^3 = 2^3\, 2^2\, 2^1\, 2^0 = 1\,0\,0\,0$). More digits can be used to give finer sampling levels. Thus, if n binary digits represent each sample, then one can have 2^n quantization levels. Fig. 13.1(g) shows a conversion process of an analog signal to digital form, and then to bar code.

Bit Error Rate (BER)

The performance of a digital communication system is measured by the probability of an error occurring in a data bit; a bit error rate (BER) equal to 0 is ideal. Define p_1 as the probability of misinterpreting a 1 bit as a 0, and p_0 as the probability of misinterpreting 0 as a 1. If the 0 or 1 bits are equally likely to be transmitted, then BER = $1/2\, p_1 + 1/2\, p_0$. In telecommunication systems, a typical acceptable BER is 10^{-9} (i.e., an average of one error every 10^9 bit).

Optical fiber devices used in any communication system, such as light sources (transmitters) photodetectors (receivers), fiber optic cables, multiplexers, demultiplexers, optiocal amplifiers, isolators, circulators, and optical switches are explained in detail throughout the book.

 (i) *Optical fiber*:
   (a) Multimode or single mode,
   (b) Core size
   (c) Refractive index profile
   (d) Attenuation
   (e) Dispersion
 (ii) *Optical transmitter*:
   (a) LED or ILD source
   (b) Operating wavelength (emission wavelength of LED or ILD)
   (c) Output power and emission pattern
   (d) Transmitter configuration
   (d) Modulation and coding
 (iiii) *Optical receiver*:
   (a) *p-i-n* or avalanche photodiode
   (b) Responsivity at the operating wavelength
   (c) Pre amplifier design (low impedance, or transimpedance front end)
   (d) Demodulation and decoding

Fig. 13.1. (g) Conversion process of an analog signal to digital and to bar code.

We may note that decision regarding the above requirements are interdependent. However, one should kept in mind that the choices are made to optimise the system performance for a particular application.

To ensure that the desired system performance can be met, normally, two types of system analysis are carried out: (i) link power budget analysis, and (ii) rise time budget analysis. The first analysis determines the power margin between the optical transmitter output and the minimum required receiver sensitivity, so that this margin may be allocated to connector, splice, and fiber losses, or to any future degradation of components. The second analysis ensures that the desired overall system performance has been met. We now examine in detail these two types of system analysis.

(i) Link Power Budget Analysis

For optimizing link power budget an optical power loss model is to be studied as shown in Fig. 13.2.

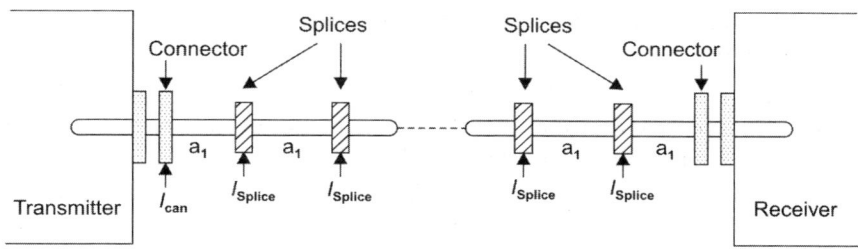

Fig. 13.2. Optical power loss model

The optical power received at the photodetector depends on the amount of optical power coupled into the fiber by the transmitter and on the losses occurring in the optical fiber as well as at the connectors and splices. This means that the power budget is derived from the sequential contributions at all the loss elements in the link. However, in addition to the link loss contributors, a safety margin of 6–8 dB is normally provided to allow for any future degradation of components and/or future addition of splices, etc. Thus, one can assume that the average power supplied by the transmitter is P_{tx}, the sensitivity of the receiver is P_{rx}, the total link loss of channel loss (including the fiber splice and connector losses is P_T, and the system's safety margin is M_s, then one finds that the following relationship should be satisfied:

$$P_{tx} = P_{rx} + C_L + M_s \qquad (13.1)$$

The channel loss C_L may be expressed as:

$$C_L = \alpha_f L + \alpha_{con} + \alpha_{splice} \qquad (13.2)$$

where α_f is the fiber loss (in dB/km), L is the link length, i.e., the transmission distance, α_{con} is the sum of losses due to all connectors in the link, and α_{splice} is the sum of losses at all the splices in the link. In Eqs (13.1) and (13.2), P_{tx} and P_{rx} are expressed in dB·m, and C_L, M_s, α_{cone} and α_{splice} are expressed in dB. One can use Eqs (13.1) and (13.2) to estimate the maximum transmission distance for a given choice of components.

(ii) Rise-Time Budget Analysis

Rise time gives important information for initial system design. Rise-time budget analysis determines the dispersion limitation of an optical fiber link. Rise time budget analysis is particularly important in the case of digital systems, where it is to be ensured that the system will be able to operate satisfactorily at the desired bit rate. One can define the rise time t_r of a linear system as the time during which the system's response increases from 10% to 90% of the maximum output value when its input changes abruptly (a step function).

We consider a simple RC circuit (Fig. 13.3) as an example of linear system. When the input voltage (V_{in}) across this circuit changes abruptly (i.e., in a step) from 0 to V_0, the output voltage (V_{out}) changes with time in accordance with the relation:

$$V_{out}(t) = V_0[1 - \exp(-t/RC)] \qquad (13.3)$$

where R is the resistance and C the capacitance of the circuit shown in Fig. 13.3. The rise time t_r is expressed as:

$$t_r = 2.2\,RC \qquad (13.4)$$

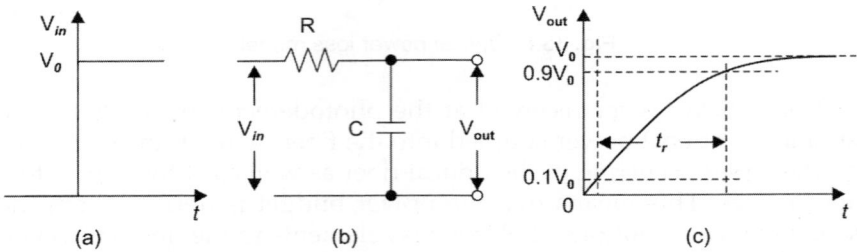

Fig. 13.3. The response of a low-pass RC filter circuit to a voltage step input V_0

Taking *Fourier transform* of Eq. (13.4), one obtains the transfer function $H(f)$ as:

$$H(f) = \frac{1}{(1 + 2\pi jf\,RC)} \qquad (13.5)$$

The 3 dB electrical bandwidth Δf for the circuit corresponds to the frequency at which $|H(f)|^2 = 1/2$ and is given by the following relation:

$$\Delta f = \frac{1}{2\pi\,RC} \qquad (13.6a)$$

Therefore, using Eqs (13.4) to (13.6a), one can relate t_r to Δf by the relation:

$$t_r = \frac{2.2}{2\pi\,\Delta f} = \frac{0.35}{\Delta f} \qquad (13.6b)$$

In Fig. 13.1(b), we have seen that in a fiber-optic communication system, there are three building blocks, and each block has its own rise time associated with it. Therefore, the total rise time of the system, t_{sys}, is obtained by taking the root-sum-square of rise time of each contributor, i.e. block. If we assume that the rise times associated with the transmitter, fiber, and receiver are respectively t_{tx}, t_f and t_{rx}, then, we can write:

$$t_{sys} = \left[t_{tx}^2 + t_f^2 + t_{rx}^2\right]^{1/2} \qquad (13.7)$$

We must note:

(i) The link components must be switched fast enough and the fiber dispersion must be low enough to meet the bandwidth

Fiber Optic Communication **673**

requirements of the application. Adequate bandwidth for a system can be assured by developing a rise time budget.

(ii) As the light sources and detectors has a finite response time to inputs. The device does not turn-on or turn-off instantaneously. Rise time and fall time determines the overall response time and hence, the resulting bandwidth.

(iii) Connectors, couplers and splices do not affect system speed, they need not be accounted in rise time budget but they appear in the link power budget.

Further, the rise time of the optical fiber include the contribution of *intermodal dispersion* ($t_{\text{intermodal}}$) and *intramodal dispersion* ($t_{\text{intramodal}}$) through the relation:

$$t_f = \left(t_{\text{intermodal}}^2 + t_{\text{intramodal}}^2\right)^{1/2} \tag{13.8}$$

In the absence of mode coupling, $t_{\text{intermodal}}$ and $t_{\text{intramodal}}$ are normally approximated by the time delays (ΔT) caused by intermodal and intramodal dispersion, respectively.

There are four basic elements that contributes to the rise time are:
- Transmitter rise-time (T_{tx})
- Group-velocity dispersion (GVD) rise time (t_{GVD})
- Modal dispersion rise time of fiber (t_{mod} or t_f)
- Receiver rise-time (t_{tx})

$$t_{\text{sys}} = \left[t_{tx}^2 + t_{\text{mod}}^2 + t_{\text{GVD}}^2 + t_{rx}^2\right]^{1/2} \tag{13.9}$$

Rise time due to modal dispersion is given as:

$$t_{\text{mod}} = \frac{440}{B_M} = \frac{440L^q}{B_0} \tag{13.10}$$

where,
B_M is bandwidth (MHz),
L is length of fiber (km),
q is a parameter ranging between 0.5 and 1, and
B_0 is bandwidth of 1 km length fiber.

Rise time due to group velocity dispersion is:

$$t_{\text{GVD}} = D^2 \sigma_\lambda^2 L^2 \tag{13.11}$$

where,
D is dispersion [ns/(nm·km)],
σ_λ is half-power spectral width of source, and
L is length of fiber

Receiver front end rise-time in nanoseconds is:

$$t_{rx} = \frac{350}{B_{rx}} \tag{13.12}$$

where, B_{rx} is 3 dB-BW of receiver (MHz).

Equation (13.9) can be written as:
$$t_{sys} = \left[t_{tx}^2 + t_{mod}^2 + t_{GVD}^2 + t_{rx}^2\right]^{1/2}$$

$$t_{sys} = \left[t_{tx}^2 + \left(\frac{440\,Lq}{B_0}\right)^2 + D^2\sigma_\lambda^2 L^2 + \left(\frac{350}{B_{rx}}\right)^2\right]^{1/2} \tag{13.13}$$

We may note that times are in nanoseconds (ns).
The system bandwidth is given by:
$$BW = \frac{0.35}{t_{sys}} \tag{13.14}$$

$$ct_{sys} = \sqrt{(1.75^2 + 3.5^2 + 3.89^2 + 1.00^2 + 1.94^2)}$$
$$t_{sys} = 5.93 \text{ ns}$$

System BW is given by:
$$BW = \frac{0.35}{t_{sys}} = \frac{0.35}{5.93 \text{ ns}} = 59 \text{ MHz} \tag{13.15}$$

We may note that that the optical power generated by the optical transmitter is generally proportional to its *inpout current*, and the *optical power received by the receiver* is proportional to the *power launched into and propagated* by the optical fiber. Finally, the output of the receiver is also proportional to its input. This means that a fiber-optic communication system can be considered to be a *band-limited linear system*, and hence Eq. (13.6(a)) is valid for this system too. Therefore, for a fiber-optic communication system, the *total rise time* t_{sys} may be expressed as:

$$t_{sys} = \frac{0.35}{\Delta f} \tag{13.16}$$

Now, the relationship between the electrical bandwidth Δf and the bit rate B depends on the digital pulse format. For the return-to-zero (RZ) format, $\Delta f = B$ and for the non-return-to-zero (NRZ) format $\Delta f = B/2$. Therefore, for digital system, t_{sys} should be below its maximum value given by:

$$t_{sys} \leq \begin{cases} \dfrac{0.35}{B} & \text{for the RZ format} \\ \dfrac{0.70}{B} & \text{for the NRZ format} \end{cases} \tag{13.17}$$

We may note that the RZ and NRZ formats, to be discussed in the next subsection, are used for signal encoding.

Line coding: There is an important criterion in the design of optical fiber link, i.e., the decision circuit in the receiver should be able to extract precise timing information from the incoming optical signal. The precise timings are desired to (i) allow the signal to be sampled by the receiver at a time when the signal-to-noise ratio is maximum, (ii) maintain proper pulse spacing, and (iii) indicate the start and end of each timing interval. We may remember that channel noise and the distortion mechanism may cause errors in the signal detection process. Obviously, one will desire for the transmitted optical signal to have an inherent error-detecting capability. There is possibility to incorporate these features into the data stream by restructuring or encoding the signal. The main function of time coding is to introduce redundancy into the data stream for the sake of minimizing errors, especially those resulting from channel interference effects.

Let us try to understand line codes. For this purpose, we want to get ourself familiarize with some commonly used terms:

(i) *Digital signal*: This comprises a series of discrete voltage pulses. An individual pulse in the total signal is termed a *signal element*. We may note that *binary* data are transmitted by encoding each data bit into signal elements. The element may be a positive or negative voltage pulse. However, when a signal consists of both positive and negative voltage pulses, it is termed as *bipolar*. When only one polarity of voltage pulse is present, the signal is called *unipolar*.

(ii) *Data rate (R)*: R is the transmission rate of data in bits per second.

(iii) *Bit duration* $(T_b = 1/R)$: T_b is the time taken by the transmitter to transmit one bit. One can use a variety of wave shapes (signal elements) to represent binary data.

(iv) *Encoding*: The mapping of binary data bits to signal elements is termed as encoding. There are three popular encoding schemes as shown in Fig. 13.4.

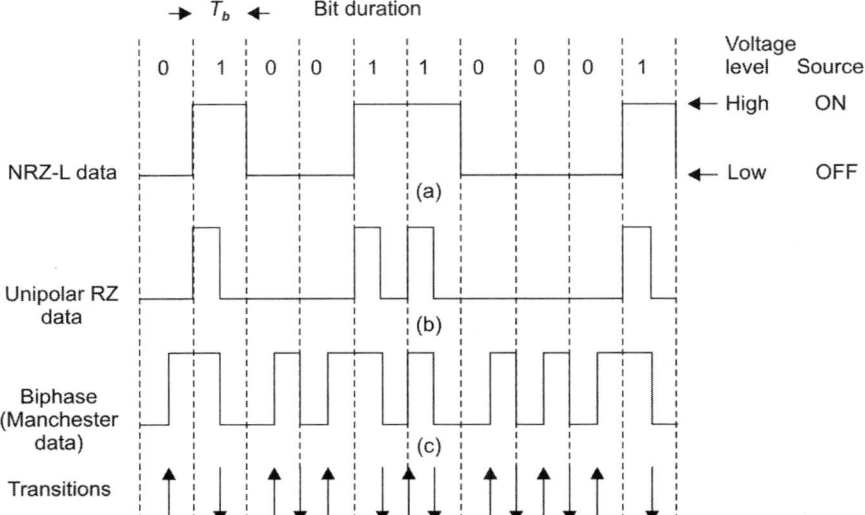

Fig. 13.4. Three popular encoding schemes: (a) NRZ, (b) RZ, and (c) Biphase

(a) *NRZ*: The simplest NRZ code is the NRZ level (or NRZ-L) as shown in Fig. 13.4(a), in which 1's or 0's of a serial data stream are represented by high voltage level and a 0 represented by low voltage level. Obviously, for a 1, there will be a light pulse filling the entire bit period, and for a 0, no light pulse will be transmitted. We may note that these codes are easier to generate and decode but do not possess an *inherent error-monitoring* capability. However, they make efficient use of bandwidth.

(b) *RZ*: This code differs from NRZ codes in that only half the bit period is used for data, while the voltage is zero in the secondhalf of the bit period. Obviously, in a unipolar RZ data format (Fig. 13.4(b)), a 1 is represented by a half-period optical pulse that occurs in the first half of the bit period and a 0 is represented by no signal during the bit period. We may note that the disadvantages of the unipolar RZ format are that it requires double the bandwidth of NRZ-L format, and a long string of 0's can cause loss of timing information.

(c) *Biphase-L*: This is a data format which possesses the virtues of easy time synchronization, no *dc* component, and some inherent facility for error detection. Biphase-L is also called as the *optical Manchester code* shown in Fig. 13.4(c). In biphase-L code, there is a transition in the middle of the bit interval, the voltage level is high for a 1 and low for a 0. In this scheme, obviously, a transition from high to low in the middle of the bit interval represents a 1 and a transition from low to high represents a 0. We may note that these codes are widely used in fiber-optic systems.

(b) Analog Systems

For long-haul communication links, digital systems with single-mode fibers are normally considered superior, even with their expensive terminal equipment for coding, multiplexing, timing, etc. The main reason for this is that an analog fiber-optic system required 20–30 dB higher signal-to-noise ratio as compared with that required for a similar digital fiber-optic system. However, for *short-haul* and *medium-haul links*, analog fiber-optic systems can be very attractive, especially for the transmission of video signals, because of their simplicity and cost effectiveness. There are several other application of analog systems. It is reported that for most applications, analog transmitters use *laser diodes*. In the design implementation of an analog system, the main parameters that need to be considered are carrier-to-noise ratio, bandwidth, and the signal distortion resulting from non-linearities in the transmission system. In such system, carrier-to-noise ratio analysis is used instead of signal-to-noise ratio, because the information signal is normally superposed on an *rf* carrier.

Figure 13.5 shows the basic elements of two types of analog fiber-optic links. In the first system (Fig. 13.5a), the optical transmitter contains either an LED or ILD as the optical source. The output intensity of the source is directly changed or modulated by the information-carrying analog signal. It is necessary to first set a bias point on the source approximately in the middle of the linear output region. The analog signal can then be transmitted using one of the several modulation techniques, the simplest one being direct intensity modulation. In this scheme, the optical output from the source is modulated by varying the current around the bias point in proportion to the level of the information signal. Thus, the signal is directly transmitted in the baseband. The modulated signal travels down the optical fiber and is demodulated at the receiver.

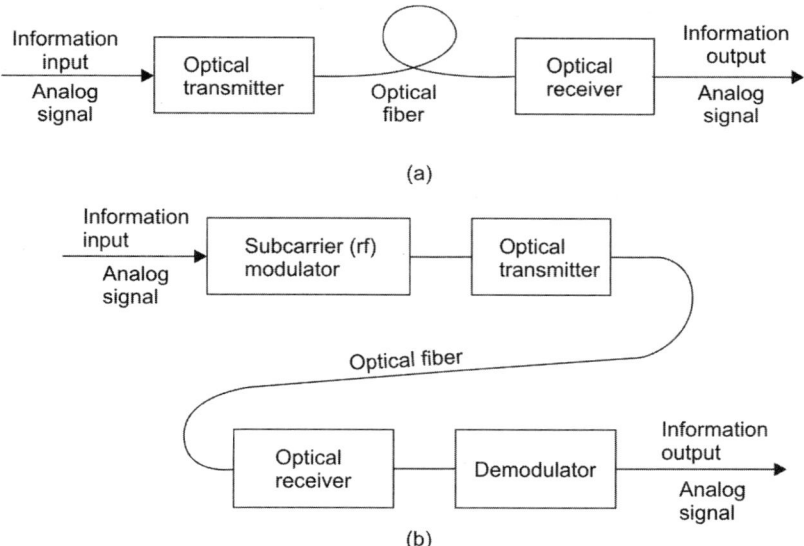

Fig. 13.5. An analog fiber-optic communication system: (a) type I—optical intensity is directly modulated by the analog signal, (b) type II—A subcarrier is modulated by the analog signal

There also exists a more efficient way of modulation in which the base-band signal is first translated into an electrical (subcarrier prior to intensity modulation as shown in Fig. 13.5(b). This is accomplished using standard techniques of amplitude modulation (AM), frequency modulation (FM), or phase modulation (PM). These modulation techniques are employed when there is a need to send multiple analog signals over the same fiber as in the case of broadband common antenna television (CATV) supetrunks.

The performance of analog systems is normally analysed by calculating the carrier-to-noise ratio (CNR) of the system. It is defined as the ratio of the rms carrier power to the rms noise power (resulting from the source, detector, amplifier, and intermodulation) at the output of the receiver, thus:

$$\text{CNR} = \frac{\text{rms carrier power}}{\text{rms noise power}} \tag{13.18}$$

For single-channel transmission the noise contributions of the source, detector, and amplifier are considered, whereas for the transmission of multiple information channels through the same fiber, the intermodulation factor is also considered. Here, we are examining a simple single-channel amplitude-modulated signal sent at baseband frequencies.

In order to determine carrier power, let us consider a laser transmitter. The optical signal variation by the source is caused by the drive current (through the source), which is a sum of the fixed bias and an analog input signal (a time-varying sinusoid), as shown in Fig. 13.6. An ILD acts as a *square-law device*, so that the envelope of the output optical power $P(t)$ has the same waveform as the input drive current. If we assume that the time-varying analog drive signal is $s(t)$, then, one may write:

$$P(t) = P_B [1 + ms(t)] \tag{13.19}$$

where P_B is the output optical power at the bias current level (I_B) and m

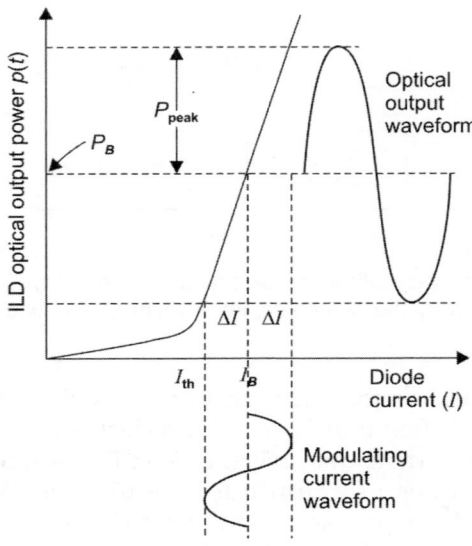

Fig. 13.6. Schematic representation of the biasing conditions of an ILD and its response to analog signal modulation

is the modulation index defined in terms of peak optical power P_{peak} as follows:

$$m = \frac{P_{\text{peak}}}{P_B} \tag{13.20}$$

For a time-varying sinusoidally received signal, the rms carrier power C at the output of the receiver (in units of A^2) is given by:

$$C = \frac{1}{2}(mRM\overline{P})^2 \tag{13.21}$$

where R is the unity gain responsivity of the photodetector, M is the gain of the APD ($M = 1$ for p-n and p-i-n photodiodes), and $\overline{P}$ is the average received optical power.

The rms noise power in Eq. (13.18) is the sum of the noise powers arising due to the source, photodetector, and pre-amplifier. The source noise, in the present case, will be given by:

$$\langle i^2_{\text{source}} \rangle = \text{RIN}(R\overline{P})^2 \, \Delta f \tag{13.22}$$

where RIN is the laser relative intensity noise measured in dB/Hz and is defined by:

$$\text{RIN} = \frac{\langle (\Delta P_L)^2 \rangle}{\overline{P}_L^2} \tag{13.23}$$

where $\overline{P}_L$ is the average laser light intensity and $\langle (DP_L)^2 \rangle$ is the mean square intensity fluctuation of the laser output. Δf is the effective noise bandwidth. This type of noise decreases as the injection current level increases.

The photodiode noise arises mainly due to shot noise and bulk dark current noise. Thus, combining the two, one obtains an expression for photodiode noise as follows:

$$\langle i^2_N \rangle = 2e\,(I_p + I_d)\,M^2 F(M)\,(\Delta f) \tag{13.24}$$

Here, $I_p = R\overline{P}$ is the primary photocurrent, I_d is the bulk dark current of the detector, M is the detector's gain with the associated noise figure F(M), and Δf is the effective noise bandwidth of the receiver. The thermal noise of the photodetector and the noise of the pre-amplifier may be combined in the expression:

$$\langle i^2_T \rangle = \frac{4kT}{R_{\text{eq}}}(\Delta f) F_t \tag{13.25}$$

where R_{eq} is the equivalent resistance of the photodiode and the pre-amplifier, and F_t is the noise factor of the pre-amplifier.

Substituting the values of the rms carrier power from Eq. (13.21) and the rms noise power [which is a sum of expressions (13.22), (13.24), and (13.25)] into Eq. (13.18), one obtains the CNR for a single-channel AM fiber-optic system as:

$$\text{CNR} = \frac{(1/2)(mRm\overline{P})^2}{\text{RIN}(R\overline{P})\,\Delta f + 2e(I_p + I_d)M^2 F(M)\,\Delta f + 4kT(\Delta f)F_t/R_{eq}} \quad (13.26)$$

Most of the general design considerations for digital fiber-optic systems described in Sec. 13.2 may be applied to analog systems as well. However, one must take extra care to ensure that the optical source and the photodetector have linear input-output characteristics, in order to avoid optical signal distortion. A careful link power budget analysis is a must because analog system require a higher SNR at the receiver than their digital counterparts. The temporal response of analog systems may be determined by rise-time calculations similar to those done for digital systems. In this case too, the maximum attainable bandwidth Δf is related to t_{sys} by Eq. (13.16).

13.3 BASIC COMMUNICATION SYSTEM

The following are basic definitions for the networks used in communication systems:

Station is a collection of devices with which users wish to communicate. such as computers, terminals, telephones, and videos. Stations are also called data terminal equipment (DTE) in network systems, and they can be connected directly to a transmission line.

Network is a group of two or more stations linked or interconnected by a transmission medium, such as a fiber optic cable or coaxial cable.

Topology is the logical manner or structure in which nodes/devices are linked together in information-transmission channels to form a network.

Switching is the transfer of information from source to destination through a series of intermediate nodes. A switch is a device that filters and forwards packets between *network segments*. Switches operate at the data link layer (layer 2) and sometimes at the network layer (layer 3) of the open system interconnection (OSI) reference model. A description of the layers is presented later in this chapter. Switches support any pocket protocol.

Routing is the process of moving a packet of data from the source to the destination by the selection of a suitable path through a network. Routing is usually performed by a dedicated device called a router.

Protocol is an agreed-upon format for transmitting data between two devices. The protocol determines the following items:
- Type of error checking to be used;
- Data compression method, if any; and
- Hand shaking, how the device indicates when finished sending a message.

A popular protocol used in optical LANs is the fiber distributed data interface (FDDI) protocol; SONET/SDH protocols are used in optical networks in metro or wider areas. Logical topologies are bound to the network protocols that direct how the data moves across a network. The Ethernet protocol is a common logical bus topology protocol.

13.4 TYPES OF TOPOLOGIES

Topology refers to the shape of a network, or the network's layout. There are different nodes in a network which communicate by the network's topology. Topologies are either physical or logical. Connections between the nodes are via optical couplers. The five most common network topologies are (i) bus topology, (ii) ring topology, (iii) star topology, (iv) mesh topology, and (v) tree topology.

(i) Bus Topology

In a bus topology, all stations are connected to a central cable which is called the bus or backbone, as shown in Fig. 13.7.

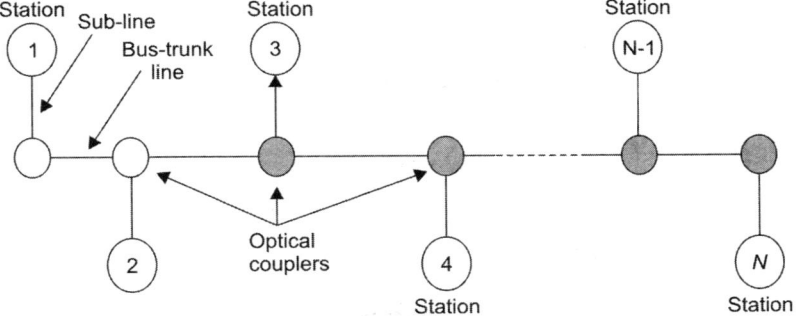

Fig. 13.7. Bus topology

(ii) Ring Topology

In a ring topology, all stations are connected to one another in the shape of a closed loop, so that each station is connected directly to two other stations on either side, as shown in Fig. 13.8.

(iii) Star Topology

In a star topology, all stations are connected to a central hub, which is an optical star coupler. Stations communicate across the network by passing data through the hub, as shown in Fig. 13.9.

(iv) Mesh Topology

In a a mesh topology, all stations are connected via many redundant interconnections, as shown in Fig. 13.10. In a true mesh topology, every station has a connection to every other station in the network.

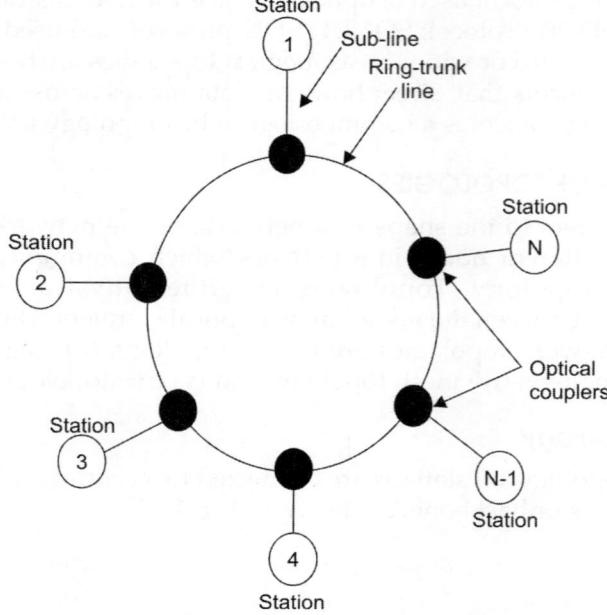

Fig. 13.8. Ring topology

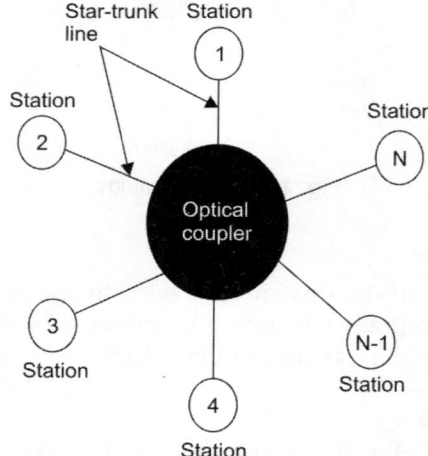

Fig. 13.9. Star topology

(v) Tree Topology

In a tree (hybrid) topology, all stations are connected by various topologies, as shown in Fig. 13.11. Groups of star-configured networks are connected to a linear bus backbone.

Fiber Optic Communication **683**

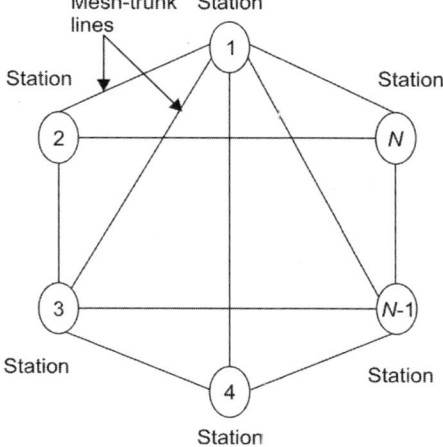

Fig. 13.10. Mesh topology

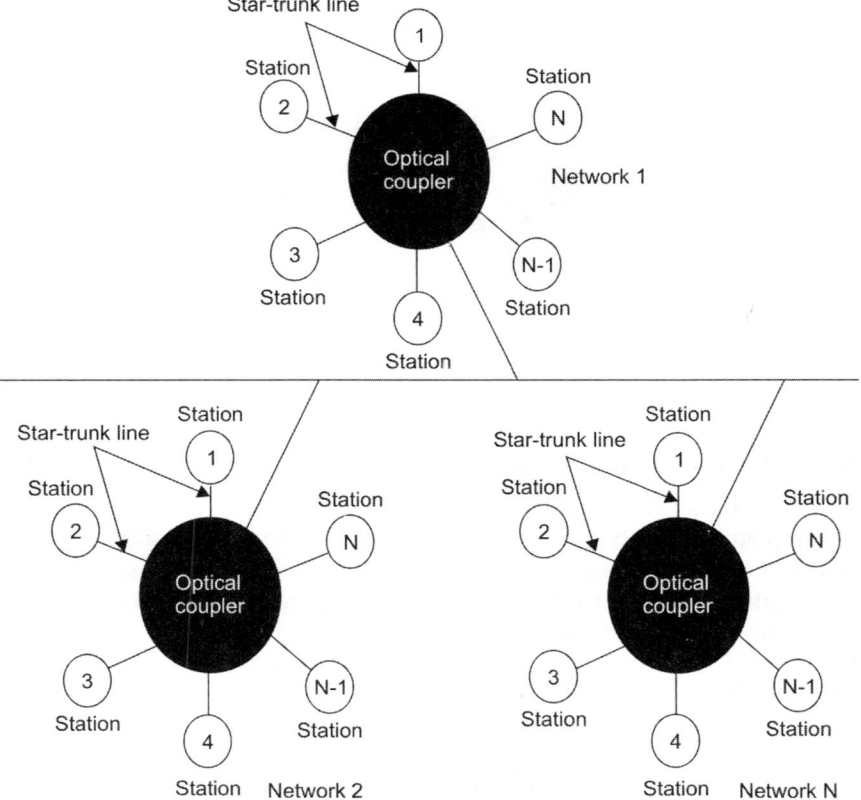

Fig. 13.11. Tree topology

13.5 TYPES OF NETWORKS

Networks are divided into five types based on the size of the area that the network covers: (i) home-area networks (HANs), (ii) local-area network, (iii) campus-area networks (CANs), (iv) metropolitan area networks (MANs), and (v) Wide area networks (WANs).

(i) Home Area Networks (HANs)

As shown in Fig. 13.12, a HAN is a network that connects the digital devices to computers, which are contained within an individual user's home.

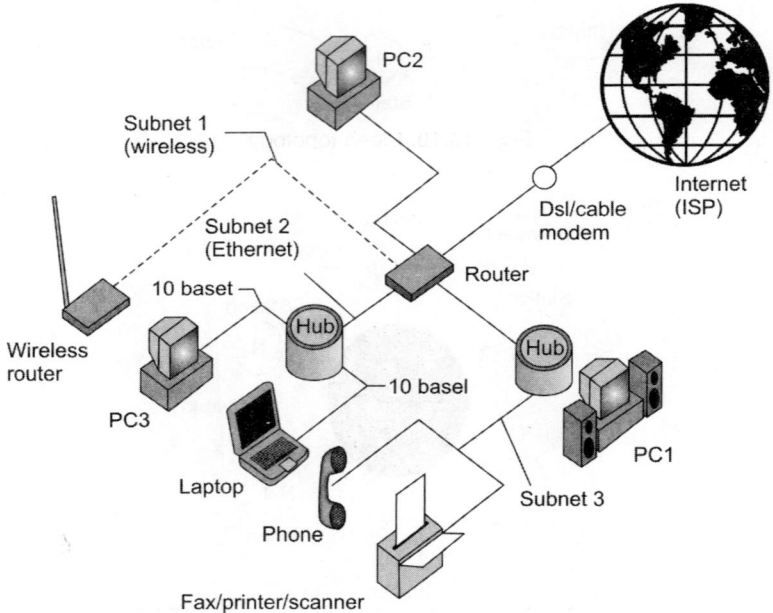

Fig. 13.12. Home-area networks (HANs)

(ii) Local Area Networks (LANs)

A LAN interconnects users in a localized area such as a department, a small group of buildings, or a factory complex. Fig. 13.13 shows a LAN connected to the internet (through a firewall). This figure shows how workstations are connected to a hub and router. The router provides connections to file servers, and printers.

(iii) Campus Area Network (CAN)

A CAN is a computer network made up of an interconnection of LANs which are located within a limited geographical area, such as a university or college campus. In the case of a university campus-based CAN, the network is likely to link a variety of campus buildings,

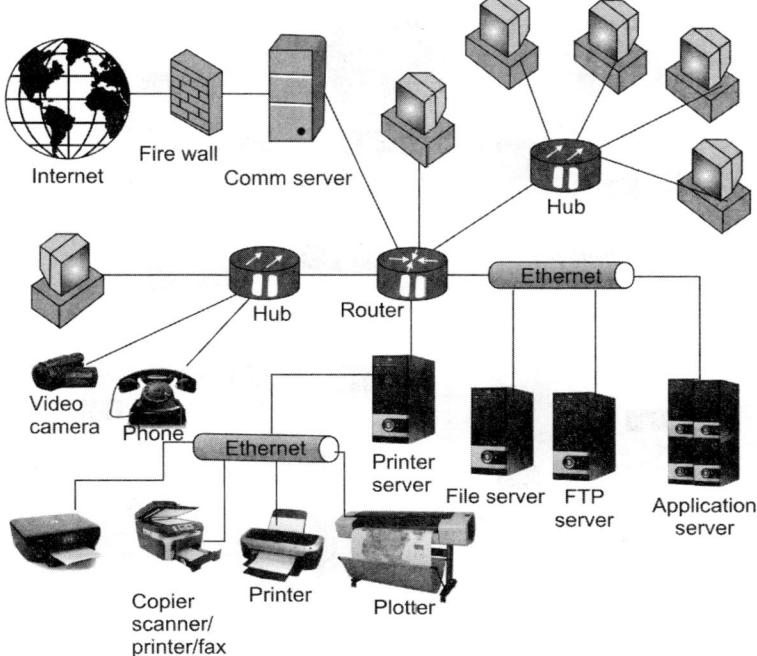

Fig. 13.13. Local-area network (LAN)

including academic departments, the university library, and student halls of residence. A CAN is larger than a LAN, but smaller than a MAN. In addition, CAN stands for corporate-area network.

(iv) Metropolitan Area Networks (MANs)

As shown in Fig. 13.14, a MAN provides interconnections within a city or in a metropolitan area surrounding a city. MANs typically use wireless infrastructure or optical fiber connections to link their sites (nodes).

For instance, a university or college may have a MAN that joins together many of their LANs, which are situated in a site whose area is less than a square kilometer. Beyond that, the MAN could have several WAN links to other universities or the internet.

Some of the technologies used for this purpose are asynchronous transfer mode (ATM) and fiber distributed data interface (FDDI). These older technologies are in the process of being displaced by ethernet-based MANs (e.g., Metro Ethernet) in most areas. MAN links between LANs have been built without cables, by using microwave, radio, or infra-red free-space optical communication links.

(v) Wide Area Networks (WANs)

A WAN covers a large geographical area connecting cities, countries and continents (Fig. 13.15). The best example of a WAN is the *internet*.

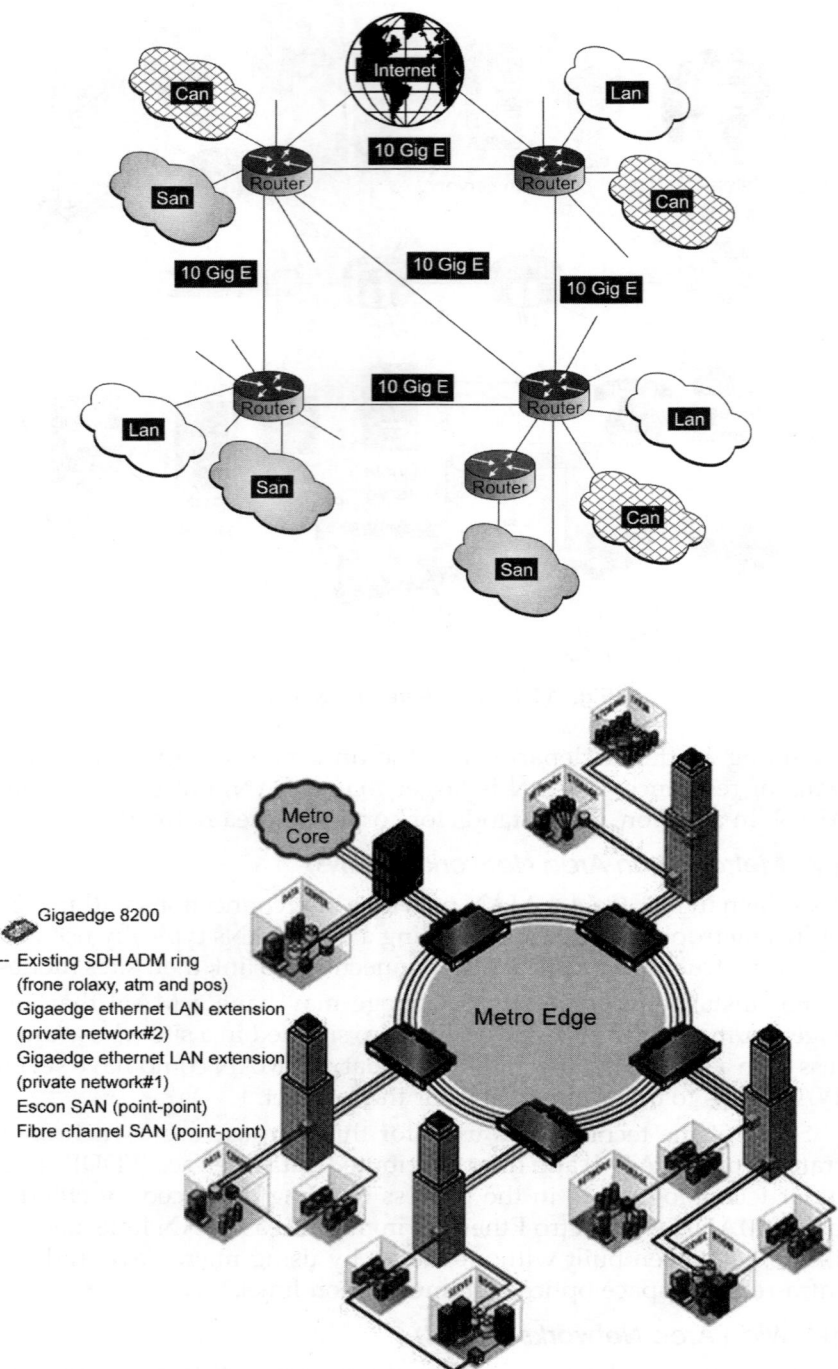

Fig. 13.14. Metropolitan area network (MAN)

WANs are used to connect LANs or MANs together, so that users and computers in one location can communicate with users and computers in other locations. Many WANs are built for one particular organization and are private. Others, built by internet service providers, provide connections from an organization's LAN to the Internet. A router connects to the LAN on one side of the line, and a hub within the WAN on the other. Network protocols, including TCP/IP, deliver transport and addressing functions. Protocols include Packet over SONET/SDH, ATM, and Frame Relay. These protocols are often used by service providers to deliver the links used in WANs. X.25 was an important early WAN protocol, and is often considered to be the grandfather of Frame Relay, as many of the underlying protocols and functions of X.25 are still in use today (with upgrade) by Frame Relay.

13.6 SUBMARINE CABLES

Oceans cover 70% of our planet, separating the continents and people. People rely on submarine cable networks for voice, data and Internet communication. Extreme demands between continents for network reliability and high capacity is achieved by using submarine networks that are reliable and well designed. Submarine cables are sometimes known as *underwater cables*. The world-leading supplier of submarine networks connected every continent from Europe to Japan, the length

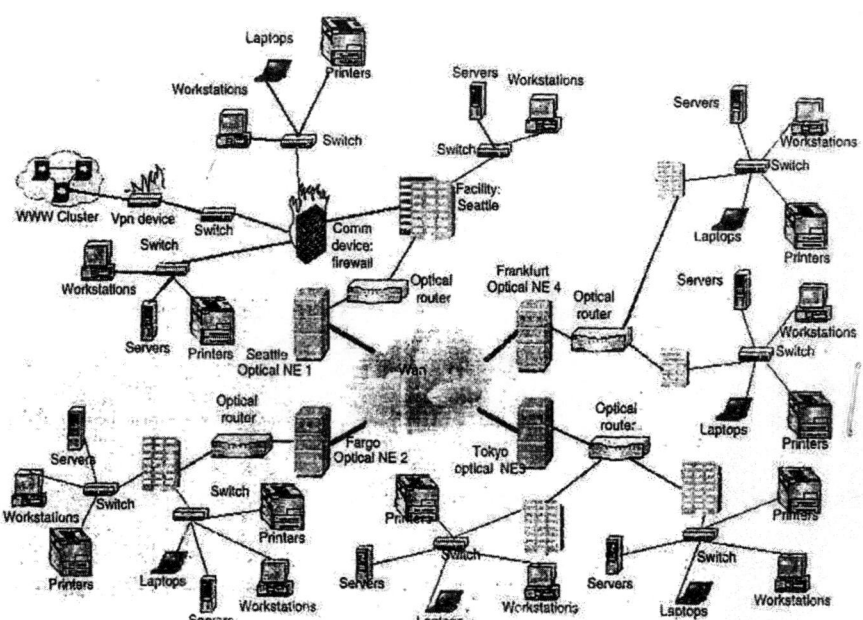

Fig. 13.15. Wide-area network (WAN)

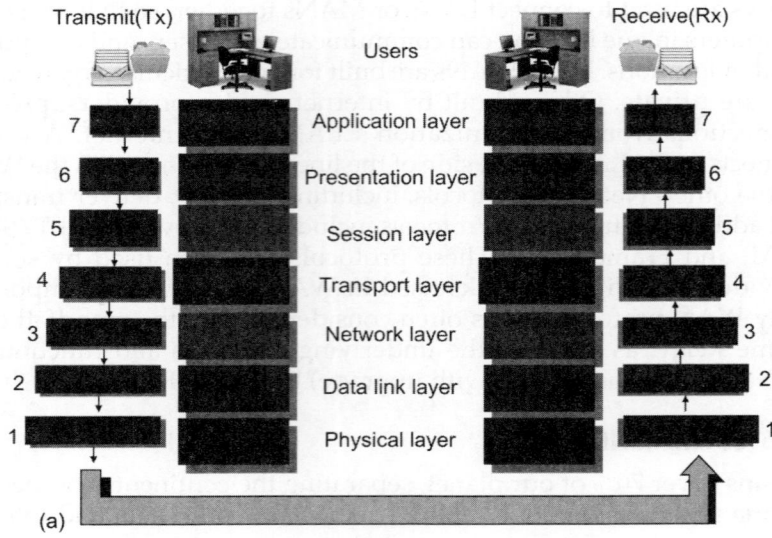

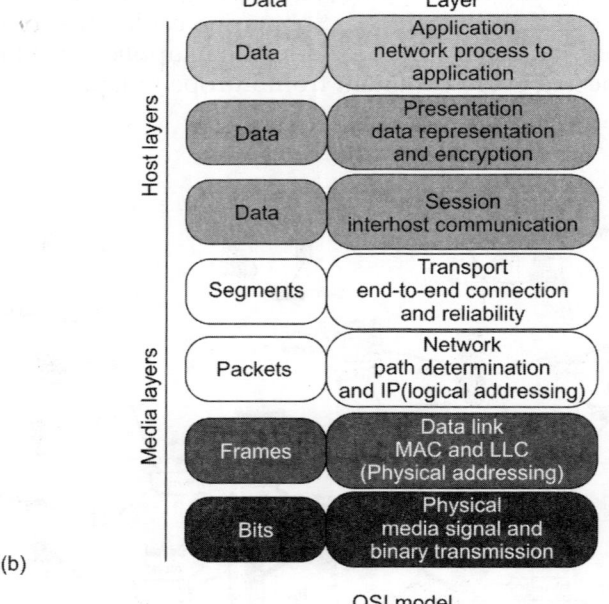

Fig. 13.16. Open system interconnection (OSI), (a) OSI and data process. (b) OSI model

of the Americas, and across the Pacific. For high capacity, dense wavelength division multiplexer (DWDM) technology is used in telecommunication systems.

13.7 OPEN SYSTEM INTERCONNECTION (OSI)

An OSI is a model that defines a networking framework used for implementing protocols in seven layers of communications. Control is passed from one layer to the next, starting at the application layer in one station, and proceeding through to the optical bottom layer. Control passes over the channel (in optical pulses) to the next station and back up the hierarchy. Fig. 13.16(a) shows the layers of OSI and data process, while Fig. 13.16(b) shows the host and media layers located in the OSI model.

(i) Physical (Layer 1)

This layer conveys the bit stream (electrical impulse, light, or radio signal) through the network at the electrical and mechanical level. This is transmission of raw data over a communication medium. It provides the hardware means of sending and receiving data on a carrier which includes defining cables, cards, and physical aspects. SONET/SDH, Fast Ethernet, RS232, and ATM are protocols with physical layer components.

(ii) Data Link (Layer 2)

Layer 2 includes transfer of data frames/packets, addressing, and error connection. At this layer, data packets are encoded and decoded into bits. Layer 2 furnishes transmission protocol knowledge and management, and handles errors in the physical layer, flow control, and frame synchronization. The society of electrical and electronic engineers formed the 802 committee which was responsible for dividing the data link layer into two sublayers: the media access control (MAC) layer and the 802.2 logical link control (LLC) layer. The MAC sublayer controls how a computer on the network gains access to the data and permission to transmit it. The LLC layer controls frame synchronization, flow control, and error checking.

(iii) Network (Layer 3)

This layer provides switching and routing functions, creating logical paths, known as virtual circuits (e.g. X.25 connection), for transmitting data from node to node. Other functions of this layer are routing and forwarding, as well as addressing, internetworking, error handling, congestion control, and packet sequencing. Routing of data packets across networks provides software interface between the physical and data link layers.

(iv) Transport (Layer 4)

This layer provides transparent transfer of data between end systems, or hosts, and is responsible for end-to-end error recovery and flow control. It ensures complete data transfer.

(v) Session (Layer 5)

This layer establishes, manages, and terminates internode connections between applications and uses standards to move data between the applications. The session layer sets up, coordinates, and terminates conversations, exchanges, and dialogues between the application at each end. It deals with session and connection coordination.

(vi) Presentation (Layer 6)

This layer involves data formatting, character conversion, security, and coding. It provides independence from differences in data representation (e.g., encryption) by translating from application to network format, and vice versa. The presentation layer transforms data into the form that the application layer can accept. This layer formats and encrypts data to be sent across a network, providing freedom from compatibility problems. It is sometimes called the *syntax layer*.

(vii) Application (Layer 7)

This layer supports application and end-user processes. Communication partners and quality of service are identified, user authentication and privacy are considered, and any constraints on data syntax are recognized. Everything at this layer is application-specific. This layer provides application services for file transfers, e-mail, network operating systems, application programs, and other network software services. Telnet and FTP are applications that exist entirely in the application level. Tiered application architectures are part of this layer.

13.8 SYSTEM ARCHITECTURES

One can classify fiber-optic communication systems into three broad categories: (i) *Point-to-point links*, (ii) *Distribution networks*, and (iii) *Local area networks* (LANs).

13.8.1 Point-to-point Links

We have mentioned in our previous discussion of digital and analog fiber-optic communication systems that these are based on essentially point to point links. Their role is to transport information from one point to another as accurately as possible. For short-haul applications (say, less than 10 km), the attenuation, dispersion, and bandwidth of optical fibers are not of major concern. In such cases, optical fibers are used primarily because of their immunity to electromagnetic interference and radio-frequency interference. However, for long-haul applications, e.g., transoceanic light wave systems, the low loss, low dispersion, and large bandwidth of optical fibers are important factors. Therefore, whenever the link length exceeds a certain value, it becomes essential to compensate for the fiber loss and/or dispersion. Such a compensation is normally carried out by *optical amplifiers*, dispersion-compensating fibers, or other means.

13.8.2 Distributed Networks

Distributed networks are preferred when data is to be transmitted to a group of subscribers. The transmission distance is relatively short (< 50 km). Examples of distributed networks are – broadcast of video channels over cable TV, telephone and FAX, commonly used topologies for distributed networks are: (i) Hub topology, (ii) Bus topology.

(i) Hub Topology

In hub topology channel distribution takes at hubs or central locations. Hub facilitates the cross-connect switched channels in electrical domain. Fig. 13.17 shows hub topology.

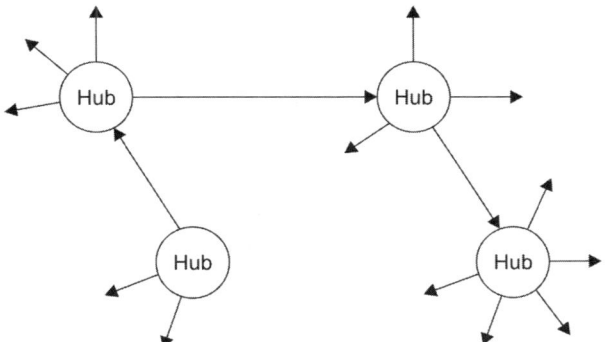

Fig. 13.17. Hub topology

(ii) Linear Bus Topology

Linear bus configuration is similar to *ethernet topology* using co-axial cable. Fig. 13.18 shows linear bus configuration.

A single fiber cable carries the multichannel optical signal throughout the area of service. Distribution is done by using optical taps which divert a small fraction of optical power to each station.

A problem with bus topology is that the signal loss increases exponentially with number of taps for stations. This limits the number

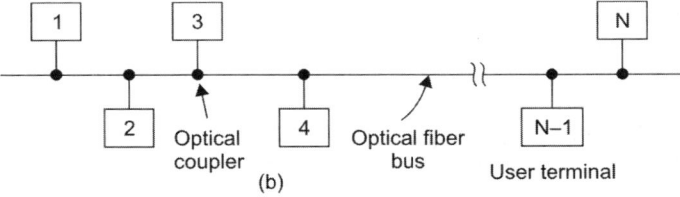

Fig. 13.18. Bus topology

of stations or subscribers that can be served by a single optical fiber bus.

Use of optical amplifiers can boost the optical power of bus and therefore, large number of stations can be connected to linear bus as long as the effect of fiber dispersion is negligible.

As compared with the coaxial cable bus, an optical fiber based bus network is more difficult to implement. The main problem is the ready availability of bidirectional optical taps which can efficiently couple optical signals into and out of the main optical fiber trunk. Access to an optical data bus is normally achieved by means of either an *active* or a *passive coupler*, as shown in Fig. 13.19.

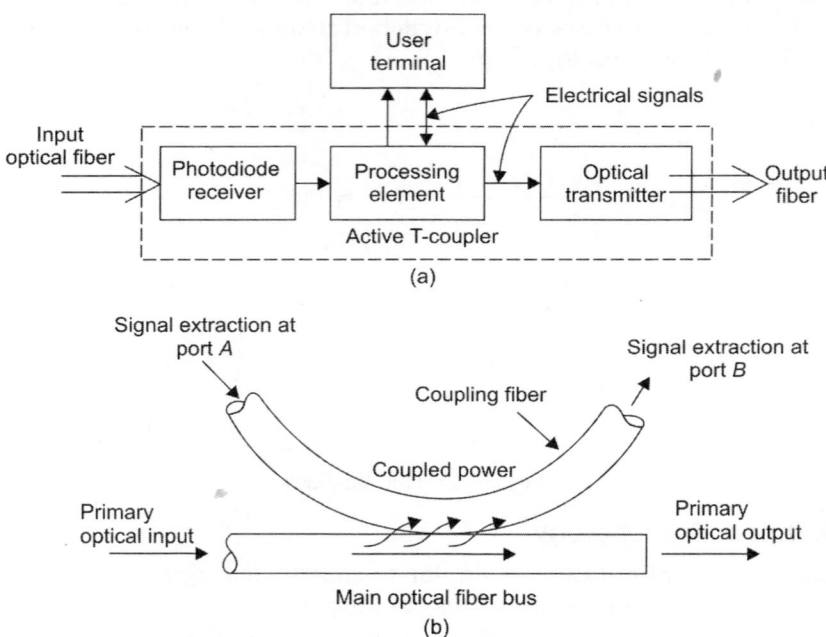

Fig. 13.19. Schematic of (a) an active T-coupler and (b) a passive T-coupler. The tilted arrows show that some optical power has been tapped off the main fiber bus into the receiver at port *B*

In the case of an active coupler, a front-end photodiode receiver converts the optical signal from the bus into an electrical signal. The processing element removes or copies a part of this signal for transmission to the user terminal and sends the remainder to optical transmitter. The latter, in turn, converts the electrical signal back into the optical bit stream, while gets coupled into the output fiber that is connected to the next terminal. The advantage of such a linear fiber bus network is that every accessing terminal acts as a repeater. Therefore, at least in principle, an active bus can accommodate an unlimited number of terminals. However, the reliability of each repeater is critical to the operation of a *single-fiber bus network*. The failure of any one repeater will stop all the traffic. This problem may be overcome by using some bypass scheme, so that if one repeater fails, the bypass ensures optical continuity from the preceding transmitter to the next terminal.

However, in the case of a *passive coupler* no repeaters are used. At each terminal node a passive coupler is used to remove a fraction of the optical signal from the main fiber bus trunk line or to inject additional optical signals into the trunk. A major problem with this type of coupler is that the *optical signal is not regenerated at each terminal node*. Therefore, optical losses at each tap coupled with the fiber losses between the taps limit the size of the network to a small number of terminals.

(iii) Local Area Networks (LAN)

Many applications of fiber optic communication technology require networks in which a large number of users within local campus are interconnected in such a way that any user can access the network randomly to transmit data to any other user. Such networks are called **local area networks (LANs)**.

Fiber optic cables are used in implementation of networks. Since the transmission distance is relatively short (less than 10 km), fiber losses are not at much concern for LAN applications. Use of fiber optic offers large bandwidth.

The commonly used topologies for LANs are: (a) Ring topology, (b) Star topology.

(a) Ring topology

In a ring topology consecutive nodes are connected by point to point links to form a closed ring. Fig. 13.20 shows ring topology.

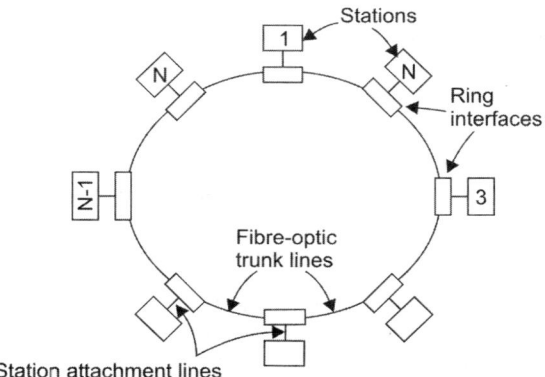

Fig. 13.20. Ring topology

Each node can transmit and receive the data by using a transmitter receiver pair. A token (predefined bit sequence) is passed around the ring. Each node monitors the bit stream to listen for its own address and to receive the data.

The use of ring topology for fiber optic LANs is known as fiber distributed data interface (FDDI). FDDI operates at 100 Mb/s by using 1.3 µm multimode fibers and LED based transmitters. It can provide backbone services e.g. interconnection of lower speed LAN.

(b) Star topology

In a star topology, all nodes are connected through point-to-point link to central node called a central nod or hub. Figure 13.21 shows star topology.

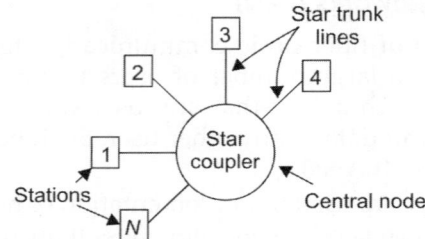

Fig. 13.21. Schematic of star LAN topology

The central node may be an active or a passive device. In an active central node, all incoming optical signals are converted into electrical signals through photodiode receivers. The electrical signal is then distributed to drive different node transmitters. One can also perform the switching operation at this node. In a star configuration with a passive central node the distribution takes place in the optical domain, through devices such as directional couplers. In this case, the input from one node is distributed to many output nodes, and hence the power transmitted to each node depends on the number of users.

The network topologies discussed above are also used by metropolitan area network (MANs), which connect users within a city or in a metropolitan area around the city, and wide area networks (WANs), which provide user interconnection over a large geographical area. We have already discussed MANs and WANs in section 5.

13.9 LINE CODING IN OPTICAL LINKS

Line coding or channel coding is a process of arranging the signal symbols in a specific pattern. Line coding introduces redundancy into the data stream for minimizing errors. There are three types of line codes used in optical fiber communication. These are:

(i) Non-Return-to-Zero (NRZ)
(ii) Return-to-Zero (RZ)
(iii) Phase-Encoded (PE)

The following properties of line codes are desirable:
(i) The line code should contain timing information.
(ii) The line code should be immune to channel noise and interference.
(iii) The line code should allow error detection and correction.

NRZ and RZ Signal Formats

The simplest method for encoding data is the unipolar *nonzero*-to-zero (NRZ) code. Unipolar means that a logic 1 is represented by a voltage or light pulse that fills an entire bit period, whereas for a logic 0 no pulse is transmitted. Although different types of NRZ codes are introduced to suit the variety of transmission requirements, but the simplest form of NRZ code is NRZ level. As stated above, it is a unipolar code, i.e., the waveform is simple *on-off* type.

When symbol "1" is to be transmitted, the signal occupies high level for full bit period. When a symbol "0" is to be transmitted, the signal has zero volts for full bit period. Fig. 13.22 shows example of NRZ-L data pattern.

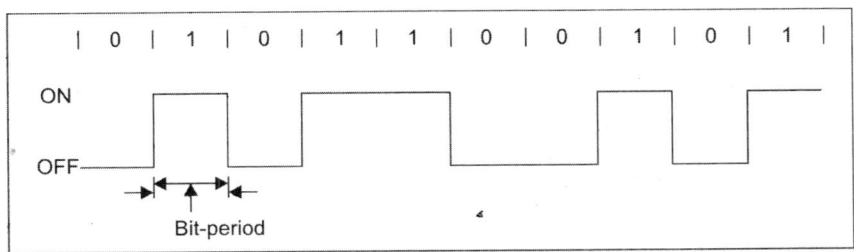

Fig. 13.22. NRZ-level data pattern

Since this process turns the light signal on and off, it is also known as *amplitude shift keying* (ASK) or *on-off keying* (OOK). If 1 and 0 pulses occur with equal probability, and if the amplitude pulse is A, then the average transmitted power for this code is $A^2/2$. In optical system one typically describes a pulse in terms of its optical power level. Thus, in this case the average power for an equal number of 1 and 0 pulses is $P/2$, where P is the peak power in a 1 pulse.

Salient features of NRZ codes
(i) Simple to generate and decode.
(ii) No timing (self-clocking) information.

(iii) No error-monitoring or correcting capabilities.
(iv) NRZ coding needs minimum BW (bandwidth)

We may note that the lack of timing capabilities in an NRZ code can lead to misinterpretation of the bit stream at the receiver.

If an adequate bandwidth margin exists, the timing problem associated with NRZ encoding can be alleviated with a *return-to-zero* (RZ) code.

In unipolar RZ data pattern a 1-bit is represented by a *half-period* in either *first* or *second half of the bit-period*. A 0-bit is represented by zero volts during the bit period. Fig. 13.23 shows RZ data pattern.

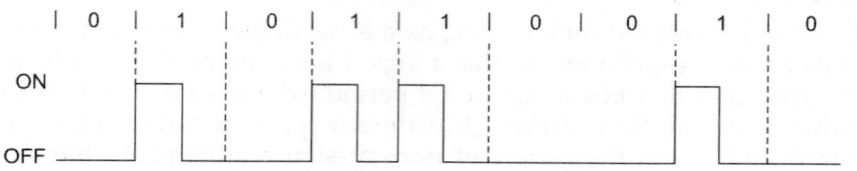

Fig. 13.23. RZ unipolar codes

Salient features of RZ codes

(i) The signal transition during high-bit period provides the timing information.
(ii) Long strings of 0 bits can cause loss of timing synchronization.
(iii) The RZ code has an amplitude transition at the beginning of each bit interval when a binary 1 is transmitted and no transition for a binary 0. Obviously, for a RZ pulse a bit occupies only part of the bit interval and returns to zero in the remainder of the bit interval. No pulse is used for a 0 bit.
(iv) Although the RZ pulse nominally occupies exactly half a bit period in electronic digital transmission systems, in an electronic communication link the RZ pulse might occupy only a fraction of a bit period. A variety of RZ formats are used for links that send data at the rates of 10 Gb/s and higher.

13.10 ERROR CONTROL OR CORRECTION

In any digital transmission system, errors are likely to occur when there is a sufficient signal-to-noise ratio to provide a low bit rate. The acceptance of a certain level of errors depends on the network user.

To control errors and to improve the reliability of communication line, first, it is necessary to be able to detect the errors and then either to correct them or restransmit the information. Error detection methods encode the information stream to have a specific pattern. If the segments in the received data stream violate this pattern, then errors have occurred.

The two basic schemes for error correction are *automatic repeat request* (ARQ) and *forward error correction* (FEC).

In *ARQ scheme*, the information word is coded with adequate redundant bits so as to enable detection of errors at the receiving end. If an error is detected, the receiver asks the sender to retransmit the particular information word.

Each retransmission adds one round trip time to latency. Therefore, ARQ techniques are not used where low latency is desirable. Fig. 13.24 shows the scheme of *ARQ error correction scheme*.

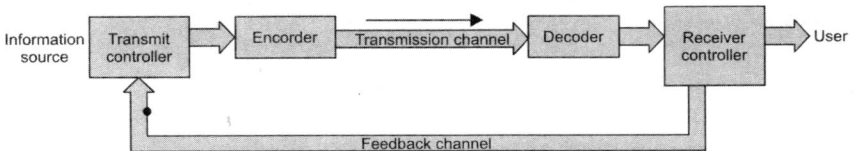

Fig. 13.24. ARQ error correction scheme

Forward Error Correction (FEC) system adds redundant information with the original information to be transmitted. The error or lost data is used reconstructed by using redundant bit. Since the redundant bits to be added are small, hence much additional BW is not required.

Most common error correcting codes are cyclic codes. Whenever, highest level of data integrity and confidentiality is needed, FEC is considered.

13.11 PERFORMANCE OF PASSIVE LINEAR OPTICAL NETWORKS

To evaluate the performance of passive linear networks, consider the fraction of optical power (F_C) lost at a particular interface or component along the transmission path, as shown in Fig. 13.25. The power ratio (A_0) over an optical fiber of length (x) will be:

$$A_0 = \frac{P_{(x)}}{P_{(0)}} = 10^{-\alpha x/10} \qquad (13.27)$$

where $P_{(x)}$ is the power received, $P_{(0)}$ is power transmitted, and a is the fiber attenuation (dB/km)..

If F_C is lost at each port of the coupler, then the connecting loss (L_C) will be:

$$L_C = 10 \log (1 - F_C) \qquad (13.28)$$

For example, if $F_C = 20\%$, then $L_C = -0.9691$ dB. The optical power gets reduced by the L_C of 1 dB at any connection junction.

The power extracted from the bus is called tap loss (L_{tap}), and is given by:

$$L_{tap} = 10 \log C_T \qquad (13.29)$$

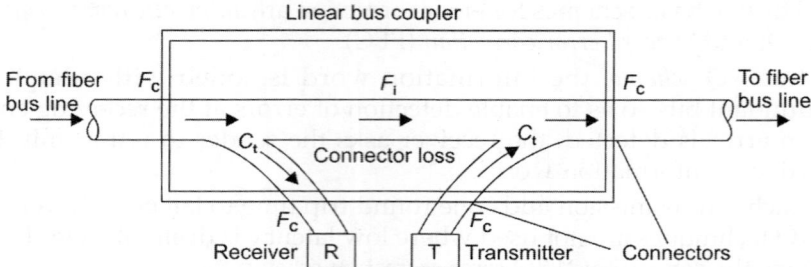

Fig. 13.25. Losses in a passive linear-bus coupler consisting of two cascaded directional couplers

where C_T is the fraction power removed from the bus and delivered to the detected port.

Then, the throughput coupling loss (L_{thru}) is given by:

$$L_{thru} = 10 \log (1 - C_T)^2 = 20 \log (1 - C_T) \qquad (13.30)$$

In addition to the losses L_C and L_{tap}, there is an intrnisic loss (L_i) associated with each bus coupler. If the fraction of power lost in the coupler is F_i, then:

$$L_i = 10 \log (1 - F_i)$$

All losses are measures in decibels (dB).

A linear bus configuration, as shown in Fig. 13.26, consisted of a number of stations (N) separated by a various lengths. For simplicity, assume a constant distance L.

The fiber attenuation between two adjacent stations is given by:

$$L_{fiber} = 10 \log A_0 = \alpha L \qquad (13.31)$$

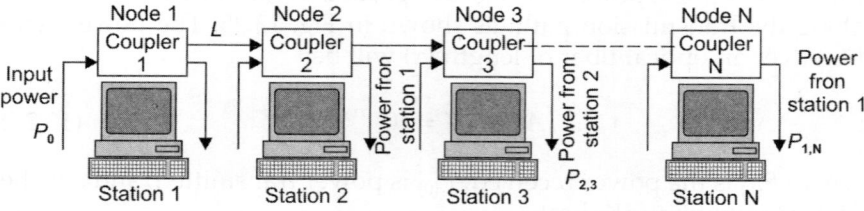

Fig. 13.26. Topology of a simplex linear bus consisting of N uniformly speed stations

13.11.1 Power Budget Calculation

To calculate the power budget of a fiber link consisting of N stations, as shown in Fig. 13.26, the fractional power losses F_c should be examined first.

13.11.2 Nearest-Distance Power Budget

If P_0 is the optical power launched from the optical source at station 1, then the optical power detected $P_{1,2}$ at the station 2 is given by:

$$P_{1,2} = A_0 C_T^2 (1-F_C)^4 (1-F_i)^2 P_0 \qquad (13.32)$$

Then the power budget (considering all losses) between stations 1 and 2 is:

$$10 \log\left(\frac{P_0}{P_{1,2}}\right) = \alpha L + 2L_{tap} + 2L_c + 2L_i \qquad (13.33)$$

13.11.3 Largest-Distance Power Budget

If P_0 is the optical power launched from the optical source at station 1, then the optical power detected $P_{1,N}$ at the station N is given by:

$$P_{1,N} = A_0^{N-1} (1-C_T)^{2(N-2)} C_T^2 (1-F_c)^{2N} (1-F_i)^N P_0 \qquad (13.34)$$

Then the power budget (considering all losses) between stations 1 and N is:

$$10 \log\left(\frac{P_0}{P_{1,N}}\right) = N(\alpha L + 2L_c + L_{thru} + L_i) - \alpha L - 2L_{thru} + 2L_{tap} \qquad (13.35)$$
$$= [\text{fibre} + \text{connector} + \text{coupler throughput} + \text{ingress/egress} + \text{coupler intrinsic}] \text{ losses}$$

As shown in Fig. 13.27, the losses (in dB) of a linear bus configuration in Fig. 13.26 increase linearly with the number of stations N.

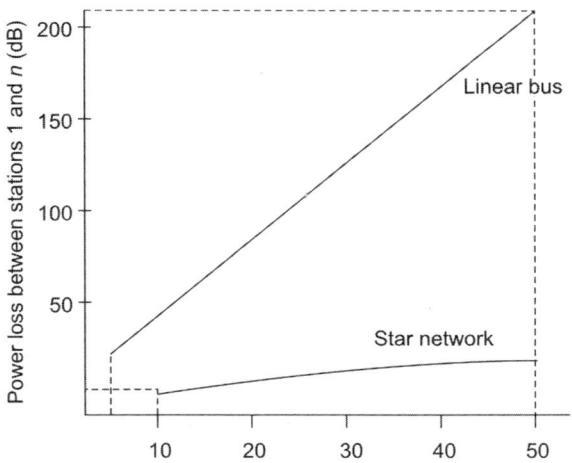

Fig. 13.27. Total loss vs number of station in linear-bus and star networks

13.12 PERFORMANCE OF STAR OPTICAL NETWORKS

The optical input power is evenly divided among the output ports in an ideal star fiber coupler, as shown in Fig. 13.28.

The total optical power loss of the coupler consists of its splitting loss and the excess loss in each path through the star configuration. The splitting loss (L_{split}) among the N stations is given by:

$$L_{\text{split}} = 10 \log \left(\frac{1}{N}\right) = 10 \log N \qquad (13.36)$$

The star fibre coupler excess loss (L_{excess}) is defined as the ratio of the single input power (P_{in}) to the total output power ($P_{\text{out},i}$) of the N stations ($i = 1....N$).

$$L_{\text{excess}} = 10 \log \left(\frac{P_{\text{in}}}{\sum_{i=1}^{N} P_{\text{out},i}} \right) \qquad (13.37)$$

The total loss within the fiber star coupler is given by:

$$\text{Total loss} = L_{\text{split}} + L_{\text{excess}} \qquad (13.38)$$

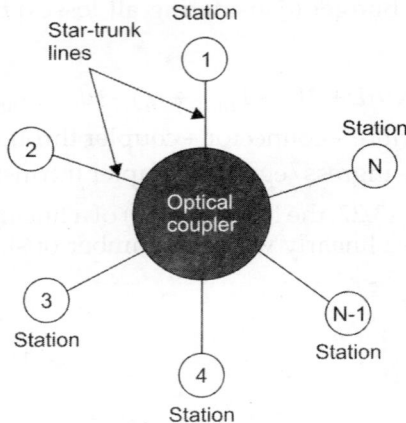

Fig. 13.28. A star coupler

The optical power balance equation between any two stations in a star network, include all losses, and is defined as:

$$P_S - P_R = L_{\text{excess}} + \alpha(2L) + 2L_c + L_{\text{split}}$$
$$= L_{\text{excess}} + \alpha(2L) + 2L_c + 10 \log N \qquad (13.39)$$

where P_S is the fiber coupled output power from source in dBm; P_R is the minimum optical power required at the receiver; L_{excess} is the star fiber coupler excess loss; L_{split} is the splitting loss; L_C is the connector loss; L is the distance from the star coupler, assuming all stations at the same distance; and α is the fiber attenuation.

As more stations are added, the loss in a star network increases much slower than the loss in a linear bus network, as shown in Fig. 13.27.

13.13 SONET AND SDH

With the advent and development of fiber optic cables and optical fiber amplifiers, the next important evolution of the digital time-division multiplexing (TDM) scheme was a standard signal format. This format is called *synchronous optical network* (SONET) in North America, and *synchronous digital hierarchy* (SDH) in other parts of the world. SONET and SDH are both optical interface standards that allow internetworking of services from multiple service providers. SONET and SDH are almost identical standards dedicated to transporting data, voice imaging, and video over optical networks. Fig. 13.29 illustrates a standard optical interface between SONET and SDH network.

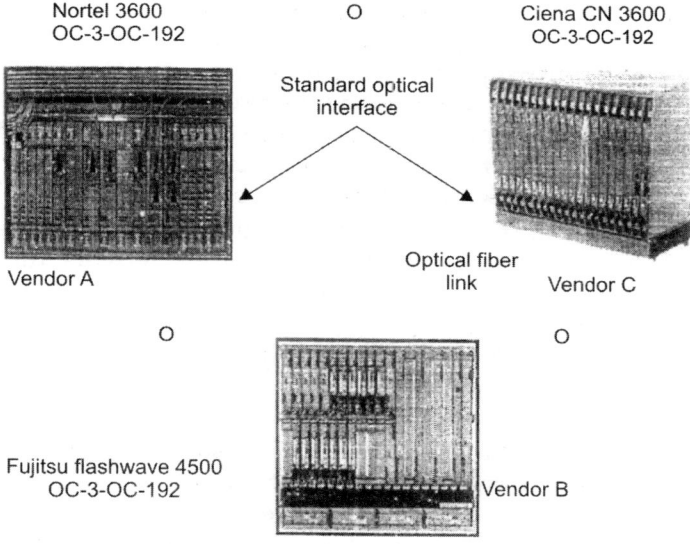

Fig. 13.29. SCNET/SDH

SONET was specified primarily by Bellcore (now Telcordia) in the late 1980s. It was submitted to the international standards bodies as a proposed international standard. After some negotiation, SDH emerged as the international standard, of which SONET can be considered a complete subset. The standards define a hierarchy of high-speed transmission rates, which currently range from 51.84 Mb/s (SDH at 155 Mb/s) up to 40 Gb/s.

The standard specifies the physical interfaces, such as the optical wavelengths, pulse shapes, and link budgets. For some of the lower rates in the hierarchy, electrical interfaces are specified as well, including line coding and electrical pulse shapes. The organization of the digital information crossing the interface is defined by specifications for frame structures, payload mappings, and overhead assignments. In addition, special signals are specified which allow communication between the two optical network elements for the purpose of operations, administration, maintenance, and provisioning (OAM&P). The OAM&P of SONET/SDH networks rely on the telecommunication management

Network architecture (TMN) as defined by the ITU. SONET/SDH network elements are managed by using translation language 1 (TL1) protocol.

Purposes and Features of SONET/SDH

There are several purposes for implementing SONET or SDH. Some of the purposes and features are as follows:

(i) *Multi-vendor networks*: The primary motivation for developing the SONET standard was to allow the deployment of multi-vendor optical networks. Prior to SONET, fiber optic transport systems could only be deployed in point-to-point configurations, resulting in a lot of unnecessary equipment to terminate the proprietary interfaces. With SONET, a completely optical transport network is envisaged whereby high capacity fibers interconnect network elements that provide access to the transport network and manage the fiber bandwidth (BW).

(ii) *Cost reduction*: It was also assumed that the resulting network would be cheaper due to the consolidation of network functions, elimination of unnecessary functions, and increased competition between vendors. Network providers feel that they no longer have to lock themselves into a single vendor's proprietary (non-standard) solution.

(iii) *Survivability and availability*: With the increasing concern over network reliability and robustness, SONET also provides the ability to build survivable networks; networks that can restore traffic within 50 ms, in the event of fiber cuts and equipment failure.

(iv) *New high-speed services*: The demand for higher bandwidth pipes continues to increase. SONET provided the opportunity to define an interface, and therefore, a network that was capable of carrying a variety of payload types with differing bandwidth requirements, including broadband payloads beyond 50 Mb/s.

(v) *Bandwidth (BW) management*: SONET manages bandwidth by introducing the concept of the payload envelope, into which all payloads are mapped as they enter the SONET network. This allows the infrastructure to be concerned only with transporting the envelops, regardless of their contents. A limited number of envelops are defined, with payload capacities ranging from 1.5 Mb/s to 10 GB/s. In addition, different envelops may be combined on the same fiber.

(vi) *Network management/Single-ended operations*: Each rate in the SONET/SDH hierarchy is an integer multiple of a basic rate. For SONET, this basic rate is 51.84 Mb/s; for SDH, it is 155.52 Mb/s. The signal format at each level is created by synchronously multiplexing the basic format. This format is called the synchronous transport signal-level 1 (STS-1) in SONET and the synchronous transport module-level (STM-1) in SDH. Synchronous multiplexing simplifies bandwidth management, allowing access to individual tributaries within the fiber signal without having to completely de-multiplex the fiber signal. The creation of an all-optical network enables management of the network bandwidth using automated techniques. Spare bandwidth on one route may be reallocated to another route, and new connections between end offices can be created quickly when required. SONET includes overhead allocations for a variety of OAM&P functionality. Examples include a data communications channel to allow network elements to be monitored from a central operations system, and integrity checks to allow single-ended performance monitoring of fiber systems and end-to-end networks.

13.14 MULTIPLEXING TERMINOLOGY AND SIGNALING HIERARCHY

13.14.1 Existing Multiplexing Terminology and Digital Signalling Hierarchy

In order to understand the role that the SONET standard plays, it is first necessary to understand what interface standards existed previously. In North America, the standard pre-SONET digital hierarchy consisted mainly of digital signals of several levels. The digital signal-level 1 (DS1), is capable of transmitting and receiving data at a bit rate of 1.544 MB/s (150 MB/s) and the DS3 has the capability of 44.736 Mb/s, as shown in Fig. 13.30. The DS2 really only exists as an intermediate step in the DS1–DS3 multiplex. The DS1 is the main interface to digital voice switches and channel banks, whereas the DS3 is the main interface to fiber optic transmission systems. The M13 multiplexer links the two together, and is named for its ability to multiplex DS1 into a DS3. This hierarchy is based on time division multiplexing (TDM).

A similar hierarchy, called electrical signal E, exists in most of the rest of the world. The 2.048 MB/s interface is the key digital switch interface, whereas most pre-standard fiber optic transmission systems carry the 139 MB/s signal. 2.048 MB/s is called E1, and the hierarchy is based on multiples of 4 EIs:

$E2 = 4 \times E1 = 8$ Mb/s
$E3 = 4 \times E2 = 34$ Mb/s
$E4 = 4 \times E3 = 140$ Mb/s
$E5 = 4 \times E4 = 565$ Mb/s

The E3 tributaries are faster than the E2 tributaries, while the E2 tributaries are faster than the E1 tributaries, and so forth. To synchronize with other tributaries, extra bits, called justification bits, are added. These tell the multiplexers which bits are data and which are spare. Multiplexers on the same level of the hierarchy remove the spare bits, and are synchronized with each other at that level only. Multiplexers on one level operate on a different timing from multiplexers on another level. For instance, the timing between primary rate muxes (which

704 Photonics | Optoelectronics

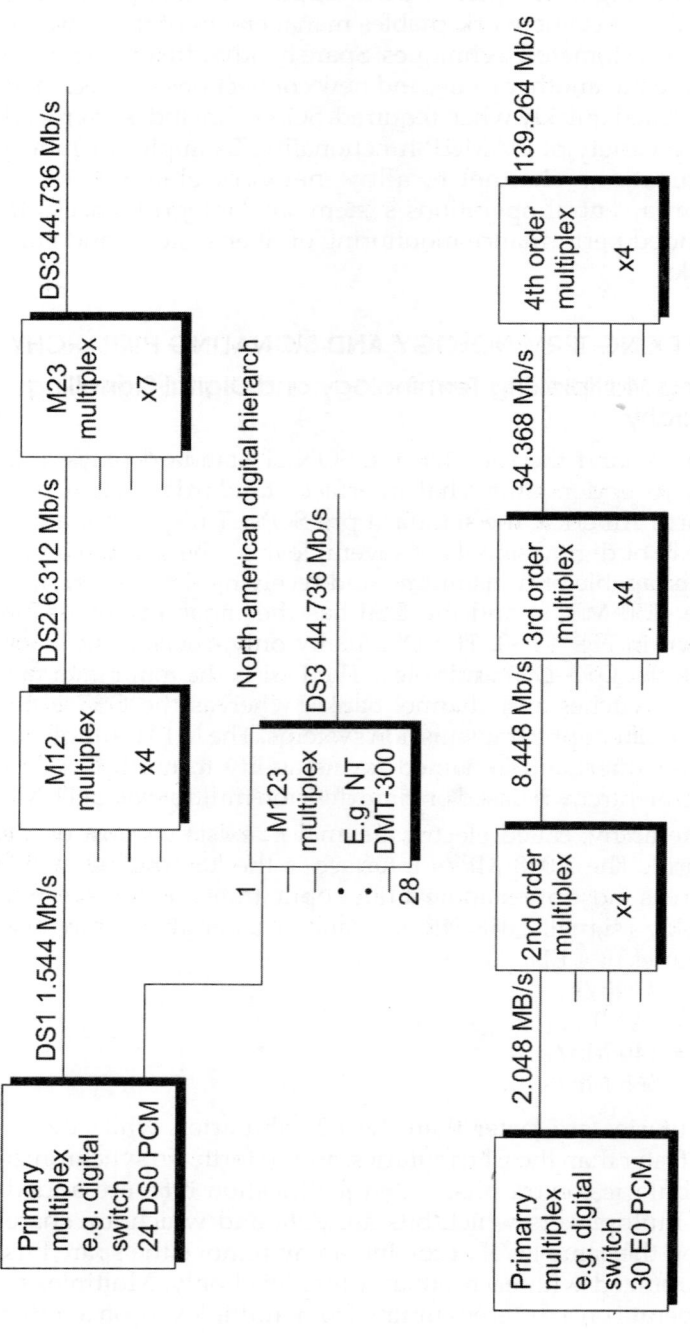

Fig. 13.30. Existing digital/electrical hierarchies

combine 30 × 64 Kb/s channels into 2.048 Mb/s E1) will be different from the timing between 8 Mbit muxes (which combine up to 4 × 2 Mb/s into 8 Mb/s.

DSI is sometimes called *transport level-1* (T1). T1 is the optimal rate for accessing low-level devices. It is a type of telephone service capable of transporting the equivalent of 24 conventional telephone lines, using only two pairs of wires. T1 uses two pairs of copper wires (four individual wires) to carry up to 24 simultaneous conversations (channels) that would normally need one pair of wires each. Each 64 Kbit/s channel can be configured to carry voice or data traffic. Most telephone companies allow customers to buy just some of these individual channels, a service called fractional T1. Typically, fractional T1 lines are sold in increments of 56 Kbps (the extra 8 Kbps per channel are used for administration purposes). One of the most common uses of a T1 line is an Internet T1. This connection is used to provide Internet access to businesses of all sizes, assisting of data at speed of 256 Kbit/s, 512 Kbit/s, 1.544 Mbit/s, and sometimes 3 Mbit/s.

13.14.2 SONET Multiplexing Terminology and Optical Signaling Hierarchy

Table 13.1 presents the optically transmitted SONET signal which is referred to as an optical carrier-level N (OC-N). The OC-N is essentially the optical equivalent of the STS-N; however, the STS-N terminology is used when referring to the SONET format. As shown in Fig. 13.31, the STS-N consists of a synchronous multiplex of N STS-1s. The STS-1 has a bit rate of 51.84 MB/s, therefore, the STS-N and OC-N have a bit rate of N times 51.84 Mb/s.

Table 13.1. SONET terminology

Optical carrier-level N (OC-N)	Optical SONET signal at N times the basic rate of 51.84 Mb/s
Synchronous transport signal-level N (STS-N)	The electrical SONET signal, or SONET format, at N times the basic rate of 51.84 Mb/s, consists of a multiplex of N STS-1s
Synchronous transport signal-level 1 (STS-1)	Electrical SONET signal at 51.84 Mb/s, also used to refer to the SONET format
Synchronous payload envelope (SPE)	In SONET, all payloads are mapped into several types: the VT SPEs carry 1.5-Mb/s payloads, the STS-1 SPE carries 50 Mb/s payloads, and the STS-Nc SPE carries 150 Mb/s and higher payloads
Virtual tributary group (VTG)	A logical grouping of VTs prior to multiplexing into the STS-1 SPE
Virtual tributary (VT)	The unit into the STS-1 SPE can be subdivided to carry payloads that require much less than 51.84 Mb/s

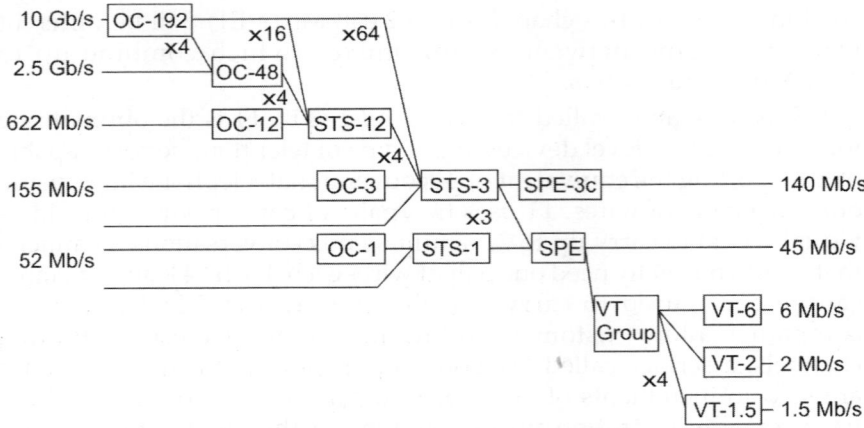

Fig. 13.31. SONET multiplexing hierachies

In SONET, all payloads are mapped into synchronous payload envelopes (SPE) at the edge of the SONET network, as shown in Fig. 13.30. The core of the SONET network transports the envelopes. Carried within the STS-1 is the STS-1 SPE, which has a payload capacity of approximately 51.85 Mb/s. N STS-1s may be concatenated to carry an STS-Nc SPE, which has a payload capacity of N × 51.84 Mb/s, as shown in Table 13.1.

13.14.3 SDH Multiplexing Terminology and Optical Signalling Hierarchy

As discussed previously, a European standard, SDH, was developed parallel to the SONET standard. SDH uses a different terminology, as shown in Table 13.2 and Fig. 13.32. Aside from the terminology, most other features of SONET can be extended to SDH.

Table 13.2. SDH terminology

Synchronous transport module-level N (STM-N)	A synchronous multiplex of N STM-1s
Synchronous transport module-level 1 (STM-1)	The basic rate (155.52 Mb/s) and format of the SDH hierarchy; also refers to the optical signal
Administrative unit group (AUG)	A logical grouping of like AUs
Administrative unit (AU)	Similar to the TU; consists of a higher order VC and a payload pointer
Tributary unit group (TUG)	A logical grouping of like TUs
Tributary unit (TU)	A logical element consisting of a lower order VC and a payload pointer
Virtual container (VC)	The SDH structure into which all payloads are mapped

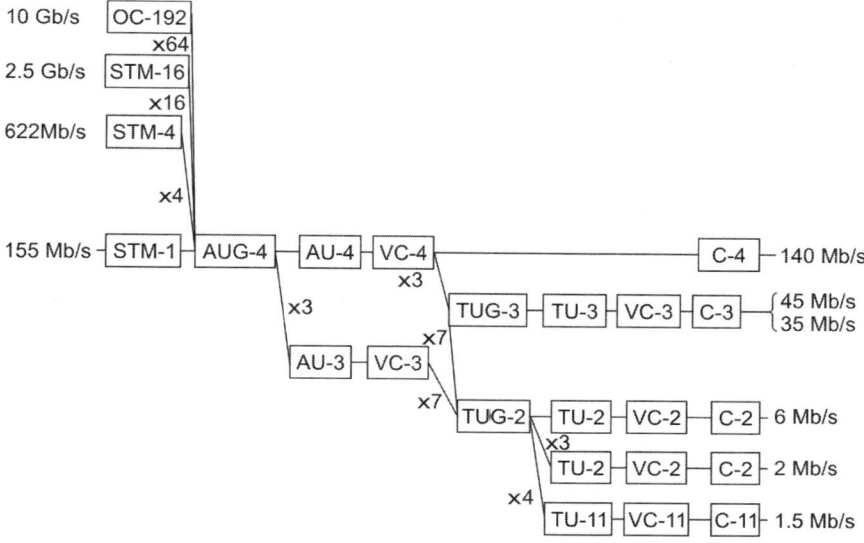

Fig. 13.32. SDH multiplexing hierarchies

13.15 SONET AND SDH TRANSMISSION RATES

Although the SONET multiplexing scheme would in theory allow any multiple of STS-1s, only certain rates are defined as standard transmission rates. Physical (photonic) interfaces are specified for these rates. Table 13.3 lists the most commonly supported SONET rates with their SDH equivalents.

Table 13.3. SONET/SDH rates

SONET	SDH	Rate (Mb/s)
OC-1	STM-0	51.84
OC-3	STM-1	155.52
OC-12	STM-4	622.08
OC-48	STM-16	2488.32
OC-192	STM-64	9953.28
OC-768	STM-256	39813.12

13.16 SONET SYSTEMS

SONET network elements are combined to create systems that are classified based on the mechanism used to provide traffic protection and survivability. Two main types of protection provides in SONET are: (i) linear, and (ii) ring.

(i) *Linear systems*: Linear systems transport traffic along a single route that may consist of one or more working fibers. A one plus one (1 + 1) system has one protection fiber and one working fiber, as shown in Fig. 13.33. The traffic is permanently bridged onto both fibers, so that the receiving end can autonomously choose the fiber that is operating better.

Fig. 13.33. Linear SONET system

(ii) *Ring system*: Ring systems transport traffic around a ring, allowing traffic to be added and dropped anywhere along the ring, as shown in Fig. 13.34. Spare capacity is allocated around the ring so that when a failure occurs at any one point, the affected traffic can be restored using the spare capacity. In a unidirectional path switched ring (UPSR), traffic added to the ring is bridged onto both directions, such that the drop node can autonomously select the better path. In a bi-directional line switched ring (BLSR), the traffic affected by a failure at any point on the ring is rerouted the other way around the ring. A protocol operating around the ring coordinates this action.

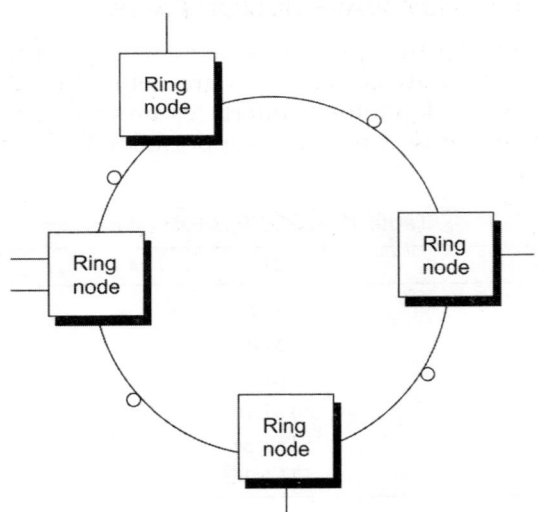

Fig. 13.34. SONET ring system

13.17 METRO AND LONG-HAUL OPTICAL NETWORKS

Metro optical networks can be thought of as consisting of *core networks* and *access* networks, as shown in Fig. 13.35. SONET/SDH metro and long-haul core networks are typically configured as point to point or ring connections that are spaced in tens or thousands of kilometers apart. The metro optical network consists of optical links between the end users and central office (CO). The ring configuration shown in Fig. 13.35 contains three or four modes.

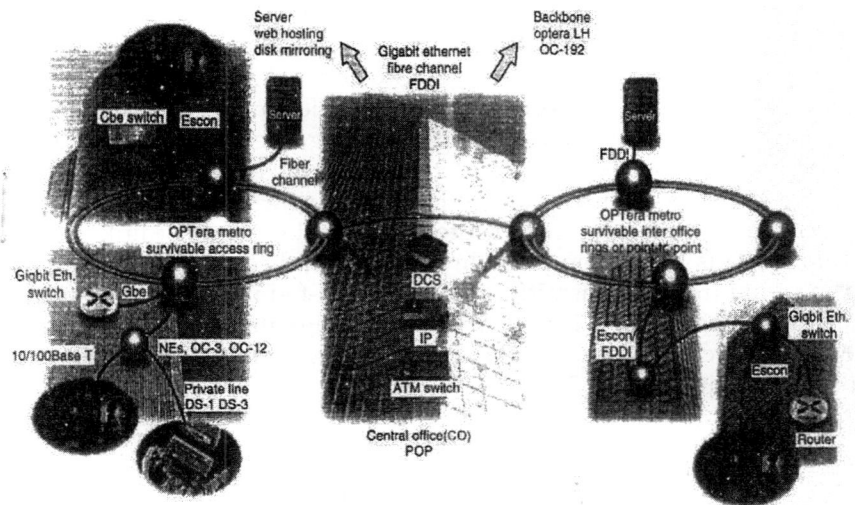

Fig. 13.35. Metro and long-haul optical networks

13.18 NETWORK CONFIGURATION

13.18.1 Automatic Protection Switching (APS)

Ring automatic protection switching (APS) provides an increased level of survivability for SONET/SDH networks by allowing as much traffic as possible to be restored, even in the event of a cable cut or node failure.

13.18.2 SONET/SDH Ring Configurations

SONET/SDH networks has two main types of ring configurations: (i) a UPSR, and (ii) a bidirectional line switched ring (BLSR).

A UPSR consists of two fibers on each span transmitting in opposite directions between adjacent nodes.

There are two types of *bidirectional line switched ring* (BLSR): (a) *two-fiber BLSR*, and (b) *four fiber BLSR*. In the two-fiber BLSR, there are two fibers transmitting in opposite directions between adjacent nodes. In the four-fiber BLSR, there are four fibers between adjacent nodes, with two fibers transmitting in one direction and other two fibers in the opposite direction. The four-fiber BLSR supports more traffic.

For a UPSR network, the selection of data path is made on a per path basic using the path layer integrity information. Thus, the ring is called path switched. In the case of a BLSR network, the decision to switch to the other path is made by the nodes adjacent to the failure using line layer integrity information. Thus, the ring is called *line switched*.

(a) Two-Fiber UPSR Configuration

Figure 13.36 shows a two-fiber UPSR network. By convention, in a unidirectional ring, the normal working traffic travels clockwise around the ring on the primary (working) path, e.g. the connection from the node 1 to node 3 uses links 1 and 2, whereas the traffic from the node 3 to node 1 traverses links 3 and 4. In a UPSR ring, the counterclockwise path is used as an alternate route for protection against link or node failures. This protection path (links 5–8) is indicated by dashed lines. The signal from a transmitting node is dual-fed into both the primary and protection fibers. This establishes a designated protection path on which traffic flow counterclockwise, namely, from node 1 to node 3 via protection links 8 and 7.

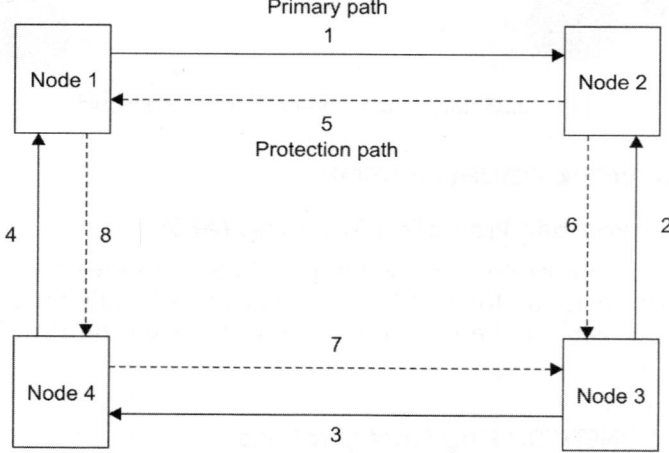

Fig. 13.36. A 2-fiber unidirectional ring with a counter-rotating protection path

Two-fiber UPSR configuration (Traffic flow): In Fig. 13.37, two signals from node 1 arrive at their destination at node 3 from opposite directions. The receiver at node 3 selects the signal from the primary (working) path. However, if the quality of received signal on the primary path is poor, then it selects the signal from the protection path. In case of any failure in the node 2 equipment or on the primary path 2, node 3 will switch to the protection path via node 4 to receive the signal from node 1.

(b) Four-Fiber BLSR Configuration

In Fig. 13.38, two primary fiber loops are used for normal bidirectional communication while the other two secondary fiber loops are standby

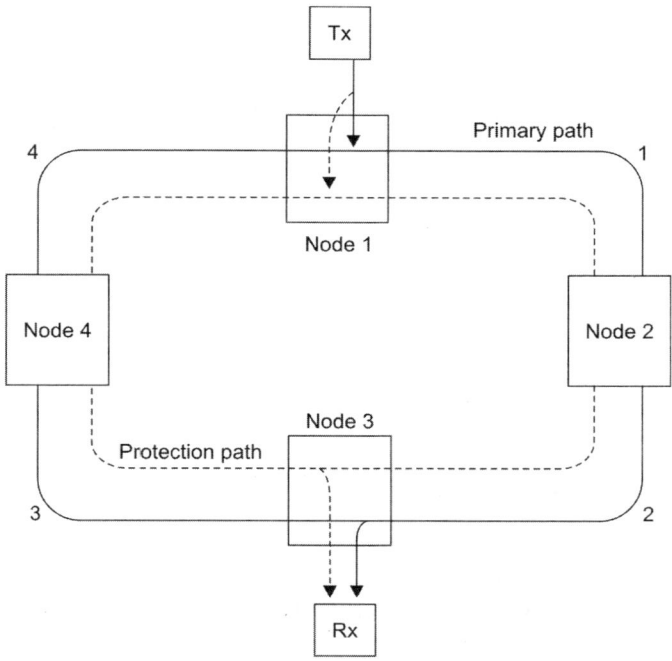

Fig. 13.37. Flow of primary and protection traffic from Node 1 to Node 3

links for protection purposes. The two primary fiber loops have fiber segments labeled 1p through 8p, which provides for an arrange-ment of 1p, 2p, 3p, and 4p in one primary loop, and 8p, 7p, 6p, and 5p in the second primary loop. The two secondary fiber loops have fiber segments labeled 1s through 8s, grouping 1s, 2s, 3s and 4s in one secondary loop, and 8s, 7s, 6s, and 5s in the second secondary loop.

Let us consider the connection from node 3 to node 1. The traffic from node 1 to node 3 flows in a clockwise direction along the links 1p and 2p. The traffic in the return path flows counterclockwise from node 3 to node 1, along links 6p and 5p. Thus, the information between node 1 and node 3 does not tie up any of the primary channel bandwidth in the other half of the ring.

Four-fiber BLSR reconfiguration (Failure 1): Consider the scenario shown in Fig. 13.39, where a transmitter or receiver circuit card used on the primary ring gails (in either node 3 or 4, in this case). In this case the affected nodes detect a lose-of signal (LOS) condition and switch both primary fibers, connecting them to the secondary protection pair. The protection between these nodes (3 or 4, in this case) now becomes part of the primary bidirectional loop.

Four-fiber BLSR reconfiguration (Failure 2): The exact same reconfiguration scenario as in failure 1 will occur when the primary fiber

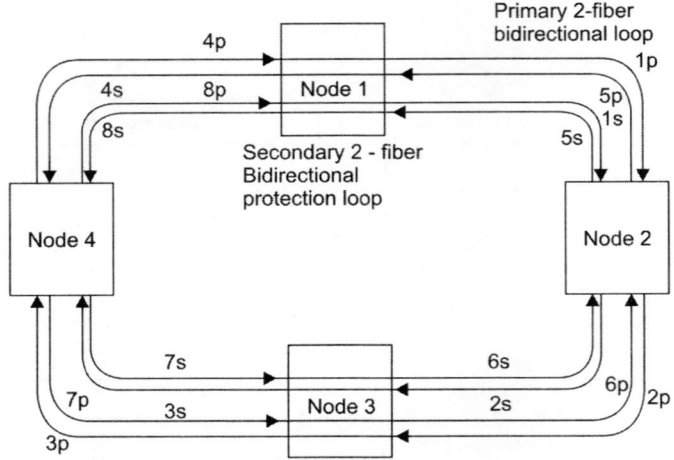

Fig. 13.38. Four-fiber bidirectional line switched ring network

connecting two nodes (in this case, nodes 3 and 4 breaks, as in Fig. 13.39. Note that in any case, the other links remain unaffected.

Four-fiber BLSR reconfiguration (Failure 3): In Fig. 13.40, consider the scenario where an entire node fails (in this case node 3), or both the primary and the protection fibers in a given span are severed, which could happen if they are in the same cable between 2 nodes (in this case nodes 3 and 4). In this scenario, the nodes on either side of the failed internal span will internally switch the primary path connection from

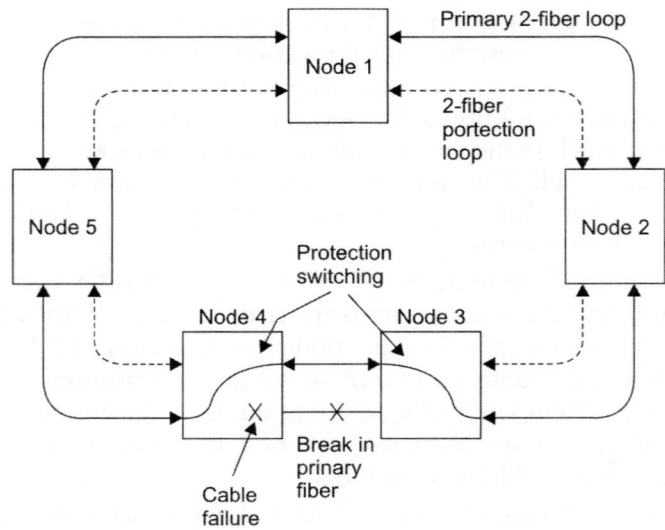

Fig. 13.39. Reconfiguration under transceiver card or line failure

their receivers and transmitters to the protection fibers, in order to loop traffic back to the previous node.

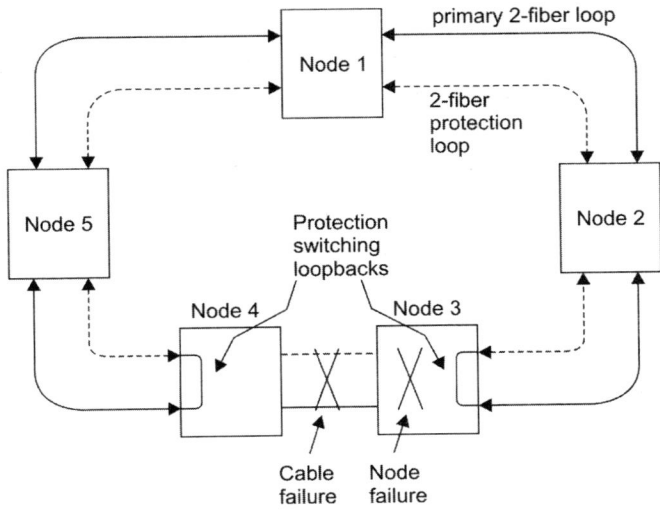

Fig. 13.40. Reconfiguration under node or fiber cable failure

(c) Generic SONET Network

SONET/SDH architecture allows the interconnections and interoperability of a variety of network configuration, as shown in Fig. 13.41. One can build point-to-point links, linear chains, UPSR, bidirectional link switched rings (BLSR), and interconnected rings. Each of the individual configuration has their own failure-recovery and protection mechanism, and SONET/SDH network management procedures.

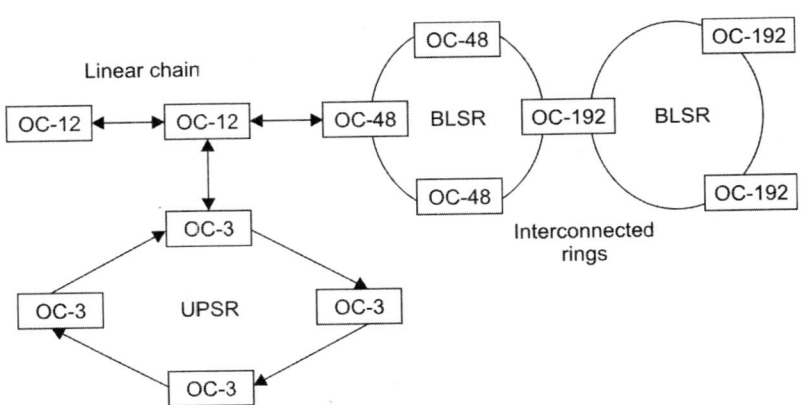

Fig. 13.41. Generic configuration of large SONET network consisting of various types of interconnected systems

(d) SONET ADM

One of the important features in SONET/SDH architecture is the add/drop multiplexer (ADM), as shown in Fig. 13.42. Various pieces of equipment are fully synchronized, and a byte-oriented multiplexer is used to add and drop sub-channels within an OC-N stream. Here, various OC-12s and OC-3s are multiplexed into OC-48 stream. Upon entering an ADM, these multiplexed sub-channels can be individually dropped by the ADM, and others can be added.

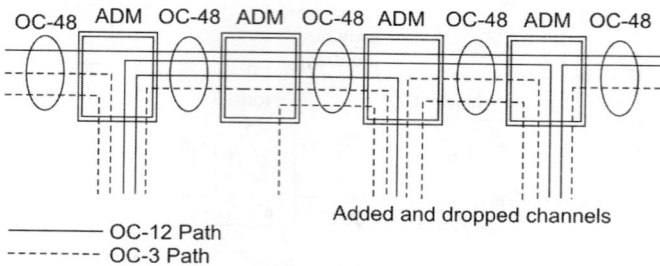

Fig. 13.42. Functional concept of an add/drop multiplexer (ADM) for SONET applications

(e) Dense WDM Deployment

SONET/SDH architecture can also be implemented with wavelengths. Figure 13.43 shows an example of dense wavelength division multiplexer (DWDM) deployment on an OC-192 trunk ring for n wavelengths (e.g., one could have $n = 16$). The different wavelength outputs from each OC-192 transmitter are first passed through the variable attenuator (VA) to equalize the out powers. These are then fed into a wavelength multiplexer, possibly amplified by a post optical amplifier, and sent out over the transmission fiber. Additional optical amplifiers might be located at intermediate points or at the receiver end.

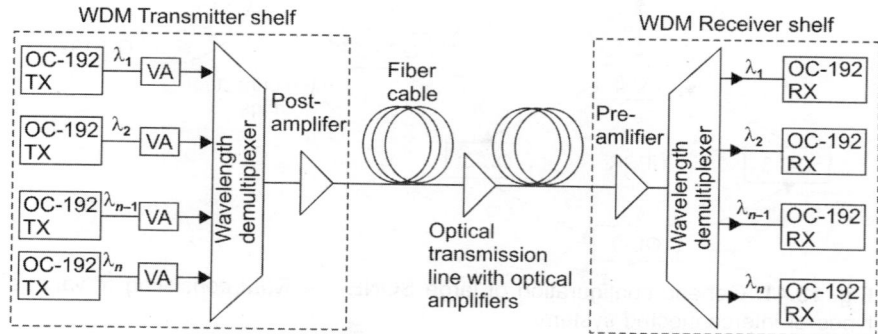

Fig. 13.43. DWD deployment of n wavelengths in an OC-192 trunk ring

13.19 NONLINEAR EFFECTS

Nonlinear phenomena in optical fiber affects the overall performance of optical fiber networks. These nonlinearities arise when high-strength optical fields from different signal wavelengths are present in an optical fiber at the same time and when these fields interact with acoustic waves and molecular vibrations.

The nonlinearities arise above a certain optical power threshold, and hence, the effect becomes negligible, once the signal has become sufficiently attenuated after travelling a distance along the fiber. Some important nonlinear effects are as follows:

  (i) Group velocity dispersion (GVD)
 (ii) Non-uniform gain for different wavelength
(iii) Polarization mode dispersion (PMD)
 (iv) Reflection from splices and connectors
  (v) Non-linear inelastic scattering processes
 (vi) Variation in reflactive index in optical fiber.

The major nonlinear processes physically affect system performance. These non-linearities are stimulated *Raman scattering*, stimulated *Brillouin scattering, self-phase modulation, cross-phase modulation,* and *four wave mixing* (FWM).

Non-linear processes are difficult to model because they depend on various factors, such as the *transmission length, cross sectional area of the fiber, transmitted power level,* etc. The problem can be understood as follows. If the signal power is assumed to be constant, the effect of a non-linear process increases with distance. However, we know that due to attenuation within the fiber, the signal power does not remain constant but decreases with distance. As a consequence, the non-linear process diminishes in magnitude. In practice, therefore, it is fairly reasonable to assume that the power is constant over a certain effective fiber length, which is less than the actual length of the fiber, and also take into account the exponential decay in power due to absorption. The effective length L_{eff} is given by:

$$L_{eff} = \frac{1}{P_0}\int_0^L P(z)\,dz = \frac{1}{P_0}\int_0^L P_0 e^{-\alpha z}\,dz = \frac{1 - e^{-\alpha L}}{\alpha} \quad (13.40)$$

where L is the actual length of the fiber, α is the attenuation coefficient, P_0 is the power at the input end of the fiber, and $P(z)$ is the power at a distance z. For large values of L, $L_{eff} \to 1/\alpha$. Thus, if we take α to be typically around 0.22 dB km^{-1} (which is equivalent to a coefficient of 0.0507 km^{-1}) at 1.55 µm, $L_{eff} \approx 20$ km. If the link incorporates optical amplifiers and the total amplified link length is L_A and the total span length of the fiber between amplifiers is L, the effective length is given approximately by the following relation:

$$L_{eff} = \frac{1-e^{-\alpha L}}{\alpha} = \frac{L}{L_A} \qquad (13.41)$$

Thus, the *total effective length decreases as the amplifier span increases.*

It has been said above that the effect of non-linear processes increases in light intensity transmitted through the fiber. This intensity, however, is inversely proportional to the cross-sectional area of the fiber core. We have seen earlier that for a single-mode fiber, this power is not distributed uniformly over the entire cross-sectional area of the core. In practice, therefore, it is convenient to use an effective cross-sectional area A_{eff}. If the mode field radius is w, $A_{eff} \approx \pi w^2$. Typical effective areas for a conventional single-mode fiber (CSF), dispersion-shifted fiber (DSF), and dispersion-compensating fiber (DCF) are 80 μm², 50 μm², and 20 μm², respectively.

13.19.1 Stimulated Raman Scattering

Stimulated Raman scattering (SRS) is the result of the inelastic scattering of a light wave (propagating through a silica-based optical fiber) by silica molecules. When a photon energy $E_1 = h\nu_1$ (where the symbols have their usual meaning) interacts with a silica moleule, some of its energy, depending on the vibrational frequency of the molecule, is absorbed by the latter and the photon is scattered. As the original photon has lost some energy, the energy of the scattered photon becomes less, say, $E_2 = h\nu_2$. This change in the frequency of the interacting photon ν_1 and ν_2 ($\nu_1 > \nu_2$) is called *Stoke's shift*. Since the light wave propagating through the fiber is a source of interacting photons, it is normally called a *pump wave*. This process generates scattered light at a wavelength longer than the pump wave ($\lambda_2 > \lambda_1$ as $\nu_2 < \nu_1$). If two or more signals at different wavelengths are simultaneously injected into the fiber, SRS can cause power to be transferred from the lower wavelength signals to the higher wavelength signals. This effect is shown in Fig. 13.44. As a consequence, SRS can severely affect the performance of a multi-channel fiber-optic communication system by transferring energy from shorter wavelength channels to other longer wavelength channels. This effect occurs in both the directions.

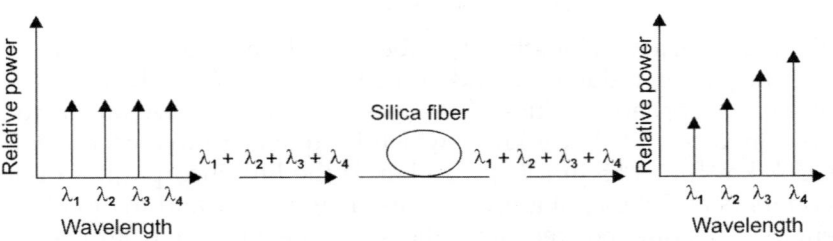

Fig. 13.44. The effect of stimulated Raman scattering (SRS)

The effect of SRS can be estimated following Buck (1995) as follows. Consider a WDM system with N equally spaced channels, 0, 1, 2, ... $(N-1)$, with a channel spacing of $\Delta\lambda_2$. With the assumptions that the same power is transmitted in all the channels, the Raman gain increases linearly, and that there is no interaction between other channels, the fraction of power coupled from channel 0 to channel i is given approximately by:

$$F(i) = g_R \frac{i\Delta\lambda_s}{\Delta\lambda_c} \frac{P_0 L_{\text{eff}}}{2 A_{\text{eff}}}$$

where g_R is the *peak Raman gain coefficient*, $\Delta\lambda_c$ is the *total channel spacing*, and the other symbols have their usual meaning. Therefore, the fraction of power coupled from channel 0 to all the other channels is given by:

$$F = \sum_{i=1}^{N-1} F(i) = \frac{g_R \Delta\lambda_s}{\Delta\lambda_c} \frac{P_0 L_{\text{eff}}}{2 A_{\text{eff}}} \frac{N(N-1)}{2} \qquad (13.42)$$

The power penalty for channel 0 is then $-10 \log_{10}(1-F)$. Thus, in order to keep this penalty below 0.5 dB, the fraction F should be less than 0.1.

SRS is not a serious problem in systems with small number of channels. However, it can create severe problems in WDM systems with large numbers of wavelength channels. In order to alleviate the effect of SRS, (i) the channels should be spaced as closely as possible, and (ii) the transmitted power level in each channel should be kept below the threshold (in other words, the distance between the optical amplifiers in the link should be reduced).

13.19.2 Stimulated Brillioun Scattering (SBS)

Stimulated Brillioun scattering (SBS) may be viewed as the scattering of a pump wave by an acoustic wave (generated by the oscillating electric field of the pump wave). This process creates a *Stokes' wave* of lower frequency, which travels in the backward direction. The Stokes' wave experiences gain at the expense of the depletion of the signal power of the forward propagating signal (i.e., the pump wave). The frequency shift due to SBS is called the Brillioun shift and is given by:

$$v_B = 2n V_A / \lambda_p \qquad (13.43)$$

where n is the mode index of the fiber, V_A is the velocity of the acoustic wave, and λ_p is the wavelength of the pump wave. If we take typical values of $n = 1.46$ for the silica fiber, $V_A = 5960$ m·s^{-1}, and $\lambda_p = 1.55$ μm, then $v_B = 11.22$ GHz. This interaction occurs over a very narrow line width of $\Delta v_B = 20$ MHz at $\lambda_p = 1.55$ μm. There are two important features of SBS: (i) it does not cause any interaction between different wavelengths, as long as the wavelength spacing is much greater than

20 MHz, and (ii) its effect can create significant distortion within a single channel, especially when the amplitude of the scattered wave is comparable with the signal power.

A simple criterion for determining the impact of SBS is to consider the SBS threshold P_{th}, which is defined as the signal power at which the backscattered power equals the fiber input power. It is given by the following approximate expression:

$$P_{th} = \frac{21bA_{eff}}{g_B L_{eff}}\left[1 + \frac{\Delta v_{source}}{\Delta v_B}\right] \tag{13.44}$$

where Δv_{source} is the line width of the source, g_B is the *Brillioun gain*, and the value of b lies between 1 and 2, depending on the relative polarizations of the pump and Stokes' waves. Thus P_{th} increases with increase in the source line width.

13.19.3 Four-Wave Mixing

In a WDM system, if three waves with angular frequencies ω_i, ω_j, and ω_k co-propagate inside a *silica fiber* simultaneously, then the non-linear susceptibility of the silica fiber generates new waves at angular frequencies $\omega_i \pm \omega_j \pm \omega_k$. This phenomenon is known as *four-wave mixing* because three waves at frequencies, ω_i, ω_j, ω_k combine to produce a fourth wave at a frequency $\omega_i \pm \omega_j \pm \omega_k$. In principle, several frequencies corresponding to the combinations of plus and minus signs are possible. However, most of them do not build up due to the lack of a *phase-matching condition*. The frequency combinations of the $\omega_i + \omega_j + \omega_k$ (with $i, j \neq k$) are often troublesome for WDM systems, as they can become phase matched when the wavelength channels are closely spaced or are spaced near the dispersion zero of the fiber. Such frequency combinations can be defined as:

$$\omega_{ijk} = \omega_i + \omega_j + \omega_k \ (i, j \neq k) \tag{13.45}$$

For N wavelength channels co-propagating through the fiber, the number of generated frequencies is given byL

$$M = \frac{N^2}{2}(N-1) \tag{13.46}$$

If the wavelength channels are equally spaced, the new waves overlap the original injected frequencies. This causes severe crosstalk and the depletion of the original signal waves, thus degrading the system performance.

In general, the penalty due to four-wave mixing can be reduced by (i) making the channel spacing unequal, (ii) increasing the channel spacing, and (iii) using a non-zero dispersion-shifted fiber instead of a dispersion-shifted fiber.

13.19.4 Self- and Cross-Phase Modulation

Self-phase modulation (SPM) arises because the refractive index n of the fiber depends on the intensity I (which is equivalent to the power per unit effective area of the fiber). The relation is as follows:

$$n = n_0 + n_2 I = n_0 + n_2 \left(\frac{P}{A_{\text{eff}}} \right) \qquad (13.47)$$

where n_0 is the ordinary refractive index of the fiber core and n_2 is the non-linear index coefficient, P is the optical power and A_{eff} is the effective area of the fiber. Depending on the dopant, the value of n_2 for a silica fiber varies from 2.2 to 3.4×10^{-8} $\mu m^2/W$.

To understand the effect of SPM, let us consider a *Gaussian pulse propagating* through a fiber with a non-linear index of refraction given by Eq. (13.47). The pulse shape is shown in Fig. 13.45. The time axis is normalized to the time parameter t_0, the pulse half-width measured at $1/e$ intensity point. As is evident from the figure, the intensity of the pulse first rises from zero to a maximum and then falls to zero again. Because the refractive index of the fiber is dependent on intensity, it will also vary with time. This variation of n with t will give rise to a temporally varying phase change in exactly the same fashion. Thus, different parts of the pulse undergo different phase shifts, which gives rise to what is known as *frequency chirping*; that is, the rising edge of the pulse shifts towards higher frequencies (red shift) and the trailing edge shifts towards lower frequencies (blue shift). This pulse chirping, in turn, enhances the group velocity dispersion (GVD) induced pulse broadening. Moreover, this effect is proportional to the transmitted signal power, and hence, SPM effects are more pronounced in long-haul systems where the transmitted powers are high.

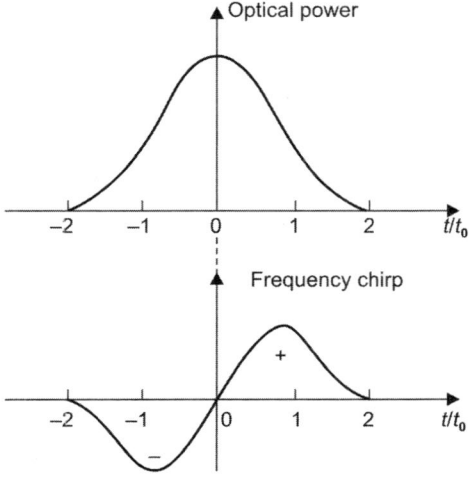

Fig. 13.45. SPM-induced frequency chirp for a Gaussian pulse

In WDM systems, the intensity-dependent refractive index of the fiber gives rise to another kind of non-linear effect, called cross-phase modulation (CPM). In this process, the power fluctuation in one channel produces phase fluctuations in other co-propagating channels. The effect can be significant if the system is using dispersion-shifted fibers and is operating above 10 Gb/s. The effect may be reduced by employing non-zero dispersion single-mode fibers.

13.20 DISPERSION

The pulse gets distorted as it travels along the fiber lengths. Pulse spreading in fiber is referred as **dispersion**. Dispersion is caused by difference in the propagation times of light rays that takes different paths during the propagation. The light pulses travelling down the fiber encounter dispersion effect because of this the pulse spreads out in time domain. Dispersion limits the information bandwidth. The distortion effects can be analyzed by studying the group velocity in guided mode.

Information Capacity Determination

Dispersion and attenuation of pulse travelling along the fiber is shown in Fig. 13.46(a).

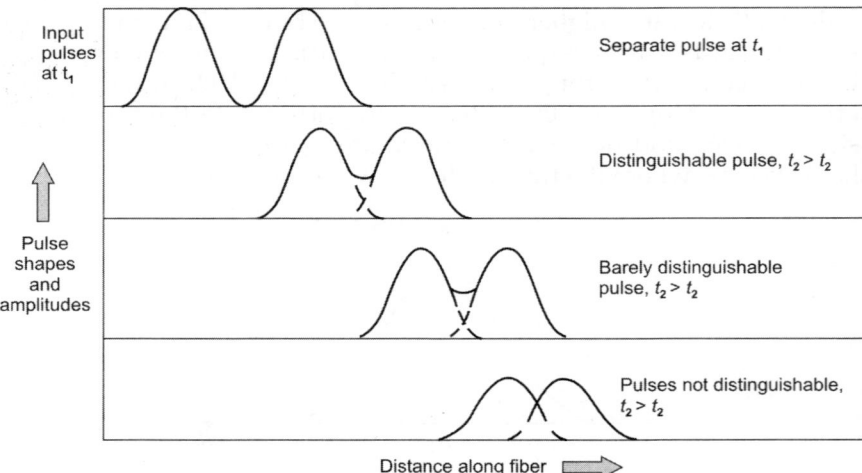

Fig. 13.46. (a) Dispersion and attenuation in fiber

Figure 13.46(a) shows, after travelling some distance, pulse starts broadening and overlap with the neighbouring pulses. At certain distance, the pulses are not even distinguishable and error will occur at receiver. Therefore, the information capacity is specified by bandwidth-distance product (MHz·km). For step index bandwidth distance produce is 20 MHz·km and for graded index, it is 2.5 GHz·km.

Fiber Optic Communication

Group Delay

Consider a fiber cable carrying optical signal equally with various modes and each mode contains all the spectral components in the wavelength band. All the spectral components travel independently and they observe different **time delay** and **group delay** in the direction of propagation. The velocity at which the energy in a pulse travels along the fiber is known as a **group velocity**. Group velocity is given by:

$$V_g = \frac{\partial \omega}{\partial \beta} \tag{13.48a}$$

Thus different frequency components in a signal will travel at different group velocities and so will arrive at their destination at different times, for digital modulation of carrier, this results in dispersion of pulse, which affects the maximum rate of modulation. Let the difference in propagation times for two side bands is $\delta\tau$.

$$\delta\tau = \frac{d\tau}{d\lambda} \times \delta\lambda \tag{13.48b}$$

where $\delta\lambda$ = Wavelength difference between upper and lower sideband (spectral width)

$\dfrac{\delta\tau}{d\lambda}$ = Dispersion coefficient (D)

Then, $D = \dfrac{1}{L} \cdot \dfrac{d\tau}{d\lambda}$ where, L is length of fiber

$D = \dfrac{d}{d\lambda}\left(\dfrac{1}{V_g}\right)$ As $\tau = \dfrac{1}{V_g}$ and considering unit length $L = 1$

$\dfrac{1}{V_g} = \dfrac{d\beta}{d\omega}$

$\dfrac{1}{V_g} = \dfrac{d\lambda}{d\omega} \times \dfrac{d\beta}{d\lambda}$

$\dfrac{1}{V_g} = \dfrac{-\lambda^2}{2\pi c} \times \dfrac{d\beta}{d\lambda}$

$\therefore \quad D = \dfrac{d}{d\lambda}\left(\dfrac{-\lambda^2}{2\pi c} \cdot \dfrac{d\beta}{d\lambda}\right) \tag{13.48c}$

Dispersion is measured in picoseconds per nanometer per kilometer.

Material dispersion (MD) occurs because the index of refraction (n) varies as a function of the optical wavelength (λ). As a consequence, since the group velocity (V_g) of a mode is a function of n, the various spectral component of a given mode will travel at different speeds, depending on λ. Obviously, *material dispersion* is, therefore, an *intermodal* dispersion effect, and is of particular importance for single-mode waveguides for LED system (since an LED has a broader output spectrum than a laser diode).

Material dispersion is also called as **chromatic dispersion**. Material dispersion exists due to change in index of refraction for different wavelengths. A light ray contains components of various wavelengths centered at wavelength λ_0. The time delay is different for different wavelength components. This results in time dispersion of pulse at the receiving end of fiber. Figure 13.46(b) shows index of refraction as a function of optical wavelength.

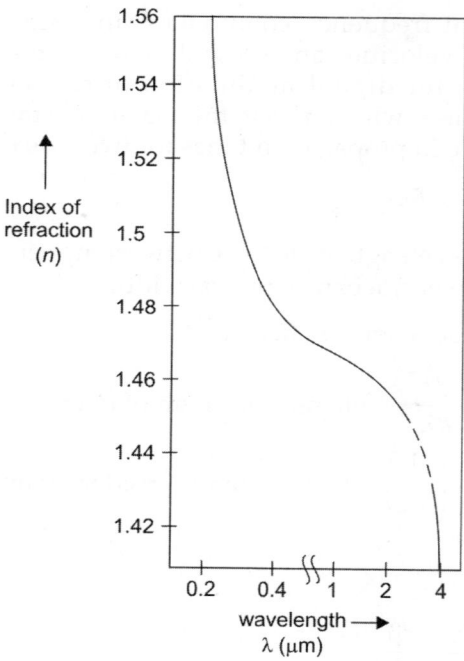

Fig. 13.46. (b) Index of refraction (n) as a function of wavelength (λ)

The material dispersion for unit length ($L = 1$) is given by:

$$D_{mat} = \frac{-\lambda}{c} \times \frac{d^2n}{d\lambda^2} \qquad (13.48d)$$

where c = Light velocity
λ = Centre wavelength

$\dfrac{d^2n}{d\lambda^2}$ = Second derivative of index of refraction with respect to wavelength

Negative sign shows that the upperside band signal (lowest wavelength) arrives before the power side band (highest wavelength).

A plot of material dispersion and wavelength is shown in Fig. 13.46(c). The unit of dispersion is: ps/nm·km.

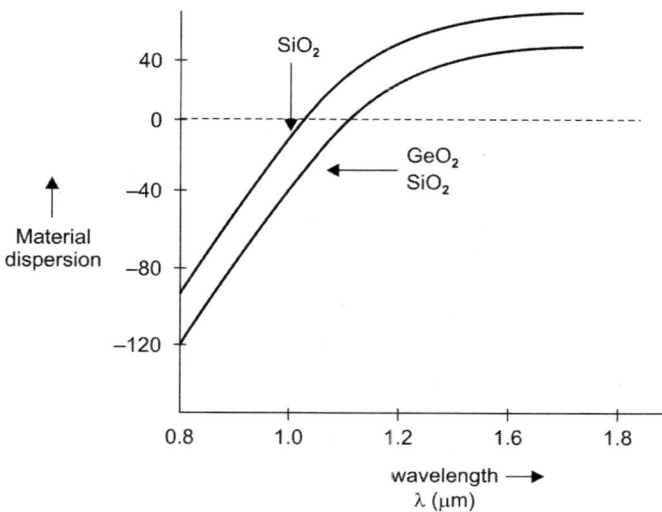

Fig. 13.46. (c) Material dispersion as a function of λ

For single mode fibers, *waveguide dispersion* (WD) is of importance and can be of the same order of magnitude as material dispersion. The pulse spread σ_{wg} occurring over a distribution of wavelengths σ_λ is obtained from the derivative of the group delay (τ_{wg}) with respect to λ:

$$\sigma_{wg} \approx \left|\frac{d\tau_{wg}}{d\lambda}\right| \sigma_\lambda = L\,|D_{wg}(\lambda)|\,\sigma_\lambda \qquad (13.48e)$$

where $D_{wg}(\lambda)$ is the *waveguide dispersion*.

(i) Waveguide dispersion is caused by the difference in the index of refraction between the core and cladding, resulting in a 'drag' effect between the core and cladding portions of the power.

(ii) Waveguide dispersion is significant only in fibers carrying fewer than 5–10 modes. Since multimode optical fibers carry hundreds of modes, they will not have observable waveguide dispersion.

(iii) The group delay (τ_{wg}) arising due to waveguide dispersion.

$$\tau_{wg} = \frac{L}{c}\left[n_2 + n_2\Delta\frac{d(kb)}{dk}\right] \qquad (13.48f)$$

where, b = Normalized propagation constant

$k = 2\pi/\lambda$ (group velocity)

Normalized frequency υ:

$$\upsilon = k\,a\left(n_1^2 - n_2^2\right)^2 = k\,a\,n_2\sqrt{2\Delta} \quad \text{(For small } \Delta\text{)}$$

$$\therefore \quad \tau_{wg} = \frac{L}{c}\left[n_2 + n_2 \Delta \frac{d(V_b)}{dV}\right] \quad (13.48g)$$

The second term $\frac{d(V_b)}{dV}$ is waveguide dispersion and is mode dependent term.

Polarization-mode dispersion (PMD): The effects of fiber birefringence on the polarization states of an optical signal are another source of pulse broadening. This is particularly critical for high-rate long haul transmission links (e.g., 10 and 40 Gb/s over tens of kilometers), i.e.:

(i) Different frequency component of a pulse acquires different polariation states (such as linear polarization and circular polarization). This results in pulse broadening is known as **Polarization-Mode Dispersion** (PMD).

(ii) PMD is the limiting factor for optical communication system at high data rates. The effects of PMD must be compensated.

A varying birefringence along its length will cause each polarization mode to travel at a slightly different velocity. The resulting difference in propagation times $\Delta\tau_{PMD}$ between the two orthogonal polarization modes will result in pulse spreading. This is *polarization-mode* dispersion (PMD). If the group velocities of two orthogonal polarization modes are V_{gx} and V_{gy}, then the *differential time delay* $\Delta\tau_{PMD}$ between the two polarization components during propagation of the pulse over a distance L is:

$$\Delta\tau_{PMD} = \left|\frac{L}{V_{gx}} - \frac{L}{V_{gy}}\right| \quad (13.49a)$$

We may note that, in contrast to chromatic dispersion, which is relatively stable phenomenon along a fiber, PMD varies randomly along a fiber. Typical values of D_{PMD} range from 0.05 to 1.0 ps/√km.

Modal Dispersion

As only a certain number of modes can propagate down the fiber, each of these modes carries the modulation signal and, each one is incident on the boundary at a different angle, they will each have their own individual propagation times. The net effect is spreading of pulse, this form of dispersion is called **modal dispersion.**

Modal dispersion takes place in multimode fibers. It is moderately present in graded index fibers and almost eliminated in single mode step index fiber.

Modal dispersion is given by:

$$\Delta t_{modal} = \frac{n_1 Z}{c}\left(\frac{\Delta}{1-\Delta}\right) \quad (13.49b)$$

where Δt_{modal} = Dispersion,
n_1 = Core refractive index,
Z = Total fiber length,
c = Velocity of light in air, and
Δ = Fractional refractive index.

Putting $\Delta = \dfrac{NA^2}{2n_1^2}$ in above equation:

$$\Delta t_{modal} = \frac{(NA)^2 Z}{2n_1 c} \quad (13.49c)$$

The modal dispersion Δt_{modal} describes the optical pulse spreading due to modal effects. Optical pulse width can be converted to electrical rise time through the relationship.

$$t_{r\,mod} = 0.44\,(\Delta t_{modal}) \quad (13.49d)$$

Signal Distortion in Single Mode Fibers

The pulse spreading σ_{wg} over range of wavelengths can be obtained from derivative of group delay with respect to λ:

$$\sigma_{wg} = \left|\frac{d\tau_{wg}}{d\lambda}\right|\sigma_\lambda = L\,|D_{wg}(\lambda)|\sigma_\lambda = \frac{V}{\lambda}\left|\frac{d\tau_{wg}}{d\lambda}\right|\sigma_\lambda \quad (13.49e)$$

$$= \frac{n_2 L \Delta \sigma_\lambda}{c\lambda}\left[V\frac{d^2(Vb)}{dv^2}\right]$$

where

$$D_{wg}(\lambda) = \frac{-n_2 \Delta}{c\lambda} V\left[\frac{d^2(Vb)}{dV^2}\right] \quad (13.49f)$$

This is the equation for waveguide dispersion for unit length.

Higher Order Dispersion

Higher order dispersive effective effects are governed by dispersion slope S.

$$S = \frac{dD}{d\lambda}$$

where, D is total dispersion.

Also,
$$S = \left(\frac{2\pi c}{\lambda^2}\right)^2 \beta_3 + \left(\frac{4\pi c}{\lambda^3}\right)\beta_2$$

where, β_2 and β_3 is second and third order dispersion parameter.

Dispersion slope S plays an important role in designing WDM system.

Dispersion Induced Limitations

The extent of pulse broadening depends on the width and the shape of input pulses. The pulse broadening is studied with the help of wave equation.

Basic Propagation Equation

The basic propagation equation which governs pulse evolution in a single mode fiber is given by:

$$\frac{\partial A}{\partial z} + \beta_1 \frac{\partial A}{\partial t} + \frac{i\beta_2}{2} \cdot \frac{\partial^2 A}{\partial t^2} - \frac{\beta_3}{6} \frac{\partial^3 A}{\partial t^3} = 0$$

where, β_1, β_2 and β_3 are different dispersion parameters.

Chirped Gaussian Pulses

A pulse is said to be **chirped** if its carrier frequency changes with time.

For a Gaussian spectrum having spectral width σ_ω, the pulse broadening factor is given by:

$$\frac{\sigma^2}{\sigma_0^2} = \left(1 + \frac{C\beta_2 L}{2\sigma_0^2}\right)^2 + (1 + V_\omega^2)\left(\frac{\beta_2 L}{2\sigma_0^2}\right)^2 + (1 + C + V_\omega^2)^2 \left(\frac{\beta_3 L}{4\sqrt{2}\,\sigma_0^3}\right)^2$$

where, $V_\omega = 2\sigma_\omega \sigma_0$

Limitations of Bit Rate

The limiting bit rate is given by:

$$4B\sigma \leq 1$$

The condition related to bit-rate distance product (BL) and dispersion (D) is given by

$$BL|D|\sigma_\lambda \leq \frac{1}{4}$$

$$BL|S|\sigma_\lambda^2 \leq \frac{1}{\sqrt{8}}$$

where, S is dispersion slope.

Limiting bit rate for a single-mode fibers as a function of fiber length for $\sigma_\lambda = 0$, 1 and 5 nm is shown in Fig. 13.46(d).

Dispersion Management

Erbium-doped fiber amplifiers have opened a new era of optical transmission technology, allowing one to use WDM or DWDM with compact as well as economical approaches. However, the price that has to be paid for this success is the *combat* with the accumulated impact of the dispersive or non-linear effects of the transmission fiber, which grows

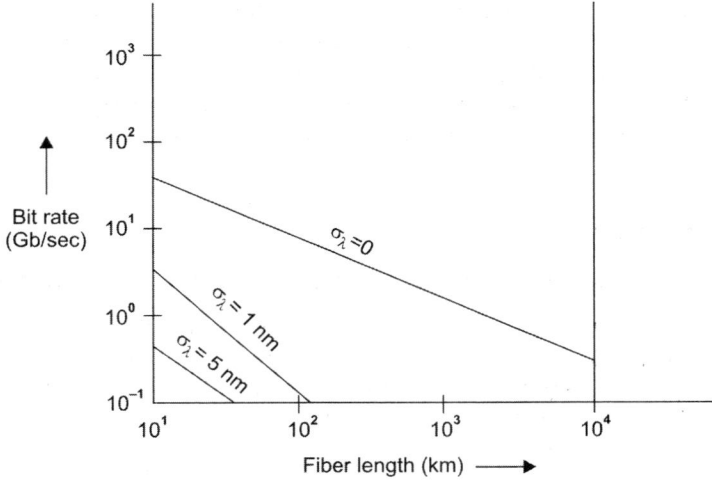

Fig. 13.46. (d) Dependence of bit rate on fiber length

with the transmission distance. Dispersion management techniques have been developed to solve this problem. The basic idea is to use two types of fibers with opposite signs of dispersion to produce a sawtooth pattern of the dispersion map as shown in Fig. 13.47. The relation for perfect dispersion compensation is:

$$D_1 L_1 + D_2 L_2 = 0 \qquad (13.50a)$$

where D_1 and D_2 are the dispersion of the two types of fibers and L_1 and L_2 are their respective lengths. A special type of fiber called the *dispersion-compensating fiber* (DCF) has been developed for this purpose. Typically, a small length of the DCF may be placed just before the optical

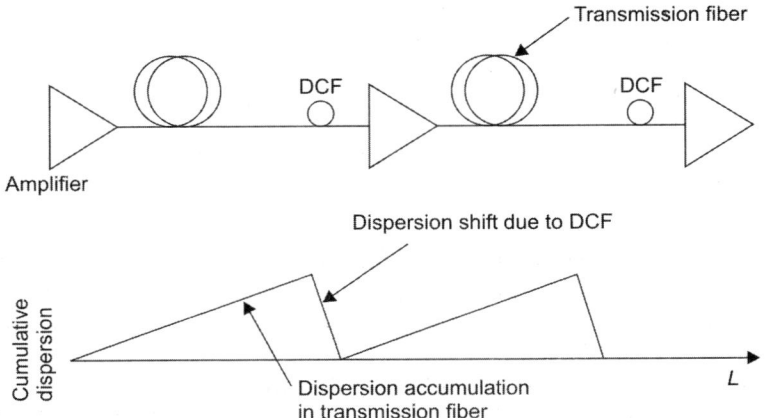

Fig. 13.47. Dispersion management using dispersion compensating fiber (DCF)

amplifier. If the transmission fiber has a low positive dispersion, the DCF should have a large negative dispersion. With this approach, the total cumulative dispersion is made zero (or small) so that the dispersion-induced penalties are negligible, but the dispersion is non-zero everywhere along the link so that the penalties due to non-linear effects are also reduced.

13.21 SOLITONS

13.21.1 Nonlinear Effects, i.e. Nonlinear Optical Susceptibility

"Soliton" refers to special kinds of waves that propagate undistorted over long distances and remain unaffected after collision with each other. Solitons exist due to *nonlinearity* and *dispersion*. Optical responses including nonlinear effects are described as:

$$P(t) = \epsilon_0 \left\{ \chi^{(1)} E(t) + \chi^{(2)} E^{(2)}(t) + \chi^{(3)} E^3(t) + ... \right\} \quad (13.50b)$$
$$\equiv P^{(1)}(t) + P^{(2)}(t) + P^{(3)}(t) + ...$$

where, we have expressed polarization $P(t)$ as a power series in the field strength $E(t)$. The quantities $\chi^{(1)}$, $\chi^{(2)}$, $\chi^{(3)}$ are known as susceptibilities; $\chi^{(1)}$ is a linear susceptibility and $\chi^{(2)}$ and $\chi^{(3)}$ are known as the second order- and third-order nonlinear susceptibilities respectively.

For a typical solid-state system $\chi^{(1)}$ is of the order of unity whereas $\chi^{(2)}$ is the order of $1/E_{at}$ and $\chi^{(3)}$ of the order of $1/E_{at}^2$, where $E_{at} = e/(4\pi \epsilon_0 a_0^2)$ is the characteristic atomic electric field strength and $a_0 = 4\pi\epsilon_0 \hbar^2/me^2$ is the Bohr radius of the hydrogen atom. Explicitly.

$$\chi^{(2)} \simeq 1.94 \times 10^{-12} \text{ m/V}$$
$$\chi^{(3)} \simeq 3.78 \times 10^{-24} \text{ m}^2/V^2$$

Formal expression for the third-order susceptibility is:

$$P_1(\omega_0 + \omega_n + \omega_m) = \epsilon_0 D \sum_{jkl} \chi^{(3)}_{ijkl}(\omega_0 + \omega_n + \omega_m, \omega_0, \omega_n, \omega_m)$$
$$\times E_j(\omega_0) E_k(\omega_n) E_l(\omega_n)$$

where i, j, k, l refer to the Cartesian components of the fields and the degeneracy factor D represents the number of distinct permutations of the frequencies $\omega_0, \omega_m, \omega_m$.

$\chi^{(j)}$ ($j = 1, 2, ...$) is the j-th order susceptibility. The linear susceptibility $\chi^{(1)}$ contributes to the linear refractive index n_0 (real and imaginary parts; the imaginary part being responsible for attenuation). The second-order susceptibility $\chi^{(2)}$ is responsible for the second harmonic generation. For SiO_2, the second-order nonlinear effect is negligible since SiO_2 has the inversion symmetry. Therefore, optical fibers normally do not show the second-order nonlinear effects.

The third-order susceptibility $\chi^{(3)}$ is responsible for the *lowest order nonlinear effects in optical fibers*. Generally, it manifests itself as the change in the refractive index with optical power or as a scattering phenomenon. It is linked with the *optical Kerr effect*, four-wave mixing, third-harmonic generation, stimulated Raman scattering etc.

Assuming linear polarization of propagating light and neglecting tensorial character of $\chi^{(3)}_{ijkl}$, one finds the following relation for the nonlinear polarization:

$$P^{NL}(\omega) = 3\,\epsilon_0\,\chi^{(3)}\,(\omega = \omega - \omega - \omega)\,|E(\omega)|^2\,E(\omega)$$

Total polarization, which consists of linear and nonlinear parts, is written as:

$$P(\omega) = \epsilon_0\,\chi^{(1)}E(\omega) + 3\,\epsilon_0\,\chi^3\,|E(\omega)|^2\,E(\omega) = \epsilon_0\,\chi_{eff}E(\omega)$$

The effective susceptibility is field dependent as:

$$\chi_{eff} = \chi^{(1)} + 3\chi^{(3)}\,|E(\omega)|^2$$

and it is linked to refractive index as:

$$n = 1 + \chi^{(3)} \equiv n_0 + n_2 I \qquad (13.50c)$$

Here I denotes the time-averaged intensity of the optical field. The main features of nonlinear effects are as follows:

13.21.2 Main Nonlinear Effects

Kerr effect: Kerr in 1875 found that a transparent liquid becomes *doubly refracting* (birefringent) when placed in a strong electric field. Generally, Kerr effect described situations where refractive index depends on electric field as:

$$n(\omega, |E^2|) = n_0(\omega) + n_2(\omega)\,|E|^2$$

Here n_2 is known as Kerr coefficient and it is related to susceptibility as:

$$n_2(\omega) = \frac{3}{4\pi}\,\chi^{(3)}_{xxx} \qquad (13.50d)$$

for a linearly polarized wave in the x direction. For silica, its value is approximately $1.3 \times 10^{-22}\ m^2/V^2$. Kerr effect originates from the *non-harmonic motion of electrons bound in molecules*. Consequently, it is fast effect, the response time of the order of 10^{-15} s.

13.21.3 Stimulated Raman Scattering

Scattering phenomena are responsible for Raman and Brillouin effects. During those scatterings, the energy of the optical field is transferred to local phonons: in Raman sactering optical phonons are generated whereas in the Brillouin scattering the acoustic phonons.

13.21.4 Derivation of the Nonlinear Schödinger Equation

Solitons in optical fibers are described by the so-called nonlinear Schrödinger (NSE) equation. In the derivation, we use the concept of the Fourier spectrum of the propagating pulse (Fig. 13.47).

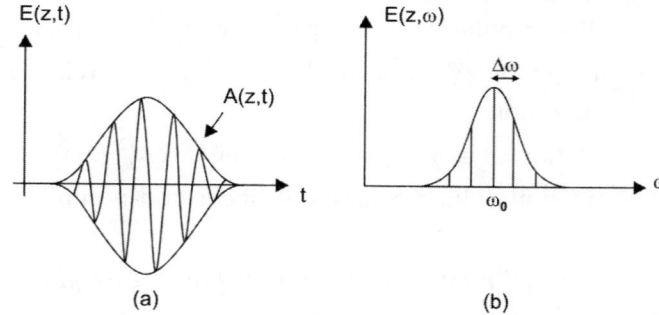

Fig. 13.47. (a) Illustration of the propagating modulated pulse, (b) and its spectrum

A medium where solitons propagate exhibits Kerr nonlinearity. In such a medium, refractive index depends on intensity of electric field $I(t)$ given by Eq. (13.50), again reproduced here:

$$n(t) = n_0 + n_2 \cdot I(t) \qquad (13.51)$$

where
$$I(t) = 2n_0\varepsilon_0 c\,|A(z,t)|^2 \qquad (13.52)$$

Here $A(z, t)$ is the slowly varying envelope connected to the optical pulse described by the optical pulse $E(z, t)$ as shown in Fig. 13.47.

$$E(z,t) = A(z,t)\,e^{i(\omega_0 t - \beta_0 z)} \qquad (13.53)$$

Fourier transform of the optical field is:

$$E(z,t) = \int_{-\infty}^{+\infty} d\omega\, \tilde{E}(z,\omega)\, e^{i(\omega t - \beta z)} \qquad (13.54)$$

where $\tilde{E}(z,\omega)$ is the Fourier spectrum of the pulse, β propagation constant and ω_0 the frequency at which the pulse spectrum is centered (also known as carrier frequency) (Fig. 13.47).

For quasi-monochromatic pulses with $\Delta\omega \equiv \omega - \omega_0$, it is useful to expand propagation constant $\beta(\omega)$ in a Taylor series:

$$\beta(\omega) = \beta_0 + \beta_1 \cdot (\omega - \omega_0) + \frac{1}{2}\beta_2 \cdot (\omega - \omega_0)^2 + \Delta\beta_{NL} \qquad (13.55)$$

where we have neglected higher order derivatives. Here $\Delta\beta_{NL} = n_2 k_0 I$ is the nonlinear contribution to the propagation constant.

Substituting expansion of Eq. (13.54) into Eq. (13.55), one obtains

$$E(z,t) = e^{-i\beta_0 z} \int_{-\infty}^{+\infty} d(\Delta\omega)\, \tilde{E}(z,\omega)\, \exp\left(i\omega t - i\beta_1 z \Delta\omega - \frac{1}{2}i\beta_2 z \Delta\omega^2 - i\cdot z \cdot \Delta\beta_{NL} \right)$$

$$= e^{-i(\omega_0 t - \beta_0 z)} \int_{-\infty}^{+\infty} d(\Delta\omega)\, \tilde{E}(z, \omega_0 + \Delta\omega)$$

$$\times \exp\left(it\Delta\omega - i\beta_1 z\Delta\omega - \frac{1}{2}i\beta_2 z\Delta\omega^2 - i\cdot z \cdot \Delta\beta_{NL} \right)$$

$$\equiv e^{i(\omega_0 t - \beta_0 z)} A(z, t)$$

where we have introduced

$$A(z, t) = \int_{-\infty}^{+\infty} d(\Delta\omega)\, \tilde{E}(z, \omega_0 + \Delta\omega)$$

$$\times \exp\left(it\Delta\omega - i\beta_1 z\Delta\omega - \frac{1}{2}i\beta_2 z\Delta\omega^2 - i\cdot z \cdot \Delta\beta_{NL} \right)$$

$$A(z, t) \equiv \int_{-\infty}^{+\infty} d(\Delta\omega)\, \tilde{E}(z, \omega_0 + \Delta\omega)\, e^{ig(z,t)} \tag{13.56}$$

Now, we will obtain a differential equation describing evolution of the amplitude $A(z, t)$ from Eq. (13.56) which is in the integral form. To do this, one needs to take partial derivatives of Eq. (13.56). One obtains:

$$\frac{\partial A(z, t)}{\partial t} = \int_{-\infty}^{+\infty} d(\Delta\omega)\, \tilde{E}(z, \omega_0 + \Delta\omega)\, i\Delta\omega\, e^{ig(z,t)}$$

$$\frac{\partial A(z, t)}{\partial t^2} = \int_{-\infty}^{+\infty} d(\Delta\omega)\, \tilde{E}(z, \omega_0 + \Delta\omega)\, (i\Delta\omega)^2\, e^{ig(z,t)}$$

$$\frac{\partial A(z, t)}{\partial z} = \int_{-\infty}^{+\infty} d(\Delta\omega)\, \tilde{E}(z, \omega_0 + \Delta\omega)\left(-i\beta_1\Delta\omega - \frac{1}{2}i\beta_2\Delta\omega^2 - i\Delta\beta_{NL} \right) e^{ig(z,t)}$$

While evaluating time derivatives, we have assumed in the above that $I(t)$ does not depend on time. Addition of the above combination of derivatives produces:

$$\frac{\partial A(z, t)}{\partial z} + \beta_1 \frac{\partial A(z, t)}{\partial z} - i\frac{1}{2}\beta_2 \frac{\partial A(z, t)}{\partial z^2}$$

$$= \int_{-\infty}^{+\infty} d(\Delta\omega)\, \tilde{E}(z, \omega_0 + \Delta\omega)$$

$$\times \left[\left(-i\beta_1\Delta\omega - \frac{1}{2}i\beta_2\Delta\omega^2 - i\cdot\Delta\beta_{NL} \right) + \beta_1 i\Delta\omega - i\frac{1}{2}\beta_2(i\Delta\omega)^2 \right]$$

The term in the bracket is, $[\ldots] = -i\cdot\Delta\beta_{NL} = in_2 k_0 I$. Thus, Eq. (15.9) gives:

$$\frac{\partial A(z, t)}{\partial z} + \beta_1 \frac{\partial A(z, t)}{\partial t} - i\frac{1}{2}\beta_2 \frac{\partial^2 A(z, t)}{\partial t^2}$$

$$= \int_{-\infty}^{+\infty} d(\Delta\omega)\, \tilde{E}(z, \omega_0 + \Delta\omega)(-in_2 k_0 I)\, e^{ig(z,t)}$$

$$= -ink_0 I \int_{-\infty}^{+\infty} d(\Delta\omega)\, \tilde{E}(z, \omega)\, e^{ig(z,t)}$$

$$= in_2 k_0 I A(z, t) \tag{13.57}$$

The final equation describing solitons is therefore:

$$\frac{\partial A(z,t)}{\partial z} + \beta_1 \frac{\partial A(z,t)}{\partial t} + i\frac{1}{2}\beta_2 \frac{\partial^2 A(z,t)}{\partial z^2}$$
$$= i\gamma |A(z,t)|^2 A(z,t) - \frac{\alpha}{2} A(z,t) \qquad (13.58)$$

where we have defined nonlinear coefficient γ as:

$$\gamma = \frac{2\pi n_2}{\lambda A_{\text{eff}}} \qquad (13.59)$$

Here A_{eff} is the effective core area.

Our interest lie here in the pulse evolution during propagation and not in the time of pulse arrival. We can, therefore, simplify the above equation by transforming it to a coordinate system which moves with group v_g. In this moving frame, new time T and new coordinate Z are:

$$Z = z \qquad (13.60)$$
$$T = t - b_1 z$$

To obtain the transformed equation, we must evaluate derivatives with respect to new variables as follows:

$$\frac{\partial A}{\partial t} = \frac{\partial A}{\partial T}\frac{\partial T}{\partial t} + \frac{\partial A}{\partial /Z}\frac{\partial Z}{\partial t} = \frac{\partial A}{\partial T}$$

since $\frac{\partial A}{\partial t} = 1$ and $\frac{\partial Z}{\partial t} = 0$. From the above one finds:

$$\frac{\partial^2 A}{\partial t^2} = \frac{\partial^2 A}{\partial T^2}$$

Using the above results, one has:

$$\frac{\partial A}{\partial z} = \frac{\partial A}{\partial T}\frac{\partial T}{\partial z} + \frac{\partial A}{\partial /Z}\frac{\partial Z}{\partial z} = -\beta_1 \frac{\partial A}{\partial T} + \frac{\partial A}{\partial /Z}$$

The last result is used in Eq. (13.58) to replace $\frac{\partial A}{\partial t}$. The transformed equation is:

$$\frac{\partial A}{\partial z} + i\frac{1}{2}\beta_2 \frac{\partial^2 A}{\partial T^2} - i\gamma |A|^2 A + \frac{1}{2}\alpha A = 0 \qquad (13.61)$$

where in the final step, we replaced Z by z. It is known as a *nonlinear Schrödinger equation* (NSE).

To further analyse NSE, we will introduce two characteristic lengths describing dispersion (L_D) and nonlinearity (L_{NL}). Those are defined as:

$$L_D = \frac{T_0^2}{|\beta_2|} = \frac{T_0^2 \, 2\pi c}{|D|\lambda^2} \qquad (13.62)$$

and

$$L_{NL} = \frac{1}{\gamma P_0} \quad (13.63)$$

where P_0 is the peak power of the slowly varying envelope $A(z, T)$ and T_0 is a temporal characteristic value of the initial pulse, which is often defined as full width half maximum (the pulse 3 dB width). Those two lengths characterize how far a pulse must propagate to show the respective effect. Physically, L_D is the propagation length at which a Gaussian pulse broadens by a factor of $\sqrt{(2)}$ due to *group velocity dispersion* (GVD).

GVD dominates pulse propagation in fibres whose length L is $L \ll L_{NL}$ and $L \geq L_D$. In such situation, the nonlinearity in NLSE can be ignored and the equation can be solved analytically. Nonlinear effects dominate in fiber where $L \ll L_D$ and $L \geq L_{NL}$. In this limit, the dispersion term can be ignored.

Typical values of parameters for solitons are summarized in Table 13.4. The numerical analysis normalize variables:

$$U = \frac{1}{\sqrt{P_0}} A \quad \text{and} \quad \tau = \frac{T}{T_0} \quad (13.64)$$

Table 13.4. Typical parameters for solitons

Parameter	Symbol	Value	Unit
Wavelength	λ	1.55	µm
Nonlinear coeff.	γ	1.3	1/km·W
GVD	β_2	15×10^{-24}	s²/km
Width parameter	T_0	100	ps
Peak power	P_0	0.15	mW
Losses	α	0.01	dB/km^{-1}
Chirp parameter	c	1.2	dimensionless
Soliton period	z_0	1047.2	km

The width parameter T_0 is related to the full-width at half-maximum (FWHM) intensity of the input pulse. Specifically:

$$T_s = 2T_0 \ln(1 + \sqrt{(2)}) \approx 1.763 T_0 \quad (13.65)$$

After simplification, Eq. (13.61) takes the form:

$$\frac{\partial U}{\partial z} - i \frac{\sin(\beta_2)}{2L_D} \frac{\partial^2 U}{\partial \tau^2} + i \frac{1}{L_{NL}} |U|^2 U + \frac{1}{2} \alpha U = 0 \quad (13.66)$$

Another normalized form of the Schrödinger equation exists in the literature. We obtain it in the lossless case, i.e. $\alpha = 0$. To derive it, normalize the z coordinate as follows:

$$\xi = \frac{z}{L_D} \qquad (13.67)$$

After a few algebraic steps, one obtains:

$$\frac{\partial U}{\partial \xi} - i \frac{\sin(\beta_2)}{2} \frac{\partial^2 U}{\partial \tau^2} + i N^2 |U|^2 U = 0 \qquad (13.68)$$

where N is known as soliton order and is defined as:

$$N^2 = \frac{L_D}{L_{NL}} = \frac{\gamma P_0 T_0^2}{|\beta_2|} \qquad (13.69)$$

The last form of the NLSE is found by introducing u as:

$$u = NU = \left(\frac{\gamma P_0 T_0^2}{|\beta_2|}\right)^{1/2} A \qquad (13.70)$$

Equation (13.68) then takes the form:

$$\frac{\partial u}{\partial \xi} - i \frac{\sin(\beta_2)}{2} \frac{\partial^2 u}{\partial \tau^2} - i |u|^2 u = 0 \qquad (13.71)$$

We will now discuss the numerical solution of the nonlinear Schrödinger equation (NSE) which describes propagation of optical solitons using the so-called *split-step Fourier method* (SSFM).

The SSFM is a numerical technique used to solve nonlinear partial differential equations like the NSE. The method relies on computing the solution in small steps and on taking into account the linear and nonlinear steps separately. The linear step (dispersion) can be made in either frequency or time domain, while the nonlinear step is made in the time domain. The method is widely used for studying nonlinear pulse propagation in optical fibers.

A nonlinear Schrödinger equation, Eq. (13.61) contains dispersive and nonlinear terms. To introduce SSFTM, write the NLSE equation in the following form:

$$\frac{\partial A(z,T)}{\partial z} = (\hat{L} + \hat{N}) A(z,t) \qquad (13.72)$$

where

$$\hat{L}A = -\frac{\alpha}{2} - \frac{i}{2}\beta_2 \frac{\partial^2 A}{\partial T^2} \qquad (13.73)$$

contains losses and dispersion in the linear medium and nonlinear term:

$$\hat{N}A = i\gamma |A|^2 A \qquad (13.74)$$

accounts for the nonlinear effects in the medium.

The basis of the SSFM is to split a propagation form z to $z + h$ (h is a small step) into two operations (assuming that they act independently):

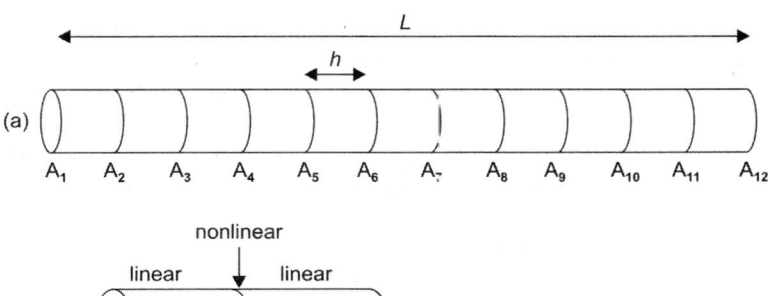
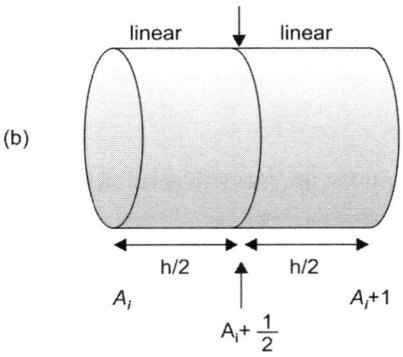

Fig. 13.48. Illustration of a split-step Fourier method (a) Division of optical fiber into N regions (here N = 11) of equal lengths (b) Illustration of operation of linear and nonlinear operations at arbitrary segments

during first step nonlinear effects are included and in the second step one accounts for linear effects (Fig. 13.48).

Formal solution of Eq. (13.72) over a small step h is thus:

$$A(z+h,t) = e^{h(\hat{L}+\hat{N})} A(z,t) \qquad (13.75)$$

In the first-order approximation, the above formula can be written as:

$$A(z+h,t) = e^{h\hat{L}} e^{h\hat{N}} A(z,t) + O(h^2) \qquad (13.76)$$

The basis of this approximation is established by Baker-Hausdorf lemma, which is:

$$e^{\hat{A}} e^{\hat{B}} = e^{\hat{A}+\hat{B}} e^{1/2[\hat{A},\hat{B}]} \qquad (13.77)$$

given that operation $\hat{A}$ and $\hat{B}$ commute with $[\hat{A}$ and $\hat{B}]$.

The basis of the method is suggested by Eq. (13.76). It tells us that $A(z+h,t)$ can be determined by applying the two operators independently. The propagation from z to $z+h$ is split into two operations: first the nonlinear step and then the linear step assuming that they act independently. If h is sufficiently small, Eq. (13.76) gives good results.

The value of step h can be determined by assuming that the maximum phase shift $\phi_{max} = g|A_p|^2 h$, where A_p is the peak value of $A(z,t)$ due to the nonlinear operator is smaller than the predefined value. Iannone showed that $\phi_{max} \leq 0.05$ rad.

For a practical implenentation of the SSFM, we need to establish practical expressions for dispersive and nonlinear terms. In the following, we will, therefore, analyse the effect of both terms independently neglecting losses.

Let us analyse the effect of the dispersive term alone. For that, we temporarily switch off the nonlinear term. After Fourier transform, the 'linear equation' becomes:

$$\frac{\partial \widetilde{A}(z,\omega)}{\partial z} = -\frac{i}{2}\omega^2 \beta_2 \widetilde{A}(z,\omega)$$

which has the solution:

$$\widetilde{A}(z,\omega) = \widetilde{A}(0,\omega) e^{-i\omega^2 \beta_2 z/2}$$

The action of the nonlinear term alone is described by the equation:

$$\frac{\partial A(z,t)}{\partial z} = i\gamma |A(z,t)|^2 A(z,t)$$

The 'natural' solution is in the time domain. It produces:

$$A(z,t) = A(0,t) e^{i\gamma |A|^2 A}$$

Split-step Fourier Transform Method

The propagation medium (say, cylindrical optical fiber) is divided into small segments, each of length h (Fig. 13.48). Further, each indi-vidual segment of length h is subdivided into two of equal lengths. The linear operator operates over each subsegment in the frequency domain, whereas the nonlinear operator operates only locally at the central point.

Operation of the linear operator $\hat{L}$, Eq. (13.73) over first subsegment is done as follows:

$$e^{h\hat{L}/2} A(z,t) = F^{-1}\left\{e^{h\hat{L}/2} F\{A(z,t\}\right\} \quad (13.78)$$

i.e., one must Fourier transform original amplitude from time domain into frequency domain, apply linear operator $\hat{L}$ and then, apply inverse Fourier transform to get the amplitude back to time domain.

The operation of nonlinear operator defined by Eq. (13.74) is as follows:

$$A_{i+1/2,L}(z,t) = A_{i+1/2,R}(z,t) e^{h\hat{N}} \quad (13.79)$$

where $A_{i+1/2,L}$ is the value of field amplitude at an infinitesimal point left from $i+\frac{1}{2}$. Finally the operation of linear operator over second subsegment of length $\frac{h}{2}$ is done exactly the same way as over the first segment.

Fiber Optic Communication

To summarize, the method over each segment of length h consists of three steps:

$$\text{Step 1} \begin{cases} \widetilde{A}_i(z, \omega) = F\{A_i(z, t)\} \\ \widetilde{A}_{i-}(z, \omega) = \widetilde{A}_i(z, \omega) \cdot \exp(-i\frac{1}{2}\omega^2 \beta_2 h) \\ A_{i-}(z, t) = F^{-1}\{\widetilde{A}_{i-}(z, \omega)\} \end{cases}$$

$$\text{Step 2} \quad A_{i+}(z, t) = A_{i-}(z, t) \cdot \exp(-i\gamma |A|^2 Ah)$$

$$\text{Step 3} \begin{cases} \widetilde{A}_{i+1}(z, \omega) = F\{A_{i+1}(z, t)\} \\ \widetilde{A}_{i+1}(z, \omega) = \widetilde{A}_{i+1}(z, \omega) \cdot \exp(-i\frac{1}{2}\omega^2 \beta_2 h) \\ A_{i-}(z, t) = F^{-1}\{\widetilde{A}_{i+}(z, \omega)\} \end{cases}$$

where F indicates Fourier transform (FT) and F^{-1} inverse FT.

Symmetrized Split-step Fourier Transform Method (SSSFM)

The simulation time of Eq. (13.76) essentially depends on the size to step h. To reduce simulation time, a new method was invented which allows us to use larger steps, h.

Mathematically, one uses the following second-order approximation:

$$A(z, +h, t) = e^{1/2h\hat{L}} \exp\left\{\int_z^{z+h} N(z')dz'\right\} e^{1/2h\hat{L}} A(z, t) + O(h^3) \quad (13.80)$$

In this approach, one assumes that the nonlinearities are distributed over h, which is more realistic. For small h, one can approximately evaluate:

$$\int_z^{z+h} \widehat{N}(z')dz' \approx \frac{h}{2}\left[\widehat{N}(z) + \widehat{N}(z+h)\right]$$

In the SSSFM algorithm Eq. (13.79) is replaced by:

$$A_{i+1/2, L}(z, t) = A_{i+1/2, R}(z, t) e^{h[\widehat{N}(z) + \widehat{N}(z+h)]} \quad (13.81)$$

The above scheme requires iteration to find A_{i+1} since it is not known at $z + \frac{1}{2}h$. Initially $\widehat{N}(z + H)$ will be assumed to be the same as $\widehat{N}(z)$.

Single Solitons

In analysis, we used fiber and pulse parameters summarized in Table 13.2. The input soliton was considered to be in the form:

$$A_{in}(\xi = 0, T) = \sqrt{A_0} \, \text{sec}\, h\left(\frac{T}{T_0}\right) \quad (13.82)$$

where N is the soliton order given by Eq. (13.69). In Fig. 13.49, we illustrated the propagation of $N = 1$ soliton without damping and in Fig. 13.50, the evolution of the same soliton with damping ($\alpha = 0.01$ dB km^{-1}).

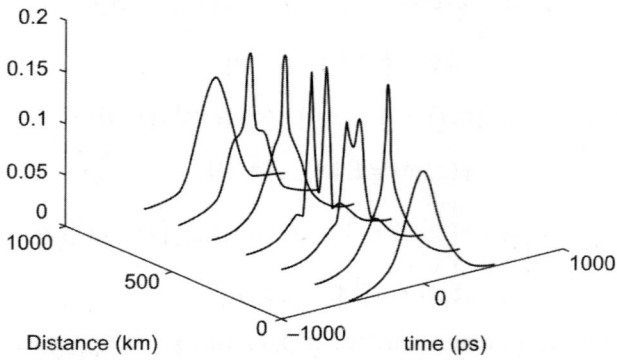

Fig. 13.49. Evolution in time over one soliton period for $N = 1$ soliton

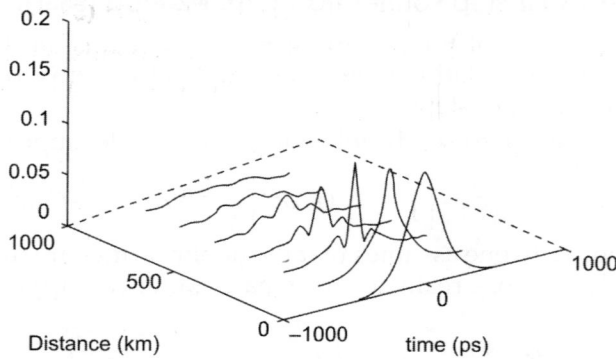

Fig. 13.50. Evolution in time over one soliton period for $N = 1$ soliton with damping

A higher-order soliton with $N = 3$ is defined as:

$$A_{in}(\xi = 0, T) = N^3 \sqrt{A_0} \operatorname{sec} h\left(\frac{T}{T_0}\right) \quad (13.83)$$

Propagation of higher-order soliton with $N = 3$ is shown in Fig. 13.51 (no damping) and in Fig. 13.52 (with damping). One can observe periodic evolution of the undamped soliton.

Chirped Solitons

Here, we analyse how chirp affects single soliton propagation. We assume the following input:

$$A_{in}(\xi = 0, T) = \sqrt{A_0} \operatorname{sec} h\left(\frac{T}{T_0}\right) \exp\left[-iC/2\left(\frac{T}{T_0}\right)^2\right] \quad (13.84)$$

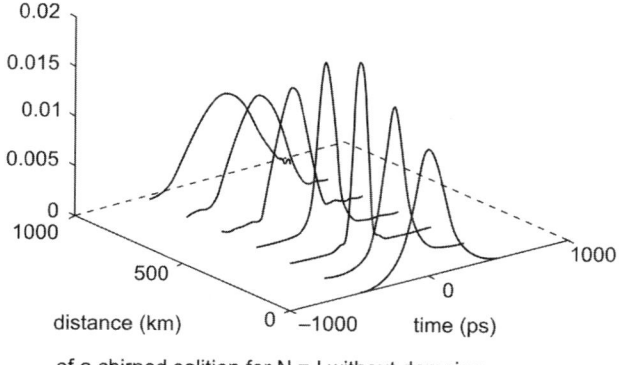

of a chirped soliton for N = I without damping

Fig. 13.51. Evolution in time over one soliton period for $N = 3$ soliton. Note soliton splitting near $z_0 = 0.5$ and its recovery beyond that point

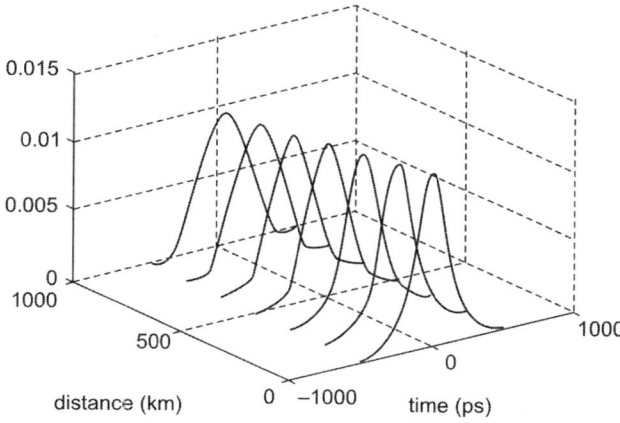

Fig. 13.52. Evolution in time over one soliton period for $N = 3$ soliton with damping

where C is the chirped parameter. Evolution of such soliton in the case of $N = 1$ and $C = 1.6$ is shown in Fig. 13.53. The pulse is initially compressed and then broadens.

Two Interacting Solitons

The effect of nonlinearity produces mutual interaction between soliton pulses, if they are launched close together. This interaction is important from a practical point of view and also from a fundamental perspective related to soliton propagation. For example, it was shown that nonlinear interaction between solitons can result in a bandwidth reduction by a factor of 10.

First, we consider an interaction between two solitons of the same strength and $N = 1$. We assume the initial pulses of the form:

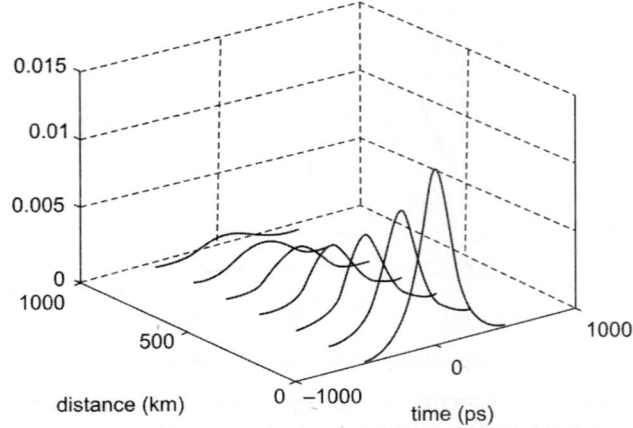

Fig. 13.53. Evolution in time of a chirped soliton for $N = 1$ without damping

$$A_{in}(\xi=0,T) = \sqrt{A_0}\,\text{sec}\,h\left(\frac{T}{T_0}-1\right) + \sqrt{A_0}\,\text{sec}\,h\left(\frac{T}{T_0}+1\right) \quad (13.85)$$

The result of two interacting solitons is shown in Fig. 13.54.

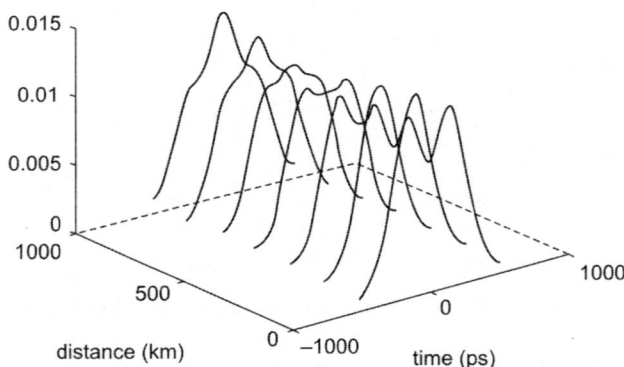

Fig. 13.54. Evolution in time of two interacting solitons without damping

An important application of solitons is in the transmission of information in optical fiber systems. Soliton pulses are stable when they propagate over long distances. Losses in fibers are an important limiting factor, so it becomes necessary to compensate periodically for fiber loss. This can be done by using EDFA.

In the transmission of information, solitons are considered like linear pulses. In a bit stream each soliton has its own bit slot and it represents logical one or zero (Fig. 13.55). Here T_B is the duration of the bit slot and it is related to bit rate B as:

$$T_B = \frac{1}{B} = 2aT_0 \quad (13.86)$$

where T_0 is the soliton width and $2a$ specifies the distance between neighbouring solitons. To prevent their interactions, the neighbouring solitons should be well separated.

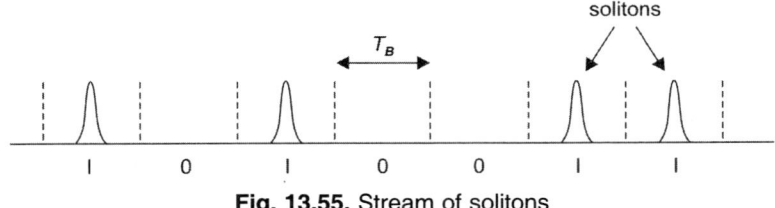

Fig. 13.55. Stream of solitons

The shape of fundamental solitons that are being experimented upon along with optical amplifiers for communication is shown in Fig. 13.56. In fact the solitons overcome the detrimental effect of chromatic dispersion, and optical amplifiers negate the attenuation. Hence, using the two together offers the promise of very high bit rate transmission with large repeaterless distances.

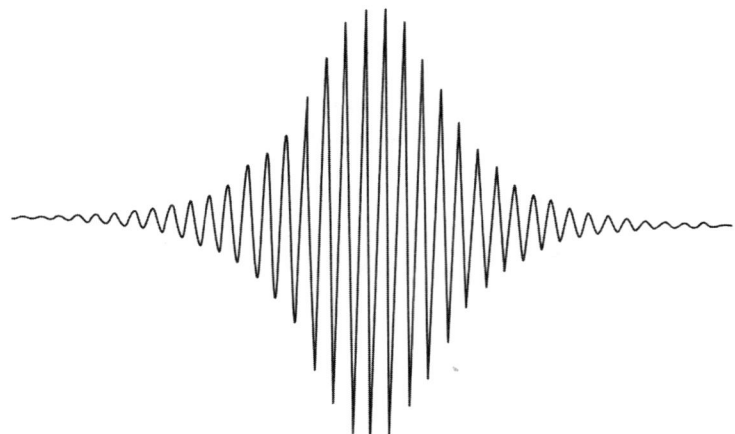

Fig. 13.56. A fundamental soliton pulse

Solitons are very well suited for *long-haul communication* because of their high information carrying capacity and the possibility of periodic amplification. However, soliton systems are still waiting for the full field deployment.

Example 1

A *p-i-n* photodiode has capacitance of 5 pf. Find the maximum value of load resistance R_L which will make the post detection bandwidth (B) of 10 MHz and also calculate the decrease in bandwidth with the same load resistance when the amplifier has a input capacitance of 5 pf.

Solution

The maximum bandwidth is given by:

$$C = \frac{1}{2\pi R_L C_d}$$

or $\quad R_L = \dfrac{1}{2\pi B C_d} = \dfrac{1}{2\pi \times 10 \times 10^6 \times 5 \times 10^{-12}} = 3.18 \times 10^3 \, \Omega$

Here
$C_d = 5 \text{ pf} = 5 \times 10^{-12} \text{ f}$
$B = 10 \times 10^6$

When the system is connected to an amplifier, whose input capacitance is 5 pf and load resistance is $3.18 \times 10^3 \, \Omega$, the bandwidth ($B'$) is:

$$B' = \frac{1}{2\pi \times 3.18 \times 10^3 \times (5+5) \times 10^{-12}} = 5 \times 10^6$$

Obviously, as soon as the detector is connected to the amplifier, the bandwidth decreases to 5 MHz.

Example 2

A digital fiber link operating at 850 nm requires a BER of 10^{-9}. Show that the quantum limit in terms of quantum efficiency is $20.7 \, h\nu/\eta$.

Solution

Probability of error:

$$P_r(0) = e^{-N} = 10^{-9}$$

Here, $\quad \lambda = 850 \text{ nm} = 850 \times 10^{-9} \text{ m}$
$\quad \quad \text{BER} = 10^{-9}$

$\therefore \quad \overline{N} = 9 \log 10 = 20.7$

where $\overline{N}$ represents electron-hole pair generated. Quantum efficiency (η), photon energy ($h\nu$) and energy received (E) are related by:

$$N = \frac{\eta E}{h\nu}$$

or $\quad E = N\dfrac{h\nu}{\eta} = 20.7 \times \dfrac{h\nu}{\eta} = 20.7 \dfrac{h\nu}{\eta}$

Example 3

A high input impedance amplifier is used in an optical fiber receiver. The effective input resistance of the amplifier is 5 mΩ matched with the bias resistance of the detector of the same value. Calculate: (i) The maximum bandwidth that may be obtained without equalization if the total capacitance $(C_T) = 5$ pf. (ii) The mean square thermal noise current per unit bandwidth generated by this high input impedance configuration when operating at 27°C. (iii) Compare the values obtained in (i) and (ii) with those values obtained when the high input amplifier is replaced by a trans-impedance amplifier having feedback resistor $R_F = 100$ kΩ and open loop gain of 400. You may assume $R_F \ll R_{TL}$ and total capacitance $(C_T) = 5$ pf.

Solution

(i) We have:
$$R_{TL} = \frac{R_b R_a}{R_b + R_a}$$

where R_b is detector's load resistance
R_a is input resistance of the amplifier

Now, $R_{TL} = \dfrac{5 \times 10^6 \times 5 \times 10^6}{5 \times 10^6 + 5 \times 10^6} = 2.5$ mΩ

Here, $R_b = 5 \times 10^6$ W, $R_a = 5 \times 10^6$ W

∴ Maximum bandwidth

$$B = \frac{1}{2\pi R_{TL} C_T} = \frac{1}{2\pi \times 2.5 \times 10^6 \times 5 \times 10^{-12}} = 1.27 \times 10^4 \text{ Hz}$$

Obviously, the maximum bandwidth that may be obtained without equalization is 1.27×10^4 Hz.

(ii) The mean thermal energy noise current per unit bandwidth for the high impedance configuration:

$$\langle i^2_{amp} \rangle = \frac{4k_B T}{R_{TL}} = \frac{4 \times 1.38 \times 10^{-23} \times 300}{2.5 \times 10^6} = 6.62 \times 10^{-27} \text{ A}^2 \text{ Hz}^{-1}$$

Here $B = 1$ Hz, $T = 273 + 27 = 300$ K

(iii) The maximum bandwidth without equalization for the transimpedance configuration:

$$B = \frac{A}{2\pi R_F C_F} = \frac{400}{2\pi \times 10^5 \times 5 \times 10^{-12}} = 1.24 \times 10^8 \text{ Hz}$$

Obviously, the bandwidth with transimpedance is 124 MHz.

Now, assuming $R_F \ll R_{TL}$, the mean square thermal noise current per unit bandwidth for the transimpedance configuration can be obtained as:

$$\langle i^2_{th} \rangle = \frac{4k_B T}{R_F} = \frac{4 \times 1.38 \times 10^{-23} \times 300}{10^5} = 16.56 \times 10^{-25} \text{ A}^2 \text{ Hz}$$

Example 4

Consider the design of a typical digital fiber-optic link which has to transmit at a data rate of 20 Mb/s with a BER of 10^{-9} using the NRZ code. The transmitter uses a GaAlAs LED emitting at 850 nm, which can couple on an average 100 µW (–10 dBm) of optical power into a fiber of core size 50 µm. The fiber cable consists of a graded-index fiber with the manufacture's specification as follows: $a_f = 2.5$ dB/km, $(\Delta T)_{mat} = 3$ ns/km, $(\Delta T)_{modal} = 1$ ns/km. A silicon p-i-n photodiode has been chosen, for detecting 850-nm optical signals, for the front end of the receiver. The detector has a sensitivity of –42 dBm in order to give the desired BER. The source along with its drive circuit has a rise time of 12 ns and the receiver has a rise time of 11 ns. The cable requires splicing every 1 km, with a loss of 0.5 dB/splice. Two connectors, one at the transmitter end and the other at the receiver end, are also required. The loss at each connector is 1 dB. It is predicted that a safety margin of 6 dB will be required. Calculate the maximum possible link length without repeaters and the total rise time of the system for accessing the feasibility of the desired system. [B.Tech.]

Solution

We have $C_L = \alpha_f L + \alpha_{con} + \alpha_{spice}$. The total channel loss C_L may be calculated as follows:

$C_L = \alpha_f L$ + (splice loss per km) × L + (loss per connector) × no. of connectors

$\quad = (2.5$ dB/km$) \times L$ (km) + $(0.5$ dB/splice$) \times (1$ splice/km$) \times L$ (km) + $(1$ dB$) \times$

$\quad = (3L + 2)$ dB

Here, $P_{tx} = -10$ dBm, $P_{rx} = -42$ dBm, and $M_s = 6$ dB. Substititing the values of P, P_{rx}, C_L, and M_S in Eq. $P_{tx} = P_{rx} + C_L + M_S$, we get:

$$-10 = -42 + (3L + 2) + 6$$

or $\quad L = 8$ km

Obviously, a maximum transmission path of 8 km is possible without repeaters.

Let us now calculate the total rise time t_{sys} using following Eqs:

$$t_{sys} = \left[t_{tx}^2 + t_f^2 + f_{rx}^2 \right]^{1/2} \tag{i}$$

and
$$t_f = \left[t_{intermodal}^2 + t_{intramodal}^2 \right]^{1/2} \tag{ii}$$

It is given that $t_{tx} = 12$ ns, $t_{rx} = 11$ ns. In the case of multimode fibers, intramodal dispersion, primarily due to material dispersion, and hence $t_{intramodal} \approx t_{mat}$:

$$t_{mat} = (3 \text{ ns/km}) \times L = (3 \text{ ns/km}) \times (8 \text{ km}) = 24 \text{ ns}$$

$t_{\text{intermodal}} = (1 \text{ ns/km}) \times L = (1 \text{ ns/km}) \times 8 \text{ (km)} = 8 \text{ ns}$

∴ $t_{\text{sys}} = [(12)^2 + (24)^2 + (8)^2 + (11)^2]^{1/2} = 30 \text{ ns}$

The maximum allowable rise time t_{sys} for our 20-Mb/s NRZ data stream from Eq. (i):

$$t_{\text{sys}} \leq \frac{0.35}{B} \text{ for the RZ format}$$

$$\frac{0.70}{B} \text{ for the NRZ format}$$

∴ $t_{\text{sys}} \leq \dfrac{0.70}{B} = \dfrac{0.70}{20 \times 10^6} \text{ s} = 35 \text{ ns}$

Since t_{sys} (= 30 ns) for the proposed link is less than the maximum allowable limit, the choice of components is adequate to meet the system design criteria.

Example 5

A transmitter has an output power of 0.1 mW. It is used with a fiber having following specifications:

NA = 0.25, attenuation of 6 dB/km, and length = 0.5 km

The link contains two connectors of 2 dB average loss. The sensitivity (minimum acceptable power) of the receiver is –35 dBm. The designer kept a 4 dB margin. Find link power budget.

Solution

Coupling loss = $-10 \log (\text{NA}^2) = -10 \log (0.25)^2 = 12 \text{ dB}$

Here, Source power, $P_S = 0.1 \text{ mW}$

$P_S = 10 \text{ dBm}$

NA = 0.25

Fiber loss = $\alpha_f L$

∴ $l_f = (6 \text{ dB/km}) (0.5 \text{ km}) = 3 \text{ dB}$

Connector loss = 2 (2 dB)

$l_c = 4 \text{ dB}$

Design margin $P_m = 4 \text{ dB}$

∴ Actual power output P_{out} = Source power − (Σ losses)

or $P_{\text{out}} = 10 \text{ dBm} - [12 \text{ dBm} + 3 + 4 + 4]$
$= -13 \text{ dBm}$

Since receiver sensitivity is given = -35 dB m, i.e. $P_{\text{min}} = -35 \text{ dB·m}$.

We note that $P_{\text{out}} > P_{\text{min}}$, i.e., the system will perform adequately over the system operating life.

Example 6

Calculate the total power that should be transmitted over 32 channels of a WDM system that are spaced 0.8 nm apart at 1.55 μm for a repeaterless distance of 80 km. Take L_{eff} = 20 km, $\Delta\lambda_c$ = 125 nm, g_R = 6 × 10⁻¹⁴ m/W, and A_{eff} = 55 μm². [B.Tech.]

Solution

Let us assume that the power penalty is to be kept below 0.5 dB, obviously, we must have $F \leq 0.1$. Therefore, from Eq.

$$F = \frac{g_R \, \Delta\lambda_s}{\Delta\lambda_c} \frac{P_0 L_{eff}}{2 A_{eff}} \frac{N(N-1)}{2}$$

where $\Delta\lambda_c$ is total channel spacing, $\Delta\lambda_s$ is channel spacing, g_R is peak Raman gain coefficient, N are equally spaced channels, 0, 1, 2, ..., $(N-1)$ in a WDM system, P_0 is power at the input end of the fiber, L_{eff} is effective length and A_{eff} is effective cross-sectional area.

We can calculate the total transmitted power P_{tot} as follows:

$$P_{tot} = NP_0 = \frac{4F \, \Delta\lambda_c \, A_{eff}}{g_R \, \Delta\lambda L_{eff} \, (N-1)}$$

or

$$P_{tot} = \frac{0.1 \, (125 \text{ nm}) \times 4 \times (55 \times 10^{-12} \text{ m}^2) \times (10^{-3} \text{ km/m})}{(6 \times 10^{-14} \text{ m/W}) \times (0.8 \text{ km}) \times (20 \text{ nm}) \times (32-1)}$$

$$= 0.0924 \text{ W} = 92.4 \text{ mW}$$

The transmitted power per channel, P_0, therefore, should be less than 2.887 mW. In this calculation, it has been assumed that there is no dispersion in the system.

Example 7

In a fiber link, the laser diode output power is 5 dBm, source-fiber coupling loss = 3 dB, connector loss = 2 dB and has 50 splices of 0.1 dB loss. If fiber attenuation loss for 100 km is 25 dB, compute the loss margin for (i) APB receiver having sensitivity −40 dBm, (ii) Hybrid PINFET high impedance receiver with sensitivity −32 dBm. [M.Tech.]

Solution

Calculations for power budget:

Source output	5 dBm
Source fiber coupling loss	3 dB
Connector loss	2 dB
Splicing loss (50 × 0.1)	5 dB
Fiber attenuation	25 dB
Total loss	35 dB

Now, available power to receiver: (5−35) dBm = −30 dBm:

(i) APD sensitivity = −40 dBm
Loss margin = [−40 − (−30)] dBm = 10 dBm
(ii) Hybrid (H) - PINFET high-impedance receiver = −3 dBm
Loss margin = [−32 − (−30)] = 2 dBm.

Example 8

A college campus CATV system uses an optical bus to distribute video signals to the subscribers. The transmitter couples 0 dB·m (1 mW) of optical power into the bus. Each receiver has a sensitivity of −40 dB·m. Each optical tap couples 5% of optical power to the subscriber and has a 0.5 dB insertion loss. How many subscribers can be added to the optical bus before the signal needs in-line amplification? [B.Tech.]

Solution

With reference to Fig. 13.57, if we neglect the optical loss within the bus (the optical fiber itself), the power available at the Nth tap is given by:

$$P_N = P_T C \,[(1 - \delta)(1 - C)^{N-1}]$$

where P_T is the transmitted power, C is the fraction of optical power coupled out at each tap, and δ is the fractional insertion loss (assumed to be the same) at each tap, and N is the number of subscribers.
Here, $P_T = 1$ mW, $P_N = -40$ dB·m $= 10^{-4}$ mW, and $C = 0.05$. δ may be calculated as follows: the insertion loss (in dB):

$$L = -10 \log_{10}(1 - \delta)$$

Here $L = 0.5$ dB, therefore $d = 0.11$. Thus, one obtains:

$$10^{-4} = 1 \times 0.05\,[(1 - 0.11)(1 - 0.05)]^{N-1}$$

This gives $N = 38$. That is, at the most 38 subscribers may be added to the bus without any in-line amplification of the signal.

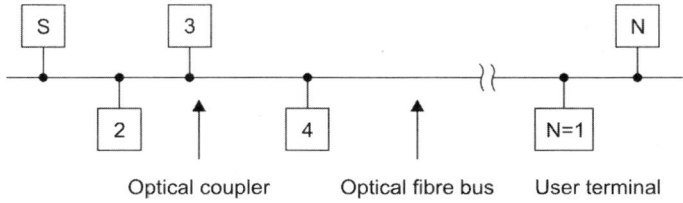

Fig. 13.57. Configuration of distribution network: bus topology

Example 9

For a multimode fiber, calculate system rise time by using following values of various parameters:
 (i) Rise time of LED with drive circuit = 15 ns
 (ii) LED spectral width = 40 nm

(iii) Material dispersion related rise time degradation = 21 ns over 6 km link
   (iv) Receiver bandwidth = 25 MHz
   (v) Model dispersion rise time = 3.9 ns [B.Tech.]

Solution
We have:
$$t_{rx} = \frac{350}{B_{rx}}$$

$$\therefore \quad t_{rx} = \frac{350}{25} = 14 \text{ ns}$$

Here, $t_{tx} = 15$ ns, $t_{mat} = 21$ ns, $t_{mod} = 3.9$ ns

Now:
$$t_{sys} = \left(\sum_{i=1}^{N} t_{ri}^2\right)^{1/2}$$

$$\therefore \quad t_{sys} = [15^2 + 21^2 + 3.9^2 + 14^2]^{1/2} = 29.6 \text{ ns}$$

Example 10
Calculate the SBS threshold power for the worst and best possible cases if the line width of the source is 100 MHz, $g_B = 4 \times 10^{-11}$ m/W, $A_{eff} = 55 \times 10^{-12}$ m^2, $L_{eff} = 20$ km, and $\lambda_p = 1.55$ μm. [B.Tech.]

Solution
We have at $\lambda_p = 1.55$ μm, $\Delta\nu_B = 20$ MHz. For the worst case, $b = 1$. Therefore, using SBS threshold power, we get:

$$P_{th} = \frac{21\, b\, A_{eff}}{g_B\, L_{eff}} \left[1 + \frac{\Delta\nu_{source}}{\Delta\nu_B}\right]$$

where $\Delta\nu_{source}$ is line width of the source, g_B is the Brillion gain, $\Delta\nu_B$ very narrow line width, A_{eff} is effective cross-sectional area and the value of b lies between 1 and 2.

$$P_{th} = \frac{21 \times 1 \times 55 \times 10^{-12}}{6 \times 10^{-11} \times 20 \times 10^3}\left[1 + \frac{100 \times 10^6}{20 \times 10^6}\right] = 8.66 \times 10^{-3} \text{ W}$$

or $\quad P_{th} = 8.66$ mW

For the best possible case, $b = 2$ and we have:
$$P_{th} = 17.32 \text{ mW}$$

Example 11
Using the following data related to fiber link, compute the system rise time and bandwidth:

Component	BW (MHz)	Rise time (t_r in ns)
Transmitter	200	1.75
LED (850 nm)	100	3.50
Fiber cable	90	3.89
Pin detector	350	1.00
Receiver	180	1.94

[B.Tech.]

Solution
We have,

$$\text{System rise time } t_{sys} = \left(\sum_{i=1}^{N} t_{ri}^2\right)^{1/2}$$

or

$$t_{sys} = \sqrt{1.75^2 + 3.5^2 + 3.89^2 + 1.00^2 + 1.94^2} = 593 \text{ ns}$$

Now, system BW is given by:

$$BW = \frac{0.35}{t_{sys}} = \frac{0.35}{5.93 \text{ ns}} = 59 \text{ MHz}$$

Example 12
For a single mode fiber, $n_2 = 1.48$ and $\Delta = 0.2\%$ operating at $\lambda = 1320$ nm, compute the waveguide dispersion if $V \cdot \frac{d^2(Vb)}{dv^2} = 0.26$.

Solution
Here $n_2 = 1.48$,
$\Delta = 0.2$, and
$\lambda = 1320$ nm.

Waveguide dispersion is given by:

$$D_{wg}(\lambda) = \frac{-n_2\Delta}{c\lambda}\left[V \frac{d^2(Vb)}{dV^2}\right] = \frac{-1.48 \times 0.2}{3 \times 10^5 \times 1320} [0.20]$$

$$= -1.943 \text{ picosec/nm} \cdot \text{km}$$

SUGGESTED READINGS
1. Agrawal GP, 'Fiber-Optic Communication Systems', 2nd ed., Wiley (1997).
2. Dutton HJR, 'Understanding Optical Communications', IBM Prentice Hall (1998).
3. Keiser G, 'Fiber Optic Communications, 3rd ed., McGraw Hill (2000).

4. Kolimbiris N, 'Fiber Optics Communications', Prentice Hall (2004).
5. Yeh C, 'Applied Photonics', Academic Press (1994).
6. Palais JC, 'Fiber Optic Communications', 4th ed., Prentice Hall (1998).
7. Kumar S and M Jamal Deen, 'Fiber Optic Communications', Wiley (2014).
8. Degiorgio V and I Critiani, 'Photonics', Springer (2014).

GLIMPSES

- *Fiber optic communication* has the advantage of high available bandwidths and hence, high data rates. Optical communication was enhanced by the invention of lasers and fibers. High cost and special advanced mechanisms to couple and direct light are some of the limitations of fiber-optic communication.
- The important components of a fiber optic communication system are: Transmitter, Information channel, and Receiver.
- System design considerations for point-to-point links are related to two important transmission parameters of optical fibers: (i) fiber attenuation, and (ii) fiber dispersion.
- To ensure that the designed system performance has been met, two types of system analysis are performed: (i) link power budget analysis, and (ii) rise time budget analysis.
  Following relation can be used to estimate maximum repeaterless transmission-distance:
  $$P_{tx} = P_{rx} + C_L + M_S$$
  where P_{tx} is average power supplied by the transmitter, P_{rx} is the sensitivity of the receiver, C_L is total link loss or channel loss (including the fiber splice and connector losses) and M_S is system safety margin.
  For the desired bit rate, the *total rise time* (t_{sys}), the digital system should be below its maximum value given by:
  $$t_{sys} \leq \begin{cases} \dfrac{0.35}{B} \text{ for RZ format} \\ \dfrac{0.70}{B} \text{ for NRZ format} \end{cases}$$
  where RZ stands for return-to-zero (RZ) format, $\Delta f = B$ and NRZ stands for the non-return-to-zero (NRZ) format $\Delta f = B/2$.
  The performance of analog systems is normally analyzed by calculating the carrier-to-noise ratio (CNR) of the system.
- *Optical sources*: *Optical transmitter* converts electrical input signal into corresponding optical signal. The optical signal is then launched into the fiber-optical source is the major component in an optical transmitter.

LED and semiconductor laser diodes (LD) are popularly used optical transmitters.

- *Attenuation*: Attenuation is a measure of decay of signal strength or loss of light power that occurs as light power propagate as light pulses through the length of the fiber.

  In optical fibers the attenuation is mainly caused by two physical factors: absorption and scattering losses. Absorption is because of fiber material and scattering due to structural imperfections within the fiber.

- *Optical receiver*: An optical system converts optical energy into electrical signal, amplify the signal and process it. The important blocks of optical receiver are: Photodetector/Front end, Amplifier/Linear channels, Signal processing circuitry/Data recovery.

- Fiber optic communication systems may be classified into three broad categories: (i) the point-to-point link, (ii) distribution networks, and (iii) local area networks.

- *Point-to-point links*: A point to point link comprises of one transmitter and a receiver system. This is the simplest form of optical communication link and it sets the basis for examining complex optical communication links. For analyzing the performance of any link, important aspects are: (i) distance of transmission, (ii) channel data rate, and (iii) bit error rate.

- *Distributed networks* are preferred when data is to be transmitted to a group of subscribers. The transmission distance is relatively short (< 50 km). Examples of distributed networks are – broadcast of video channels over cable TV, telephone and FAX, commonly used topologies for distributed network are: Hub topology, Bus topology.

- *Local Area Networks (LAN)*: Several applications of fiber optic communication technology require networks in which a large number of users within local campus are interconnected in such a way that any user can access the network randomly to transmit data to any other user. The commonly used topologies for LANs are: (i) Star topology, (ii) Ring topology.

- In *Ring topology*, consecutive nodes are connected by point-to-point links to form a closed link.

- In *Star topology*, all nodes are connected through point-to-point link to central node called a hub.

- WDM: Optical signals of different wavelength (1300–1600 nm) can propagate without interfering with each other. The scheme of combining a number of wavelengths over a single fiber is called *wavelength division multiplexing* (WDM).

- *Star coupler* is mainly used for combining optical powers from N-inputs and divide them equally at M-output ports.

- *Optical amplifier*: An optical amplifier is nothing but a laser without feedback. Optical gain is achieved when the amplifier is pumped optically or electrically to achieve population inversion. Optical amplification depends on: (i) frequency (or wavelength) of incident signal, and (ii) local beam intensity.

  Optical amplifiers are mainly classified into two types: (i) Semiconductor Optical Amplifiers (SOA), and (ii) Doped Fiber Amplifier (DFA).
- There are two types of optical nonlinear effects that place limitations on system performance particularly at high transmitted power levels or at high bit rates exceeding 10 Gb/s: (i) non-linear inelastic process, e.g. SRS and SBS, and (ii) non-linear effects arising from the intensity-independent variation in the refractive index of fiber core, e.g., SPM, CPM, and FWM.
- *Material dispersion* occurs because of refraction varies as a function of the optical wavelength. Dispersion defines pulse spread as a function of wavelength and is measured in picoseconds per kilometer per nanometer [ps/(nm·km)]. Dispersion in long haul systems can be compensated using two types of fibers with opposite signs of dispersion.
- *'Soliton'* refers to special kinds of waves that can propagate undistorted over long distances and remain unaffected after collisions with each other. In an optical communication system, solitons are very narrow, high frequency optical pulses that retain their shape through the balancing pulse dispersion with the nonlinear properties of an optical fiber.

  Solitons are very narrow optical pulses with high peak powers that retain their shapes as they propagate along the fiber. Owing to GVD, the pulse propagating through the fiber gets broadened. In the anomalous regime (where chromatic dispersion is positive, say, above 1.32 µm in silica-based optical fibers), SPM causes the pulse to narrow, thereby partly compensating the chromatic dispersion. If the relative effects of SPM and GVD for an appropriate pulse shape are controlled properly, the compression of the pulse resulting from SPM can exactly balance the broadening of the pulse due to GVD. Therefore, the pulse shape either does not change or changes periodically as the pulse propagates down the fiber. The family of pulses that do not undergo any change in shape are called *fundamental solitons* and the pulses that undergo periodic changes are known as higher order solitons.

  Fundamental solitons along with optical amplifiers offer the promise of very high bit rate transmission over large repeaterless distance.

REVIEW QUESTIONS

1. Draw a relevent diagram showing signal path through optical data link via transmitter, fiber and receiver and describe the nature of the signal wave forms.
2. Give the block diagram of a receiver showing different types of noise generated and derive the expression for each type of noise.
3. Using steady-state amplifier equations and appropriate approximations, derive an analytical expression for an optimum length of fiber which gives maximum amplification.
4. Explain link power budget and rise-time budget analyses? Perform these analyses for a fiber-optic link that uses a conventional single-mode fiber that has been upgraded by DCF loops and optical amplifiers.
5. Mention different types of a system architectures and discuss them. Suggest the application of each of these topologies.
6. Explain non-linear effects observed in optical fibers. Also explain, why do these effects become pronounced at high power levels?
7. Write and define the networks used in communication systems.
8. How many types of topologies are there? Explain each topology.
9. How many types of networks are there based on the size of the area that the network covers? Explain briefly each of them.
10. Explain, how passive linear optical networks perform?
11. Explain the performance of star optical networks.
12. What are transmission links in an optical communication system. Explain each of them.
13. Define and explain SONET and SDH. What do you understand by SONET and SDH transmission rates?
14. What is four-fiber BLSR configuration? Explain briefly.
15. What is FWM? Mention negative effects of FWM in WDM systems. Can it be used in a beneficial way? If yes, how?
16. Define SPM and CPM and explain how SPM can be used to produce fundamental solitons? Write unique properties of solitons.

PROBLEMS

1. Calculate the minimum power required to maintain a bit error of 10^{-6} achieved with a photodiode detector with a responsivity 0.4 A/W. Assuming that the signal-to-noise ratio is limited by thermal noise with 50 W load, 400 K noise temperature and a 10 MHz noise bandwidth. [Ans. P_{min} = 1.587 µW]

    [Hint. we have:

$$20 \log \frac{\langle i_s \rangle}{\langle i_n \rangle} = 19.6 \text{ dB}$$

$$\therefore \frac{\langle i_s \rangle}{\langle i_N \rangle} = 9.55$$

Since $\langle i_N \rangle = \sqrt{\langle i_N \rangle^2}$

Now, $\langle i_s \rangle = 9.55 \langle i_N \rangle^2 = 9.55 \sqrt{\frac{4kTB}{R}}$

$$= 9.55 \sqrt{\frac{4 \times 1.38 \times 10^{-23} \times 400 \times 10^7}{50}}$$

$$= 6.35 \times 10^{-7} = 635 \text{ NA}$$

Now, $P_{min} = \frac{\langle i_s \rangle}{R} = \frac{635 \times 10^{-9}}{0.4} = 1.587 \times 10^{-6} \text{ W} = 1.587 \text{ μW}]$

2. Design an optical fiber link for transmitting 15 Mb/s of data for a distance of 4 km with BER of 10^{-9}.

   [**Hint.** Bandwidth × length = 15 Mb/s × 4 km = (60 Mb/s) km

   (i) *Optical source*: LED at 820 nm is suitable for short distances. LED generates 120 dB·m optical power.

   (ii) *Optical detector*: PIN-FET optical detector is reliable and has 50 dB·m sensitivity.

   (iii) *Optical fiber*: Step index multimode fiber can be selected. This fiber has bandwidth length product of 100 (Mb/s) km.

   Link power budget: We assume:
   Splicing loss = 0.5 dB/slice
   Connector loss = 1.5 dB
   System link power margin, $P_m = 8$ dB
   Fiber attenuation, $\alpha_f = 6$ dB/km
   Actual total loss $= (2 \times l_c) + \alpha_f L + P_m$
   or $P_T = (2 \times 1.5) + (6 \times 4) + 8 = 35$ dB
   Maximum permissible system loss
   P_{max} = Optical source output power − Optical receiver sensitivity
   $= -10$ dB·m $-(-50$ dB·m$) = 40$ dB·m

   We see that actual losses in the system are less than the allowable loss, hence the system is functional.

3. A type-I intensity-modulated fiber-optic link employs a laser transmitter which couples a mean optical power of 0 dB·m into a multi-mode optical fiber cable. The cable exhibits an attenuation of 3.0 dB/km with splice losses estimated at 0.5 dB/km. A connector at the receiver end shows a loss of another 1.5 dB. The *p-i-n* photodiode receiver has a sensitivity of −25 dB·m for a CNR of −50 dB with a modulation index of 0.5. A safety margin of 7 dB is required. The

rise times of ILD and p-i-n diode are 1 ns and 5 ns, respectively, and the intermodal and intramodal rise times of the fiber cable are 9 ns/km and 2 ns/km, respectively. (i) What is the maximum possible link length without repeaters? (ii) What is the maximum permitted 3-dB bandwidth of the system? [B.Tech.]

[Hint:
  (i) *Link power budget*: The mean optical power coupled into the fiber cable by the laser transmitter (P_{tx}) = 0 dB·m, the mean optical power required at the p-i-n receiver (P_{rx}) = –25 dB·m, and the total system margin $(P_{tx} - P_{rx})$ = 25 dB.

  Let us assume that the repeaterless link length is L. Then, using following Eq.
  $$C_L = \alpha_f L + \alpha_{con} + \alpha_{splice}$$
  where α_f is the fiber loss in (dB/km), L is link length, α_{con} is the sum of the losses at all the connectors in the link, and α_{splice} is the sum of the losses at all the splices in the link. C_L, α_{con} and α_{splice} are expressed in dB, the total channel loss C_L may be calculated as follows:

  C_L = (attenuation/km) × L + (splice loss/km) × L + connector loss
  = (3 dB/km) × L + (0.5 dB/km) × L + 1.5 dB
  = (3.5 L + 1.5) dB

  Therefore, from Eq. $P_{tx} = P_{rx} + C_L + M_S$
  or $P_{tx} - P_{rx} = C_L + M_S$, we obtain
  25 dB = [(3.5 L + 1.5) + 7] dB

  or $$L = \frac{16.5}{3.5} = 4.7 \text{ km}$$

  (ii) *Rise-time budget*:
  $$t_f^2 = [(9 \text{ ns/km} \times 4.7 \text{ km})^2 + (2 \text{ ns/km} \times 4.7 \text{ km})^2]$$
  $$= 1877.65 \text{ ns}^2$$
  $$t_{sys} = (t_{tx}^2 + t_f^2 + t_{rx}^2)^{1/2} = \{(1 \text{ ns})^2 + 1.877.65 \text{ ns}^2 + (5 \text{ ns})^2\}^{1/2}$$
  $$= 43.63$$

  The system bandwidth:
  $$\Delta f = \frac{0.35}{t_{sys}} = \frac{0.35}{43.6 \times 10^{-9}} \text{ Hz} = 8 \times 10^6 \text{ Hz} = 8 \text{ MHz}$$

  Thus, the proposed link length without repeaters is 4.7 km with a 3-dB bandwidth of 8 MHz.]

SHORT ANSWER QUESTIONS

1. Write the names of important parts of an optical transmitter?
Ans. (i) Interface circuit, (ii) Source driver circuit, (iii) An optical source.

2. **What is an optical fiber receiver?**

Ans. An optical fiber receiver is an electro-optic device that accepts optical signals from an optical fiber and converts them into electrical signals.

3. **What is an optical detector?**

Ans. An optical detector is a device that converts optical signal into electrical signal. The electrical signal is proportional to the intensity of the optical radiation.

4. **What are the basic applications of optical amplifiers (OAs)?**

Ans. OAs can be used to boost signal power after multiplexing, or before demultiplexing, or at any point in modern optical networks. OAs are ideal for metro and long-haul dense wavelength multiplexing (DWDM) as well as single wavelength applications.

5. **What is Wavelength Division Multiplexing (WDM)?**

Ans. WDM is an optical technology that permits several wavelengths to be coupled into the same fiber cable, effectively increasing the aggregation bandwidth per fiber cable.

6. **What is an optical receiver?**

Ans. Optical receivers are an essential part of a communication system. An optical receiver converts an optical signal, transmitted through an optical fiber cable into an electrical signal suitable for a receiving device installed at the other end of the communication system. The conversion process in the receiver is performed by two essential parts: (i) a detector, and (ii) an electronic signal processor. The detector converts the optical signal into an electrical signal. The electronic signal processor converts the raw detector signal into a form decipherable by the receiving device, such as a telephone, camera, or scanner.

7. **What are the important parameters which define the performance of an optical receiver?**

Ans. (i) Receiver sensitivity, (ii) Bandwidth, and (iii) Dynamic range.

8. **How the signal is carried in fiber optic cable?**

Ans. Through photons. Photon is a quanta of energy ($E = h\upsilon$). Photon energy depend only on the frequency υ.

9. **Why the optical fiber cables are not used for point to point transmiossion?**

Ans. We know that transmission through metal wires requires simple circuitry. Point to point transmission from one device to another using metal wire is much simple than using optical fibers as they don't require the use of transmitter and receiver. However, optical fibers are preferred when data rates becomes too high.

10. **What is the speed of laboratory optical fiber local area networks (LANs)?**

Ans. Hundreds of megabits per second. Optical fibers and LANs are the hottest communication technologies now. The main problem in optical fibers is the implementation of couplers and extra cost of optical fiber equipment. We know that metal wires and coaxial cables can operate from 1 Mb/s to 10 Mb/s. Optical fiber connected Lans which were tested, were operating at 100 Mb/s and up and so far not exceeded far from this value.

11. What is dispersion?

Ans. Dispersion is caused by the expansion of light pulses as they travel through optical components. This occurs because the speed of light through the optical medium is dependent on the wavelength, the propagation mode, and the optical properties of the material along the light path. Dispersion is given by:

$$D = \frac{1}{L}\frac{d\tau_g}{d\lambda} = \frac{d}{d\lambda}\left(\frac{1}{V_g}\right) = \frac{2\pi c}{\lambda^2}\beta_2$$

The factor $\beta_2 = d^2b/d\omega^2$ is the GVD parameter, which determines how much a light pulse broadens as it travels along an optical fiber. Dispersion defines the pulse spread as a function of wavelength and is measured in picoseconds per kilometer per nanometer [ps/(nm·km)].

12. What is the speed of laboratory fiber optic local area networks?

Ans. Hundreds of megabits per second. Presently, fiber optics and LANs are the hottest communication technologies. The main problem in the optical fibers is the implementation of couplers and the extra cost of fiber optic equipment. Metal wires and coaxial cables can operate from 1 Mb/s to 10 Mb/s. Optical fibers connected LANs which were tested operating at 100 Mb/s and up but so far not exceeded this figure.

13. What are solitons?

Ans. Solitons are very narrow optical pulses with high peak powers that retain their shapes as they propagate along the fiber.

Group velocity dispersion (GVD) causes most pulses to broaden in time as they propagate through an optical fiber. However, a soliton takes advantage of nonlinear effects in silica, particularly, *self phase modulation* (SPM), resulting from Kerr nonlinearity, to overcome the pulse-broadening effects of GVD.

The term *"soliton"* refers to special kind of waves that can propagate undistorted over long distances and remain unaffected after collisions with each other. In an optical communication system, solitons are very narrow, high-intensity optical pulses that retain their shape through the interaction of balancing pulse dispersion with the nonlinear properties of an optical fiber.

14. **What stands FTTB?**

Ans. FTTB means *"Fiber to the building"* and refers to installing optical fiber from the telephone company central office to a specific building such as a business or apartment house.

15. **What stands FITH?**

Ans. FITH means *"Fiber to the Home"* is network technology that deploys fiber optic cable directly to the home or business to deliver voice, video or data services.

16. **What stands FTTC?**

Ans. FTTC stands for "Fiber to the Curb" and refers to the installation and use of optical fiber cable directly to the curbs near homes or any business environment as replacement for Plain Old Telephone Service (POTS).

17. **What is SONET?**

Ans. SONET is synchronous optical networking. General SONET equipment uses one wavelength to carry an OC level, which can be divided into time slots for individual circuits. In Europe and Asia, SONET is known as SDH (Synchronous Digital Hierarchy).

18. **What is HFC?**

Ans. Hybrid Fiber Coax (HFC) is a way of delivering video, voice telephony, data and other interactive services over coaxial and fiber optic cables.

An HFC network consists of a headend office, distribution center, fiber nodes, and network interface units.

19. **What is Dense Wavelength Division Multiplexing (DWDM)?**

Ans. DWDM is an optical technology used to increase bandwidth over existing fiber-optic backbones.

DWDM works by combining and transmitting multiple signals simultaneously at differen wavelengths on the same fibers.

20. **Write the names of different components of an optical fiber communication system.**

Ans. (i) Optical transmitter, (ii) Fiber channel, (iii) Optical amplifier, (iv) Optical detector.

21. **What makes optical fibers immune to EMI?**

Ans. Optical fibers transmit signals in a light rather than electric current. This is the main reason that optical fibers are immune to EMI, which is a type of noise that originates with one of the properties of electromagnetism. We know that EMI is induced when magnetic field lines are cut across a conductor.

22. **What is topological reconfiguration?**

Ans. Topological reconfiguration relies on the capability of the transmission subsystem to establish an alternate data path in case of failure by changing its basic topology. This reconfigured topology should allow the live stations to operate normally despite the fault condition. We may note that this is very effective for protection against fiber breaks and applies specifically to an active structure.

23. Which method permits large number of independent, selectable channels to exist on a single fiber.

Ans. Frequency division multiplexing (FDM). Using FDM, multiple frequencies of light, i.e. multiple channels can coexist on a single optical fiber.

MULTIPLE CHOICE QUESTIONS

1. One aspect of SONET that has allowed it to survive during a time of tremendous changes in network capacity needs is its:
   (a) functionality   (b) capability
   (c) versatility   (d) scalability
2. The key requirement/requirements needed in analyzing a link is/are
   (a) data rate   (b) distance of transmission
   (c) bit-error rate   (d) All of the above
3. The bit duration of a 2.5-Gbits/s signal is:
   (a) 2.5 ns   (b) 1 ns
   (c) 0.4 ns   (d) 0.1 ns
4. For a passive star network, the total optical power supplied by the central node is 1 mW and that received at the terminal nodes is 0.1 µW. If the fractional insertion loss at each coupler is 0.05, what is the number of subscribers (nodes)?
   (a) 50   (b) 100
   (c) 250   (d) 500
5. Which of the following is a non-linear inelastic process?
   (a) SRS   (b) SPM
   (c) CPM   (d) FWM
6. Which of the following non-linear effects arise from the intensity-dependent variation of the refractive index of a fiber?
   (a) SPM   (b) CPM
   (c) FWM   (d) All of these
7. Assuming the fiber link span to be large, what will be its effective length if the attenuation per unit length is taken to be 0.20 dB/km?
   (a) 95.6 km   (b) 50.2 km
   (c) 22.2 km   (d) 10 km
8. A typical standard dispersion-shifted fiber has a dispersion parameter of 16 ps·nm^{-1}·km^{-1}. The repeaterless fiber length is 50 km. What should be the dispersion parameter of a DCF of length

2 km in order to compensate for the accumulated dispersion of the DSF for a repeaterless distance?
(a) 180 ps·nm^{-1}·km^{-1}
(b) –80 ps·nm^{-1}·km^{-1}
(c) –200 ps·nm^{-1}·km^{-1}
(d) –400 ps·nm^{-1}·km^{-1}

9. A fiber-optic link of length 50 km has two splices each exhibiting a loss of 1 dB. The fiber itself has a rated 0.2 dB/km loss. If the minimum power required to run a photodetector is 20 nW, what power must be supplied by the optical source?
(a) 20 nW
(b) 0.317 µW
(c) 0.317 mW
(d) 0.90 mW

10. A fiber used in a typical link has $(\Delta T)_{mat} = 3$ ns/km and $(\Delta T)_{modal} = 1$ ns/km. The link length is 8 km. The rise times of the transmitter and receiver are 12 ns and 11 ns, respectively. The total rise time of the system is:
(a) 30 ns
(b) 35 ns
(c) 23 ns
(d) 57 ns

11. The channel loss C_L may be expressed by the following equation (symbols have usual meanings):
(a) $C_L = \alpha_f L + \alpha_{con} + \alpha_{splice}$
(b) $C_L = \alpha_f - \alpha_{con} + \alpha_{splice}$
(c) $C_L = \alpha_f L - \alpha_{con} - \alpha_{splice}$
(d) $C_L = \alpha_f L + \alpha_{con} - \alpha_{splice}$

12. For a fiber-optic communication system, the total rise time t_{sys} in terms of electrical bandwidth Δf may be written as:
(a) $t_{sys} = \dfrac{0.65}{\Delta f}$
(b) $t_{sys} = \dfrac{0.35}{\Delta f}$
(c) $t_{sys} = \dfrac{1}{\Delta f}$
(d) $t_{sys} = \dfrac{0.1}{\Delta f}$

13. Which of the following relation for t_{sys} for digital system is correct?

(a) $t_{sys} \leq \begin{cases} \dfrac{0.65}{B} \text{ for RZ format} \\ \dfrac{0.75}{B} \text{ for NRZ format} \end{cases}$

(b) $t_{sys} \leq \begin{cases} \dfrac{0.35}{B} \text{ for RZ format} \\ \dfrac{0.70}{B} \text{ for NRZ format} \end{cases}$

(c) $t_{sys} \leq \begin{cases} \dfrac{0.70}{B} \text{ for RZ format} \\ \dfrac{0.35}{B} \text{ for NRZ format} \end{cases}$

(d) $t_{sys} \leq \begin{cases} \dfrac{0.15}{B} \text{ for RZ format} \\ \dfrac{0.65}{B} \text{ for NRZ format} \end{cases}$

14. If the number of wavelength channels in a WDM system is 10, what will be number of frequencies created by four-wave mixing?
    (a) 300 (b) 1000
    (c) 450 (d) 900

15. Which of the following process is used to compensate for the GVD-induced dispersion in a soliton?
    (a) FWM (b) SPM
    (c) CPM (d) All of the above

16. The carrier-to-noise ratio (CNR) for the performance of analog system is defined as:
    (a) $CNR = \dfrac{\text{rms carrier power}}{\text{rms noise power}}$
    (b) $CNR = \dfrac{\text{carrier frequency}}{\text{noise frequency}}$
    (c) $CNR = \dfrac{\text{carrier power}}{\text{noise power}}$
    (d) None of the above

17. For N wavelength channels copropagating through the optical fiber, the number of generated frequencies is:
    (a) $M = \dfrac{N^2}{2}$
    (b) $M = \dfrac{N}{2}(N-1)$
    (c) $M = \dfrac{N^2}{2}(N-1)$
    (d) $M = N(N-1)^2$

18. If D_1 and D_2 are the dispersion parameters of the two types of fibers and L_1 and L_2 are their respective lengths, then the condition for perfect dispersion compensation is:
    (a) $D_1L_1 + D_2L_2 = 1$
    (b) $D_1L_1 + D_2L_2 = 100$
    (c) $D_1L_1 + D_2L_2 = 0$
    (d) $D_1L_1 + D_2L_2 = 1000$

ANSWERS

1. (d) 2. (d) 3. (c) 4. (d) 5. (a) 6. (d) 7. (c)
8. (d) 9. (b) 10. (a) 11. (a) 12. (b) 13. (b) 14. (d)
15. (c) 16. (a) 17. (c) 18. (c)

14

Optical Fiber Sensors

14.1 INTRODUCTION

Optical fibers has found major applications in *sensor technology* due to their inherent advantages, e.g., immunity to electromagnetic interference or radio frequency interference, electrical isolation, small size, chemical passivity, low weight and ability to optical fibers to interface with a wide variety of measurands. Development in fiber optics due to rapid progress in the field of optical fiber communication industry, along with the progress in the field of optoelectronics have enabled optical fiber sensor technology to attain an ideal potential for several industrial applications.

Basically, fiber optic sensor is a *transducer* which can convert various input variables (physical quantity into an electrical signal in a measurable form.

The application of fiber optic sensors are increased due to their low cost and improved quality compared to traditional sensors. Optical fiber sensors find many applications such as measuring and sensing various parameters like, pressure, temperature, electric field, magnetic field, current, humidity, acoustic vibration, flow rate, liquid level and so on. Temperature and pressure changes can give rise to the change in effective length of the fiber and this change in length produces a small phase change. When the fiber is long, even this small phase change gets considerably magnified over the whole length of the fiber and provides a measurable amount of phase change. Since this phase change is due to change in temperature or pressure, so it automatically measures temperature or pressure respectively.

The major advantages of using optical fibers in sensing applications are:

    (i) They are light in weight and small in size.
    (ii) They have good geometrical flexibility.
    (iii) First optical sensors are free from the risk of fire or sparks since, they are made of silica.

(iv) They are electrically passive, i.e., they are immune to electro-magnetic interference and moreover do not distort the surrounding electrical and magnetic fields.
(v) The chemical and environmental ruggedness is more.
(vi) Fiber optic sensors have large bandwidth and also they are highly sensitive.

The fiber optic sensors are non-contact and generally high accuracy devices and systems. In optical fiber sensors, the optical wave is the information carrier and sensor.

Whenever we use fiber optic sensors for measuring a physical parameter, any one of the characteristics like amplitude, intensity, phase, polarisation, frequency and direction of propagation of the wave gets modulated by the measured quantity.

In the present chapter, we discusses fiber optic sensors, their classification, their configuration and their applications.

14.2 FIBER-OPTIC SENSOR (FOS)

FOS is a transducer which uses light guided within an optical fiber to detect any external physical, chemical, biomedical or any other parameter, i.e., it can convert various input variables (physical quantity) into an electrical signal in a measurable form.

Figure 14.1 shows a generalized configuration as a block diagram. We see that it consists of an optoelectronic source, optical fiber(s), a modulating element, an optoelectronic detector, a signal processor, and a read out device.

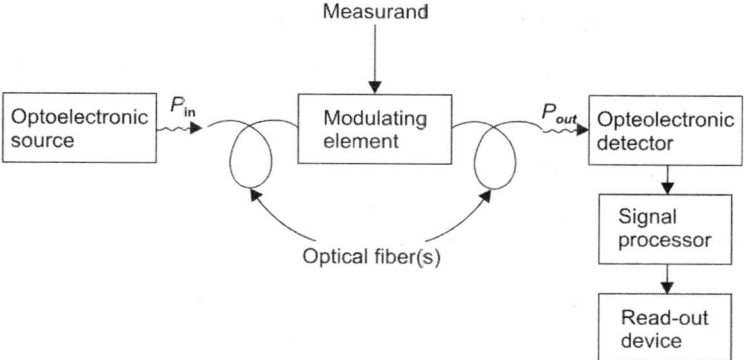

Fig. 14.1. Block diagram of a fiber-optic system. P_{in} and P_{out} are input and output optical powers respectively

We now consider a simple FOS in which the measurand (e.g., displacement, pressure, force, temperature, etc.) modulates the intensity of light propagating through the optical fiber and modulating element combination. However, the modulated light changes the detector output, which can be further processed and calibrated to give the value of measurand. These have been variety of schemes suggested for this modulation. From Fig. 14.1, we see that the signal S developed by the detector, to a good approximation can be given by the following relation:

$$S \approx P(\lambda)\, \eta\, T(\lambda, l)\, M(I, \phi\ or\ \lambda)\, R(\lambda) \qquad (14.1)$$

where $P(\lambda)$ is the power furnished by the optoelectronic source as a function of λ, η is the coupling efficiency of the input/output fiber(s) with the modulating element, $T(\lambda, l)$ is the transmission efficiency of the optical fiber(s), which will depend on the wavelength λ and length of the fiber l, $M(I, \phi\ or\ \lambda)$ is the response of the modulating element, which may modulate the intensity (I), phase (ϕ), or spectral distribution (λ) (we will restrict to only intensity modulation), and $R(\lambda)$ is the responsivity of the photodetector. We assume that (i) the system is using a source that provides almost monochromatic light, (ii) the optical fiber is a single mode fiber, i.e. LP_{01} (linearly polarized) mode is propagating through the fiber (however, if multimode fiber is used, all the modes are excited uniformly), and (iii) the response of the modulating element and that of the detector are almost linear.

Equation (14.1) providesus a basis for exploring possible methods by optimizing the design of FOSs. We may note that the choice of components is largely governed by the selection of modulation scheme rather than the measurand, because of the same measurand can be sensed using various modulating mechanisms.

14.3 CLASSIFICATION OF FIBER-OPTIC SENSORS

One can classify fiber-optic sensors in two ways. The first way is to put them into the two types:

14.3.1 Extrinsic or Hybrid Fiber-optic Sensors

In these sensors (also called passive sensor), the interaction between the light and the measurand (the quantity under measurement) takes place outside the fiber. The light from an optical source is launched into the fiber (Fig. 14.1) and is guided to a point where the measurement is performed. At this point, the light is allowed to exit the fiber and get modulated by the measurand in a separate zone prior being relaunched into the same or a different fiber. The devices working on this principle are named as extrinsic sensors.

In extrinsic sensors, the fiber acts merely as a waveguide. This type of sensors have sensor head, and the sensed optical signal is transferred to the measurement point with low attenuation and enhanced mechanical stability for signal processing.

Extrinsic sensors can be used for measurement of voltages, current, temperature, pressure, force, displacement, etc.

14.3.2 Intrinsic Sensors

In these sensors measurand acts directly on the fiber itself and produces a change in the transmission characteristics. The light launched into the fiber gets modulated in response to the measurand whilst still being guided in the fiber.

Liquid level sensors and Faraday sensors or gyroscope are of this type.

The other more logical way to classify optical fiber sensors is based on the modulation scheme that is used in making the sensor. Accordingly, one can place them into following three groups:

(i) *Intensify-modulated sensors* in which the intensity of light launched into the fiber is changed either intrinsically or extrinsically by the measurand.

(ii) *Phase-modulated sensors* in which the phase of monochromatic light propagating through the fiber is changed (normally intrinsically) by the measurand.

(iii) *Spectrally modulated sensors* in which the wavelength of light is changed (normally extrinsically) by the measurand.

Several mechanisms of fiber-optic sensing based on these modulation schemes have been suggested, but we will restrict to the study of a few representative ones to get an insight into the favourable and unfavourable features of devices of these types.

14.4 INTENSITY-MODULATED SENSORS

Various schemes have been suggested for intensity modulation. Among these, one important scheme involves the displacement of one fiber relative to the other by measurand. One fiber is kept fixed and light is launched through it, and a second fiber is made to undergo either longitudinal (or axial) displacement, or lateral (or transverse) displacement, or an angular displacement with respect to first fiber, by the measurand (Fig. 14.2).

The coupling efficiency η of Eq. (14.1) in this case will depend on several factors. In fact, for the three types of displacements shown in Fig. 14.2(a), and 14.2(c), the coupling efficiencies for two-compatible multimode fibers are given by Eqs (14.2), (14.3) and (14.4), respectively. Thus, the coupling efficiency η_{long} for longitudinal displacement Δx between two fibers is given by:

$$\eta_{\text{long}} \approx \frac{16 k^2}{(1+k)^4} \left[1 - \frac{\Delta x (\text{NA})}{4an} \right] \qquad (14.2)$$

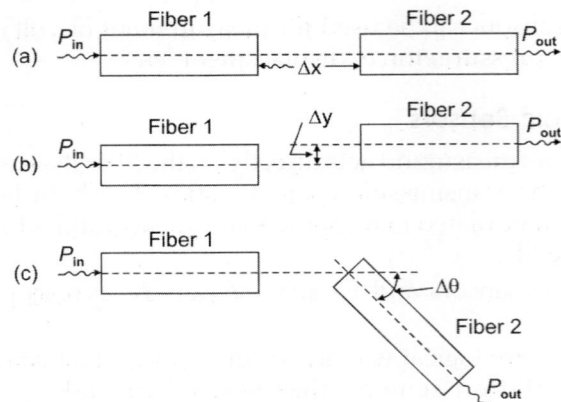

Fig. 14.2. Schematic of intensity modulated sensor. (a) Longitudinal (or axial), (b) lateral (or transverse), and (c) angular displacement of fiber 2 with respect to fiber 1 caused by the measurand

Here $k = n_1/n$, n_1 is the refractive index of the core of the fiber, n is the refractive index of the medium surrounding the fiber (e.g., air), a is the core radius, and NA is the numerical aperture of the fibers. The coupling efficiency η_{lat} for a lateral offset Δy between the axes of the two fibers is given by:

$$\eta_{lat} \approx \frac{16\, k^2}{(1+k)^4} \frac{2}{\pi} \left[\cos^{-1}\left(\frac{\Delta y}{2a}\right) - \left(\frac{\Delta y}{2a}\right) \left\{ 1 - \left(\frac{\Delta y}{2a}\right)^2 \right\}^{1/2} \right] \quad (14.3)$$

Similarly, the coupling efficiency η_{ang} for an angular displacement $\Delta\theta$ between the axes of the two fibers is given by:

$$\eta_{lat} \approx \frac{16\, k^2}{(1+k)^4} \left[1 - \frac{n\Delta\theta}{\pi(NA)} \right] \quad (14.4)$$

With a specific configuration of the sensing device (i.e., with an LED or ILD of specific wavelength, fibers of fixed length, diameter, etc., and a photodiode combination), the terms $P(\lambda)$, $T(\lambda, l)$, and $R(\lambda)$ in Eq. (14.1) may be taken to be constant. With a specific medium (e.g., dry air or an index-matched fluid) between the two fiber ends, $k = n_1/n$ may also be considered constant. Thus, using Eqs (14.1) and (14.2), the signal developed by the detector (normalized to that corresponding to $\Delta x = 0$, $\Delta y = 0$, and $\Delta\theta = 0$) for longitudinal displacement Δx, keeping $\Delta\theta = 0$, $\Delta y = 0$, may be written as:

$$S_{long} \text{ (normalized)} = 1 - \frac{\Delta x\, (NA)}{4\, an} \quad (14.5)$$

Similarly, the signals developed by the detector for lateral displacement Δy (normalized to that for $\Delta y = 0$) keeping $\Delta x = 0$ and $\Delta\theta = 0$ and the angular displacement $\Delta\theta$ (normalized to that for $\Delta\theta = 0$), keeping $\Delta x = 0$ and $\Delta y = 0$ are given, respectively, by the following expressions:

$$S_{\text{lat}} \text{ (normalized)} = \frac{2}{\pi}\left[\cos^{-1}\left(\frac{\Delta y}{2a}\right) - \left(\frac{\Delta y}{2a}\right)\left\{1 - \left(\frac{\Delta y}{2a}\right)^2\right\}\right] \quad (14.6)$$

and
$$S_{\text{lat}} \text{ (normalized)} = 1 - \frac{n\Delta\theta}{\pi\,(\text{NA})} \quad (14.7)$$

Here, $\cos^{-1}(\Delta y/2a)$ in Eq. (14.6) and $\Delta\theta$ in Eq. (14.7) are expressed in radians.

Clearly, any parameter that can be transformed into any one of these three types of displacements (e.g., longitudinal, lateral, or angular) can be measured with the sensor.

The design of a *differential fiber-optic sensor* is shown in Fig. 14.3(a). We can see that herein, light from an optical source is launched into the transmitting fiber. Two separate compatible fibers of equal length are placed with respect to the transmitting fiber such that both these fibers receive equal amount of power in the equilibrium position. The output of these two fibers is detected by two identical detectors. When the transmitting fiber is displaced by the measured through a distance d, the area of overlap of receiving fiber 1 increases and that of receiving fiber 2 decreases. In other words, the lateral offset (Δy_1) between the axes of the transmitting fiber and receiving fiber 1 decreases, whereas, the offset (Δy_2) between the axes of the transmitting fiber and receiving fiber 2 increases, as shown in Figs 14.3(a) and (b). In fact, $\Delta y_1 = a - d$ and $\Delta y_2 = a + d$, where a is the radius of each fiber. Consequently, the coupling coefficiencies η_1 and η_2 of receiving fibers 1 and 2 with respect to the transmitting fiber increase and decrease, respectively. Thus, in turn, causes the outputs V_1 and V_2 of the two detectors to vary accordingly. If we take $V_1 - V_2 = V_c$, it can be shown that the normalized value of V_c may be expressed as:

$$(V_c)_{\text{normalized}} = \frac{V_c}{(V_c)_{\text{max}}} = \frac{\eta_1 - \eta_2}{(\eta_1 - \eta_2)_{\text{max}}} \quad (14.8)$$

Figure 14.4 shows an example of *intrinsic sensor*. Herein, the optical source launches light into a single multimode or monomode fiber and its output is detected by the detector. The measurand causes the tapered teeth (or some other device) to produce microbending of the optical fiber. This, in turn, results in the loss of higher modes. As the microbending increases, the power received by the detector decreases.

One can show that the loss is maximum for periodic microbending, with the bend pitch Λ given by:

$$\Lambda = \frac{a\pi an}{\text{NA}} \quad (14.9)$$

768 Photonics | Optoelectronics

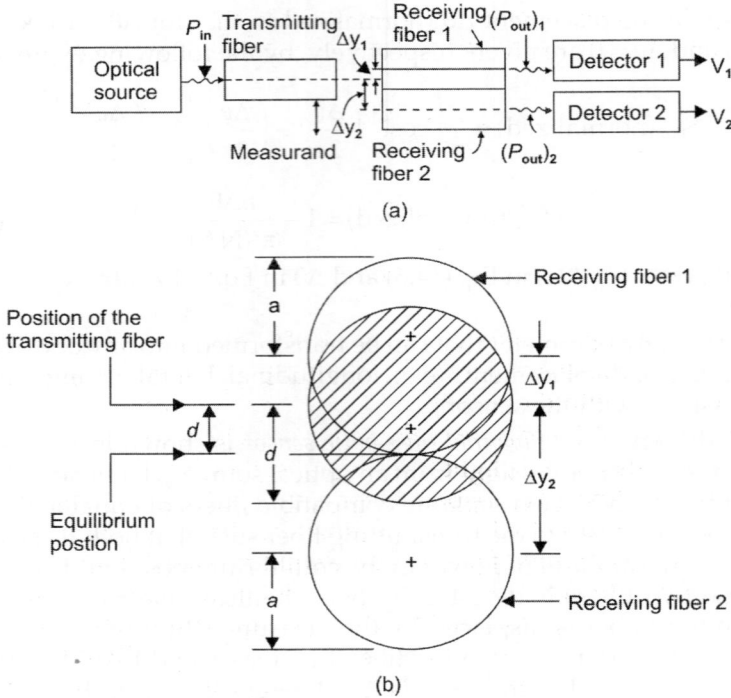

Fig. 14.3. (a) Design concept of a *differential sensor*. P_{in} is the input optical power, and $(P_{out})_1$ and $(P_{out})_2$ are the output powers of receiving fibers 1 and 2, (b) Overlap of the cross-sectional areas of the transmitting fiber and receiving fibers 1 and 2 in the displaced position is shown

where α is the profile parameter of the core refractive index of the fiber, a is the core radius, n is the refractive index of the fiber core, and NA is the numerical aperture of the fiber.

Intensity-modulated sensors offer the virtues of simplicity of construction, reliability, low cost, and compatibility to multimode fiber technology. However, the signal developed by the detector depends on a large number of factors, some of which may not be under full control. Therefore, absolute measurements with these sensors may not be possible. For good performance, they need some of referencing. In fact such sensors are most suitable for *switching applications*, and applications in which digital (on-off) or modulation frequency encoders are used.

14.5 PHASE-MODULATED SENSORS

The measurand in phase modulated sensors (e.g., temperature, pressure, strain, magnetic field, etc.) causes the phase modulation of an optical wave guided by the fiber. Usually, this scheme requires mono-

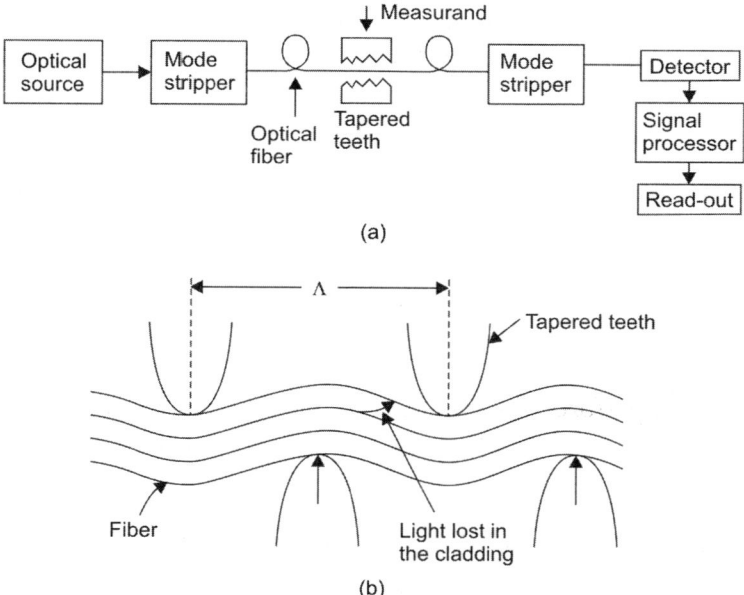

Fig. 14.4. (a) Schematic of an intrinsic-type microbend sensor. (b) Details of the loss mechanism

chromatic sources and single-mode fibers. In order to visualize how phase modulation may be achieved, consider monochromatic light of free-space wavelength λ propagating through a single-mode fiber of length L. The propagation constant β (i.e., phase change per unit length) of such a wave in the fiber will be given by:

$$\beta = \frac{2\pi}{\lambda} \eta_{\text{eff}} \qquad (14.10)$$

where η_{eff} is the effective index of the LP_{01} mode supported by the fiber. Thus, the total phase change after propagating through a length L of the fiber will be given by:

$$\phi = \beta L \qquad (14.11)$$

$$= \frac{2\pi}{\lambda} \eta_{\text{eff}} L \qquad (14.12)$$

A small change $\Delta\phi$ in ϕ can then be described by:

$$\Delta\phi = \beta \Delta L + L \Delta \beta \qquad (14.13)$$

where $\beta \Delta L$ is the phase change due to the change in length ΔL of the fiber caused by the measurand (e.g. axial strain), and $L\Delta\beta$ is the phase change caused by the change $\Delta\beta$ in the propagation constant (due to the strain in the fiber). If we assume that the change in the fiber diameter due to strain is negligible and that the difference in the core and cladding refractive indices is very small, one may write to a good approximation $n_{eff} = n$, where n is the core index of the fiber. Hence $\beta = (2\pi/\lambda)n$ and $\Delta\beta = (2\pi/\lambda)\Delta n$.

Substituting the values of β and $\Delta\beta$ in Eq. (14.13), one obtains:

$$\Delta\phi = \frac{2\pi}{\lambda} n\Delta L + \frac{2\pi}{\lambda} L\Delta n = \frac{2\pi}{\lambda}(n\Delta L + L\Delta n) \qquad (14.14)$$

and

$$\phi = \frac{2\pi}{\lambda} nL$$

Therefore

$$\frac{\Delta\phi}{\phi} = \frac{\Delta L}{L} + \frac{\Delta n}{n} \qquad (14.15)$$

One can perform similar calculations for phase modulation due to pressure and strain. Now, we discuss a few representative sensors of this type.

(a) Fiber-optic Mach–Zehnder Interferometric Sensor

Figure 14.5 shows the basic configuration of an all fiber Mach–Zehnder interferometric sensor. We see that light from a laser source is split by a 3 dB coupler and sent equally to the sensing fiber and reference fiber arms (both are single-mode fibers). The outputs of the two fibers are combined at the second 3 dB coupler. The sensing arm is in direct contact with the measurand, whereas, the reference arm is in shielded from external perturbation. Here, the measurand acts on the sensing arm and changes the phase of the light wave either by changing its length, its refractive index, or both.

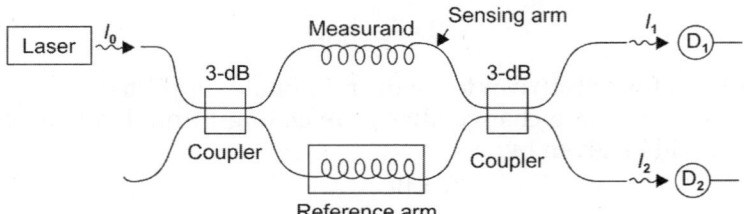

Fig. 14.5. Schematic of fiber-optic Mach-Zehnder interferometric sensor

As the reference arm is not affected by the measurand, there appears a phase difference $\Delta\phi$ between the two waves arriving at the second coupler. The intensity I_1 and I_2 of light arriving at the two output ports of the second coupler will depend on this phase difference between the two waves emerging from the sensing and reference arms.

Let us assume that the coupling coefficients of the two couplers are same, i.e., $k_1 = k_2 = k = 0.5$, and the fibers exhibit negligible loss, then one can show that:

$$I_1 = (I_0/2)(1 + \cos \Delta\phi) = I_0 \cos^2(\Delta\phi/2) \qquad (14.16)$$
and
$$I_2 = (I_0/2)(1 - \cos \Delta\phi) = I_0 \cos^2(\Delta\phi/2) \qquad (14.17)$$

where $I_0 = I_1 + I_2$ is the input intensity.

If the sensing and reference arms introduce identical phase shifts, $\Delta\phi$ will be zero and $I_1 = I_0$ and $I_2 = 0$; that is, the entire launched power appears at output port 1. On the other hand, if $\Delta\phi = \pi$, $I_1 = 0$, and $I_2 = I_0$, the entire launched power appears at output port 2. For other values of $\Delta\phi$, the power will be divided into the two ports according to Eqs (14.16) and (14.17).

However, in practice, the sensor is operated at the quadrature point corresponding to $\Delta\phi = \pi/2$. Thus a fixed bias of $\pi/2$ is induced and the phase change θ caused by the measurand is related to the output intensity. One may, therefore, write:

$$\Delta\phi = \pi/2 + \theta$$

If θ is very small, substituting this value of $\Delta\phi$ in Eq. (14.17), one obtains:

$$I_2 = (I_0/2)[1 - \cos(\pi/2 + \theta)]$$
$$= (I_0/2)[1 + \sin\theta] = (I_0/2)(1 + \theta) \quad (\because \sin\theta \approx \theta) \qquad (14.18)$$
Similarly, $I_1 = (I_0/2)(1 - \theta)$ \hfill (14.19)

Clearly, for small values of θ, I_1 and I_2 vary almost linearly with the phase change θ caused by the measurand.

(b) Fiber-optic Gyroscope

This is essentially a rotation sensor (Fig. 14.6). As shown in Fig. 14.6, it consists of a loop of single mode fiber (usually a single polarization fiber), a laser (coherent source), and an optical detector. The three devices are connected via 3-dB directional couplers (DCs).

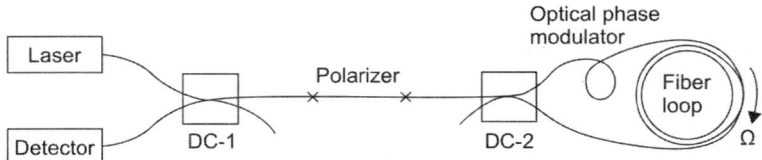

Fig. 14.6. Schematic of a fiber-optic gyroscope

A light beam from a laser is split by DC-2 into two equal halves and fed simultaneously into the two ends of the fiber loop and thus two beams propagate through the loop in opposite direction (i.e., one beam in clockwise and other beam in anticlockwise direction). When these beams emerge at their respective ends, they recombine at the same coupler and interfere at the detector via DC-1.

This arrangement may be thought of as a special form of the Mach-Zehnder interferometer, in which the two arms lie within the same single fiber, but the two beams travel in opposite directions. If we assume that the couplers are lossless and the fiber loop is stationary, the counter-propagating beams will take the same time to emerge, resulting in all the power returning towards DC-1.

Let us suppose that the entire arrangement is rotating clockwise at an angular velocity Ω. Then the clockwise travelling beam will view the end of fiber receding from it as it travels, and hence, it will have to travel further to emerge. Conversely, the anticlockwise travelling beam will see its corresponding end approaching, and hence, it will have to travel a smaller path to emerge. Consequently, there is a relative phase shift between the two beams when they emerge at the respective ends. This causes a corresponding shift in the interference pattern formed at the detector. This phenomenon of phase shift is shown as the *Sagnac effect*. This effect can, therefore, be used to measure the rotational speed Ω.

One can show that the phase difference between the two counter-propagating beams, when the fiber loop is rotating, is given by:

$$\Delta\phi = \frac{8\pi NA\Omega}{c\lambda} \qquad (14.20)$$

where N is the number of turns in the loop, A is the area of a single turn, c is the speed of light in free space, and λ is the free-space wavelength of light.

The arrangement shown in Fig. 14.6 is called the *minimum configuration design*. The function of the polarizer and the phase modulator is to ensure that the detection bias is maintained at the point of maximum sensitivity. Fiber-opticl gyroscope having configuration described above has several applications ranging from *ballistic missiles*, through vehicle navigation systems, to industrial machinery.

14.6 SPECTRALLY MODULATED SENSORS

In spectrally modulated sensors, the wavelength of light is modulated by the measurand. We will describe two important techniques out of several suggested schemes of wavelength modulation:

(a) Fiber-optic Fluorescence Temperature Sensors

Fluorescence is a type of luminescence in which the emission of electromagnetic radiation ceases as soon as excitation ceases. The radiation emitted is usually, but not necessarily, light. Excitation is commonly by ionizing radiation or by electromagnetic radiation of different wavelength from that which is emitted. Normally the emitted radiation is of longer wavelength than the incident electromagnetic radiation

(Stoke's law), but it can be shorter if the substance is not initially in the ground state. The intensity of fluorescence, in most phosphors is reported to varies with temperature. This clearly reveals that by selecting an appropriate fluorescent material, a temperature sensor based on this technique can be designed.

A sensor developed by ASEA, a Swedish company is shown in Fig. 14.7.

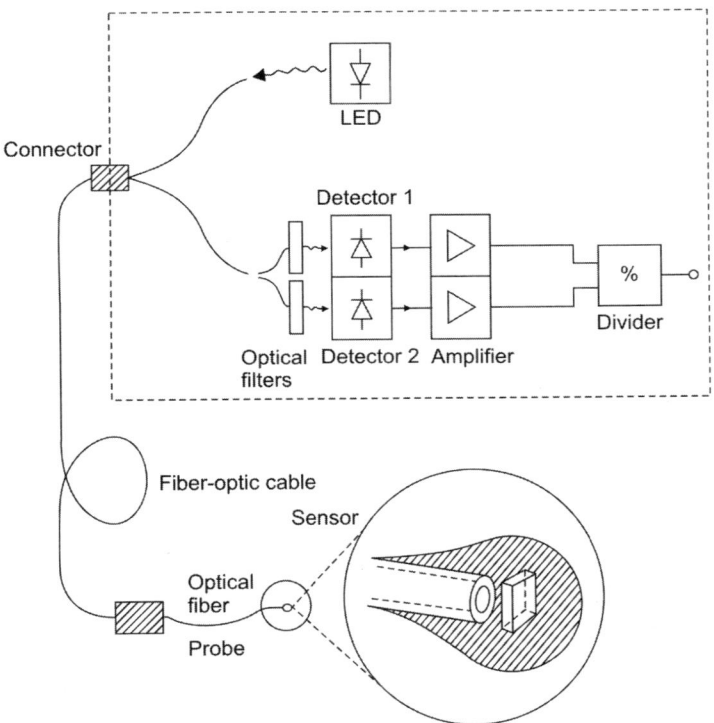

Fig. 14.7. Schematic of a fiber-optic fluorescence temperature sensor, ASEA model 1010

This device utilizes a small crystal of GaAs sandwiched between GaAlAs layers in the sensor head. This is made to fluoresce by absorption of light emitted by a GaAs LED. As the temperature at the sensor head is increased, the emission band broadens and shifts towards the longer wavelength side. Two narrow bandpass filters are used to select two portions of the emission spectrum, and the intensity at each band is measured by respective detectors (photodiodes), their ratio obtained and correlated with temperature.

Another commercial device based on fluorescence is shown in Fig. 14.8. It uses a UV source for excitation and a rare-earth phosphor in the sensor head. The change in the ratio of the emission at two

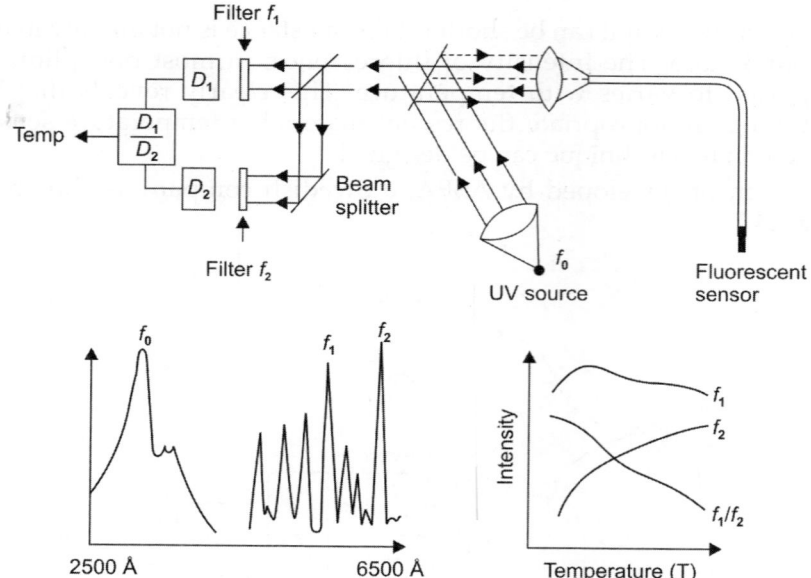

Fig. 14.8. Schematic of a 'Luxtron' fluorescent temperature sensor

different wavelength (λ_1 and λ_2) emitted by the sensor is calibrated as a function of temperature.

In sensors of this type, another channel is required for referencing, because there are several factors other than temperature that may cause the fluorescence intensity to vary. Consequently, a sensing technique based on the measurement of the lifetime of fluorescence has been developed, which has found wider application. Fig. 14.9 shows the fundamental principle of this technique. The fluorescent material is excited by a pulse of suitable wavelength. The intensity rises exponentially with time. After the cessation of excitation, the intensity I falls again exponentially with time t and may be expressed by the following relation:

$$I(t) = I_0 \exp(-t/\tau) \qquad (14.21)$$

where I_0 is the maximum intensity at time $t = 0$ and τ is the time constant, i.e., the time required for the intensity to decay to I_0/e. This constant is also called the *lifetime of fluorescence*. The lifetime is an intrinsic parameter of the fluorescent material, and hence is not affected by the change in fluorescence intensity. With such a system, any configuration of the probe can be used depending on the design requirement. Moreover, there is the choice of using the most appropriate material to meet the requirement of particular application.

(b) Fiber-Bragg Grating (FBG) Sensors

An FBG is written into a segment of a Ge-doped single-mode fiber in which refractive index of the core is made to vary periodically by allowing exposure to a spatial pattern of ultraviolet (UV) light. When FBG is illuminated by an optical source having a broad spectral width, a Bragg wavelength λ_B, satisfying the following condition, is refracted by it:

$$\lambda_B = 2n_{\text{eff}} \Lambda \tag{14.22}$$

where Λ is the spatial periof of index modulation and n_{eff} is the effective mode index.

Such gratings can be used for sensing strain, temperature, pressure, etc. Indeed, the FBG central wavelength λ_B varies with changes in these parameters and the corresponding wavelength shifts are given as follows:

(a) For a longitudinal strain $\Delta\varepsilon$ applied to an FBG, the shift in λ_B is given by:

$$(\Delta\lambda_B)_{\text{strain}} = \lambda_B (1 - \rho_a) \Delta\varepsilon \tag{14.23}$$

where ρ_a is the photoelastic coefficient of the fiber.

(b) For a temperature change of ΔT, the corresponding shift in λ_B is given by:

$$(\Delta\lambda_B)_{\text{temp}} = \lambda_B (1 + \xi) \Delta T \tag{14.24}$$

where ξ is the thermo-optic coefficient of the fiber.

(c) For a pressure change of ΔP, the corresponding shift in λ_B is given by:

$$(\Delta\lambda_B)_{\text{pressure}} = \lambda_B \left[\frac{1}{\Lambda} \frac{\partial \Lambda}{\partial P} + \frac{1}{n_{\text{eff}}} \frac{\partial n_{\text{eff}}}{\partial P} \right] \Delta P \tag{14.25}$$

In addition to these parameters, FBG sensors have found an important application in *'fiber-optic smart structures'* for monitoring strain distributions.

14.7 DISTRIBUTED FIBER-OPTIC SENSORS (DFOSs)

The fiber-optic sensors discussed so far provide a single measurand value averaged over a defined range. These are generally referred to as "point" sensors, even though the length of the sensing fiber over which the averaging is done in some cases may be quite large.

Whenever, there are a number of points where the parameter of interest is to monitor, there exist following two solutions:

(i) The point sensors may be arranged in a desired network or array configuration and their outputs may be multiplexed into an optical fiber telemetry system using common multiplexing techniques such as time-division multiplexing (TDM), frequency-division multiplexing (FDM), or wavelength-division multiplexing (WDM). Such a system is called a *quasi-distributed system*. The limitation of such a system is that the measurand only at a finite number of predetermined locations, where the point sensors have been placed. However, this solution is expensive and generally broadly inadequate.

(ii) It is possible to use a continuous length of a suitably configured optical fiber and determine the value of the desired measurand continuously as a function of the position of the fiber. Such a system is called a *fully distributed system*. Such systems are normally implemented as intrinsic sensors in which the optical fiber is the sensor as well as the transmission medium. Now onwards, we call them simply distributed sensors.

Distributed sensors have enormous possibilities for industrial applications, e.g., (i) monitoring of strain distributions in large structures such as buildings, bridges, dams, ships, aircrafts, spacecrafts, etc., (ii) monitoring of temperature profiles in boilers, power transformers, generators, furnaces, etc., and (iii) mapping of electric- and magnetic-field distributions and the intrusion alarm system for homes and industrial machines.

In distributed sensing, the monitoring of the measurand along the contour of the optical fiber requires some means of identifying the signal originating from a given section of the fiber. There are several methods by which this can be achieved. We discuss a commonly used technique of optical time domain reflectometry (OTDR). Herein, a pulsed signal is transmitted into one end of the fiber and the backscattered signals from different parts of the fiber are recovered at the same fiber end. Let us see how this method may be used for sensing.

A distributed sensor based on monitoring backscattered power with OTDR is shown in Fig. 14.9. An optical pulse of high intensity is launched into the fiber through a DC. As it propagates, it is backscattered due to Rayleigh scattering. The backscattered signal is coupled to the detector again through the DC, processed, and finally read out or recorded. A plot of backscattered power as a function of time/distance is also shown in Fig. 14.9. If the fiber is homogenous and subject to a uniform environ-ment, the backscattered power is given by the following relation:

$$P(l) = \frac{1}{2} P_0 W S \alpha_s (l) V_g \exp\left\{-\int_0^l 2\alpha(z) dz\right\} \quad (14.26)$$

where P_0 is the power launched into the fiber, W is the pulse width, $P(l)$ is the backscattered power coupled to the detector as a function of length l of the fiber, where $l = ct/2n$ is the location of the forward-travelling pulse at the time of generation of the detected backscattered signal.

Thus, OTDR can sense the parameter that can change the total attenuation coefficient α, keeping the scattering coefficient α_s and the capture fraction S constant. Thus, Eq. (14.26) will be modified to:

$$P(l) = A \exp\left\{-\int_0^l \alpha(z) dz\right\} \quad (14.27)$$

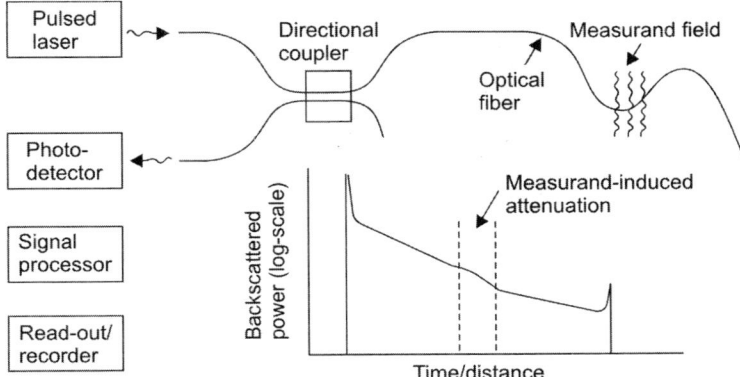

Fig. 14.9. Schematic of a distributed sensor based on monitoring attenuation with OTDR

where A is a constant. Alternatively, it is also possible to sense parameters that can change the scattering coefficient α_s if α and S are kept constant. Then Eq. (14.26) will take the form:

$$P(l) = A'\alpha_s(l) \exp(-B'l) \qquad (14.28)$$

where A' and B' are constants.

Clearly, there are many parameters, e.g., pressure, strain, temperature, etc., that can cause α or α_s to vary, and hence they can easily be measured or monitored by this technique.

14.8 FIBER-OPTIC SMART STRUCTURES

DFOSs can be embedded in composites or building materials, e.g., concrete, to create a 'smart structure' or 'smart skin'. The aim is to create a structural element that can monitor the internal conditions of the component throughout its life. Fig. 14.10(a) shows a conceptual fiber-optic smart structure system. We may note that fiber-optic sensors corresponding to environmental effects such as strain, temperature, pressure, etc., to be monitored, are embedded into the composite panel. For this purpose, either multiplexed or distributed sensors may be used, the outputs of which are sent to the signal processor having optical as well as electronic components. The processed information is conveyed to the control system, which may be aircraft engine performance or flight control or health or damage assessment. One can also use optical fibers to control actuators.

Figure 14.10(b) shows the relevant technologies associated with fiber-optic smart structures and skins.

The prime issues list below need consideration:
(i) Embedding fibers into composite materials, which means the selection of appropriate coatings compatible with the composites,

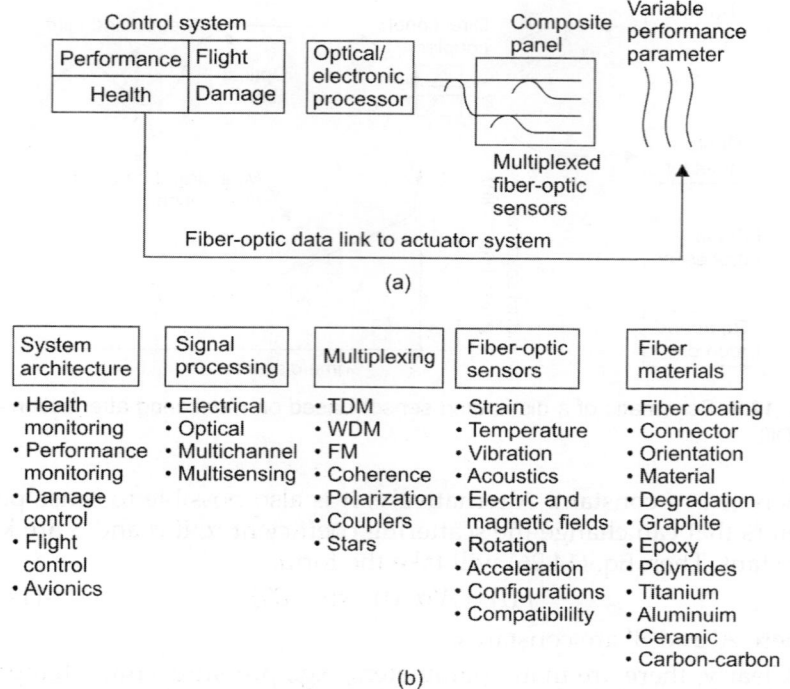

Fig. 14.10. (a) Schematic of a fiber-optic smart structure (b) Technologies associated with 'smart structures' and 'smart skins'

orientation of the fiber in the material, and the means to access the ends of optical fibers so that they can be connected to other parts or to fiber-optic links outside the structure.

(ii) Selecting the technique of fiber-optic sensing from the available ones. Usually, interferometric sensors are employed, as they are highly sensitive, accurate, and compatible with multiplexing techniques. FBG sensors are also strong candidates for smart skins. Distributed sensors can also serve thepurpose in a better way).

(iii) Designing the signal processing unit, which may be used to processw the outputs of embedded sensors and also monitor the performance of the finished structure throughout its life.

Some typical applications of fiber-optic smart structures are shown in Fig. 14.11. Figure 14.11(a) shows a space-based habitat having arrays of fiber-optic sensors arranged to find the location and extent of impact damage by monitoring strain distribution, acoustics, or fiber breakage. Several other sensors can also be used to measure, say, leakage, radiation dose, etc. Fig. 14.11(b) illustrates the use of DFOSs in monitoring the performance of pressurized tanks.

Optical Fiber Sensors 779

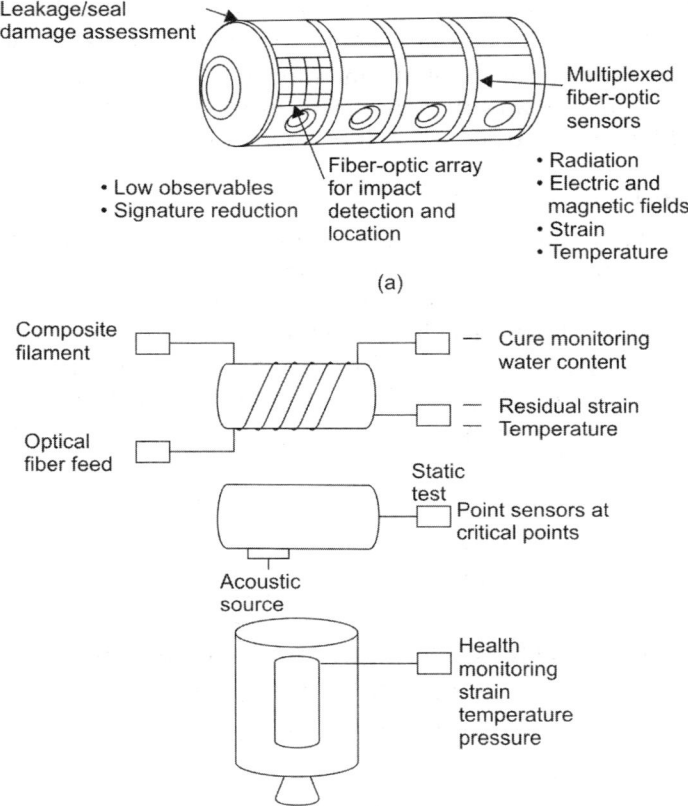

Fig. 14.11. Schematic of fiber-optic smart structures technology for (a) space-based habitat, and (b) pressurized tanks

Truely, fiber-optic smart structures have the potential to revolutionize the field of future composite materials and intelligent structures.

14.9 INDUSTRIAL APPLICATIONS OF FIBER-OPTIC SENSORS

Fiber-optic sensors possess the virtues of excellent sensitivity, dynamic range, low cost, high reliability, immunity to electromagnetic interference, electromagnetic pulses, and radio-frequency interference, small size, and low weight. Almost all the parameters needed for industrial process control, e.g., temperature, pressure, strain, fluid level, flow rate, displacement/position, vibration, pH, electric and magnetic fields, voltage and current, etc., can be measured or monitored by FOSs. A number of commercial devices are now available, e.g., Herga pressure mats (by Herga Ltd) based on distributed microbend losses are being used to protect personnel working close to robots or to detect collisions of remotely controlled vehicles. The distributed cryogenic leak detection system, fully distributed temperature sensor system, DTS-II (by York

Ltd.), distributed cable strain monitor (by NTT), fiber-optic hydrophone array (by Plessey Ltd.), etc. are available in the market and used.

The performance of the above mentioned sensors has been demonstrated to be more than sufficient for industrial applications. We expect that all-fiber smart structures will be soon a reality.

Example 1

Light of wavelength $\lambda = 0.633$ µm is propagating through a single mode silica-based optical fiber. The measurand is temperature, which changes at the rate of 10^{-5} °C^{-1}. The nominal refractive index of the core is $n = 1.45$ and the fractional change in the length of the fiber per degree change in temperature is 5.1×10^{-7} °C^{-1}. Show that the phase change per unit length per degree rise in the temperature of the fiber is 106.5 rad m^{-1} °C^{-1}. Comment on your result. [B.Tech.]

Solution

We have:

$$\frac{\Delta \phi}{\Delta T} = \frac{2\pi}{\lambda} \left(n \frac{\Delta L}{\Delta T} + L \frac{\Delta n}{\Delta T} \right)$$

Now, the phase change per degree rise in temperature is given by:

$$\frac{1}{L} \frac{\Delta \phi}{\Delta T} = \frac{2\pi}{L} \left[\frac{n}{L} \frac{\Delta L}{\Delta T} + \frac{\Delta n}{\Delta T} \right]$$

Substituting the values, we get:

$$\frac{1}{L} \frac{\Delta \phi}{\Delta T} = \frac{2\pi}{0.633 \times 10^{-6}} [1.45 \times 5.1 \times 10^{-7} + 10^{-5}] = 106.5 \text{ rad m}^{-1} \text{°C}^{-1}$$

Here

$$n = 1.45$$

$$\frac{\Delta n}{\Delta T} = 10^{-5} \text{ °C}^{-1}$$

$$\frac{1}{L} \frac{\Delta L}{\Delta T} = 5.1 \times 10^{-7} \text{ °C}^{-1}$$

We see that the amplitude is large.

Example 2

The diameter of a circular coil of a fiber-optic gyroscope is 0.1 m and the total length of the optical fiber in the coil is 500 m. Considering that, it is operating at $\lambda = 0.85$ µm, show that the phase shift corresponding to earth's rotational speed, $\Omega = 7.3 \times 10^{-5}$ rad s^{-1} is 8.99×10^{-5} rad. [B.Tech.]

Solution
We have:
$$\Delta\phi = \frac{2\pi L D \Omega}{c\lambda}$$

Substituting the values, one obtains:
$$\Delta\phi = \frac{2\pi \times 500 \times 0.1 \times 7.3 \times 10^{-5}}{3 \times 10^8 \times 0.85 \times 10^{-6}} = 8.99 \times 10^{-5} \text{ rad}$$

Here $L = 500$ m
$D = 0.1$ m
$\Omega = 7.3 \times 10^{-5}$ rad s^{-1}
$\lambda = 0.85 \times 10^{-6}$ m
$c = 3 \times 10^8$ ms^{-1}

The value is quite small but can be easily measured.

SUGGESTED READINGS
1. Grattan KTV and Meggitt BT (Eds.), Optical Fiber Sensor Technology, Chapman & Hall (1998).
2. Udd E, Fiber Optic Smart Structures, John Wiley (1995).
3. Keiser G, Optical Fiber Communications, McGraw Hill, 4th ed. (2011).
4. Agrawal, GP, Fiber-Optic Communication Systems, 3rd ed., John Wiley (2002).

GLIMPSES
- Fiber optic sensor (FOS) is a transducer which can convert various input variables (physical quantity) into an electrical signal in a measureable form. Basically, FOS is a device that uses light guided within an optical fiber to detect any external physical, chemical, or any other parameter.
- FOS are of two types: (i) extrinsic sensors, and (ii) intrinsic sensors.
- In extrinsic sensors, light from an optical source is launched into the fiber and guided to a point where the measurement is to be performed, i.e., the interaction between the light and the measurand (the quantity under measurement) takes place outside the fiber and fiber acts merely as a waveguide. Here the light gets modulated by the measurand and is relaunched into the same (or other) fiber.
- In intrinsic sensors, the measurand acts directly on the fiber itself and produces a change in the transmission characteristics, i.e., the light launched into the fiber gets modulated in response to the measurand whilst still being guided in the fiber.

- FOSs can also be classified based on the scheme of modulation as used for making the sensor: (i) intensity-modulated sensors, (ii) phase-modulated sensors, and (iii) spectrally modulated sensors.
- One can achieve the *intensity modulation* in a number of ways. Again, there are extrinsic and intrinsic type of mechanisms for this kind of modulation.
- Phase modulation gives more accurate results, but it is more difficult to implement. One can use Mach–Zehnder type of sensors for the measurement of pressure and strain. One can use fiber-optic gyroscopes for rotation sensing as well as some other parameters.
- One can achieve spectral modulation in several ways but this scheme has been used prominently for making fiber-optic fluorescent thermometers.
- In addition to *point sensors*, there are several application areas where distributed sensing is required. In implementing DFOSs, OTDR is the main technique used. Using these sensors, temperature, pressure, strain, etc. can be sensed along a given contour.
- To make smart skins and smart structures, point or distributed FOSs can be embedded in composite materials.
- Using FOSs or DFOSs, almost any industrial parameter, e.g. temperature, pressure, displacement/position, liquid level, pH, etc. can be measured.
- The optical sensors are non-contact and generally high accuracy devices and systems. In optical fiber sensors, the optical wave is the information carrier and sensor.
- Whenever fiber optic sensors are used for measuring a physical parameter, any one of the characteristics like amplitude, intensity, etc. is measured.
- Optical fiber sensors have advantages due to their small size, flexibility, and high-temperature resistance for cure monitoring. In addition, many kinds of physical values such as temperature, pressure, strain, refractive index, and molecular structure can be monitored.

REVIEW QUESTIONS

1. What are the parameters that can be measured using fiber optic sensors?
2. List few physical measurands and their modulation effects in fiber optic sensors.
3. How are fiber optic sensors classified? What are extrinsic and intrinsic sensors?
4. Explain displacement sensor function based on (a) intensity measurement and diffraction pattern.

5. Explain the working of a pressure sensor based on (a) phase modulation, (b) photoelastic effect.
6. Give examples of extrinsic and intrinsic sensors.
7. Mention the different field areas where fiber optic sensors are used.
8. What are the advantages of fiber optic sensors over traditional sensors? Explain microbending and photoelastic pressure sensors.
9. Consider the fiber-optic Mach–Zehnder interferometric (MZI) pressure sensor, and show that the phase change ($\Delta\phi$) per unit length in the sensing arm due to change in pressure (ΔP) is approximately given by:

$$\frac{\Delta\phi}{L\Delta P} = \frac{2\pi}{\lambda}\left[n\frac{\Delta L}{L\Delta P} = \frac{\Delta n}{\Delta P}\right]$$

where the symbols have usual meanings. [B.Tech.]

10. Briefly explain the different pressure sensors with their principle.
11. Show that the shift $\Delta\lambda_B$ in the Bragg wavelength λ_B due to pressure change ΔP for an FBG sensor may be given by:

$$\frac{\Delta\lambda_B}{\Delta P} = \lambda_B \frac{1}{\Lambda}\frac{\partial\Lambda}{\partial P} + \frac{1}{n_{eff}} + \frac{1}{n_{eff}}\frac{\partial n_{eff}}{\partial P}$$

where n_{eff} is the effective mode index and Λ is the spatial period of index modulation.

12. What is fluorescence? Describe a fiber-optic fluorescence temperature sensor.
13. Describe a fiber-optic gyroscope.
14. Explain fiber-optic smart structures? Mention possible application of such structures.

PROBLEMS

1. A fiber optic gyroscope has a circular coil of diameter 0.12 m. The total length of the fiber used in the coil is 400 m. Considering that it is operating at $\lambda = 0.633$, calculate the phase shift corresponding to angular speed of 5×10^{-4} rad s^{-1}. [**Ans.** 0.066 rad]
2. The identical fibers employed in the angular displacement sensor have core index 1.46 and cladding index 1.45 considering that the range of angular deviation to be detected varies from 0° to 10°, calculate (a) the range of S_{ang} (normalized), (b) the range of loss, and (c) the minimum power that should be launched by the source if the photodetector has a sensitivity of –30 dB·m Assume that loss in fibers is negligible.
[**Ans.** (a) 1 – 0.673, (b) 0 to 2.027 dB, (c) 1.6 µW]

SHORT ANSWER QUESTIONS

1. What is a fiber optic sensor?

Ans. Fiber optic sensor is a transducer which can convert various input variables (physical quantity) into an electrical signal in a measurable form.

2. How many types of fiber optic sensors are there?

Ans. Two types: (i) Extrinsic or passive fiber optic sensors, and (ii) Intrinsic or active sensors.

3. Mention the advantages of optical sensors in sensing applications.

Ans. (i) They are light in weight and small in size. (ii) They have good geometrical flexibility. (iii) They are electrically passive. (iv) The chemical and environmental ruggedness is more.

4. Write the characteristics of the wave that may get modulated in fiber optic sensor?

Ans. The wave characteristics like amplitude or intensity, phase polarisation, frequency and direction of propagation may be modulated.

5. What are interferometric sensors?

Ans. These are fiber optic sensors based on the principle of interference pattern produced due to phase shift on superimposing a reference beam and a sensing beam.

6. What are displacement sensors?

Ans. These sensors are used to measure the displacement based on the intensity modulation of the beam.

7. What is a pressure sensor?

Ans. This sensor is used to measure the pressure acting over a fiber. These sensors are mainly based on the principle of interference pattern due to phase shift.

8. What are the different field areas where pressure sensors are used?

Ans. These are used in field areas like aerospace, oceanography, meterology, pressure control, energy exploration and hydrology.

9. What is the principle of a photoelastic sensor?

Ans. Photoelastic sensor is based on photoelastic effect.

10. What are intensity modulated sensors?

Ans. In these sensors the intensity of light launched into the fiber is changed either intrinsically or extrinsically by the measurand.

11. What are phase modulated sensors?

Ans. These are the sensors in which the phase of monochromatic light propagating through the fiber is changed (normally intrinsically) by the measurand.

Optical Fiber Sensors

12. What are spectrally modulated sensors?
Ans. These are the sensors in which the wavelength of light is changed (normally extrinsically) by the measurand.
13. What is a fiber-optic Gyroscope?
Ans. This is essentially a rotation sensor. It consists of a loop of a single-mode fiber (preferably a single polarization fiber), a coherent source of light (e.g., a laser), and a detector.
14. What is a photoelastic pressure sensor?
Ans. This is a multimode optical fiber sensor which is capable of measuring pressure.
15. What are the special features of optical fiber sensors from industrial applications point of view.
Ans. Fiber optic sensors possess the virtues of excellent sensitivity, dynamic range, low cost, high reliability, immunity to electromagnetic interference, electromagnetic pulses, and radio-frequency interference, small size, and low weight industrial process control.

MULTIPLE CHOICE QUESTIONS

1. Which of the following is used as a prime means of measurement by fiber-optic sensors?
   (a) Gravitational field   (b) Electric field
   (c) Magnetic field   (d) Optical field
2. Fiber optical sensor can convert various input variables (physical quantity) into an electrical signal in a measurable form and hence it is a:
   (a) galvanometer   (b) ammeter
   (c) transducer   (d) spectrometer
3. In extrinsic sensors, the interaction between the light and the measurand (the quantity under measurement) takes place:
   (a) inside the fiber
   (b) outside the fiber
   (c) sometimes inside and sometimes outside the fiber
   (d) None of the above
4. Which of the following measurands can change the refractive index of a silica based fiber?
   (a) Acoustic wave   (b) Temperature
   (c) Pressure   (d) All of the above
5. Which of the following measurands cannot be measured by a microbend sensor?
   (a) Electric current   (b) Pressure
   (c) Displacement   (d) Temperature

6. In intrinsic sensors, the measurand acts directly on the fiber itself and produces a change in the:
   (a) absorption characteristics
   (b) reflection characteristics
   (c) transmission characteristics
   (d) all of the above
7. Interferometric sensors are based on the principle of the .... produced due to a phase shift on superimposing a reference beam and a sensing beam.
   (a) diffraction pattern   (b) polarization
   (c) Raman effect   (d) interference pattern
8. Microbending sensors are based on measurement of the light losses caused by
   (a) controlled microbending
   (b) uncontrolled microbending
   (c) interference
   (d) None of the above
9. In photoelastic pressure sensor, the phenomena based on which effect is used to measure the pressure?
   (a) Raman effect   (b) Photoelastic effect
   (c) Photoelectric effect   (d) None of the above
10. Which one of the following is an example of an intensity-modulated sensor?
    (a) A fiber optic gyroscope
    (b) A Mach-Zehnder interferometer
    (c) A sensor based on the relative displacement of two fibers
    (d) None of the above
11. The fiber-optic sensors which are used to measure the displacement based on the intensity modulation of the transmitted beam is called:
    (a) interferometric sensors   (b) displacement sensors
    (c) pressure sensors   (d) photoelastic sensors
12. A fiber-optic smart structure can be used for:
    (a) mapping of thermal profiles
    (b) monitoring internal strain(s)
    (c) monitoring the structural integrity of the completed component
    (d) All of the above

13. In principle DFOSs are based on the following property (properties) of a silica fiber:
    (a) scattering of light
    (b) absorption of light
    (c) attenuation of light
    (d) All of the above
14. Fiber Bragg grating cannot be used for the measurement of:
    (a) temperature
    (b) liquid level
    (c) pressure
    (d) strain

ANSWERS

1. (d)   2. (c)   3. (b)   4. (d)   5. (a)   6. (c)   7. (d)
8. (a)   9. (b)   10. (a)   11. (b)   13. (d)   13. (d)   14. (b)

15

Laser Based Systems

15.1 INTRODUCTION

In Chapter 3, we have studied about semiconductor lasers. Besides semiconductor lasers, there are several other kinds of lasers, e.g. solid state laser, gas laser, dye laser, excimer lasers, liquid lasers or dye lasers, chemical lasers, colour centre lasers, UV and X-ray lasers, etc. In solid state lasers active medium is an insulating dielectric solid. These solids can be crystalline or amorphous. These lasers are used to generate a wide range of wavelengths from the vacuum ultraviolet to midinfrared. They are capable of generating high peak powers (~ 10^{14} W) because of long life metastable states which allow higher energy storage compared to other active media. The common examples of solid state lasers are ruby, which is a chromium doped Al_2O_3, titanium doped Al_2O_3 (Ti:Al_2O_3), neodymium-doped yttrium aluminium garnet (Nd:YAG) and neodymium-doped glasses (Nd:glass).

In gas lasers, gas or gas mixture is capable of withstanding a large amount of power. The pointing stability and high optical quality beams are their great assets. Common examples are copper-vapour lasers, helium neon (He-Ne) laser, helium-cadmium laser and carbon dioxide (CO_2) laser.

The free electron lasers (FEL) are operated in far infrared and ultra-violet (UV) (below 100 nm) where atomic or molecular lasers are not readily available; and for large average power and high efficiency systems. The disadvantages of FEL lasers are their complexity and the cost of a particle accelerator.

The important characteristics of laser dye (liquid) media are their broad wavelength tunability, wide spectral coverage and practical simplicity. Hundreds of dyes are reported to have lasting action. The examples of various dye classes are oligophenylenes, coumarins, ranthenes, merocyanines and cyanines with spectral emission from 300 nm to more than 1100 nm.

15.2 SOLID STATE LASERS

In a solid state laser the active medium is a solid material instead of a gas or a liquid. All solid state lasers are optically pumped, which means that excitation of the active medium is achieved by the absorption of light. Solid state laser operation may be either *continuous wave* or *pulsed*.

Figures 15.1(a) and (b) shows a solid-state laser employing optical pumping.

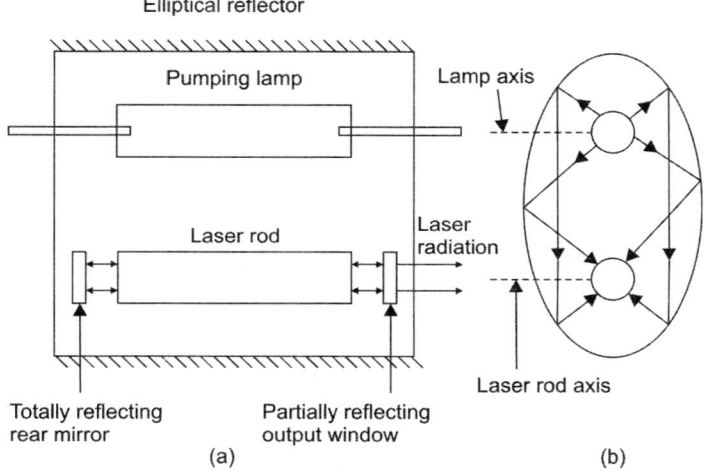

Fig. 15.1. (a) Schematic of solid-state laser with elliptical reflector, (b) focusing of pumping light on the laser rod

A linear pumping source and the laser rod are placed parallel to each other within an elliptical reflector. The pumping lamp is placed along one principal axis (focus) and the laser rod along the other. The configuration achieves a higher concentration of light flux from the pumping lamp onto the laser rod. The lamps employed for optical pumping are basically discharge tubes, whose configuration may be linear, π-shaped, or helical as shown in Fig. 15.2. Linear and π-shaped discharge tubes are suitable for systems shown in Fig. 15.1. If a helical lamp is employed, the laser rod is placed along the axis of the helix, and the entire system is kept within a cylindrical reflector.

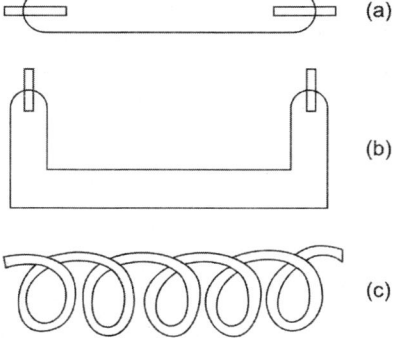

Fig. 15.2. Schematic of lamps for optical pumping: (a) linear, (b) π-shaped, and (c) helical

Normally, solid state lasers operate in the pulsed mode, and therefore, use pulsed power sources for the supply of power to the flash tube. Figure 15.3 shows a generalized block diagram of pulsed power supply. We can see that it consists of a current source, a rectifier, a control circuit, and a flash tube circuit. However, the actual circuit configuration usually depend on the objective sought.

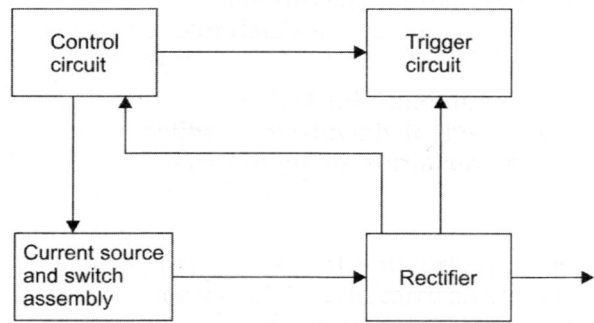

Fig. 15.3. A generalized block diagram of a pulsed power supply

As stated earlier, there are large number of solid state materials seem to be promising for the use as an active medium for solid state lasers, but only ruby and Ne-doped hosts have been developed commercially. Here, we discuss only these two laser systems:

15.2.1 Ruby Laser

Historically, ruby laser is the first laser, demonstrated by Maiman of Hughes Research Laboratory in July 1960. It is a doped insulator, optically pumped laser. In this laser, ruby is the active medium, which is an aluminium oxide, doped with chromium.* The crystal (host lattice) doesn't take part directly into lasing action. The properties of dopant ion are different from its properties in free state. When the ion is in the gaseous state, many of its energy levels are degenerate (degenerate means having same energies), but when it is lattice field, degeneracy is partly removed. The host lattice determines the absorption and emission properties of dopant. The chromium ion doped with aluminium oxide, absorbs strongly blue and green light of visible spectrum, and therefore, ruby has a beautiful red colour.

The crystal is taken in cylindrical shape having length of a few cm and diameter of a few mm. The ends of the rod are polished and made flat and parallel. One end is fully silvered and other is partially silvered. A xenon flash lamp is used to create population inversion.

---

* An addition of impurity is called doping, impurity is called *dopant* and crystal is called *host lattice*.

The lamp is pulsed with duration of millisecond. The heat generated by flash lamp and nonradiative transitions is removed by cooling system. In Fig. 15.4(a), cooling system is not shown. Many electrons go to $^4F_1$ and $^4F_1$ levels [Fig. 15.4(b)]. These levels decay rapidly into ^2E level. These are nonradiative decays. ^2E is the metastable state (life time is about 3 ms). The later transition occurs at the wavelength of 694.3 nm between ^2E and ground level. Ruby may be considered a three-level laser. We may briefly summarize the working of a ruby laser:

(a) High-voltage electricity causes the quartz flash tube to emit an intense burst of light, exciting some of the atoms in the ruby crystal to higher energy levels.

(b) At a specific energy level, some atoms emit photons. At first, the photons are emitted in all directions. Photons from one atom stimulate emission of photons from other atoms and the light intensity is rapidly amplified.

(c) The photons leave through the partially silvered mirror at one end. This is laser beam.

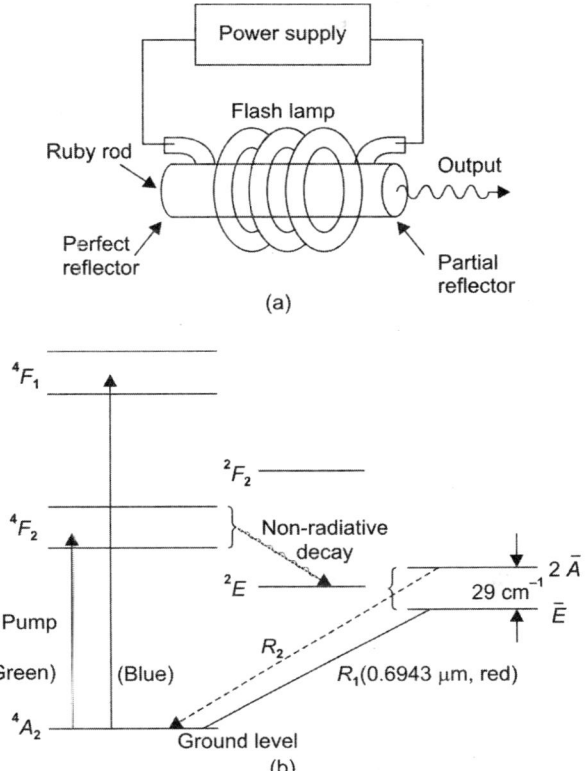

Fig. 15.4. Ruby laser (a) construction, (b) energy level diagram (not to the scale)

15.2.2 Nd³⁺: YAG Laser

Pure YAG, yttrium aluminium garnet ($Y_3Al_5O_{12}$) is a synthetic, crystal-line material of the garnet group. $Y_3Al_5O_{12}$ is an optically isotropic crystal with a cubic structure. YAG is commonly used as a host material in various laser active materials. Due to similar size, yttrium ions can be replaced with rare-earth laser active ions without strongly affecting the lattice structure. $Y_3Al_3O_{12}$ is doped with trivalent neodymium (Nd^{3+}) ions (about 0.73% by weight) to make it a laser active material. Nd^{3+} ions serve as active atoms. Nd^{3+} introduce well-defined energy level in YAG as shown in Fig. 15.5.

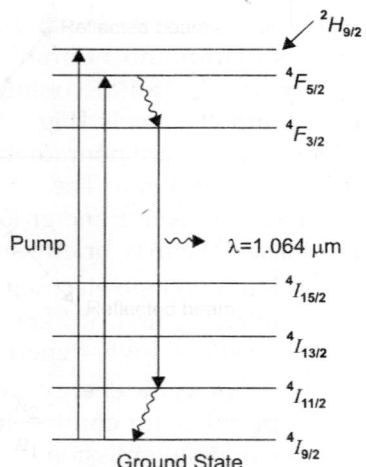

Fig. 15.5. Simplified energy-level diagram of Nd^{3+} in YAG (Nd-YAG laser). The typical pumping route and subsequent decays are shown by appropriate arrows in figure

Absorption of pump frequencies raises the atoms to higher levels: the $^4F_{3/2}$ level which forms the upper lasing level. The dominant laser transition is from $^4F_{3/2}$ (at 11.507 cm⁻¹) to $^4I_{11/2}$ (at 2110 cm⁻¹), giving a spectral line at $\lambda_0 = 1.064$ µm. The lower laser level $^4I_{11/2}$ decays non-radiatively. The Nd^{3+}: YAG laser may be considered a four-level laser. It can be operated on power levels of upto kilowatts and can be directly Q-switched with Cr^{4+}: YAG. The energy levels of the Nd^{3+} ion are responsible for the fluorescent properties, i.e. active particles in the amplifying process.

In this case, the optical gain is much greater than that of ruby. This causes the laser threshold to be very low and also makes CW operation much easier. As the absorption bands are narrow, krypton gas, whose spectrum matches the pumping bands, is normally used in the pumping lamp.

The cylindrical crystal forms the laser cavity and has reflective ends, one coated so that it is 100% reflective, and the other is either sufficiently reflective or coated to allow only part of the amplified light to pass through feedback so that oscillation may occur as shown in Fig. 15.6

Nd^{3+}-doped glass also make useful laser systems. However, because of the amorphous nature of glass, the absorption bands are much broader than those of YAG. This leads to wider fluorescent lines. Second, glass is a poor thermal conductor, and hence, it is difficult to remove this waste heat, which in turn limits the repetition rate of glass lasers.

15.2.3 Applications of Nd:YAG Lasers

This laser is used in medical field as a means of correcting posterior capsular opacification (after cataract). This laser is used for peripheral

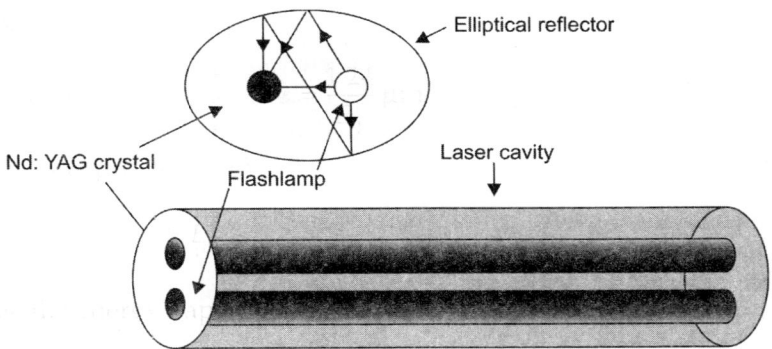

Fig. 15.6. Schematic of Nd:Yag laser

iridotomy in patients with acute angle closure glaucoma, where it has superseded surgical iridectomy. Frequency-doubled Nd:YAG laser (532 nm) is used for pan-retinal photocoagulation (in place of argon laser) in patients with diabetic retinopathy.

Nd:YAG lasers are also used extensively in the field of cosmetic medicine for laser hair removal and the treatment of minor vascular defects, e.g., spider veins on the face and legs.

These lasers are used for soft tissue surgeries in the oral cavity, e.g. gingivectomy, periodontal sulcular debridgement, frenectomy, biopsy, and coagulation of graft donor sites.

These lasers are used in manufacturing as a means of engraving, etching, or marking a variety of metals and plastics. These lasers are widely used in manufacturing for cutting and welding steel and super alloys. Super alloy drilling (for gas turbine parts) typically uses these lasers. These lasers are also employed to make subsurface markings in transparent materials, e.g., glass or acrylic glass.

These lasers can also be used for flow visualization techniques in fluid dynamics, e.g., particle image velocimetry or induced fluorescence.

Q-Switching

This is sometimes known as pulse formation, is a technique by which a laser can be made to produce a pulsed output beam. The technique permits the production of light pulses with extremely high (gigawatt) peak power, much higher than would be produced by the same laser if it were operating in a continuous wave mode. As compared to mode locking technique, this is another technique for pulse generation with lasers. Q switching leads to much lower pulse repetition rates, much higher pulse energies, and much longer pulse durations. Both techniques are also applied sometimes at once. Details are given in later sections.

15.3 GAS LASERS

In gas lasers, an electric current is discharged through a gas to produce light. Although, a number of gases have been shown to exhibit laser action, only a few have been exploited commercially. Of them, helium-neon (He-Ne) argon ion, and CO_2 lasers have been extensively studied and used. A typical gas laser is shown in Fig. 15.7. The pumping is normally achieved through the electrical discharge between a pair of electrodes. The discharge occurs along the axis of the laser cavity. In this case, some volume of the gas is not utilized. In order to achieve uniform excitation of a large volume of the gas, transverse discharge is employed. Excitation of gas lasers can also be carried out by electron beams. The operation of CW gas lasers often requires rectified ac, while pulsed lasers use power supplies similar to those used in solid-state lasers.

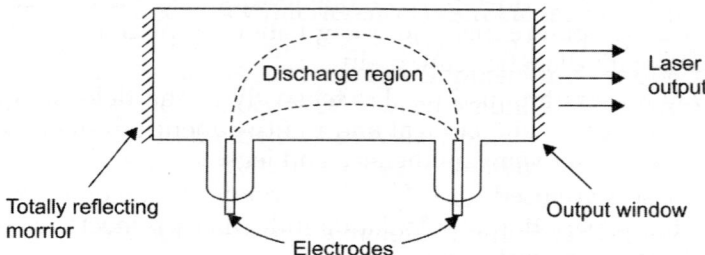

Fig. 15.7. Schematic of a typical gas laser

Advantages of gas lasers are as follows:
- High volume of active material.
- Active material is comparatively inexpensive.
- There is very little possibility to damage the active material.
- Heat can be removed very quickly from the cavity.

The different types of gas lasers are as follows:
- He-Ne gas laser
- Argon gas laser
- Krypton gas laser
- Xenon gas laser
- Nitrogen gas laser
- CO_2 gas laser
- Ion gas laser

15.3.1 Helium-Neon (He-Ne) Gas Laser

This is the most common and best known laser. The first gas laser has reported by Javan et al. in 1961. It emitted continuously at 1.15 μm (infrared) transition of neon. Soon after that in 1962, White and Rigden reported (red) transitions at 632.8 nm. The active medium is a gas mixture of helium and neon in 10:1. It is electric discharge laser. A brief construction is shown in Fig. 15.8.

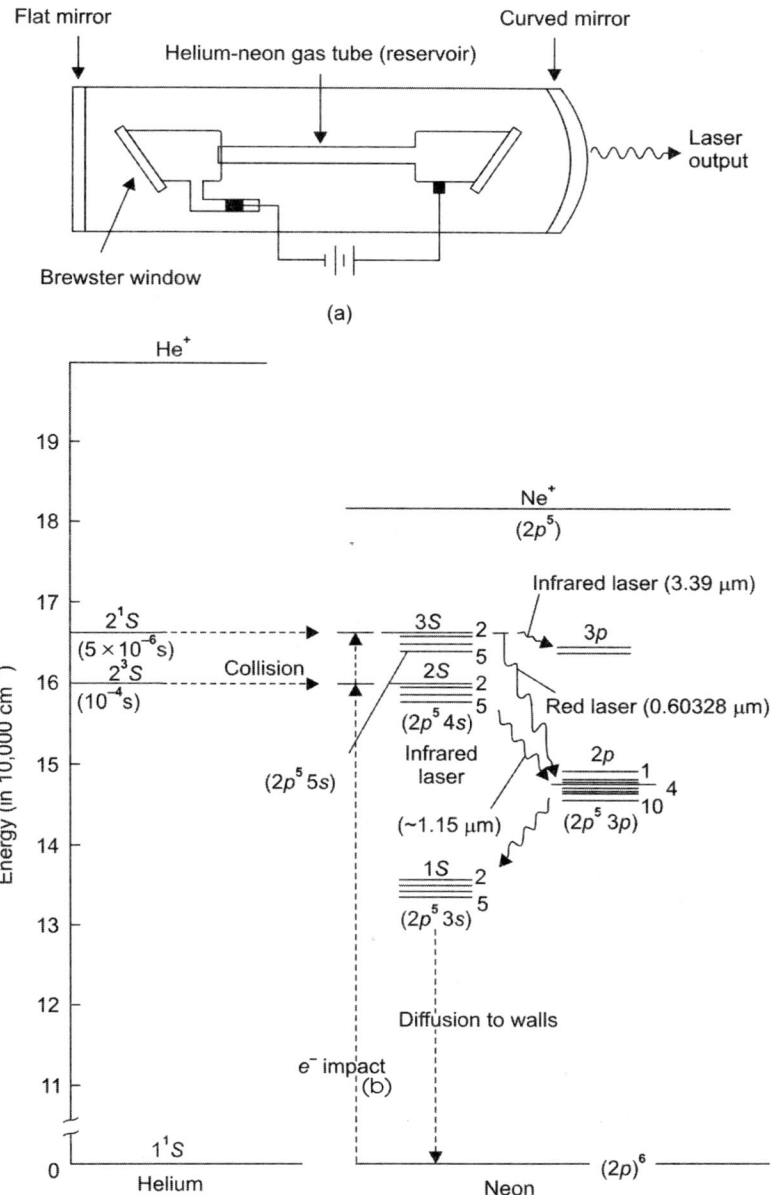

Fig. 15.8. He-Ne laser. (a) Schematic diagram of He-Ne laser, (b) Energy level diagram exhibiting the dominating excitation paths for the red and infrared laser transitions

The discharge tube is 10 to 30 cm long with *Brewster-angle windows* at the end to obtain linearly polarized output radiation. Mirrors are fitted as shown in Fig. 15.8(a). One mirror is flat and other is curved. Laser transitions occur in neon energy levels. The helium atoms provide an efficient excitation mechanism for neon atoms. The electric discharge excites helium atoms to their metastable states. Two energy states of neon have the same energy as that of two metastable states of helium. Collisions occur between helium and neon atoms. In these collisions, energy is transferred from helium to neon atoms. These collisions are called *resonant collisions*. After transferring energy to neon atoms, helium atoms return to ground state and neon atom go to excited state. The energy levels of neon are shown in Fig. 15.8(b). The strongest lasing transitions of neon are 3.39 µm, 1.15 µm and 0.6328 µm. The lowest two excited states of helium (2^3s and 2^1s) are metastable. The lasing occurs at 3s → 2p, 3p, 2s → 2p and 3s → 3p, 3s → 2p transitions also lase in the visible region. By using a prism in the optical cavity, one can select a particular transition in the range 543–633 nm region; (543, 594, 612 or 633 nm) red, orange or yellow green.

When the excited He atoms collide with the Ne atoms, the energy is exchanged and the Ne atoms are pumped to the respective levels. The atoms at the Ne 3S level eventually decay to the 2p level as a result of stimulated emission, and a spectral line of $\lambda = 0.6328$ µm. However, stimulated emission tends to occur between the 3S and 3p levels emitting light at $\lambda = 3.39$ µm. Thus, laser action would normally take place at 3.39 µm (infrared) instead of the desired 0.6328 µm (red). This problem is overcome by attenuating the 3.39 µm line in the cavity. Normally, the output power of He-Ne lasers is in the range 0.5–5 mW.

Specifications of He-Ne lasers are summarized in Table 15.1.

Table 15.1. He-Ne lasers	
Wavelength	632.8 nm
Output power	0.5–50 mW
Beam diameter	0.5–2.0 mm
Beam divergence	0.5–3 m Rad
Coherence length	0.1–2 m
Power stability	5% Hr
Life time	> 20,000 hours

He-Ne gas lasers are mostly used in:
- pumping of dye lasers
- measurement of air pollution (LIDAR)
- scientific research

15.3.2 Carbon Dioxide (CO_2) Gas Laser

This gas laser is fundamentally different from other lasers. It is a molecular gas laser.

The overall efficiency of most of the gas lasers is below 1%; but that of CO_2 laser is 20%. The unusual high efficiency of this laser was first recognized by CKN Patel (in 1964).

The construction of CO_2 laser is similar to He-Ne laser. It is a discharge tube filled with a mixture of carbon dioxide, nitrogen and helium gases in the ratio 1:4:5. The energy spectrum of molecular gas is more complex than the atomic gas. The molecule has energy levels due to vibrations and rotations of the atoms with the electronic energy levels of atoms. The carbon dioxide molecule has three vibrational modes (Fig. 15.9) and for each mode there are rotational modes.

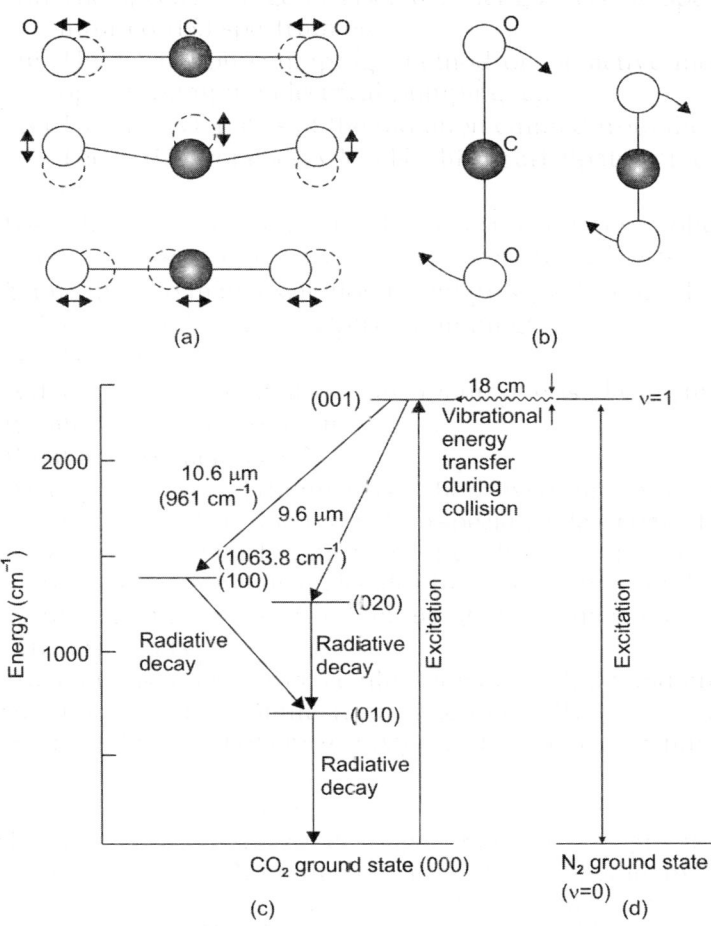

Fig. 15.9. Vibrational modes of CO_2 molecule: (a) symmetric stretch mode, bending mode, asymmetric stretch mode. (b) Rotational modes of CO_2 molecule. (c) Some of the low-lying vibrational levels of the CO_2 molecule, which includes lower and upper levels for 10.6 µm and 9.6 µm laser transitions. (d) Ground state (v = 0) and the first excited state (v = 1) of the N_2 molecule, which plays an active and important role in the selective excitation of the (001) CO_2 level

The CO_2 molecule vibrates in three modes, namely, symmetric stretch mode, bending mode and asymmetric stretch mode. Each mode is associated with a set of energy levels. For the symmetric stretch mode, they are denoted by $(n00)$, where n is an integer; for bending mode $(0n0)$ and for asymmetric stretch mode $(00n)$. The first excited asymmetric stretch state (001) is almost exactly equal to lowest vibrational level of nitrogen. The trick of energy transfer by resonant collision, used in He-Ne laser, is used here also. The nitrogen is excited electrically and it delivers the energy to CO_2 (001) state by way of collisions. If we attempt to populate (001) state of CO_2 directly by electron collisions, most of the energy goes to low-lying (010) and (020) states, Therefore, nitrogen is used as vehicle. The lasing occurs between (001) → (100) and (001) → (020), at 10.6 µm and 9.6 µm respectively. The vibrational-vibrational energy transfer pumping process results in high efficiency and enormous output power, increasing utility of this laser. Improvement in heat removal from the laser discharge is the main purpose of adding helium in the mixture. It also helps in speeding up the transition from (100) to ground state (000) to maintain population inversion. Laser oscillations can occur at 10.6 µm but the gain is stronger at 10.6 µm.

Output spectrum is in the infrared (IR) spectrum: 9– 11 µm. Continuous power outputs upto 25 kW are obtainable. This laser is widely favoured for materials processing applications, e.g. cutting, welding and annealing.

CO_2 laser, unlike most other gas lasers, has an appreciably high efficiency ~10 to 15%. To reach the high powers required for these lasers, cavity lengths can stretch to 2 to 3 meters or even more.

15.4 DYE OR LIQUID LASERS

These lasers use organic dye molecules as the active lasing medium. Dye lasers can lase in a wide frequency range, i.e., they are frequency tunable. The spectral range of dyes covers infrared, visible and ultraviolet light. In fact, these lasers can be tuned to give a continuously variable output over a wide range of wavelengths. Pumping is by another pulsed/continuous laser, or by pulsed lamp (CW argon or pulsed nitrogen or by flash tubes). Fig. 15.10 shows a tunable dye laser.

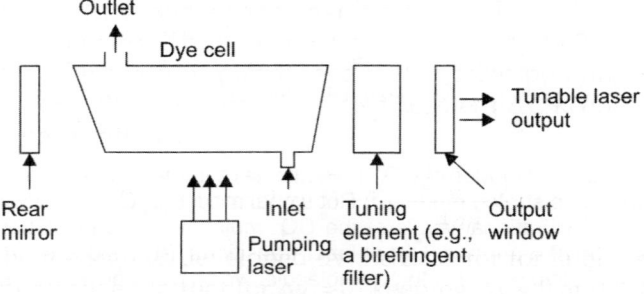

Fig. 15.10. A dye laser

Dye lasers cover the spectral range from about 0.42 to about 0.80 µm. The lasing ranges of some dyes are as follows. Carbostyril, 0.419–0.485 µm; coumarin, 0.435–0.565 µm; rhodamine, 0.540–0.635 µm; oxazine, 0.695–0.801 µm; and so on. For CW operation, the dye is dissolved into a suitable solvent, e.g. ethylene glycol, and then circulated through the dye cell. It is excited by another laser or some other source. The tuning is achieved by rotating the birefringent filter placed inside the optical cavity.

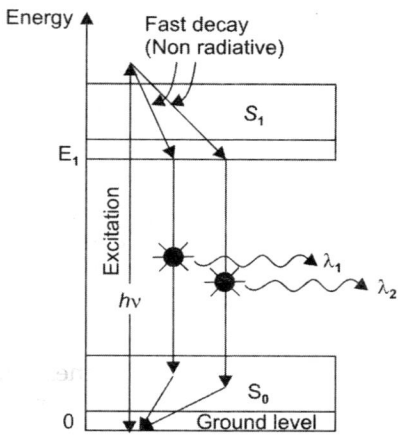

Fig. 15.11. Simplified energy level diagram of dye laser

The output of a dye laser is always a coherent radiation tunable over a specific spectrum region, determined by the dye material.

The energy level diagram of dye molecules in a solvent is very complex. Fig. 15.11 shows a simplified energy level diagram of a dye laser.

The width of each energy band is of the order of tenths of eV. The thermal energy of vibration is of the order of $1/40$ eV. Obviously, the bottom of each energy level is filled.

In a dye laser, the liquid dye is inside a transparent container, and the optical pump energy is coming through the walls of the container or the liquid dye is flowing through a special nozzle, and the optical pump energy is shining on it while it flows out of the nozzle (Fig. 15.12).

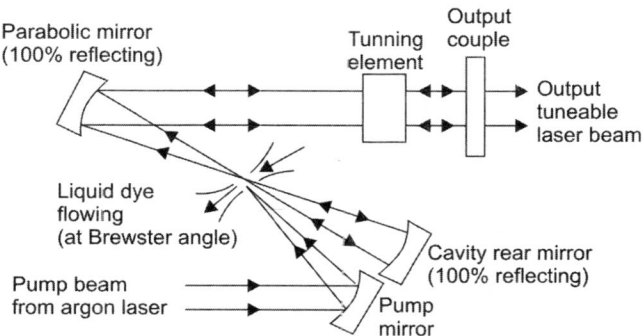

Fig. 15.12. Tunable dye laser with flowing dye

Advantages

(i) Liquid is homogeneous by nature, and there is no difficulty of manufacturing homogeneous perfect solid with no defects.

(ii) In dye lasers, it is relatively easy to change the type of liquid used as an active medium. Thus, changing the range of wavelength of the emitted radiation.
(iii) The liquid carry with it the heat evolved during the lasing process, so cooling the dye laser is simple. In dye lasers active medium is replaced continuously.
(iv) The output radiation of a dye laser have very narrow line width and very short pulses.

Disadvantages

(i) The maintainance of dye laser is complicated due to use of liquid as the active medium by most dye lasers.
(ii) The excitation in these lasers is done by another laser, which complicates the system.
(iii) Dye life time is short and further the dye quality degrades with time, and need to be changed.
(iv) Continuing operating expenses.
(v) Volatile solvents.
(vi) Potentially toxic (poisonous) chemicals.
(vii) Hazardous wast disposal.

Applications

(i) The main use of CW dye laser is to reduce the bandwidth, i.e., line width of e.m. radiation.

One can pump a dye laser with another laser, and the output line width from the dye laser will be upto 1% of the pump laser, while maintaining about 70% of the original energy. The result is a *tunable* laser with a very narrow line width. Line width of 100 Hz at a wavelength of 10^{15} Hz, i.e., a ratio of $1:10^{13}$ can be achieved.

(ii) One can easily get very short pulses out of dye lasers (~ pulses of 10^{-14} sec).
(iii) In medicine dye lasers are used for destroying tumors which have selective wavelength dependent absorption, photo-dynamic therapy (PDT) and for destroying kidney stones by shock waves created by the short pulses.

15.5 CHEMICAL LASERS

In chemical lasers pump energy comes from a chemical reaction between two atoms. In these lasers, pollution inversion results from a chemical reaction. The high energy density that can be released in exothermic chemical reactions has led to the interest in converting this energy to coherent optical energy. If one could use the energy of such chemical reactions to obtain a population inversion directly, one could built compact and powerful lasers. Such lasers may also prove to be a powerful tool in the study of kinetic of chemical reactions.

Most chemical lasers are based on hydrogen halides. The most common is hydrogen fluoride (HF). The emitted radiation is in IR, with a few lines in the spectrum range: 2.6–3.0 mm.

In 1969, Cool and Stephens developed a purely chemical laser based on following chemical reactions:

$$F_2 + NO_2 \rightarrow NOF + F$$
$$F + D_2 \rightarrow DF + D$$
$$D + F_2 \rightarrow DF + F$$
$$(DF)^* + CO_2 \rightarrow DF + CO_2$$

In the first chemical reaction F_2 is dissociated to give free fluoride atoms which on mixing with D_2 yield vibrationally excited DF. The excited DF transfers its energy to CO_2 and laser action results at wavelength 10.6 microns. The power output from this laser is about 500–600 W.

Schematic drawing of a chemical laser structure is shown in Fig. 15.13.

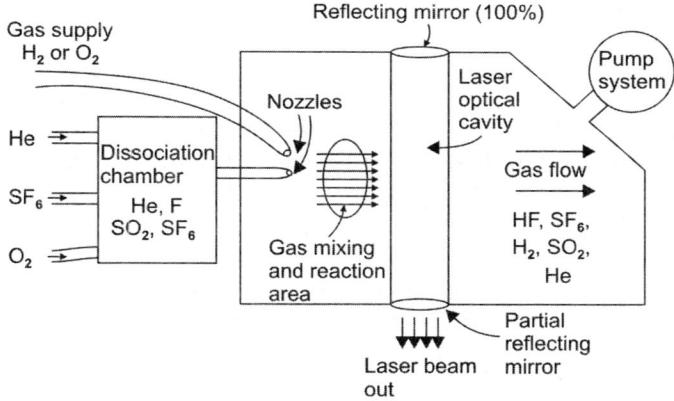

Fig. 15.13. Simplified structure of the chemical laser

The gases are injected into the laser through pipes with pinholes at their ends.

The design of the pinholes is critical to avoid thermodynamic equilibrium of the gas. The gas flows rapidly out of the pinholes and creates a turbulent flow. This results in excited hydrogen-halide molecule. The excited gas enters the laser optical cavity at right angle of the laser optical axis.

Advantages

(i) The source of energy is conveniently stored in chemical lasers (gas balloons).
(ii) Very high output power.

The atmosphere is more transparent to the emitted spectrum out of DF lasers than for HF lasers, so the DF laser is more developed, although its efficiency is lower, and the price of the deuterium isotope is higher.

Disadvantages

(i) Fluorine is a very reactive gas.
(ii) Hydrogen gas can explore easily.

In a commercial chemical laser, high voltage of about 8,000 V is applied to the electrodes of the laser tube. Some lasers use ultraviolet (UV) radiation prior to the electric discharge to preionize the gas and increase the efficiency of the chemical reaction. The chemical reaction between free fluorine and hydrogen releases a large amount of heat while creating the molecule HF, which is in an excited state.

The efficiency of the electrical input versus the laser output shows that one can get more than 100%, because of the chemical energy released by the reaction between the free fluorine and hydrogen. In chemical lasers, the electrical efficiency is less than 1%, while the chemical efficiency is about 20%.

Applications

Most of the applications of chemical lasers are military applications. *Mild Infra-Red Advanced Chemical Laser* (MIRACL) is the well known chemical laser. It is designed to destroy enemy missiles in the air. It was the first megawatt; continuous wave (CW), chemical laser built and was operated first in 1980. This laser can emit a continuous power of up to 2 megawatts, for a short time up to a maximum of 70 seconds. The clear aperture of the special telescope used to direct this laser is 1.5 meters, with computer automatic tracking of the target.

15.6 COLOR CENTER LASER

When crystals of alkali halides are exposed to high energy radiation, e.g., x-rays or electrons, *point defects* are created within the crystal. These point defects add more energy levels to the atoms in the crystal similar to impurity energy levels in semiconductors. These extra energy levels can cause optical absorption at specific wavelengths, thus adding color to the transparent alkali halides. After these colors gave the name *color center lasers* to these lasers.

There is a defect called F-center in crystals. This defect in the crystal causes local region with extra positive charge. This region can be regarded as a "nucleus" around which electrons can assemble similar to hydrogen atom. The electron is bound to a positive halogen vacancy. The excited electron energy states in the lattice are strongly coupled to lattice vibrations. Thus, all electronic states are broadband, resulting in broad absorption and emission bands.

The energy levels of the color center are wide and occupy bands, because of the interaction neighbouring atoms i.e. crystal vibrations. Absorption bands are not the same as emission bands (Fig. 15.14).

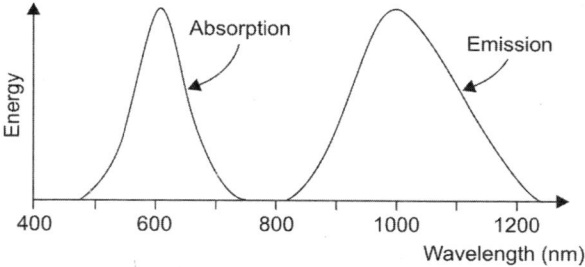

Fig. 15.14. Absorption and emission spectrum bands of F center in KCl crystal

Tunable color center lasers in alkali halide crystals can in principle cover the spectrum range from 0.6–4 μm. However, there are problems with shelf life of these lasers and with their stability during operation. Color center lasers operate at liquid nitrogen temperature (77 K). The main advantage of color center laser is its *single frequency purity*. In single mode continuous wave (CW) operation, linewidth below 4 kHz have been achieved.

Color center laser is a *solid state laser*. It is optically pumped usually by another laser which emits in the absorption spectrum of the color center. Since the energy levels are not discrete but band, it is a *tunable laser*, and the emitted wavelength can be controlled.

Applications

(i) *Basic research*: Spectroscopy of atoms and molecules (because of the narrow bandwidth of the emitted wavelength, and the broad range of tunability).

(ii) *Laser chemistry*: To initiate chemical reaction by selective excitation of specific levels of atoms and molecules.

15.7 SPECIAL LASERS

Some lasers are based on different physical principles than the "standard" lasers that were described so far. The nonstandard features can be the pump energy is of special form such as in free electron laser (FEL), the wavelength is so special such as x-ray laser or the active medium takes a special form as in fiber laser.

In recent years, possibility of laser without inversion (LWI) i.e. lasers in which lasing action can occur in a special case without achieving population inversion and atom laser which is related to Bose–Einstein condensation (BEC) at very low temperatures are being explored.

Free Electron Laser (FEL)

Gas lasers, or solid-state lasers, emit electromagnetic radiation at specific wavelengths, which correspond to specific transitions between energy levels in the active medium of the laser.

FEL is a device that can emit high power electromagnetic radiation at any wavelength. The emitted wavelength depends on the design of the laser, and not on the properties of the active medium. The efficiency FELs can be very high, upto 65%.

The electromagnetic radiation in a FEL is created as a result of the interaction of three factors:
   (i) Beam of high energy electrons, which move at a speed close to the speed of light.
   (ii) Beam of electromagnetic radiation moving in the same direction as the beam of electrons.
   (iii) Magnetic field which is arranged in periodic manner in space. Such magnetic field is formed by a series of magnets where each neighbour is opposite in direction. Such arrangement is called Wiggler or Undulator.

The accelerated electrons can transfer part of their kinetic energy to the oscillating electric field of electromagnetic radiation. This occurs when the direction of movement of the electron is in the direction of the oscillating electric field. The electromagnetic radiation is a transverse wave, in which the electric field is oscillating in a direction perpendicular to the direction of propagation of the beam. Thus, the electromagnetic beam must be linearly polarized (oscillate only in one plane) in the plane of oscillation of the electrons, which is perpendicular to the plane of periodic magnetic field.

The velocity of the electrons (v) along the periodic magnetic field (B) must fit the periodicity of the wiggler. The requirement can be understood by looking at the Lorenz force which describes the force acting on a moving charged particle:

$$F = q\,(v \times B)$$

The magnetic field (B) is perpendicular to the direction of movement of the electrons (v). Thus, it changes the direction of movement in a direction that is both, perpendicular to the magnetic field and to the direction of movement of the electron (vector product). Since the magnetic field is periodic, so is the change in the direction of movement of the electrons. The result is a curvilinear trajectory of the electrons which causes them to lose energy to the electromagnetic radiation (accelerated electric charges emit electromagnetic radiation). The emitted wavelength out of a free electron laser is determined by the period of the wiggler, and the energy of the electrons.

Most of the applications of free electron laser are military applications. Today, the main applications are for medical purposes, because of the tunability of the wavelength, and the adaptability to the wavelength specific interaction of the electromagnetic radiation with the biological tissue. To operate a free electron laser, an electron beam is accelerated to very high velocities close to the speed of light (relativistic electrons). For this, very high current (~ thousands of amperes) and very high voltage (~ thousands of volts) are used.

The main disadvantages of the free electron laser are the *big dimensions* because of the electron accelerator, *high cost* and *hazards* of x-rays created by the accelerated electrons.

15.8 X-RAY LASER

Theoretically, since the physical nature of electromagnetic waves is the same with no dependence on their wavelength, it is logical to think that it is possible to create laser at any wavelength. In practice, there is a problem creating a laser that operates in the short wavelength of x-rays. The lasing process depends on the properties of the medium in which lasing occurs. The requirements from the active medium depend on the wavelengths which are needed to be created:

- In the visible spectrum, and in the near infrared (NIR) spectrum, the radiation is emitted as a result of electronic transitions between outer electrons energy levels of the atoms or molecules.
- To create radiation (photons) in the x-rays spectrum a much higher energy is needed. Such energy can come from transitions from the outer energy levels into inner energy levels. Exciting these energy levels requires much more energy for pumping, and the lifetime of the excited states is very short (of the order of pico-seconds).

To create a laser in the x-rays spectrum region requires a large amount of pump energy, in a very short time.

Lasers operating in the x-rays spectrum range can be used as a sophisticated weapon system. One of the most promising energy sources for exciting an x-ray laser is nuclear energy.

One application is to use a nuclear explosion to pump a large number of x-ray lasers. A simplified sketch of such a system is shown in Fig. 15.15.

The large amount of energy created in the nuclear explosion evaporates the active medium and creates plasma (state of matter in which the electrons and the ions are separated from each other). Before the plasma has time to epand, some free electrons are captured in the inner shells, because they are attracted to the charge nuclei. Their binding energy is released in the form of x-rays. Since at the plasma state, there are almost no electrons occupying the inner levels, the situation is of population inversion. These transitions which emit

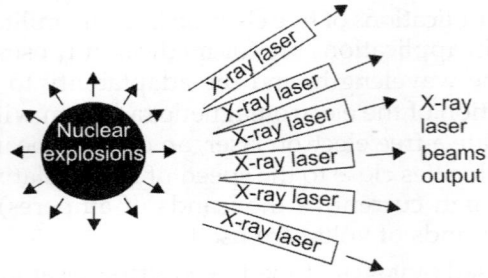

Fig. 15.15. Nuclear explosions to pump x-ray lasers

x-rays have a very high gain. Thus, even one pass through the medium is enough to create a lot of x-ray laser energy, and no mirrors are required. The beam shape out of x-ray laser depends on the geometry of the plasma and excitation.

Scientists are trying to develop γ-ray laser (*Graser*). A graser may be possible if the ways can be found to achieve *Mossbauer transitions* and population inversion simultaneously in nuclear isomers. These lasers are of military importance and can find several applications in diverse fields.

15.9 Q-SWITCHING

This is a means of generating short, high peak-power pulses with relatively low-power pump sources. In a Q-switched laser, the loss of the cavity is maintained at a high level unit a large population inversion is achieved. At such time, the loss is rapidly decreased so that the inversion is well above its new threshold value, resulting in a short, high power pulse. Q-switching relies on the fact that the lifetime of the population inversion is much longer than the output pulse width. The gain medium is, therefore, able to store energy, which can be quickly released in the form of optical pulse. The cavity loss is used to control the performance of the laser.

Lasers have been divided into three groups in accordance with their mode of oscillation:
  (i) *Continuous lasers*: These lasers emit continuous high beam with constant power. Obviously, such devices require continuous steady-state pumping of the active medium.
  (ii) *Pulsed lasers with free oscillations*: The emission in these lasers takes the form of periodic light pulses. These devices require pulse operation of the pumping system. The system achieves population inversion of the lasing levels periodically and for short periods only.
  (iii) *Pulsed lasers with controlled losses*: In these lasers, the concentration of energy reaches a maximum, and hence, they give rise to giant

pulses of short duration. This peak power is of the order of 10^8 W or even more.

The giant-pulse mode of oscillation is realized by controlling the losses of inside the optical cavity. This controlled is achieved through Q-switch. Fig. 15.16 shows some examples of Q-switching.

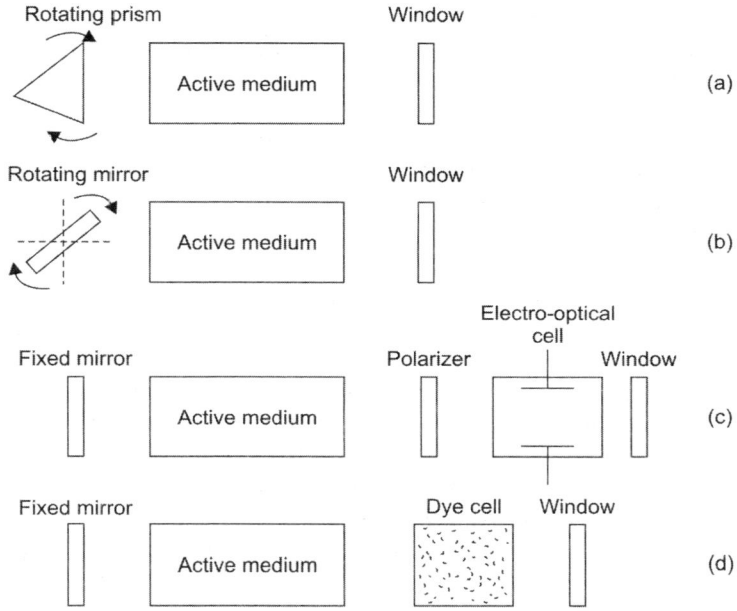

Fig. 15.16. Commonly used Q-switches: (a) rotating prism device, (b) rotating mirror device, (c) electro-optical switch based on the Kerr effect, and (d) dye cell

Conventional Q-switches are located within the laser cavity and control the amount of cavity loss. The first devices used to Q-switch a laser were mechanical switches, such as rotating mirrors. Their relatively slow speed has led to their replacement by acousto-optic and electro-optic Q-switches, except in cases where optical damage limits the use of the alternative technologies.

In Q-switched mode, i.e. giant pulse mode, the active medium is excited without feedback by blocking the reflection from one of the end mirrors of the optical cavity. The end mirror is then allowed suddenly to reflect, employing either a mechanical or an electro-optical switch. The applied feedback from the mirror causes a rapid population inversion of lasing levels, which results in a quite high peak power output pulse. The duration of the light pulse may be ~ 0.1 µs.

A mechanically driven device, e.g. a rotating deviation prism or a mirror, or a passive device such as an electro-optical cell or a dye cell may be used as an *optical switch*. In a *rotating prism optical switch*

(Fig. 15.16(a)), the totally reflecting end mirror of the cavity is replaced by a deviating prism. This prism is rotated by a synchronous motor at a high speed (~ 30,000 revolutions/min). When this optical switch is employed, the pulses of the flash lamp are electronically controlled and synchronized with the rotation of the optical switch.

In the rotating mirror optical switch, one of the end mirrors is rotated as shown in Fig. 15.16(b). In the electro-optical switch based on the Kerr effect, the light leaving the optical cavity passes through the polarizer and a Kerr cell to a partially reflecting mirror (i.e. window) [Fig. 15.16(c)]. When an appropriate voltage is applied to the capacitor plates of the Kerr cell, the material (e.g. nitrobenzene) inside it becomes birefringent. By appropriate variation of the voltage, the Kerr cell either blocks or transmits the polarized beam. The pumping system and terminal voltage of the Kerr cell should be controlled by an electronic unit.

Another *passive form* of the Q-switch uses a cell containing an organic dye [(Fig. 15.16(d)]. Initially, the light output of the laser is absorbed by the dye, preventing reflection from the window mirror, until the dye is bleached when a relatively high intensity has been reached. At this instance, reflection from the window mirror is possible, which results in a rapid increase in cavity gain. This causes rapid depopulation, and a very high peak power pulse may be obtained.

We may note that Q-switching are of two types:

(i) *Active Q-switching*: In active Q-switching, the Q-switch is an externally controlled variable attenuator. This may be a mechanical device, e.g., shutter, chopper wheel or spinning mirror placed within the cavity, or it may be modulator such as an acousto-optic device or an electro-optic device – a Pockels cell or Kerr cell. The reduction of losses and increase of Q is triggered by an external event, typically an electrical signal. Clearly, the pulse repetition rate can therefore, be controlled externally.

(ii) *Passive Q-switching*: In passive Q-switching, the Q-switch is a saturable absorber, a material whose transmission increases when the intensity of light exceeds some threshold. The material may be an ion-doped crystal, e.g. Cr:YAG, which is used for Q-switching of Nd:YAG lasers, a bleachable dye, or a passive semiconductor device. However, initially, the loss of the absorber is high, but still low enough to permit some lasing once a large amount of energy is stored in the gain medium. With the increase in laser power, it saturates the absorber, i.e. rapidly reduces the resonator loss, so that the power can increase even faster. Ideally, this brings the absorber into a state with low losses to permit efficient extraction of the stored energy by the laser pulse. After the pulse, the absorber recovers to its high-loss state prior to the gain medium is entirely replenished. The pulse repetition rate can only indirectly be controlled, e.g., by varying the pump power of laser and the amount of saturable absorber in the cavity. Direct control of the repetition rate can be achieved by making use of a pulsed pump source as well as passive Q-switching.

Q-switching laser: This is a laser to which the technique of active or passive Q-switching is applied, so that it emits energetic pulses. These lasers find applications in material processing, e.g., cutting, drilling, laser marking, pumping non-linear frequency conversion devices, range finding, and remote sensing.

Q-switched lasers can be pumped either continuously or with pulses, e.g., from discharge lamps. For continuous pumping, a gain medium having long upper state lifetime to reach a high enough stored energy rather than losing the energy as fluorescence is required. However, the saturation energy should not be low, as this could lead to excessive gain, so that the premature onset of lasing is more difficult to suppress. The latter problem can occur particularly for fiber lasers.

The most common type of Q-switching laser is the actively *Q-switched solid state bulk laser*. For wavelength in the 1 µm spectral region, a laser based on a neodymium-doped laser crystal, e.g., ND:YAG, $Nd:YVO_4$, or Nd:YLF is used. A typical Q-switched laser, e.g., ND:YAG laser, with a resonator length of 10 cm can produce light pulses of several tens of nanoseconds duration. Even when the average power is below 1 W, the peak power can be many kilowatts.

Q-switched lasers are often used in applications where high laser intensities in nanosecond pulses, e.g., dentistry, metal cutting or pulsed holography. These lasers can also be used for measurement purposes, e.g. distance measurement by measuring the time it takes for the pulse to get to some target and the reflected light to get back to sender.

Q-switched lasers also used to remove tattoos. These lasers are used to shatter tattoo pigment into particles that are cleared by the body's lymphatic system.

15.10 MODE LOCKING

Mode locking refers to situation where the phases of several longitudinal cavity modes are fixed (or locked) with respect to each other such that the coherent addition of the electric fields adds constructively for a short period of time. This allows the generation of periodic train of high peak power, ultrashort pulses.

Mode locking is a technique in optics by which a laser can be made to produce pulses of light of extremely short duration (ultrashort pulses), of the order of picoseconds (10^{-12} s) or femtoseconds (10^{-15} s) or attoseconds (10^{-18} s). The lasers are then called mode-locked lasers.

The basis of technique is to induce a fixed phase relationship between the modes of the laser's resonant cavity. The laser is then said to be phase-locked or mode-locked. Interference between these modes causes the laser light to be produced as a train of pulses. Depending on the properties of the laser, these pulses may be of extremely brief duration, as short as a few demtoseconds.

In general, mode locking will occur if the net gain for a mode-locked train of pulses is greater than the net gain for any other combination of cavity modes.

Active mode (AM) locking: In AM mode locking, the loss of some element in the laser cavity is modulated at the round trip frequency. This modulates the amplitude of the optical field and generates sidebands that are resonant with cavity modes. With the proper phase relationship between these modes, most of the light is incident on the modulator during its minimum loss. This combination of modes sees a lower loss than any other combination of modes and is, therefore, favoured. The same result occurs if the gain of the cavity is modulated. Gain modulation through modulation of the pump source is known as *synchronous* pumping. The minimum pulse width obtainable by active mode locking is often linked by the handwidth of the action modulation.

FM mode locking: In FM mode locking, the optical length of the laser cavity (physical length or refractive index) is modulated at the round-trip cavity frequency. As in the case of AM mode locking, the modulation generates sidebands that are resonant with cavity modes.

Passive mode locking: Passive mode locking can occur when a laser cavity contains a nonlinear optical element, such as a saturable absorber. In this case, the more intense the light incident on the saturable absorber, the less the total absorption. The total loss of the cavity is, therefore, minimized by putting all of the energy into the short pulses. This is, essentially, self induced AM mode locking.

Theory

We know that an ideal homogeneously broadened laser can oscillate at a single frequency. However, all practical lasers are inhomogeneously broadened, and hence, they may oscillate at a number of frequencies, which are separated by:

$$\omega_q - \omega_{q-1} = \frac{\pi c}{l} \equiv \omega \tag{15.1}$$

where c is the speed of light in free space and l is the length of the gain medium as well as the distance between the mirrors of the cavity. We have assumed $n = 1$, i.e., the refractive index of the gain medium is 1 (as in the case of He-Ne laser).

Now, the total optical electric field resulting from such multimode oscillation at a particular point may be expressed by:

$$e(t) = \sum_n E_n \exp\left[i\{(\omega_0 + n\omega)t + \phi_n\}\right] \tag{15.2}$$

where E_n is the amplifier of the nth mode, which is oscillating at an angular frequency of $(\omega_0 + n\omega)$; ω_0 is arbitrarily taken as the reference frequency. ϕ_n is the phase of the nth mode. One can easily prove that $e(t)$ is periodic in $T = 2\pi/\omega = 2l/c$, which is the round trip transit time inside the resonator. Thus, we have:

$$e(t+T) = \sum_n E_n \exp\left[i\left\{(\omega_0 + n\omega)\left(t + \frac{2\pi}{\omega}\right) + \phi_n\right\}\right]$$

$$= \sum_n E_n \exp[i\{(\omega_0 + n\omega)t + \phi_n\}] \exp\left[i\left\{2\pi\left(\frac{\omega_0}{\omega} + n\right)\right\}\right] \quad (15.3)$$

Since ω_0 is a reference frequency, one can take it to be $\omega_0 = m\pi c/l$; m is an integer and $\omega = \pi c/l$ [from Eq. (15.1), the ratio of ω_0/ω is an integer m. Therefore, we have:

$$\exp[i\{2\pi(m+n)\}] = 1$$

This reduces Eq. (15.3) to:

$$e(t+T) = \sum_n E_n \exp[i\{(\omega_0 + n\omega)t + \phi_n\}] = e(t) \quad (15.4)$$

In order that $e(t)$ maintains a periodic nature, the phases ϕ_n should be fixed. However, in many lasers, the phases ϕ_n vary randomly with time. This causes the laser output power to fluctuate randomly, thus reducing the possibility of its application in cases where *temporal coherence* is an important consideration.

There are two ways in which the laser can be made coherent. These are: (i) to allow the laser to oscillate only at a single frequency and (ii) to force the phases ϕ_n of the modes to maintain their relative values. The second method is called *mode-locking*, and this causes the oscillation intensity to consist of a periodic train with a period $T = 2l/c = 2\pi/\omega$.

In order to make things simple, let us lock each mode to a common origin of time and take each phase ϕ_n to be zero. Further, assume that there are N oscillating modes with equal amplitudes E_n, and that $E_n = 1$. Substituting these parameters in Eq. (15.2), we obtain:

$$e(t) = \sum_{-(N-1)/2}^{(N-1)/2} e^{i(\omega_0 + n\omega)t} \quad (15.5)$$

$$= e^{i\omega_0 t} \frac{\sin(N\omega t/2)}{\sin(\omega t/2)} \quad (15.6)$$

The average laser output power:

$$P(t) \propto e(t)e^*(t)$$

where $e^*(t)$ is the complex conjugate of $e(t)$:

or

$$P(t) \propto \frac{\sin(N\omega t/2)}{\sin(\omega t/2)} \quad (15.7)$$

Some evident features of $P(t)$ are as follows: (i) the peak power P_{peak} is equal to n times the average power P_{av}, where N is the number of modes locked together; (ii) the peak amplitude of the field is equal to N times the amplitude of a single mode; (iii) the individual pulse width τ_p, defined as the time from the peak to the first zero, can be estimated by the relation:

$$(P_{peak})\tau_p \approx (P_{av})T = (P_{av})\left(\frac{2\pi}{\omega}\right) = (P_{av})\left(\frac{2l}{c}\right)$$

or

$$(NP_{av})\tau_p = (P_{av})\frac{2l}{c}$$

or

$$\tau_p = \frac{2l}{cN} \tag{15.8}$$

But the number of oscillating modes N is approximately given by the ratio of the transition line shape width $\Delta\omega$ to the frequency spacing ω between the modes; i.e.:

$$N = \frac{\Delta\omega}{\omega} = \frac{\Delta\omega}{\pi c/l} \tag{15.9}$$

[where we have used Eq. (15.1).

From Eqs (15.8) and (15.9), one obtains:

$$\tau_p = \frac{2l}{c}\frac{(\pi c/l)}{\Delta\omega} = \frac{2\pi}{\Delta\omega} = \frac{1}{\Delta\nu} \tag{15.10}$$

Equation (15.10) shows that the width of the mode-locked pulse is inversely proportional to the gain line width. This time corresponds to a frequency exactly equal to the mode spacing of laser, $\Delta\nu = \dfrac{1}{\tau_p}$.

Applications

(i) *Photochemistry*: One can use the short pulses to probe the process of the reaction at a very high temporal resolution, allowing the detection of short lived intermediate molecules. This method is particularly useful in biochemistry, where it is used to analyze details of protein folding and function.
(ii) *Nuclear fusion*
(iii) *Nonlinear optics*, such as second harmonic generation, parametric down-conversion, optical parametric oscillators and generation of terahertz radiation.

15.11 APPLICATION OF LASERS

The applications of lasers are so vast in different fields due to two special features, which are not available in light from ordinary sources:

(i) *Narrow bandwidth*: Which is the same thing as high monochromaticity or high temporal coherence.

(ii) *Narrow angular spread*: Which is same thing as high directionality (hence, high intensity) or large spatial coherence.

In the past three decades, there has been a tremendous increase in the applications of lasers in various fields, e.g., medicine, surgery, communication, industry, defence, science and technology, environmental monitoring, etc.

Here, few examples of major laser based applications are briefly described.

(i) Welding and Cutting

(a) The highly collimated beam of laser can be further focussed to a microscopic dot of extremely high energy density for welding and cutting.

(b) The automobile industry makes extensive use of CO_2 lasers with powers upto several kW for computer-controlled welding on auto-assembly lines.

(c) CO_2 lasers find interesting application in the welding of stainless steel handles on copper cooking pots. A nearly impossible task for conventional welding because of the great difference in thermal conductivities between stainless steel and copper is done so quickly by the laser that the material conductivities are irrelevant.

(ii) Material Preparation and Processing

Laser beams are being used to melt materials at very high melting point. They are successfully used for producing alloys, intermetallics and other useful materials. Lasers are also used to change optical properties, e.g. refractive index, dielectric constant, transparency, etc. of materials without destroying them.

The use of lasers in industrial processing of materials is increasing rapidly. Material processing with high-power lasers include cutting, welding, drilling, marking, surface modification, prototyping/manufacturing, etc. These are, in fact, examples of the peaceful use of directed energy application. When an intense laser beam strikes a target, a part of it is reflected and the remainder is absorbed. The absorbed energy heats the surface. This heating can be very rapid, and its extent can be controlled for different applications. Some important industrial lasers and their potential applications are summarized in Table 15.2.

(iii) Lasers in Scientific Research and Investigations

Lasers have been used several times to repeat Michelson-Morley experiment with 50 times more accuracy than it was initially done in 1881. Raman effect in solids, liquid or gases can be recorded using laser source and infrared recorders without any delay. These studies made using laser have provided informations about the energy states of

Table 15.2. Some important industrial lasers and their applications

Type of laser	Applications
Solid-state lasers	
Nd^{3+}:YAG	• Light to heavy duty industrial drilling, cutting, welding, marking, etc.
Nd^{3+}:glass	
Ho^{3+}:YLF	• Industrial pollution monitoring, wireless initiation of thermal batteries, explosives, propellants, etc.
Er^{3+}:YLF	
Gas lasers	
CO_2	• Light to heavy duty industrial drilling, cutting, welding,
N_2	surface modification, etc.
Argon ion	• Industrial wire stripping
Semiconductor lasers	
Laser diodes	• Fiber-optic communication, compact disc drivers, laser printers, bar code scanners, optoelectronic devices
Other lasers	
Dye lasers	• R&D, medical diagnostics
Excimer laser (a combination of two gases: rare gas + halogen)	• Optical lithography and stereolithography, precision micro-matching, polishing, etc.
Chemical lasers	• Light to heavy duty jobs for material processing
Free-electron lasers	• Mainly for directed energy applications
X-ray lasers	

molecules not hitherto studies. Laser beam screening is used to get maps of distant object, moon and detail profile of celestial objects. Lasers are used successfully to measure distances and velocities with extremely high accuracy. Lasers are also used to detect internal stress and defects in materials employing non-destructive techniques, to detect underground shocks, to measure drift of continents and so on.

Lasers are widely used in the study of atoms and molecules. There is a separate branch of spectroscopy called *laser spectroscopy*. A laser is an ideal source for selective excitations of atoms and molecules for studying absorption and emission properties due to their high radiant power and narrow spectral width. A *laser microscope* is a very powerful tool for studying materials available in very small quantities, i.e., of the order of micrograms or even less.

One can use a laser microprobe in two ways for spectral analysis:

(i) In the first method, the energy of the laser beam is used for simultaneously evaporating the sample and also exciting the vapour simultaneously and finally causing emission.

(ii) In the second method, a fairly powerful Q-switched mode of operation is desired (Fig. 15.17). A high-speed spectrograph or spectrometer is employed to record the emission spectrum.

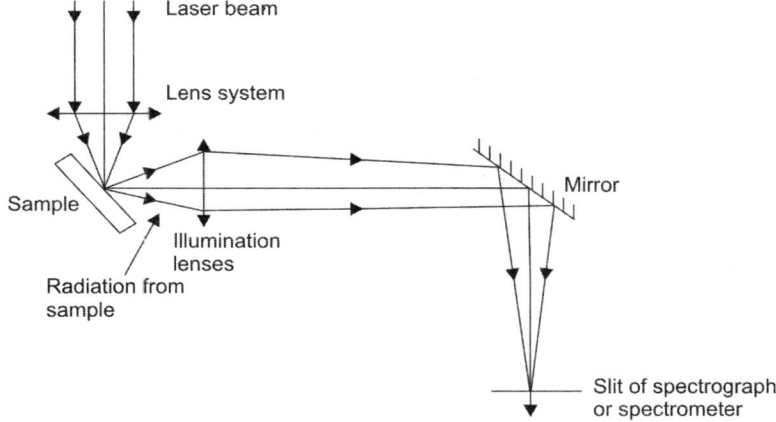

Fig. 15.17. Schematic of optics involved in laser-induced emission analysis

A spark gap is formed by two pointed carbon electrodes for the application of second method. As shown in Fig. 15.18, the center of the spark gap coincides with the optic axis of the laser beam. The sample is kept about 1–2 km below the center of the spark gap. 1–5 kV dc voltage is applied to the electrodes. As soon as the laser pulse irradiates the sample, it produces a strongly ionized vapour.

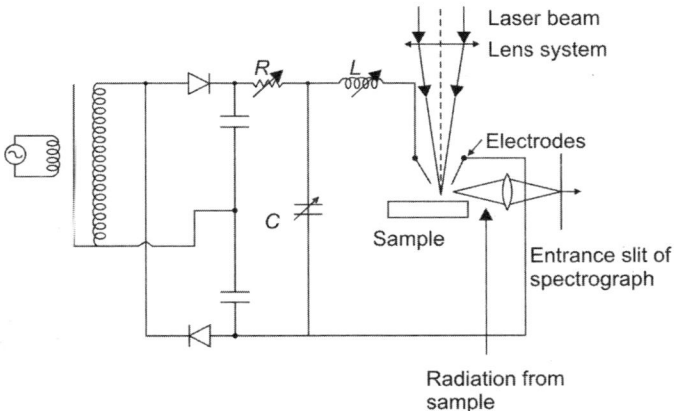

Fig. 15.18. Schematic of laser microanalyser with cross-excitation

This vapour renders the spark gap conducting and the capacitor C discharges, which gives rise to the spark. The excitation of the vapour is caused by the spark. The value of the damping resistance R is maintained at such a value that no spark occurs following the discharge of the capacitor C. The characteristics of the spectrum can be modified with the help of the inductor L within the discharge circuit. The amount of sample necessary for this technique is of the order of 0.1 µg.

These two methods are generally employed for microanalysis or local analysis. These methods can also be used for macroanalysis, i.e. one can study the composition of a large homogeneous sample. There are two ways: (i) one way is to focus an energetic laser beam onto a large area, i.e., with a diameter of several tenths of a millimeter, and (ii) the second way is to scan the specimen with a sequence of laser pulses. In the latter case, one has to employ two *servometers* for moving the sample in two mutually perpendicular directions. These servometers are controlled electronically.

Number of other systems have been developed for measurements of different kinds in various fields of scientific investigation.

Optical tweezers (combination of laser and microscope) are used to trap and manipulate very small particles ranging in size from 25 nm to 10 µm using a laser. When combined with another laser, usually a UV laser, particles can be cut, moved and fused with other particles. This technique of very minute and accurate manipulation of particles has enormous applications in biological sciences and already laser tweezers are being commonly used in this area.

Lasers are useful for excitation of molecules to specific levels, and examination of the emitted radiation. They can be used for measurements of the relaxation time of specific excited levels of molecules.

Lasers are used to disrupt chemical bonds in molecules in specific region.

(iv) Lasers in Medical Uses

Lasers have become an indispensible tool in medical diagnosis and surgery. Lasers are used for photocoagulation of the retina to halt the retinal hamorrhaging and for taking of retinal tears. High power lasers are used after cataract surgery if the supportive membrane surrounding the implanted lens gets milky. Photodisruption of the membrane often can cause it to draw back like shade, almost instantly restoring vision. A focussed laser can act as extremely sharp scalpel for extremely delicate surgery, counterizing as it cuts (cauterizing refers to long stainding medical practices of using a hot instrument or a high frequency electrical probe to single the tissue around an incision, sealing off tiny blood vessel to stop bleeding). The cauterizing action is particularly important for surgical procedures in blood-rich tissues, e.g., lever.

For the purpose of diagnosis the laser-induced fluorescence method is usually preferred, e.g., it is possible to distinguish between normal and diseases tissues by correlating their spectral features with other pathological data. Doctors distinguish between normal, benign and malignant human breast tissues with the help of this technique.

The technique has also been used by the doctors for the detection of lung cancer. A chemical called *haematoporphyrin derivative* (or Hpd) is introduced into the patient's body. The chemical concentrates in the cancerous cells. The suspected areas are illuminated by light of 0.40 µm (blue laser), which is the absorption region of Hpd. The cancerous cells containing this chemical fluoresce in the range 0.60–0.70 µm and the process reveal their presence.

One can also analyse cells and their contents for the diagnosis of hereditary diseases using laser microfluorimetry. Cellular structures as small as 0.3 µm and cellular processes as fast as 0.2 ns can easily be recorded by using technique.

The use of lasers in medical surgery is increasing tremendously. The principal uses of lasers are to precisely cut, cauterize, and damage or destroy affected areas/tissues. The use of lasers provides the following major advantages over conventional surgery:

  (i) Surgery can be performed under a microscope so that the affected areas can be located and treated accurately.
 (ii) Sterilization is not required because no mechanical instruments are used.
(iii) There is less danger of haemorrhage and also less post-operative pain because photocoagulation is done by the laser.
(iv) Fiber-optic endoscopes can be used to locate, diagnose, and treat inaccessible areas of the body.
 (v) Lasers can be used to weld blood vessels and, hence, reduce the number of sutures.
(vi) Surgeries can be controlled by computers.

Table 15.3 summarizes some typical applications of lasers in surgery.

Table 15.3. Lasers in surgery

Type of laser	Applications in surgery
Solid-state lasers	
Nd^{3+}:YAG laser	• Eye surgery, photocoagulation, spinal surgery, brain surgery, plastic surgery
Ruby laser	• Hair removal (cosmetics)
Gas lasers	
He-Ne laser	• Laser doppler velocity meter
Argon ion laser	• Eye surgery (removal of cataract), photocoagulation, angioplasty, brain surgery
CO_2 laser	• Removal of cancerous growth, lesions, dermatology, spinal surgery, skin resurfacing (cosmetics)
Other lasers	
Metal-vapour lasers	• Fluorimetry (Hpd)
Excimer lasers	• Eye surgery, angioplasty
Dye lasers	• Removal of benign pigmented marks
Semiconductor lasers	
Diode lasers	• Cosmetics: hair removal and teeth whitening

(v) Laser Printers

Lasers are used in a major type of computer printers. Now-a-days, the laser printer has become a dominant mode of printing in offices, laser printers employ a semiconductor laser and xerography principle. The laser is focussed and scanned across a photoactive selenium coated drum where it produces a charge pattern which mirrors the material to be printed. This drum, then holds the particles of the toner to transfer to paper which is rolled over the drum in the presence of heat. The images are carried from the drum to paper by means of electrostatic photocopying. The laser for this application is the AlGaAs laser at 760 nm wavelength, just into the infrared.

A small plastic disc, that is compact disc (CD) can store large amount of data (MB to GB) in digital form. A laser is used in recording and reading this data, e.g., audio and video, CD, DVD etc.

(vi) Laser in Communications

Optical fiber cables are a major mode of communication partly because multiple signals can be sent with high quality and low loss by light propagating along the optical fibers. The light signal can be modulated with information to be sent by either LED or lasers. The lasers have significant advantages because the light frequency is so high and the intensity can be rapidly altered to encode very complex signals. Lasers are nearly monochromatic and this allows the pulse shape to be maintained over long distances. Obviously, if a better pulse shape can be maintained, then the communication can be sent at higher rates without overlap of the pulses.

(vii) Laser in Electronics Industry

The capability of coherent-infrared, visible or ultraviolet radiation to focus to beam diameters below 100 μm is used by electronics industry for drilling, trimming and marking of materials, e.g. Xe laser which operates at 537 nm is used in resistor trimming, a CO_2 laser at 10.6 μm is used for marking. Other applications in electronic industry are microlithography, deposition and etching of thin films.

(viii) Holography

In production and reading of holograms, laser is used. The holographic technique is used in testing of stresses and structural defects in materials. The holograms are produced with and without stress and then compared. Using holographic principle, in medical research three dimensional view of living cells is obtained.

(ix) Remote Sensing Using Light Detection and Ranging

One can measure the distance of a remote object by measuring time t

taken by a short laser pulse to reach the object and get reflected back to the observer. Then the distance d of the object will be:

$$d = \frac{ct}{2} \quad (15.11)$$

where c is the velocity of light in vacuum (3×10^8 m·s^{-1}). We can see that this technique is similar to the radar technique using radio-frequency waves, and therefore, it is called an *optical radar* or *laser radar* or *LIDAR* (light detection and ranging). Lidar is used to determine the distance between points on the earth and the moon to an accuracy of a few inches. Fig. 15.19(a) shows a lidar system. The major components of lidar system are: (i) a pulsed laser, (ii) a photodetector, with a timing circuit, and (iii) a collecting and focusing telescope [Fig. 15.19(b)] to collect the reflected light.

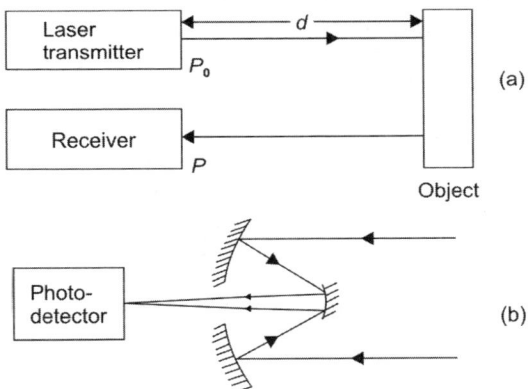

Fig. 15.19. (a) Schematic of LIDAR system (b) collecting and focusing telescope

Power received (P) at the telescope may be given by Beer's law:

$$P = P_0 \exp(-\alpha \, 2d) \quad (15.12)$$

where α is the coefficient of attenuation per unit length and P_0 is the power transmitted. Attenuation of the transmitted signal may be caused due to absorption, and scattering of light by the medium between the object and laser.

Lidar has several applications, all of which derive from high directionality and high power of laser radiation. These applications include monitoring of volcanic aerosols that may affect global climate, detection and characterization of fog layers, monitoring of atmospheric pollution, laser based weaponry.

Lidar is also used in *airborne laser bathymeters*, used to measure the depth of water or the location of any submerged object, e.g., a typical airborne laser bathymeter may be mounted on a fixed-wing aircraft or a helicopter. Suppose, it employs a 3-MW peak power Q-switched

Nd^{3+}:YAG laser with 4-ns pulses at the rate of 200 Hz. Let us assume that it is emitting a fundamental wavelength $\lambda_1 = 1.064$ μm (IR) and also its second harmonic at $\lambda_2 = 0.532$ μm (green). The IR pulse at λ_1 is reflected back from the sea surface, while the green pulse at λ_2 penetrates the sea water and is received by the airborne sensor after getting reflected by the bottom of the sea or any submerged object. The difference Δt in the arrival times of the two pulses will give the depth of the sea bottom or that of the submerged object. Now, if we assume that the refractive index of sea water is approximately 1.33 and that the submerged object is located at a depth of 50 m from the surface of the sea, then the time difference between the two pulses will be:

$$\Delta t = \frac{n \times 2 \times 50 \text{ m}}{\text{cm·s}^{-1}} = \frac{1.33 \times 2 \times 50 \text{ m}}{3 \times 10^8 \text{ m·s}^{-1}} = 0.44 \times 10^{-6} \text{ s} = 0.44 \text{ μs}$$

Obviously, this is indeed a long duration as compared to the duration of the pulse, which is only 4 ns.

(x) Military Applications

Lasers find several applications in defence, e.g. laser range finder, target destination, laser weapons, laser blinding for man and sensitive equipment, etc. Laser can be used in battlefield for target range finding, target destination and tracking, and guidance. Almost all battlefield tanks and armoured combat vehicles utilize the services of a laser range finder. Normally, these devices use the Q-switched Nd^{3+}:YAG laser. Laser systems which can serve as EOCM devices have been developed. The main aim of these devices is to temporarily or even permanently disable electro-optic devices/sensors used by the enemy in the battlefield.

Laser weapons or direct energy weapons (DEWs) is also an area of interest in defence. These weapons are nothing but lasers which have high enough power and appropriate control mechanisms to disable the guidance system and warheads, or trigger an explosion of the fuel or warhead, or cause temporary or permanent damage to the target. Lasers are also used in *electro-optic counter measures* (EOCMs) in the battlefield. The aim of this system is to make ineffective similar systems used by the enemy on ground, air, or at sea, ensuring at the same time reliable and uninterrupted functioning of the system.

Example 1

Calculate the relative population in the laser transition levels in a Ruby laser in thermal equilibrium ($T = 300$ K). The wavelength $\lambda = 6943$ Å.

Solution

$$\frac{N_1}{N_2} = e^{\frac{E_2 - E_1}{k_B T}} = e^{\frac{h\nu}{k_B T}}$$

Now, $\upsilon\lambda = c \Rightarrow \upsilon \dfrac{c}{\lambda} = \dfrac{3 \times 10^8}{6943 \times 10^{-10}} = 4.32 \times 10^{14}$ Hz

$$\dfrac{N_1}{N_2} = e^{\left[\dfrac{6.63 \times 10^{-34} \times 4.32 \times 10^{14}}{1.38 \times 10^{-23} \times 300}\right]} = e^{69.14} = 1.064 \times 10^{30}$$

Thus, the ratio of population is 1.064×10^{30}.

Example 2

Evaluate the wavelength of radiation given out by a laser with $E_2 - E_1 = 3$ eV.

Solution

$$E = h\nu$$
$$E = E_2 - E_1 = h\nu$$
$$3\text{eV} = 6.63 \times 10^{-34} \lambda\text{s} \times 3 \times 10^8 \text{ m/s}$$
$$\lambda = \dfrac{6.63 \times 10^{-34} \times 3 \times 10^8}{3 \times 1.6 \times 10^{-19}} = 4.41 \times 10^{-7} \text{ m}$$

Thus, the wavelength of radiation is 4.41×10^{-7} m.

Example 3

A He-Ne laser emits light of a wavelength 632.8 nm and has an output power of 2.3 mW. How many photons are emitted each minute by this laser when operated?

Solution

$$\text{Power} = n\, h\nu$$
$$2.3 \times 10^{-3} = n \dfrac{hc}{\lambda}$$
$$n = \dfrac{2.3 \times 10^{-3} \times 632.8 \times 10^{-9}}{6.63 \times 10^{-34} \times 3 \times 10^8} = 73.17 \times 10^{14} \text{ photons/s}$$
$$= 4.38 \times 10^{17} \text{ photons/min}$$

Thus, there are 4.39×10^{17} photons emitted per min.

Example 4

For a pulsed ruby laser, calculate the energy in the pulse required to obtain threshold inversion. Assume that the density of Cr^{3+} atoms per cm^3 is 1.9×10^{-19}. The flash lamp for pumping produces a pulse of light of duration $\tau_f = 0.5$ ms. The lifetime of spontaneous emission $\tau_{sp} = 3$ ms; the average absorption coefficient over the blue and green bands is $\overline{\alpha(\nu)} = 2$ cm^{-1}; the average pump frequency absorbed by ruby,

$\bar{v} \approx 5.45 \times 10^{14}$ Hz. (b) Calculate the threshold electrical energy input to the flash lamp per cm² of the ruby crystal surface. Assume that the efficiency of conversion from electrical energy to optical energy is 50%; the fraction of total optical output that is usefully absorbed by the active medium (ruby) is 15%; the fraction of the lamp light focused by the optical system onto the laser rod is 20%. [B.Tech.]

Solution

(a) For the sake of simplicity, let us assume the shape of the pulse to be rectangular. If we assume that a pulse of duration τ_f produces an optical flux of $W(v)$ watts per unit area per unit frequency v at the surface of the ruby crystal, then the amount of energy absorbed by the crystal per unit volume is given by:

$$\tau_f \int_0^\infty W(v)\,\alpha(v)\,dv$$

where $\alpha(v)$ is the absorption coefficient of the crystal. Now, we assume that the absorption quantum efficiency (that is, the probability that of the absorption of a pump photon at a frequency v results in transferring one atom into the upper laser level, say, $\bar{E}$) is $\eta(v)$, then the number of atoms pumped into level 2 (i.e., level $\bar{E}$) per unit volume will be given by:

$$N_2 = \tau_f \int_0^\infty \frac{W(v)\,\alpha(v)\,\eta(v)}{hv}\,dv \qquad \text{(i)}$$

As the lifetime of spontaneous emission, $\tau_{sp} = 3$ ms, of level 2 is much longer than the flash duration of 0.5 ms, the spontaneous decay out of level 2 during the time of flash may be neglected. Therefore, N_2 may be taken as the population of level 2 (i.e. $\bar{E}$) after the flash. Now, if the useful absorption is limited to a narrow spectral region Δv, one may approximate Eq. (i) by:

$$N_2 = \frac{\tau_f\,\overline{W(v)}\,\overline{\alpha(v)}\,\overline{\eta(v)}\,\overline{\Delta v}}{hv} \qquad \text{(ii)}$$

where the bars represent average values over the useful absorption region of width $\overline{\Delta v}$.

As ruby is a *three-level laser*, the populations N_1 and N_2 at levels 1 and 2, respectively, should satisfy the condition:

$$N_1 + N_2 = N_0 \qquad \text{(iii)}$$

where N_0 is the density of the active atoms (Cr^{3+}). Of course, here we are assuming that the population N_3 at level 3 is negligible because of the very fast transition rate out of level 3. If the pumping level is high enough so that the population at level 2 becomes:

$$N_2 = N_1 = N_0/2 \qquad \text{(iv)}$$

the optical gain will be zero. This means that roughly half of the chromium atoms must be raised to level $\bar{E}$ to achieve transparency (zero gain) on the R_1 line. Further pumping will yield gain and oscillation if appropriate feedback is supplied.

Using the values of various parameters, one obtains:
$$N_2 \approx \frac{N_0}{2} = \frac{1.9}{2} \times 10^{19} = 9.5 \times 10^{18} \text{ cm}^{-3}$$

Taking $\bar{v} = 5.45 \times 10^{14}$ Hz and $\bar{\eta}(v) \approx 1$, we obtain [from Eq. (ii)], the following pump energy that must fall on each cm² of the ruby crystal surface in order to achieve threshold inversion:

$$\overline{W}(v)\overline{\Delta v \tau}_f = \frac{N_2 h \bar{v}}{\bar{\alpha}(v)\bar{\eta}(v)}$$

$$= \frac{9.5 \times 10^{18} \times 6.6 \times 10^{-34} \times 5.45 \times 10^{14}}{2 \times 1} = 1.7 \text{ J cm}^{-2}$$

(b) From the value given, the threshold energy input to the flash lamp per cm² of the ruby surface will be:

$$\frac{1.7 \text{ J cm}^{-2}}{0.15 \times 0.20 \times 0.5} \approx 113 \text{ J cm}^{-2}$$

We may note that these are rough estimates and give the order of magnitude of the powers involved in laser pumping.

Example 5
Calculate the wavelength of emission from GaAs material whose bandgap $E_g = 1.44$ eV.

Solution
$$\lambda = \frac{ch}{E_g} = \frac{3 \times 10^8 \times 6.63 \times 10^{-34}}{1.44 \times 1.6 \times 10^{-19}} = 8628 \times 10^{-10} \text{ m} = 8628 \text{ Å}$$

Thus, the wavelength of emission is 8628 Å.

Example 6
When a beam of light traversing from one medium to another is split into two beams at the interface as shown in Fig. 15.20. A part of the beam is reflected back into the same medium at an angle of incidence θ_i, and the remaining part is refracted at an angle θ_r. According to Snell's law:

$$n_1 \sin \theta_i = n_2 \sin \theta_r$$

where n_1 and n_2 are the refractive indices of the two media. When the incident beam is unpolarized, the reflected and refracted beams become partially polarized. The refracted beam tends to be polarized in the plane of incidence while the reflected beam is polarized in a plane normal to it. However, if the angle of incidence θ_i is such that:

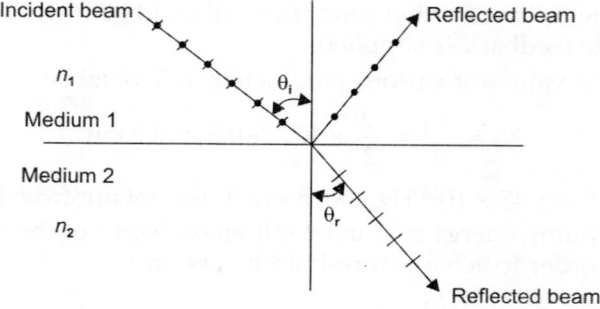

Fig. 15.20

$$\tan \theta_i = \tan \theta_B = \frac{n_2}{n_1}$$

then the refracted and reflected beams become perfectly polarized. In this special case, the angle of incidence $\theta_i = \theta_B$ is referred to as the *Brewster angle*, and the reflected and refracted beams are at right angles to each other.

Example 7

Whether it is possible to achieve the condition of polarization selectivity of laser emission employing the above phenomenon? Explain how?
[B.Tech.]

Solution

If the end face of the laser rod (in the case of a solid-state laser) or the gas discharge tube (in the case of a gas laser) is tilted such that the normal to the end face and the optic axis OO' are at the Brewster angle corresponding to the refractive index of the material of the end face window, the emitted laser radiation will be plane-polarized. The relevant configuration is as shown in Fig. 15.21.

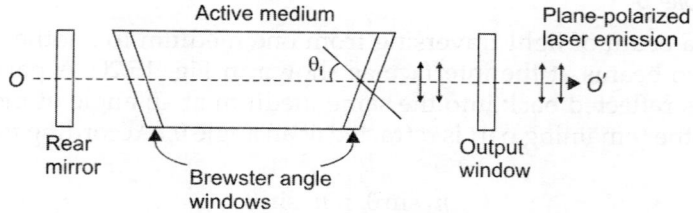

Fig. 15.21

Example 8

The InGaAsP diode laser has peak emission at wavelength $\lambda = 1.55\ \mu m$. Calculate its energy gap in eV.

Solution

We know that
$$\lambda = \frac{1.24}{E_g} \, \mu m$$

where E_g is expressed in eV.

$$E_g = \frac{1.24}{\lambda(\mu m)} = \frac{1.24}{1.55} = 0.8 \, eV$$

Thus, the energy gap is 0.8 eV.

Example 9

A laser beam of wavelength 740 nm has coherence time 4×10^{-5} sec. Calculate the order of magnitude of its coherence length, spectral half-width and purity factor.

Solution

Here, coherence length = $L = \tau_c = 4 \times 10^{-5} \times 3 \times 10^8 = 12 \times 10^3$ m.

Spectral half width $\Delta\lambda = \dfrac{\lambda^2}{L} = \dfrac{(740 \times 10^{-9})^2}{12 \times 10} = 4.5 \times 10^{-17}$ m

Purity factor $Q = \dfrac{\lambda}{\Delta\lambda} = \dfrac{740 \times 10^{-9}}{4.5 \times 10^{-17}} = 1.6 \times 10^{10}$

Thus, the coherence length is 12×10^3 m, spectral half width is 4.5×10^{-17} m and purity factor is 1.6×10^{10}.

SUGGESTED READINGS

1. Senior JM, 'Optical Fiber Communications: Principles and Practice', Pearson (3rd ed., 2009).
2. Quimby RS, 'Photonics and Lasers: An Introduction', Wiley (2006).
3. Gupta MC and Ballato J (Eds.), "The Handbook of Photonics", CRC Press (2007).
4. Davis, CC, 'Lasers and Electro-Optics: Fundamentals and Engineering, Cambridge University Press (2002).
5. Liu JM, 'Photonic Devices', Cambridge University Press (2005).
6. Chuang, JM, 'Physics of Photonic Devices', Wiley (2005).
7. Siegman AJ, 'Lasers', University Science Books (1986).
8. Silfvast WT, 'Laser Fundamentals', 2nd ed., Cambridge (2004).
9. Weber MJ (Ed.), 'Handbook of Laser Science and Technology', CRC (1995).
10. Weber MJ (Ed.), 'Handbook of Lasers', CRC (2003).

GLIMPSES

- Broadly there are six different types of lasers; solid state laser, gas laser, dye laser, semiconductor laser, UV and X-ray laser and free electron laser. On the basis of the state of active medium, lasers may be placed under four broad categories, namely, (i) solid state lasers, (ii) gas lasers, (iii) dye lasers, and (iv) semiconductor lasers.
- In *solid-state lasers* active medium is an insulating dielectric solid. These solids can be crystalline or amorphous. These lasers are used to generate a wide range of wavelengths from the vacuum UV to mid IR. They are capable of high peak powers (~ 10^{14} W) because of long life metastable states which allow higher energy storage compared to other active media. The common examples of solid state lasers are ruby (chromium doped Al_2O_3), titanium doped Al_2O_3 (Ti:Al_2O_3), neodymium-doped aluminium garnet (Nd:YAG) and neodymium-doped glasses (Nd:glass), etc.

  In *gas lasers*, gas or gas mixture is capable of withstanding a large amount of power. The pointing stability and high optical quality beams are their greatest assets. Common examples are He-Ne laser, helium-cadmium laser, copper vapour laser and CO_2 laser.

  The important characteristics of laser *die media* are their broad wavelength tunability, wide spectral coverage and practical simplicity. Hundreds of dyes are reported to have lasing action. The examples of various dye classes are Oligophenylenes, Coumarins, Xanthenes, Merocyanines and Cyanines with spectral emission from 300 nm to more than 1100 nm.

  The *free electron lasers* (FEL) are operated in far infrared and UV (below 100 nm) where atomic and molecular lasers are not readily available; and for large average power and high efficiency systems. The disadvantage of FEL are greater complexity and the cost of particle accelerator.

  Semiconductor lasers have some special features, e.g. compactness, high efficiency, capability for high speed direct modulation, wide emission spectrum and high reliability. Their main disadvantage is that they are sensitive to temperature.
- According to the mode of oscillation, one can categorize lasers in three groups: (i) continuous wave (CW) lasers, (ii) pulsed lasers with free oscillations, and (iii) Q-switched lasers.
- Lasers can be made *coherent* either through single-mode operation or through mode-locking.
- Lasers are put to numerous applications in the different fields ranging from medicine in military, and from ground based equipment to satellites in the sky due to their two special feathers: (i) narrow bandwidth, and (ii) narrow angular spread.

REVIEW QUESTIONS

1. How many different types of lasers are there? Suggest at least three ways of classifying lasers with examples in each category.
2. Explain the working of He-Ne lasers. Clearly explain the role of Ne. Mention its uses.
3. What is ruby? Describe the construction and working of a ruby laser. Why He-Ne laser is superior to ruby laser?
4. What do you mean by pumping? Describe in brief the various types of pumping.
5. What are liquid lasers and excimer lasers. Briefly explain their principles.
6. Describe methods of Q-switching lasers. Derive an expression for the maximum number of photons within the cavity of Q-switched laser.
7. What do you understand by Mode-locking? Suggest a method for mode-locking lasers. Derive an expression for an individual temporal pulse width of a mode locked laser.
8. Mention some important applications of lasers.
9. What is laser range finder? Describe its generalized configuration. Mention its civilian as well as military applications.

PROBLEMS

1. Calculate the population of two states in He-Ne laser that produces light of wavelengths 700 nm at 27° C.   [**Ans.** $N_2/N_1 = 5.9 \times 10^{-29}$]
2. The ratio of population inversion of two energy levels out of which upper one corresponds to a metastable state is 1.059×10^{-30}. Show that the wavelength of light emitted at 330 K is 632 nm.
3. Consider a pulsed ruby laser with the following parameters: $N_0 = 1.5 \times 10^{19}$ atoms of Cr^{3+} per cm^3, $\tau_f = 0.4$, $\tau_{sp} = 3$ ms, $\bar{\alpha}(v) = 2$ cm^{-1}, $\bar{\eta}(v) = 0.95$ and $\bar{v} = 5.1 \times 10^{14}$ Hz. Calculate the pump energy that must fall on each cm^2 of the ruby crystal surface to achieve threshold inversion.   [B.Tech.] [**Ans.** 1.328 J cm^{-2}]
4. Estimate the critical population inversion $(N_2 - N_1)_{th}$ for a He-Ne laser emitting at 0.6328 μm. Assume that $\tau_{sp} = 0.1$ μs, length of the cavity = 10 cm, $R_1 = 1$, $R_2 = 0.99$, and the total loss coefficient $\alpha = 0$. The Doppler-broadened width of the laser transition, $\Delta v \approx 10^9$ Hz.
   [B.Tech.] [**Ans.** 6.28×10^8 cm^{-3}]
5. An injection laser has an active cavity with losses of 30 cm^{-1} and the reflectivity of each cleaved laser facet is 30%. Determine the laser gain coefficient for the cavity when it has a length of 600 μm.
   [**Ans.** 50 cm^{-1}]

[**Hint.** The threshold gain per unit length is given by:

$$g = \alpha + \frac{1}{L} \ln \frac{1}{\gamma} = 30 + \frac{1}{0.06} \ln \left(\frac{1}{0.3}\right) = 50 \text{ cm}^{-1}$$

(Here $r_1 = r_2 = r$)
The threshold gain per unit length is equivalent to the laser gain coefficient for the active cavity, which is 50 cm^{-1}].

6. Compare the ratio of the threshold current densities at 20°C and 80°C for an AlGaAs injection laser with $T_0 = 160$ K and the similar ratio for an InGaAsP device with $T_0 = 55$ K. [**Ans.** 1.46, 1.98]

[**Hint.** The threshold current density:

$$J \propto \exp\left(\frac{T}{T_0}\right)$$

where T is the device absolute temperature and T_0 is the threshold temperature coefficient which is a characteristic temperature describing the quality of the material, but which is also affected by the source of the device.

For AlGaAs device:

$$J_1 (20°C) \propto \exp \frac{293}{160} = 6.24$$

$$J_2 (80°C) \propto \exp \frac{353}{160} = 9.08$$

$$\therefore \quad \frac{J_2 (80°C)}{J_1 (20°C)} = \frac{9.08}{6.24} = 1.46$$

For InGaAsP device:

$$J_1 (20°C) \propto \exp \frac{293}{55} = 205.88$$

$$J_2 (80°C) \propto \exp \frac{353}{55} = 612.89$$

Hence the ratio of current densities

$$\frac{J_2 (80°C)}{J_2 (20°C)} = \frac{612.89}{205.88} = 2.98].$$

7. Calculate the pulse width τ_p and the spatial length L_p of mode-locked pulses for the following cases: (a) a He-Ne laser for which $\Delta v = 1.5 \times 10^9$ Hz and (b) a ruby laser for which $\Delta v = 6 \times 10^{10}$ Hz. Comment on the results. [B.Tech.]

[**Hint.**
 (a) $\tau_p = 1/(\Delta v) = 1/(1.5 \times 10^9)$ s $= 0.66$ ns
 $L_p = c\tau_p = 3 \times 10^8 \times 0.66 \times 10^{-9}$ m $= 19.8$ cm

(b) $\tau_p = 1.66 \times 10^{-11}$ s = 16.6 ps
$L_p = 3 \times 10^8 \times 1.66 \times 10^{-11} = 5 \times 10^{-3}$ m = 5 mm

The results simply indicate that enormous number of photons can be packed into a very small space L_p occupied by these pulses.]

SHORT ANSWER QUESTIONS

1. In how many groups lasers can be divided?

Ans. (i) The state of matter of the *active* medium: solid, liquid, gas, or plasma.
 (ii) The spectral range of laser wavelength: visible spectrum, infrared (IR) spectrum etc.
 (iii) The excitation (pumping) method of the active medium: optical pumping electrical pumping, etc.
 (iv) The characteristics of the radiation emitted from the laser.
 (v) The number of energy levels which participate in the lasing process.

Basically, there are four broad categories of lasers: solid state lasers, gas lasers, dye lasers, and semiconductor lasers.

2. Name the two main excitation techniques used for gas lasers.

Ans. (i) Electric discharge, (ii) Optical pumping.

3. What is a dye laser?

Ans. A dye laser is a laser that uses an organic dye as a lasing medium usually as a liquid solution.

4. What are excimer lasers?

Ans. An excimer (excited dimmer) is a short-lived dimeric or heterodimeric molecule formed from two species, at least one of which is in an excited state. Excimers are often diatomic and are formed between two atoms or molecules that would not bond if both were in ground stage. The lifetime of an excimer is very short ~ nanoseconds.

Excimer gas lasers, produce ultraviolet (UV) light and are used in semiconductor manufacturing and in LASIK eye surgery, e.g. F_2 (157 nm), ArF (193 nm), KrCl (222 nm), KrF (248 nm), XeCl (308 nm), XeF (351 nm).

5. What are the free electron lasers?

Ans. These lasers have the ability to generate wavelengths from the microwave to the X-ray region. These lasers operate by having an electron beam is an optical cavity pass through a wiggler magnetic field. The change in direction exerted by the magnetic field on the electrons causes them to emit photons.

6. What is a gas laser?

Ans. A gas laser is a laser in which an electric current is discharged through a gas to produce light. Examples are: He-Ne gas laser.

7. Mention few advantages of gas lasers.

Ans. (i) High volume of active material.

(ii) Active material is relatively cheaper.

(iii) Almost impossible to damage the active material.

(iv) Heat can be removed quickly from the cavity.

8. What is an ion laser?

Ans. An ion laser is a gas laser which uses an ionized gas as its lasing medium. Like other gas lasers, ion lasers feature a sealed cavity containing the laser medium and mirrors forming a Fabry-Perot resonator.

9. What are metal vapour lasers?

Ans. In these lasers, the active medium is a vapour consisting of metal atoms. There are: (a) *neutral metal vapour* lasers which include: (i) copper vapour laser (CVL), and (ii) gold vapour laser (GVL); (b) *ionized metal vapour* lasers which include: (i) He-Cd laser.

10. What is mode locking?

Ans. Mode locking is a technique in optics by which a laser can be made to produce pulses of light of extremely short duration (ultrashort pulses), of the order of 10^{-12} s (picoseconds), or femoseconds (10^{-15} s) or attoseconds (10^{-18} s). The lasers are then termed as mode-lock lasers.

MULTIPLE CHOICE QUESTIONS

1. The most important characteristic of a laser is:
    (a) polarization
    (b) coherence
    (c) high intensity
    (d) directionality

2. The intensity of a laser beam does not decrease with distance in accordance with the inverse square law because:
    (a) the laser light is monochromatic
    (b) the laser light is very intense
    (c) the laser light is very directional
    (d) the laser light obeys Planck's law

3. For a laser beam $\lambda = 4400$ Å and coherence time $= 4 \times 10^{-5}$ s, the coherence length will be:
    (a) 12 km
    (b) 1.2 km
    (c) 0.12 km
    (d) 0.012 km

4. In a laser beam minimum angular divergence depends on:
    (a) wavelength λ only
    (b) diameter of mirror D only
    (c) both λ and D
    (d) alignment of mirrors only

5. The wavelengths produced by a He-Ne laser correspond to transition in:
   (a) both helium and neon
   (b) helium
   (c) neon
   (d) neither helium nor neon
6. The term 'population inversion' means:
   (a) population of the ionised state is maximum
   (b) population of the lowest state is maximum
   (c) population of the lower level is more than that of the higher level
   (d) population of the higher level is more than that of the lower level
7. A laser operates at a frequency of 3×10^{14} Hz and has an aperture of 10^{-2} m. The angular spread will be:
   (a) 10^{-2} rad
   (b) 10^{-3} rad
   (c) 10^{-4} rad
   (d) 10^{-5} rad
8. The function of He atoms in the He-Ne laser is:
   (a) to quench the neon atoms
   (b) to provide energy to the neon atoms
   (c) to make neon atoms inactive
   (d) None of the above
9. Which of the following characteristics is not associated with a laser light?
   (a) Coherence
   (b) Brightness
   (c) Polarization
   (d) Birefringes
10. In ruby laser, the rod is surrounded by a helical photographic flash lamp filled with:
    (a) chromium
    (b) aluminium
    (c) xenon
    (d) neon
11. CO_2 lasers can be used as:
    (a) light to heavy duty industrial jobs
    (b) medical surgery
    (c) military defence
    (d) All of the above
12. Laser range finders use the following lasers:
    (a) Excimer lasers
    (b) Semiconductor lasers
    (c) He-Ne lasers
    (d) Q-switched Nd^{3+}:YAG lasers
13. Directed energy weapons (DEWs) can accomplish the following tasks in the battlefield:
    (a) Disable EOCM of the enemy
    (b) Trigger an explosion of the fuel or warhead
    (c) Cause a temporary or permanent damage to the target
    (d) All of the above

14. The Nd^{3+}:YAG laser is a:
    (a) solid-state laser
    (b) gas laser
    (c) dye laser
    (d) semiconductor laser
15. The CO_2 laser is:
    (a) an atomic laser
    (b) a molecular laser
    (c) an ion laser
    (d) an excimer laser
16. The He-Ne laser emits the following wavelength:
    (a) 0.6943 μm
    (b) 1.064 μm
    (c) 0.6328 μm
    (d) 10.6 μm
17. For an air-glass interface, what is the Brewster angle? (Assume that the refractive index for glass is 1.5).
    (a) 30°
    (b) 49.4°
    (c) 56.3°
    (d) 90°
18. Which of the following modes of operation should be used to achieve laser pulses of very high power?
    (a) Continuous-wave operation
    (b) Pulsed operation with free oscillation
    (c) Q-switched operation
    (d) All of these
19. What is the temporal width of a mode-locked rhodamine 6G dye laser emitting at 0.6 μm if $\Delta v = 10^{13}$ Hz?
    (a) 6.6×10^{-10} s
    (b) 3×10^{-11} s
    (c) 5.3×10^{-13} s
    (d) 10^{-13} s
20. What is the spatial length of a mode-locked Nd^{3+}:glass laser if $\Delta v = 3 \times 10^{12}$ Hz?
    (a) 0.1 mm
    (b) 5.3 mm
    (c) 4 m
    (d) 7 m

ANSWERS

1. (b) 2. (c) 3. (a) 4. (c) 5. (c) 6. (d) 7. (c)
8. (b) 9. (d) 10. (c) 11. (d) 12. (d) 13. (d) 14. (a)
15. (b) 16. (c) 17. (c) 18. (c) 19. (d) 20. (a)

16

Metamaterials

16.1 INTRODUCTION

Metamaterials are artificially created structures with predefined electromagnetic properties. They are fabricated from identical elements (atoms) which form one-, two- or three-dimensional structures. They resemble natural solid state structures. Metamaterials typically form a periodic arrangement of artificial elements designed to achieve new properties usually not seen in nature. In a sense, they are composed of elements in the same way as matter consists of atoms.

Metamaterials are characterized and defined by their response to electromagnetic wave. Optical properties of such materials are determined by an effective permittivity ε_{eff} and permeability μ_{eff} valid on a length scale greater than the size of the constituent units. In order to introduce such a description, one requires that the size of artificial inclusions characterized by d be much smaller than wavelength λ, i.e. $d \ll \lambda$.

The name *meta* originates from Greek, μετα and means 'beyond'. Main characteristics of metamaterials (MM) are:
- man-made,
- have properties not found in nature,
- have rationally designed properties,
- are constructed by placing inclusions at desired locations.

With modern fabrication techniques, it is possible to create structure which are much smaller than the wavelength of visible light. An example of the unusual properties that such structures could have is the *negative refractive index*. Although the problem has a long history (for example, Mandelstam in 1945 discussed negative refraction and negative group velocity), it was only in 1967 when Victor Veselago indicated the possibility of materials having simultaneously negative μ and ε. He demonstrated the technological potential that materials with a negative refractive index could have for application in *imaging*. It recently made its way to optics, mostly due to rapid progress in *nanofabrication*.

The most well-known property of metamaterials is the *negative index of refraction* (NIM–negative index materials). Another popular name used is *left-handed material*.

The situation is illustrated in Fig. 16.1. For a longer wavelength, one cannot sense the properties of individual constituent atoms, shown on the left, whereas the shorter wavelength, on the order of the distance between 'atoms', can effectively be used to determine some of the atom's properties (like their locations). In metamaterials, we deal with the situation on the left.

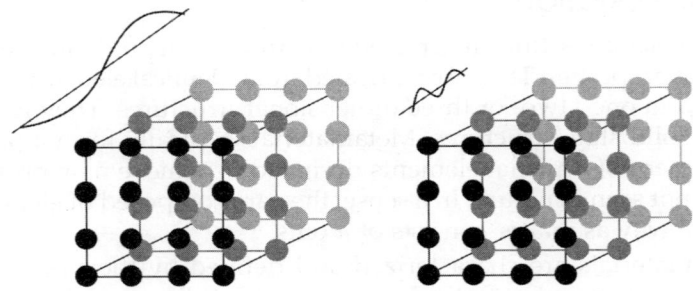

Fig. 16.1. Schemes of the elementary cells and the wavelength of external electromagnetic wave in two extreme cases

Metamaterials can have controlled magnetic and electric responses over a broad range of frequencies. Those response depend on the properties of individual elements (atoms). In the long-wavelength limit, where $a \ll \lambda$, where a is the characteristic dimension and λ is the wavelength of electromagnetic wave, one should perform some sort of averaging procedure to determine effective parameters of MM. In the end, it is possible to achieve the conditions where $\varepsilon_{eff} < 0$ and $\mu_{eff} < 0$.

A diagram which illustrates the classification of metamaterials is shown in Fig. 16.2. In general negative index materials do not exist in Nature; the rare exception is bismuth, which when placed in a waveguide shows a negative refractive index at a wavelength of $\lambda = 60$ μm. There are no known naturally occurring NIMs in the optical range. However, artificially designed materials (metamaterials) can act as NIM. One should, however, notice that the occurrence of NIM needs both negative $\varepsilon_{eff} < 0$ and $\mu_{eff} < 0$ over the same frequency range.

Metamaterials can open new avenues to achieve unprecedented physical properties and functionality unattainable with naturally existing materials. Optical NIMs promise to create entirely new prospects for controlling and mainpulating light, optical sensing and nanoscale imaging and photolithography.

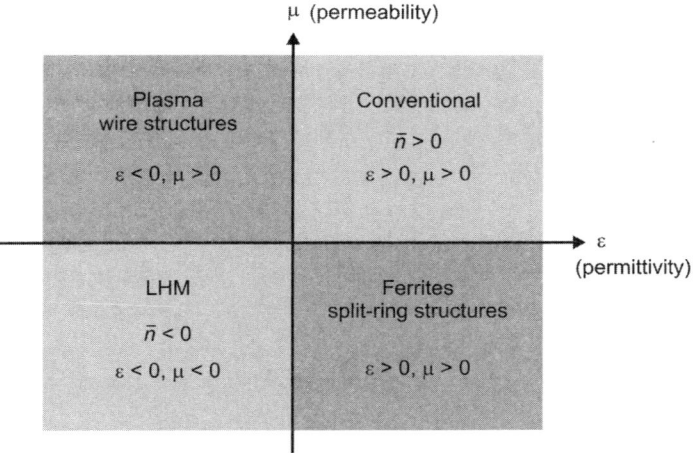

Fig. 16.2. Classification of materials according to the sign of electric permittivity and magnetic permeability

16.2 PREHISTORY OF METAMATERIALS

(i) **Artificial dielectrics:** The artificial dielectric material is usually defined as a composite reproducing on a much larger scale, processes occurring in the molecules of usual dielectric. This involved arranging metallic elements in a three-dimensional (3D) array or lattice structure to stimulate the crystalline lattices of dielectric materials. Such an array responds to radio waves just as a molecular lattice respond to light waves: Free electrons in the metal elements flow back and forth under the action of the alternating electric field. Metal elements called also as lattice inclusions or lattice particles become oscillating dipoles similar to the oscillating molecular dipoles of a natural dielectric.

Artificial dielectrics are not necessarily regular lattices. They can be randomly mixtures. Even the nonuniform concentration of particles is sometimes allowed, which offers unusual properties of such composite media. When the concentration of particles exceeds the so-called percolation threshold (particles touch one another and/or the capacitive coupling between adjacent particles is very strong) artificial dielectrics in the same low frequency range become artificial conductors with complex conductivity. Their conductivity can be engineered (i.e. controlled by the design parameters) and is in principle tunable magnetically or electrically. Artificial conductors find applications in electromechanical devices, fuel cells, and other techniques where controlled heating by electric current is desired.

(ii) **Artificial magnetics:** Magnetism without magnetic constituents has been known since 1940s due to works of Schelkunoff and Friis that suggested so-called split-ring resonators (SRRs) (Fig. 16.3).

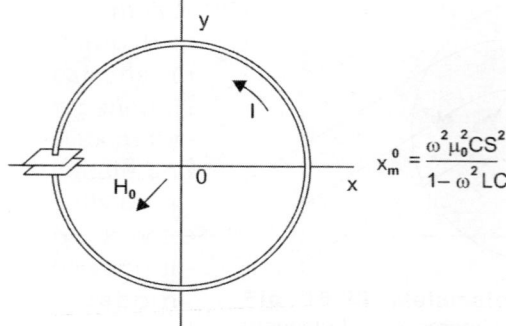

Fig. 16.3. SRSs in the 1950s

We may note that "artificial magnetism" also happens in ordinary structures, like wet snow. The loop-forming parts of liquid water cause diamagnetic behaviour. However, in lattices of SRRs, the artificial magnetism is significantly enhanced in the resonant frequency range (and it is paramagnetic at lower frequencies).

Particles with metal loops of various shapes were studied in the 1980s and in combination with other shapes in 1990s, especially in connection with artificial bianisotropic materials for microwave applications. Polarisabilities of these bianisotropic particles in magnetic and electric fields were studied analytically, numerically and experimentally. Such double SRRs could be used to create artificial magnetics without chirality. In 1997, one finds probably the first experimental demonstration of negative permeability at artificial microwave materials.

The design with strong capacitive coupling between loops suggested by Pendry and coworkers in 1998 seems to be more appropriate for the artificial magnetism.

The strong coupling of two loops allowed one to obtain the magnetic resonance at lower frequencies. This means that the resonant frequency is low enough to consider the lattice of SRRs as a continuous medium. Because of its planar structure SRRs (Fig. 16.4(a)) are perhaps very practical ways of creating artificial magnetism at microwaves. Fig. 16.4(b) shows swiss roll metal scatterers seems to be more efficient as magnetic resonators but work at considerably lower frequencies. The amplitude and frequency bandwidth of magnetic response can be enhanced by using very densely packed stacks of split rings, called metasolenoids.

(iii) **Artificial plasma:** Artificial plasma, i.e., a medium with negative permittivity is presently termed wire medium (Fig. 16.5). Usually this is a square lattice of thin parallel wires which can be considered at microwaves as perfectly conducting ones. This is called as simple wire medium. Recent studies revealed that these phenomena arise due to spatial dispersion. When the wave propagates normally with respect to the wires, the spatial dispersion does not arise. Then, the effective

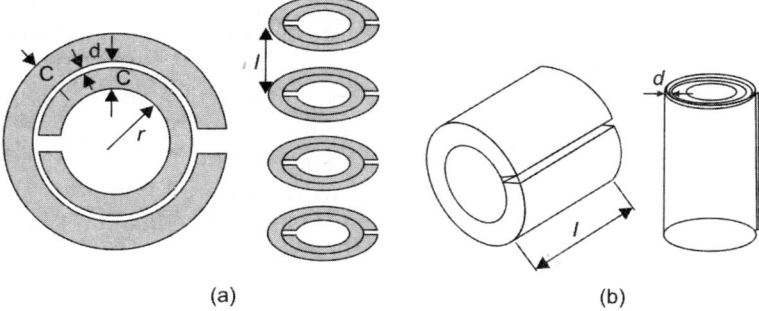

Fig. 16.4. (a) SRRs in the 1990s, and (b) swiss rolls

permittivity of the wire medium obeys the so-called *Drude model* of electric (nonmagnetized) plasma.

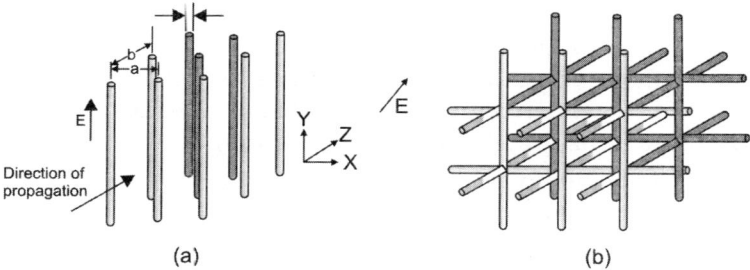

Fig. 16.5. (a) Simple wire media in the 1960s; and (b) Triple wire media in the 1960s

The lattice of parallel wires was considered as a kind of artificial dielectric because it was invented and practically used for applications in microwave lenses. In 1970s tunable lattices of wires in which PiN diodes were inserted in order to switch the negative effective permittivity of the lattice to the positive one were created and produced by industry.

The term wire media appeared after theoretically revealing the effects of spatial dispersion in recent work. Experimental verification of some of these effects also appeared in the literature. We may note that the array of parallel swiss rolls mentioned above behaves like a wire medium (of thick wires) with respect to the magnetic field of propagating wave.

(iv) Backward-waves in bulk media: In isotropic media, the absolute value of the wave vector is fully determined by the frequency. Therefore, the group velocity:

$$V_g = \frac{d\omega(k)}{dk} \hat{k} \text{ or } V_g = \frac{d\omega}{dk} \qquad (16.1)$$

is directed along k or opposite to it, depending on the sign of the derivative $d\omega/dk$. The case of negative sign corresponds to the negative dispersion. According to Mandelshtam, in case of negative dispersion, the wave in the medium is backward and the negative refraction should occur at an interface with such a medium. Mandelshtam presented a physical example of a 3D structure supporting backward electromagnetic waves. It was an inhomogeneous material with permittivity periodically varying in space. Basically, this work predicted the negative refraction in *photonic crystals* later rediscovered.

16.3 ELECTROMAGNETIC METAMATERIALS

Electromagnetic metamaterials (EM) are usually defined as artificial effectively homogeneous structures with specific properties, which cannot be observed in natural materials. A classical example of such a metamaterial is a structure exhibiting simultaneously negative values of the dielectric permittivity ε and magnetic permeability μ ($\varepsilon<0, \mu<0$). These materials represent a new paradigm in electronics and photonics. Very often, the concept of left-handedness is used for structures with backward EM waves in contrast to conventional materials with forward EM waves where the electric field, the magnetic field, and the propagation vector form the right handed (rH trial). In contrast to conventional periodic structures that are usually operated in the Bragg regime (period $\approx \lambda_g/2$, where λ_g is guided wavelength) for filtering, metamaterials are long wavelength structures (average distance between unit elements for atoms $<< \lambda_g$). Since they are not readily available naturally, they are engineered under the form of 1D, 2D, or 3D structures.

Metamaterials may be engineered using resonant particles. However, resonant metamaterials are narroband or lossy. In contrast, transmission lime (TL) metamaterials, which are constituted of patterned lumped elements (inductances and capacitances). (We may note that TL metamaterials may be understood as the limiting case of resonant metamaterials with strongly coupled particles) TL metamaterials are non-resonant structures exhibiting low loss, broad bandwidth and dispersion engineering capability. The most general, efficient, and applied TL metamaterials to date are the so-called composite right-/left-handed (CRLH) TL metamaterials. While they always exhibit a filtering response, and while in 1D configurations, they may exhibit an appearance similar to conventional filters. TL metamaterials are substantially distinct from conventional filters. They are designed for phase specifications and include exclusively subwavelength, typically identical, reactive elements, while conventional filters follow magnitude specifications based on insertion loss prototyping functions and do not exhibit media (i.e. continuous macroscopic)—constitutive parameters.

Metamaterials are usually obtained by combining materials with different properties, in the form of metallic/dielectric inclusions periodically or randomly distributed in a host material. The degrees of freedom available in the frame of complex artificial materials allow engineering unusual combinations of inclusions and host materials that may give rise to new properties, not actually available in natural materials.

TL approach gives an efficient design tool for microwave applications providing a correct description of physical properties of metamaterials. A conventional TL with a positive phase velocity behaves as a right-handed transmission line (RHTL). An artificial RHTL can be formed as a ladder network of capacitors connected in shunt- and series-connected inductors. Fig. 16.6(a) shows the unit cell of RHTL. The dual TL can be designed as a ladder network of inductors connected in shunt- and series-connected capacitors as in the unit cell Fig. 16.6(b). This line has a negative phase velocity and is referred to as the left handed transmission like. A backward-wave propagates along the LHTL, which can be considered as a 1D metamaterial. A more general model of a LHTL is a composite right-/left-handed structure, which includes RH effects. In most practical applications, the influence of RH effects is negligibly small and many interesting features can be observed when a combination RH and LHLTs is used. The most important feature of the LH and RHTLs is that their dispersion characteristics.

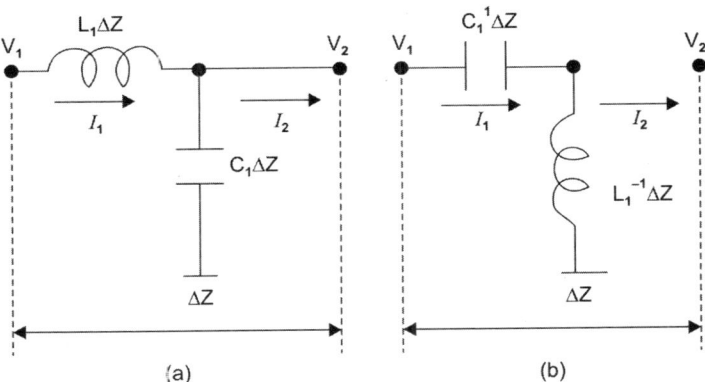

Fig. 16.6. Unit cells of (a) RHTL, and (b) LHTL

16.4 GENERAL EQUATIONS FOR THE RH AND TLs

One can describe the homogeneous RHTL presented as cascade connection of unit cells (Fig. 16.6(a)) by the telegraph equations:

$$\frac{\partial V}{\partial z} = -L_1 \frac{\partial I}{\partial \pi}$$

$$\frac{\partial I}{\partial z} = -C_1 \frac{\partial V}{\partial \pi} \qquad (16.2)$$

where L_1 and C_1 are the inductance and capacitance per unit length respectively, V is the voltage and I is the current.

For sinusoidal wave, $\frac{\partial}{\partial \pi} = i\omega$ and the wave equations look as:

$$\frac{\partial^2 V}{\partial z^2} = -\omega^2 L_1 C_1 V \qquad (16.3)$$

$$\frac{\partial^2 I}{\partial z^2} = -\omega^2 L_1 C_1 I \qquad (16.4)$$

The solution to Eqns. (16.3) and (16.4) are:

$$V = V_0 \exp[i(\omega t - kz)] \qquad (16.5)$$
$$I = I_0 \exp[i(\omega t - kz)] \qquad (16.6)$$

where the wave number $k = k_R$, is defined as:

$$k_R = \omega \sqrt{L_1 C_1} > 0 \qquad (16.7)$$

Phase and group velocities are both positive:

$$V_{Ph} = \frac{\omega}{k_R} = \frac{1}{\sqrt{L_1 C_1}} > 0 \qquad (16.8)$$

$$V_g = \left(\frac{\partial k_R}{\partial \omega}\right)^{-1} = V_{Ph} = \frac{1}{\sqrt{L_1 C_1}} > 0 \qquad (16.9)$$

Therefore, the forward-wave propagation in RHTL. In line with Eqn. (16.7), the dispersion law is linear. One can define the characteristic impedance as:

$$Z_0 = \frac{V}{I} = \sqrt{\frac{L_1}{C_1}} \qquad (16.10)$$

For a homogeneous perfect LHTL formed as a cascaded connection of unit cells (Fig. 16.6(b)), the telegraph equations in the case of sinusoidal waves look as:

$$\frac{\partial V}{\partial z} = -\frac{1}{i\omega}\left(\frac{1}{C}\right)_1 I \qquad (16.11)$$

$$\frac{\partial I}{\partial z} = -\frac{1}{i\omega}\left(\frac{1}{C}\right)_1 V \qquad (16.12)$$

where $(1/C)_1$ and $(1/L)_1$ are the inverse capacitance and the inverse inductance per unit length. One can write the wave equations as:

$$\frac{\partial^2 V}{\partial z^2} = -\frac{1}{\omega^2}\left(\frac{1}{L}\right)_1 \left(\frac{1}{C}\right)_1 V \qquad (16.13)$$

$$\frac{\partial^2 I}{\partial z^2} = -\frac{1}{\omega^2}\left(\frac{1}{L}\right)_1\left(\frac{1}{C}\right)_1 I \qquad (16.14)$$

The solution to Eqns. (16.13) and (16.14) are the same as in Eqns. (16.5) and (16.6), with the wave number defined as:

$$\frac{1}{k_L} = \frac{1}{2}\sqrt{\left(\frac{1}{L}\right)_1\left(\frac{1}{C}\right)_1} < 0 \qquad (16.15)$$

In this case the phase velocity is negative:

$$V_{Ph} = \omega^2 \frac{1}{\sqrt{\left(\frac{1}{L}\right)_1\left(\frac{1}{C}\right)_1}} < 0 \qquad (16.16)$$

whereas the group velocity is positive:

$$V_g = \left(\frac{\partial k_L}{\partial \omega}\right)^{-1} = \omega^2 \frac{1}{\sqrt{\left(\frac{1}{L}\right)_1\left(\frac{1}{C}\right)_1}} > 0 \qquad (16.17)$$

Obviously, the backward wave propagates in the LHTL. In line with Eqn. (16.15), the wave number (propagation constant) is inversely proportional to the frequency and the dispersion law is nonlinear.

$$Z_0 = \sqrt{\frac{(1/C)_s}{(1/L)_1}} \qquad (16.18)$$

Equation (16.2) through (16.18) describe the homogeneous, infinitely long TLs with distributed parameters. One can define a section of such a line of length l by the electrical length θ as:

$$\theta_{R,L} = k_{R,L} \cdot l \qquad (16.19)$$

In accordance with Eqn. (16.7), the frequency dependence of the electrical length of a section of RHTL is:

$$\theta_R(\omega) = \theta_{OR} \frac{\omega}{\omega_0} \qquad (16.20)$$

where $\theta_{OR} = k_R l > 0$ is the electrical length at the frequency ω_0. In the case of section of LHTL, the electrical length can be found from Eqn. (16.13) and the frequency dependence is expressed as:

$$\theta_L(\omega) = \theta_{OL} \frac{\omega}{\omega_0} \qquad (16.21)$$

where $\theta_L(\omega) = \theta_{OL} \frac{\omega_0}{\omega}$ is the electrical length at the frequency ω_0.

In practice, one can compose the artificial RH and LHTLs as a periodical structure containing one inductive and one capacitive component in the unit cell of the length l defined by a real length of the lumped components (Fig. 16.7). In this case, we have to use the translation symmetry for a description of the I–V wave propagating along the one-dimensional structure. The telegraph equations for the RHTL can be expressed as:

$$V_{n+1} - V_n = -i\omega L_0 I_{n+1} \tag{16.22}$$

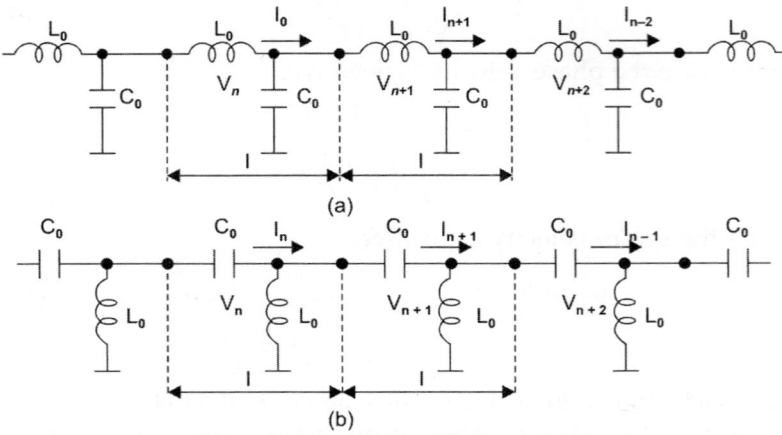

Fig. 16.7. Artificial lumped element TLs: (a) RHTL and (b) LHTL

$$I_{n+1} - I_n = -i\omega C_0 V_n \tag{16.23}$$

The voltage and current are:

$$V_n = V_0 e^{-in\theta}, \quad I_n = I_0 e^{-in\theta} \tag{16.24}$$

with
$$\theta = kl, \, n = 1, 2, 3, \ldots$$

Substituting Eqn. (16.24) into Eqns. (16.22) and (16.23), one obtains:

$$\sin^2\left(\frac{\theta}{2}\right) = \frac{1}{4}\omega^2 L_0 C_0 = \frac{\omega^2}{\omega_c^2} \tag{16.25}$$

where the cut-off frequency:

$$\omega_c = \frac{2}{\sqrt{L_0 C_0}} \tag{16.26}$$

For $\omega > \omega_c$, θ is an imaginary quantity and the wave attenuates: the higher the frequency, the more is attenuation. In the low frequency limit, i.e. $\omega \ll \omega_c$, we have:

$$\frac{\omega}{\omega_c} = \pm\sin\left(\frac{\theta}{2}\right) \approx \pm\frac{\theta}{2} \tag{16.27}$$

and
$$k = \frac{2\omega}{l\omega_c} = \omega\sqrt{L_1 C_1} \tag{16.28}$$

with $L_1 = L_0/l$ and $C_1 = C_0/l$. The artificial lumped element RHTL behaves at $\omega \ll \omega_c$ as an infinitely long, perfect TL with the linear dispersion law. In a wide frequency range, the artificial periodic RHTL (Fig. 16.7(a)) is considered as a low pass lumped element TL.

The same consideration of the artificial LHTL: (Fig. 10.7(b)) gives the dispersion relation.

$$\sin^2\frac{\theta}{2} = \frac{1}{4}\frac{1}{\omega^2 L_0 C_0} = \frac{\omega_c^2}{\omega^2} \tag{16.29}$$

with the cut-off frequency:

$$\omega_c = \frac{1}{2\sqrt{L_0 C_0}} \tag{16.30}$$

For $\omega < \omega_c$, θ is an imaginary quantity and the wave attenuates: the lower the frequency, the more is the attenuation. In the high frequency limit ($\omega \gg \omega_c$), one finds:

$$\frac{\omega_c}{\omega} = \pm \sin\left(\frac{\theta}{2}\right) \approx \pm\frac{\theta}{2} \tag{16.31}$$

and

$$k = \frac{-1}{\omega}\sqrt{\left(\frac{1}{L}\right)_1 \left(\frac{1}{C}\right)_1} \tag{16.32}$$

with $\left(\frac{1}{L}\right)_1 = \left(\frac{1}{L_0}\right)/l$ and $\left(\frac{1}{C}\right)_1 = \left(\frac{1}{C_0}\right)/l$

We find that the artificial lumped element LHTL (Fig. 16.7(b)) behaves as an infinitely long, perfect with k inversely proportional to ω. The artificial periodic LHTL is considered as a high-pass lumped element TL. Fig. 16.8 shows the dispersion characteristics of RH and LHTLs. In further consideration the frequency range is limited by the inequalities $\omega \ll \omega_c$ for the RHTL and $\omega \gg \omega_c$ for the LHTL.

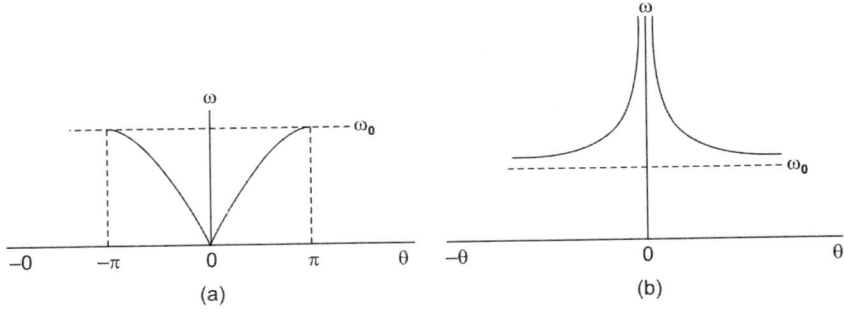

Fig. 16.8. Dispersion characteristics of (a) RH lumped element TL and (b) LH lumped element TL

Microwave Phase Shifters Based on Switchable LH and RHTLs

The transmission type phase shifter is a lossless two-port providing a change in a phase response of the EM wave under the control signal (current or voltage). The digital phase shifters using switchable channels are well known: the EM waves propagate in turn along two channels formed by TL sections of different electrical lengths.

Using specific dispersion properties of RH and LHTLs makes it possible to combine the benefits of switchable channel phase shifters of both kinds. Using cascade connection of RH and LHTL sections, one can obtain a similar slope of the phase response of two TLs of different electrical lengths. Obviously, in principle, it is possible to design a controllable phase shifter based on switchable metamaterial TLs, providing a flat differential phase response.

Figure 16.9 illustrates the operational principle of a digital phase shifter using switchable RH and LHTL sections.

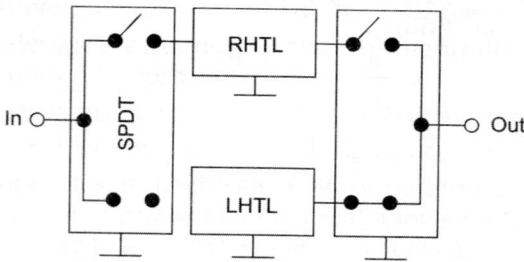

Fig. 16.9. Structure of a digital phase shifter based on switchable TL sections of different electrical lengths

In one state, the signal goes through the RHTL section with a negative phase response ϕ_1, whereas in another state, it propagates through the LHTL section with a positive phase response ϕ_2. The differential phase response (phase shift), $\Delta\phi = \phi_1 - \phi_2$, is obtained by switching the signal path using two SDPT switches.

It is reasonable to form the phase shifter containing switchable RH and LHTL unit cells as T- or TT networks as shown in Fig. 16.10. For different bits of a digital N-bit phase shifter giving the phase shift $\Delta\phi_m$, the equivalent electrical length θ, of the both RHTL and LHTL sections should be chosen as follows:

$$|\phi_m| = \Delta\phi_m/2 \qquad (16.33)$$

where
$$\Delta\phi_m = \Delta\phi_{LH} - \Delta\phi_{RH} \qquad (16.34)$$

and $m = 1, 2, ...N$ is the bit number.

In general, one can describe the TL section by the ABCD matrix:

$$\begin{bmatrix} A & B \\ C & D \end{bmatrix}_{TL} = \begin{bmatrix} \cos\theta & iZ_0 \sin\theta \\ i\sin\theta/Z_0 & \cos\theta \end{bmatrix} \qquad (16.35)$$

Taking into account the symmetry of the section of a homogeneous TL, it is possible to replace it by the symmetric lumped element T- or π-circuits.

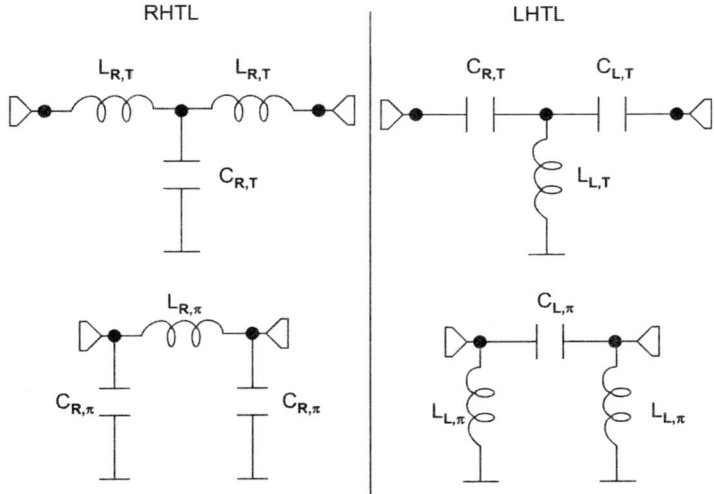

Fig. 16.10. Lumped-element equivalent circuits of the RH and LHTL sections

One obtains the dispersion equation as:

$$\phi_{RH}(\omega) = -\arccos(A)_{T,\pi} = -\arccos(1 - \omega^2 L_R C_R) \qquad (16.36)$$

$$\phi_{LH}(\omega) = -\arccos(A)_{T,\pi} = \arccos\frac{1}{\omega^2 L_L C_L} \qquad (16.37)$$

on differentiation Eqs (16.36) and (16.37), one obtains the main characteristic of interest, i.e. slope parameter of phase characteristics:

$$\frac{d\phi_{RH}}{d\omega} = \frac{\sqrt{2L_R C_R}}{\sqrt{1-(\omega^2 L_R C_R/2)}} \qquad (16.38)$$

$$\frac{d\phi_{LH}}{d\omega} = \frac{\sqrt{2L_L C_L}}{\sqrt{\omega^2}}\frac{1}{\sqrt{1-(1/2\omega^2 L_L C_L)}} \qquad (16.39)$$

The phase response of a TL section relates to the electrical length as $\phi(\omega) = -\theta(\omega)$. For the maximum available digital phase shift, $\Delta\phi = \phi_{LH} - \theta_{RH} = 180°$ ($m = 1$), the electrical lengths of both RH and LH channels are $|\phi_0| = 90°$, and consequently $\omega_0^2 L_R C_R = \omega_0^2 L_L C_L = 1$. At the central frequency of the operational bandwidth ω_0:

$$\left.\frac{d\phi_{RH}}{d\omega}\right|_{\omega=\omega_0} = \left.\frac{d\phi_{LH}}{d\omega}\right|_{\omega=\omega_0} = 2\sqrt{L_{R,L} C_{R,L}} \qquad (16.40)$$

The equality is valid at the central frequency only. The frequency dependence of the slope is different for LH and RHTLs (Fig. 16.11).

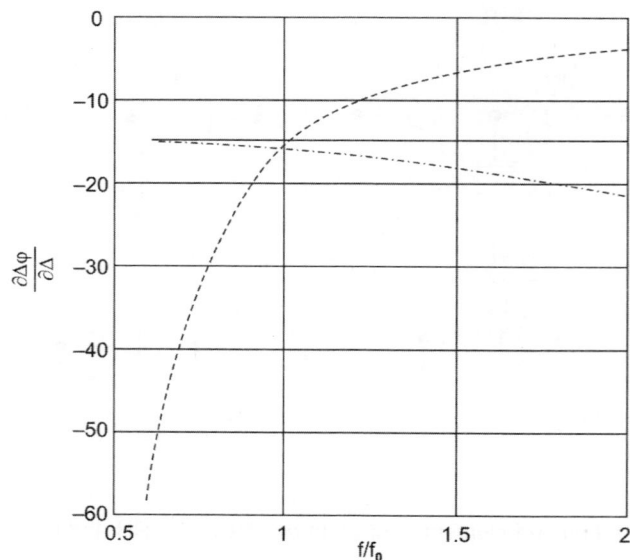

Fig. 16.11. Frequency dependence of derivative $\dfrac{\partial(\Delta\varphi)}{\partial\omega}$ in deg for the natural TL section (solid line), T-section of lumped element RHTL (dash-dotted line), and T-section of lumped LHTL (dashed line); the electrical length $|\theta|_0 = 45°$ at $f/f_0 = 1$

For $m > 1$, the terms in Eqns. (16.38) and (16.39), $\omega_0^2 L_R C_R / 2 \ll 1$ and $\dfrac{1}{2}\omega_0^2 L_L C_L / 2 \ll 1$, both decreases rapidly when m arises. Thus, one can simplify Eqns. (16.38) and (16.39) for $m > 1$:

$$\frac{d\phi_{RH}}{d\omega} = -\sqrt{2L_R C_R} \qquad (16.41)$$

$$\frac{d\phi_{LH}}{d\omega} = -\frac{\sqrt{2L_L C_L}}{\omega^2} \qquad (16.42)$$

Figure 16.12 shows the frequency dependencies for the slope parameter of the distributed RHTL section for the natural TL section, lumped element section of RHTL, the lumped element section of LHTL with electrical length $\phi_0 = 45°$ ($m = 2$) at $f/f_0 = 1$. The difference in the slope parameters of the LH and RHTL sections is remarkably pronounced at a lower frequency range ($\omega \ll \omega_0$) and is less at higher frequencies, though rises at $\omega \gg \omega_0$. Concluding, one finds that for a design of the broadband phase shifter on switchable RH and LHTL sections, it is reasonable to use single cells with a small value of the equivalent electrical length. Thus, for a lower m, the RH and LH branches should be designed as a cascaded connection of RH and LHTL single cells (T or π) having a small electrical length.

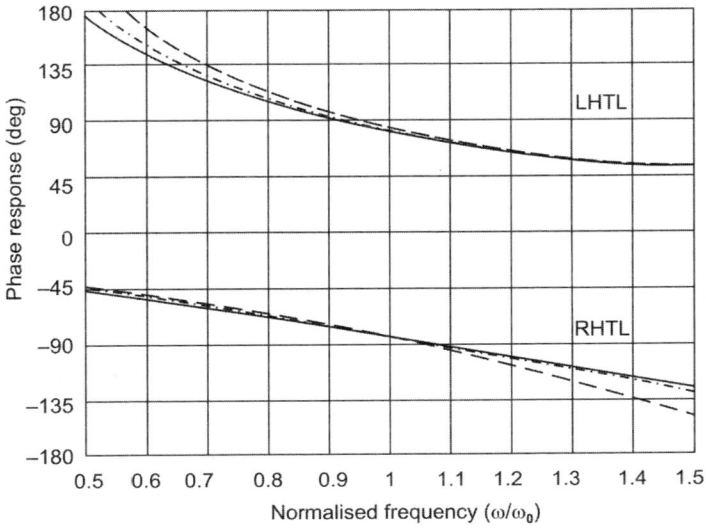

Fig. 16.12. Phase characteristics vs normalised frequency for different numbers of LC section: single LC section, dashed lines; two LC sections, dash-dotted lines; and ideal TL section, solid lines

Fig. 16.12 presents the theoretical phase response of the RH and LHTL sections for different numbers of LC sections (single cell in Fig. 16.10) providing a phase shift of 180° while switching the RH and LH channels.

When the characteristic impedance of both the TLs Z_0 is the same, the perfect matching of the network shown in Fig. 16.9 is provided in any frequency range for both states if Z_0 is equal to the port impedance.

We have seen that combination of the RH and LHTL sections is useful for applications in microwave technique. A new approach to designing digital phase shifters that are free of spurious response resonators with enlarged functionality resulted in exceptional performance of the devices. The combination of the RH and LHTL section can also be used for improving the characteristics of directional couplers, power splitters, matching circuits, etc. Modern multilayer technology makes it possible to design fully integrated miniature microwave devices on the RH/LHTLs.

16.5 METAMATERIAL LENSES AND SUPERLENSES

The concept of a perfect lens that is able to restore both the propagating and evanescent spectra of a source at the image is a product of the theoretical work of Veselago, who in 1968 consolidated earlier thoughts about the focussing of rays without aberration through a LH metamaterial slab by way of negative refraction, and Pendry, who is 2000 proposed the conditions that would enable Veselago's slab lens to transmit near-field information by way of the apparent amplification of evanescent waves. Such a lens is called Veselago-Pendry superlens, for it is able to exceed the performance of conventional lenses, which are constrained by the classical diffraction limit to produce images whose resolutions are limited to the order of $\lambda/2$ of illumination. These superlenses have many potential applications in biomedical, microelectronics, and defense related fields in subdiffraction microscopy, lithography, tomography, and sensing, but their design is complicated by the strict requirement for isotropic $\mu = -\mu_0$ and $\varepsilon = -\varepsilon_0$. These necessary conditions ensure a refractive index of $n = -1$ while guaranteeing matching to free space. Furthermore, these lenses require extremely low-loss characteristics and must be adequately thin.

The first successful attempt at superlensing of Veselago-Pendry type, and indeed, the first demonstration of subdiffraction imaging using metamaterials employed negative-refractive index transmission line (NRI-TL) metamaterials, which consist of fine TL grid loaded with inductors and capacitors to control their effective-medium response. This effort began with the realisation of planar NRI-TL lenses based on discrete-lumped loading, followed by the development of a rigorous theory that enabled the design of a planar NRI-TL superlens, which successfully verified the phenomenon of Veselago-Pendry superlensing, albeit in a planar microstrip environment. The need to extend the benefits of the NRI-TL method to free-space realisation led to volumetric topologies that, although polarisation-specific, could be easily constructed, and one of these topologies was shown to exhibit true free-space superlensing of the Veselago-Pendry type. Nowadays, fully isotropic 3D metamaterials based on NRI-TL techniques have been proposed, and it seems that topological simplications, along with ongoing developments in fabrication techniques, should facilitate their realization and suggest practical applications in near future.

16.6 ELECTROMAGNETIC AND PHOTONIC CRYSTALS

Electromagnetic crystals are artificial periodical structures operating at frequencies where the wavelength is comparable with the characteristic period of the structure. In the optical frequency range, such structures are called *photonic crystals*. The inherent feature of such media is the existence of frequency bands inside which wave propagation is not allowed. These bands are called bandgaps, and therefore, these crystals are sometimes, also called *electromagnetic bandgap* (EBG) or *photonic bandgap* (PBG) structures. The bandgaps originate physically from the spatial (Bloch) resonances of the crystal and strongly depend on the direction of wave propagation. This reveals that electromagnetic crystals are media with spatial dispersion.

16.7 HYPERLENS

Hyperlens is one of the extraordinary applications of metamaterials; utilizing hyperlens, a metamaterial with permittivities in opposite signs in two orthogonal directions of cylindrical coordinates, to overcome the resolution limit of optical imaging. A hyperlens can be a simple additional part to any existing microscopes that will remarkably improve the resolution.

An EM wave has two distinct properties, the angular frequency (ω) and the wave vector k that is the inverse of wavelength (λ). There is a special relationship between the two and this dispersion in a medium is represented by the fllowing in k-space:

$$\frac{k_r^2}{\varepsilon_\theta} + \frac{k_\theta^2}{\varepsilon_r} = \frac{\omega^2}{c^2} \qquad (16.43)$$

where k_r and k_θ are electric permittivities in radial and tangential directions, respectively. ω is the angular frequency and c is the speed

Equation (16.43) is written in cylindrical coordinator for convenience. If a medium exhibits isotropic electric response ($\varepsilon_r = \varepsilon_0 = \varepsilon$) in all directions (i.e. an isotropic medium), Eqn. (16.43) can be graphically represented with a circle (Fig. 16.13)(a) solid). However, for imaging application, this circular dispersion is undesirable because accessible wave vectors (k_q) are limited to the refractive index of the medium, $n = \sqrt{\varepsilon}$.

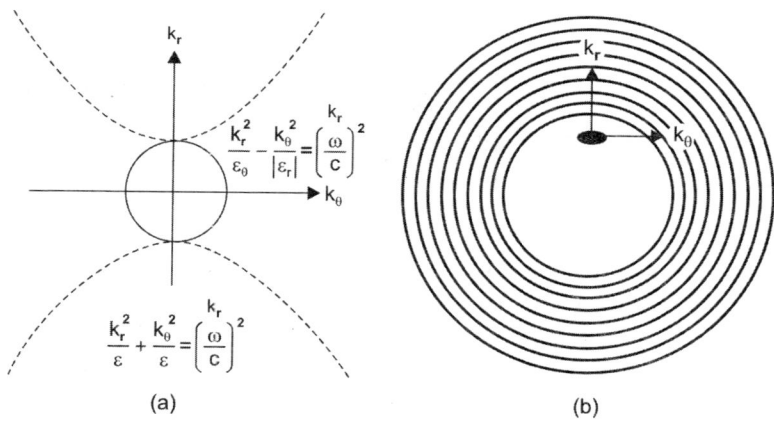

Fig. 16.13. Principle of hyperlens. (a) Solid circle; dispersion isofrequency curve of EM wave in isotropic medium with permittivity ε in cylindrical coordinates. Dotted curve; hyperbolic dispersion in an anisotropic medium when ε_r and ε_θ have opposite signs. (b) Suggested realisable cylindrical hyperlens structure. Multilayer and concentric layer of alternating metal and dielectric layers make up an anisotropic metamaterial

This is why higher index materials are sought to be used in immersion microscopy; the larger the k_θ collected, the better the resolution. We know that current immersion methods have their limitations simply because there are not many materials available in nature with high values of n. However, a material with anisotropic electrical response ($\varepsilon_r \neq \varepsilon_\theta$) can indeed support the propagation of large wave vectors. This means that if ε_r can be engineered to be negative and ε_θ positive, there is a way to harness arbitrarily k_θ. In such a case, dispersion relation reads as:

$$\frac{k_r^2}{\varepsilon_\theta} - \frac{k_\theta^2}{|\varepsilon_r|} = \frac{\omega^2}{c^2} \qquad (16.44)$$

which represents a hyperbola. With a hyperbolic dispersion (Fig. 16.12(a)), theoretically, there is no limit to the maximum k_θ that can be supported by the material. This device was named 'hyperlens'.

Materials of such characteristics are not found in nature. However, a hyperlens can be engineered as a metamaterial. Fig. 16.13(b) shows one example of such materials that consists of alternating layers of a metal and a dielectric, whose cylindrical configuration is similar to earlier reported cylindrical superlens. Since in visible frequency, metal tend to have negative and dielectrics have positive permittivities, a metamaterial that contains both metals and dielectrics can be designed to have opposite signs of permittivities in two orthogonal directions using the following equations:

$$\varepsilon_\theta = \frac{\varepsilon_m + \varepsilon_d}{2} \qquad (16.45)$$

$$\varepsilon_r = \frac{2\varepsilon_m \varepsilon_d}{\varepsilon_m + \varepsilon_d} \qquad (16.46)$$

where ε_m and ε_d are permittivities of the metal and the dielectric layers, respectively. Assuming the layer thickness is much smaller than the working wavelength, this cylindrical anisotropy can give rise to a hyperbolic dispersion behaving as one bulk matter, or as an effective matter. We may note that, in addition to the anisotropy, it is the cylindrical geometry that provides the imaging capability that a stack of flat layers cannot. When a small object is placed at the centre of the cylindrical hyperlens (Fig. 16.13(b)) and illuminated, the object scatters the light and generates a wideband of k_θ, which would normally be evanescent in common dielectrics, can now propagate along the radial direction, and because of the curvature, the k_θ shrinks due to the conversation of angular momentum:

$$m = k_\theta r \qquad (16.47)$$

where m is angular momentum mode number and r is the radius.

Since m is constant for a given mode, the k_θ has to become smaller as r becomes larger, which simply means a magnification process.

We may note that the hyperlens design is not limited to the cylindrical geometry that has a drawback of being able to magnify in only one direction. A spherical hyperlens would be more applicable in real life as a true two-dimensional (2D) nanoimaging device.

16.8 CLOAKING

One set of novel and exciting applications of metamaterials deals with making a given object essentially 'invisible' to the impinging electromagnetic wave. This application of metamaterials, associating with it the term "cloaking". Studies have shown that covering a dielectric or conducting object with a plasmonic material or metamaterial with permittivity lower than one of the background may realize the desired cloaking in the range of frequency of interest. This effect relies on a scattering cancellation, for which the wave scattered from the cloak may cancel the one from the object to be cloaked, leaving an external observer with a very low residual scattering that makes the system practically invisible around the design frequency. Besides this, there are three other cloaking techniques, e.g., the possibility of including anomalous localised resonances which may isolate and cloak a given object, use of transmission line circuits to match a given network surrounding an object with the background material. It is expected that metamaterial cloaking may become of major interest for various exciting applications in various scenarios, including *camouflaging* and non-invasive probing, and sensing in medical and biological applications.

16.9 METAMATERIAL TRANSMISSION LINES

These are 1D artificial propagating media that exhibit either a single-negative (SNG) or a double negative (DNG, or left-handed, LH) behaviour. In the first case, signal propagation is not allowed in the frequency region of interest (even though the medium is transparent at other frequencies) since the effective permeability or permittivity are negative in that region. Conversely, LH metamaterial transmission lines exhibit backward-wave propagation in the LH band (due to the coexistence of negative effect permittivity and permeability in that band) and forward-wave propagation in a certain frequency interval at higher frequencies. Purely left-handed (PLH) lines (i.e., those lines exhibiting only a backward-wave transmission band) cannot be implemented in practice due to the presence of parasitic elements. This is why, LH lines actually exhibit a composite behaviour, and for this reason, these lines are called composite right/left-handed (CRLH) lines. Two main approaches can be used for the implementation of metamaterial transmission lines: (i) the nonresonant type approach, where a host line is periodically loaded with series capacitors and/or shunt capacitors, and (ii) the resonant-type approach, where the host transmission lines are loaded with split ring resonators (SRRs) or complimentary split ring resonators, (CSRRs). SRRs have been widely used to design and fabricate left-handed materials (LHMs), because SRPs provide the required μ-negative (MNG) property, whereas wise medium provides ε negative (ENG) property.

CRLH transmission line (TL) metamaterials which exhibit both a LH frequency band and RH frequency band with an infinite wavelength tran-sition frequency when a structure resonances are balanced, represent a new paradigm in electromagnetic theory and engineering. In their metamaterial (effectively homogeneous) range, CRLH structures may be conveniently described as ideal uniform TLs. When a broadband description is required, to account for bandage; anisotropy; stopband; and diffraction effect, CRLH structures may be analyzed using Bloch-Floquet TL analysis (1D, 2D or 3D). The fundamental parameters of CRLH metamaterials are their dispersion relation and their Bloch impedance, and their operation bandwidth is determined by the lumped LC elements of their unit cell. Useful CRLH metamaterial concepts for antennas include the infinite-wavelength power divider, which may be used as a compact and low-loss series feeding network, multiband capabilities (dual-band for the C-CRLH case, tri-band for the DL-CRLH case, and quad-band for the extended CRLH case). CRLH metamaterials have lead to several breakthrough in both 1D and 2D leaky wave antenna and reflector applications.

16.10 CONTROLLABLE METAMATERIALS

The concept of controllability appears with the comparison of the behaviour of two complementary passive metallic bidimensional photonic bandgap (2D PBG). The first one, which consists of a square lattice of continuous metallic rods, is well-known for having a wide forbidden band from 0Hz to its common frequency called plasmon-like band. This last one depends on the geometrical parameters of the structure. In this forbidden band, the effective permittivity is negative. The second one is a square lattice of discontinuous metallic rods. In this case, a valence (or transmission) band occurred instead of the plasmon-like band at low frequency. The effective permittivity becomes positive in this allowed band.

Controllable photonic bandgap (CPBG) materials are used to realize planar and conformable antennas for communication at 12 GHz and in the GSM-UMTs. The main advantage of these CPBG materials is their capability to realize multifunction smart antennas. The main drawback of these materials is their thickness, which is of the order of several $\lambda/4$, where λ is the operating wavelength.

16.11 OPTOELECTRONIC CONTROL OF METAMATERIALS

While metamaterials have been utilized to create materials with unprecedented properties such as invisibility cloaks, negative refraction, and perfect lenses, metamaterials alone suffer important drawbacks. The resonant nature of most metamaterial design to date leads to significant frequency dispersion, resulting in narrow banwidth may be desirable, while for others a broader bandwidth is required, which present distinct challenges. Truly speaking, the usable range of any given design may be limited to about 5%–10% compared to its centre frequency ω_0. The narrowband resonant response and large metamaterial absorption severely restricts several potential applications. Similarly, the interesting and often desirable physical properties of many natural materials are also subject to limitations. However, composites that result from the combination of natural materials and metamaterials (usually termed hybrid metamaterials) have the potential to address deficiencies of either system, and are relatively unexplored so far.

Hybrid metamaterials form when the properties of a natural material, such as the dynamic photoconductivity of a semiconductor, strongly couples with the resonance of a metamaterial element. The resulting hybrid material will have passive properties (e.g., negative electric response, negative index, or gradient index) that can be selected by the controlled design and patterning of the metamaterial elements, and dynamic properties (photoconductivity, nonlinearity, gain etc.) that result from strategic incorporation of natural materials. Metamaterials, thus not only extend and complement natural material response, but they can also yield tunable properties, enabling them to actively and individually control electric and magnetic fields. Coupled with their ability to yield nearly any EM response from radio frequencies to near optical frequencies underscores their importance and substantiates their use for future state-of-the-art devices operating in any technologically relevant spectral range.

Metamaterials hybrid structures that can be controlled optoelectronically may be fabricated on, or constructed with, various semiconductors, e.g. semiinsulating gallium arsenide (SI-GaAs), GaAs:ErAs nanoisland superlattices, low temperature grown GaAs, silicon-on sapphire (SOS), silicon-on-insulator, and a thin n-doped GaAs layer, etc.

For optical control of metamaterials, the carrier generation efficiency must be considered. The number of carriers generated per cm^3 is given by:

$$n = 6.25 \times 10^{18} \frac{I_0 \, (I-R) \cdot \tau}{Adh\omega} \qquad (16.48)$$

where, I_0 is the average power in watts, R is the reflectivity, $\hbar\omega$ is the photon energy in eV, A is the area of the laser excitation and d is either the penetration depth or sample thickness in the appropriate limit.

The expression is valid for continuous wave (CW) and pulse excitation.

We may note that amplitude or frequency control of the metamaterial resonance may be possible through the dependence of epsilon infinity (of the substrate material) or interband transitions. The frequency dependence of the metamaterial electric response can be approximately modelled by the *Brude-Lorentz model*:

$$\overline{\varepsilon}(\omega) = \varepsilon_\infty + \frac{\omega_p^2}{\omega_0^2 - \omega^2 - i\gamma\omega} \quad (16.49)$$

where, ε_∞ is the real part of dielectric constant at $\omega = \infty$, ω_0 is the central frequency of the oscillator and γ is damping.

Under optical excitation, one can increase ε_∞ thereby decreasing the frequency at which $\varepsilon_1(\omega) = 0$ on the high-frequency side of the resonance. In such cases, no additional loss is introduced at the metamaterial resonance frequency. In this way, we can extend the range of optical control of metamaterials to the mid-IR and perhaps near-IR regimes; although high powers are essential, significant heating of the substrate is problematic, and relatively small changes are expected.

Active control of metamaterial devices at THz frequencies has been achieved by biasing *n*-GaAs near the gaps of split rings, which effectively tuned the strength of the resonance of the metamaterials. While active control required arrays of conducting wires, it did not affect the oscillator strength of the response, lending potential flexibility to future devices. Dynamical control of metamaterial devices at THz frequencies was realized through photoexcitation of carriers in the substrate of the SRR. This demonstration of optical control of metamaterials allows for fast switching times that are not realizaable by electronic means.

16.12 EXTRAORDINARY TRANSMISSION

The experimental discovery of unexpectedly high levels of transmission of light through perforated metal screens initiated interest on the problem of transmission of EM waves through electrically small holes practiced in opaque slabs. This phenomenon was termed extraordinary transmission because classical Bethe's theory for small holes predicted much lower power transmission per hole. Extraordinary transmission phenomenon can be classified under following three different categories:

(i) Transmission through periodic (or quasiperiodic) arrays of apertures in opaque screens. These apertures can be either infinitely long slits or finite size holes of various shapes.
(ii) Transmission through a single aperture around the surface of the screen is periodically structured (with corrugations, e.g.).
(iii) Transmission through one, two, or three holes without periodic patterns around them.

The phenomenon of extraordinary transmission of EM waves through metal screens periodically performated with small holes or narrow slits can be explained in terms of impedance matching concepts. This point of view directly leads to equivalent circuit models that are very useful to make qualitative and quantitative predictions. The circuit models only require the determination of a reduced set of parameters that can be used for obtaining the entire transmission spectrum in the frequency range of interest.

16.13 PHOTONIC CRYSTALS (PhCs)

Although fundamentally photonic crystals are not really subwavelength structures, and are not considered as metamaterials but some negative refraction is possible in these dielectric structures without a negative refractive index or "backward wave" effect, if they contain periodic variations of the scale of wavelength of the electromagnetic radiation. These materials can also be obtained using self-organisation principles. This effect in photonic crystals comes from complex Bragg scattering effects, which differ from effects in left-handed metamaterials.

There are numerous opportunities for PhCs to bring decisive improvements in active systems. In addition to their applications to individual device, PhCs obviously offer the possibility of revisiting integration in optics. Given the present thrust for integration and for PICs, complex sources combining passive and active devices have a good chance to be the first integrated systems that will exploit PhCs. One advantage of PhCs is simply due to the fact that they impose the polarisation state of emitting devices (typically TE).

It seems that the use of PhCs in practical systems does not rest so much on their bandgap properties. At present, no specific direction of application can be mentioned, future information technologies will undoubtedly benefit from today's research on semiconductor based PhCs. Photonics is a rich and diversified field, where all opportunities to combine light and matter are properly scrutinised.

Photonic metamaterials open up a way to overcome the constraints set by ordinary materials. The basic idea is to create an artificial crystals with deep subwavelength periods. Analogous to an ordinary optical material, such a photonic metamaterial can be treated as an effective medium that is characterized by effective material parameters $\varepsilon(\omega)$ and $\mu(\omega)$. However, the proper design of the elementary building blocks ("artificial atoms") of the photonic metamaterial allows for a non-vanishing magnetic response and even $\mu < 0$ at optical frequencies—despite the fact that constituent materials of photonic metamaterials and nonmagnetic.

16.14 FABRICATION OF METAMATERIALS

Lithographic tools have played a pivotal role in the fabrication of metamaterials. As the dimensions and wavelengths have shrunk, the technology of printed circuit boards has been replaced by other lithographic processes, such as optical and electron-beam lithography (EBL). All three technologies share a common base—the use of *resist*. Resist refers to the patterning placed upon a material that can then be transferred into the material by the application of ink or dye. In other words, covering a selected area of material (e.g., the copper in circuit boards, dielectrics, or semiconductor materials) with resist allows it to be processed in a different manner from that used on the rest of the material. Pattern transfer is achieved by first applying resist to the whole surface of a piece of material or substrate and then selectively removing some of the resist. A further processing step, which may be deposition of another material or etching of the existing material, transfers the pattern itself into the material.

The resist is usually a polymer and the selective removal of the resist can be achieved by exposing it to either ultraviolet (UV) light or electrons (X-rays and ions are also possible). Resist comes in either a positive or a negative form: positive organic resists consists of polymers, the chains of which are broken during the exposure process; conversely, in negative organic resists, the exposure to UV light or electrons causes the formation of additional cross-links between the polymer chains. Examples of positive and negative commercially available resists are polymethyl methacrylate (PMMA) and AR-N-a novolak resin-based resist, respectively.

Once the resist has been developed, the next stage of processing can begin, which involves either an etching step onto previously defined metallisation or selective deposition of metal through holes developed in the resist layer.

Following methods of fabrication are used:
  (A) **Lithography:**
      A.1 Optical lithography
      A.2 Electron-beam lithography
      A.3 Lift-off
      A.4 Multiple layer structures
      A.5 Arbitrary multilayer structures
      A.6 Nanoimprint lithography
  (B) Focussed ion beam etching.
  (C) SRR-based metamaterials in different metallisations.

The approaches include high-precision "top-down" deterministic ones that are typically specified, in the first place, as a set of pattern which are transferred either directly or via a mask to be unstructured materials followed by deposition and/or etching.

16.15 HOW TO CREATE METAMATERIALS?

16.15.1 Metamaterials with Negative Effective Permittivity in the Microwave Regime

At optical freqnencies metals are characterized by an electric permittivity that varies with frequency according to *Drude relation*:

$$\varepsilon(\omega) = \varepsilon_0 \left[1 - \frac{\omega_p^2}{\omega(\omega + i\gamma)}\right] \quad (16.50)$$

Here $\omega_p^2 = \frac{Ne^2}{m\varepsilon_0}$ in the plasma frequency, i.e., the frequency with which the plasma consisting of free electrons oscillates in the presence of an external electric field. Typical values for ω_p are in the ultraviolet regime. The other symbols are: N is the electron density; e is charge of an electron and m its mass. The parameter g describes damping and its value, for example for copper is $\gamma \approx 4 \times 10^{13}$ rad s^{-1}.

In the limit when $\gamma = 0$, from Eq. (16.50), it follows that $\varepsilon < 0$ for $\omega < \omega_p$; i.e., the medium is characterized by a negative permittivity. Considering typical values of ω_p, the resulting range of negative values of ε is in the ultraviolet regime. Unfortunately, in this frequency range $\omega \ll \gamma$, and as a result losses dominate the behaviour of ε. Thus, using metals to achieve negative ε over this frequency range will be impractical (high losses) and the propagation of light will be mainly evanescent.

To achieve negative ε at microwave frequencies, Pendry and coworkers in 1998 proposed to use periodic structure consisting of long thin metallic wires of radius r arranged on a horizontal plane (xy), (Fig. 16.14). The unit cell of this periodic structure is a square whose sides have length equal to a.

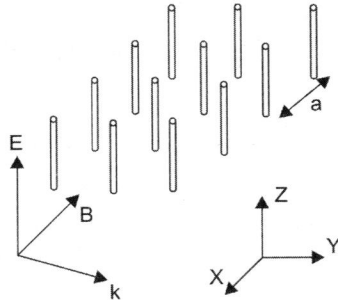

Fig. 16.14. Schematic illustration of a periodic arrangement of infinitely long thin wires along z-direction used in the creation of an effective plasma medium at microwave frequencies

When electric field $E = E_0 e^{-i(\omega t - kz)}z$ is incident on this structure, it forces free electrons to move inside the wires in the direction of the incident field. Effective electron density N_{eff} of such structures which participate in plasma oscillations is $N_{\text{eff}} = N\frac{\pi r^2}{a^2}$ (with N being the electron density inside each wire, r the radius of a wire and a distance between wires), which is significantly smaller compared to N, thus reducing the effective plasma frequency. For example, for a wire with radius $r = 1$ μm and wire spacing $a = 5$ mm, one finds that $N_{\text{eff}} \approx 1.3 \times 10^{-7}$ N; i.e. the effective electron density of the new medium is reduced by seven orders of magnitude compared to that of the free electron gas inside an isolated wire.

Additionally, in such engineered structures, the effective mass m_{eff} of the electrons is significantly larger compared to that of a free electron. To determine effective mass of an electron in this wired medium, we use the classifical equation of motion of a moving electron:

$$d(mv)/dt = e[E + v \times B] \rightarrow \frac{d}{dt}[mv + eA] = -e\nabla(\varphi - v \cdot A) \quad (16.51)$$

where e is the charge of an electron and v the velocity of an electron. One also has $B = \nabla \times A$ and $E = -\nabla\varphi - \partial A/\partial t$. We divide the xy plane into circles of radius R_c, centred at each wire and having area equal to that of the square unit cell; i.e. $R_c = a/\sqrt{\pi}$. Furthermore, we assume that the wires are sufficiently apart from each other so that the magnetic field inside each circle arises only from the current I that flows perpendicularly to the centre of the circle, and that the field at the circumference of each circle vanishes; i.e. $H(R_c) = 0$. Magnetic field intensity at a distance R from each wire is given by:

$$H = \frac{I}{2\pi R}\left(1 - \frac{R^2}{R_c^2}\right) \quad (16.52)$$

Magnetic field H associated with a vector potential. A according to the relation $H = \mu_0^{-1} \nabla \times A$ gives the following expression for the vector potential A:

$$A = \frac{\mu_0 I}{2\pi}\left(\ln(R_c/R) + \frac{R^2 - R_c^2}{2R_c^2}\right)\hat{z} \quad (16.53)$$

where $\hat{z}$ is a unit vector along the z-direction. It has been assumed that $A(R \geq R_c) = 0$. For distances very close to the wires, i.e. for $R \rightarrow r \ll R_c$, Eq. (16.53) gives:

$$A \approx \frac{\mu_0 I}{2\pi}\ln(a/r)\hat{z} \quad (16.54)$$

Since both v and A point along $\hat{z}$, the right hand side of the second part of Eq. (16.51) will vanish (for a given R); i.e. the so-called 'conjugate momentum' $(mv + eA)$ of an electron will be *conserved along the z-direction*. As a result, an electron will be moving in the above engineered medium with an effective mass, $m_{\text{eff}} = eA(r)/v$. Realizing that the current I can be simply re-expressed as $I = -(\pi r^2)(Nev)$, we finally obtain using Eq. (16.64) the effective mass of a moving electron inside our effective medium $m_{\text{eff}} = 0.5 \times \mu_0 Ne^2 r^2 \ln(a/r)$. Thus, for copper wires of radius $r = 1$ μm, being separated by $a = 5$ mm, we obtain: $m_{\text{eff}} \approx 1.3 \times 10^4$ m; i.e. the effective mass of an electron in our engineered medium is increased by more than four orders of magnitude. This, combined with the fact that the effective electron density is reduced by approximately seven orders of magnitude, leads to an effective plasma frequency that is in the microwave regime:

$$\omega_p^2 = \frac{N_{\text{eff}} e^2}{m_{\text{eff}} \varepsilon_0} = 5.1 \times 10^{10} \text{ [rad s}^{-1}\text{]}^2 \rightarrow \nu_p = \omega_p / 2\pi = 8.2 \text{ GHz} \quad (16.55)$$

We must note that, based on Eq. (16.55), the calculated wavelength, $\lambda_p = c/\nu_p$, which corresponds to our medium's effective plasma frequency, turns out to be considerably larger compared to the periodicity of the structure ($\lambda_p \approx 7a$), justifying the description of the periodic structure as an effective medium. Therefore, we are able to construct an engineered medium that can exhibit a *negative* electric permittivity in the *microwave* regime with reasonably low losses.

16.15.2 Magnetic Properties: Split-ring Resonators

In the previous section, we examined how to construct an artificial structure possessing negative effective permittivity ε. Here, we will examine, how to create negative effective permeability μ in the micro-wave regime using the so-called *split-ring resonators* (SRR). The perspective view of SRR is shown in Fig. 16.15. The structure consists of two cut cylinders. The model used to represent SRR is illustrated in Fig. 16.16. Here, we will describe description of split-ring resonators (SRR) based on equivalent circuit model approach. The structure under consideration is shown in Fig. 16.16.

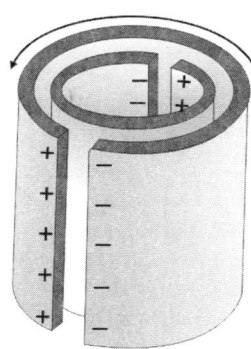

Fig. 16.15. Perspective view of split-ring resonator

We will establish the equivalent circuit model of this structure and show that it can produce negative effective permittivity μ. The unit cell shown above can be modelled using equivalent circuit model shown in Fig. 16.17.

Apply a time-varying external magnetic field H_0 in the y-direction. It will induce a current I flowing in each SRR unit. From Faraday's law, the voltage V due to the external field H_0 is:

$$V = i\omega\mu_0 \pi r^2 H_0 \quad (16.56)$$

where r is the radius of the ring. Along the y-direction, SRR loops form a column (stack) which behaves like a solenoid. In each loop, there is a current I flowing. Neglect the fringing effects, i.e., spreading of the magnetic field lines. The magnetic flux in such a column is thus:

$$\Phi = \pi r^2 \mu_0 \frac{I}{l} \quad (16.57)$$

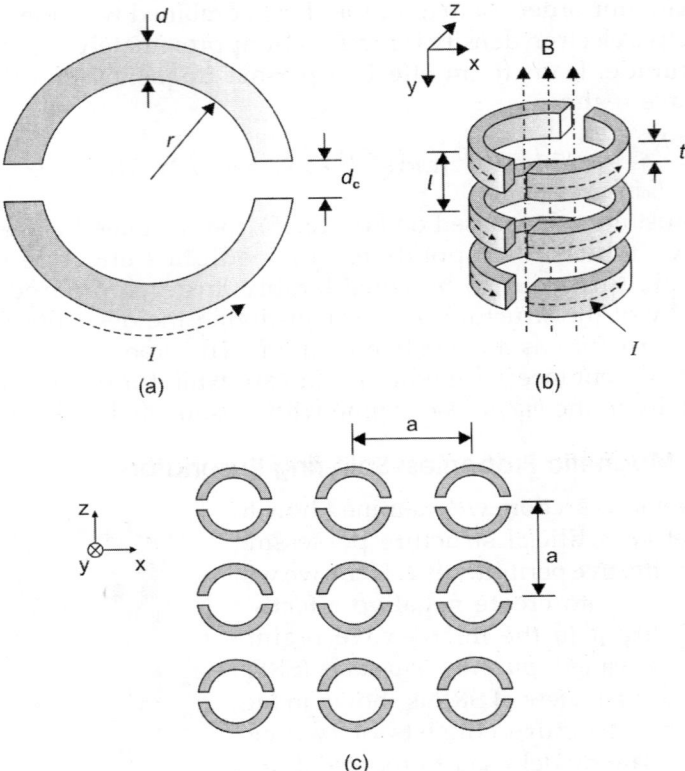

(a)

(b)

(c)

Fig. 16.16. Arrangement of SRR in space

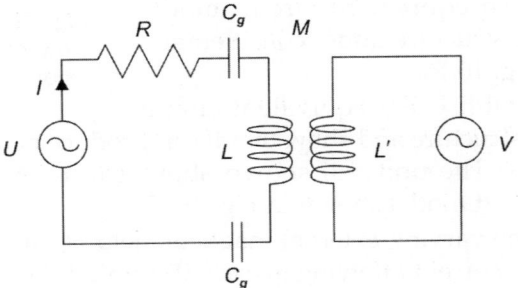

Fig. 16.17. Equivalent circuit of SRR

where l is the separation between loops. The inductance L appearing in the circuit model is:

$$L = \mu_0 \pi \frac{r^2}{l} \qquad (16.58)$$

The inductance L is defined per ring of an infinite column of rings.

Let L' be the total inductance of the SRR units in the other column (excluding column represented by inductance L). Coupling between L and L' is represeted by the mutual inductance M. For very long columns in y-direction, the mutual inductance M is:

$$M = \frac{\Phi_L}{I} = \lim_{n \to \infty} \frac{\pi r^2}{na^2} \frac{\phi_d}{I} = \lim_{n \to \infty} \frac{\pi r^2}{na^2}(n-1)L = \frac{\pi r^2}{a^2} L + F \cdot L \quad (16.59)$$

Here Φ_L is the flux of the depolarization field located in the interior of L, $\Phi = L \cdot I$ is the depolarization field generated by one column of the rings, $\phi_d = (n-1)\Phi = (n-1)L \cdot I$ is the flux of the total depolarization field, and $F = \frac{\pi r^2}{a^2}$ is the fractional volume of the periodic unit cell in the xz plane occupied by the interior of the SRR.

Applying Kirchoff's voltage drop law for a loop in the equivalent circuit shown in Fig. 16.17 gives:

$$V = R \cdot I + \frac{1}{-i\omega C} + (-i\omega L)I - (-i\omega M)I \quad (16.60)$$

where $C = \frac{1}{2}C_g$ is the total capacitance in the loop.

From the previous equations:

$$V = i\omega\mu_0\pi r^2 H_0 = i\omega L \, I \, H_0 \quad (16.61)$$

From Eq. (16.60), one has:

$$V = \left(R - \frac{1}{i\omega C} - i\omega L + i\omega M\right) I$$

Solving for current:

$$I = \frac{V}{R - \dfrac{1}{i\omega C} - i\omega L + i\omega M} = \frac{i\omega L I H_0}{R - \dfrac{1}{i\omega C} - i\omega L + i\omega M \cdot L}$$

$$= \frac{LH_0}{\dfrac{R}{i\omega L} + \dfrac{1}{\omega^2 LC} - 1 + F} = \frac{-LH_0}{(1-F) - \dfrac{1}{\omega^2 LC} + i\dfrac{R}{\omega L}} \quad (16.62)$$

Magnetic dipole moment per unit volume of the material is:

$$M = \frac{\pi r^2}{la^2} I$$

with the current I inferred from the above equation to be:

$$I = -\frac{H_0 l}{(1-F) - 1/(\omega^2 LC) + iR/(\omega L)} \quad (16.63)$$

As a result, the (relative) effective magnetic permeability associated with this medium will be (in the direction, x, that the incident magnetic field is polarized):

$$\mu_r = \frac{B/\mu_0}{B/\mu_0 - M_d} = 1 - \frac{F}{1 - 1/(\omega^2 LC) + iR/(\omega L)} \quad (16.64)$$

From Eq. (16.64), we can see that μ assumes negative values in the range: $1/\sqrt{LC} < \omega_p < 1/\sqrt{KC(1-F)}$, where $\omega_{mo} = 1/\sqrt{LC}$ is the resonance frequency of the Lorentzian variation of the medium's magnetic permeability, and $\omega_{mp} = 1/\sqrt{LC(1-F)}$ is the corresponding plasma frequency (where Re{u} = 0). Crucially, we note that the resonant wavelength (λ_{m0}) of the structure depends entirely on the rings' effective inductance (L) and capacitance (C), and can, therefore, be made considerably larger than the periodicity (a) of the structure, thereby fully justifying its description as an effective medium.

Plot of real and imaginary parts of effective magnetic permeability is shown in Fig. 16.18.

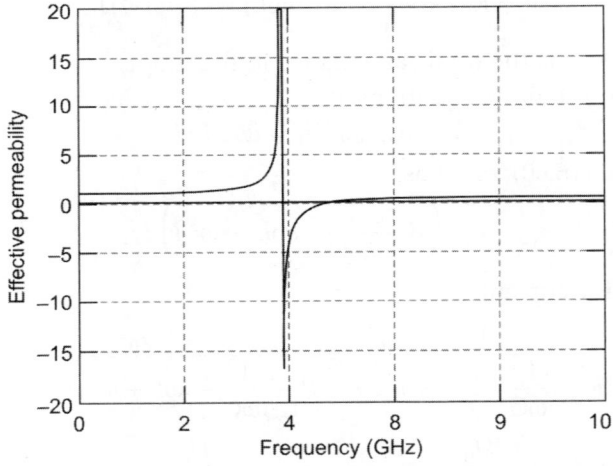

Fig. 16.18. Real and imaginary parts of magnetic permeability

The combination of long wires and SRR in a unit cell is shown in Fig. 16.19. The effect of both results in a negative effec-tive permittivity and permeability over the same frequency band.

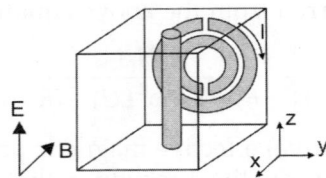

Fig. 16.19. Elementary cell of metamaterial formed by SRR and thin wire

Figure 16.20 shows schematic plots of effective permittivity (ε_{eff}) and perme-ability (μ_{eff}).

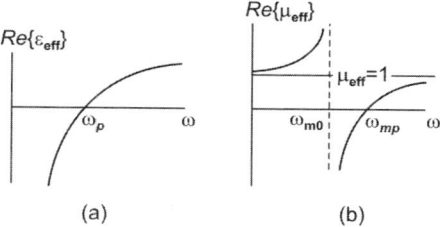

Fig. 16.20. (a) Permittivity of wire medium demonstrating plasma-like frequency dependent permittivity and (b)Frequency response of effective permeability

16.16 SOME APPLICATIONS OF METAMATERIALS

Materials with such unusual properties allow for unusual applications. Here, just explore on a few interesting possibilities. We start with the possibility of creating a so-called *'perfect lens'*.

16.16.1 Perfect Lenses

Until recently, it was thought that the manipulation of light is limited by the fundamental law of diffraction to a relatively long wavelengths (say around 0.5 λ). The sub-wavelength details are carried by evanescent harmonics which decay exponentially and are also subject to noise. A conventional lens only collects the propagating waves. The evanescent waves are lost due to their decay.

A planar slab consisting of NIM with sufficient thickness can act as a lens, known as Vaselago lens. Such a slab lens with refractive index $n = -1$ placed in vacuum (Fig. 16.21) can resolve details of an object with subwavelength precision.

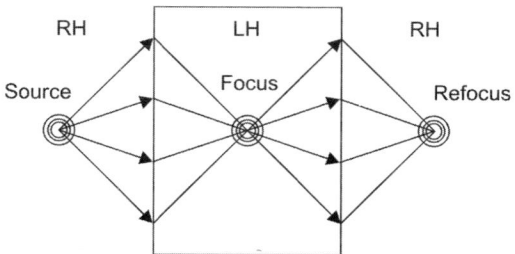

Fig. 16.21. Double focussing of a source with a planar double negative (left-handed) metamaterials slab surrounded by regular (right-handed) dielectrics

To understand the problem, consider a slab of thickness d and refractive index of, e.g. $n = -1$, surrounded by air. It will bring all rays emanating from a source to a double focus: first, at a point inside the

NIM slab, at a distance $s = l < d$, where l is the distance of the source from the slab, and second at a point outside the slab, at a distance $d - l$. Hence, such a slab acts like a lens, and is able to bring the rays radiated by a source to a focus outside the slab, without reflections occurring at the media interfaces because the $n = -1$ slab is impedance-matched to free space.

Evanescent waves are associated with the high spatial frequencies of the electromagnetic waves creates by source. They carry fine (subwavelength) features of the source. Therefore, a NIM slab can, in principle, enable us to obtain the image of an object with 'perfect' resolution, containing all the subwavelength features of an object and overcoming the usual diffraction limitations that characterize conventional lenses.

16.16.2 Stopped Light in Metamaterials

For decades, scientists maintained that optical data cannot be stored statically and must be processed and switched on the fly. The reason for this conclusion was that stopping and storing an optical signal by dramatically reducing the speed of light itself was thought to be unfeasible.

At present, some of the most successful slow-light designs based on photonic-crystals (PhCs) or coupled-resonator optical waveguides (CROWs), can indeed slow down light efficiently by a factor of only 40; otherwise, large group-velocity-dispersion and attenuation-dispersion occur: i.e., the guided light pulses broaden and the attainable bandwidth is severely restricted.

Tsakmakidis and Coworkers in 2007 proposed a new method that can allow for a true stopping of light in NIM. The stopping of light in the proposed configuration is associated with a negative Goos-Hanchen (G-H) phase shift, a lateral displacement of light ray when it is totally reflected at the interface of two different dielectric media. Classical G-H shift between dielectrics having positive refractive indices is illustrated in Fig. 16.22.

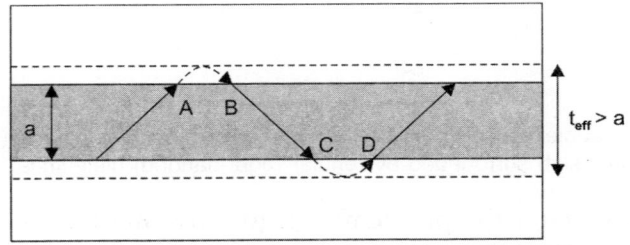

Fig. 16.22. Classical G-H shift in regular dielectrics

In a structure, where the central layer is formed by NIM, the G-H shift is reversed as it is illustrated in Fig. 16.23.

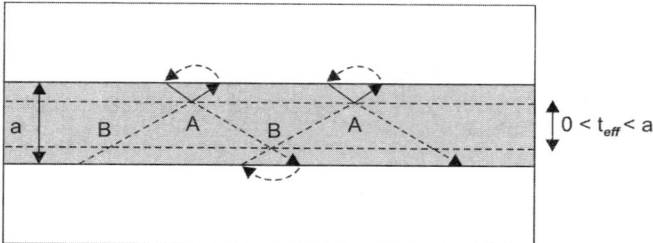

Fig. 16.23. G-H shift in NIM forming central layer

To more precisely understand the manner in which light is decelerated in this structure, let us imagine a ray of light propagating in a zig-zag fashion along a waveguide with a negative-index (left-handed) core. The ray experiences negative Goos-Hänchen lateral displacement each time it strikes the interfaces of the core with the positive-index (right-handed) claddings (Fig. 16.23). Accordingly, the cross points of the incident and reflected rays will sit inside the left-handed core and the effective thickness of the guide will be smaller than its natural thickness. It is reasonable to expect that by gradually reducing the core physical thickness, the effective thickness of the guide will eventually vanish. Obviously, beyond that point the ray will not be able to propagate further down, and will effectively be trapped inside the negative index metamaterial (NIM) heterostructure (Fig. 16.24).

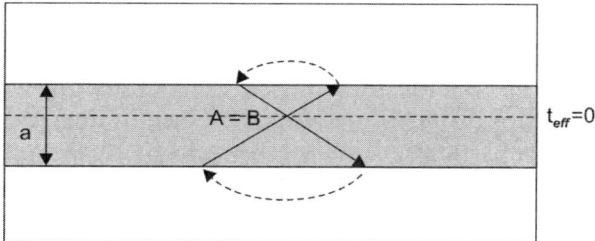

Fig. 16.24. G-H shifts in NIM at critical thickness

Tsakmakidis and coworkers further proposed stopping a light pulse by varying the thickness of the waveguide core to the point where the cycle-averaged power flow in the core and the cladding becomes comparable. At the degeneracy point, where the magnitudes of these powers become equal, the total time-averaged power flow directed along the central axis of the core vanishes. At this point the group (or energy) velocity goes to zero and the path of the light ray forms a double light cone (optical clepsydra) where the negative GH lateral shift experienced by the ray is equal to its positive lateral displacement as it travels across the core. Adiabatically reducing the thickness of the NIM core layer may thus, in principle, enable complete trapping of a range of light rays, each corresponding to a different frequency contained within a guided wavepacket.

This ability of metamaterial-based heterostructures to dramatically decelerate or even *completely stop* light under realistic experimental conditions, has recently led to a series of experimental works that have reported an observation of so-called '*trapped rainbow*' light-stopping in metamaterial waveguides.

16.16.3 Cloaking (Invisibility)

Unusual light-bending properties constitute a rather generic feature of metamaterials. Part of the excitement surrounding these materials is that they could be engineered to 'cloak' objects from electromagnetic radiation such as light: that is, make them seem invisible at specific frequencies. A metamaterial 'invisibility' cloak can be designed such that it does not reflect waves back nor scatter them in other directions. Several methods were proposed to make extended bodies invisible, such as those based on cancellation of scattering or on coordinate transformations. The method suggested by Pendry and coworkers in 2006 relies on controlling the paths of electromagnetic waves. It was applied to a spherical volume and uses coordinate transformation that expels the paths of electromagnetic waves (rays) from a spherical volume, squeezing them into a spherical shell around the volume that is to be cloaked, thereby, making it invisible to incident radiation (Fig. 16.25). The light rays smoothly avoid the cloaked object and flow around it like a fluid. They appear to have properties of the free space when observed externally. The rays make a detour around the hidden part of the device. On the left side, they must arrive at the same time as if they were propagating through empty space. Since in the cloaking region, they travel longer distances, their phase velocity must exceed c. This is in principle possible for a specific frequency).

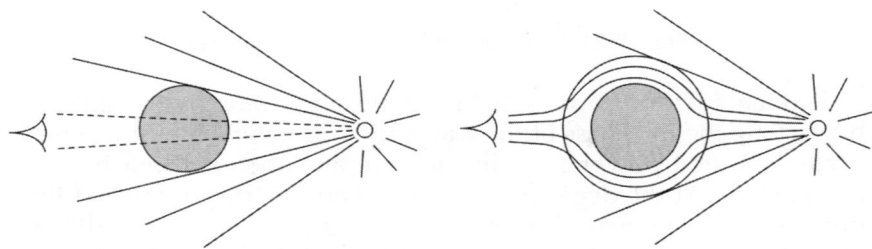

Fig. 16.25. The rays go around the inner object and then go to the eye. An observer at the left of the cloak would see the point source

To introduce the relevant transformation, one starts with a mathematical point (placing it at the origin of a coordinate system), which is obviously invisibly small. To hide an extended object, say a sphere of radius R_1, we transform the mathematical point at the origin into a sphere of radius R_1 and the vicinity of the sphere into another sphere of radius $R_2 > R_1$. The transformation which does this is:

$$r' = R_1 + \frac{R_2 - R_1}{R_2} r, \quad \theta' = \theta, \quad \phi' = \phi \tag{16.65}$$

Maxwell's equations are from-invariant to coordinate transformations like the one above. Only the components of μ and ε are effected by the transformation. They become spatially varying and anisotropic tensors. Their forms were determined by Pendry and coworkers and for $R_1 < r < R_2$ are:

$$\varepsilon'_{r'} = \mu'_{r'} = \frac{R_2}{R_2 - R_1} \frac{(r' - R_1)^2}{r'} \tag{16.66a}$$

$$\varepsilon'_{\theta'} = \mu'_{\theta'} = \frac{R_2}{R_2 - R_1} \tag{16.66b}$$

$$\varepsilon'_{\phi'} = \mu'_{\phi'} = \frac{R_2}{R_2 - R_1} \tag{16.66c}$$

Inside the sphere with radius R_1, the permittivity and permeability can be arbitrary, whereas the space between the spheres is filled with a material having the permittivity and permeability tensors determined by Eq. (16.66). As a result, any object inside the sphere with radius $r < R_1$ is concealed. These conclusions follow from exact manipulations of Maxwell's equations and are not restricted to a ray approximation.

16.16.4 Optical Black Holes

By extending the concept behind the invisibility cloak, one can speculate on the possibility of creating an optical analogue of a black hole. In this concept, by proper design of a new class of metamaterials, one can expect to concentrate and trap light waves, similarly to what can happen in a 'black hole' (Fig. 16.26). In such a system, light will be permanently trapped. Such a black hole design has been proposed by Narimanov and Kildishev in 2009. Numerical simulations showed a highly efficient light absorption. The electromagnetic black hole was built recently and it operates at microwave frequencies. The structure is composed of 60 concentric layers, and each layer is a thin printed circuit board etched with a number of subwavelength unit structures on one side and coated with 0.018 mm thick copper on the other side. The permittivity changes radially in the shell of the microwave black hole, and hence, the unit cells are identical in each layer but have different sizes in adjacent layers. The structure can efficiently absorb electromagnetic waves coming from all directions owing to the local control of electromagnetic fields. It is expected that such a microwave black hole could find important applications in solar-light harvesting, thermal emitting, and cross-talk reduction in microwave circuits and devices. These specially designed analogues of black holes could also be used for controlling, slowing and trapping electromagnetic waves, as well as for investigating some of the more exotic physics associated with celestial mechanics.

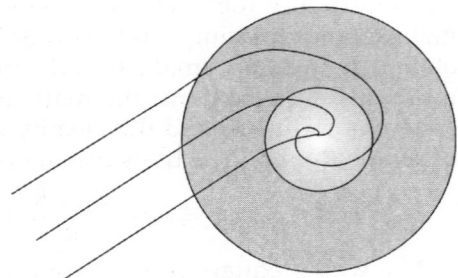

Fig. 16.26. Metamaterial black hole fabricated from properly crafted gradient-index material

16.17 METAMATERIALS WITH AN ACTIVE ELEMENT

Now, we briefly discuss the problem of losses in metamaterials and how to compensate for them.

As mentioned before, metamaterials show large losses which at present are orders of magnitude too large for practical applications and are considered as an important factor limiting practical applications of metamaterials. For example, detailed analytical studies show that losses limit the superresolution of a theoretical superlens. There was some controversy whether loss elimination can be feasible, but it is possible to completely eliminate losses in metamaterials.

Very recently, several computational and experimental works have demonstrated that optical losses can be fully overcome in realistic negative-refractive-index metamaterials. The specific loss-free design is consisted of two metallic films perforated with small rectangular holes (fishnets), and with an active medium (laser dye) spacer between two films. Additionally, several reports speculated about possible compensation for losses in metamaterials by introducing a gain element. For example, Wegener and coworkers in 2008 formulated a simple model where gain is represented by a fermionic two-level system which is coupled via a local-field to a single bosonic resonance representing the plasmonic resonance of the metamaterial. Also Fang and coworkers in 2009 described a model where the gain system is modelled by a generic four-level atomic system. They conducted numerical analysis using the FDTD technique. Gain material was introduced in the gap region of the split-ring resonators (SRR). The system had a magnetic resonance frequency at 100 THz.

GLIMPSIS

- Metamaterials are artificial engineered materials that can provide unusual electromagnetic properties not readily available in nature, represent a new paradigm in electronics and photonics.

- One of the most important properties of metamaterials is the feasibility of left-handed materials (LHMs) from the combination of electric conductors. These LHMs are characterised by their negative values of electric conductors. These LHMs are characterised by their negative values of electric permittivity ($\varepsilon < 0$) and magnetic permeability ($\mu < 0$); and hence, left-handed materials (LHMs) are also referred to as double negative (DNG) materials. Several applications arise from the use of such LHMs, like the negative refraction index, or the feasibility of LH transmission lines.
- The beauty of metamaterial concept is the simple epiphany that we do not have to rearrange atoms- or subatomic particles therein – in order to engineer a new matter; instead, we fabricate artificial structure with the unit components much smaller than the wavelength of interacting electromagnetic (EM) waves. The collective response of the bulk can result in novel EM properties that are not observed in nature.
- Metamaterials are beginning to transform optics and microwave technology. The versatile properties of these materials can be tailoerd according to practical needs and desires.
- Perhaps the best-known metamaterials are the materials used in the pioneering demonstrations of negative refraction or invisibility cloaking of microwaves, or for negative refraction of near-visible light. These materials consist of metallic cells that are smaller than the relevant electromagnetic wavelength. Each cell acts like an artificial atom that can be tuned by changing the shape and the dimensions of the metallic structure.
- Different approaches are used for a description of the fundamental EM properties of metamaterials and the practical realization of these materials as well. Among them, the transmission line (TL) approach gives an efficient design tool for microwave applications providing a correct description of physical properties of metamaterials.
- In contrast to conventional periodic swtructures that are usually operated in the Bragg regime (period $\approx \lambda_g/2$, where λ_g is the guided wavelength) for filtering, metamaterials are long-wavelength structures (average distance between unit elements or atoms $\ll \lambda_g$). Since they are not readily available naturally, they are engineered under the form of 1D, 2D or 3D structures.
- Composite right/left-handed (CRLH) metamaterials have lead to several breakthroughs in both 1D and 2D leaky wave antenna and reflector applications.
- An interesting aspect of using metamaterials in radiating systems is the possibility to create highly directive beams from simple sources placed inside planar layered structures.

- Electronically controllable photonic bandgap (CPBG) materials are used to realize planar and comfortable antennas for communication at 12 GHz and in the GSM-UMTs band. The main advantage of these CPBG materials is their capability to realize multifunction smart antennas. The main drawback of these materials is their thickness, which is of the order of several $\lambda/3$, where λ is the operating wavelength.

SUGGESTED READINGS

1. Pendry, JB, *Phys. Rev. Lett.*, 85, 3966 (2000).
2. Veselago VG, *Sov. Phys.* Uspekhi, 10, 509 (1968).
3. Zhu X and Obtsu M (eds.), Near-Field Optics: Principles and Applications, World Scientific, Singapore (1999).
4. Caloz C and Itoh T, Electromagnetic Metamaterials: Transmission Line Theory and Microwave Applications, John Wiley (2006).
5. Eleftheriades E and Balmain KG, Negative-Refraction Metamaterials: Fundamental Principle and Applications, Wiley, New York (2006).
6. Engheta N and Ziolkowski R (eds.), Metamaterial Physics and Engineering Explorations, John Wiley, New York (2006).
7. Alitala P and Tretyakov S, Metamaterials, 1(2), 81–88 (2007).
8. Capolino F (ed.), Theory and Phenomena of Metamaterials, CRC Press, Taylor and Francis Group (2009).
9. Cui TJ, Smith DR, Liu R (eds.), Metamaterials: Theory, Design and Applications, Springer (2010).
10. Cai W and Shalaev V, Optical metamaterials: Fundamentals and Applications, Springer (2010).
11. Kakani SL and Kakani A, Engineering Materials, New Age International Publishers, New Delhi (2014).

REVIEW QUESTIONS

1. What are metamaterials? How they differ from normal materials?
2. What are electromagnetic and photonic crystals? Explain the concept of negative refraction and backward wave effects.
3. Write the properties of materials with extreme anisotropy.
4. What is super resolution effect? Explain the concept of super-resolution imaging with hyperlens.
5. Give a brief account of metamaterials for cloaking applications.
6. What are electromagnetic bandgaps? How they are used to microwave circuit design?
7. Why metamaterials are also referred as double negative (BNG) materials.

8. Give a brief account of microwave phase shifters and filters based on a combination of left-handed and right-handed transmission lines.
9. Give a brief account composite right/left-handed (CRLH) transmission line (TL) metamaterials. How these materials are substantially different from filters?
10. Give a brief account of metamaterial antenna applications.
11. Briefly explain the important applications of metamaterials.

SHORT ANSWER QUESTIONS

1. What are metamaterials?

Ans. These are artificial engineered materials that can provide unusual electromagnetic properties. One of their most important properties is the feasibility of left-handed materials (LHMs) from the combination of electric conductors. These LHMs are characterized by their negative values of electric permittivity ($\varepsilon < 0$) and magnetic permeability ($\mu < 0$); and hence, LHMs are also referred to as *double negative* materials. Several applications arise from the use of such LHMs, like negative refraction index, or the feasibility of LH transmission lines.

2. What are electromagnetic and photonic crystals?

Ans. Electromagnetic crystals are artificial periodical structures operating at frequencies where the wavelength is comparable with the characteristic period of the structure. In the optical frequency range, such structures are referred as photonic crystals. The inherent feature of such media is the existence of frequency bands inside which wave propagation is not allowed. These bands are also called band gaps, and therefore, these crystals are sometimes also called *electromagnetic band gap* (EBG) or *photonic band gap* (PBG) structures.

3. Mention the properties of materials with extreme anisotropy?

Ans. These materials have remarkable properties. All the extraordinary waves supported by the media and propagating waves. Thus, theoretically, it is possible to transport an arbitrary field distribution with suitable polarisation through such material with no loss of resolution. Lens formed by slabs of such materials provide a unique opportunity to transmit the near-field with super-resolution. At the front interface with free space, the evanescent waves are transformed into poropagating waves, which prevents their decay and preserves the subwavelength information. The propagating waves travel through the slab and reproduce the image at the back interface of the lens. This phenomenon is known as *'canalisation'*. This regime is particularly appropriate for the transmission of images with super-resolution over significant distances in terms of wavelength. We may note that this effect cannot be achieved using other available imaging techniques.

4. How the innovation of metamaterials opened wide the door in the field of electromagnetics?

Ans. The beauty of metamaterial concept is the simply epiphany that we do not have to rearrange atoms- or subatomic particles therein—in order to engineer a new matter; instead, we fabricate artificial structure with the unit components much smaller than the wavelength of the interacting electromagnetic (EM) waves. The collective respopnse of the bulk can result in novel EM properties that are not observed in nature.

5. What is a hyperlens?

Ans. A hyperlens utilizes unusual characteristic of an anisotropic metamaterial to carry large wave vectors into the far field. Concentric multilayers of metal and dielectrics make up this metamaterial structure. The metamaterial lens was designed such that (i) large wave vectors can be collected and (ii) to gradually convert them into smaller wave vectors. These small wave vectors are then captured by a conventional microscope that directly delivers the magnified image of nanoscale objects to our eyes. A hyperlens can be a simple additional part fo any existing microscopes that will remarkably improve the resolution. Hyperlensing requires no postimaging processes and is a direct far-field imaging technique.

6. What is metamaterial cloaking?

Ans. One set of novel and exciting applications of metamaterials deals with making a given object essentially "invisible" to the impinging electromagnetic wave. This is termed as metamaterial cloaking. Metamaterial cloaking may become of major interest for various exciting applications in various scenarios, including camouflaging and noninvasive probing, and sensing in medical and biological applications.

7. What are metamaterial transmission lines?

Ans. Metamaterial transmission lines are one-dimensional (1D) artificial propagating media that exhibit either a single-negative (SNG) or a double-negative (DNG or left-handed, LH) behaviour. In the first case, signal propagation is not allowed in the frequency region of interest (even though the medium is transparent at other frequencies) since the effective permeability and permittivity are negative in that region. Conversely, LH metamaterial transmission lines exhibit backward wave propagation in the LH band (due to the coexistence of negative effect permittivity and permeability in that band) and forward wave propagation in a certain frequency interval at higher frequencies. Purely left-handed (PLH) lines (i.e. those lines exhibiting only a backward wave transmission band) cannot be implemented in practice due to the presence of parasitic elements. Hence, LH lines actually exhibit a composite behaviour, and for this reason, these lines are called composite light/left-handed (CRLH) lines. Two main approaches can be used for the implementation of metamaterial transmission lines: (i) the nonresonant-type approach, where a host line is periodically loaded with series capacitors and/or shunt capacitors, and the resonant-type approach, where the host transmission lines are loaded with SRRs or CSRRs.

8. Explain the concept of microwave phase shifters and filters based on a combination of left-handed (LH) and right-handed (RH) transmission lines.

Ans. Different approaches are used for a description of the fundamental FM properties of metamaterials and the practical realization of these materials as well. Among them, the transmission line (TL) approach gives an efficient design tool for microwave applications providing a correct description of physical properties of metamaterials. A conventional TL with a positive phase velocity behaves as right-handed transmission line (RHTL). An artificial RHTL can be formed as a ladder network of capacitors connected in shunt- and series-connected inductors. A combination of the RH and LHTL sections is useful for applications in microwave techniques. A new approach to designing digital phase shifters that are free of spurious response resonators with enlarged functionality resulted in exceptional performance of the devices. One can also use the combination of RH and LHTL sections for improving the characteristics of directional couplers, power splitters, matching circuits, etc. Modern multilayer technology makes it possible to design fully integrated miniature microwave devices on the RH/LHTLs.

9. What is difference between conventional periodic structures and metamaterials?

Ans. Metamaterials are broadly defined as effectively homogeneous structures exhibiting unusual electromagnetic properties not readily available in nature, represent a new paradigm in electronics and photonics. In contrast to conventional periodic structures that are usually operated in the Bragg regime (period $\approx \lambda_g/2$, where λ_g is the guided wavelength) for filtering, metamaterials are long-wavelength structures (average distance between unit elements or atoms $<< \lambda_g$). Since they are not readily available naturally, they are engineered under the form of 1D, 2D or 3D structures. The most general, efficient, and applied TL metamaterials to date are the so-called composite right/left-handed (CRLH) TL metamaterials. The fundamental parameters of CRLH metamaterials are their dispersion relation and their Bloch impedance, and their operation bandwidth is determined by the lumped LC elements of their unit cell.

10. What is active and dynamical control of metamaterial devices at THz frequencies?

Ans. Active control is achieved by biasing n-GaAs near the gaps of split rings, which effectively tuned the strength of the resonance of the metamaterial. Active control required arrays of conducting wires, but it did not affect the oscillator strength of the response, lending potential flexibility to future devices. Dynamical control is realized through photoexcitation of carriers in the substrate of the split ring resonator (SRR). This demonstration of the optical control allows for fast switching times that are not realizable by electronic means.

11. Is ruby glass a matameterial?

Ans. Yes, ruby glass contains nanoscale gold colloids that tender the glass neither golden nor transparent, but ruby, depending on the size and concentration of the gold droplets. The colour originates from a resonance of the surface plasmons on the metallic droplets. Metamaterials per se are nothing new: what is new is the degree of control over the structures in the material that generate the desired properties.

FILL IN THE BLANKS

1. Electromagnetic crystals are artificial .... structures operating at frequencies where the wavelength is comparable with the .... period of the structure. [periodic, characteristic]
2. When the usual (forward) wave propagates in the isotropic medium, its wave vector and its group velocity vector are .... In the backward-wave, these two vectors are by definition .... [parallel, antiparallel]
3. Lens based imaging devices such as conventional microscopes can not provide resolution better than ...., where λ is the wavelength of radiation. This restriction is known as the .... limit and holds irrespectively of the frequency of operation—from microwave frequencies up to the visible range. [$\lambda/2$, diffraction]
4. When a multilayer array of plasmonic nanosphere is placed near an electric dipole, on the other side, the optical near field is .... due to the excitation of .... modes in the layers. [enhanced, resonance]
5. The smallest resolvable feature (Δ) of any optical system can be approximated by the diffraction limit:

$$\Delta = \frac{\lambda}{2(NA)}$$

where λ is the wavelength of the light in the free space and NA = $n \sin \theta$ is the ...., where n is the refractive index of medium and θ is the light collection angle of the optics. [numerical aperture]

6. A hyperlens utilizes unusual characteristic of an .... metamaterial to carry large wave vectors into the far field. [anisotropic]
7. Metamaterials are materials with electromagnetic properties that originate from human-made .... structures. [subwavelength]
8. Perhaps the best known metamaterials are the materials used in the pioneering demonstrations of .... refraction or invisibility .... of microwaves, or for .... refraction of near visible light. [negative, cloaking, negative]
9. One of the most important properties of metamaterials is the feasibility of .... materials from the combination of electric conductors. [left-handed (LH]
10. In the contrast to conventional periodic structures that are usually operated in the Bragg regime (period $\approx \lambda_g/2$, where λ_g is the guided wavelength) for filtering, metamaterials are .... wavelength structures (average distance between unit elements or atoms << ....). [long, λ]

MULTIPLE CHOICE QUESTIONS

1. Electromagnetic crystals are media with:
   (a) spatial dispersion (b) interference
   (c) polarisation (d) None of these
2. Lens-based imaging devices such as conventional microscopes cannot provide resolution better than:
   (a) λ (b) $\lambda/2$
   (c) λ^2 (d) λ^3
3. Hyperlen is a metamaterial having signs of permittivities:
   (a) same
   (b) opposite
   (c) one zero and other positive
   (d) None of these
4. Metamaterials are materials with electromagnetic properties that originate from human-made:
   (a) subwavelength structures
   (b) interference
   (c) diffraction
   (d) polarisation

5. Ruby glass is an example of:
   (a) semiconductor
   (b) metal
   (c) insulator
   (d) metamaterial
6. Electromagnetic bandgaps (EBGs) are periodic structures that are able to inhibit signal propagation at certain:
   (a) frequencies
   (b) directions
   (c) frequencies and directions
   (d) None of these
7. Left-handed materials (LHMs) are also referred to as:
   (a) singly negative materials
   (b) double negative materials
   (c) positive materials
   (d) None of these
8. Metamaterials are not readily available naturally, they are engineered under the form of:
   (a) 1D structure only
   (b) 2D structure only
   (c) 3D structure only
   (d) 1D, 2D or 3D structures
9. The proper design of the elementary building blocks (artificial atoms) of the photonic metamaterial allows for a nonvanishing magnetic response and m = 0 at optical frequencies–despite the fact that constituent materials of the photonic metamaterial are:
   (a) magnetic
   (b) nonmagnetic
   (c) superconducting
   (d) None of these

ANSWERS

1. (a) 2. (b) 3. (b) 4. (a) 5. (d) 6. (c) 7. (b)
8. (d) 9. (b)

Index

A

Acceptance
  angle 119
  cone 119
Acousto-optic
  effect 511
  modulators 511
  switches 600
  tunable filters 638
Active
  glass fibers 195
  medium 264
Amplifier noise 531
Analog 665
  systems 676
Antireflection coating 33
Artificial plasma 836
Arrayed waveguide gratings 628
Attenuation 148
Avalanche
  breakdown 469
  multiplication 362
  photo-diodes 361, 474
Azimuthual eigenvalue 72

B

Band gap 283
Bandwidth 339
Barrier potential 258
Bessel function of first kind 73
BER 481
Birefringence 539
Birefringent crystals 13
Bit error rate 669
Bit-rate transparency 453

Boundary conditions 20
Bragg
  diffraction 511
  mirrors 34
  modulator 513
Brewster's angle 12
Butt-joint connectors 218
Bus topology 681

C

Campus area network 684
Chalcogenide glass fibers 195
Chemical lasers 800
Chirped solitons 739
Cladding 116
Cloaking 851, 866
Coarse wavelength division
      multiplexing 613
Coherence 268
Color center laser 802
Common fiber connector assembly 217
Compound semiconductors 328
Confinement factors 68
Connectors and splices 211
CO_2 gas laser 796
Couplers 614
Cylindrical waveguides 70

D

Dark current 331, 347
Decibel 124
Dense wavelength division
      multiplexing 613

Depletion layer photo current 472
Depletion region 451
Diffraction 8
Diffusion current 461
Digital 665
Direct band gap 259
Direct band gap semiconductors 426
Directionality 268
Dispersion 720
Distributed fiber optic sensors 775
Double heterostructure laser 273
Dye or liquid lasers 798

E

Ecology 1
Einstein's relation 264
Electric boundary conditions 20
Electromagnetic metamaterials 838
Electromagnetic waves 17, 54
Electronic signal processor 449
Electro-optic
    effect 503, 514
    switches 592
Energy bands 255
Epoxyless connector 218
Erbium doped fiber optical
    amplifiers 549
Expanded beam connectors 219
External modulation 502
Extraordinary ray 14

F

Fabry-Perot cavity amplifiers 526
Femtochemistry 2
Fiber
    connectors 213
    fabrication 153
    grating tunable filters 637
    light wave 118
    losses 147
    materials 194
    splicing 223
    strength and durability 152
    waveguide 118
Fiber Bragg grating 630
    sensor 774
Fiber optic
    connector types 215
    connectors 152
    gyroscope 771

Mach-Zehnder interferometer
    sensors 770
    sensors 763
    smart structures 777
Fiber to fiber misalignment losses 207
Fractional modal power
    distribution 79
Free electron laser 804
Frequency division multiplexing 612
Fusion splices 223

G

Gain saturation 530
Gas lasers 793
Generation-recombination current 461
Generic SONET network 713
Glass fibers 194
Graded index fibers 82
GRINSCH laser 275
Group velocity 721

H

Half wave voltage 510
Halide glass fibers 195
Hankel function 74
Helium-neon gas laser 794
Heterojunction 283
Holography 818
Home area networks 684
Homogeneous p-n junction 282
Hot-melt connector 217
Hub topology 691
Hyperlens 848

I

Imaging 1
Indirect band gap 260
Infrared emitting diode 381
Inhomogeneous medium 56
Injection electroluminescence 378
In-line optical amplifiers 564
Intensity modulated sensors 765
Intermediate band solar cells 433
Intermodal dispersion 673
Internal
    modulation 502
    reflection 10
Intrinsic
    material 257
    sensors 765
Isotropic medium 56

Index

J
Jacket or sheat 117
Junction
 capacitance 348
 capacitance effects 360

K
Kerr effect 503
Key-lock mechanical fiber optic splices 226

L
Laser
 components 264
 diode 388
 oscillations and resonant modes 270
Leakage current 461
LED indicator circuits 380
Left handed material 834
Light
 emitting diodes 374
 guide 118
Linear
 electro-optic modulator 503
 optical amplifiers 584
Link power budget analysis 670
Liquid grating tunable filters 638
Liquid phase melting techniques 153, 196
Local area networks 684, 693
Loss due to Fresnel reflection 207

M
Mach-Zehnder interferometer 632
Magnetic boundary conditions 21
Majority carriers 258
Mass action law 258
Material dispersion 721
Maxwell's equations 18, 54
Mechanical splices 224
3D-MEMS based optical switches 604
Mesh topology 681
Metal semiconductor metal
 detectors 465
 photodiode 368
Metamaterials 833
Metamaterial lenses and superlenses 847
Metastable state 265

Metro and long haul optical network 708
Metropolitan area network 685
Microelectromechanical systems 601, 606
Minority carriers 258
Modal dispersion 724
Mode locking 809
Modern fiber optic processes 156
Modes of a symmetric step index fiber 62
Monochromaticity 268
Multicolour LEDs 381
Multifiber connectors 219
Multimode
 graded index fiber 133
 step index fiber 131
Multi junctions 432
Multiple
 quantum well laser 275
 splices 227
Multi segment LED displays 381
Multiplexers and demultiplexers 624

N
Nd^{3+}:YAG laser 792
Negative refractive index 833
Neodymium doped optical amplifiers 559
Noise
 current 348
 equivalent power 348
 equivalent ratio 340
Nonlinear effects 2
NRZ and RZ signal formats 695
Numerical aperature 119

O
Optical
 absorption 452
 amplifier 523
 black holes 867
 communication 663
 coupler 329
 fiber 116
 fiber amplification 162
 fiber cables 150, 202
 gain 286
 interconnects 4
 nonlinearities in fibers 164

Optoelectronic
  devices 328
  integrated circuits 328
Optoisolator 329
Optomechanical switches 581, 585
Outside vapour phase oxidation
    process 198

P

Perfect lenses 863
Phase modulated sensors 768
Photoconductor 335
Photoconductive effect 330
Photoconductive mode 331
Photoconductivity 331
Photodetector noise 476
Photodetectors 333, 453
Photodiode 342
Photoemissive effect 330
Photonic band gap fibers 86
Photonic crystal fibers 86
Photonic crystals 855
Photonic sensors 3
Photons 15
Phototransistors 402
Photovolatic
  devices 420
  mode 332
Photovoltaics 464
p-i-n photodiode 354, 451, 470
p-n photodiodes 467
Planar optical waveguide 60
Planar waveguide optical
    amplifiers 564
Plane polarized wave 26
Plasma activated chemical vapour
    deposition method 202
Plastic optical fibers 195
Point to point WDM links 639
Polarization 8, 159
Polarizer 13
Population inversion 266
Postamplifier 565
Poynting theorem 35
Praseodymium doped fluoride optical
    amplifiers 559
Preamplifier 565
Pulse amplification 538

Q

Q-switching 793, 806
Quanta 15

Quantum
  communiction 4
  dot lasers 276
  dots and multi-junctions 432
  efficiency 337, 346, 456
  information processing 4
  nature of light 15
  transition 261
  well lasers 274

R

Radiation mode 62
Raman fiber optical
    amplifiers 548, 560
Raman-Nath
  diffraction 511
  modulator 512
Rare earth doped fiber amplifiers 548
Ray propagation in graded index
    fibers 139
Receiver sensitivity 452
Refractive index profile 129
Response time 348, 459, 473
Responsivity 338, 457
Reverse breakdown voltage 348
Ring topology 681, 683
Rise power budget analysis 671
RMS pulse width 145
Ruby laser 790

S

Schottky photodiodes 367
Screen-printed monocrystalline silicon
    cells 437
SDH 703
Semiconductor
  lasers 271
  optical amplifiers 524, 532
  photo-diode 450
Semiconductors 254
Sensitivity 461
Shot noise 478
Simulations 4
Single mode step index fiber 130
Single solitons 737
Snell's law 9
Snug tube splice 224
Solid state lasers 789
Soliton in optical fiber 165

Solitons 728
SONET 705
SONET/SDH ring configuration 709
Space charge region 259
Split ring resonators 859
Spontaneous emission 261
Star coupler 619
Star topology 681
Stimulated
    absorption 261
    Brillouin scattering 717
    emission 261
    Raman scattering 715, 729
Superlattice 367
Switch logic 598

T

Table-type mechanical fiber optic
    splices 227
TE polarization 28
Telluride-based erbium doped fiber
    optical amplifiers 559
Thermal noise 478
Thermo-optic
    switches 595
    tunable filters 638
Thin film filters 631
Thulium doped optical amplifiers 559
Time division multiplexing 612
TM polarization 30

Total internal reflection 118
Transit time 336
Transverse electro-optic
    modulator 509
Transverse magnetic modes 60
Travelling wave amplifiers 526
Tree topology 682
Tunable
    optical filters 635
    sources 633
Tunneling current 461

V

Vapour axial deposition process 200
Vapour phase deposition
    techniques 155, 196
Vertical cavity surface emitting 278
V-parameter 125

W

Wave particle duality 16
Wavelength division multiplexing
    technology 607, 663
Wavelength routed networks 639
WDM star, rings and meshes 641
Wide area networks 685

X

X-ray laser 805